Universitext

For other titles in this series, go to
www.springer.com/series/223

W.A. Coppel

Number Theory

An Introduction to Mathematics

Second Edition

 Springer

W.A. Coppel
3 Jansz Crescent
2603 Griffith
Australia

ISBN 978-0-387-89485-0   e-ISBN 978-0-387-89486-7
DOI 10.1007/978-0-387-89486-7
Springer Dordrecht Heidelberg London New York

Library of Congress Control Number: 2009931687

Mathematics Subject Classification (2000): 11-xx, 05B20, 33E05

Springer is part of Springer Science+Business Media (www.springer.com)

For Jonathan, Nicholas, Philip and Stephen

Contents

Preface to the Second Edition

Undergraduate courses in mathematics are commonly of two types. On the one hand there are courses in subjects, such as linear algebra or real analysis, with which it is considered that every student of mathematics should be acquainted. On the other hand there are courses given by lecturers in their own areas of specialization, which are intended to serve as a preparation for research. There are, I believe, several reasons why students need more than this.

First, although the vast extent of mathematics today makes it impossible for any individual to have a deep knowledge of more than a small part, it is important to have some understanding and appreciation of the work of others. Indeed the sometimes surprising interrelationships and analogies between different branches of mathematics are both the basis for many of its applications and the stimulus for further development. Secondly, different branches of mathematics appeal in different ways and require different talents. It is unlikely that all students at one university will have the same interests and aptitudes as their lecturers. Rather, they will only discover what their own interests and aptitudes are by being exposed to a broader range. Thirdly, many students of mathematics will become, not professional mathematicians, but scientists, engineers or schoolteachers. It is useful for them to have a clear understanding of the nature and extent of mathematics, and it is in the interests of mathematicians that there should be a body of people in the community who have this understanding.

The present book attempts to provide such an understanding of the nature and extent of mathematics. The connecting theme is the theory of numbers, at first sight one of the most abstruse and irrelevant branches of mathematics. Yet by exploring its many connections with other branches, we may obtain a broad picture. The topics chosen are not trivial and demand some effort on the part of the reader. As Euclid already said, there is no royal road. In general I have concentrated attention on those hard-won results which illuminate a wide area. If I am accused of picking the eyes out of some subjects, I have no defence except to say "But what beautiful eyes!"

The book is divided into two parts. Part A, which deals with elementary number theory, should be accessible to a first-year undergraduate. To provide a foundation for subsequent work, Chapter I contains the definitions and basic properties of various mathematical structures. However, the reader may simply skim through this chapter and refer back to it later as required. Chapter V, on Hadamard's determinant problem, shows that elementary number theory may have unexpected applications.

Part B, which is more advanced, is intended to provide an undergraduate with some idea of the scope of mathematics today. The chapters in this part are largely independent, except that Chapter X depends on Chapter IX and Chapter XIII on Chapter XII.

Although much of the content of the book is common to any introductory work on number theory, I wish to draw attention to the discussion here of quadratic fields and elliptic curves. These are quite special cases of algebraic number fields and algebraic curves, and it may be asked why one should restrict attention to these special cases when the general cases are now well understood and may even be developed in parallel. My answers are as follows. First, to treat the general cases in full rigour requires a commitment of time which many will be unable to afford. Secondly, these special cases are those most commonly encountered and more constructive methods are available for them than for the general cases. There is yet another reason. Sometimes in mathematics a generalization is so simple and far-reaching that the special case is more fully understood as an instance of the generalization. For the topics mentioned, however, the generalization is more complex and is, in my view, more fully understood as a development from the special case.

At the end of each chapter of the book I have added a list of selected references, which will enable readers to travel further in their own chosen directions. Since the literature is voluminous, any such selection must be somewhat arbitrary, but I hope that mine may be found interesting and useful.

The computer revolution has made possible calculations on a scale and with a speed undreamt of a century ago. One consequence has been a considerable increase in 'experimental mathematics'—the search for patterns. This book, on the other hand, is devoted to 'theoretical mathematics'—the explanation of patterns. I do not wish to conceal the fact that the former usually precedes the latter. Nor do I wish to conceal the fact that some of the results here have been proved by the greatest minds of the past only after years of labour, and that their proofs have later been improved and simplified by many other mathematicians. Once obtained, however, a good proof organizes and provides understanding for a mass of computational data. Often it also suggests further developments.

The present book may indeed be viewed as a 'treasury of proofs'. We concentrate attention on this aspect of mathematics, not only because it is a distinctive feature of the subject, but also because we consider its exposition is better suited to a book than to a blackboard or a computer screen. In keeping with this approach, the proofs themselves have been chosen with some care and I hope that a few may be of interest even to those who are no longer students. Proofs which depend on general principles have been given preference over proofs which offer no particular insight.

Mathematics is a part of civilization and an achievement in which human beings may take some pride. It is not the possession of any one national, political or religious group and any attempt to make it so is ultimately destructive. At the present time there are strong pressures to make academic studies more 'relevant'. At the same time, however, staff at some universities are assessed by 'citation counts' and people are paid for giving lectures on chaos, for example, that are demonstrably rubbish.

The theory of numbers provides ample evidence that topics pursued for their own intrinsic interest can later find significant applications. I do not contend that curiosity has been the only driving force. More mundane motives, such as ambition or the necessity of earning a living, have also played a role. It is also true that mathematics pursued for the sake of applications has been of benefit to subjects such as number theory; there is a two-way trade. However, it shows a dangerous ignorance of history and of human nature to promote utility at the expense of spirit.

This book has its origin in a course of lectures which I gave at the Victoria University of Wellington, New Zealand, in 1975. The demands of my own research have hitherto prevented me from completing it, although I have continued to collect material. If it succeeds at all in conveying some idea of the power and beauty of mathematics, the labour of writing it will have been well worthwhile.

As with a previous book, I have to thank Helge Tverberg, who has read most of the manuscript and made many useful suggestions.

The first Phalanger Press edition of this book appeared in 2002. A revised edition, which was reissued by Springer in 2006, contained a number of changes. I removed an error in the statement and proof of Proposition II.12 and filled a gap in the proof of Proposition III.12. The statements of the Weil conjectures in Chapter IX and of a result of Heath-Brown in Chapter X were modified, following comments by J.-P. Serre. I also corrected a few misprints, made many small expository changes and expanded the index.

In the present edition I have made some more expository changes and have added a few references at the end of some chapters to take account of recent developments. For more detailed information the Internet has the advantage over a book. The reader is referred to the American Mathematical Society's MathSciNet (www.ams.org/mathscinet) and to The Number Theory Web maintained by Keith Matthews (www.maths.uq.edu.au/~krm/).

I am grateful to Springer for undertaking the commercial publication of my book and hope you will be also. Many of those who have contributed to the production of this new softcover edition are unknown to me, but among those who are I wish to thank especially Alicia de los Reyes and my sons Nicholas and Philip.

W.A. Coppel
May, 2009
Canberra, Australia

I

The Expanding Universe of Numbers

For many people, numbers must seem to be the essence of mathematics. *Number theory*, which is the subject of this book, is primarily concerned with the properties of one particular type of number, the 'whole numbers' or *integers*. However, there are many other types, such as complex numbers and *p*-adic numbers. Somewhat surprisingly, a knowledge of these other types turns out to be necessary for any deeper understanding of the integers.

In this introductory chapter we describe several such types (but defer the study of *p*-adic numbers to Chapter VI). *To embark on number theory proper the reader may proceed to Chapter II now* and refer back to the present chapter, via the Index, only as occasion demands.

When one studies the properties of various types of number, one becomes aware of formal similarities between different types. Instead of repeating the derivations of properties for each individual case, it is more economical – and sometimes actually clearer – to study their common algebraic structure. This algebraic structure may be shared by objects which one would not even consider as numbers.

There is a pedagogic difficulty here. Usually a property is discovered in one context and only later is it realized that it has wider validity. It may be more digestible to prove a result in the context of number theory and then simply point out its wider range of validity. Since this is a book on number theory, and many properties were first discovered in this context, we feel free to adopt this approach. However, to make the statements of such generalizations intelligible, in the latter part of this chapter we describe several basic algebraic structures. We do not attempt to study these structures in depth, but restrict attention to the simplest properties which throw light on the work of later chapters.

0 Sets, Relations and Mappings

The label '0' given to this section may be interpreted to stand for 'Optional'. We collect here some definitions of a logical nature which have become part of the common language of mathematics. Those who are not already familiar with this language, and who are repelled by its abstraction, should consult this section only when the need arises.

W.A. Coppel, *Number Theory: An Introduction to Mathematics*, Universitext,
DOI: 10.1007/978-0-387-89486-7_1, © Springer Science + Business Media, LLC 2009

We will not formally define a *set*, but will simply say that it is a collection of objects, which are called its *elements*. We write $a \in A$ if a is an element of the set A and $a \notin A$ if it is not.

A set may be specified by listing its elements. For example, $A = \{a, b, c\}$ is the set whose elements are a, b, c. A set may also be specified by characterizing its elements. For example,

$$A = \{x \in \mathbb{R} : x^2 < 2\}$$

is the set of all real numbers x such that $x^2 < 2$.

If two sets A, B have precisely the same elements, we say that they are *equal* and write $A = B$. (If A and B are not equal, we write $A \neq B$.) For example,

$$\{x \in \mathbb{R} : x^2 = 1\} = \{1, -1\}.$$

Just as it is convenient to admit 0 as a number, so it is convenient to admit the *empty set* $\emptyset$, which has no elements, as a set.

If every element of a set A is also an element of a set B we say that A is a *subset* of B, or that A is *included* in B, or that B *contains* A, and we write $A \subseteq B$. We say that A is a *proper* subset of B, and write $A \subset B$, if $A \subseteq B$ and $A \neq B$.

Thus $\emptyset \subseteq A$ for every set A and $\emptyset \subset A$ if $A \neq \emptyset$. Set inclusion has the following obvious properties:

(i) $A \subseteq A$;
(ii) *if $A \subseteq B$ and $B \subseteq A$, then $A = B$*;
(iii) *if $A \subseteq B$ and $B \subseteq C$, then $A \subseteq C$*.

For any sets A, B, the set whose elements are the elements of A or B (or both) is called the *union* or 'join' of A and B and is denoted by $A \cup B$:

$$A \cup B = \{x : x \in A \text{ or } x \in B\}.$$

The set whose elements are the common elements of A and B is called the *intersection* or 'meet' of A and B and is denoted by $A \cap B$:

$$A \cap B = \{x : x \in A \text{ and } x \in B\}.$$

If $A \cap B = \emptyset$, the sets A and B are said to be *disjoint*.

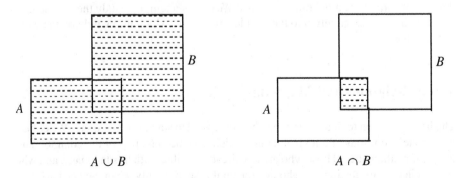

Fig. 1. Union and Intersection.

It is easily seen that union and intersection have the following algebraic properties:

$$A \cup A = A, \quad A \cap A = A,$$
$$A \cup B = B \cup A, \quad A \cap B = B \cap A,$$
$$(A \cup B) \cup C = A \cup (B \cup C), \quad (A \cap B) \cap C = A \cap (B \cap C),$$
$$(A \cup B) \cap C = (A \cap C) \cup (B \cap C), \quad (A \cap B) \cup C = (A \cup C) \cap (B \cup C).$$

Set inclusion could have been defined in terms of either union or intersection, since $A \subseteq B$ is the same as $A \cup B = B$ and also the same as $A \cap B = A$.

For any sets A, B, the set of all elements of B which are not also elements of A is called the *difference* of B from A and is denoted by $B \backslash A$:

$$B \backslash A = \{x : x \in B \text{ and } x \notin A\}.$$

It is easily seen that

$$C \backslash (A \cup B) = (C \backslash A) \cap (C \backslash B),$$
$$C \backslash (A \cap B) = (C \backslash A) \cup (C \backslash B).$$

An important special case is where all sets under consideration are subsets of a given universal set X. For any $A \subseteq X$, we have

$$\emptyset \cup A = A, \quad \emptyset \cap A = \emptyset,$$
$$X \cup A = X, \quad X \cap A = A.$$

The set $X \backslash A$ is said to be the *complement* of A (in X) and may be denoted by A^c for fixed X. Evidently

$$\emptyset^c = X, \quad X^c = \emptyset,$$
$$A \cup A^c = X, \quad A \cap A^c = \emptyset,$$
$$(A^c)^c = A.$$

By taking $C = X$ in the previous relations for differences, we obtain 'De Morgan's laws':

$$(A \cup B)^c = A^c \cap B^c, \ (A \cap B)^c = A^c \cup B^c.$$

Since $A \cap B = (A^c \cup B^c)^c$, set intersection can be defined in terms of unions and complements. Alternatively, since $A \cup B = (A^c \cap B^c)^c$, set union can be defined in terms of intersections and complements.

For any sets A, B, the set of all ordered pairs (a, b) with $a \in A$ and $b \in B$ is called the (*Cartesian*) *product* of A by B and is denoted by $A \times B$.

Similarly one can define the product of more than two sets. We mention only one special case. For any positive integer n, we write A^n instead of $A \times \cdots \times A$ for the set of all (ordered) *n-tuples* $(a_1, \ldots, a_n)$ with $a_j \in A$ $(1 \le j \le n)$. We call a_j the *j*-th *coordinate* of the *n*-tuple.

A *binary relation* on a set A is just a subset R of the product set $A \times A$. For any $a, b \in A$, we write $a R b$ if $(a, b) \in R$. A binary relation R on a set A is said to be

reflexive if $a\,R\,a$ for every $a \in A$;
symmetric if $b\,R\,a$ whenever $a\,R\,b$;
transitive if $a\,R\,c$ whenever $a\,R\,b$ and $b\,R\,c$.

It is said to be an *equivalence relation* if it is reflexive, symmetric and transitive.

If R is an equivalence relation on a set A and $a \in A$, the *equivalence class* R_a of a is the set of all $x \in A$ such that $x\,R\,a$. Since R is reflexive, $a \in R_a$. Since R is symmetric, $b \in R_a$ implies $a \in R_b$. Since R is transitive, $b \in R_a$ implies $R_b \subseteq R_a$. It follows that, for all $a, b \in A$, either $R_a = R_b$ or $R_a \cap R_b = \emptyset$.

A *partition* $\mathscr{C}$ of a set A is a collection of nonempty subsets of A such that each element of A is an element of exactly one of the subsets in $\mathscr{C}$.

Thus the distinct equivalence classes corresponding to a given equivalence relation on a set A form a partition of A. It is not difficult to see that, conversely, if $\mathscr{C}$ is a partition of A, then an equivalence relation R is defined on A by taking R to be the set of all $(a, b) \in A \times A$ for which a and b are elements of the same subset in the collection $\mathscr{C}$.

Let A and B be nonempty sets. A *mapping* f of A into B is a subset of $A \times B$ with the property that, for each $a \in A$, there is a unique $b \in B$ such that $(a, b) \in f$. We write $f(a) = b$ if $(a, b) \in f$, and say that b is the *image* of a under f or that b is the *value* of f at a. We express that f is a mapping of A into B by writing $f : A \to B$ and we put

$$f(A) = \{f(a) : a \in A\}.$$

The term *function* is often used instead of 'mapping', especially when A and B are sets of real or complex numbers, and 'mapping' itself is often abbreviated to *map*.

If f is a mapping of A into B, and if A' is a nonempty subset of A, then the *restriction* of f to A' is the set of all $(a, b) \in f$ with $a \in A'$.

The *identity map* i_A of a nonempty set A into itself is the set of all ordered pairs (a, a) with $a \in A$.

If f is a mapping of A into B, and g a mapping of B into C, then the *composite mapping* $g \circ f$ of A into C is the set of all ordered pairs (a, c), where $c = g(b)$ and $b = f(a)$. Composition of mappings is associative, i.e. if h is a mapping of C into D, then

$$(h \circ g) \circ f = h \circ (g \circ f).$$

The identity map has the obvious properties $f \circ i_A = f$ and $i_B \circ f = f$.

Let A, B be nonempty sets and $f : A \to B$ a mapping of A into B. The mapping f is said to be 'one-to-one' or *injective* if, for each $b \in B$, there exists at most one $a \in A$ such that $(a, b) \in f$. The mapping f is said to be 'onto' or *surjective* if, for each $b \in B$, there exists at least one $a \in A$ such that $(a, b) \in f$. If f is both injective and surjective, then it is said to be *bijective* or a 'one-to-one correspondence'. The nouns *injection, surjection* and *bijection* are also used instead of the corresponding adjectives.

It is not difficult to see that f is injective if and only if there exists a mapping $g : B \to A$ such that $g \circ f = i_A$, and surjective if and only if there exists a mapping $h : B \to A$ such that $f \circ h = i_B$. Furthermore, if f is bijective, then g and h are unique and equal. Thus, for any bijective map $f: A \to B$, there is a unique *inverse* map $f^{-1}: B \to A$ such that $f^{-1} \circ f = i_A$ and $f \circ f^{-1} = i_B$.

If $f: A \to B$ and $g: B \to C$ are both bijective maps, then $g \circ f: A \to C$ is also bijective and

$$(g \circ f)^{-1} = f^{-1} \circ g^{-1}.$$

1 Natural Numbers

The natural numbers are the numbers usually denoted by $1, 2, 3, 4, 5, \ldots$. However, other notations are also used, e.g. for the chapters of this book. Although one notation may have considerable practical advantages over another, it is the properties of the natural numbers which are basic.

The following system of axioms for the natural numbers was essentially given by Dedekind (1888), although it is usually attributed to Peano (1889):

The natural numbers are the elements of a set $\mathbb{N}$*, with a distinguished element* 1 (*one*) *and map* $S: \mathbb{N} \to \mathbb{N}$*, such that*

(N1) *S is injective*, i.e. *if* $m, n \in \mathbb{N}$ *and* $m \neq n$*, then* $S(m) \neq S(n)$;
(N2) $1 \notin S(\mathbb{N})$;
(N3) *if* $M \subseteq \mathbb{N}$*,* $1 \in M$ *and* $S(M) \subseteq M$*, then* $M = \mathbb{N}$.

The element $S(n)$ of $\mathbb{N}$ is called the *successor* of n. The axioms are satisfied by $\{1, 2, 3, \ldots\}$ if we take $S(n)$ to be the element immediately following the element n.

It follows readily from the axioms that 1 is the only element of $\mathbb{N}$ which is not in $S(\mathbb{N})$. For, if $M = S(\mathbb{N}) \cup \{1\}$, then $M \subseteq \mathbb{N}$, $1 \in M$ and $S(M) \subseteq M$. Hence, by **(N3)**, $M = \mathbb{N}$.

It also follows from the axioms that $S(n) \neq n$ for every $n \in \mathbb{N}$. For let M be the set of all $n \in \mathbb{N}$ such that $S(n) \neq n$. By **(N2)**, $1 \in M$. If $n \in M$ and $n' = S(n)$ then, by **(N1)**, $S(n') \neq n'$. Thus $S(M) \subseteq M$ and hence, by **(N3)**, $M = \mathbb{N}$.

The axioms **(N1)**–**(N3)** actually determine $\mathbb{N}$ up to 'isomorphism'. We will deduce this as a corollary of the following general *recursion theorem*:

Proposition 1 *Given a set A, an element a_1 of A and a map $T: A \to A$, there exists exactly one map $\varphi: \mathbb{N} \to A$ such that $\varphi(1) = a_1$ and*

$$\varphi(S(n)) = T\varphi(n) \quad \text{for every } n \in \mathbb{N}.$$

Proof We show first that there is at most one map with the required properties. Let φ_1 and φ_2 be two such maps, and let M be the set of all $n \in \mathbb{N}$ such that

$$\varphi_1(n) = \varphi_2(n).$$

Evidently $1 \in M$. If $n \in M$, then also $S(n) \in M$, since

$$\varphi_1(S(n)) = T\varphi_1(n) = T\varphi_2(n) = \varphi_2(S(n)).$$

Hence, by **(N3)**, $M = \mathbb{N}$. That is, $\varphi_1 = \varphi_2$.

We now show that there exists such a map φ. Let $\mathscr{C}$ be the collection of all subsets C of $\mathbb{N} \times A$ such that $(1, a_1) \in C$ and such that if $(n, a) \in C$, then also $(S(n), T(a)) \in C$. The collection $\mathscr{C}$ is not empty, since it contains $\mathbb{N} \times A$. Moreover, since every set in $\mathscr{C}$ contains $(1, a_1)$, the intersection D of all sets $C \in \mathscr{C}$ is not empty. It is easily seen that actually $D \in \mathscr{C}$. By its definition, however, no proper subset of D is in $\mathscr{C}$.

Let M be the set of all $n \in \mathbb{N}$ such that $(n, a) \in D$ for exactly one $a \in A$ and, for any $n \in M$, define $\varphi(n)$ to be the unique $a \in A$ such that $(n, a) \in D$. If $M = \mathbb{N}$, then $\varphi(1) = a_1$ and $\varphi(S(n)) = T\varphi(n)$ for all $n \in \mathbb{N}$. Thus we need only show that $M = \mathbb{N}$. As usual, we do this by showing that $1 \in M$ and that $n \in M$ implies $S(n) \in M$.

We have $(1, a_1) \in D$. Assume $(1, a') \in D$ for some $a' \neq a_1$. If $D' = D \backslash \{(1, a')\}$, then $(1, a_1) \in D'$. Moreover, if $(n, a) \in D'$ then $(S(n), T(a)) \in D'$, since $(S(n), T(a)) \in D$ and $(S(n), T(a)) \neq (1, a')$. Hence $D' \in \mathscr{C}$. But this is a contradiction, since D' is a proper subset of D. We conclude that $1 \in M$.

Suppose now that $n \in M$ and let a be the unique element of A such that $(n, a) \in D$. Then $(S(n), T(a)) \in D$, since $D \in \mathscr{C}$. Assume that $(S(n), a'') \in D$ for some $a'' \neq T(a)$ and put $D'' = D \backslash \{(S(n), a'')\}$. Then $(S(n), T(a)) \in D''$ and $(1, a_1) \in D''$. For any $(m, b) \in D''$ we have $(S(m), T(b)) \in D$. If $(S(m), T(b)) = (S(n), a'')$, then $S(m) = S(n)$ and $T(b) = a'' \neq T(a)$, which implies $m = n$ and $b \neq a$. Thus D contains both (n, b) and (n, a), which contradicts $n \in M$. Hence $(S(m), T(b)) \neq (S(n), a'')$, and so $(S(m), T(b)) \in D''$. But then $D'' \in \mathscr{C}$, which is also a contradiction, since D'' is a proper subset of D. We conclude that $S(n) \in M$. $\square$

Corollary 2 *If the axioms* (N1)–(N3) *are also satisfied by a set* $\mathbb{N}'$ *wth element* $1'$ *and map* $S': \mathbb{N}' \to \mathbb{N}'$, *then there exists a bijective map* φ *of* $\mathbb{N}$ *onto* $\mathbb{N}'$ *such that* $\varphi(1) = 1'$ *and*

$$\varphi(S(n)) = S'\varphi(n) \quad \text{for every } n \in \mathbb{N}.$$

Proof By taking $A = \mathbb{N}'$, $a_1 = 1'$ and $T = S'$ in Proposition 1, we see that there exists a unique map $\varphi: \mathbb{N} \to \mathbb{N}'$ such that $\varphi(1) = 1'$ and

$$\varphi(S(n)) = S'\varphi(n) \quad \text{for every } n \in \mathbb{N}.$$

By interchanging $\mathbb{N}$ and $\mathbb{N}'$, we see also that there exists a unique map $\psi: \mathbb{N}' \to \mathbb{N}$ such that $\psi(1') = 1$ and

$$\psi(S'(n')) = S\psi(n') \quad \text{for every } n' \in \mathbb{N}'.$$

The composite map $\chi = \psi \circ \varphi$ of $\mathbb{N}$ into $\mathbb{N}$ has the properties $\chi(1) = 1$ and $\chi(S(n)) = S\chi(n)$ for every $n \in \mathbb{N}$. But, by Proposition 1 again, χ is uniquely determined by these properties. Hence $\psi \circ \varphi$ is the identity map on $\mathbb{N}$, and similarly $\varphi \circ \psi$ is the identity map on $\mathbb{N}'$. Consequently φ is a bijection. $\square$

We can also use Proposition 1 to define addition and multiplication of natural numbers. By Proposition 1, for each $m \in \mathbb{N}$ there exists a unique map $s_m: \mathbb{N} \to \mathbb{N}$ such that

$$s_m(1) = S(m), \quad s_m(S(n)) = Ss_m(n) \quad \text{for every } n \in \mathbb{N}.$$

We define the *sum* of m and n to be

$$m + n = s_m(n).$$

It is not difficult to deduce from this definition and the axioms **(N1)–(N3)** the usual rules for *addition: for all $a, b, c \in \mathbb{N}$,*

(A1) if $a + c = b + c$, then $a = b$; (cancellation law)
(A2) $a + b = b + a$; (commutative law)
(A3) $(a + b) + c = a + (b + c)$. (associative law)

By way of example, we prove the cancellation law. Let M be the set of all $c \in \mathbb{N}$ such that $a + c = b + c$ only if $a = b$. Then $1 \in M$, since $s_a(1) = s_b(1)$ implies $S(a) = S(b)$ and hence $a = b$. Suppose $c \in M$. If $a + S(c) = b + S(c)$, i.e. $s_a(S(c)) = s_b(S(c))$, then $Ss_a(c) = Ss_b(c)$ and hence, by **(N1)**, $s_a(c) = s_b(c)$. Since $c \in M$, this implies $a = b$. Thus also $S(c) \in M$. Hence, by **(N3)**, $M = \mathbb{N}$.

We now show that

$$m + n \neq n \quad \text{for all } m, n \in \mathbb{N}.$$

For a given $m \in \mathbb{N}$, let M be the set of all $n \in \mathbb{N}$ such that $m + n \neq n$. Then $1 \in M$ since, by **(N2)**, $s_m(1) = S(m) \neq 1$. If $n \in M$, then $s_m(n) \neq n$ and hence, by **(N1)**,

$$s_m(S(n)) = Ss_m(n) \neq S(n).$$

Hence, by **(N3)**, $M = \mathbb{N}$.

By Proposition 1 again, for each $m \in \mathbb{N}$ there exists a unique map $p_m : \mathbb{N} \to \mathbb{N}$ such that

$$p_m(1) = m,$$
$$p_m(S(n)) = s_m(p_m(n)) \quad \text{for every } n \in \mathbb{N}.$$

We define the *product* of m and n to be

$$m \cdot n = p_m(n).$$

From this definition and the axioms **(N1)–(N3)** we may similarly deduce the usual rules for *multiplication: for all $a, b, c \in \mathbb{N}$,*

(M1) if $a \cdot c = b \cdot c$, then $a = b$; (cancellation law)
(M2) $a \cdot b = b \cdot a$; (commutative law)
(M3) $(a \cdot b) \cdot c = a \cdot (b \cdot c)$; (associative law)
(M4) $a \cdot 1 = a$. (identity element)

Furthermore, addition and multiplication are connected by

(AM1) $a \cdot (b + c) = (a \cdot b) + (a \cdot c)$. (distributive law)

As customary, we will often omit the dot when writing products and we will give multiplication precedence over addition. With these conventions the distributive law becomes simply

$$a(b + c) = ab + ac.$$

We show next how a relation of order may be defined on the set $\mathbb{N}$. For any $m, n \in \mathbb{N}$, we say that m *is less than* n, and write $m < n$, if

$$m + m' = n \quad \text{for some } m' \in \mathbb{N}.$$

Evidently $m < S(m)$ for every $m \in \mathbb{N}$, since $S(m) = m + 1$. Also, if $m < n$, then either $S(m) = n$ or $S(m) < n$. For suppose $m + m' = n$. If $m' = 1$, then $S(m) = n$. If $m' \neq 1$, then $m' = m'' + 1$ for some $m'' \in \mathbb{N}$ and

$$S(m) + m'' = (m + 1) + m'' = m + (1 + m'') = m + m' = n.$$

Again, if $n \neq 1$, then $1 < n$, since the set consisting of 1 and all $n \in \mathbb{N}$ such that $1 < n$ contains 1 and contains $S(n)$ if it contains n.

It will now be shown that the relation '$<$' induces a *total order* on $\mathbb{N}$, which is compatible with both addition and multiplication: *for all* $a, b, c \in \mathbb{N}$,

(O1) *if* $a < b$ *and* $b < c$, *then* $a < c$; (transitive law)

(O2) *one and only one of the following alternatives holds*:

$$a < b, a = b, b < a; \quad \text{(law of trichotomy)}$$

(O3) $a + c < b + c$ *if and only if* $a < b$;

(O4) $ac < bc$ *if and only if* $a < b$.

The relation **(O1)** follows directly from the associative law for addition. We now prove **(O2)**. If $a < b$ then, for some $a' \in \mathbb{N}$,

$$b = a + a' = a' + a \neq a.$$

Together with **(O1)**, this shows that at most one of the three alternatives in **(O2)** holds.

For a given $a \in \mathbb{N}$, let M be the set of all $b \in \mathbb{N}$ such that at least one of the three alternatives in **(O2)** holds. Then $1 \in M$, since $1 < a$ if $a \neq 1$. Suppose now that $b \in M$. If $a = b$, then $a < S(b)$. If $a < b$, then again $a < S(b)$, by **(O1)**. If $b < a$, then either $S(b) = a$ or $S(b) < a$. Hence also $S(b) \in M$. Consequently, by **(N3)**, $M = \mathbb{N}$. This completes the proof of **(O2)**.

It follows from the associative and commutative laws for addition that, if $a < b$, then $a + c < b + c$. On the other hand, by using also the cancellation law we see that if $a + c < b + c$, then $a < b$.

It follows from the distributive law that, if $a < b$, then $ac < bc$. Finally, suppose $ac < bc$. Then $a \neq b$ and hence, by **(O2)**, either $a < b$ or $b < a$. Since $b < a$ would imply $bc < ac$, by what we have just proved, we must actually have $a < b$.

The law of trichotomy **(O2)** implies that, for given $m, n \in \mathbb{N}$, the equation

$$m + x = n$$

has a solution $x \in \mathbb{N}$ only if $m < n$.

As customary, we write $a \leq b$ to denote either $a < b$ or $a = b$. Also, it is sometimes convenient to write $b > a$ instead of $a < b$, and $b \geq a$ instead of $a \leq b$.

A subset M of $\mathbb{N}$ is said to have a *least element* m' if $m' \in M$ and $m' \leq m$ for every $m \in M$. The least element m' is uniquely determined, if it exists, by **(O2)**. By what we have already proved, 1 is the least element of $\mathbb{N}$.

Proposition 3 *Any nonempty subset M of* $\mathbb{N}$ *has a least element.*

Proof Assume that some nonempty subset M of $\mathbb{N}$ does not have a least element. Then $1 \notin M$, since 1 is the least element of $\mathbb{N}$. Let L be the set of all $l \in \mathbb{N}$ such that $l < m$ for every $m \in M$. Then L and M are disjoint and $1 \in L$. If $l \in L$, then $S(l) \leq m$ for every $m \in M$. Since M does not have a least element, it follows that $S(l) \notin M$. Thus $S(l) < m$ for every $m \in M$, and so $S(l) \in L$. Hence, by **(N3)**, $L = \mathbb{N}$. Since $L \cap M = \emptyset$, this is a contradiction. $\square$

The method of *proof by induction* is a direct consequence of the axioms defining $\mathbb{N}$. Suppose that with each $n \in \mathbb{N}$ there is associated a proposition P_n. To show that P_n is true for every $n \in \mathbb{N}$, we need only show that P_1 is true and that P_{n+1} is true if P_n is true.

Proposition 3 provides an alternative approach. To show that P_n is true for every $n \in \mathbb{N}$, we need only show that if P_m is false for some m, then P_l is false for some $l < m$. For then the set of all $n \in \mathbb{N}$ for which P_n is false has no least element and consequently is empty.

For any $n \in \mathbb{N}$, we denote by I_n the set of all $m \in \mathbb{N}$ such that $m \leq n$. Thus $I_1 = \{1\}$ and $S(n) \notin I_n$. It is easily seen that

$$I_{S(n)} = I_n \cup \{S(n)\}.$$

Also, for any $p \in I_{S(n)}$, there exists a bijective map f_p of I_n onto $I_{S(n)} \backslash \{p\}$. For, if $p = S(n)$ we can take f_p to be the identity map on I_n, and if $p \in I_n$ we can take f_p to be the map defined by

$$f_p(p) = S(n), \quad f_p(m) = m \quad \text{if } m \in I_n \backslash \{p\}.$$

Proposition 4 *For any* $m, n \in \mathbb{N}$, *if a map* $f : I_m \to I_n$ *is injective and* $f(I_m) \neq I_n$, *then* $m < n$.

Proof The result certainly holds when $m = 1$, since $I_1 = \{1\}$. Let M be the set of all $m \in \mathbb{N}$ for which the result holds. We need only show that if $m \in M$, then also $S(m) \in M$.

Let $f : I_{S(m)} \to I_n$ be an injective map such that $f(I_{S(m)}) \neq I_n$ and choose $p \in I_n \backslash f(I_{S(m)})$. The restriction g of f to I_m is also injective and $g(I_m) \neq I_n$. Since $m \in M$, it follows that $m < n$. Assume $S(m) = n$. Then there exists a bijective map g_p of $I_{S(m)} \backslash \{p\}$ onto I_m. The composite map $h = g_p \circ f$ maps $I_{S(m)}$ into I_m and is injective. Since $m \in M$, we must have $h(I_m) = I_m$. But, since $h(S(m)) \in I_m$ and h is injective, this is a contradiction. Hence $S(m) < n$ and, since this holds for every f, $S(m) \in M$. $\square$

Proposition 5 *For any* $m, n \in \mathbb{N}$, *if a map* $f : I_m \to I_n$ *is not injective and* $f(I_m) = I_n$, *then* $m > n$.

Proof The result holds vacuously when $m = 1$, since any map $f : I_1 \to I_n$ is injective. Let M be the set of all $m \in \mathbb{N}$ for which the result holds. We need only show that if $m \in M$, then also $S(m) \in M$.

Let $f: I_{S(m)} \to I_n$ be a map such that $f(I_{S(m)}) = I_n$ which is not injective. Then there exist $p, q \in I_{S(m)}$ with $p \neq q$ and $f(p) = f(q)$. We may choose the notation so that $q \in I_m$. If f_p is a bijective map of I_m onto $I_{S(m)} \setminus \{p\}$, then the composite map $h = f \circ f_p$ maps I_m onto I_n. If it is not injective then $m > n$, since $m \in M$, and hence also $S(m) > n$. If h is injective, then it is bijective and has a bijective inverse $h^{-1}: I_n \to I_m$. Since $h^{-1}(I_n)$ is a proper subset of $I_{S(m)}$, it follows from Proposition 4 that $n < S(m)$. Hence $S(m) \in M$. $\square$

Propositions 4 and 5 immediately imply

Corollary 6 *For any $n \in \mathbb{N}$, a map $f: I_n \to I_n$ is injective if and only if it is surjective.*

Corollary 7 *If a map $f: I_m \to I_n$ is bijective, then $m = n$.*

Proof By Proposition 4, $m < S(n)$, i.e. $m \leq n$. Replacing f by f^{-1}, we obtain in the same way $n \leq m$. Hence $m = n$. $\square$

A set E is said to be *finite* if there exists a bijective map $f: E \to I_n$ for some $n \in \mathbb{N}$. Then n is uniquely determined, by Corollary 7. We call it the *cardinality* of E and denote it by $\#(E)$.

It is readily shown that if E is a finite set and F a proper subset of E, then F is also finite and $\#(F) < \#(E)$. Again, if E and F are disjoint finite sets, then their union $E \cup F$ is also finite and $\#(E \cup F) = \#(E) + \#(F)$. Furthermore, for any finite sets E and F, the product set $E \times F$ is also finite and $\#(E \times F) = \#(E) \cdot \#(F)$.

Corollary 6 implies that, for any finite set E, a map $f: E \to E$ is injective if and only if it is surjective. This is a precise statement of the so-called *pigeonhole principle*.

A set E is said to be *countably infinite* if there exists a bijective map $f: E \to \mathbb{N}$. Any countably infinite set may be bijectively mapped onto a proper subset F, since $\mathbb{N}$ is bijectively mapped onto a proper subset by the successor map S. Thus a map $f: E \to E$ of an infinite set E may be injective, but not surjective. It may also be surjective, but not injective; an example is the map $f: \mathbb{N} \to \mathbb{N}$ defined by $f(1) = 1$ and, for $n \neq 1$, $f(n) = m$ if $S(m) = n$.

2 Integers and Rational Numbers

The concept of number will now be extended. The natural numbers $1, 2, 3, \ldots$ suffice for counting purposes, but for bank balance purposes we require the larger set $\ldots, -2, -1, 0, 1, 2, \ldots$ of integers. (From this point of view, -2 is not so 'unnatural'.) An important reason for extending the concept of number is the greater freedom it gives us. In the realm of natural numbers the equation $a + x = b$ has a solution if and only if $b > a$; in the extended realm of integers it will always have a solution.

Rather than introduce a new set of axioms for the integers, we will define them in terms of natural numbers. Intuitively, an integer is the difference $m - n$ of two natural numbers m, n, with addition and multiplication defined by

$$(m - n) + (p - q) = (m + p) - (n + q),$$
$$(m - n) \cdot (p - q) = (mp + nq) - (mq + np).$$

However, two other natural numbers m', n' may have the same difference as m, n, and anyway what does $m - n$ mean if $m < n$? To make things precise, we proceed in the following way.

Consider the set $\mathbb{N} \times \mathbb{N}$ of all ordered pairs of natural numbers. For any two such ordered pairs, (m, n) and (m', n'), we write

$$(m, n) \sim (m', n') \quad \text{if } m + n' = m' + n.$$

We will show that this is an *equivalence relation*. It follows at once from the definition that $(m, n) \sim (m, n)$ (reflexive law) and that $(m, n) \sim (m', n')$ implies $(m', n') \sim (m, n)$ (symmetric law). It remains to prove the transitive law:

$$(m, n) \sim (m', n') \text{ and } (m', n') \sim (m'', n'') \text{ imply } (m, n) \sim (m'', n'').$$

This follows from the commutative, associative and cancellation laws for addition in $\mathbb{N}$. For we have

$$m + n' = m' + n, \quad m' + n'' = m'' + n',$$

and hence

$$(m + n') + n'' = (m' + n) + n'' = (m' + n'') + n = (m'' + n') + n.$$

Thus

$$(m + n'') + n' = (m'' + n) + n',$$

and so $m + n'' = m'' + n$.

The equivalence class containing $(1, 1)$ evidently consists of all pairs (m, n) with $m = n$.

We define an *integer* to be an equivalence class of ordered pairs of natural numbers and, as is now customary, we denote the set of all integers by $\mathbb{Z}$.

Addition of integers is defined componentwise:

$$(m, n) + (p, q) = (m + p, n + q).$$

To justify this definition we must show that it does not depend on the choice of representatives within an equivalence class, i.e. that

$$(m, n) \sim (m', n') \text{ and } (p, q) \sim (p', q') \text{ imply } (m + p, n + q) \sim (m' + p', n' + q').$$

However, if

$$m + n' = m' + n, \quad p + q' = p' + q,$$

then

$$(m + p) + (n' + q') = (m + n') + (p + q')$$
$$= (m' + n) + (p' + q) = (m' + p') + (n + q).$$

It follows at once from the corresponding properties of natural numbers that, also in $\mathbb{Z}$, addition satisfies the commutative law **(A2)** and the associative law **(A3)**. Moreover, the equivalence class 0 (*zero*) containing $(1,1)$ is an *identity element* for addition:

(A4) $a + 0 = a$ *for every* a.

Furthermore, the equivalence class containing (n, m) is an *additive inverse* for the equivalence containing (m, n):

(A5) *for each* a, *there exists* $-a$ *such that* $a + (-a) = 0$.

From these properties we can now obtain

Proposition 8 *For all* $a, b \in \mathbb{Z}$, *the equation* $a + x = b$ *has a unique solution* $x \in \mathbb{Z}$.

Proof It is clear that $x = (-a) + b$ is a solution. Moreover, this solution is unique, since if $a + x = a + x'$ then, by adding $-a$ to both sides, we obtain $x = x'$. $\square$

Proposition 8 shows that the cancellation law **(A1)** is a consequence of **(A2)**–**(A5)**. It also immediately implies

Corollary 9 *For each* $a \in \mathbb{Z}$, 0 *is the only element such that* $a + 0 = a$, $-a$ *is uniquely determined by* a, *and* $a = -(-a)$.

As usual, we will henceforth write $b - a$ instead of $b + (-a)$.
Multiplication of integers is defined by

$$(m, n) \cdot (p, q) = (mp + nq, mq + np).$$

To justify this definition we must show that $(m, n) \sim (m', n')$ and $(p, q) \sim (p', q')$ imply

$$(mp + nq, mq + np) \sim (m'p' + n'q', m'q' + n'p').$$

From $m + n' = m' + n$, by multiplying by p and q we obtain

$$mp + n'p = m'p + np,$$
$$m'q + nq = mq + n'q,$$

and from $p + q' = p' + q$, by multiplying by m' and n' we obtain

$$m'p + m'q' = m'p' + m'q,$$
$$n'p' + n'q = n'p + n'q'.$$

Adding these four equations and cancelling the terms common to both sides, we get

$$(mp + nq) + (m'q' + n'p') = (m'p' + n'q') + (mq + np),$$

as required.

It is easily verified that, also in $\mathbb{Z}$, multiplication satisfies the commutative law **(M2)** and the associative law **(M3)**. Moreover, the distributive law **(AM1)** holds and, if 1 is the equivalence class containing $(1 + 1, 1)$, then **(M4)** also holds. (In practice it does not cause confusion to denote identity elements of $\mathbb{N}$ and $\mathbb{Z}$ by the same symbol.)

Proposition 10 *For every* $a \in \mathbb{Z}$, $a \cdot 0 = 0$.

Proof We have

$$a \cdot 0 = a \cdot (0 + 0) = a \cdot 0 + a \cdot 0.$$

Adding $-(a \cdot 0)$ to both sides, we obtain the result. □

Proposition 10 could also have been derived directly from the definitions, but we prefer to view it as a consequence of the properties which have been labelled.

Corollary 11 *For all* $a, b \in \mathbb{Z}$,

$$a(-b) = -(ab), \ (-a)(-b) = ab.$$

Proof The first relation follows from

$$ab + a(-b) = a \cdot 0 = 0,$$

and the second relation follows from the first, since $c = -(-c)$. □

By the definitions of 0 and 1 we also have

(AM2) $1 \neq 0$.

(In fact $1 = 0$ would imply $a = 0$ for every a, since $a \cdot 1 = a$ and $a \cdot 0 = 0$.)

We will say that an integer a is *positive* if it is represented by an ordered pair (m, n) with $n < m$. This definition does not depend on the choice of representative. For if $n < m$ and $m + n' = m' + n$, then $m + n' < m' + m$ and hence $n' < m'$.

We will denote by P the set of all positive integers. The law of trichotomy **(O2)** for natural numbers immediately implies

(P1) *for every* a, *one and only one of the following alternatives holds*:

$$a \in P, \quad a = 0, \quad -a \in P.$$

We say that an integer is *negative* if it has the form $-a$, where $a \in P$, and we denote by $-P$ the set of all negative integers. Since $a = -(-a)$, **(P1)** says that $\mathbb{Z}$ is the disjoint union of the sets P, $\{0\}$ and $-P$.

From the property **(O3)** of natural numbers we immediately obtain

(P2) *if* $a \in P$ *and* $b \in P$, *then* $a + b \in P$.

Furthermore, we have

(P3) *if* $a \in P$ *and* $b \in P$, *then* $a \cdot b \in P$.

To prove this we need only show that if m, n, p, q are natural numbers such that $n < m$ and $q < p$, then

$$mq + np < mp + nq.$$

Since $q < p$, there exists a natural number q' such that $q + q' = p$. But then $nq' < mq'$, since $n < m$, and hence

$$mq + np = (m + n)q + nq' < (m + n)q + mq' = mp + nq.$$

We may write **(P2)** and **(P3)** symbolically in the form

$$P + P \subseteq P, \quad P \cdot P \subseteq P.$$

We now show that there are no *divisors of zero* in $\mathbb{Z}$:

Proposition 12 *If $a \neq 0$ and $b \neq 0$, then $ab \neq 0$.*

Proof By **(P1)**, either a or $-a$ is positive, and either b or $-b$ is positive. If $a \in P$ and $b \in P$ then $ab \in P$, by **(P3)**, and hence $ab \neq 0$, by **(P1)**. If $a \in P$ and $-b \in P$, then $a(-b) \in P$. Hence $ab = -(a(-b)) \in -P$ and $ab \neq 0$. Similarly if $-a \in P$ and $b \in P$. Finally, if $-a \in P$ and $-b \in P$, then $ab = (-a)(-b) \in P$ and again $ab \neq 0$. $\qquad\qquad\square$

The proof of Proposition 12 also shows that any nonzero square is positive:

Proposition 13 *If $a \neq 0$, then $a^2 := aa \in P$.*

It follows that $1 \in P$, since $1 \neq 0$ and $1^2 = 1$.

The set P of positive integers induces an order relation in $\mathbb{Z}$. Write

$$a < b \quad \text{if } b - a \in P,$$

so that $a \in P$ if and only if $0 < a$. From this definition and the properties of P it follows that the order properties **(O1)**–**(O3)** hold also in $\mathbb{Z}$, and that **(O4)** holds in the modified form:

(O4)′ *if $0 < c$, then $ac < bc$ if and only if $a < b$.*

We now show that we can represent any $a \in \mathbb{Z}$ in the form $a = b - c$, where $b, c \in P$. In fact, if $a = 0$, we can take $b = 1$ and $c = 1$; if $a \in P$, we can take $b = a + 1$ and $c = 1$; and if $-a \in P$, we can take $b = 1$ and $c = 1 - a$.

An element a of $\mathbb{Z}$ is said to be a *lower bound* for a subset X of $\mathbb{Z}$ if $a \leq x$ for every $x \in X$. Proposition 3 immediately implies that if a subset of $\mathbb{Z}$ has a lower bound, then it has a least element.

For any $n \in \mathbb{N}$, let n' be the integer represented by $(n + 1, 1)$. Then $n' \in P$. We are going to study the map $n \to n'$ of $\mathbb{N}$ into P. The map is injective, since $n' = m'$ implies $n = m$. It is also surjective, since if $a \in P$ is represented by (m, n), where $n < m$, then it is also represented by $(p + 1, 1)$, where $p \in \mathbb{N}$ satisfies $n + p = m$. It is easily verified that the map preserves sums and products:

$$(m + n)' = m' + n', \quad (mn)' = m'n'.$$

Since $1' = 1$, it follows that $S(n)' = n' + 1$. Furthermore, we have

$$m' < n' \quad \text{if and only if } m < n.$$

Thus the map $n \to n'$ establishes an 'isomorphism' of $\mathbb{N}$ with P. In other words, P is a copy of $\mathbb{N}$ situated within $\mathbb{Z}$. By identifying n with n', we may regard $\mathbb{N}$ itself as a subset of $\mathbb{Z}$ (and stop talking about P). Then 'natural number' is the same as 'positive integer' and any integer is the difference of two natural numbers.

Number theory, in its most basic form, is the study of the properties of the set $\mathbb{Z}$ of integers. It will be considered in some detail in later chapters of this book, but to relieve the abstraction of the preceding discussion we consider here the *division algorithm*:

Proposition 14 *For any integers a, b with $a > 0$, there exist unique integers q, r such that*

$$b = qa + r, \quad 0 \le r < a.$$

Proof We consider first uniqueness. Suppose

$$qa + r = q'a + r', \quad 0 \le r, r' < a.$$

If $r < r'$, then from

$$(q - q')a = r' - r,$$

we obtain first $q > q'$ and then $r' - r \ge a$, which is a contradiction. If $r' < r$, we obtain a contradiction similarly. Hence $r = r'$, which implies $q = q'$.

We consider next existence. Let S be the set of all integers $y \ge 0$ which can be represented in the form $y = b - xa$ for some $x \in \mathbb{Z}$. The set S is not empty, since it contains $b - 0$ if $b \ge 0$ and $b - ba$ if $b < 0$. Hence S contains a least element r. Then $b = qa + r$, where $q, r \in \mathbb{Z}$ and $r \ge 0$. Since $r - a = b - (q + 1)a$ and r is the least element in S, we must also have $r < a$. ∎

The concept of number will now be further extended to include 'fractions' or 'rational numbers'. For measuring lengths the integers do not suffice, since the length of a given segment may not be an exact multiple of the chosen unit of length. Similarly for measuring weights, if we find that three identical coins balance five of the chosen unit weights, then we ascribe to each coin the weight 5/3. In the realm of integers the equation $ax = b$ frequently has no solution; in the extended realm of rational numbers it will always have a solution if $a \ne 0$.

Intuitively, a rational number is the ratio or 'quotient' a/b of two integers a, b, where $b \ne 0$, with addition and multiplication defined by

$$a/b + c/d = (ad + cb)/bd,$$
$$a/b \cdot c/d = ac/bd.$$

However, two other integers a', b' may have the same ratio as a, b, and anyway what does a/b mean? To make things precise, we proceed in much the same way as before.

Put $\mathbb{Z}^\times = \mathbb{Z} \backslash \{0\}$ and consider the set $\mathbb{Z} \times \mathbb{Z}^\times$ of all ordered pairs (a, b) with $a \in \mathbb{Z}$ and $b \in \mathbb{Z}^\times$. For any two such ordered pairs, (a, b) and (a', b'), we write

$$(a, b) \sim (a', b') \quad \text{if } ab' = a'b.$$

To show that this is an equivalence relation it is again enough to verify that $(a, b) \sim (a', b')$ and $(a', b') \sim (a'', b'')$ imply $(a, b) \sim (a'', b'')$. The same calculation as before, with addition replaced by multiplication, shows that $(ab'')b' = (a''b)b'$. Since $b' \ne 0$, it follows that $ab'' = a''b$.

The equivalence class containing $(0, 1)$ evidently consists of all pairs $(0, b)$ with $b \ne 0$, and the equivalence class containing $(1, 1)$ consists of all pairs (b, b) with $b \ne 0$.

We define a *rational number* to be an equivalence class of elements of $\mathbb{Z} \times \mathbb{Z}^\times$ and, as is now customary, we denote the set of all rational numbers by $\mathbb{Q}$.

Addition of rational numbers is defined by

$$(a, b) + (c, d) = (ad + cb, bd),$$

where $bd \neq 0$ since $b \neq 0$ and $d \neq 0$. To justify the definition we must show that

$$(a, b) \sim (a', b') \text{ and } (c, d) \sim (c', d') \text{ imply } (ad + cb, bd) \sim (a'd' + c'b', b'd').$$

But if $ab' = a'b$ and $cd' = c'd$, then

$$(ad + cb)(b'd') = (ab')(dd') + (cd')(bb')$$
$$= (a'b)(dd') + (c'd)(bb') = (a'd' + c'b')(bd).$$

It is easily verified that, also in $\mathbb{Q}$, addition satisfies the commutative law **(A2)** and the associative law **(A3)**. Moreover **(A4)** and **(A5)** also hold, the equivalence class 0 containing $(0, 1)$ being an identity element for addition and the equivalence class containing $(-b, c)$ being the additive inverse of the equivalence class containing (b, c).

Multiplication of rational numbers is defined componentwise:

$$(a, b) \cdot (c, d) = (ac, bd).$$

To justify the definition we must show that

$$(a, b) \sim (a', b') \text{ and } (c, d) \sim (c', d') \text{ imply } (ac, bd) \sim (a'c', b'd').$$

But if $ab' = a'b$ and $cd' = c'd$, then

$$(ac)(b'd') = (ab')(cd') = (a'b)(c'd) = (a'c')(bd).$$

It is easily verified that, also in $\mathbb{Q}$, multiplication satisfies the commutative law **(M2)** and the associative law **(M3)**. Moreover **(M4)** also holds, the equivalence class 1 containing $(1, 1)$ being an identity element for multiplication. Furthermore, addition and multiplication are connected by the distributive law **(AM1)**, and **(AM2)** also holds since $(0, 1)$ is not equivalent to $(1, 1)$.

Unlike the situation for $\mathbb{Z}$, however, every nonzero element of $\mathbb{Q}$ has a *multiplicative inverse*:

(M5) *for each $a \neq 0$, there exists a^{-1} such that $aa^{-1} = 1$.*

In fact, if a is represented by (b, c), then a^{-1} is represented by (c, b).

It follows that, for all $a, b \in \mathbb{Q}$ with $a \neq 0$, the equation $ax = b$ has a unique solution $x \in \mathbb{Q}$, namely $x = a^{-1}b$. Hence, if $a \neq 0$, then 1 is the only solution of $ax = a$, a^{-1} is uniquely determined by a, and $a = (a^{-1})^{-1}$.

We will say that a rational number a is *positive* if it is represented by an ordered pair (b, c) of integers for which $bc > 0$. This definition does not depend on the choice of representative. For suppose $0 < bc$ and $bc' = b'c$. Then $bc' \neq 0$, since $b \neq 0$ and $c' \neq 0$, and hence $0 < (bc')^2$. Since $(bc')^2 = (bc)(b'c')$ and $0 < bc$, it follows that $0 < b'c'$.

Our previous use of P having been abandoned in favour of $\mathbb{N}$, we will now denote by P the set of all positive rational numbers and by $-P$ the set of all rational numbers

$-a$, where $a \in P$. From the corresponding result for $\mathbb{Z}$, it follows that **(P1)** continues to hold in $\mathbb{Q}$. We will show that **(P2)** and **(P3)** also hold.

To see that the sum of two positive rational numbers is again positive, we observe that if a, b, c, d are integers such that $0 < ab$ and $0 < cd$, then also

$$0 < (ab)d^2 + (cd)b^2 = (ad + cb)(bd).$$

To see that the product of two positive rational numbers is again positive, we observe that if a, b, c, d are integers such that $0 < ab$ and $0 < cd$, then also

$$0 < (ab)(cd) = (ac)(bd).$$

Since **(P1)–(P3)** all hold, it follows as before that Propositions 12 and 13 also hold in $\mathbb{Q}$. Hence $1 \in P$ and **(O4)$'$** now implies that $a^{-1} \in P$ if $a \in P$. If $a, b \in P$ and $a < b$, then $b^{-1} < a^{-1}$, since $bb^{-1} = 1 = aa^{-1} < ba^{-1}$.

The set P of positive elements now induces an order relation on $\mathbb{Q}$. We write $a < b$ if $b - a \in P$, so that $a \in P$ if and only if $0 < a$. Then the order relations **(O1)–(O3)** and **(O4)$'$** continue to hold in $\mathbb{Q}$.

Unlike the situation for $\mathbb{Z}$, however, the ordering of $\mathbb{Q}$ is *dense*, i.e. if $a, b \in \mathbb{Q}$ and $a < b$, then there exists $c \in \mathbb{Q}$ such that $a < c < b$. For example, we can take c to be the solution of $(1 + 1)c = a + b$.

Let $\mathbb{Z}'$ denote the set of all rational numbers a' which can be represented by $(a, 1)$ for some $a \in \mathbb{Z}$. For every $c \in \mathbb{Q}$, there exist $a', b' \in \mathbb{Z}'$ with $b' \neq 0$ such that $c = a'b'^{-1}$. In fact, if c is represented by (a, b), we can take a' to be represented by $(a, 1)$ and b' by $(b, 1)$. Instead of $c = a'b'^{-1}$, we also write $c = a'/b'$.

For any $a \in \mathbb{Z}$, let a' be the rational number represented by $(a, 1)$. The map $a \to a'$ of $\mathbb{Z}$ into $\mathbb{Z}'$ is clearly bijective. Moreover, it preserves sums and products:

$$(a + b)' = a' + b', \quad (ab)' = a'b'.$$

Furthermore,

$$a' < b' \quad \text{if and only if } a < b.$$

Thus the map $a \to a'$ establishes an 'isomorphism' of $\mathbb{Z}$ with $\mathbb{Z}'$, and $\mathbb{Z}'$ is a copy of $\mathbb{Z}$ situated within $\mathbb{Q}$. By identifying a with a', we may regard $\mathbb{Z}$ itself as a subset of $\mathbb{Q}$. Then any rational number is the ratio of two integers.

By way of illustration, we show that if a and b are positive rational numbers, then there exists a positive integer l such that $la > b$. For if $a = m/n$ and $b = p/q$, where m, n, p, q are positive integers, then

$$(np + 1)a > pm \geq p \geq b.$$

3 Real Numbers

It was discovered by the ancient Greeks that even rational numbers do not suffice for the measurement of lengths. If x is the length of the hypotenuse of a right-angled triangle whose other two sides have unit length then, by Pythagoras' theorem, $x^2 = 2$.

But it was proved, probably by a disciple of Pythagoras, that there is no rational number x such that $x^2 = 2$. (A more general result is proved in Book X, Proposition 9 of Euclid's *Elements*.) We give here a somewhat different proof from the classical one.

Assume that such a rational number x exists. Since x may be replaced by $-x$, we may suppose that $x = m/n$, where $m, n \in \mathbb{N}$. Then $m^2 = 2n^2$. Among all pairs m, n of positive integers with this property, there exists one for which n is least. If we put

$$p = 2n - m, \quad q = m - n,$$

then p and q are positive integers, since clearly $n < m < 2n$. But

$$p^2 = 4n^2 - 4mn + m^2 = 2(m^2 - 2mn + n^2) = 2q^2.$$

Since $q < n$, this contradicts the minimality of n.

If we think of the rational numbers as measuring distances of points on a line from a given origin O on the line (with distances on one side of O positive and distances on the other side negative), this means that, even though a dense set of points is obtained in this way, not all points of the line are accounted for. In order to fill in the gaps the concept of number will now be extended from 'rational number' to 'real number'.

It is possible to define real numbers as infinite decimal expansions, the rational numbers being those whose decimal expansions are eventually periodic. However, the choice of base 10 is arbitrary and carrying through this approach is awkward.

There are two other commonly used approaches, one based on *order* and the other on *distance*. The first was proposed by Dedekind (1872), the second by Méray (1869) and Cantor (1872). We will follow Dedekind's approach, since it is conceptually simpler. However, the second method is also important and in a sense more general. In Chapter VI we will use it to extend the rational numbers to the *p-adic numbers*.

It is convenient to carry out Dedekind's construction in two stages. We will first define 'cuts' (which are just the positive real numbers), and then pass from cuts to arbitrary real numbers in the same way that we passed from the natural numbers to the integers.

Intuitively, a cut is the set of all rational numbers which represent points of the line between the origin O and some other point. More formally, we define a *cut* to be a nonempty proper subset A of the set P of all positive rational numbers such that

(i) *if $a \in A, b \in P$ and $b < a$, then $b \in A$;*
(ii) *if $a \in A$, then there exists $a' \in A$ such that $a < a'$.*

For example, the set I of all positive rational numbers $a < 1$ is a cut. Similarly, the set T of all positive rational numbers a such that $a^2 < 2$ is a cut. We will denote the set of all cuts by $\mathscr{P}$.

For any $A, B \in \mathscr{P}$ we write $A < B$ if A is a proper subset of B. We will show that this induces a *total order* on $\mathscr{P}$.

It is clear that if $A < B$ and $B < C$, then $A < C$. It remains to show that, for any $A, B \in \mathscr{P}$, one and only one of the following alternatives holds:

$$A < B, \quad A = B, \quad B < A.$$

It is obvious from the definition by set inclusion that at most one holds. Now suppose that neither $A < B$ nor $A = B$. Then there exists $a \in A \backslash B$. It follows from (i), applied to B, that every $b \in B$ satisfies $b < a$ and then from (i), applied to A, that $b \in A$. Thus $B < A$.

Let $\mathscr{S}$ be any nonempty collection of cuts. A cut B is said to be an *upper bound* for $\mathscr{S}$ if $A \leq B$ for every $A \in \mathscr{S}$, and a *lower bound* for $\mathscr{S}$ if $B \leq A$ for every $A \in \mathscr{S}$. An upper bound for $\mathscr{S}$ is said to be a *least upper bound* or *supremum* for $\mathscr{S}$ if it is a lower bound for the collection of all upper bounds. Similarly, a lower bound for $\mathscr{S}$ is said to be a *greatest lower bound* or *infimum* for $\mathscr{S}$ if it is an upper bound for the collection of all lower bounds. Clearly, $\mathscr{S}$ has at most one supremum and at most one infimum.

The set $\mathscr{P}$ has the following basic property:

(P4) *if a nonempty subset $\mathscr{S}$ has an upper bound, then it has a least upper bound.*

Proof Let C be the union of all sets $A \in \mathscr{S}$. By hypothesis there exists a cut B such that $A \subseteq B$ for every $A \in \mathscr{S}$. Since $C \subseteq B$ for any such B, and $A \subseteq C$ for every $A \in \mathscr{S}$, we need only show that C is a cut.

Evidently C is a nonempty proper subset of P, since $B \neq P$. Suppose $c \in C$. Then $c \in A$ for some $A \in \mathscr{S}$. If $d \in P$ and $d < c$, then $d \in A$, since A is a cut. Furthermore $c < a'$ for some $a' \in A$. Since $A \subseteq C$, this proves that C is a cut. $\qquad\square$

In the set P of positive rational numbers, the subset T of all $x \in P$ such that $x^2 < 2$ has an upper bound, but no least upper bound. Thus **(P4)** shows that there is a difference between the total order on P and that on $\mathscr{P}$.

We now define addition of cuts. For any $A, B \in \mathscr{P}$, let $A + B$ denote the set of all rational numbers $a + b$, with $a \in A$ and $b \in B$. We will show that also $A + B \in \mathscr{P}$. Evidently $A + B$ is a nonempty subset of P. It is also a proper subset. For choose $c \in P \backslash A$ and $d \in P \backslash B$. Then, by (i), $a < c$ for all $a \in A$ and $b < d$ for all $b \in B$. Since $a + b < c + d$ for all $a \in A$ and $b \in B$, it follows that $c + d \notin A + B$.

Suppose now that $a \in A$, $b \in B$ and that $c \in P$ satisfies $c < a + b$. If $c > b$, then $c = b + d$ for some $d \in P$, and $d < a$. Hence, by (i), $d \in A$ and $c = d + b \in A + B$. Similarly, $c \in A + B$ if $c > a$. Finally, if $c \leq a$ and $c \leq b$, choose $e \in P$ so that $e < c$. Then $e \in A$ and $c = e + f$ for some $f \in P$. Then $f \in B$, since $f < c$, and $c = e + f \in A + B$.

Thus $A + B$ has the property (i). It is trivial that $A + B$ also has the property (ii), since if $a \in A$ and $b \in B$, there exists $a' \in A$ such that $a < a'$ and then $a + b < a' + b$. This completes the proof that $A + B$ is a cut.

It follows at once from the corresponding properties of rational numbers that addition of cuts satisfies the commutative law **(A2)** and the associative law **(A3)**.

We consider next the connection between addition and order.

Lemma 15 *For any cut A and any $c \in P$, there exists $a \in A$ such that $a + c \notin A$.*

Proof If $c \notin A$, then $a + c \notin A$ for every $a \in A$, since $c < a + c$. Thus we may suppose $c \in A$. Choose $b \in P \backslash A$. For some positive integer n we have $b < nc$ and hence $nc \notin A$. If n is the least positive integer such that $nc \notin A$, then $n > 1$ and $(n - 1)c \in A$. Consequently we can take $a = (n - 1)c$. $\qquad\square$

Proposition 16 *For any cuts A, B, there exists a cut C such that $A + C = B$ if and only if $A < B$.*

Proof We prove the necessity of the condition by showing that $A < A + C$ for any cuts A, C. If $a \in A$ and $c \in C$, then $a < a + c$. Since $A + C$ is a cut, it follows that $a \in A + C$. Consequently $A \leq A + C$, and Lemma 15 implies that $A \neq A + C$.

Suppose now that A and B are cuts such that $A < B$, and let C be the set of all $c \in P$ such that $c + d \in B$ for some $d \in P \backslash A$. We are going to show that C is a cut and that $A + C = B$.

The set C is not empty. For choose $b \in B \backslash A$ and then $b' \in B$ with $b < b'$. Then $b' = b + c'$ for some $c' \in P$, which implies $c' \in C$. On the other hand, $C \leq B$, since $c + d \in B$ and $d \in P$ imply $c \in B$. Thus C is a proper subset of P.

Suppose $c \in C$, $p \in P$ and $p < c$. We have $c + d \in B$ for some $d \in P \backslash A$ and $c = p + e$ for some $e \in P$. Since $d + e \in P \backslash A$ and $p + (d + e) = c + d \in B$, it follows that $p \in C$.

Suppose now that $c \in C$, so that $c + d \in B$ for some $d \in P \backslash A$. Choose $b \in B$ so that $c + d < b$. Then $b = c + d + e$ for some $e \in P$. If we put $c' = c + e$, then $c < c'$. Moreover $c' \in C$, since $c' + d = b$. This completes the proof that C is a cut.

Suppose $a \in A$ and $c \in C$. Then $c + d \in B$ for some $d \in P \backslash A$. Hence $a < d$. It follows that $a + c < c + d$, and so $a + c \in B$. Thus $A + C \leq B$.

It remains to show that $B \leq A + C$. Pick any $b \in B$. If $b \in A$, then also $b \in A + C$, since $A < A + C$. Thus we now assume $b \notin A$. Choose $b' \in B$ with $b < b'$. Then $b' = b + d$ for some $d \in P$. By Lemma 15, there exists $a \in A$ such that $a + d \notin A$. Moreover $a < b$, since $b \notin A$, and hence $b = a + c$ for some $c \in P$. Since $c + (a + d) = b + d = b'$, it follows that $c \in C$. Thus $b \in A + C$ and $B \leq A + C$. $\square$

We can now show that addition of cuts satisfies the order relation (**O3**). Suppose first that $A < B$. Then, by Proposition 16, there exists a cut D such that $A + D = B$. Hence, for any cut C,

$$A + C < (A + C) + D = B + C.$$

Suppose next that $A + C < B + C$. Then $A \neq B$. Since $B < A$ would imply $B + C < A + C$, by what we have just proved, it follows from the law of trichotomy that $A < B$.

From (**O3**) and the law of trichotomy, it follows that addition of cuts satisfies the cancellation law (**A1**).

We next define multiplication of cuts. For any A, $B \in \mathscr{P}$, let AB denote the set of all rational numbers ab, with $a \in A$ and $b \in B$. In the same way as for $A + B$, it may be shown that $AB \in \mathscr{P}$. We note only that if $a \in A$, $b \in B$ and $c < ab$, then $b^{-1}c < a$. Hence $b^{-1}c \in A$ and $c = (b^{-1}c)b \in AB$.

It follows from the corresponding properties of rational numbers that multiplication of cuts satisfies the commutative law (**M2**) and the associative law (**M3**). Moreover (**M4**) holds, the identity element for multiplication being the cut I consisting of all positive rational numbers less than 1.

We now show that the distributive law (**AM1**) also holds. The distributive law for rational numbers shows at once that

$$A(B + C) \leq AB + AC.$$

It remains to show that $a_1b + a_2c \in A(B + C)$ if $a_1, a_2 \in A$, $b \in B$ and $c \in C$. But

$$a_1b + a_2c \leq a_2(b + c) \quad \text{if } a_1 \leq a_2,$$

and

$$a_1b + a_2c \leq a_1(b + c) \quad \text{if } a_2 \leq a_1.$$

In either event it follows that $a_1b + a_2c \in A(B + C)$.

We can now show that multiplication of cuts satisfies the order relation **(O4)**. If $A < B$, then there exists a cut D such that $A + D = B$ and hence $AC < AC + DC = BC$. Conversely, suppose $AC < BC$. Then $A \neq B$. Since $B < A$ would imply $BC < AC$, it follows that $A < B$.

From **(O4)** and the law of trichotomy **(O2)** it follows that multiplication of cuts satisfies the cancellation law **(M1)**.

We next prove the existence of multiplicative inverses. The proof will use the following multiplicative analogue of Lemma 15:

Lemma 17 *For any cut A and any $c \in P$ with $c > 1$, there exists $a \in A$ such that $ac \notin A$.*

Proof Choose any $b \in A$. We may suppose $bc \in A$, since otherwise we can take $a = b$. Since $b < bc$, we have $bc = b + d$ for some $d \in P$. By Lemma 15 we can choose $a \in A$ so that $a + d \notin A$. Since $b + d \in A$, it follows that $b + d < a + d$, and so $b < a$. Hence $ab^{-1} > 1$ and

$$a + d < a + (ab^{-1})d = ab^{-1}(b + d) = ac.$$

Since $a + d \notin A$, it follows that $ac \notin A$. $\qquad\qquad\qquad\square$

Proposition 18 *For any $A \in \mathscr{P}$, there exists $A^{-1} \in \mathscr{P}$ such that $AA^{-1} = I$.*

Proof Let A^{-1} be the set of all $b \in P$ such that $b < c^{-1}$ for some $c \in P \backslash A$. It is easily verified that A^{-1} is a cut. We note only that $a^{-1} \notin A^{-1}$ if $a \in A$ and that, if $b < c^{-1}$, then also $b < d^{-1}$ for some $d > c$.

We now show that $AA^{-1} = I$. If $a \in A$ and $b \in A^{-1}$ then $ab < 1$, since $a \geq b^{-1}$ would imply $a > c$ for some $c \in P \backslash A$. Thus $AA^{-1} \leq I$. On the other hand, if $0 < d < 1$ then, by Lemma 17, there exists $a \in A$ such that $ad^{-1} \notin A$. Choose $a' \in A$ so that $a < a'$, and put $b = (a')^{-1}d$. Then $b < a^{-1}d$. Since $a^{-1}d = (ad^{-1})^{-1}$, it follows that $b \in A^{-1}$ and consequently $d = a'b \in AA^{-1}$. Thus $I \leq AA^{-1}$. $\qquad\square$

For any positive rational number a, the set A_a consisting of all positive rational numbers c such that $c < a$ is a cut. The map $a \rightarrow A_a$ of P into $\mathscr{P}$ is injective and preserves sums and products:

$$A_{a+b} = A_a + A_b, \quad A_{ab} = A_a A_b.$$

Moreover, $A_a < A_b$ if and only if $a < b$.

By identifying a with A_a we may regard P as a subset of $\mathscr{P}$. It is a proper subset, since **(P4)** does not hold in P.

This completes the first stage of Dedekind's construction. In the second stage we pass from cuts to real numbers. Intuitively, a real number is the difference of two cuts. We will deal with the second stage rather briefly since, as has been said, it is completely analogous to the passage from the natural numbers to the integers.

On the set $\mathscr{P} \times \mathscr{P}$ of all ordered pairs of cuts an equivalence relation is defined by

$$(A, B) \sim (A', B') \quad \text{if } A + B' = A' + B.$$

We define a *real number* to be an equivalence class of ordered pairs of cuts and, as is now customary, we denote the set of all real numbers by $\mathbb{R}$.

Addition and multiplication are unambiguously defined by

$$(A, B) + (C, D) = (A + C, B + D),$$
$$(A, B) \cdot (C, D) = (AC + BD, AD + BC).$$

They obey the laws **(A2)–(A5)**, **(M2)–(M5)** and **(AM1)–(AM2)**.

A real number represented by (A, B) is said to be *positive* if $B < A$. If we denote by $\mathscr{P}'$ the set of all positive real numbers, then **(P1)–(P3)** continue to hold with $\mathscr{P}'$ in place of P. An order relation, satisfying **(O1)–(O3)**, is induced on $\mathbb{R}$ by writing $a < b$ if $b - a \in \mathscr{P}'$. Moreover, any $a \in \mathbb{R}$ may be written in the form $a = b - c$, where $b, c \in \mathscr{P}'$. It is easily seen that $\mathscr{P}$ is isomorphic with $\mathscr{P}'$. By identifying $\mathscr{P}$ with $\mathscr{P}'$, we may regard both $\mathscr{P}$ and $\mathbb{Q}$ as subsets of $\mathbb{R}$. An element of $\mathbb{R} \backslash \mathbb{Q}$ is said to be an *irrational* real number.

Upper and lower bounds, and suprema and infima, may be defined for subsets of $\mathbb{R}$ in the same way as for subsets of $\mathscr{P}$. Moreover, the least upper bound property **(P4)** continues to hold in $\mathbb{R}$. By applying **(P4)** to the subset $-\mathscr{S} = \{-a : a \in \mathscr{S}\}$ we see that if a nonempty subset $\mathscr{S}$ of $\mathbb{R}$ has a lower bound, then it has a greatest lower bound.

The least upper bound property implies the so-called *Archimedean property*:

Proposition 19 *For any positive real numbers a, b, there exists a positive integer n such that $na > b$.*

Proof Assume, on the contrary, that $na \le b$ for every $n \in \mathbb{N}$. Then b is an upper bound for the set $\{na : n \in \mathbb{N}\}$. Let c be a least upper bound for this set. From $na \le c$ for every $n \in \mathbb{N}$ we obtain $(n + 1)a \le c$ for every $n \in \mathbb{N}$. But this implies $na \le c - a$ for every $n \in \mathbb{N}$. Since $c - a < c$ and c is a least upper bound, we have a contradiction. $\square$

Proposition 20 *For any real numbers a, b with $a < b$, there exists a rational number c such that $a < c < b$.*

Proof Suppose first that $a \ge 0$. By Proposition 19 there exists a positive integer n such that $n(b - a) > 1$. Then $b > a + n^{-1}$. There exists also a positive integer m such that $mn^{-1} > a$. If m is the least such positive integer, then $(m - 1)n^{-1} \le a$ and hence $mn^{-1} \le a + n^{-1} < b$. Thus we can take $c = mn^{-1}$.

If $a < 0$ and $b > 0$ we can take $c = 0$. If $a < 0$ and $b \le 0$, then $-b < d < -a$ for some rational d and we can take $c = -d$. $\square$

Proposition 21 *For any positive real number a, there exists a unique positive real number b such that $b^2 = a$.*

Proof Let S be the set of all positive real numbers x such that $x^2 \leq a$. The set S is not empty, since it contains a if $a \leq 1$ and 1 if $a > 1$. If $y > 0$ and $y^2 > a$, then y is an upper bound for S. In particular, $1 + a$ is an upper bound for S. Let b be the least upper bound for S. Then $b^2 = a$, since $b^2 < a$ would imply $(b + 1/n)^2 < a$ for sufficiently large $n > 0$ and $b^2 > a$ would imply $(b - 1/n)^2 > a$ for sufficiently large $n > 0$. Finally, if $c^2 = a$ and $c > 0$, then $c = b$, since

$$(c - b)(c + b) = c^2 - b^2 = 0. \qquad \square$$

The unique positive real number b in the statement of Proposition 21 is said to be a *square root* of a and is denoted by $\sqrt{a}$ or $a^{1/2}$. In the same way it may be shown that, for any positive real number a and any positive integer n, there exists a unique positive real number b such that $b^n = a$, where $b^n = b \cdots b$ (n times). We say that b is an *n-th root* of a and write $b = \sqrt[n]{a}$ or $a^{1/n}$.

A set is said to be a *field* if two binary operations, addition and multiplication, are defined on it with the properties **(A2)–(A5)**, **(M2)–(M5)** and **(AM1)–(AM2)**. A field is said to be *ordered* if it contains a subset P of 'positive' elements with the properties **(P1)–(P3)**. An ordered field is said to be *complete* if, with the order induced by P, it has the property **(P4)**.

Propositions 19–21 hold in any complete ordered field, since only the above properties were used in their proofs. By construction, the set $\mathbb{R}$ of all real numbers is a complete ordered field. In fact, any complete ordered field F is isomorphic to $\mathbb{R}$, i.e. there exists a bijective map $\varphi : F \to \mathbb{R}$ such that, for all $a, b \in F$,

$$\varphi(a + b) = \varphi(a) + \varphi(b),$$
$$\varphi(ab) = \varphi(a)\varphi(b),$$

and $\varphi(a) > 0$ if and only if $a \in P$. We sketch the proof.

Let e be the identity element for multiplication in F and, for any positive integer n, let $ne = e + \cdots + e$ (n summands). Since F is ordered, ne is positive and so has a multiplicative inverse. For any rational number m/n, where $m, n \in \mathbb{Z}$ and $n > 0$, write $(m/n)e = m(ne)^{-1}$ if $m > 0$, $= -(-m)(ne)^{-1}$ if $m < 0$, and $= 0$ if $m = 0$. The elements $(m/n)e$ form a subfield of F isomorphic to $\mathbb{Q}$ and we define $\varphi((m/n)e) = m/n$. For any $a \in F$, we define $\varphi(a)$ to be the least upper bound of all rational numbers m/n such that $(m/n)e \leq a$. One verifies first that the map $\varphi : F \to \mathbb{R}$ is bijective and that $\varphi(a) < \varphi(b)$ if and only if $a < b$. One then deduces that φ preserves sums and products.

Actually, any bijective map $\varphi : F \to \mathbb{R}$ which preserves sums and products is also order-preserving. For, by Proposition 21, $b > a$ if and only if $b - a = c^2$ for some $c \neq 0$, and then

$$\varphi(b) - \varphi(a) = \varphi(b - a) = \varphi(c^2) = \varphi(c)^2 > 0.$$

Those whose primary interest lies in real analysis may *define* $\mathbb{R}$ to be a complete ordered field and omit the tour through $\mathbb{N}, \mathbb{Z}, \mathbb{Q}$ and $\mathscr{P}$. That is, one takes as axioms the 14 properties above which define a complete ordered field and simply assumes that they are consistent.

The notion of convergence can be defined in any totally ordered set. A sequence $\{a_n\}$ is said to *converge*, with *limit* l, if for any l', l'' such that $l' < l < l''$, there exists a positive integer $N = N(l', l'')$ such that

$$l' < a_n < l'' \quad \text{for every } n \geq N.$$

The limit l of the convergent sequence $\{a_n\}$ is clearly uniquely determined; we write

$$\lim_{n \to \infty} a_n = l,$$

or $a_n \to l$ as $n \to \infty$.

It is easily seen that any convergent sequence is *bounded*, i.e. it has an upper bound and a lower bound. A trivial example of a convergent sequence is the *constant* sequence $\{a_n\}$, where $a_n = a$ for every n; its limit is again a.

In the set $\mathbb{R}$ of real numbers, or in any totally ordered set in which each bounded sequence has a least upper bound and a greatest lower bound, the definition of convergence can be reformulated. For, let $\{a_n\}$ be a bounded sequence. Then, for any positive integer m, the subsequence $\{a_n\}_{n \geq m}$ has a greatest lower bound b_m and a least upper bound c_m:

$$b_m = \inf_{n \geq m} a_n, \quad c_m = \sup_{n \geq m} a_n.$$

The sequences $\{b_m\}_{m \geq 1}$ and $\{c_m\}_{m \geq 1}$ are also bounded and, for any positive integer m,

$$b_m \leq b_{m+1} \leq c_{m+1} \leq c_m.$$

If we define the *lower limit* and *upper limit* of the sequence $\{a_n\}$ by

$$\underline{\lim_{n \to \infty}} a_n := \sup_{m \geq 1} b_m, \quad \overline{\lim_{n \to \infty}} a_n := \inf_{m \geq 1} c_m,$$

then $\underline{\lim}_{n \to \infty} a_n \leq \overline{\lim}_{n \to \infty} a_n$, and it is readily shown that $\lim_{n \to \infty} a_n = l$ if and only if

$$\underline{\lim_{n \to \infty}} a_n = l = \overline{\lim_{n \to \infty}} a_n.$$

A sequence $\{a_n\}$ is said to be *nondecreasing* if $a_n \leq a_{n+1}$ for every n and *nonincreasing* if $a_{n+1} \leq a_n$ for every n. It is said to be *monotonic* if it is either nondecreasing or nonincreasing.

Proposition 22 *Any bounded monotonic sequence of real numbers is convergent.*

Proof Let $\{a_n\}$ be a bounded monotonic sequence and suppose, for definiteness, that it is nondecreasing: $a_1 \leq a_2 \leq a_3 \leq \cdots$. In this case, in the notation used above we have $b_m = a_m$ and $c_m = c_1$ for every m. Hence

$$\underline{\lim_{n \to \infty}} a_n = \sup_{m \geq 1} a_m = c_1 = \overline{\lim_{n \to \infty}} a_n. \qquad \square$$

Proposition 22 may be applied to the centuries-old algorithm for calculating square roots, which is commonly used today in pocket calculators. Take any real number $a > 1$ and put

$$x_1 = (1 + a)/2.$$

Then $x_1 > 1$ and $x_1^2 > a$, since $(a - 1)^2 > 0$. Define the sequence $\{x_n\}$ recursively by

$$x_{n+1} = (x_n + a/x_n)/2 \quad (n \geq 1).$$

It is easily verified that if $x_n > 1$ and $x_n^2 > a$, then $x_{n+1} > 1, x_{n+1}^2 > a$ and $x_{n+1} < x_n$. Since the inequalities hold for $n = 1$, it follows that they hold for all n. Thus the sequence $\{x_n\}$ is nonincreasing and bounded, and therefore convergent. If $x_n \to b$, then $a/x_n \to a/b$ and $x_{n+1} \to b$. Hence $b = (b+a/b)/2$, which simplifies to $b^2 = a$.

We consider now sequences of real numbers which are not necessarily monotonic.

Lemma 23 *Any sequence $\{a_n\}$ of real numbers has a monotonic subsequence.*

Proof Let M be the set of all positive integers m such that $a_m \geq a_n$ for every $n > m$. If M contains infinitely many positive integers $m_1 < m_2 < \cdots$, then $\{a_{m_k}\}$ is a nonincreasing subsequence of $\{a_n\}$. If M is empty or finite, there is a positive integer n_1 such that no positive integer $n \geq n_1$ is in M. Then $a_{n_2} > a_{n_1}$ for some $n_2 > n_1$, $a_{n_3} > a_{n_2}$ for some $n_3 > n_2$, and so on. Thus $\{a_{n_k}\}$ is a nondecreasing subsequence of $\{a_n\}$. $\square$

It is clear from the proof that Lemma 23 also holds for sequences of elements of any totally ordered set. In the case of $\mathbb{R}$, however, it follows at once from Lemma 23 and Proposition 22 that

Proposition 24 *Any bounded sequence of real numbers has a convergent subsequence.*

Proposition 24 is often called the Bolzano–Weierstrass theorem. It was stated by Bolzano (c. 1830) in work which remained unpublished until a century later. It became generally known through the lectures of Weierstrass (c. 1874).

A sequence $\{a_n\}$ of real numbers is said to be a *fundamental sequence*, or 'Cauchy sequence', if for each $\varepsilon > 0$ there exists a positive integer $N = N(\varepsilon)$ such that

$$-\varepsilon < a_p - a_q < \varepsilon \quad \text{for all } p, q \geq N.$$

Any fundamental sequence $\{a_n\}$ is bounded, since any finite set is bounded and

$$a_N - \varepsilon < a_p < a_N + \varepsilon \quad \text{for } p \geq N.$$

Also, any convergent sequence is a fundamental sequence. For suppose $a_n \to l$ as $n \to \infty$. Then, for any $\varepsilon > 0$, there exists a positive integer N such that

$$l - \varepsilon/2 < a_n < l + \varepsilon/2 \quad \text{for every } n \geq N.$$

It follows that

$$-\varepsilon < a_p - a_q < \varepsilon \quad \text{for } p \geq q \geq N.$$

The definitions of convergent sequence and fundamental sequence, and the preceding result that 'convergent' implies 'fundamental', hold also for sequences of rational numbers, and even for sequences with elements from any ordered field. However, for sequences of real numbers there is a converse result:

Proposition 25 *Any fundamental sequence of real numbers is convergent.*

Proof If $\{a_n\}$ is a fundamental sequence of real numbers, then $\{a_n\}$ is bounded and, for any $\varepsilon > 0$, there exists a positive integer $m = m(\varepsilon)$ such that

$$-\varepsilon/2 < a_p - a_q < \varepsilon/2 \quad \text{for all } p, q \geq m.$$

But, by Proposition 24, the sequence $\{a_n\}$ has a convergent subsequence $\{a_{n_k}\}$. If l is the limit of this subsequence, then there exists a positive integer $N \geq m$ such that

$$l - \varepsilon/2 < a_{n_k} < l + \varepsilon/2 \quad \text{for } n_k \geq N.$$

It follows that

$$l - \varepsilon < a_n < l + \varepsilon \quad \text{for } n \geq N.$$

Thus the sequence $\{a_n\}$ converges with limit l. $\square$

Proposition 25 was known to Bolzano (1817) and was clearly stated in the influential *Cours d'analyse* of Cauchy (1821). However, a rigorous proof was impossible until the real numbers themselves had been precisely defined.

The Méray–Cantor method of constructing the real numbers from the rationals is based on Proposition 25. We define two fundamental sequences $\{a_n\}$ and $\{a_n'\}$ of rational numbers to be equivalent if $a_n - a_n' \to 0$ as $n \to \infty$. This is indeed an equivalence relation, and we define a real number to be an equivalence class of fundamental sequences. The set of all real numbers acquires the structure of a field if addition and multiplication are defined by

$$\{a_n\} + \{b_n\} = \{a_n + b_n\}, \quad \{a_n\} \cdot \{b_n\} = \{a_n b_n\}.$$

It acquires the structure of a complete ordered field if the fundamental sequence $\{a_n\}$ is said to be positive when it has a positive lower bound. The field $\mathbb{Q}$ of rational numbers may be regarded as a subfield of the field thus constructed by identifying the rational number a with the equivalence class containing the constant sequence $\{a_n\}$, where $a_n = a$ for every n.

It is not difficult to show that an ordered field is complete if every bounded monotonic sequence is convergent, or if every bounded sequence has a convergent subsequence. In this sense, Propositions 22 and 24 state equivalent forms for the least upper bound property. This is not true, however, for Proposition 25. An ordered field need not have the least upper bound property, even though every fundamental sequence is convergent. It is true, however, that an ordered field has the least upper bound property if and only if it has the Archimedean property (Proposition 19) *and* every fundamental sequence is convergent.

In a course of real analysis one would now define continuity and prove those properties of continuous functions which, in the 18th century, were assumed as 'geometrically obvious'. For example, for given $a, b \in \mathbb{R}$ with $a < b$, let $I = [a, b]$ be the *interval* consisting of all $x \in \mathbb{R}$ such that $a \leq x \leq b$. If $f : I \to \mathbb{R}$ is continuous, then it attains its supremum, i.e. there exists $c \in I$ such that $f(x) \leq f(c)$ for every $x \in I$. Also, if $f(a)f(b) < 0$, then $f(d) = 0$ for some $d \in I$ (the intermediate-value theorem). Real analysis is not our primary concern, however, and we do not feel obliged to establish even those properties which we may later use.

4 Metric Spaces

The notion of convergence is meaningful not only for points on a line, but also for points in space, where there is no natural relation of order. We now reformulate our previous definition, so as to make it more generally applicable.

The *absolute value* $|a|$ of a real number a is defined by

$$|a| = a \quad \text{if } a \geq 0,$$
$$|a| = -a \quad \text{if } a < 0.$$

It is easily seen that absolute values have the following properties:

$$|0| = 0, |a| > 0 \quad \text{if } a \neq 0;$$
$$|a| = |-a|;$$
$$|a + b| \leq |a| + |b|.$$

The first two properties follow at once from the definition. To prove the third, we observe first that $a + b \leq |a| + |b|$, since $a \leq |a|$ and $b \leq |b|$. Replacing a by $-a$ and b by $-b$, we obtain also $-(a + b) \leq |a| + |b|$. But $|a + b|$ is either $a + b$ or $-(a + b)$.

The *distance* between two real numbers a and b is defined to be the real number

$$d(a, b) = |a - b|.$$

From the preceding properties of absolute values we obtain their counterparts for distances:

(D1) $d(a, a) = 0, d(a, b) > 0$ *if* $a \neq b$;
(D2) $d(a, b) = d(b, a)$;
(D3) $d(a, b) \leq d(a, c) + d(c, b)$.

The third property is known as the *triangle inequality*, since it may be interpreted as saying that, in any triangle, the length of one side does not exceed the sum of the lengths of the other two.

Fréchet (1906) recognized these three properties as the essential characteristics of any measure of distance and introduced the following general concept. A set E is a *metric space* if with each ordered pair (a, b) of elements of E there is associated a real number $d(a, b)$, so that the properties **(D1)**–**(D3)** hold for all $a, b, c \in E$.

We note first some simple consequences of these properties. For all $a, b, a', b' \in E$ we have

$$|d(a, b) - d(a', b')| \leq d(a, a') + d(b, b') \qquad (*)$$

since, by **(D2)** and **(D3)**,

$$d(a, b) \leq d(a, a') + d(a', b') + d(b, b'),$$
$$d(a', b') \leq d(a, a') + d(a, b) + d(b, b').$$

Taking $b = b'$ in $(*)$, we obtain from **(D1)**,

$$|d(a, b) - d(a', b)| \leq d(a, a'). \qquad (**)$$

In any metric space there is a natural *topology*. A subset G of a metric space E is *open* if for each $x \in G$ there is a positive real number $\delta = \delta(x)$ such that G also contains the whole open ball $B_\delta(x) = \{y \in E : d(x, y) < \delta\}$. A set $F \subseteq E$ is *closed* if its complement $E \backslash F$ is open.

For any set $A \subseteq E$, its *closure* $\bar{A}$ is the intersection of all closed sets containing it, and its *interior* int A is the union of all open sets contained in it.

A subset F of E is *connected* if it is not contained in the union of two open subsets of E whose intersections with F are disjoint and nonempty. A subset F of E is (se-quentially) *compact* if every sequence of elements of F has a subsequence converging to an element of F (and *locally compact* if this holds for every bounded sequence of elements of F).

A map $f : X \to Y$ from one metric space X to another metric space Y is *contin-uous* if, for each open subset G of Y, the set of all $x \in X$ such that $f(x) \in G$ is an open subset of X. The two properties stated at the end of §3 admit far-reaching gen-eralizations for continuous maps between subsets of metric spaces, namely that under a continuous map the image of a compact set is again compact, and the image of a connected set is again connected.

There are many examples of metric spaces:

(i) Let $E = \mathbb{R}^n$ be the set of all n-tuples $a = (\alpha_1, \ldots, \alpha_n)$ of real numbers and define

$$d(b, c) = |b - c|,$$

where $b - c = (\beta_1 - \gamma_1, \ldots, \beta_n - \gamma_n)$ if $b = (\beta_1, \ldots, \beta_n)$ and $c = (\gamma_1, \ldots, \gamma_n)$, and

$$|a| = \max_{1 \le j \le n} |\alpha_j|.$$

Alternatively, one can replace the *norm* $|a|$ by either

$$|a|_1 = \sum_{j=1}^n |\alpha_j|$$

or

$$|a|_2 = \left(\sum_{j=1}^n |\alpha_j|^2 \right)^{1/2}.$$

In the latter case, $d(b, c)$ is the *Euclidean distance* between b and c. The triangle in-equality in this case follows from the *Cauchy–Schwarz inequality*: for any real numbers $\beta_j, \gamma_j (j = 1, \ldots, n)$

$$\left(\sum_{j=1}^n \beta_j \gamma_j \right)^2 \le \left(\sum_{j=1}^n \beta_j^2 \right) \left(\sum_{j=1}^n \gamma_j^2 \right).$$

(ii) Let $E = \mathbb{F}_2^n$ be the set of all n-tuples $a = (\alpha_1, \ldots, \alpha_n)$, where $\alpha_j = 0$ or 1 for each j, and define the *Hamming distance* $d(b, c)$ between $b = (\beta_1, \ldots, \beta_n)$ and $c = (\gamma_1, \ldots, \gamma_n)$ to be the number of j such that $\beta_j \ne \gamma_j$. This metric space plays a basic role in the theory of *error-correcting codes*.

(iii) Let $E = \mathscr{C}(I)$ be the set of all continuous functions $f: I \to \mathbb{R}$, where

$$I = [a, b] = \{x \in \mathbb{R} : a \leq x \leq b\}$$

is an interval of $\mathbb{R}$, and define $d(g, h) = |g - h|$, where

$$|f| = \sup_{a \leq x \leq b} |f(x)|.$$

(A well-known property of continuous functions ensures that f is bounded on I.) Alternatively, one can replace the *norm* $|f|$ by either

$$|f|_1 = \int_a^b |f(x)| dx$$

or

$$|f|_2 = \left(\int_a^b |f(x)|^2 dx \right)^{1/2}.$$

(iv) Let $E = \mathscr{C}(\mathbb{R})$ be the set of all continuous functions $f: \mathbb{R} \to \mathbb{R}$ and define

$$d(g, h) = \sum_{N \geq 1} d_N(g, h)/2^N[1 + d_N(g, h)],$$

where $d_N(g, h) = \sup_{|x| \leq N} |g(x) - h(x)|$. The triangle inequality **(D3)** follows from the inequality

$$|\alpha + \beta|/[1 + |\alpha + \beta|] \leq |\alpha|/[1 + |\alpha|] + |\beta|/[1 + |\beta|]$$

for arbitrary real numbers α, β.

The metric here has the property that $d(f_n, f) \to 0$ if and only if $f_n(x) \to f(x)$ uniformly on every bounded subinterval of $\mathbb{R}$. It may be noted that, even though E is a vector space, the metric is not derived from a norm since, if $\lambda \in \mathbb{R}$, one may have $d(\lambda g, \lambda h) \neq |\lambda| d(g, h)$.

(v) Let E be the set of all measurable functions $f: I \to \mathbb{R}$, where $I = [a, b]$ is an interval of $\mathbb{R}$, and define

$$d(g, h) = \int_a^b |g(x) - h(x)|(1 + |g(x) - h(x)|)^{-1} dx.$$

In order to obtain **(D1)**, we identify functions which take the same value at all points of I, except for a set of measure zero.

Convergence with respect to this metric coincides with *convergence in measure*, which plays a role in the theory of probability.

(vi) Let $E = \mathbb{F}_2^\infty$ be the set of all infinite sequences $a = (\alpha_1, \alpha_2, \ldots)$, where $\alpha_j = 0$ or 1 for every j, and define $d(a, a) = 0$, $d(a, b) = 2^{-k}$ if $a \neq b$, where $b = (\beta_1, \beta_2, \ldots)$ and k is the least positive integer such that $\alpha_k \neq \beta_k$.

Here the triangle inequality holds in the stronger form

$$d(a, b) \leq \max[d(a, c), d(c, b)].$$

This metric space plays a basic role in the theory of *dynamical systems*.

(vii) A connected *graph* can be given the structure of a metric space by defining the distance between two vertices to be the number of edges on the shortest path joining them.

Let E be an arbitrary metric space and $\{a_n\}$ a sequence of elements of E. The sequence $\{a_n\}$ is said to *converge*, with *limit* $a \in E$, if

$$d(a_n, a) \to 0 \quad \text{as } n \to \infty,$$

i.e. if for each real $\varepsilon > 0$ there is a corresponding positive integer $N = N(\varepsilon)$ such that $d(a_n, a) < \varepsilon$ for every $n \geq N$.

The limit a is uniquely determined, since if also $d(a_n, a') \to 0$, then

$$d(a, a') \leq d(a_n, a) + d(a_n, a'),$$

and the right side can be made arbitrarily small by taking n sufficiently large. We write

$$\lim_{n \to \infty} a_n = a,$$

or $a_n \to a$ as $n \to \infty$. If the sequence $\{a_n\}$ has limit a, then so also does any (infinite) subsequence.

If $a_n \to a$ and $b_n \to b$, then $d(a_n, b_n) \to d(a, b)$, as one sees by taking $a' = a_n$ and $b' = b_n$ in (∗).

The sequence $\{a_n\}$ is said to be a *fundamental sequence*, or 'Cauchy sequence', if for each real $\varepsilon > 0$ there is a corresponding positive integer $N = N(\varepsilon)$ such that $d(a_m, a_n) < \varepsilon$ for all $m, n \geq N$.

If $\{a_n\}$ and $\{b_n\}$ are fundamental sequences then, by (∗), the sequence $\{d(a_n, b_n)\}$ of real numbers is a fundamental sequence, and therefore convergent.

A set $S \subseteq E$ is said to be *bounded* if the set of all real numbers $d(a, b)$ with $a, b \in S$ is a bounded subset of $\mathbb{R}$.

Any fundamental sequence $\{a_n\}$ is bounded, since if

$$d(a_m, a_n) < 1 \quad \text{for all } m, n \geq N,$$

then

$$d(a_m, a_n) < 1 + \delta \quad \text{for all } m, n \in \mathbb{N},$$

where $\delta = \max_{1 \leq j < k \leq N} d(a_j, a_k)$.

Furthermore, any convergent sequence $\{a_n\}$ is a fundamental sequence, as one sees by taking $a = \lim_{n \to \infty} a_n$ in the inequality

$$d(a_m, a_n) \leq d(a_m, a) + d(a_n, a).$$

A metric space is said to be *complete* if, conversely, every fundamental sequence is convergent.

By generalizing the Méray–Cantor method of extending the rational numbers to the real numbers, Hausdorff (1913) showed that any metric space can be embedded in a complete metric space. To state his result precisely, we introduce some definitions.

A subset F of a metric space E is said to be *dense* in E if, for each $a \in E$ and each real $\varepsilon > 0$, there exists some $b \in F$ such that $d(a, b) < \varepsilon$.

A map σ from one metric space E to another metric space E' is necessarily injective if it is distance-preserving, i.e. if

$$d'(\sigma(a), \sigma(b)) = d(a, b) \quad \text{for all } a, b \in E.$$

If the map σ is also surjective, then it is said to be an *isometry* and the metric spaces E and E' are said to be *isometric*.

A metric space $\bar{E}$ is said to be a *completion* of a metric space E if $\bar{E}$ is complete and E is isometric to a dense subset of $\bar{E}$. It is easily seen that any two completions of a given metric space are isometric.

Hausdorff's result says that *any metric space E has a completion $\bar{E}$*. We sketch the proof. Define two fundamental sequences $\{a_n\}$ and $\{a_n'\}$ in E to be equivalent if

$$\lim_{n \to \infty} d(a_n, a_n') = 0.$$

It is easily shown that this is indeed an equivalence relation. Moreover, if the fundamental sequences $\{a_n\}, \{b_n\}$ are equivalent to the fundamental sequences $\{a_n'\}, \{b_n'\}$ respectively, then

$$\lim_{n \to \infty} d(a_n, b_n) = \lim_{n \to \infty} d(a_n', b_n').$$

We can give the set $\bar{E}$ of all equivalence classes of fundamental sequences the structure of a metric space by defining

$$\bar{d}(\{a_n\}, \{b_n\}) = \lim_{n \to \infty} d(a_n, b_n).$$

For each $a \in E$, let $\bar{a}$ be the equivalence class in $\bar{E}$ which contains the fundamental sequence $\{a_n\}$ such that $a_n = a$ for every n. Since

$$\bar{d}(\bar{a}, \bar{b}) = d(a, b) \quad \text{for all } a, b \in E,$$

E is isometric to the set $E' = \{\bar{a} : a \in E\}$. It is not difficult to show that E' is dense in $\bar{E}$ and that $\bar{E}$ is complete.

Which of the previous examples of metric spaces are complete? In example (i), the completeness of $\mathbb{R}^n$ with respect to the first definition of distance follows directly from the completeness of $\mathbb{R}$. It is also complete with respect to the two alternative definitions of distance, since a sequence which converges with respect to one of the three metrics also converges with respect to the other two. Indeed it is easily shown that, for every $a \in \mathbb{R}^n$,

$$|a| \leq |a|_2 \leq |a|_1$$

and

$$|a|_1 \leq n^{1/2}|a|_2, \quad |a|_2 \leq n^{1/2}|a|.$$

In example (ii), the completeness of $\mathbb{F}_2^n$ is trivial, since any fundamental sequence is ultimately constant.

In example (iii), the completeness of $\mathscr{C}(I)$ with respect to the first definition of distance follows from the completeness of $\mathbb{R}$ and the fact that the limit of a uniformly convergent sequence of continuous functions is again a continuous function.

However, $\mathscr{C}(I)$ is not complete with respect to either of the two alternative definitions of distance. It is possible also for a sequence to converge with respect to the two alternative definitions of distance, but not with respect to the first definition. Similarly, a sequence may converge in the first alternative metric, but not even be a fundamental sequence in the second.

The completions of the metric space $\mathscr{C}(I)$ with respect to the two alternative metrics may actually be identified with spaces of functions. The completion for the first alternative metric is the set $L(I)$ of all *Lebesgue measurable* functions $f: I \to \mathbb{R}$ such that

$$\int_a^b |f(x)|dx < \infty,$$

functions which take the same value at all points of I, except for a set of measure zero, being identified. The completion $L^2(I)$ for the second alternative metric is obtained by replacing $\int_a^b |f(x)|dx$ by $\int_a^b |f(x)|^2 dx$ in this statement.

It may be shown that the metric spaces of examples (iv)–(vi) are all complete. In example (vi), the strong triangle inequality implies that $\{a_n\}$ is a fundamental sequence if (and only if) $d(a_{n+1}, a_n) \to 0$ as $n \to \infty$.

Let E be an arbitrary metric space and $f: E \to E$ a map of E into itself. A point $\bar{x} \in E$ is said to be a *fixed point* of f if $f(\bar{x}) = \bar{x}$. A useful property of complete metric spaces is the following *contraction principle*, which was first established in the present generality by Banach (1922), but was previously known in more concrete situations.

Proposition 26 *Let E be a complete metric space and let $f: E \to E$ be a map of E into itself. If there exists a real number θ, with $0 < \theta < 1$, such that*

$$d(f(x'), f(x'')) \leq \theta d(x', x'') \quad \text{for all } x', x'' \in E,$$

then the map f has a unique fixed point $\bar{x} \in E$.

Proof It is clear that there is at most one fixed point, since $0 \leq d(x', x'') \leq \theta d(x', x'')$ implies $x' = x''$. To prove that a fixed point exists we use the *method of successive approximations*.

Choose any $x_0 \in E$ and define the sequence $\{x_n\}$ recursively by

$$x_n = f(x_{n-1}) \quad (n \geq 1).$$

For any $k \geq 1$ we have

$$d(x_{k+1}, x_k) = d(f(x_k), f(x_{k-1})) \leq \theta d(x_k, x_{k-1}).$$

Applying this k times, we obtain

$$d(x_{k+1}, x_k) \leq \theta^k d(x_1, x_0).$$

Consequently, if $n > m \geq 0$,

$$\begin{aligned} d(x_n, x_m) &\leq d(x_n, x_{n-1}) + d(x_{n-1}, x_{n-2}) + \cdots + d(x_{m+1}, x_m) \\ &\leq (\theta^{n-1} + \theta^{n-2} + \cdots + \theta^m) d(x_1, x_0) \\ &\leq \theta^m (1 - \theta)^{-1} d(x_1, x_0), \end{aligned}$$

since $0 < \theta < 1$. It follows that $\{x_n\}$ is a fundamental sequence and so a convergent sequence, since E is complete. If $\bar{x} = \lim_{n \to \infty} x_n$, then

$$\begin{aligned} d(f(\bar{x}), \bar{x}) &\leq d(f(\bar{x}), x_{n+1}) + d(x_{n+1}, \bar{x}) \\ &\leq \theta d(\bar{x}, x_n) + d(\bar{x}, x_{n+1}). \end{aligned}$$

Since the right side can be made less than any given positive real number by taking n large enough, we must have $f(\bar{x}) = \bar{x}$. The proof shows also that, for any $m \geq 0$,

$$d(\bar{x}, x_m) \leq \theta^m (1 - \theta)^{-1} d(x_1, x_0). \qquad \square$$

The contraction principle is surprisingly powerful, considering the simplicity of its proof. We give two significant applications: an inverse function theorem and an existence theorem for ordinary differential equations. In both cases we will use the notion of differentiability for functions of several real variables. The unambitious reader may simply take $n = 1$ in the following discussion (so that 'invertible' means 'nonzero'). Functions of several variables are important, however, and it is remarkable that the proper definition of differentiability in this case was first given by Stolz (1887).

A map $\varphi : U \to \mathbb{R}^m$, where $U \subseteq \mathbb{R}^n$ is a *neighbourhood* of $x_0 \in \mathbb{R}^n$ (i.e., U contains some open ball $\{x \in \mathbb{R}^n : |x - x_0| < \rho\}$), is said to be *differentiable* at x_0 if there exists a linear map $A : \mathbb{R}^n \to \mathbb{R}^m$ such that

$$|\varphi(x) - \varphi(x_0) - A(x - x_0)| / |x - x_0| \to 0 \quad \text{as } |x - x_0| \to 0.$$

(The inequalities between the various norms show that it is immaterial which norm is used.) The linear map A, which is then uniquely determined, is called the *derivative* of φ at x_0 and will be denoted by $\varphi'(x_0)$.

This definition is a natural generalization of the usual definition when $m = n = 1$, since it says that the difference $\varphi(x_0 + h) - \varphi(x_0)$ admits the linear approximation Ah for $|h| \to 0$.

Evidently, if φ_1 and φ_2 are differentiable at x_0, then so also is $\varphi = \varphi_1 + \varphi_2$ and

$$\varphi'(x_0) = \varphi_1'(x_0) + \varphi_2'(x_0).$$

It also follows directly from the definition that derivatives satisfy the *chain rule*: If $\varphi : U \to \mathbb{R}^m$, where U is a neighbourhood of $x_0 \in \mathbb{R}^n$, is differentiable at x_0, and if $\psi : V \to \mathbb{R}^l$, where V is a neighbourhood of $y_0 = \varphi(x_0) \in \mathbb{R}^m$, is differentiable at y_0, then the composite map $\chi = \psi \circ \varphi : U \to \mathbb{R}^l$ is differentiable at x_0 and

$$\chi'(x_0) = \psi'(y_0)\varphi'(x_0),$$

the right side being the composite linear map.

We will also use the notion of norm of a linear map. If $A: \mathbb{R}^n \to \mathbb{R}^m$ is a linear map, its *norm* $|A|$ is defined by

$$|A| = \sup_{|x| \le 1} |Ax|.$$

Evidently

$$|A_1 + A_2| \le |A_1| + |A_2|.$$

Furthermore, if $B: \mathbb{R}^m \to \mathbb{R}^l$ is another linear map, then

$$|BA| \le |B||A|.$$

Hence, if $m = n$ and $|A| < 1$, then the linear map $I - A$ is invertible, its inverse being given by the geometric series

$$(I - A)^{-1} = I + A + A^2 + \cdots.$$

It follows that for any invertible linear map $A: \mathbb{R}^n \to \mathbb{R}^n$, if $B: \mathbb{R}^n \to \mathbb{R}^n$ is a linear map such that $|B - A| < |A^{-1}|^{-1}$, then B is also invertible and $|B^{-1} - A^{-1}| \to 0$ as $|B - A| \to 0$.

If $\varphi: U \to \mathbb{R}^m$ is differentiable at $x_0 \in \mathbb{R}^n$, then it is also continuous at x_0, since

$$|\varphi(x) - \varphi(x_0)| \le |\varphi(x) - \varphi(x_0) - \varphi'(x_0)(x - x_0)| + |\varphi'(x_0)||x - x_0|.$$

We say that φ is *continuously differentiable* in U if it is differentiable at each point of U and if the derivative $\varphi'(x)$ is a continuous function of x in U. The *inverse function theorem* says:

Proposition 27 *Let U_0 be a neighbourhood of $x_0 \in \mathbb{R}^n$ and let $\varphi: U_0 \to \mathbb{R}^n$ be a continuously differentiable map for which $\varphi'(x_0)$ is invertible.*
  Then, for some $\delta > 0$, the ball $U = \{x \in \mathbb{R}^n : |x - x_0| < \delta\}$ is contained in U_0 and

  (i) *the restriction of φ to U is injective;*
 (ii) *$V := \varphi(U)$ is open, i.e. if $\eta \in V$, then V contains all $y \in \mathbb{R}^n$ near η;*
(iii) *the inverse map $\psi: V \to U$ is also continuously differentiable and, if $y = \varphi(x)$, then $\psi'(y)$ is the inverse of $\varphi'(x)$.*

Proof To simplify notation, assume $x_0 = \varphi(x_0) = 0$ and write $A = \varphi'(0)$. For any $y \in \mathbb{R}^n$, put

$$f_y(x) = x + A^{-1}[y - \varphi(x)].$$

Evidently x is a fixed point of f_y if and only if $\varphi(x) = y$. The map f_y is also continuously differentiable and

$$f_y'(x) = I - A^{-1}\varphi'(x) = A^{-1}[A - \varphi'(x)].$$

Since $\varphi'(x)$ is continuous, we can choose $\delta > 0$ so that the ball $U = \{x \in \mathbb{R}^n : |x| < \delta\}$ is contained in U_0 and

$$|f_y'(x)| \le 1/2 \quad \text{for } x \in U.$$

If $x_1, x_2 \in U$, then

$$|f_y(x_2) - f_y(x_1)| = \left| \int_0^1 f'((1-t)x_1 + tx_2)(x_2 - x_1)dt \right|$$
$$\leq |x_2 - x_1|/2.$$

It follows that f_y has at most one fixed point in U. Since this holds for arbitrary $y \in \mathbb{R}^n$, the restriction of φ to U is injective.

Suppose next that $\eta = \varphi(\xi)$ for some $\xi \in U$. We wish to show that, if y is near η, the map f_y has a fixed point near ξ.

Choose $r = r(\xi) > 0$ so that the closed ball $B_r = \{x \in \mathbb{R}^n : |x - \xi| \leq r\}$ is contained in U, and fix $y \in \mathbb{R}^n$ so that $|y - \eta| < r/2|A^{-1}|$. Then

$$|f_y(\xi) - \xi| = |A^{-1}(y - \eta)|$$
$$\leq |A^{-1}||y - \eta| < r/2.$$

Hence if $|x - \xi| \leq r$, then

$$|f_y(x) - \xi| \leq |f_y(x) - f_y(\xi)| + |f_y(\xi) - \xi|$$
$$\leq |x - \xi|/2 + r/2 \leq r.$$

Thus $f_y(B_r) \subseteq B_r$. Also, if $x_1, x_2 \in B_r$, then

$$|f_y(x_2) - f_y(x_1)| \leq |x_2 - x_1|/2.$$

But B_r is a complete metric space, with the same metric as $\mathbb{R}^n$, since it is a closed subset (if $x_n \in B_r$ and $x_n \to x$ in $\mathbb{R}^n$, then also $x \in B_r$). Consequently, by the contraction principle (Proposition 26), f_y has a fixed point $x \in B_r$. Then $\varphi(x) = y$, which proves (ii).

Suppose now that $y, \eta \in V$. Then $y = \varphi(x), \eta = \varphi(\xi)$ for unique $x, \xi \in U$. Since

$$|f_y(x) - f_y(\xi)| \leq |x - \xi|/2$$

and

$$f_y(x) - f_y(\xi) = x - \xi - A^{-1}(y - \eta),$$

we have

$$|A^{-1}(y - \eta)| \geq |x - \xi|/2.$$

Thus

$$|x - \xi| \leq 2|A^{-1}||y - \eta|.$$

If $F = \varphi'(\xi)$ and $G = F^{-1}$, then

$$\psi(y) - \psi(\eta) - G(y - \eta) = x - \xi - G(y - \eta)$$
$$= -G[\varphi(x) - \varphi(\xi) - F(x - \xi)].$$

Hence

$$|\psi(y) - \psi(\eta) - G(y - \eta)|/|y - \eta| \le 2|A^{-1}||G||\varphi(x) - \varphi(\xi) - F(x - \xi)|/|x - \xi|.$$

If $|y - \eta| \to 0$, then $|x - \xi| \to 0$ and the right side tends to 0. Consequently ψ is differentiable at η and $\psi'(\eta) = G = F^{-1}$.

Thus ψ is differentiable in U and, *a fortiori*, continuous. In fact ψ is continuously differentiable, since F is a continuous function of ξ (by hypothesis), since $\xi = \psi(\eta)$ is a continuous function of η, and since F^{-1} is a continuous function of F. □

To bring out the meaning of Proposition 27 we add some remarks:

(i) The invertibility of $\varphi'(x_0)$ is necessary for the existence of a differentiable inverse map, but not for the existence of a continuous inverse map. For example, the continuously differentiable map $\varphi: \mathbb{R} \to \mathbb{R}$ defined by $\varphi(x) = x^3$ is bijective and has the continuous inverse $\psi(y) = y^{1/3}$, although $\varphi'(0) = 0$.

(ii) The hypothesis that φ is *continuously* differentiable cannot be totally dispensed with. For example, the map $\varphi: \mathbb{R} \to \mathbb{R}$ defined by

$$\varphi(x) = x + x^2 \sin(1/x) \quad \text{if } x \ne 0, \varphi(0) = 0,$$

is everywhere differentiable and $\varphi'(0) \ne 0$, but φ is not injective in any neighbourhood of 0.

(iii) The inverse map may not be defined throughout U_0. For example, the map $\varphi: \mathbb{R}^2 \to \mathbb{R}^2$ defined by

$$\varphi_1(x_1, x_2) = x_1^2 - x_2^2, \quad \varphi_2(x_1, x_2) = 2x_1 x_2,$$

is everywhere continuously differentiable and has an invertible derivative at every point except the origin. Thus the hypotheses of Proposition 27 are satisfied in any connected open set $U_0 \subseteq \mathbb{R}^2$ which does not contain the origin, and yet $\varphi(1, 1) = \varphi(-1, -1)$.

It was first shown by Cauchy (c. 1844) that, under quite general conditions, an ordinary differential equation has local solutions. The method of successive approximations (i.e., the contraction principle) was used for this purpose by Picard (1890):

Proposition 28 *Let $t_0 \in \mathbb{R}, \xi_0 \in \mathbb{R}^n$ and let U be a neighbourhood of (t_0, ξ_0) in $\mathbb{R} \times \mathbb{R}^n$. If $\varphi: U \to \mathbb{R}^n$ is a continuous map with a derivative φ' with respect to x that is continuous in U, then the differential equation*

$$dx/dt = \varphi(t, x) \tag{1}$$

has a unique solution $x(t)$ which satisfies the initial condition

$$x(t_0) = \xi_0 \tag{2}$$

and is defined in some interval $|t - t_0| \le \delta$, where $\delta > 0$.

Proof If $x(t)$ is a solution of the differential equation (1) which satisfies the initial condition (2), then by integration we get

$$x(t_0) = \xi_0 + \int_{t_0}^{t} \varphi[\tau, x(\tau)]d\tau.$$

Conversely, if a *continuous* function $x(t)$ satisfies this relation then, since φ is continuous, $x(t)$ is actually differentiable and is a solution of (1) that satisfies (2). Hence we need only show that the map $\mathscr{F}$ defined by

$$(\mathscr{F}x)(t) = \xi_0 + \int_{t_0}^{t} \varphi[\tau, x(\tau)]d\tau$$

has a unique fixed point in the space of continuous functions.

There exist positive constants M, L such that

$$|\varphi(t, \xi)| \le M, \quad |\varphi'(t, \xi)| \le L$$

for all (t, ξ) in a neighbourhood of (t_0, ξ_0), which we may take to be U. If $(t, \xi_1) \in U$ and $(t, \xi_2) \in U$, then

$$|\varphi(t, \xi_2) - \varphi(t, \xi_1)| = \left| \int_0^1 \varphi'(t, (1 - u)\xi_1 + u\xi_2)(\xi_2 - \xi_1)du \right|$$

$$\le L|\xi_2 - \xi_1|.$$

Choose $\delta > 0$ so that the box $|t - t_0| \le \delta, |\xi - \xi_0| \le M\delta$ is contained in U and also $L\delta < 1$. Take $I = [t_0 - \delta, t_0 + \delta]$ and let $\mathscr{C}(I)$ be the complete metric space of all continuous functions $x: I \to \mathbb{R}^n$ with the distance function

$$d(x_1, x_2) = \sup_{t \in I} |x_1(t) - x_2(t)|.$$

The constant function $x_0(t) = \xi_0$ is certainly in $\mathscr{C}(I)$. Let E be the subset of all $x \in \mathscr{C}(I)$ such that $x(t_0) = \xi_0$ and $d(x, x_0) \le M\delta$. Evidently if $x_n \in E$ and $x_n \to x$ in $\mathscr{C}(I)$, then $x \in E$. Hence E is also a complete metric space with the same metric. Moreover $\mathscr{F}(E) \subseteq E$, since if $x \in E$ then $(\mathscr{F}x)(t_0) = \xi_0$ and, for all $t \in I$,

$$|(\mathscr{F}x)(t) - \xi_0| = \left| \int_{t_0}^{t} \varphi[\tau, x(\tau)]d\tau \right| \le M\delta.$$

Furthermore, if $x_1, x_2 \in E$, then $d(\mathscr{F}x_1, \mathscr{F}x_2) \le L\delta d(x_1, x_2)$, since for all $t \in I$,

$$|(\mathscr{F}x_1)(t) - (\mathscr{F}x_2)(t)| = \left| \int_{t_0}^{t} \{\varphi[\tau, x_1(\tau)] - \varphi[\tau, x_2(\tau)]\}d\tau \right|$$

$$\le L\delta\, d(x_1, x_2).$$

Since $L\delta < 1$, the result now follows from Proposition 26. $\qquad\qquad\square$

Proposition 28 only guarantees the local existence of solutions, but this is in the nature of things. For example, if $n = 1$, the unique solution of the differential equation

$$dx/dt = x^2$$

such that $x(t_0) = \xi_0 > 0$ is given by

$$x(t) = \{1 - (t - t_0)\xi_0\}^{-1}\xi_0.$$

Thus the solution is defined only for $t < t_0 + \xi_0^{-1}$, even though the differential equation itself has exemplary behaviour everywhere.

To illustrate Proposition 28, take $n = 1$ and let $E(t)$ be the solution of the (linear) differential equation

$$dx/dt = x \tag{3}$$

which satisfies the initial condition $E(0) = 1$. Then $E(t)$ is defined for $|t| < R$, for some $R > 0$. If $|\tau| < R/2$ and $x_1(t) = E(t + \tau)$, then $x_1(t)$ is the solution of the differential equation (3) which satisfies the initial condition $x_1(0) = E(\tau)$. But $x_2(t) = E(\tau)E(t)$ satisfies the same differential equation and the same initial condition. Hence we must have $x_1(t) = x_2(t)$ for $|t| < R/2$, i.e.

$$E(t + \tau) = E(t)E(\tau). \tag{4}$$

In particular,

$$E(t)E(-t) = 1, \quad E(2t) = E(t)^2.$$

The last relation may be used to extend the definition of $E(t)$, so that it is continuously differentiable and a solution of (3) also for $|t| < 2R$. It follows that the solution $E(t)$ is defined for all $t \in \mathbb{R}$ and satisfies the *addition theorem* (4) for all $t, \tau \in \mathbb{R}$.

It is instructive to carry through the method of successive approximations explicitly in this case. If we take $x_0(t)$ to be the constant 1, then

$$x_1(t) = 1 + \int_0^t x_0(\tau)d\tau = 1 + t,$$

$$x_2(t) = 1 + \int_0^t x_1(\tau)d\tau = 1 + t + t^2/2,$$

$$\cdots.$$

By induction we obtain, for every $n \geq 1$,

$$x_n(t) = 1 + t + t^2/2! + \cdots + t^n/n!.$$

Since $x_n(t) \to E(t)$ as $n \to \infty$, we obtain for the solution $E(t)$ the infinite series representation

$$E(t) = 1 + t + t^2/2! + t^3/3! + \cdots,$$

valid actually for every $t \in \mathbb{R}$. In particular,

$$e := E(1) = 1 + 1 + 1/2! + 1/3! + \cdots = 2.7182818\ldots.$$

Of course $E(t) = e^t$ is the *exponential function*. We will now adopt the usual notation, but we remark that the definition of e^t as a solution of a differential equation provides a meaning for irrational t, as well as a simple proof of both the addition theorem and the exponential series.

The power series for e^t shows that

$$e^t > 1 + t > 1 \quad \text{for every } t > 0.$$

Since $e^{-t} = (e^t)^{-1}$, it follows that $0 < e^t < 1$ for every $t < 0$. Thus $e^t > 0$ for all $t \in \mathbb{R}$. Hence, by (3), e^t is a strictly increasing function. But $e^t \to +\infty$ as $t \to +\infty$ and $e^t \to 0$ as $t \to -\infty$. Consequently, since it is certainly continuous, the exponential function maps the real line $\mathbb{R}$ bijectively onto the positive half-line $\mathbb{R}_+ = \{x \in R: x > 0\}$. For any $x > 0$, the unique $t \in \mathbb{R}$ such that $e^t = x$ is denoted by $\ln x$ (the *natural logarithm* of x) or simply $\log x$.

5 Complex Numbers

By extending the rational numbers to the real numbers, we ensured that every positive number had a square root. By further extending the real numbers to the complex numbers, we will now ensure that all numbers have square roots.

The first use of complex numbers, by Cardano (1545), may have had its origin in the solution of cubic, rather than quadratic, equations. The cubic polynomial

$$f(x) = x^3 - 3px - 2q$$

has three real roots if $d := q^2 - p^3 < 0$ since then, for large $X > 0$,

$$f(-X) < 0, \quad f(-p^{1/2}) > 0, \quad f(p^{1/2}) < 0, \quad f(X) > 0.$$

Cardano's formula for the three roots,

$$f(x) = \sqrt[3]{(q + \sqrt{d})} + \sqrt[3]{(q - \sqrt{d})},$$

gives real values, even though d is negative, because the two summands are conjugate complex numbers. This was explicitly stated by Bombelli (1572). It is a curious fact, first proved by Hölder (1891), that if a cubic equation has three distinct real roots, then it is impossible to represent these roots solely by real radicals.

Intuitively, complex numbers are expressions of the form $a + ib$, where a and b are real numbers and $i^2 = -1$. But what is i? Hamilton (1835) defined complex numbers as ordered pairs of real numbers, with appropriate rules for addition and multiplication. Although this approach is similar to that already used in this chapter, and actually was its first appearance, we now choose a different method.

We define a *complex number* to be a 2×2 matrix of the form

$$A = \begin{pmatrix} a & b \\ -b & a \end{pmatrix},$$

where a and b are real numbers. The set of all complex numbers is customarily denoted by $\mathbb{C}$. We may define addition and multiplication in $\mathbb{C}$ to be matrix addition and multiplication, since $\mathbb{C}$ is closed under these operations: if

$$B = \begin{pmatrix} c & d \\ -d & c \end{pmatrix},$$

then

$$A + B = \begin{pmatrix} a+c & b+d \\ -(b+d) & a+c \end{pmatrix}, \quad AB = \begin{pmatrix} ac-bd & ad+bc \\ -(ad+bc) & ac-bd \end{pmatrix}.$$

Furthermore $\mathbb{C}$ contains

$$0 = \begin{pmatrix} 0 & 0 \\ 0 & 0 \end{pmatrix}, \quad 1 = \begin{pmatrix} 1 & 0 \\ 0 & 1 \end{pmatrix},$$

and $A \in \mathbb{C}$ implies $-A \in \mathbb{C}$.

It follows from the properties of matrix addition and multiplication that addition and multiplication of complex numbers have the properties (A2)–(A5), (M2)–(M4) and (AM1)–(AM2), with 0 and 1 as identity elements for addition and multiplication respectively. The property (M5) also holds, since if a and b are not both zero, and if

$$a' = a/(a^2 + b^2), \quad b' = -b/(a^2 + b^2),$$

then

$$A^{-1} = \begin{pmatrix} a' & b' \\ -b' & a' \end{pmatrix}$$

is a multiplicative inverse of A. Thus $\mathbb{C}$ satisfies the axioms for a *field*.

The set $\mathbb{C}$ also contains the matrix

$$i = \begin{pmatrix} 0 & 1 \\ -1 & 0 \end{pmatrix},$$

for which $i^2 = -1$, and any $A \in \mathbb{C}$ can be represented in the form

$$A = a1 + bi,$$

where $a, b \in \mathbb{R}$. The multiples $a1$, where $a \in \mathbb{R}$, form a subfield of $\mathbb{C}$ isomorphic to the real field $\mathbb{R}$. By identifying the real number a with the complex number $a1$, we may regard $\mathbb{R}$ itself as contained in $\mathbb{C}$.

Thus we will now stop using matrices and use only the fact that $\mathbb{C}$ is a field containing $\mathbb{R}$ such that every $z \in \mathbb{C}$ can be represented in the form

$$z = x + iy,$$

where $x, y \in \mathbb{R}$ and $i \in \mathbb{C}$ satisfies $i^2 = -1$. The representation is necessarily unique, since $i \notin \mathbb{R}$. We call x and y the *real* and *imaginary parts* of z and denote them by $\mathscr{R}z$

and $\mathscr{I}z$ respectively. Complex numbers of the form iy, where $y \in \mathbb{R}$, are said to be *pure imaginary*.

It is worth noting that $\mathbb{C}$ cannot be given the structure of an *ordered* field, since in an ordered field any nonzero square is positive, whereas $i^2 + 1^2 = (-1) + 1 = 0$.

It is often suggestive to regard complex numbers as points of a plane, the complex number $z = x + iy$ being the point with coordinates (x, y) in some chosen system of rectangular coordinates.

The *complex conjugate* of the complex number $z = x + iy$, where $x, y \in \mathbb{R}$, is the complex number $\bar{z} = x - iy$. In the geometrical representation of complex numbers, $\bar{z}$ is the reflection of z in the x-axis. From the definition we at once obtain

$$\mathscr{R}z = (z + \bar{z})/2, \quad \mathscr{I}z = (z - \bar{z})/2i.$$

It is easily seen also that

$$\overline{z + w} = \bar{z} + \bar{w}, \quad \overline{zw} = \bar{z}\bar{w}, \quad \bar{\bar{z}} = z.$$

Moreover, $\bar{z} = z$ if and only if $z \in \mathbb{R}$. Thus the map $z \to \bar{z}$ is an 'involutory automorphism' of the field $\mathbb{C}$, with the subfield $\mathbb{R}$ as its set of fixed points. It follows that $\overline{-z} = -\bar{z}$.

If $z = x + iy$, where $x, y \in \mathbb{R}$, then

$$z\bar{z} = (x + iy)(x - iy) = x^2 + y^2.$$

Hence $z\bar{z}$ is a positive real number for any nonzero $z \in \mathbb{C}$. The *absolute value* $|z|$ of the complex number z is defined by

$$|0| = 0, \quad |z| = \sqrt{(z\bar{z})} \text{ if } z \neq 0,$$

(with the positive value for the square root). This agrees with the definition in §3 if $z = x$ is a positive real number.

It follows at once from the definition that $|\bar{z}| = |z|$ for every $z \in \mathbb{C}$, and $z^{-1} = \bar{z}/|z|^2$ if $z \neq 0$.

Absolute values have the following properties: *for all $z, w \in \mathbb{C}$,*

(i) $|0| = 0, |z| > 0$ *if $z \neq 0$;*
(ii) $|zw| = |z||w|$;
(iii) $|z + w| \leq |z| + |w|$.

The first property follows at once from the definition. To prove (ii), observe that both sides are non-negative and that

$$|zw|^2 = zw\overline{zw} = zw\bar{z}\bar{w} = z\bar{z}w\bar{w} = |z|^2|w|^2.$$

To prove (iii), we first evaluate $|z + w|^2$:

$$|z + w|^2 = (z + w)(\bar{z} + \bar{w}) = z\bar{z} + (z\bar{w} + w\bar{z}) + w\bar{w} = |z|^2 + 2\mathscr{R}(z\bar{w}) + |w|^2.$$

Since $\mathscr{R}(z\bar{w}) \leq |z\bar{w}| = |z||w|$, this yields

$$|z + w|^2 \le |z|^2 + 2|z||w| + |w|^2 = (|z| + |w|)^2,$$

and (iii) follows by taking square roots.

Several other properties are consequences of these three, although they may also be verified directly. By taking $z = w = 1$ in (ii) and using (i), we obtain $|1| = 1$. By taking $z = w = -1$ in (ii) and using (i), we now obtain $|-1| = 1$. Taking $w = -1$ and $w = z^{-1}$ in (ii), we further obtain

$$|-z| = |z|, \quad |z^{-1}| = |z|^{-1} \quad \text{if } z \ne 0.$$

Again, by replacing z by $z - w$ in (iii), we obtain

$$||z| - |w|| \le |z - w|.$$

This shows that $|z|$ is a continuous function of z. In fact $\mathbb{C}$ is a metric space, with the metric $d(z, w) = |z - w|$. By considering real and imaginary parts separately, one verifies that this metric space is complete, i.e. every fundamental sequence is convergent, and that the Bolzano–Weierstrass property continues to hold, i.e. any bounded sequence of complex numbers has a convergent subsequence.

It will now be shown that any complex number has a square root. If $w = u + iv$ and $z = x + iy$, then $z^2 = w$ is equivalent to

$$x^2 - y^2 = u, \quad 2xy = v.$$

Since

$$(x^2 + y^2)^2 = (x^2 - y^2)^2 + (2xy)^2,$$

these equations imply

$$x^2 + y^2 = \sqrt{(u^2 + v^2)}.$$

Hence

$$x^2 = \{u + \sqrt{(u^2 + v^2)}\}/2.$$

Since the right side is positive if $v \ne 0$, x is then uniquely determined apart from sign and $y = v/2x$ is uniquely determined by x. If $v = 0$, then $x = \pm\sqrt{u}$ and $y = 0$ when $u > 0$; $x = 0$ and $y = \pm\sqrt{(-u)}$ when $u < 0$, and $x = y = 0$ when $u = 0$.

It follows that any quadratic polynomial

$$q(z) = az^2 + bz + c,$$

where $a, b, c \in \mathbb{C}$ and $a \ne 0$, has two complex roots, given by the well-known formula

$$z = \{-b \pm \sqrt{(b^2 - 4ac)}\}/2a.$$

However, much more is true. The so-called *fundamental theorem of algebra* asserts that any polynomial

$$f(z) = a_0 z^n + a_1 z^{n-1} + \cdots + a_n,$$

where $a_0, a_1, \ldots, a_n \in \mathbb{C}, n \geq 1$ and $a_0 \neq 0$, has a complex root. Thus by adjoining to the real field $\mathbb{R}$ a root of the polynomial $z^2 + 1$ we ensure that every non-constant polynomial has a root. Today the fundamental theorem of algebra is considered to belong to analysis, rather than to algebra. It is useful to retain the name, however, as a reminder that our own pronouncements may seem equally quaint in the future.

Our proof of the theorem will use the fact that any polynomial is differentiable, since sums and products of differentiable functions are again differentiable, and hence also continuous. We first prove

Proposition 29 *Let $G \subseteq \mathbb{C}$ be an open set and E a proper subset (possibly empty) of G such that each point of G has a neighbourhood containing at most one point of E. If $f : G \to \mathbb{C}$ is a continuous map which at every point of $G \backslash E$ is differentiable and has a nonzero derivative, then $f(G)$ is an open subset of $\mathbb{C}$.*

Proof Evidently $G \backslash E$ is an open set. We show first that $f(G \backslash E)$ is also an open set. Let $\zeta \in G \backslash E$. Then f is differentiable at ζ and $\rho = |f'(\zeta)| > 0$. We can choose $\delta > 0$ so that the closed disc $B = \{z \in \mathbb{C} : |z - \zeta| \leq \delta\}$ contains no point of E, is contained in G and

$$|f(z) - f(\zeta)| \geq \rho |z - \zeta|/2 \quad \text{for every } z \in B.$$

In particular, if $S = \{z \in \mathbb{C} : |z - \zeta| = \delta\}$ is the boundary of B, then

$$|f(z) - f(\zeta)| \geq \rho\delta/2 \quad \text{for every } z \in S.$$

Choose $w \in \mathbb{C}$ so that $|w - f(\zeta)| < \rho\delta/4$ and consider the minimum in the compact set B of the continuous real-valued function $\phi(z) = |f(z) - w|$. On the boundary S we have

$$\phi(z) \geq |f(z) - f(\zeta)| - |f(\zeta) - w| \geq \rho\delta/2 - \rho\delta/4 = \rho\delta/4.$$

Since $\phi(\zeta) < \rho\delta/4$, it follows that ϕ attains its minimum value in B at an interior point z_0. Since $z_0 \notin E$, we can take

$$z = z_0 - h[f'(z_0)]^{-1}\{f(z_0) - w\},$$

where $h > 0$ is so small that $|z - \zeta| < \delta$. Then

$$f(z) - w = f(z_0) - w + f'(z_0)(z - z_0) + o(h) = (1 - h)\{f(z_0) - w\} + o(h).$$

If $f(z_0) \neq w$ then, for sufficiently small $h > 0$,

$$|f(z) - w| \leq (1 - h/2)|f(z_0) - w| < |f(z_0) - w|,$$

which contradicts the definition of z_0. We conclude that $f(z_0) = w$. Thus $f(G \backslash E)$ contains not only $f(\zeta)$, but also an open disc $\{w \in \mathbb{C} : |w - f(\zeta)| < \rho\delta/4\}$ surrounding it. Since this holds for every $\zeta \in G \backslash E$, it follows that $f(G \backslash E)$ is an open set.

Now let $\zeta \in E$ and assume that $f(G)$ does not contain any open neighbourhood of $\omega := f(\zeta)$. Then $f(z) \neq \omega$ for every $z \in G \backslash E$. Choose $\delta > 0$ so small that the closed disc $B = \{z \in \mathbb{C}: |z - \zeta| \le \delta\}$ is contained in G and contains no point of E except ζ. If $S = \{z \in \mathbb{C}: |z - \zeta| = \delta\}$ is the boundary of B, there exists an open disc U with centre ω that contains no point of $f(S)$. It follows that if $A = \{z \in \mathbb{C}: 0 < |z - \zeta| < \delta\}$ is the annulus $B \setminus (S \cup \{\zeta\})$, then $U \setminus \{\omega\}$ is the union of the disjoint nonempty open sets $U \cap \{\mathbb{C} \setminus f(B)\}$ and $U \cap f(A)$. Since $U \setminus \{\omega\}$ is a connected set (because it is *path-connected*), this is a contradiction.   □

From Proposition 29 we readily obtain

Theorem 30 *If*

$$f(z) = z^n + a_1 z^{n-1} + \cdots + a_n$$

is a polynomial of degree $n \ge 1$ with complex coefficients $a_1, \ldots, a_n$, then $f(\zeta) = 0$ for some $\zeta \in \mathbb{C}$.

Proof Since

$$f(z)/z^n = 1 + a_1/z + \cdots + a_n/z^n \to 1 \quad \text{as } |z| \to \infty,$$

we can choose $R > 0$ so large that

$$|f(z)| > |f(0)| \quad \text{for all } z \in \mathbb{C} \text{ such that } |z| = R.$$

Since the closed disc $D = \{z \in \mathbb{C}: |z| \le R\}$ is compact, the continuous function $|f(z)|$ assumes its minimum value in D at a point ζ in the interior $G = \{z \in \mathbb{C}: |z| < R\}$. The function $f(z)$ is differentiable in G and the set E of all points of G at which the derivative $f'(z)$ vanishes is finite. (In fact E contains at most $n - 1$ points, by Proposition II.15.) Hence, by Proposition 29, $f(G)$ is an open subset of $\mathbb{C}$. Since $|f(z)| \ge |f(\zeta)|$ for all $z \in G$, this implies $f(\zeta) = 0$.   □

The first 'proof' of the fundamental theorem of algebra was given by d'Alembert (1746). Assuming the convergence of what are now called Puiseux expansions, he showed that if a polynomial assumes a value $w \ne 0$, then it also assumes a value w' such that $|w'| < |w|$. A much simpler way of reaching this conclusion, which required only the existence of k-th roots of complex numbers, was given by Argand (1814). Cauchy (1820) gave a similar proof and, with latter-day rigour, it is still reproduced in textbooks. The proof we have given rests on the same general principle, but uses neither the existence of k-th roots nor the continuity of the derivative. These may be called *differential calculus proofs*.

The basis for an *algebraic proof* was given by Euler (1749). His proof was completed by Lagrange (1772) and then simplified by Laplace (1795). The algebraic proof starts from the facts that $\mathbb{R}$ is an ordered field, that any positive element of $\mathbb{R}$ has a square root in $\mathbb{R}$ and that any polynomial of odd degree with coefficients from $\mathbb{R}$ has a root in $\mathbb{R}$. It then shows that any polynomial of degree $n \ge 1$ with coefficients from $\mathbb{C} = \mathbb{R}(i)$, where $i^2 = -1$, has a root in $\mathbb{C}$ by using induction on the highest power of 2 which divides n.

Gauss (1799) objected to this proof, because it assumed that there were 'roots' and then proved that these roots were complex numbers. The difficulty disappears if one uses the result, due to Kronecker (1887), that a polynomial with coefficients from an arbitrary field K decomposes into linear factors in a field L which is a finite extension of K. This general result, which is not difficult to prove, is actually all that is required for many of the previous applications of the fundamental theorem of algebra.

It is often said that the first rigorous proof of the fundamental theorem of algebra was given by Gauss (1799). Like d'Alembert, however, Gauss assumed properties of algebraic curves which were unknown at the time. The gaps in this proof of Gauss were filled by Ostrowski (1920).

There are also *topological proofs* of the fundamental theorem of algebra, e.g. using the notion of topological degree. This type of proof is intuitively appealing, but not so easy to make rigorous. Finally, there are *complex analysis proofs*, which depend ultimately on Cauchy's theorem on complex line integrals. (The latter proofs are more closely related to either the differential calculus proofs or the topological proofs than they seem to be at first sight.)

The *exponential function* e^z may be defined, for any complex value of z, as the sum of the everywhere convergent power series

$$\sum_{n \geq 0} z^n/n! = 1 + z + z^2/2! + z^3/3! + \cdots .$$

It is easily verified that $w(z) = e^z$ is a solution of the differential equation $dw/dz = w$ satisfying the initial condition $w(0) = 1$.

For any $\zeta \in \mathbb{C}$, put $\varphi(z) = e^{\zeta - z} e^z$. Differentiating by the product rule, we obtain

$$\varphi'(z) = -e^{\zeta - z} e^z + e^{\zeta - z} e^z = 0.$$

Since this holds for all $z \in \mathbb{C}$, $\varphi(z)$ is a constant. Thus $\varphi(z) = \varphi(0) = e^\zeta$. Replacing ζ by $\zeta + z$, we obtain the *addition theorem* for the exponential function:

$$e^\zeta e^z = e^{\zeta + z} \quad \text{for all } z, \zeta \in \mathbb{C}.$$

In particular, $e^{-z} e^z = 1$ and hence $e^z \neq 0$ for every $z \in \mathbb{C}$.

The power series for e^z shows that, for any real y, e^{-iy} is the complex conjugate of e^{iy} and hence

$$|e^{iy}|^2 = e^{iy} e^{-iy} = 1.$$

It follows that, for all real x, y,

$$|e^{x+iy}| = |e^x||e^{iy}| = e^x.$$

The *trigonometric functions* $\cos z$ and $\sin z$ may be defined, for any complex value of z, by the formulas of Euler (1740):

$$\cos z = (e^{iz} + e^{-iz})/2, \quad \sin z = (e^{iz} - e^{-iz})/2i.$$

It follows at once that

$$e^{iz} = \cos z + i \sin z,$$
$$\cos 0 = 1, \quad \sin 0 = 0,$$
$$\cos(-z) = \cos z, \quad \sin(-z) = -\sin z,$$

and the relation $e^{iz}e^{-iz} = 1$ implies that

$$\cos^2 z + \sin^2 z = 1.$$

From the power series for e^z we obtain, for every $z \in \mathbb{C}$,

$$\cos z = \sum_{n \geq 0}(-1)^n z^{2n}/(2n)! = 1 - z^2/2! + z^4/4! - \cdots,$$

$$\sin z = \sum_{n \geq 0}(-1)^n z^{2n+1}/(2n+1)! = z - z^3/3! + z^5/5! - \cdots.$$

From the differential equation we obtain, for every $z \in \mathbb{C}$,

$$d(\cos z)/dz = -\sin z, \quad d(\sin z)/dz = \cos z.$$

From the addition theorem we obtain, for all $z, \zeta \in \mathbb{C}$,

$$\cos(z + \zeta) = \cos z \cos \zeta - \sin z \sin \zeta,$$
$$\sin(z + \zeta) = \sin z \cos \zeta + \cos z \sin \zeta.$$

We now consider periodicity properties. By the addition theorem for the exponential function, $e^{z+h} = e^z$ if and only if $e^h = 1$. Thus the exponential function has period h if and only if $e^h = 1$. Since $e^h = 1$ implies $h = ix$ for some real x, and since $\cos x$ and $\sin x$ are real for real x, the periods correspond to those real values of x for which

$$\cos x = 1, \quad \sin x = 0.$$

In fact, the second relation follows from the first, since $\cos^2 x + \sin^2 x = 1$.

By bracketing the power series for $\cos x$ in the form

$$\cos x = (1 - x^2/2! + x^4/4!) - (1 - x^2/7 \cdot 8)x^6/6! - (1 - x^2/11 \cdot 12)x^{10}/10! - \cdots$$

and taking $x = 2$, we see that $\cos 2 < 0$. Since $\cos 0 = 1$ and $\cos x$ is a continuous function of x, there is a least positive value ξ of x such that $\cos \xi = 0$. Then $\sin^2 \xi = 1$. In fact $\sin \xi = 1$, since $\sin 0 = 0$ and $\sin' x = \cos x > 0$ for $0 \leq x < \xi$. Thus

$$0 < \sin x < 1 \quad \text{for } 0 < x < \xi$$

and

$$e^{i\xi} = \cos \xi + i \sin \xi = i.$$

As usual, we now write $\pi = 2\xi$. From $e^{\pi i/2} = i$, we obtain

$$e^{2\pi i} = i^4 = (-1)^2 = 1.$$

Thus the exponential function has period $2\pi i$. It follows that it also has period $2n\pi i$, for every $n \in \mathbb{Z}$. We will show that there are no other periods.

Suppose $e^{ix'} = 1$ for some $x' \in \mathbb{R}$ and choose $n \in \mathbb{Z}$ so that $n \leq x'/2\pi < n+1$. If $x = x' - 2n\pi$, then $e^{ix} = 1$ and $0 \leq x < 2\pi$. If $x \neq 0$, then $0 < x/4 < \pi/2$ and hence $0 < \sin x/4 < 1$. Thus $e^{ix/4} \neq \pm 1, \pm i$. But this is a contradiction, since

$$(e^{ix/4})^4 = e^{ix} = 1.$$

We show next that the map $x \to e^{ix}$ maps the interval $0 \leq x < 2\pi$ bijectively onto the *unit circle*, i.e. the set of all complex numbers w such that $|w| = 1$. We already know that $|e^{ix}| = 1$ if $x \in \mathbb{R}$. If $e^{ix} = e^{ix'}$, where $0 \leq x \leq x' < 2\pi$, then $e^{i(x'-x)} = 1$. Since $0 \leq x' - x < 2\pi$, this implies $x' = x$.

It remains to show that if $u, v \in \mathbb{R}$ and $u^2 + v^2 = 1$, then

$$u = \cos x, \quad v = \sin x$$

for some x such that $0 \leq x < 2\pi$. If $u, v > 0$, then also $u, v < 1$. Hence $u = \cos x$ for some x such that $0 < x < \pi/2$. It follows that $v = \sin x$, since $\sin^2 x = 1 - u^2 = v^2$ and $\sin x > 0$. The other possible sign combinations for u, v may be reduced to the case $u, v > 0$ by means of the relations

$$\sin(x + \pi/2) = \cos x, \quad \cos(x + \pi/2) = -\sin x.$$

If z is any nonzero complex number, then $r = |z| > 0$ and $|z/r| = 1$. It follows that any nonzero complex number z can be uniquely expressed in the form

$$z = re^{i\theta},$$

where r, θ are real numbers such that $r > 0$ and $0 \leq \theta < 2\pi$. We call these r, θ the *polar coordinates* of z and θ the *argument* of z. If $z = x + iy$, where $x, y \in \mathbb{R}$, then $r = \sqrt{(x^2 + y^2)}$ and

$$x = r\cos\theta, \quad y = r\sin\theta.$$

Hence, in the geometrical representation of complex numbers by points of a plane, r is the distance of z from O and θ measures the angle between the positive x-axis and the ray $\overrightarrow{Oz}$.

We now show that the exponential function assumes every nonzero complex value w. Since $|w| > 0$, we have $|w| = e^x$ for some $x \in \mathbb{R}$. If $w' = w/|w|$, then $|w'| = 1$ and so $w' = e^{iy}$ for some $y \in \mathbb{R}$. Consequently,

$$w = |w|w' = e^x e^{iy} = e^{x+iy}.$$

It follows that, for any positive integer n, a nonzero complex number w has n distinct n-th roots. In fact, if $w = e^z$, then w has the distinct n-th roots

$$\zeta_k = \zeta\omega^k \, (k = 0, 1, \ldots, n-1),$$

where $\zeta = e^{z/n}$ and $\omega = e^{2\pi i/n}$. In the geometrical representation of complex numbers by points of a plane, the n-th roots of w are the vertices of an n-sided regular polygon.

It remains to show that π has its usual geometric significance. Since the continuously differentiable function $z(t) = e^{it}$ describes the unit circle as t increases from 0 to 2π, the length of the unit circle is

$$L = \int_0^{2\pi} |z'(t)| dt.$$

But $|z'(t)| = 1$, since $z'(t) = ie^{it}$, and hence $L = 2\pi$.

In a course of complex analysis one would now define complex line integrals, prove Cauchy's theorem and deduce its numerous consequences. The miracle is that, if $D = \{z \in \mathbb{C}: |z| < \rho\}$ is a disc with centre the origin, then any differentiable function $f: D \to \mathbb{C}$ can be represented by a *power series*,

$$f(z) = c_0 + c_1 z + c_2 z^2 + \cdots,$$

which is convergent for $|z| < \rho$. It follows that, if f vanishes at a sequence of distinct points converging to 0, then it vanishes everywhere. This is the basis for *analytic continuation*.

A complex-valued function f is said to be *holomorphic* at $a \in \mathbb{C}$ if, in some neighbourhood of a, it can be represented as the sum of a convergent power series (its 'Taylor' series):

$$f(z) = c_0 + c_1 (z - a) + c_2 (z - a)^2 + \cdots.$$

It is said to be *meromorphic* at $a \in \mathbb{C}$ if, for some integer n, it can be represented near a as the sum of a convergent series (its 'Laurent' series):

$$f(z) = c_0 (z - a)^{-n} + c_1 (z - a)^{-n+1} + c_2 (z - a)^{-n+2} + \cdots.$$

If $c_0 \neq 0$, then $(z - a)f'(z)/f(z) \to -n$ as $z \to a$. If also $n > 0$ we say that a is a *pole* of f of *order n* with *residue* c_{n-1}. If $n = 1$, the residue is c_0 and the pole is said to be *simple*.

Let G be a nonempty connected open subset of $\mathbb{C}$. From what has been said, if $f: G \to \mathbb{C}$ is differentiable throughout G, then it is also holomorphic throughout G. If f_1 and f_2 are holomorphic throughout G and f_2 is not identically zero, then the quotient $f = f_1/f_2$ is meromorphic throughout G. Conversely, it may be shown that if f is meromorphic throughout G, then $f = f_1/f_2$ for some functions f_1, f_2 which are holomorphic throughout G.

The behaviour of many functions is best understood by studying them in the complex domain, as the exponential and trigonometric functions already illustrate. Complex numbers, when they first appeared, were called 'impossible' numbers. They are now indispensable.

6 Quaternions and Octonions

Quaternions were invented by Hamilton (1843) in order to be able to 'multiply' points of 3-dimensional space, in the same way that complex numbers enable one to multiply points of a plane. The definition of quaternions adopted here will be analogous to our definition of complex numbers.

We define a *quaternion* to be a 2×2 matrix of the form

$$A = \begin{pmatrix} a & b \\ -\bar{b} & \bar{a} \end{pmatrix},$$

where a and b are complex numbers and the bar denotes complex conjugation. The set of all quaternions will be denoted by $\mathbb{H}$. We may define addition and multiplication in $\mathbb{H}$ to be matrix addition and multiplication, since $\mathbb{H}$ is closed under these operations. Furthermore $\mathbb{H}$ contains

$$0 = \begin{pmatrix} 0 & 0 \\ 0 & 0 \end{pmatrix}, \quad 1 = \begin{pmatrix} 1 & 0 \\ 0 & 1 \end{pmatrix},$$

and $A \in \mathbb{H}$ implies $-A \in \mathbb{H}$.

It follows from the properties of matrix addition and multiplication that addition and multiplication of quaternions have the properties **(A2)**–**(A5)** and **(M3)**–**(M4)**, with 0 and 1 as identity elements for addition and multiplication respectively. However, **(M2)** no longer holds, since multiplication is not always commutative. For example,

$$\begin{pmatrix} 0 & 1 \\ -1 & 0 \end{pmatrix} \begin{pmatrix} 0 & i \\ i & 0 \end{pmatrix} \neq \begin{pmatrix} 0 & i \\ i & 0 \end{pmatrix} \begin{pmatrix} 0 & 1 \\ -1 & 0 \end{pmatrix}.$$

On the other hand, there are now two distributive laws: *for all $A, B, C \in \mathbb{H}$,*

$$A(B + C) = AB + AC, \quad (B + C)A = BA + CA.$$

It is easily seen that $A \in \mathbb{H}$ is in the *centre* of $\mathbb{H}$, i.e. $AB = BA$ for every $B \in \mathbb{H}$, if and only if $A = \lambda 1$ for some real number λ. Since the map $\lambda \to \lambda 1$ preserves sums and products, we can regard $\mathbb{R}$ as contained in $\mathbb{H}$ by identifying the real number λ with the quaternion $\lambda 1$.

We define the *conjugate* of the quaternion

$$A = \begin{pmatrix} a & b \\ -\bar{b} & \bar{a} \end{pmatrix},$$

to be the quaternion

$$\bar{A} = \begin{pmatrix} \bar{a} & -b \\ \bar{b} & a \end{pmatrix}.$$

It is easily verified that

$$\overline{A + B} = \bar{A} + \bar{B}, \quad \overline{AB} = \bar{B}\bar{A}, \quad \bar{\bar{A}} = A.$$

Furthermore,

$$\bar{A}A = A\bar{A} = n(A), \quad A + \bar{A} = t(A),$$

where the *norm* $n(A)$ and *trace* $t(A)$ are both real:

$$n(A) = a\bar{a} + b\bar{b}, \quad t(A) = a + \bar{a}.$$

Moreover, $n(A) > 0$ if $A \neq 0$. It follows that any quaternion $A \neq 0$ has a multiplicative inverse: if $A^{-1} = n(A)^{-1}\bar{A}$, then

$$A^{-1}A = AA^{-1} = 1.$$

Norms and traces have the following properties: *for all A, $B \in \mathbb{H}$,*

$$t(\bar{A}) = t(A),$$
$$n(\bar{A}) = n(A),$$
$$t(A + B) = t(A) + t(B),$$
$$n(AB) = n(A)n(B).$$

Only the last property is not immediately obvious, and it can be proved in one line:

$$n(AB) = \overline{AB}AB = \bar{B}\bar{A}AB = n(A)\bar{B}B = n(A)n(B).$$

Furthermore, for any $A \in \mathbb{H}$ we have

$$A^2 - t(A)A + n(A) = 0,$$

since the left side can be written in the form $A^2 - (A + \bar{A})A + \bar{A}A$. (The relation is actually just a special case of the 'Cayley–Hamilton theorem' of linear algebra.) It follows that the quadratic polynomial $x^2 + 1$ has not two, but infinitely many quaternionic roots.

If we put

$$I = \begin{pmatrix} 0 & 1 \\ -1 & 0 \end{pmatrix}, \quad J = \begin{pmatrix} 0 & i \\ i & 0 \end{pmatrix}, \quad K = \begin{pmatrix} i & 0 \\ 0 & -i \end{pmatrix},$$

then

$$I^2 = J^2 = K^2 = -1,$$
$$IJ = K = -JI, \quad JK = I = -KJ, \quad KI = J = -IK.$$

Moreover, any quaternion A can be uniquely represented in the form

$$A = \alpha_0 + \alpha_1 I + \alpha_2 J + \alpha_3 K,$$

where $\alpha_0, \ldots, \alpha_3 \in \mathbb{R}$. In fact this is equivalent to the previous representation with

$$a = \alpha_0 + i\alpha_3, \quad b = \alpha_1 + i\alpha_2.$$

The corresponding representation of the conjugate quaternion is

$$\bar{A} = \alpha_0 - \alpha_1 I - \alpha_2 J - \alpha_3 K.$$

Hence $\bar{A} = A$ if and only if $\alpha_1 = \alpha_2 = \alpha_3 = 0$ and $\bar{A} = -A$ if and only if $\alpha_0 = 0$.

A quaternion A is said to be *pure* if $\bar{A} = -A$. Thus any quaternion can be uniquely represented as the sum of a real number and a pure quaternion.

It follows from the multiplication table for the units I, J, K that $A = \alpha_0 + \alpha_1 I + \alpha_2 J + \alpha_3 K$ has norm

$$n(A) = \alpha_0^2 + \alpha_1^2 + \alpha_2^2 + \alpha_3^2.$$

Consequently the relation $n(A)n(B) = n(AB)$ may be written in the form

$$(\alpha_0^2 + \alpha_1^2 + \alpha_2^2 + \alpha_3^2)(\beta_0^2 + \beta_1^2 + \beta_2^2 + \beta_3^2) = \gamma_0^2 + \gamma_1^2 + \gamma_2^2 + \gamma_3^2,$$

where

$$\gamma_0 = \alpha_0\beta_0 - \alpha_1\beta_1 - \alpha_2\beta_2 - \alpha_3\beta_3,$$
$$\gamma_1 = \alpha_0\beta_1 + \alpha_1\beta_0 + \alpha_2\beta_3 - \alpha_3\beta_2,$$
$$\gamma_2 = \alpha_0\beta_2 - \alpha_1\beta_3 + \alpha_2\beta_0 + \alpha_3\beta_1,$$
$$\gamma_3 = \alpha_0\beta_3 + \alpha_1\beta_2 - \alpha_2\beta_1 + \alpha_3\beta_0.$$

This '4-squares identity' was already known to Euler (1770).

An important application of quaternions is to the parametrization of rotations in 3-dimensional space. In describing this application it will be convenient to denote quaternions now by lower case letters. In particular, we will write i, j, k in place of I, J, K.

Let u be a quaternion with norm $n(u) = 1$, and consider the mapping $T: \mathbb{H} \to \mathbb{H}$ defined by

$$Tx = uxu^{-1}.$$

Evidently

$$T(x + y) = Tx + Ty,$$
$$T(xy) = (Tx)(Ty),$$
$$T(\lambda x) = \lambda Tx \quad \text{if } \lambda \in \mathbb{R}.$$

Moreover, since $u^{-1} = \bar{u}$,

$$T\bar{x} = \overline{Tx}.$$

It follows that

$$n(Tx) = n(x),$$

since

$$n(Tx) = Tx\overline{Tx} = TxT\bar{x} = T(x\bar{x}) = n(x)T1 = n(x).$$

Furthermore, T maps pure quaternions into pure quaternions, since $\bar{x} = -x$ implies

$$\overline{Tx} = T\bar{x} = -Tx.$$

If we write

$$x = \xi_1 i + \xi_2 j + \xi_3 k,$$

then

$$Tx = y = \eta_1 i + \eta_2 j + \eta_3 k,$$

where $\eta_\mu = \sum_{v=1}^3 \beta_{\mu v} \xi_v$ for some $\beta_{\mu v} \in \mathbb{R}$. Since

$$\eta_1^2 + \eta_2^2 + \eta_3^2 = \xi_1^2 + \xi_2^2 + \xi_3^2,$$

the matrix $V = (\beta_{\mu v})$ is *orthogonal*: $V^{-1} = V^t$.

Thus with every quaternion u with norm 1 there is associated a 3×3 orthogonal matrix $V = (\beta_{\mu v})$. Explicitly, if

$$u = \alpha_0 + \alpha_1 i + \alpha_2 j + \alpha_3 k,$$

where

$$\alpha_0^2 + \alpha_1^2 + \alpha_2^2 + \alpha_3^2 = 1,$$

then

$$\beta_{11} = \alpha_0^2 + \alpha_1^2 - \alpha_2^2 - \alpha_3^2, \quad \beta_{12} = 2(\alpha_1\alpha_2 - \alpha_0\alpha_3), \quad \beta_{13} = 2(\alpha_1\alpha_3 + \alpha_0\alpha_2),$$
$$\beta_{21} = 2(\alpha_1\alpha_2 + \alpha_0\alpha_3), \quad \beta_{22} = \alpha_0^2 - \alpha_1^2 + \alpha_2^2 - \alpha_3^2, \quad \beta_{23} = 2(\alpha_2\alpha_3 - \alpha_0\alpha_1),$$
$$\beta_{31} = 2(\alpha_1\alpha_3 - \alpha_0\alpha_2), \quad \beta_{32} = 2(\alpha_2\alpha_3 + \alpha_0\alpha_1), \quad \beta_{33} = \alpha_0^2 - \alpha_1^2 - \alpha_2^2 + \alpha_3^2.$$

This parametrization of orthogonal transformations was first discovered by Euler(1770).

We now consider the dependence of V on u, and consequently write $V(u)$ in place of V. Since

$$u_1 u_2 x (u_1 u_2)^{-1} = u_1 (u_2 x u_2^{-1}) u_1^{-1},$$

we have

$$V(u_1 u_2) = V(u_1) V(u_2).$$

Thus the map $u \rightarrow V(u)$ is a 'homomorphism' of the multiplicative group of all quaternions of norm 1 into the group of all 3×3 real orthogonal matrices. In particular, $V(\bar{u}) = V(u)^{-1}$.

We show next that two quaternions u_1, u_2 of norm 1 yield the same orthogonal matrix if and only if $u_2 = \pm u_1$. Put $u = u_2^{-1} u_1$. Then $u_1 x u_1^{-1} = u_2 x u_2^{-1}$ if and only if $ux = xu$. This holds for every pure quaternion x if and only if u is real, i.e. if and only if $u = \pm 1$, since $n(u) = 1$.

The question arises whether all 3×3 orthogonal matrices may be represented in the above way. It follows readily from the preceding formulas for $\beta_{\mu v}$ that the orthogonal matrix $-I$ cannot be so represented. Consequently, if an orthogonal matrix V is represented, then $-V$ is not. On the other hand, suppose u is a pure quaternion, so that $\alpha_0 = 0$. Then $ux + xu = ux + \bar{x}\bar{u}$ is real, and given by

$$ux + xu = -2(\alpha_1\xi_1 + \alpha_2\xi_2 + \alpha_3\xi_3) = 2\langle\bar{u}, x\rangle,$$

with the notation of §10 for inner products in $\mathbb{R}^3$. It follows that

$$y = ux\bar{u} = 2\langle\bar{u}, x\rangle\bar{u} - x.$$

But the mapping $x \rightarrow x - 2\langle\bar{u}, x\rangle\bar{u}$ is a *reflection* in the plane orthogonal to the unit vector u. Hence, for every reflection R, $-R$ is represented. It may be shown that every orthogonal transformation of $\mathbb{R}^3$ is a product of reflections. (Indeed, this is a special case of a more general result which will be proved in Proposition 17 of Chapter VII.) It follows that an orthogonal matrix V is represented if and only if V is a product of an even number of reflections (or, equivalently, if and only if V has determinant 1, as defined in Chapter V, §1).

Since, by our initial definition of quaternions, the quaternions of norm 1 are just the 2×2 unitary matrices with determinant 1, our results may be summed up (cf. Chapter X, §8) by saying that there is a homomorphism of the *special unitary group* $SU_2(\mathbb{C})$ onto the *special orthogonal group* $SO_3(\mathbb{R})$, with kernel $\{\pm I\}$. (Here 'special' signifies 'determinant 1'.)

Since the quaternions of norm 1 may be identified with the points of the unit sphere S^3 in $\mathbb{R}^4$ it follows that, as a topological space, $SO_3(\mathbb{R})$ is homeomorphic to S^3 with antipodal points identified, i.e. to the projective space $P^3(\mathbb{R})$. Similarly (cf. Chapter X, §8), the topological group $SU_2(\mathbb{C})$ is the *simply-connected covering space* of the topological group $SO_3(\mathbb{R})$.

Again, by considering the map $T: \mathbb{H} \rightarrow \mathbb{H}$ defined by $Tx = vxu^{-1}$, where u, v are quaternions with norm 1, it may be seen that that there is a homomorphism of the direct product $SU_2(\mathbb{C}) \times SU_2(\mathbb{C})$ onto the special orthogonal group $SO_4(\mathbb{R})$ of 4×4 real orthogonal matrices with determinant 1, the kernel being $\{\pm(I, I)\}$.

Almost immediately after Hamilton's invention of quaternions Graves (1844), in a letter to Hamilton, and Cayley (1845) invented 'octonions', also known as 'octaves' or 'Cayley numbers'. We define an *octonion* to be an ordered pair (a_1, a_2) of quaternions, with addition and multiplication defined by

$$(a_1, a_2) + (b_1, b_2) = (a_1 + b_1, a_2 + b_2),$$
$$(a_1, a_2) \cdot (b_1, b_2) = (a_1b_1 - \bar{b}_2a_2, b_2a_1 + a_2\bar{b}_1).$$

Then the set $\mathbb{O}$ of all octonions is a commutative group under addition, i.e. the laws (A2)–(A5) hold, with $0 = (0, 0)$ as identity element, and multiplication is both left and right distributive with respect to addition. The octonion $1 = (1, 0)$ is a two-sided identity element for multiplication, and the octonion $\varepsilon = (0, 1)$ has the property $\varepsilon^2 = -1$.

It is easily seen that $\alpha \in \mathbb{O}$ is in the *centre* of $\mathbb{O}$, i.e. $\alpha\beta = \beta\alpha$ for every $\beta \in \mathbb{O}$, if and only if $\alpha = (c, 0)$ for some $c \in \mathbb{R}$.

Since the map $a \rightarrow (a, 0)$ preserves sums and products, we may regard $\mathbb{H}$ as contained in $\mathbb{O}$ by identifying the quaternion a with the octonion $(a, 0)$. This shows that multiplication of octonions is in general not commutative. It is also in general not even associative; for example,

$$(ij)\varepsilon = k\varepsilon = (0, k), \quad i(j\varepsilon) = i(0, j) = (0, -k).$$

It is for this reason that we defined octonions as ordered pairs, rather than as matrices. It should be mentioned, however, that we could have used precisely the same construction to define complex numbers as ordered pairs of real numbers, and quaternions as ordered pairs of complex numbers, but the verification of the associative law for multiplication would then have been more laborious.

Although multiplication is non-associative, $\mathbb{O}$ does inherit some other properties from $\mathbb{H}$. If we define the *conjugate* of the octonion $\alpha = (a_1, a_2)$ to be the octonion $\bar{\alpha} = (\overline{a_1}, -a_2)$, then it is easily verified that

$$\overline{\alpha + \beta} = \bar{\alpha} + \bar{\beta}, \quad \overline{\alpha\beta} = \bar{\beta}\bar{\alpha}, \quad \bar{\bar{\alpha}} = \alpha.$$

Furthermore,

$$\alpha\bar{\alpha} = \bar{\alpha}\alpha = n(\alpha),$$

where the *norm* $n(\alpha) = a_1\overline{a_1} + a_2\overline{a_2}$ is real. Moreover $n(\alpha) > 0$ if $\alpha \neq 0$, and $n(\bar{\alpha}) = n(\alpha)$.

It will now be shown that if $\alpha, \beta \in \mathbb{O}$ and $\alpha \neq 0$, then the equation

$$\zeta\alpha = \beta$$

has a unique solution $\zeta \in \mathbb{O}$. Writing $\alpha = (a_1, a_2)$, $\beta = (b_1, b_2)$ and $\zeta = (x_1, x_2)$, we have to solve the simultaneous quaternionic equations

$$x_1 a_1 - \overline{a_2} x_2 = b_1,$$
$$a_2 x_1 + x_2 \overline{a_1} = b_2.$$

If we multiply the second equation on the right by a_1 and replace $x_1 a_1$ by its value from the first equation, we get

$$n(\alpha)x_2 = b_2 a_1 - a_2 b_1.$$

Similarly, if we multiply the first equation on the right by $\overline{a_1}$ and replace $x_2\overline{a_1}$ by its value from the second equation, we get

$$n(\alpha)x_1 = b_1\overline{a_1} + \overline{a_2}b_2.$$

It follows that the equation $\zeta\alpha = \beta$ has the unique solution

$$\zeta = n(\alpha)^{-1}\beta\bar{\alpha}.$$

Since the equation $\alpha\eta = \beta$ is equivalent to $\bar{\eta}\bar{\alpha} = \bar{\beta}$, it has the unique solution $\eta = n(\alpha)^{-1}\bar{\alpha}\beta$. Thus $\mathbb{O}$ is a *division algebra*. It should be noted that, since $\mathbb{O}$ is non-associative, it is not enough to verify that every nonzero element has a multiplicative inverse.

It follows from the preceding discussion that, *for all* $\alpha, \beta \in \mathbb{O}$,

$$(\beta\bar{\alpha})\alpha = n(\alpha)\beta = \alpha(\bar{\alpha}\beta).$$

Consequently the norm is multiplicative: *for all* $\alpha, \beta \in \mathbb{O}$,

$$n(\alpha\beta) = n(\alpha)n(\beta).$$

For, putting $\gamma = \alpha\beta$, we have

$$n(\gamma)\bar{\alpha} = (\bar{\alpha}\gamma)\bar{\gamma} = (\bar{\alpha}(\alpha\beta))\bar{\gamma} = n(\alpha)\beta\bar{\gamma} = n(\alpha)\beta(\bar{\beta}\bar{\alpha}) = n(\alpha)n(\beta)\bar{\alpha}.$$

This establishes the result when $\alpha \neq 0$, and when $\alpha = 0$ it is obvious.

Every $\alpha \in \mathbb{O}$ has a unique representation $\alpha = a_1 + a_2\varepsilon$, where $a_1, a_2 \in \mathbb{H}$, and hence a unique representation

$$\alpha = c_0 + c_1 i + c_2 j + c_3 k + c_4\varepsilon + c_5 i\varepsilon + c_6 j\varepsilon + c_7 k\varepsilon,$$

where $c_0, \ldots, c_7 \in \mathbb{R}$. Since $\bar{\alpha} = \bar{a_1} - a_2\varepsilon$ and $n(\alpha) = a_1\bar{a_1} + a_2\bar{a_2}$, it follows that

$$\bar{\alpha} = c_0 - c_1 i - c_2 j - c_3 k - c_4\varepsilon - c_5 i\varepsilon - c_6 j\varepsilon - c_7 k\varepsilon$$

and

$$n(\alpha) = c_0^2 + \cdots + c_7^2.$$

Consequently the relation $n(\alpha)n(\beta) = n(\alpha\beta)$ may be written in the form

$$(c_0^2 + \cdots + c_7^2)(d_0^2 + \cdots + d_7^2) = e_0^2 + \cdots + e_7^2,$$

where $e_i = \sum_{j=0}^{7}\sum_{k=0}^{7} \rho_{ijk} c_j d_k$ for some real constants ρ_{ijk} which do not depend on the c's and d's. An '8-squares identity' of this type was first found by Degen (1818).

7 Groups

A nonempty set G is said to be a *group* if a binary operation φ, i.e. a mapping $\varphi: G \times G \to G$, is defined with the properties

(i) $\varphi(\varphi(a, b), c) = \varphi(a, \varphi(b, c))$ for all $a, b, c \in G$; (associative law)
(ii) there exists $e \in G$ such that $\varphi(e, a) = a$ for every $a \in G$; (identity element)
(iii) for each $a \in G$, there exists $a^{-1} \in G$ such that $\varphi(a^{-1}, a) = e$.(inverse elements)

If, in addition,

(iv) $\varphi(a, b) = \varphi(b, a)$ for all $a, b \in G$,(commutative law)

then the group G is said to be *commutative* or *abelian*.

For example, the set $\mathbb{Z}$ of all integers is a commutative group under addition, i.e. with $\varphi(a, b) = a + b$, with 0 as identity element and $-a$ as the inverse of a. Similarly, the set $\mathbb{Q}^\times$ of all nonzero rational numbers is a commutative group under multiplication, i.e. with $\varphi(a, b) = ab$, with 1 as identity element and a^{-1} as the inverse of a.

We now give an example of a noncommutative group. The set $\mathscr{S}_A$ of all bijective maps $f: A \to A$ of a nonempty set A to itself is a group under composition, i.e. with $\varphi(a, b) = a \circ b$, with the identity map i_A as identity element and the inverse map f^{-1} as the inverse of f. If A contains at least 3 elements, then $\mathscr{S}_A$ is a noncommutative group. For suppose a, b, c are distinct elements of A, let $f : A \to A$ be the bijective map defined by

$$f(a) = b, \quad f(b) = a, \quad f(x) = x \quad \text{if } x \neq a, b,$$

and let $g : A \to A$ be the bijective map defined by

$$g(a) = c, \quad g(c) = a, \quad g(x) = x \quad \text{if } x \neq a, c.$$

Then $f \circ g \neq g \circ f$, since $(f \circ g)(a) = c$ and $(g \circ f)(a) = b$.

For arbitrary groups, instead of $\varphi(a, b)$ we usually write $a \cdot b$ or simply ab. For commutative groups, instead of $\varphi(a, b)$ we often write $a + b$.

Since, by the associative law,

$$(ab)c = a(bc),$$

we will usually dispense with brackets.

We now derive some simple properties possessed by all groups. By (iii) we have $a^{-1}a = e$. In fact also $aa^{-1} = e$. This may be seen by multiplying on the left, by the inverse of a^{-1}, the relation

$$a^{-1}aa^{-1} = ea^{-1} = a^{-1}.$$

By (ii) we have $ea = a$. It now follows that also $ae = a$, since

$$ae = aa^{-1}a = ea.$$

For all elements a, b of the group G, the equation $ax = b$ has the solution $x = a^{-1}b$ and the equation $ya = b$ has the solution $y = ba^{-1}$. Moreover, these solutions are unique. For from $ax = ax'$ we obtain $x = x'$ by multiplying on the left by a^{-1}, and from $ya = y'a$ we obtain $y = y'$ by multiplying on the right by a^{-1}.

In particular, the identity element e is unique, since it is the solution of $ea = a$, and the inverse a^{-1} of a is unique, since it is the solution of $a^{-1}a = e$. It follows that the inverse of a^{-1} is a and the inverse of ab is $b^{-1}a^{-1}$.

As the preceding argument suggests, in the definition of a group we could have replaced left identity and left inverse by right identity and right inverse, i.e. we could have required $ae = a$ and $aa^{-1} = e$, instead of $ea = a$ and $a^{-1}a = e$. (However, left identity and right inverse, or right identity and left inverse, would not give the same result.)

If H, K are nonempty subsets of a group G, we denote by HK the subset of G consisting of all elements hk, where $h \in H$ and $k \in K$. If L is also a nonempty subset of G, then evidently

$$(HK)L = H(KL).$$

A subset H of a group G is said to be a *subgroup* of G if it is a group under the same group operation as G itself. A nonempty subset H is a subgroup of G if and only if $a, b \in H$ implies $ab^{-1} \in H$. Indeed the necessity of the condition is obvious. It is also sufficient, since it implies first $e = aa^{-1} \in H$ and then $b^{-1} = eb^{-1} \in H$. (The associative law in H is inherited from G.)

We now show that a nonempty *finite* subset H of a group G is a subgroup of G if it is closed under multiplication only. For, if $a \in H$, then $ha \in H$ for all $h \in H$. Since H is finite and the mapping $h \to ha$ of H into itself is injective, it is also surjective by the pigeonhole principle (Corollary I.6). Hence $ha = a$ for some $h \in H$, which shows that H contains the identity element of G. It now further follows that $ha = e$ for some $h \in H$, which shows that H is also closed under inversion.

A group is said to be *finite* if it contains only finitely many elements and to be of *order n* if it contains exactly n elements.

In order to give an important example of a subgroup we digress briefly. Let n be a positive integer and let A be the set $\{1, 2, \ldots, n\}$ with the elements in their natural order. Since we regard A as ordered, a bijective map $\alpha : A \to A$ will be called a *permutation*. The set of all permutations of A is a group under composition, the *symmetric group* $\mathscr{S}_n$.

Suppose now that $n > 1$. An inversion of order induced by the permutation α is a pair (i, j) with $i < j$ for which $\alpha(i) > \alpha(j)$. The permutation α is said to be *even* or *odd* according as the total number of inversions of order is even or odd. For example, the permutation $\{1, 2, 3, 4, 5\} \to \{3, 5, 4, 1, 2\}$ is odd, since there are $2 + 3 + 2 = 7$ inversions of order.

The *sign* of the permutation α is defined by

$$\mathrm{sgn}(\alpha) = 1 \text{ or } -1 \text{ according as } \alpha \text{ is even or odd.}$$

Evidently we can write

$$\mathrm{sgn}(\alpha) = \prod_{1 \le i < j \le n} \{\alpha(j) - \alpha(i)\}/(j - i),$$

from which it follows that

$$\mathrm{sgn}(\alpha\beta) = \mathrm{sgn}(\alpha)\mathrm{sgn}(\beta).$$

Since the sign of the identity permutation is 1, this implies

$$\mathrm{sgn}(\alpha^{-1}) = \mathrm{sgn}(\alpha).$$

Thus $\mathrm{sgn}(\rho^{-1}\alpha\rho) = \mathrm{sgn}(\alpha)$ for any permutation ρ of A, and so $\mathrm{sgn}(\alpha)$ is actually independent of the ordering of A.

Since the product of two even permutations is again an even permutation, the even permutations form a subgroup of $\mathscr{S}_n$, the *alternating group* $\mathscr{A}_n$. The order of $\mathscr{A}_n$ is $n!/2$. For let τ be the permutation $\{1, 2, 3, \ldots, n\} \to \{2, 1, 3, \ldots, n\}$. Since there is only one inversion of order, τ is odd. Since $\tau\tau$ is the identity permutation, a permutation is odd if and only if it has the form $\alpha\tau$, where α is even. Hence the number of odd permutations is equal to the number of even permutations.

It may be mentioned that the sign of a permutation can also be determined without actually counting the total number of inversions. In fact any $\alpha \in \mathscr{S}_n$ may be written as a product of v disjoint cycles, and α is even or odd according as $n - v$ is even or odd.

We now return to the main story. Let H be a subgroup of an arbitrary group G and let a, b be elements of G. We write $a \sim_r b$ if $ba^{-1} \in H$. We will show that this is an equivalence relation.

The relation is certainly reflexive, since $e \in H$. It is also symmetric, since if $c = ba^{-1} \in H$, then $c^{-1} = ab^{-1} \in H$. Furthermore it is transitive, since if $ba^{-1} \in H$ and $cb^{-1} \in H$, then also $ca^{-1} = (cb^{-1})(ba^{-1}) \in H$.

The equivalence class which contains a is the set Ha of all elements ha, where $h \in H$. We call any such equivalence class a *right coset* of the subgroup H, and any element of a given coset is said to be a *representative* of that coset.

It follows from the remarks in §0 about arbitrary equivalence relations that, for any two cosets Ha and Ha', either $Ha = Ha'$ or $Ha \cap Ha' = \emptyset$. Moreover, the distinct right cosets form a partition of G.

If H is a subgroup of a finite group G, then H is also finite and the number of distinct right cosets is finite. Moreover, each right coset Ha contains the same number of elements as H, since the mapping $h \to ha$ of H to Ha is bijective. It follows that the order of the subgroup H divides the order of the whole group G, a result usually known as *Lagrange's theorem*. The quotient of the orders, i.e. the number of distinct cosets, is called the *index* of H in G.

Suppose again that H is a subgroup of an arbitrary group G and that $a, b \in G$. By writing $a \sim_l b$ if $a^{-1}b \in H$, we obtain another equivalence relation. The equivalence class which contains a is now the set aH of all elements ah, where $h \in H$. We call any such equivalence class a *left coset* of the subgroup H. Again, two left cosets either coincide or are disjoint, and the distinct left cosets form a partition of G.

When are the two partitions, into left cosets and into right cosets, the same? Evidently $Ha = aH$ for every $a \in G$ if and only if $a^{-1}Ha = H$ for every $a \in G$ or, since a may be replaced by a^{-1}, if and only if $a^{-1}ha \in H$ for every $h \in H$ and every $a \in G$. A subgroup H which satisfies this condition is said to be 'invariant' or *normal*.

Any group G obviously has two normal subgroups, namely G itself and the subset $\{e\}$ which contains only the identity element. A group G is said to be *simple* if it has no other normal subgroups and if these two are distinct (i.e., G contains more than one element).

We now show that if H is a normal subgroup of a group G, then the collection of all cosets of H can be given the structure of a group. Since $Ha = aH$ and $HH = H$, we have

$$(Ha)(Hb) = H(Ha)b = Hab.$$

Thus if we define the product $Ha \cdot Hb$ of the cosets Ha and Hb to be the coset Hab, the definition does not depend on the choice of coset representatives. Clearly multiplication of cosets is associative, the coset $H = He$ is an identity element and the coset Ha^{-1} is an inverse of the coset Ha. The new group thus constructed is called the *factor group* or *quotient group* of G by the normal subgroup H, and is denoted by G/H.

A mapping $f : G \to G'$ of a group G into a group G' is said to be a (group) *homomorphism* if

$$f(ab) = f(a)f(b) \quad \text{for all } a, b \in G.$$

By taking $a = b = e$, we see that this implies that $f(e) = e'$ is the identity element of G'. By taking $b = a^{-1}$, it now follows that $f(a^{-1})$ is the inverse of $f(a)$ in G'. Since the subset $f(G)$ of G' is closed under both multiplication and inversion, it is a subgroup of G'.

If $g: G' \to G''$ is a homomorphism of the group G' into a group G'', then the composite map $g \circ f: G \to G''$ is also a homomorphism.

The *kernel* of the homomorphism f is defined to be the set N of all $a \in G$ such that $f(a) = e'$ is the identity element of G'. The kernel is a subgroup of G, since if $a \in N$ and $b \in N$, then $ab \in N$ and $a^{-1} \in N$. Moreover, it is a normal subgroup, since $a \in N$ and $c \in G$ imply $c^{-1}ac \in N$.

For any $a \in G$, put $a' = f(a) \in G'$. The coset Na is the set of all $x \in G$ such that $f(x) = a'$, and the map $Na \to a'$ is a bijection from the collection of all cosets of N to $f(G)$. Since f is a homomorphism, Nab is mapped to $a'b'$. Hence the map $Na \to a'$ is a homomorphism of the factor group G/N to $f(G)$.

A mapping $f: G \to G'$ of a group G into a group G' is said to be a (group) *isomorphism* if it is both bijective and a homomorphism. The inverse mapping $f^{-1}: G' \to G$ is then also an isomorphism. (An *automorphism* of a group G is an isomorphism of G with itself.)

Thus we have shown that, if $f: G \to G'$ is a homomorphism of a group G into a group G', with kernel N, then the factor group G/N is isomorphic to $f(G)$.

Suppose now that G is an arbitrary group and a any element of G. We have already defined a^{-1}, the inverse of a. We now inductively define a^n, for any integer n, by putting

$$a^0 = e, \qquad a^1 = a,$$
$$a^n = a(a^{n-1}), \quad a^{-n} = a^{-1}(a^{-1})^{n-1} \quad \text{if } n > 1.$$

It is readily verified that, for all $m, n \in \mathbb{Z}$,

$$a^m a^n = a^{m+n}, \quad (a^m)^n = a^{mn}.$$

The set $\langle a \rangle = \{a^n : n \in \mathbb{Z}\}$ is a commutative subgroup of G, the *cyclic subgroup generated by* a. Evidently $\langle a \rangle$ contains a and is contained in every subgroup of G which contains a.

If we regard $\mathbb{Z}$ as a group under addition, then the mapping $n \to a^n$ is a homomorphism of $\mathbb{Z}$ onto $\langle a \rangle$. Consequently $\langle a \rangle$ is isomorphic to the factor group $\mathbb{Z}/N$, where N is the subgroup of $\mathbb{Z}$ consisting of all integers n such that $a^n = e$. Evidently $0 \in N$, and $n \in \mathbb{N}$ implies $-n \in N$. Thus either $N = \{0\}$ or N contains a positive integer. In the latter case, let s be the least positive integer in N. By Proposition 14, for any integer n there exist integers q, r such that

$$n = qs + r, \quad 0 \le r < s.$$

If $n \in N$, then also $r = n - qs \in N$ and hence $r = 0$, by the definition of s. It follows that $N = s\mathbb{Z}$ is the subgroup of $\mathbb{Z}$ consisting of all multiples of s. Thus either $\langle a \rangle$ is isomorphic to $\mathbb{Z}$, and is an infinite group, or $\langle a \rangle$ is isomorphic to the factor group $\mathbb{Z}/s\mathbb{Z}$, and is a finite group of order s. We say that the element a itself is of *infinite order* if $\langle a \rangle$ is infinite and of *order* s if $\langle a \rangle$ is of order s.

It is easily seen that in a *commutative* group the set of all elements of finite order is a subgroup, called its *torsion subgroup*.

If S is any nonempty subset of a group G, then the set $\langle S \rangle$ of all finite products $a_1^{\varepsilon_1} a_2^{\varepsilon_1} \cdots a_n^{\varepsilon_n}$, where $n \in \mathbb{N}$, $a_j \in S$ and $\varepsilon_j = \pm 1$, is a subgroup of G, called the subgroup *generated* by S. Clearly $S \subseteq \langle S \rangle$ and $\langle S \rangle$ is contained in every subgroup of G which contains S.

Two elements a, b of a group G are said to be *conjugate* if $b = x^{-1}ax$ for some $x \in G$. It is easy to see that conjugacy is an equivalence relation. For $a = a^{-1}aa$, if $b = x^{-1}ax$ then $a = (x^{-1})^{-1}bx^{-1}$, and $b = x^{-1}ax, c = y^{-1}by$ together imply $c = (xy)^{-1}axy$. Consequently G may be partitioned into *conjugacy classes*, so that two elements of G are conjugate if and only if they belong to the same conjugacy class.

For any element a of a group G, the set N_a of all elements of G which commute with a,

$$N_a = \{x \in G : xa = ax\},$$

is closed under multiplication and inversion. Thus N_a is a subgroup of G, called the *centralizer* of a in G.

If y and z lie in the same right coset of N_a, so that $z = xy$ for some $x \in N_a$, then $zy^{-1}a = azy^{-1}$ and hence $y^{-1}ay = z^{-1}az$. Conversely, if $y^{-1}ay = z^{-1}az$, then y and z lie in the same right coset of N_a. If G is finite, it follows that the number of elements in the conjugacy class containing a is equal to the number of right cosets of the subgroup N_a, i.e. to the *index* of the subgroup N_a in G, and hence it divides the order of G.

To conclude, we mention a simple way of creating new groups from given ones. Let G, G' be groups and let $G \times G'$ be the set of all ordered pairs (a, a') with $a \in G$ and $a' \in G'$. Then $G \times G'$ acquires the structure of a group if we define the product $(a, a') \cdot (b, b')$ of (a, a') and (b, b') to be $(ab, a'b')$. Multiplication is clearly associative, (e, e') is an identity element and (a^{-1}, a'^{-1}) is an inverse for (a, a'). The group thus constructed is called the *direct product* of G and G', and is again denoted by $G \times G'$.

8 Rings and Fields

A nonempty set R is said to be a *ring* if two binary operations, $+$ (addition) and $\cdot$ (multiplication), are defined with the properties

(i) R is a commutative group under addition, with 0 (*zero*) as identity element and $-a$ as inverse of a;
(ii) multiplication is associative: $(ab)c = a(bc)$ for all $a, b, c \in R$;
(iii) there exists an identity element 1 (*one*) for multiplication: $a1 = a = 1a$ for every $a \in R$;
(iv) addition and multiplication are connected by the two distributive laws:

$$(a + b)c = (ac) + (bc), \quad c(a + b) = (ca) + (cb) \quad \text{for all } a, b, c \in R.$$

The elements 0 and 1 are necessarily uniquely determined. If, in addition, multiplication is commutative:

$$ab = ba \quad \text{for all } a, b \in R,$$

then R is said to be a *commutative* ring. In a commutative ring either one of the two distributive laws implies the other.

It may seem inconsistent to require that addition is commutative, but not multiplication. However, the commutative law for addition is actually a consequence of the other axioms for a ring. For, by the first distributive law we have

$$(a + b)(1 + 1) = a(1 + 1) + b(1 + 1) = a + a + b + b,$$

and by the second distributive law

$$(a + b)(1 + 1) = (a + b)1 + (a + b)1 = a + b + a + b.$$

Since a ring is a group under addition, by comparing these two relations we obtain first

$$a + a + b = a + b + a$$

and then $a + b = b + a$.

As examples, the set $\mathbb{Z}$ of all integers is a commutative ring, with the usual definitions of addition and multiplication, whereas if $n > 1$, the set $M_n(\mathbb{Z})$ of all $n \times n$ matrices with entries from $\mathbb{Z}$ is a noncommutative ring, with the usual definitions of matrix addition and multiplication.

A very different example is the collection $\mathscr{P}(X)$ of all subsets of a given set X. If we define the sum $A + B$ of two subsets A, B of X to be their *symmetric difference*, i.e. the set of all elements of X which are in either A or B, but not in both:

$$A + B = (A \cup B) \backslash (A \cap B) = (A \cup B) \cap (A^c \cup B^c),$$

and the product AB to be the set of all elements of X which are in both A and B:

$$AB = A \cap B,$$

it is not difficult to verify that $\mathscr{P}(X)$ is a commutative ring, with the empty set $\emptyset$ as identity element for addition and the whole set X as identity element for multiplication. For every $A \in \mathscr{P}(X)$, we also have

$$A + A = \emptyset, \quad AA = A.$$

The set operations are in turn determined by the ring operations:

$$A \cup B = A + B + AB, \quad A \cap B = AB, \quad A^c = A + X.$$

A ring R is said to be a *Boolean ring* if $aa = a$ for every $a \in R$. It follows that $a + a = 0$ for every $a \in R$, since

$$a + a = (a + a)(a + a) = a + a + a + a.$$

Moreover, a Boolean ring is commutative, since

$$a + b = (a + b)(a + b) = a + b + ab + ba$$

and $ba = -ba$, by what we have already proved.

For an arbitrary set X, any nonempty subset of $\mathscr{P}(X)$ which is closed under union, intersection and complementation can be given the structure of a Boolean ring in the manner just described. It was proved by Stone (1936) that every Boolean ring may be obtained in this way. Thus the algebraic laws of set theory may be replaced by the more familiar laws of algebra and all such laws are consequences of a small number among them.

We now return to arbitrary rings. In the same way as for $\mathbb{Z}$, in any ring R we have

$$a0 = 0 = 0a \quad \text{for every } a$$

and

$$(-a)b = -(ab) = a(-b) \quad \text{for all } a, b.$$

It follows that R contains only one element if $1 = 0$. We will say that the ring R is 'trivial' in this case.

Suppose R is a nontrivial ring. Then, viewing R as a group under addition, the cyclic subgroup $\langle 1 \rangle$ is either infinite, and isomorphic to $\mathbb{Z}/0\mathbb{Z}$, or finite of order s, and isomorphic to $\mathbb{Z}/s\mathbb{Z}$ for some positive integer s. The ring R is said to have *characteristic* 0 in the first case and *characteristic* s in the second case.

For any positive integer n, write

$$na := a + \cdots + a \quad (n \text{ summands}).$$

If R has characteristic $s > 0$, then $sa = 0$ for every $a \in R$, since

$$sa = (1 + \cdots + 1)a = 0a = 0.$$

On the other hand, $n1 \neq 0$ for every positive integer $n < s$, by the definition of characteristic.

An element a of a nontrivial ring R is said to be 'invertible' or a *unit* if there exists an element a^{-1} such that

$$a^{-1}a = 1 = aa^{-1}.$$

The element a^{-1} is then uniquely determined and is called the *inverse* of a. For example, 1 is a unit and is its own inverse. If a is a unit, then a^{-1} is also a unit and its inverse is a. If a and b are units, then ab is also a unit and its inverse is $b^{-1}a^{-1}$. It follows that the set $R^{\times}$ of all units is a group under multiplication.

A nontrivial ring R in which every nonzero element is invertible is said to be a *division ring*. Thus all nonzero elements of a division ring form a group under multiplication, the *multiplicative group* of the division ring. A *field* is a commutative division ring.

A nontrivial commutative ring R is said to be an *integral domain* if it has no 'divisors of zero', i.e. if $a \neq 0$ and $b \neq 0$ imply $ab \neq 0$. A division ring also has no divisors of zero, since if $a \neq 0$ and $b \neq 0$, then $a^{-1}ab = b \neq 0$, and hence $ab \neq 0$.

As examples, the set $\mathbb{Q}$ of rational numbers, the set $\mathbb{R}$ of real numbers and the set $\mathbb{C}$ of complex numbers are all fields, with the usual definitions of addition and multiplication. The set $\mathbb{H}$ of quaternions is a division ring, and the set $\mathbb{Z}$ of integers is an integral domain, but neither is a field.

In a ring with no divisors of zero, the additive order of any nonzero element a is the same as the additive order of 1, since $ma = (m1)a = 0$ if and only if $m1 = 0$. Furthermore, the characteristic of such a ring is either 0 or a prime number. For assume $n = lm$, where l and m are positive integers less than n. If $n1 = 0$, then

$$(l1)(m1) = n1 = 0.$$

Since there are no divisors of zero, either $l1 = 0$ or $m1 = 0$, and hence the characteristic cannot be n.

A subset S of a ring R is said to be a (two-sided) *ideal* if it is a subgroup of R under addition and if, for every $a \in S$ and $c \in R$, both $ac \in S$ and $ca \in S$.

Any ring R has two obvious ideals, namely R itself and the subset $\{0\}$. It is said to be *simple* if it has no other ideals and is nontrivial.

Any division ring is simple. For if an ideal S of a division ring R contains $a \neq 0$, then for every $c \in R$ we have $c = (ca^{-1})a \in S$.

Conversely, if a *commutative* ring R is simple, then it is a field. For, if a is any nonzero element of R, the set

$$S_a = \{xa : x \in R\}$$

is an ideal (since R is commutative). Since S_a contains $1a = a \neq 0$, we must have $S_a = R$. Hence $1 = xa$ for some $x \in R$. Thus every nonzero element of R is invertible.

If R is a commutative ring and $a_1, \ldots, a_m \in R$, then the set S consisting of all elements $x_1 a_1 + \cdots + x_m a_m$, where $x_j \in R$ $(1 \leq j \leq m)$, is clearly an ideal of R, the ideal *generated* by $a_1, \ldots, a_m$. An ideal of this type is said to be *finitely generated*.

We now show that if S is an ideal of the ring R, then the set $\mathscr{S}$ of all cosets $S + a$ of S can be given the structure of a ring. The ring R is a commutative group under addition. Hence, as we saw in §7, $\mathscr{S}$ acquires the structure of a (commutative) group under addition if we define the sum of $S+a$ and $S+b$ to be $S+(a+b)$. If $x = s+a$ and $x' = s' + b$ for some $s, s' \in S$, then $xx' = s'' + ab$, where $s'' = ss' + as' + sb$. Since S is an ideal, $s'' \in S$. Thus without ambiguity we may define the product of the cosets $S + a$ and $S + b$ to be the coset $S + ab$. Evidently multiplication is associative, $S + 1$ is an identity element for multiplication and both distributive laws hold. The new ring thus constructed is called the *quotient ring* of R by the ideal S, and is denoted by R/S.

A mapping $f : R \rightarrow R'$ of a ring R into a ring R' is said to be a (ring) *homomorphism* if, for all $a, b \in R$,

$$f(a + b) = f(a) + f(b), \quad f(ab) = f(a)f(b),$$

and if $f(1) = 1'$ is the identity element for multiplication in R'.

The *kernel* of the homomorphism f is the set N of all $a \in R$ such that $f(a) = 0'$ is the identity element for addition in R'. The kernel is an ideal of R, since it is a subgroup under addition and since $a \in N$, $c \in R$ imply $ac \in N$ and $ca \in N$.

For any $a \in R$, put $a' = f(a) \in R'$. The coset $N + a$ is the set of all $x \in R$ such that $f(x) = a'$, and the map $N+a \rightarrow a'$ is a bijection from the collection of all cosets of N to $f(R)$. Since f is a homomorphism, $N + (a + b)$ is mapped to $a' + b'$ and $N + ab$ is mapped to $a'b'$. Hence the map $N + a \rightarrow a'$ is also a homomorphism of the quotient ring R/N into $f(R)$.

A mapping $f : R \rightarrow R'$ of a ring R into a ring R' is said to be a (ring) *isomorphism* if it is both bijective and a homomorphism. The inverse mapping $f^{-1}: R' \rightarrow R$ is then also an isomorphism. (An *automorphism* of a ring R is an isomorphism of R with itself.)

Thus we have shown that, if $f : R \rightarrow R'$ is a homomorphism of a ring R into a ring R', with kernel N, then the quotient ring R/N is isomorphic to $f(R)$.

An ideal M of a ring R is said to be *maximal* if $M \neq R$ and if there are no ideals S such that $M \subset S \subset R$.

Let M be an ideal of the ring R. If S is an ideal of R which contains M, then the set S' of all cosets $M + a$ with $a \in S$ is an ideal of R/M. Conversely, if S' is an ideal of R/M, then the set S of all $a \in R$ such that $M + a \in S'$ is an ideal of R which contains M. It follows that M is a maximal ideal of R if and only if R/M is simple. Hence an ideal M of a commutative ring R is maximal if and only if the quotient ring R/M is a field.

To conclude, we mention a simple way of creating new rings from given ones. Let R, R' be rings and let $R \times R'$ be the set of all ordered pairs (a, a') with $a \in R$ and $a' \in R'$. As we saw in the previous section, $R \times R'$ acquires the structure of a (commutative) group under addition if we define the sum $(a, a') + (b, b')$ of (a, a') and (b, b') to be $(a + b, a' + b')$. If we define their product $(a, a') \cdot (b, b')$ to be $(ab, a'b')$, then $R \times R'$ becomes a ring, with $(0, 0')$ as identity element for addition and $(1, 1')$ as identity element for multiplication. The ring thus constructed is called the *direct sum* of R and R', and is denoted by $R \oplus R'$.

9 Vector Spaces and Associative Algebras

Although we assume some knowledge of linear algebra, it may be useful to place the basic definitions and results in the context of the preceding sections. A set V is said to be a *vector space* over a division ring D if it is a commutative group under an operation $+$ (addition) and there exists a map $\varphi : D \times V \rightarrow V$ (multiplication by a scalar) such that, if $\varphi(\alpha, v)$ is denoted by αv then, for all $\alpha, \beta \in D$ and all $v, w \in V$,

(i) $\alpha(v + w) = \alpha v + \alpha w$,
(ii) $(\alpha + \beta)v = \alpha v + \beta v$,
(iii) $(\alpha\beta)v = \alpha(\beta v)$,
(iv) $1v = v$,

where 1 is the identity element for multiplication in D. The elements of V will be called *vectors* and the elements of D *scalars*.

For example, for any positive integer n, the set D^n of all n-tuples of elements of the division ring D is a vector space over D if addition and multiplication by a scalar are defined by

$$(\alpha_1, \ldots, \alpha_n) + (\beta_1, \ldots, \beta_n) = (\alpha_1 + \beta_1, \ldots, \alpha_n + \beta_n),$$
$$\alpha(\alpha_1, \ldots, \alpha_n) = (\alpha\alpha_1, \ldots, \alpha\alpha_n).$$

The special cases $D = \mathbb{R}$ and $D = \mathbb{C}$ have many applications.

As another example, the set $\mathscr{C}(I)$ of all continuous functions $f : I \to \mathbb{R}$, where I is an interval of the real line, is a vector space over the field $\mathbb{R}$ of real numbers if addition and multiplication by a scalar are defined, for every $t \in I$, by

$$(f + g)(t) = f(t) + g(t),$$
$$(\alpha f)(t) = \alpha f(t).$$

Let V be an arbitrary vector space over a division ring D. If O is the identity element of V with respect to addition, then

$$\alpha O = O \quad \text{for every } \alpha \in D,$$

since $\alpha O = \alpha(O + O) = \alpha O + \alpha O$. Similarly, if 0 is the identity element of D with respect to addition, then

$$0v = O \quad \text{for every } v \in V,$$

since $0v = (0 + 0)v = 0v + 0v$. Furthermore,

$$(-\alpha)v = -(\alpha v) \quad \text{for all } \alpha \in D \text{ and } v \in V,$$

since $O = 0v = (\alpha + (-\alpha))v = \alpha v + (-\alpha)v$, and

$$\alpha v \neq O \quad \text{if } \alpha \neq 0 \text{ and } v \neq O,$$

since $\alpha^{-1}(\alpha v) = (\alpha^{-1}\alpha)v = 1v = v$.

From now on we will denote the zero elements of D and V by the same symbol 0. This is easier on the eye and in practice is not confusing.

A subset U of a vector space V is said to be a *subspace* of V if it is a vector space under the same operations as V itself. It is easily seen that a nonempty subset U is a subspace of V if (and only if) it is closed under addition and multiplication by a scalar. For then, if $u \in U$, also $-u = (-1)u \in U$, and so U is an additive subgroup of V. The other requirements for a vector space are simply inherited from V.

For example, if $1 \leq m < n$, the set of all $(\alpha_1, \ldots, \alpha_n) \in D^n$ with $\alpha_1 = \cdots = \alpha_m = 0$ is a subspace of D^n. Also, the set $\mathscr{C}^1(I)$ of all continuously differentiable functions $f : I \to \mathbb{R}$ is a subspace of $\mathscr{C}(I)$. Two obvious subspaces of any vector space V are V itself and the subset $\{0\}$ which contains only the zero vector.

If U_1 and U_2 are subspaces of a vector space V, then their *intersection* $U_1 \cap U_2$, which necessarily contains 0, is again a subspace of V. The *sum* $U_1 + U_2$, consisting of all vectors $u_1 + u_2$ with $u_1 \in U_1$ and $u_2 \in U_2$, is also a subspace of V. Evidently $U_1 + U_2$ contains U_1 and U_2 and is contained in every subspace of V which contains both U_1 and U_2. If $U_1 \cap U_2 = \{0\}$, the sum $U_1 + U_2$ is said to be *direct*, and is denoted by $U_1 \oplus U_2$, since it may be identified with the set of all ordered pairs (u_1, u_2), where $u_1 \in U_1$ and $u_2 \in U_2$.

Let V be an arbitrary vector space over a division ring D and let $\{v_1, \ldots, v_m\}$ be a finite subset of V. A vector v in V is said to be a *linear combination* of $v_1, \ldots, v_m$ if

$$v = \alpha_1 v_1 + \cdots + \alpha_m v_m$$

for some $\alpha_1, \ldots, \alpha_m \in D$. The *coefficients* $\alpha_1, \ldots, \alpha_m$ need not be uniquely determined. Evidently a vector v is a linear combination of $v_1, \ldots, v_m$ if it is a linear combination of some proper subset, since we can add the remaining vectors with zero coefficients.

If S is any nonempty subset of V, then the set $\langle S \rangle$ of all vectors in V which are linear combinations of finitely many elements of S is a subspace of V, the subspace 'spanned' or *generated* by S. Clearly $S \subseteq \langle S \rangle$ and $\langle S \rangle$ is contained in every subspace of V which contains S.

A finite subset $\{v_1, \ldots, v_m\}$ of V is said to be *linearly dependent* (over D) if there exist $\alpha_1, \ldots, \alpha_m \in D$, not all zero, such that

$$\alpha_1 v_1 + \cdots + \alpha_m v_m = 0,$$

and is said to be *linearly independent* otherwise.

For example, in $\mathbb{R}^3$ the vectors

$$v_1 = (1, 0, 1), \quad v_2 = (1, 1, 0), \quad v_3 = (1, 1/2, 1/2)$$

are linearly dependent, since $v_1 + v_2 - 2v_3 = 0$. On the other hand, the vectors

$$e_1 = (1, 0, 0), \quad e_2 = (0, 1, 0), \quad e_3 = (0, 0, 1)$$

are linearly independent, since $\alpha_1 e_1 + \alpha_2 e_2 + \alpha_3 e_3 = (\alpha_1, \alpha_2, \alpha_3)$, and this is 0 only if $\alpha_1 = \alpha_2 = \alpha_3 = 0$.

In any vector space V, the set $\{v\}$ containing the single vector v is linearly independent if $v \neq 0$ and linearly dependent if $v = 0$. If $v_1, \ldots, v_m$ are linearly independent, then any vector $v \in \langle v_1, \ldots, v_m \rangle$ has a unique representation as a linear combination of $v_1, \ldots, v_m$, since if

$$\alpha_1 v_1 + \cdots + \alpha_m v_m = \beta_1 v_1 + \cdots + \beta_m v_m,$$

then

$$(\alpha_1 - \beta_1) v_1 + \cdots + (\alpha_m - \beta_m) v_m = 0$$

and hence

$$\alpha_1 - \beta_1 = \cdots = \alpha_m - \beta_m = 0.$$

Evidently the vectors $v_1, \ldots, v_m$ are linearly dependent if some proper subset is linearly dependent. Hence any nonempty subset of a linearly independent set is again linearly independent.

A subset S of a vector space V is said to be a *basis* for V if S is linearly independent and $\langle S \rangle = V$. In the previous example, the vectors e_1, e_2, e_3 are a basis for $\mathbb{R}^3$, since they are not only linearly independent but also generate $\mathbb{R}^3$.

Any nontrivial finitely generated vector space has a basis. In fact if a vector space V is generated by a finite subset T, then V has a basis $B \subseteq T$. Moreover, any linearly independent subset of V is also finite and its cardinality does not exceed that of T. It follows that any two bases contain the same number of elements.

If V has a basis containing n elements, we say V has *dimension n* and we write
$\dim V = n$. We say that V has infinite dimension if it is not finitely generated, and has
dimension 0 if it contains only the vector 0.

For example, the field $\mathbb{C}$ of complex numbers may be regarded as a 2-dimensional
vector space over the field $\mathbb{R}$ of real numbers, with basis $\{1, i\}$.

Again, D^n has dimension n as a vector space over the division ring D, since it has
the basis

$$e_1 = (1, 0, \ldots, 0), \quad e_2 = (0, 1, \ldots, 0), \ldots, \quad e_n = (0, 0, \ldots, 1).$$

On the other hand, the real vector space $\mathscr{C}(I)$ of all continuous functions $f : I \to \mathbb{R}$
has infinite dimension if the interval I contains more than one point since, for any
positive integer n, the real polynomials of degree less than n form an n-dimensional
subspace.

The first of these examples is readily generalized. If E and F are fields with
$F \subseteq E$, we can regard E as a vector space over F. If this vector space is finite-
dimensional, we say that E is a *finite extension* of F and define the *degree* of E over
F to be the dimension $[E : F]$ of this vector space.

Any subspace U of a finite-dimensional vector space V is again finite-dimensional.
Moreover, $\dim U \le \dim V$, with equality only if $U = V$. If U_1 and U_2 are subspaces
of V, then

$$\dim(U_1 + U_2) + \dim(U_1 \cap U_2) = \dim U_1 + \dim U_2.$$

Let V and W be vector spaces over the same division ring D. A map $T : V \to W$
is said to be *linear*, or a *linear transformation*, or a 'vector space homomorphism', if
for all $v, v' \in V$ and every $\alpha \in D$,

$$T(v + v') = Tv + Tv', \quad T(\alpha v) = \alpha(Tv).$$

Since the first condition implies that T is a homomorphism of the additive group of V
into the additive group of W, it follows that $T0 = 0$ and $T(-v) = -Tv$.

For example, if (τ_{jk}) is an $m \times n$ matrix with entries from the division ring D, then
the map $T : D^m \to D^n$ defined by

$$T(\alpha_1, \ldots, \alpha_m) = (\beta_1, \ldots, \beta_n),$$

where

$$\beta_k = \alpha_1 \tau_{1k} + \cdots + \alpha_m \tau_{mk} \quad (1 \le k \le n),$$

is linear. It is easily seen that every linear map of D^m into D^n may be obtained in this
way.

As another example, if $\mathscr{C}^1(I)$ is the real vector space of all continuously differentiable functions $f : I \to R$, then the map $T : \mathscr{C}^1(I) \to \mathscr{C}(I)$ defined by $Tf = f'$
(the derivative of f) is linear.

Let U, V, W be vector spaces over the same division ring D. If $T : V \to W$ and
$S : U \to V$ are linear maps, then the composite map $T \circ S : U \to W$ is again
linear. For linear maps it is customary to write TS instead of $T \circ S$. The identity map

$I : V \to V$ defined by $Iv = v$ for every $v \in V$ is clearly linear. If a linear map $T : V \to W$ is bijective, then its inverse map $T^{-1} : W \to V$ is again linear.

If $T : V \to W$ is a linear map, then the set N of all $v \in V$ such that $Tv = 0$ is a subspace of V, called the *nullspace* or *kernel* of T. Since $Tv = Tv'$ if and only if $T(v - v') = 0$, the map T is injective if and only if its kernel is $\{0\}$, i.e. when T is *nonsingular*.

For any subspace U of V, its image $TU = \{Tv : v \in U\}$ is a subspace of W. In particular, TV is a subspace of W, called the *range* of T. Thus the map T is surjective if and only if its range is W.

If V is finite-dimensional, then the range R of T is also finite-dimensional and

$$\dim R = \dim V - \dim N,$$

(since $R \approx V/N$). The dimensions of R and N are called respectively the *rank* and *nullity* of T. It follows that, if $\dim V = \dim W$, then T is injective if and only if it is surjective.

Two vector spaces V, W over the same division ring D are said to be *isomorphic* if there exists a bijective linear map $T : V \to W$. As an example, if V is an n-dimensional vector space over the division ring D, then V is isomorphic to D^n. For if $v_1, \ldots, v_n$ is a basis for V and if $v = \alpha_1 v_1 + \cdots + \alpha_n v_n$ is an arbitrary element of V, the map $v \to (\alpha_1, \ldots, \alpha_n)$ is linear and bijective.

Thus there is essentially only one vector space of given finite dimension over a given division ring. However, vector spaces do not always present themselves in the concrete form D^n. An example is the set of solutions of a system of homogeneous linear equations with real coefficients. Hence, even if one is only interested in the finite-dimensional case, it is still desirable to be acquainted with the abstract definition of a vector space.

Let V and W be vector spaces over the same division ring D. We can define the *sum* $S + T$ of two linear maps $S : V \to W$ and $T : V \to W$ by

$$(S + T)v = Sv + Tv.$$

This is again a linear map, and it is easily seen that with this definition of addition the set of all linear maps of V into W is a commutative group. If D is a field, i.e. if multiplication in D is commutative, then for any $\alpha \in D$ the map αT defined by

$$(\alpha T)v = \alpha(Tv)$$

is again linear, and with these definitions of addition and multiplication by a scalar the set of all linear maps of V into W is a vector space over D. (If the division ring D is not a field, it is necessary to consider 'right' vector spaces over D, as well as 'left' ones.)

If $V = W$, then the *product* TS is also defined and it is easily verified that the set of all linear maps of V into itself is a ring, with the identity map I as identity element for multiplication. The bijective linear maps of V to itself are the units of this ring and thus form a group under multiplication, the *general linear group* $GL(V)$.

Similarly to the direct product of two groups and the direct sum of two rings, one may define the *tensor product* $V \otimes V'$ of two vector spaces V, V' and the *Kronecker product* $T \otimes T'$ of two linear maps $T : V \to W$ and $T' : V' \to W'$.

The *centre* of a ring R is the set of all $c \in R$ such that $ac = ca$ for every $a \in R$. An *associative algebra* A over a field F is a ring containing F in its centre. On account of the ring structure, we can regard A as a vector space over F. The associative algebra is said to be *finite-dimensional* if it is finite-dimensional as a vector space over F.

For example, the set $M_n(F)$ of all $n \times n$ matrices with entries from the field F is a finite-dimensional associative algebra, with the usual definitions of addition and multiplication, and with $\alpha \in F$ identified with the matrix αI.

More generally, if D is a division ring containing F in its centre, then the set $M_n(D)$ of all $n \times n$ matrices with entries from D is an associative algebra over F. It is finite-dimensional if D itself is finite dimensional over F.

By the definition for rings, an associative algebra A is *simple* if $A \neq \{0\}$ and A has no ideals except $\{0\}$ and A. It is not difficult to show that, for any division ring D containing F in its centre, the associative algebra $M_n(D)$ is simple. It was proved by Wedderburn (1908) that any finite-dimensional simple associative algebra has the form $M_n(D)$, where D is a division ring containing F in its centre and of finite dimension over F.

If $F = \mathbb{C}$, the fundamental theorem of algebra implies that $\mathbb{C}$ is the only such D. If $F = \mathbb{R}$, there are three choices for D, by the following theorem of Frobenius (1878):

Proposition 31 *If a division ring D contains the real field $\mathbb{R}$ in its centre and is of finite dimension as a vector space over $\mathbb{R}$, then D is isomorphic to $\mathbb{R}, \mathbb{C}$ or $\mathbb{H}$.*

Proof Suppose first that D is a field and $D \neq \mathbb{R}$. If $a \in D\backslash\mathbb{R}$ then, since D is finite-dimensional over $\mathbb{R}$, a is a root of a monic polynomial with real coefficients, which we may assume to be of minimal degree. Since $a \notin \mathbb{R}$, the degree is not 1 and the fundamental theorem of algebra implies that it must be 2. Thus

$$a^2 - 2\lambda a + \mu = 0$$

for some $\lambda, \mu \in \mathbb{R}$ with $\lambda^2 < \mu$. Then $\mu - \lambda^2 = \rho^2$ for some nonzero $\rho \in \mathbb{R}$ and $i = (a - \lambda)/\rho$ satisfies $i^2 = -1$. Thus D contains the field $\mathbb{R}(i) = \mathbb{R} + i\mathbb{R}$. But, since D is a field, the only $x \in D$ such that $x^2 = -1$ are i and $-i$. Hence the preceding argument shows that actually $D = \mathbb{R}(i)$. Thus D is isomorphic to the field $\mathbb{C}$ of complex numbers.

Suppose now that D is not commutative. Let a be an element of D which is not in the centre of D, and let M be an $\mathbb{R}$-subspace of D of maximal dimension which is commutative and which contains both a and the centre of D. If $x \in D$ commutes with every element of M, then $x \in M$. Hence M is a maximal commutative subset of D. It follows that if $x \in M$ and $x \neq 0$ then also $x^{-1} \in M$, since $xy = yx$ for all $y \in M$ implies $yx^{-1} = x^{-1}y$ for all $y \in M$. Similarly $x, x' \in M$ implies $xx' \in M$. Thus M is a field which properly contains $\mathbb{R}$. Hence, by the first part of the proof, M is isomorphic to $\mathbb{C}$. Thus $M = \mathbb{R}(i)$, where $i^2 = -1$, $[M : \mathbb{R}] = 2$ and $\mathbb{R}$ is the centre of D.

If $x \in D\backslash M$, then $b = (x + ixi)/2$ satisfies

$$bi = (xi - ix)/2 = -ib \neq 0.$$

Hence $b \in D\backslash M$ and $b^2 i = ib^2$. But, in the same way as before, $N = \mathbb{R} + \mathbb{R}b$ is a maximal subfield of D containing b and $\mathbb{R}$, and $N = \mathbb{R}(j)$, where $j^2 = -1$. Thus $b^2 = \alpha + \beta b$, where $\alpha, \beta \in \mathbb{R}$. In fact, since $b^2 i = ib^2$, we must have $\beta = 0$. Similarly

$j = \gamma + \delta b$, where $\gamma, \delta \in \mathbb{R}$ and $\delta \neq 0$. Since $j^2 = \gamma^2 + 2\gamma\,\delta b + \delta^2 a = -1$, we must have $\gamma = 0$. Thus $j = \delta b$ and $ji = -ij$.

If we put $k = ij$, it now follows that

$$k^2 = -1, \quad jk = i = -kj, \quad ki = j = -ik.$$

Since no $\mathbb{R}$-linear combination of $1, i, j$ has these properties, the elements $1, i, j, k$ are $\mathbb{R}$-linearly independent. But, by Proposition 32 below, $[D : M] = [M : \mathbb{R}] = 2$. Hence $[D : \mathbb{R}] = 4$ and $1, i, j, k$ are a basis for D over $\mathbb{R}$. Thus D is isomorphic to the division ring $\mathbb{H}$ of quaternions. $\qquad\square$

To complete the proof of Proposition 31 we now prove

Proposition 32 *Let D be a division ring which, as a vector space over its centre C, has finite dimension $[D : C]$. If M is a maximal subfield of D, then $[D : M] = [M : C]$.*

Proof Put $n = [D : C]$ and let $e_1, \ldots, e_n$ be a basis for D as a vector space over C. Obviously we may suppose $n > 1$. We show first that if $a_1, \ldots, a_n$ are elements of D such that

$$a_1 x e_1 + \cdots + a_n x e_n = 0 \quad \text{for every } x \in D,$$

then $a_1 = \cdots = a_n = 0$. Assume that there exists such a set $\{a_1, \ldots, a_n\}$ with not all elements zero and choose one with the minimal number of nonzero elements. We may suppose the notation chosen so that $a_i \neq 0$ for $i \leq r$ and $a_i = 0$ for $i > r$ and, by multiplying on the left by a_1^{-1}, we may further suppose that $a_1 = 1$. For any $y \in D$ we have

$$a_1 y x e_1 + \cdots + a_n y x e_n = 0 = y(a_1 x e_1 + \cdots + a_n x e_n)$$

and hence

$$(a_1 y - ya_1)x e_1 + \cdots + (a_n y - ya_n)x e_n = 0.$$

Since $a_i y = ya_i$ for $i = 1$ and for $i > r$, our choice of $\{a_1, \ldots, a_n\}$ implies that $a_i y = ya_i$ for all i. Since this holds for every $y \in D$, it follows that $a_i \in C$ for all i. But this is a contradiction, since $e_1, \ldots, e_n$ is a basis for D over C and $a_1 e_1 + \cdots + a_n e_n = 0$.

The map $T_{jk} : D \to D$ defined by $T_{jk}x = e_j x e_k$ is a linear transformation of D as a vector space over C. By what we have just proved, the n^2 linear maps T_{jk} ($j, k = 1, \ldots, n$) are linearly independent over C. Consequently every linear transformation of D as a vector space over C is a C-linear combination of the maps T_{jk}.

Suppose now that $T : D \to D$ is a linear transformation of D as a vector space over M. Since $C \subseteq M$, T is also a linear transformation of D as a vector space over C and hence has the form

$$Tx = a_1 x e_1 + \cdots + a_n x e_n$$

for some $a_1, \ldots, a_n \in D$. But $T(bx) = b(Tx)$ for all $b \in M$ and $x \in D$. Hence

$$(a_1 b - ba_1)x e_1 + \cdots + (a_n b - ba_n)x e_n = 0 \quad \text{for every } x \in D,$$

which implies $a_i b = b a_i$ $(i = 1, \ldots, n)$. Since this holds for all $b \in M$ and M is a maximal subfield of D, it follows that $a_i \in M$ $(i = 1, \ldots, n)$.

Let $\mathscr{T}$ denote the set of all linear transformations of D as a vector space over M. By what we have already proved, every $T \in \mathscr{T}$ is an M-linear combination of the maps $T_1, \ldots, T_n$, where $T_i x = x e_i$ $(i = 1, \ldots, n)$, and the maps $T_1, \ldots, T_n$ are linearly independent over M. Consequently the dimension of $\mathscr{T}$ as a vector space over M is n. But $\mathscr{T}$ has dimension $[D : M]^2$ as a vector space over M. Hence $[D : M]^2 = n$. Since $n = [D : M][M : C]$, it follows that $[D : M] = [M : C]$. $\qquad\square$

10 Inner Product Spaces

Let F denote either the real field $\mathbb{R}$ or the complex field $\mathbb{C}$. A vector space V over F is said to be an *inner product space* if there exists a map $(u, v) \to \langle u, v \rangle$ of $V \times V$ into F such that for every $\alpha \in F$ and all $u, u', v \in V$,

(i) $\langle \alpha u, v \rangle = \alpha \langle u, v \rangle$,
(ii) $\langle u + u', v \rangle = \langle u, v \rangle + \langle u', v \rangle$,
(iii) $\langle v, u \rangle = \overline{\langle u, v \rangle}$,
(iv) $\langle u, u \rangle > 0$ if $u \neq O$.

If $F = \mathbb{R}$, then (iii) simply says that $\langle v, u \rangle = \langle u, v \rangle$, since a real number is its own complex conjugate. The restriction $u \neq O$ is necessary in (iv), since (i) and (iii) imply that

$$\langle u, O \rangle = \langle O, v \rangle = 0 \quad \text{for all } u, v \in V.$$

It follows from (ii) and (iii) that

$$\langle u, v + v' \rangle = \langle u, v \rangle + \langle u, v' \rangle \quad \text{for all } u, v, v' \in V,$$

and from (i) and (iii) that

$$\langle u, \alpha v \rangle = \bar{\alpha} \langle u, v \rangle \quad \text{for every } \alpha \in F \text{ and all } u, v \in V.$$

The standard example of an inner product space is the vector space F^n, with the inner product of $x = (\xi_1, \ldots, \xi_n)$ and $y = (\eta_1, \ldots, \eta_n)$ defined by

$$\langle x, y \rangle = \xi_1 \bar{\eta}_1 + \cdots + \xi_n \bar{\eta}_n.$$

Another example is the vector space $\mathscr{C}(I)$ of all continuous functions $f : I \to F$, where $I = [a, b]$ is a compact subinterval of $\mathbb{R}$, with the inner product of f and g defined by

$$\langle f, g \rangle = \int_a^b f(t)\overline{g(t)}dt.$$

In an arbitrary inner product space V we define the *norm* $\|v\|$ of a vector $v \in V$ by

$$\|v\| = \langle v, v \rangle^{1/2}.$$

Thus $\|v\| \geq 0$, with equality if and only if $v = O$. Evidently

$$\|\alpha v\| = |\alpha| \|v\| \quad \text{for all } \alpha \in F \text{ and } v \in V.$$

Inner products and norms are connected by *Schwarz's inequality*:

$$|\langle u, v \rangle| \leq \|u\| \|v\| \quad \text{for all } u, v \in V,$$

with equality if and only if u and v are linearly dependent. For the proof we may suppose that u and v are linearly independent, since it is easily seen that equality holds if $u = \lambda v$ or $v = \lambda u$ for some $\lambda \in F$. Then, for all $\alpha, \beta \in F$, not both 0,

$$0 < \langle \alpha u + \beta v, \alpha u + \beta v \rangle = |\alpha|^2 \langle u, u \rangle + \alpha \bar{\beta} \langle u, v \rangle + \bar{\alpha} \beta \overline{\langle u, v \rangle} + |\beta|^2 \langle v, v \rangle.$$

If we choose $\alpha = \langle v, v \rangle$ and $\beta = -\langle u, v \rangle$, this takes the form

$$0 < \|u\|^2 \|v\|^4 - 2\|v\|^2 |\langle u, v \rangle|^2 + |\langle u, v \rangle|^2 \|v\|^2 = \{\|u\|^2 \|v\|^2 - |\langle u, v \rangle|^2\} \|v\|^2.$$

Hence

$$|\langle u, v \rangle|^2 < \|u\|^2 \|v\|^2,$$

as we wished to show. We follow common practice by naming the inequality after Schwarz (1885), but (cf. §4) it had already been proved for $\mathbb{R}^n$ by Cauchy (1821) and for $\mathscr{C}(I)$ by Bunyakovskii (1859).

It follows from Schwarz's inequality that

$$\|u + v\|^2 = \|u\|^2 + 2\mathscr{R}\langle u, v \rangle + \|v\|^2$$
$$\leq \|u\|^2 + 2|\langle u, v \rangle| + \|v\|^2 \leq \{\|u\| + \|v\|\}^2.$$

Thus

$$\|u + v\| \leq \|u\| + \|v\| \quad \text{for all } u, v \in V,$$

with strict inequality if u and v are linearly independent.

It now follows that V acquires the structure of a metric space if we define the distance between u and v by

$$d(u, v) = \|u - v\|.$$

In the case $V = \mathbb{R}^n$ this is the *Euclidean distance*

$$d(x, y) = \left(\sum_{j=1}^{n} |\xi_j - \eta_j|^2 \right)^{1/2},$$

and in the case $V = \mathscr{C}(I)$ it is the L^2-*norm*

$$d(f, g) = \left(\int_a^b |f(t) - g(t)|^2 dt \right)^{1/2}.$$

The norm in any inner product space V satisfies the *parallelogram law*:

$$\|u + v\|^2 + \|u - v\|^2 = 2\|u\|^2 + 2\|v\|^2 \quad \text{for all } u, v \in V.$$

This may be immediately verified by substituting $\|w\|^2 = \langle w, w \rangle$ throughout and using the linearity of the inner product. The geometrical interpretation is that in any parallelogram the sum of the squares of the lengths of the two diagonals is equal to the sum of the squares of the lengths of all four sides.

It may be shown that any normed vector space which satisfies the parallelogram law can be given the structure of an inner product space by defining

$$\langle u, v \rangle = \{\|u + v\|^2 - \|u - v\|^2\}/4 \quad \text{if } F = \mathbb{R},$$
$$= \{\|u + v\|^2 - \|u - v\|^2 + i\|u + iv\|^2 - i\|u - iv\|^2\}/4 \quad \text{if } F = \mathbb{C}.$$

(Cf. the argument for $F = \mathbb{Q}$ in §4 of Chapter XIII.)

In an arbitrary inner product space V a vector u is said to be 'perpendicular' or *orthogonal* to a vector v if $\langle u, v \rangle = 0$. The relation is symmetric, since $\langle u, v \rangle = 0$ implies $\langle v, u \rangle = 0$. For orthogonal vectors u, v, the *law of Pythagoras* holds:

$$\|u + v\|^2 = \|u\|^2 + \|v\|^2.$$

More generally, a subset E of V is said to be *orthogonal* if $\langle u, v \rangle = 0$ for all $u, v \in E$ with $u \neq v$. It is said to be *orthonormal* if, in addition, $\langle u, u \rangle = 1$ for every $u \in E$. An orthogonal set which does not contain O may be converted into an orthonormal set by replacing each $u \in E$ by $u/\|u\|$.

For example, if $V = F^n$, then the basis vectors

$$e_1 = (1, 0, \ldots, 0), \quad e_2 = (0, 1, \ldots, 0), \ldots, \quad e_n = (0, 0, \ldots, 1)$$

form an orthonormal set. It is easily verified also that, if $I = [0, 1]$, then in $\mathscr{C}(I)$ the functions $e_n(t) = e^{2\pi i n t}$ $(n \in \mathbb{Z})$ form an orthonormal set.

Let $\{e_1, \ldots, e_m\}$ be *any* orthonormal set in the inner product space V and let U be the vector subspace generated by $e_1, \ldots, e_m$. Then the norm of a vector $u = \alpha_1 e_1 + \cdots + \alpha_m e_m \in U$ is given by

$$\|u\|^2 = |\alpha_1|^2 + \cdots + |\alpha_m|^2,$$

which shows that $e_1, \ldots, e_m$ are linearly independent.

To find the *best approximation* in U to a given vector $v \in V$, put

$$w = \gamma_1 e_1 + \cdots + \gamma_m e_m,$$

where

$$\gamma_j = \langle v, e_j \rangle \quad (j = 1, \ldots, m).$$

Then $\langle w, e_j \rangle = \langle v, e_j \rangle (j = 1, \ldots, m)$ and hence $\langle v - w, w \rangle = 0$. Consequently, by the law of Pythagoras,

$$\|v\|^2 = \|v - w\|^2 + \|w\|^2.$$

Since $\|w\|^2 = |\gamma_1|^2 + \cdots + |\gamma_m|^2$, this yields *Bessel's inequality*:

$$|\langle v, e_1 \rangle|^2 + \cdots + |\langle v, e_m \rangle|^2 \leq \|v\|^2,$$

with strict inequality if $v \notin U$. For any $u \in U$, we also have $\langle v - w, w - u \rangle = 0$ and so, by Pythagoras again,

$$\|v - u\|^2 = \|v - w\|^2 + \|w - u\|^2.$$

This shows that w is the unique nearest point of U to v.

From any linearly independent set of vectors $v_1, \ldots, v_m$ we can inductively construct an orthonormal set $e_1, \ldots, e_m$ such that $e_1, \ldots, e_k$ span the same vector subspace as $v_1, \ldots, v_k$ for $1 \leq k \leq m$. We begin by taking $e_1 = v_1/\|v_1\|$. Now suppose $e_1, \ldots, e_k$ have been determined. If

$$w = v_{k+1} - \langle v_{k+1}, e_1 \rangle e_1 - \cdots - \langle v_{k+1}, e_k \rangle e_k,$$

then $\langle w, e_j \rangle = 0 \ (j = 1, \ldots, k)$. Moreover $w \neq O$, since w is a linear combination of $v_1, \ldots, v_{k+1}$ in which the coefficient of v_{k+1} is 1. By taking $e_{k+1} = w/\|w\|$, we obtain an orthonormal set $e_1, \ldots, e_{k+1}$ spanning the same linear subspace as $v_1, \ldots, v_{k+1}$. This construction is known as *Schmidt's orthogonalization process*, because of its use by E. Schmidt (1907) in his treatment of linear integral equations. The (normalized) Legendre polynomials are obtained by applying the process to the linearly independent functions $1, t, t^2, \ldots$ in the space $\mathscr{C}(I)$, where $I = [-1, 1]$.

It follows that any finite-dimensional inner product space V has an orthonormal basis $e_1, \ldots, e_n$ and that

$$\|v\|^2 = \sum_{j=1}^{n} |\langle v, e_j \rangle|^2 \quad \text{for every } v \in V.$$

In an infinite-dimensional inner product space V an orthonormal set E may even be uncountably infinite. However, for a given $v \in V$, there are at most countably many vectors $e \in E$ for which $\langle v, e \rangle \neq 0$. For if $\{e_1, \ldots, e_m\}$ is any finite subset of E then, by Bessel's inequality,

$$\sum_{j=1}^{m} |\langle v, e_j \rangle|^2 \leq \|v\|^2$$

and so, for each $n \in \mathbb{N}$, there are at most $n^2 - 1$ vectors $e \in E$ for which $|\langle v, e \rangle| > \|v\|/n$.

If the vector subspace U of all finite linear combinations of elements of E is dense in V then, by the best approximation property of finite orthonormal sets, *Parseval's equality* holds:

$$\sum_{e \in E} |\langle v, e \rangle|^2 = \|v\|^2 \quad \text{for every } v \in V.$$

Parseval's equality holds for the inner product space $\mathscr{C}(I)$, where $I = [0, 1]$, and the orthonormal set $E = \{e^{2\pi i n t} : n \in \mathbb{Z}\}$ since, by *Weierstrass's approximation theorem* (see the references in §6 of Chapter XI), every $f \in \mathscr{C}(I)$ is the uniform limit of a sequence of *trigonometric polynomials*. The result in this case was formally derived by Parseval (1805).

An *almost periodic function*, in the sense of Bohr (1925), is a function $f : \mathbb{R} \to \mathbb{C}$ which can be uniformly approximated on $\mathbb{R}$ by *generalized trigonometric polynomials*

$$\sum_{j=1}^{m} c_j e^{i\lambda_j t},$$

where $c_j \in \mathbb{C}$ and $\lambda_j \in \mathbb{R}$ $(j = 1, \ldots, m)$. For any almost periodic functions f, g, the limit

$$\langle f, g \rangle = \lim_{T \to \infty} (1/2T) \int_{-T}^{T} f(t) \overline{g(t)} dt$$

exists. The set $\mathscr{B}$ of all almost periodic functions acquires in this way the structure of an inner product space. The set $E = \{e^{i\lambda t} : \lambda \in \mathbb{R}\}$ is an uncountable orthonormal set and Parseval's equality holds for this set.

A finite-dimensional inner product space is necessarily complete as a metric space, i.e., every fundamental sequence converges. However, an infinite-dimensional inner product space need not be complete, as $\mathscr{C}(I)$ already illustrates. An inner product space which is complete is said to be a *Hilbert space*.

The case considered by Hilbert (1906) was the vector space ℓ^2 of all infinite sequences $x = (\xi_1, \xi_2, \ldots)$ of complex numbers such that $\sum_{k \geq 1} |\xi_k|^2 < \infty$, with

$$\langle x, y \rangle = \sum_{k \geq 1} \xi_k \bar{\eta}_k.$$

Another example is the vector space $L^2(I)$, where $I = [0, 1]$, of all (equivalence classes of) Lebesgue measurable functions $f : I \to \mathbb{C}$ such that $\int_0^1 |f(t)^2 dt < \infty$, with

$$\langle f, g \rangle = \int_0^1 f(t) \overline{g(t)} dt.$$

With any $f \in L^2(I)$ we can associate a sequence $\hat{f} \in \ell^2$, consisting of the inner products $\langle f, e_n \rangle$, where $e_n(t) = e^{2\pi i n t} (n \in \mathbb{Z})$, in some fixed order. The map $\mathscr{F} : L^2(I) \to \ell^2$ thus defined is linear and, by Parseval's equality,

$$\|\mathscr{F}f\| = \|f\|.$$

In fact $\mathscr{F}$ is an *isometry* since, by the *theorem of Riesz–Fischer* (1907), it is bijective.

11 Further Remarks

A vast fund of information about numbers in different cultures is contained in Menninger [52]. A good popular book is Dantzig [18].

The algebra of sets was created by Boole (1847), who used the symbols $+$ and $\cdot$ instead of $\cup$ and $\cap$, as is now customary. His ideas were further developed, with applications to logic and probability theory, in Boole [10]. A simple system of axioms for

Boolean algebra was given by Huntingdon [39]. For an introduction to Stone's representation theorem, referred to in §8, see Stone [69]; there are proofs in Halmos [30] and Sikorski [66]. For applications of Boolean algebras to switching circuits see, for example, Rudeanu [62]. Boolean algebra is studied in the more general context of lattice theory in Birkhoff [6].

Dedekind's axioms for $\mathbb{N}$ may be found on p. 67 of [19], which contains also his earlier construction of the real numbers from the rationals by means of cuts. Some interesting comments on the axioms **(N1)–(N3)** are contained in Henkin [34]. Starting from these axioms, Landau [47] gives a detailed derivation of the basic properties of $\mathbb{N}, \mathbb{Q}, \mathbb{R}$ and $\mathbb{C}$.

The argument used to extend $\mathbb{N}$ to $\mathbb{Z}$ shows that any commutative *semigroup* satisfying the cancellation law may be embedded in a commutative *group*. The argument used to extend $\mathbb{Z}$ to $\mathbb{Q}$ shows that any commutative *ring* without divisors of zero may be embedded in a *field*.

An example of an ordered field which does not have the Archimedean property, although every fundamental sequence is (trivially) convergent, is the field $*\mathbb{R}$ of hyperreal numbers, constructed by Abraham Robinson (1961). Hyperreal numbers are studied in Stroyan and Luxemburg [70].

The 'arithmetization of analysis' had a gradual evolution, which is traced in Chapitre VI (by Dugac) of Dieudonné *et al.* [22]. A modern text on real analysis is Rudin [63]. In Lemma 7 of Chapter VI we will show that all norms on $\mathbb{R}^n$ are equivalent.

The contraction principle (Proposition 26) has been used to prove the *central limit theorem* of probability theory by Hamedani and Walter [32]. Bessaga (1959) has proved a *converse* of the contraction principle: Let E be an arbitrary set, $f : E \rightarrow E$ a map of E to itself and θ a real number such that $0 < \theta < 1$. If each iterate $f^n (n \in \mathbb{N})$ has at most one fixed point and if some iterate has a fixed point, then a complete metric d can be defined on E such that $\mathrm{d}(f(x'), f(x'')) \leq \theta \mathrm{d}(x', x'')$ for all $x', x'' \in E$. A short proof is given by Jachymski [40].

There are other important fixed point theorems besides Proposition 26. *Brouwer's fixed point theorem* states that, if $B = \{x \in \mathbb{R}^n : |x| \leq 1\}$ is the n-dimensional closed unit ball, every continuous map $f : B \rightarrow B$ has a fixed point. For an elementary proof, see Kulpa [44]. The *Lefschetz fixed point theorem* requires a knowledge of algebraic topology, even for its statement. Fixed point theorems are extensively treated in Dugundji and Granas [23] (and in A. Granas and J. Dugundji, *Fixed Point Theory*, Springer-Verlag, New York, 2003).

For a more detailed discussion of differentiability for functions of several variables see, for example, Fleming [26] and Dieudonné [21]. The inverse function theorem (Proposition 27) is a local result. Some global results are given by Atkinson [5] and Chichilnisky [14]. For a holomorphic version of Proposition 28 and for the simple way in which higher-order equations may be replaced by systems of first-order equations see, e.g., Coddington and Levinson [16].

The formula for the roots of a cubic was first published by Cardano [12], but it was discovered by del Ferro and again by Tartaglia, who accused Cardano of breaking a pledge of secrecy. Cardano is judged less harshly by historians today than previously. His book, which contained developments of his own and also the formula for the roots of a quartic discovered by his pupil Ferrari, was the most significant Western contribution to mathematics for more than a thousand years.

Proposition 29 still holds, but is more difficult to prove, if in its statement "has a nonzero derivative" is replaced by "which is not constant". Read [57] shows that the basic results of complex analysis may be deduced from this stronger form of Proposition 29 without the use of complex integration.

A field F is said to be *algebraically closed* if every polynomial of positive degree with coefficients from F has a root in F. Thus the 'fundamental theorem of algebra' says that the field $\mathbb{C}$ of complex numbers is algebraically closed. The proofs of this theorem due to Argand–Cauchy and Euler–Lagrange–Laplace are given in Chapter 4 (by Remmert) of Ebbinghaus *et al.* [24]. As shown on p. 77 of [24], the latter method provides, in particular, a simple proof for the existence of n-th roots.

Wall [72] gives a proof of the fundamental theorem of algebra, based on the notion of topological degree, and Ahlfors [1] gives the most common complex analysis proof, based on Liouville's theorem that a function holomorphic in the whole complex plane is bounded only if it is a constant. A form of Liouville's theorem is easily deduced from Proposition 29: if the power series

$$p(z) = a_0 + a_1 z + a_2 z^2 + \cdots$$

converges and $|p(z)|$ is bounded for all $z \in \mathbb{C}$, then $a_n = 0$ for every $n \geq 1$.

The representation of trigonometric functions by complex exponentials appears in §138 of Euler [25]. The various algebraic formulas involving trigonometric functions, such as

$$\cos 3x = 4 \cos^3 x - 3 \cos x,$$

are easily established by means of this representation and the addition theorem for the exponential function.

Some texts on complex analysis are Ahlfors [1], Caratheodory [11] and Narasimhan [56].

The 19th century literature on quaternions is surveyed in Rothe [59]. Although Hamilton hoped that quaternions would prove as useful as complex numbers, a quaternionic analysis analogous to complex analysis was first developed by Fueter (1935). A good account is given by Sudbery [71].

One significant contribution of quaternions was indirect. After Hamilton had shown the way, other 'hypercomplex' number systems were constructed, which led eventually to the structure theory of associative algebras discussed below.

It is not difficult to show that any *automorphism* of $\mathbb{H}$, i.e. any bijective map $T : \mathbb{H} \to \mathbb{H}$ such that

$$T(x + y) = Tx + Ty, \quad T(xy) = (Tx)(Ty) \quad \text{for all } x, y \in \mathbb{H},$$

has the form $Tx = uxu^{-1}$ for some quaternion u with norm 1.

For octonions and their uses, see van der Blij [8] and Springer and Veldkamp [67]. The group of all automorphisms of the algebra $\mathbb{O}$ is the exceptional simple Lie group G_2. The other four exceptional simple Lie groups are also all related to $\mathbb{O}$ in some way.

Of wider significance are the associative algebras introduced in 1878 by Clifford [15] (pp. 266–276) as a common generalization of quaternions and Grassmann algebra. *Clifford algebras* were used by Lipschitz (1886) to represent orthogonal transformations in n-dimensional space. There is an extensive discussion of Clifford algebras in Deheuvels [20]. For their applications in physics, see Salingaros and Wene [64].

Proposition 32 has many uses. The proof given here is extracted from Nagahara and Tominaga [55].

It was proved by both Kervaire (1958) and Milnor (1958) that if a division algebra A (not necessarily associative) contains the real field $\mathbb{R}$ in its centre and is of finite dimension as a vector space over $\mathbb{R}$, then this dimension must be 1,2,4 or 8 (but the algebra need not be isomorphic to $\mathbb{R}, \mathbb{C}, \mathbb{H}$ or $\mathbb{O}$). All known proofs use deep results from algebraic topology, which was first applied to the problem by H. Hopf (1940). For more information about the proof, see Chapter 11 (by Hirzebruch) of Ebbinghaus *et al.* [24].

When is the product of two sums of squares again a sum of squares? To make the question precise, call a triple (r, s, t) of positive integers 'admissible' if there exist real numbers $\rho_{ijk} (1 \leq i \leq t, 1 \leq j \leq r, 1 \leq k \leq s)$ such that, for every $x = (\xi_1, \ldots, \xi_r) \in \mathbb{R}^r$ and every $y = (\eta_1, \ldots, \eta_s) \in \mathbb{R}^s$,

$$(\xi_1^2 + \cdots + \xi_r^2)(\eta_1^2 + \cdots + \eta_s^2) = \zeta_1^2 + \cdots + \zeta_t^2,$$

where

$$\zeta_i = \sum_{j=1}^{r} \sum_{k=1}^{s} \rho_{ijk} \xi_j \eta_k.$$

The question then becomes, which triples (r, s, t) are admissible? It is obvious that $(1, 1, 1)$ is admissible and the relation $n(x)n(y) = n(xy)$ for the norms of complex numbers, quaternions and octonions shows that (t, t, t) is admissible also for $t = 2, 4, 8$. It was proved by Hurwitz (1898) that (t, t, t) is admissible for no other values of t. A survey of the general problem is given by Shapiro [65].

General introductions to algebra are provided by Birkhoff and MacLane [7] and Herstein [35]. More extended treatments are given in Jacobson [41] and Lang [48].

The theory of groups is treated in M. Hall [29] and Rotman [60]. An especially significant class of groups is studied in Humphreys [38].

If H is a subgroup of a finite group G, then it is possible to choose a system of left coset representatives of H which is also a system of right coset representatives. This interesting, but not very useful, fact belongs to combinatorics rather than to group theory. We mention it because it was the motivation for the theorem of P. Hall (1935) on *systems of distinct representatives*, also known as the 'marriage theorem'. Further developments are described in Mirsky [53]. For quantitative versions, with applications to operations research, see Ford and Fulkerson [27].

The theory of rings separates into two parts. Noncommutative ring theory, which now incorporates the structure theory of associative algebras, is studied in the books of Herstein [36], Kasch [42] and Lam [46]. Commutative ring theory, which grew out of algebraic number theory and algebraic geometry, is studied in Atiyah and Macdonald [4] and Kunz [45].

Field theory was established as an independent subject of study in 1910 by Steinitz [68]. The books of Jacobson [41] and Lang [48] treat also the more recent theory of ordered fields, due to Artin and Schreier (1927).

Fields and groups are connected with one another by *Galois theory*. This subject has its origin in attempts to solve polynomial equations 'by radicals'. The founder of the subject is really Lagrange (1770/1). By developing his ideas, Ruffini (1799) and Abel (1826) showed that polynomial equations of degree greater than 4 cannot, in general, be solved by radicals. Abel (1829) later showed that polynomial equations *can* be solved by radicals if their 'Galois group' is commutative. In honour of this result, commutative groups are often called *abelian*.

Galois (1831, published posthumously in 1846) introduced the concept of normal subgroup and stated a necessary and sufficient condition for a polynomial equation to be solvable by radicals. The significance of Galois theory today lies not in this result, despite its historical importance, but in the much broader 'fundamental theorem of Galois theory'. In the form given it by Dedekind (1894) and Artin (1944), this establishes a correspondence between extension fields and groups of automorphisms, and provides a framework for the solution of a number of algebraic problems.

Morandi [54] and Rotman [61] give modern accounts of Galois theory. The historical development is traced in Kiernan [43]. In recent years attention has focussed on the problem of determining which finite groups occur as Galois groups over a given field; for an introductory account, see Matzat [51].

Some texts on linear algebra and matrix theory are Halmos [31], Horn and Johnson [37], Mal'cev [50] and Gantmacher [28].

The older literature on associative algebras is surveyed in Cartan [13]. The texts on noncommutative rings cited above give modern introductions.

A vast number of characterizations of inner product spaces, in addition to the parallelogram law, is given in Amir [3]. The theory of Hilbert space is treated in the books of Riesz and Sz.-Nagy [58] and Akhiezer and Glazman [2]. For its roots in the theory of integral equations, see Hellinger and Toeplitz [33]. Almost periodic functions are discussed from different points of view in Bohr [9], Corduneanu [17] and Maak [49]. The convergence of Fourier series is treated in Zygmund [73], for example.

12 Selected References

[1] L.V. Ahlfors, *Complex analysis*, 3rd ed., McGraw-Hill, New York, 1978.

[2] N.I. Akhiezer and I.M. Glazman, *Theory of linear operators in Hilbert space*, English transl. by E.R. Dawson based on 3rd Russian ed., Pitman, London, 1981.

[3] D. Amir, *Characterizations of inner product spaces*, Birkhäuser, Basel, 1986.

[4] M.F. Atiyah and I.G. Macdonald, *Introduction to commutative algebra*, Addison-Wesley, Reading, Mass., 1969.

[5] F.V. Atkinson, The reversibility of a differentiable mapping, *Canad. Math. Bull.* **4** (1961), 161–181.

[6] G. Birkhoff, *Lattice theory*, corrected reprint of 3rd ed., American Mathematical Society, Providence, R.I., 1979.

[7] G. Birkhoff and S. MacLane, *A survey of modern algebra*, 3rd ed., Macmillan, New York, 1965.

[8] F. van der Blij, History of the octaves, *Simon Stevin* **34** (1961), 106–125.

[9] H. Bohr, *Almost periodic functions*, English transl. by H. Cohn and F. Steinhardt, Chelsea, New York, 1947.

[10] G. Boole, *An investigation of the laws of thought, on which are founded the mathematical theories of logic and probability*, reprinted, Dover, New York, 1957. [Original edition, 1854]

[11] C. Caratheodory, *Theory of functions of a complex variable*, English transl. by F. Steinhardt, 2 vols., 2nd ed., Chelsea, New York, 1958/1960.

[12] G. Cardano, *The great art or the rules of algebra*, English transl. by T.R. Witmer, M.I.T. Press, Cambridge, Mass., 1968. [Latin original, 1545]

[13] E. Cartan, Nombres complexes, *Encyclopédie des sciences mathématiques, Tome I, Fasc. 4, Art. I.5*, Gauthier-Villars, Paris, 1908. [Reprinted in *Oeuvres complètes, Partie II, Vol. 1*, pp. 107–246.]

[14] G. Chichilnisky, Topology and invertible maps, *Adv. inAppl. Math.* **21** (1998), 113–123.

[15] W.K. Clifford, *Mathematical Papers*, reprinted, Chelsea, New York, 1968.

[16] E.A. Coddington and N. Levinson, *Theory of ordinary differential equations*, McGraw-Hill, New York, 1955.

[17] C. Corduneanu, *Almost periodic functions*, English transl. by G. Berstein and E. Tomer, Interscience, New York, 1968.

[18] T. Dantzig, *Number: The language of science*, 4th ed., Pi Press, Indianapolis, IN, 2005.

[19] R. Dedekind, *Essays on the theory of numbers*, English transl. by W.W. Beman, reprinted, Dover, New York, 1963.

[20] R. Deheuvels, *Formes quadratiques et groupes classiques*, Presses Universitaires de France, Paris, 1981.

[21] J. Dieudonné, *Foundations of modern analysis*, enlarged reprint, Academic Press, New York, 1969.

[22] J. Dieudonné et al., *Abrégé d'histoire des mathématiques 1700–1900*, reprinted, Hermann, Paris, 1996.

[23] J. Dugundji and A. Granas, *Fixed point theory I*, PWN, Warsaw, 1982.

[24] H.-D. Ebbinghaus et al., *Numbers*, English transl. of 2nd German ed. by H.L.S. Orde, Springer-Verlag, New York, 1990.

[25] L. Euler, *Introduction to analysis of the infinite, Book I*, English transl. by J.D. Blanton, Springer-Verlag, New York, 1988.

[26] W. Fleming, *Functions of several variables*, 2nd ed., Springer-Verlag, New York, 1977.

[27] L.R. Ford Jr. and D.R. Fulkerson, *Flows in networks*, Princeton University Press, Princeton, N.J., 1962.

[28] F.R. Gantmacher, *The theory of matrices*, English transl. by K.A. Hirsch, 2 vols., Chelsea, New York, 1959.

[29] M. Hall, *The theory of groups*, reprinted, Chelsea, New York, 1976.

[30] P.R. Halmos, *Lectures on Boolean algebras*, Van Nostrand, Princeton, N.J., 1963.

[31] P.R. Halmos, *Finite-dimensional vector spaces*, 2nd ed., reprinted, Springer-Verlag, New York, 1974.

[32] G.G. Hamedani and G.G. Walter, A fixed point theorem and its application to the central limit theorem, *Arch. Math.* **43** (1984), 258–264.

[33] E. Hellinger and O. Toeplitz, *Integralgleichungen und Gleichungen mit unendlichvielen Unbekannten*, reprinted, Chelsea, New York, 1953. [Original edition, 1928]

[34] L. Henkin, On mathematical induction, *Amer. Math. Monthly* **67** (1960), 323–338.

[35] I.N. Herstein, *Topics in algebra*, reprinted, Wiley, London, 1976.

[36] I.N Herstein, *Noncommutative rings*, reprinted, Mathematical Association of America, Washington, D.C., 1994.

[37] R.A. Horn and C.A. Johnson, *Matrix analysis*, corrected reprint, Cambridge University Press, 1990.

[38] J.E. Humphreys, *Reflection groups and Coxeter groups*, Cambridge University Press, 1990.

[39] E.V. Huntingdon, Boolean algebra: A correction, *Trans. Amer. Math. Soc.* **35** (1933), 557–558.

[40] J. Jachymski, A short proof of the converse to the contraction principle and some related results, *Topol. Methods Nonlinear Anal.* **15** (2000), 179–186.

[41] N. Jacobson, *Basic Algebra I,II*, 2nd ed., Freeman, New York, 1985/1989.

[42] F. Kasch, *Modules and rings*, English transl. by D.A.R. Wallace, Academic Press, London, 1982.

[43] B.M. Kiernan, The development of Galois theory from Lagrange to Artin, *Arch. Hist. Exact Sci.* **8** (1971), 40–154.

[44] W. Kulpa, The Poincaré–Miranda theorem, *Amer. Math. Monthly* **104** (1997), 545–550.

[45] E. Kunz, *Introduction to commutative algebra and algebraic geometry*, English transl. by M. Ackerman, Birkhäuser, Boston, Mass., 1985.

[46] T.Y. Lam, *A first course in noncommutative rings*, Springer-Verlag, New York, 1991.

[47] E. Landau, *Foundations of analysis*, English transl. by F. Steinhardt, 3rd ed., Chelsea, New York, 1966. [German original, 1930]

[48] S. Lang, *Algebra*, corrected reprint of 3rd ed., Addison-Wesley, Reading, Mass., 1994.

[49] W. Maak, *Fastperiodische Funktionen*, Springer-Verlag, Berlin, 1950.

[50] A.I. Mal'cev, *Foundations of linear algebra*, English transl. by T.C. Brown, Freeman, San Francisco, 1963.

[51] B.H. Matzat, Über das Umkehrproblem der Galoisschen Theorie, *Jahresber. Deutsch. Math.-Verein.* **90** (1988), 155–183.

[52] K. Menninger, *Number words and number symbols*, English transl. by P. Broneer, M.I.T. Press, Cambridge, Mass., 1969.

[53] L. Mirsky, *Transversal theory*, Academic Press, London, 1971.

[54] P. Morandi, *Field and Galois theory*, Springer, New York, 1996.

[55] T. Nagahara and H. Tominaga, Elementary proofs of a theorem of Wedderburn and a theorem of Jacobson, *Abh. Math. Sem. Univ. Hamburg* **41** (1974), 72–74.

[56] R. Narasimhan, *Complex analysis in one variable*, Birkhäuser, Boston, Mass., 1985.

[57] A.H. Read, Higher derivatives of analytic functions from the standpoint of functional analysis, *J. London Math. Soc.* **36** (1961), 345–352.

[58] F. Riesz and B. Sz.-Nagy, *Functional analysis*, English transl. by L.F. Boron of 2nd French ed., F. Ungar, New York, 1955.

[59] H. Rothe, Systeme geometrischer Analyse, *Encyklopädie der Mathematischen Wissenschaften* III 1.2, pp. 1277–1423, Teubner, Leipzig, 1914–1931.

[60] J.J. Rotman, *An introduction to the theory of groups*, 4th ed., Springer-Verlag, New York, 1995.

[61] J. Rotman, *Galois theory*, 2nd ed., Springer-Verlag, New York, 1998.

[62] S. Rudeanu, *Boolean functions and equations*, North-Holland, Amsterdam, 1974.

[63] W. Rudin, *Principles of mathematical analysis*, 3rd ed., McGraw-Hill, New York, 1976.

[64] N.A. Salingaros and G.P. Wene, The Clifford algebra of differential forms, *Acta Appl. Math.* **4** (1985), 271–292.

[65] D.B. Shapiro, Products of sums of squares, *Exposition. Math.* **2** (1984), 235–261.

[66] R. Sikorski, *Boolean algebras*, 3rd ed., Springer-Verlag, New York, 1969.

[67] T.A. Springer and F.D. Veldkamp, *Octonions, Jordan algebras, and exceptional groups*, Springer, Berlin, 2000.

[68] E. Steinitz, *Algebraische Theorie der Körper*, reprinted, Chelsea, New York, 1950.

[69] M.H. Stone, The representation of Boolean algebras, *Bull. Amer. Math. Soc.* **44** (1938), 807–816.

[70] K.D. Stroyan and W.A.J. Luxemburg, *Introduction to the theory of infinitesimals*, Academic Press, New York, 1976.

[71] A. Sudbery, Quaternionic analysis, *Math. Proc. Cambridge Philos. Soc.* **85** (1979), 199–225.

[72] C.T.C. Wall, *A geometric introduction to topology*, reprinted, Dover, New York, 1993.

[73] A. Zygmund, *Trigonometric series*, 3rd ed., Cambridge University Press, 2003.

Additional References

J.C. Baez, The octonions, *Bull. Amer. Math. Soc.* (*N.S.*) **39** (2002), 145–205.

J.H. Conway and D.A. Smith, *On quaternions and octonions: their geometry, arithmetic and symmetry*, A.K. Peters, Natick, Mass., 2003.

II

Divisibility

1 Greatest Common Divisors

In the set $\mathbb{N}$ of all positive integers we can perform two basic operations: addition and multiplication. In this chapter we will be primarily concerned with the second operation.

Multiplication has the following properties:

(**M1**) *if* $ab = ac$, *then* $b = c$;   (cancellation law)
(**M2**) $ab = ba$ *for all* a, b;   (commutative law)
(**M3**) $(ab)c = a(bc)$ *for all* a, b, c;   (associative law)
(**M4**) $1a = a$ *for all* a.   (identity element)

For any $a, b \in \mathbb{N}$ we say that b *divides* a, or that b is a *factor* of a, or that a is a *multiple* of b if $a = ba'$ for some $a' \in \mathbb{N}$. We write $b|a$ if b divides a and $b \nmid a$ if b does not divide a. For example, $2|6$, since $6 = 2 \times 3$, but $4 \nmid 6$. (We sometimes use $\times$ instead of $\cdot$ for the product of positive integers.) The following properties of divisibility follow at once from the definition:

(i) $a|a$ *and* $1|a$ *for every* a;
(ii) *if* $b|a$ *and* $c|b$, *then* $c|a$;
(iii) *if* $b|a$, *then* $b|ac$ *for every* c;
(iv) $bc|ac$ *if and only if* $b|a$;
(v) *if* $b|a$ *and* $a|b$, *then* $b = a$.

For any $a, b \in \mathbb{N}$ we say that d is a *common divisor* of a and b if $d|a$ and $d|b$. We say that a common divisor d of a and b is a *greatest common divisor* if every common divisor of a and b divides d. The greatest common divisor of a and b is uniquely determined, if it exists, and will be denoted by (a, b).

The greatest common divisor of a and b is indeed the *numerically* greatest common divisor. However, it is preferable not to define greatest common divisors in this way, since the concept is then available for algebraic structures in which there is no relation of magnitude and only the operation of multiplication is defined.

W.A. Coppel, *Number Theory: An Introduction to Mathematics*, Universitext,
DOI: 10.1007/978-0-387-89486-7_2, © Springer Science + Business Media, LLC 2009

Proposition 1 *Any $a, b \in \mathbb{N}$ have a greatest common divisor (a, b).*

Proof Without loss of generality we may suppose $a \geq b$. If b divides a, then $(a, b) = b$. Assume that there exists a pair a, b without greatest common divisor and choose one for which a is a minimum. Then $1 < b < a$, since b does not divide a. Since also $1 \leq a - b < a$, the pair $a - b, b$ has a greatest common divisor d. Since any common divisor of a and b divides $a - b$, and since d divides $(a - b) + b = a$, it follows that d is a greatest common divisor of a and b. But this is a contradiction. $\square$

The proof of Proposition 1 uses not only the multiplicative structure of the set $\mathbb{N}$, but also its ordering and additive structure. To see that there is a reason for this, consider the set S of all positive integers of the form $4k + 1$. The set S is closed under multiplication, since

$$(4j + 1)(4k + 1) = 4(4jk + j + k) + 1,$$

and we can define divisibility and greatest common divisors in S by simply replacing $\mathbb{N}$ by S in our previous definitions. However, although the elements 693 and 189 of S have the common divisors 9 and 21, they have no greatest common divisor according to this definition.

In the following discussion we use the result of Proposition 1, but make no further appeal to either addition or order.

For any $a, b \in \mathbb{N}$ we say that h is a *common multiple* of a and b if $a|h$ and $b|h$. We say that a common multiple h of a and b is a *least common multiple* if h divides every common multiple of a and b. The least common multiple of a and b is uniquely determined, if it exists, and will be denoted by $[a, b]$.

It is evident that, for every a,

$$(a, 1) = 1, \quad [a, 1] = a,$$
$$(a, a) = a = [a, a].$$

Proposition 2 *Any $a, b \in \mathbb{N}$ have a least common multiple $[a, b]$. Moreover,*

$$(a, b)[a, b] = ab.$$

Furthermore, for all $a, b, c \in \mathbb{N}$,

$$(ac, bc) = (a, b)c, \qquad\qquad [ac, bc] = [a, b]c,$$
$$([a, b], [a, c]) = [a, (b, c)], \quad [(a, b), (a, c)] = (a, [b, c]).$$

Proof We show first that $(ac, bc) = (a, b)c$. Put $d = (a, b)$. Clearly cd is a common divisor of ac and bc, and so $(ac, bc) = qcd$ for some $q \in \mathbb{N}$. Thus $ac = qcda'$, $bc = qcdb'$ for some $a', b' \in \mathbb{N}$. It follows that $a = qda'$, $b = qdb'$. Thus qd is a common divisor of a and b. Hence qd divides d, which implies $q = 1$.

If g is any common multiple of a and b, then ab divides ga and gb, and hence ab also divides (ga, gb). But, by what we have just proved,

$$(ga, gb) = (a, b)g = dg.$$

Hence $h := ab/d$ divides g. Since h is clearly a common multiple of a and b, it follows that $h = [a, b]$. Replacing a, b by ac, bc, we now obtain

$$[ac, bc] = acbc/(ac, bc) = abc/(a, b) = hc.$$

If we put

$$A = ([a, b], [a, c]), \quad B = [a, (b, c)],$$

then by what we have already proved,

$$A = (ab/(a, b), ac/(a, c)),$$
$$B = a(b, c)/(a, (b, c)) = (ab/(a, (b, c)), ac/(a, (b, c))).$$

Since any common divisor of $ab/(a, b)$ and $ac/(a, c)$ is also a common divisor of $ab/(a, (b, c))$ and $ac/(a, (b, c))$, it follows that A divides B. On the other hand, a divides A, since a divides $[a, b]$ and $[a, c]$, and similarly (b, c) divides A. Hence B divides A. Thus $B = A$.

The remaining statement of the proposition is proved in the same way, with greatest common divisors and least common multiples interchanged. □

The last two statements of Proposition 2 are referred to as the distributive laws, since if the greatest common divisor and least common multiple of a and b are denoted by $a \wedge b$ and $a \vee b$ respectively, they take the form

$$(a \vee b) \wedge (a \vee c) = a \vee (b \wedge c), \quad (a \wedge b) \vee (a \wedge c) = a \wedge (b \vee c).$$

Properties (i), (ii) and (v) at the beginning of the section say that divisibility is a *partial ordering* of the set $\mathbb{N}$ with 1 as least element. The existence of greatest common divisors and least common multiples says that $\mathbb{N}$ is a *lattice* with respect to this partial ordering. The distributive laws say that $\mathbb{N}$ is actually a *distributive* lattice.

We say that $a, b \in \mathbb{N}$ are *relatively prime*, or *coprime*, if $(a, b) = 1$. Divisibility properties in this case are much simpler:

Proposition 3 *For any $a, b, c \in \mathbb{N}$ with $(a, b) = 1$,*

 (i) *if $a|c$ and $b|c$, then $ab|c$;*
 (ii) *if $a|bc$, then $a|c$;*
 (iii) *$(a, bc) = (a, c)$;*
 (iv) *if also $(a, c) = 1$, then $(a, bc) = 1$;*
 (v) *$(a^m, b^n) = 1$ for all $m, n \geq 1$.*

Proof To prove (i), note that $[a, b]$ divides c and $[a, b] = ab$. To prove (ii), note that a divides $(ac, bc) = (a, b)c = c$. To prove (iii), note that any common divisor of a and bc divides c, by (ii). Obviously (iii) implies (iv), and (v) follows by induction. □

Proposition 4 *If $a, b \in \mathbb{N}$ and $(a, b) = 1$, then any divisor of ab can be uniquely expressed in the form de, where $d|a$ and $e|b$. Conversely, any product of this form is a divisor of ab.*

Proof The proof is based on Proposition 3. Suppose c divides ab and put $d = (a, c)$, $e = (b, c)$. Then $(d, e) = 1$ and hence de divides c. If $a = da'$ and $c = dc'$, then $(a', c') = 1$ and $e|c'$. On the other hand, $c'|a'b$ and hence $c'|b$. Since $e = (b, c)$, it follows that $c' = e$ and $c = de$.

Suppose $de = d'e'$, where d, d' divide a and e, e' divide b. Then $d|d'$, since $(d, e') = 1$, and similarly $d'|d$, since $(d', e) = 1$. Hence $d' = d$ and $e' = e$.

The final statement of the proposition is obvious. □

It follows from Proposition 4 that if $c^n = ab$, where $(a, b) = 1$, then $a = d^n$ and $b = e^n$ for some $d, e \in \mathbb{N}$.

The greatest common divisor and least common multiple of any finite set of elements of $\mathbb{N}$ may be defined in the same way as for sets of two elements. By induction we easily obtain:

Proposition 5 *Any* $a_1, \ldots, a_n \in \mathbb{N}$ *have a greatest common divisor* $(a_1, \ldots, a_n)$ *and a least common multiple* $[a_1, \ldots, a_n]$. *Moreover,*

(i) $(a_1, a_2, \ldots, a_n) = (a_1, (a_2, \ldots, a_n))$, $[a_1, a_2, \ldots, a_n] = [a_1, [a_2, \ldots, a_n]]$;
(ii) $(a_1 c, \ldots, a_n c) = (a_1, \ldots, a_n)c$, $[a_1 c, \ldots, a_n c] = [a_1, \ldots, a_n]c$;
(iii) $(a_1, \ldots, a_n) = a/[a/a_1, \ldots, a/a_n]$, $[a_1, \ldots, a_n] = a/(a/a_1, \ldots, a/a_n)$, *where* $a = a_1 \cdots a_n$.

We can use the distributive laws to show that

$$([a, b], [a, c], [b, c]) = [(a, b), (a, c), (b, c)].$$

In fact the left side is equal to $\{a \vee (b \wedge c)\} \wedge (b \vee c)$, whereas the right side is equal to

$$(b \wedge c) \vee \{a \wedge (b \vee c)\} = \{(b \wedge c) \vee a\} \wedge \{(b \wedge c) \vee (b \vee c)\}$$
$$= \{a \vee (b \wedge c)\} \wedge (b \vee c).$$

If

$$a = (a_1, \ldots, a_m), \quad b = (b_1, \ldots, b_n),$$

then ab is the greatest common divisor of all products $a_j b_k$, since $(a_j b_1, \ldots, a_j b_n) = a_j b$ and $(a_1 b, \ldots, a_m b) = ab$.

Similarly, if

$$a = [a_1, \ldots, a_m], \quad b = [b_1, \ldots, b_n],$$

then ab is the least common multiple of all products $a_j b_k$.

It is easily shown by induction that if $(a_i, a_j) = 1$ for $1 \leq i < j \leq m$, then

$$(a_1 \cdots a_m, c) = (a_1, c) \cdots (a_m, c), [a_1 \cdots a_m, c] = [a_1, \ldots, a_m, c].$$

Proposition 6 *If* $a \in \mathbb{N}$ *has two factorizations*

$$a = b_1 \cdots b_m = c_1 \cdots c_n,$$

then these factorizations have a common refinement, i.e. there exist $d_{jk} \in \mathbb{N}$ $(1 \leq j \leq m, 1 \leq k \leq n)$ *such that*

$$b_j = \prod_{k=1}^{n} d_{jk}, \quad c_k = \prod_{j=1}^{m} d_{jk}.$$

Proof We show first that if $a = a_1 \cdots a_n$ and $d|a$, then $d = d_1 \cdots d_n$, where $d_i|a_i$ ($1 \le i \le n$). We may suppose that $n > 1$ and that the assertion holds for products of less than n elements of $\mathbb{N}$. Put $a' = a_1 \cdots a_{n-1}$ and $d' = (a', d)$. Then $d' = d_1 \cdots d_{n-1}$, where $d_i|a_i (1 \le i < n)$. Moreover $a'' = a'/d'$ and $d'' = d/d'$ are coprime. Since $d'' = d/d'$ divides $a''a_n = a/d'$, the greatest common divisor $a_n = (a_n a'', a_n d'')$ is divisible by d''. Thus we can take $d_n = d''$.

We return now to the proposition. Since $c_1| \prod_j b_j$, we can write $c_1 = \prod_j d_{j1}$, where $d_{j1}|b_j$. Put $b'_j = b_j/d_{j1}$. Then

$$\prod_j b'_j = a/c_1 = c_2 \cdots c_n.$$

Hence we can write $c_2 = \prod_j d_{j2}$, where $d_{j2}|b'_j$. Proceeding in this way, we obtain factorizations $c_k = \prod_j d_{jk}$ such that $\prod_k d_{jk}$ divides b_j. In fact, since

$$\prod_{j,k} d_{jk} = a = \prod_j b_j,$$

we must have $b_j = \prod_k d_{jk}$. $\qquad\square$

Instead of defining divisibility and greatest common divisors in the set $\mathbb{N}$ of all positive integers, we can define them in the set $\mathbb{Z}$ of all integers by simply replacing $\mathbb{N}$ by $\mathbb{Z}$ in the previous definitions. The properties (i)–(v) at the beginning of this section continue to hold, provided that in (iv) we require $c \ne 0$ and in (v) we alter the conclusion to $b = \pm a$. We now list some additional properties:

(i)′ $a|0$ *for every* a;
(ii)′ *if* $0|a$, *then* $a = 0$;
(iii)′ *if* $c|a$ *and* $c|b$, *then* $c|ax + by$ *for all* x, y.

Greatest common divisors and least common multiples still exist, but uniqueness holds only up to sign. With this understanding, Propositions 2–4 continue to hold, and so also do Propositions 5 and 6 if we require $a \ne 0$. It is evident that, for every a,

$$(a, 0) = a, \quad [a, 0] = 0.$$

More generally, we can define divisibility in any *integral domain*, i.e. a commutative ring in which $a \ne 0$ and $b \ne 0$ together imply $ab \ne 0$. The properties (i)–(v) at the beginning of the section continue to hold, provided that in (iv) we require $c \ne 0$ and in (v) we alter the conclusion to $b = ua$, where u is a *unit*, i.e. $u|1$. The properties (i)′–(iii)′ above also remain valid.

We define a *GCD domain* to be an integral domain in which any pair of elements has a greatest common divisor. This implies that any pair of elements also has a least common multiple. Uniqueness now holds only up to unit multiples. With this understanding Propositions 2–6 continue to hold in any GCD domain in the same way as for $\mathbb{Z}$.

An important example, which we will consider in Section 3, of a GCD domain other than $\mathbb{Z}$ is the *polynomial ring* $K[t]$, consisting of all polynomials in t with coefficients from an arbitrary field K. The units in this case are the nonzero elements of K.

Another example, which we will meet in §4 of Chapter VI, is the valuation ring R of a non-archimedean valued field. In this case, for any $a, b \in R$, either $a|b$ or $b|a$ and so (a, b) is either a or b.

In the same way that the ring $\mathbb{Z}$ of integers may be embedded in the field $\mathbb{Q}$ of rational numbers, any integral domain R may be embedded in a field K, its *field of fractions*, so that any nonzero $c \in K$ has the form $c = ab^{-1}$, where $a, b \in R$ and $b \neq 0$. If R is a GCD domain we can further require $(a, b) = 1$, and a, b are then uniquely determined apart from a common unit multiple. The field of fractions of the polynomial ring $K[t]$ is the field $K(t)$ of *rational functions*.

In our discussion of divisibility so far we have avoided all mention of prime numbers. A positive integer $a \neq 1$ is said to be *prime* if 1 and a are its only positive divisors, and otherwise is said to be *composite*.

For example, 2, 3 and 5 are primes, but $4 = 2 \times 2$ and $6 = 2 \times 3$ are composite. The significance of the primes is that, as far as multiplication is concerned, they are the 'atoms' and the composite integers are the 'molecules'. This is made precise in the following so-called *fundamental theorem of arithmetic*:

Proposition 7 *If $a \in \mathbb{N}$ and $a \neq 1$, then a can be represented as a product of finitely many primes. Moreover, the representation is unique, except for the order of the factors.*

Proof Assume, on the contrary, that some composite $a_1 \in \mathbb{N}$ is not a product of finitely many primes. Since a_1 is composite, it has a factorization $a_1 = a_2 b_2$, where $a_2, b_2 \in \mathbb{N}$ and $a_2, b_2 \neq 1$. At least one of a_2, b_2 must be composite and not a product of finitely many primes, and we may choose the notation so that a_2 has these properties. The preceding argument can now be repeated with a_2 in place of a_1. Proceeding in this way, we obtain an infinite sequence (a_k) of positive integers such that a_{k+1} divides a_k and $a_{k+1} \neq a_k$ for each $k \geq 1$. But then the sequence (a_k) has no least element, which contradicts Proposition I.3.

Suppose now that

$$a = p_1 \cdots p_m = q_1 \cdots q_n$$

are two representations of a as a product of primes. Then, by Proposition 6, there exist $d_{jk} \in \mathbb{N}$ ($1 \leq j \leq m, 1 \leq k \leq n$) such that

$$p_j = \prod_{k=1}^{n} d_{jk}, \quad q_k = \prod_{j=1}^{m} d_{jk}.$$

Since p_1 is a prime, we must have $d_{1k_1} = p_1$ for some $k_1 \in \{1, \ldots, n\}$, and since q_{k_1} is a prime, we must have $q_{k_1} = d_{1k_1} = p_1$. The same argument can now be applied to

$$a' = \prod_{j \neq 1} p_j = \prod_{k \neq k_1} q_k.$$

It follows that $m = n$ and $q_1, \ldots, q_n$ is a permutation of $p_1, \ldots, p_m$. $\square$

It should be noted that factorization into primes would not be unique if we admitted 1 as a prime. The fundamental theorem of arithmetic may be reformulated in the following way: any $a \in \mathbb{N}$ can be uniquely represented in the form

$$a = \prod_p p^{\alpha_p},$$

where p runs through the primes and the α_p are non-negative integers, only finitely many of which are nonzero. It is easily seen that if $b \in \mathbb{N}$ has the analogous representation

$$b = \prod_p p^{\beta_p},$$

then $b|a$ if and only if $\beta_p \leq \alpha_p$ for all p. It follows that the greatest common divisor and least common multiple of a and b have the representations

$$(a, b) = \prod_p p^{\gamma_p}, \quad [a, b] = \prod_p p^{\delta_p},$$

where

$$\gamma_p = \min\{\alpha_p, \beta_p\}, \quad \delta_p = \max\{\alpha_p, \beta_p\}.$$

The fundamental theorem of arithmetic extends at once from $\mathbb{N}$ to $\mathbb{Q}$: any nonzero rational number a can be uniquely represented in the form

$$a = u \prod_p p^{\alpha_p},$$

where $u = \pm 1$ is a unit, p runs through the primes and the α_p are integers (not necessarily non-negative), only finitely many of which are nonzero.

The following property of primes was already established in Euclid's *Elements* (Book VII, Proposition 30):

Proposition 8 *If p is a prime and $p|bc$, then $p|b$ or $p|c$.*

Proof If p does not divide b, we must have $(p, b) = 1$. But then p divides c, by Proposition 3(ii). ☐

The property in Proposition 8 actually characterizes primes. For if a is composite, then $a = bc$, where $b, c \neq 1$. Thus $a|bc$, but $a \nmid b$ and $a \nmid c$.

We consider finally the extension of these notions to an arbitrary integral domain R. For any nonzero $a, b \in R$, we say that a divisor b of a is a *proper divisor* if a does not divide b (i.e., if a and b do not differ only by a unit factor). We say that $p \in R$ is *irreducible* if p is neither zero nor a unit and if every proper divisor of p is a unit. We say that $p \in R$ is *prime* if p is neither zero nor a unit and if $p|bc$ implies $p|b$ or $p|c$.

By what we have just said, the notions of 'prime' and 'irreducible' coincide if $R = \mathbb{Z}$, and the same argument applies if R is any GCD domain. However, in an arbitrary integral domain R, although any prime element is irreducible, an irreducible element need not be prime. (For example, in the integral domain R consisting of all complex numbers of the form $a + b\sqrt{-5}$, where $a, b \in \mathbb{Z}$, it may be seen that

$6 = 2 \times 3 = (1 + \sqrt{-5})(1 - \sqrt{-5})$ has two essentially distinct factorizations into irreducibles, and thus none of these irreducibles is prime.)

The proof of Proposition 7 shows that, in an arbitrary integral domain R, every element which is neither zero nor a unit can be represented as a product of finitely many irreducible elements if and only if the following *chain condition* is satisfied:

(#) *there exists no infinite sequence* (a_n) *of elements of* R *such that* a_{n+1} *is a proper divisor of* a_n *for every* n.

Furthermore, the representation is *essentially unique* (i.e. unique except for the order of the factors and for multiplying them by units) if and only if R is also a GCD domain.

An integral domain R is said to be *factorial* (or a 'unique factorization domain') if the 'fundamental theorem of arithmetic' holds in R, i.e. if every element which is neither zero nor a unit has such an essentially unique representation as a product of finitely many irreducibles. By the above remarks, an integral domain R is factorial if and only if it is a GCD domain satisfying the chain condition (#).

For future use, we define an element of a factorial domain to be *square-free* if it is neither zero nor a unit and if, in its representation as a product of irreducibles, no factor is repeated. In particular, a positive integer is square-free if and only if it is a nonempty product of distinct primes.

2 The Bézout Identity

If a, b are arbitrary integers with $a \neq 0$, then there exist unique integers q, r such that

$$b = qa + r, \quad 0 \le r < |a|.$$

In fact qa is the greatest multiple of a which does not exceed b. The integers q and r are called the *quotient* and *remainder* in the 'division' of b by a.

(For $a > 0$ this was proved in Proposition I.14. It follows that if a and n are positive integers, any positive integer b less than a^n has a unique representation 'to the base a':

$$b = b_0 + b_1 a + \cdots + b_{n-1} a^{n-1},$$

where $0 \le b_j < a$ for all j. In fact b_{n-1} is the quotient in the division of b by a^{n-1}, b_{n-2} is the quotient in the division of the remainder by a^{n-2}, and so on.)

If a, b are arbitrary integers with $a \neq 0$, then there exist also integers q, r such that

$$b = qa + r, \quad |r| \le |a|/2.$$

In fact qa is the nearest multiple of a to b. Thus q and r are not uniquely determined if b is midway between two consecutive multiples of a.

Both these *division algorithms* have their uses. We will be impartial and merely use the fact that

$$b = qa + r, \quad |r| < |a|.$$

An *ideal* in the commutative ring $\mathbb{Z}$ of all integers is defined to be a nonempty subset J such that if $a, b \in J$ and $x, y \in \mathbb{Z}$, then also $ax + by \in J$.

For example, if $a_1, \ldots, a_n$ are given elements of $\mathbb{Z}$, then the set of all linear combinations $a_1 x_1 + \cdots + a_n x_n$ with $x_1, \ldots, x_n \in \mathbb{Z}$ is an ideal, the ideal *generated* by $a_1, \ldots, a_n$. An ideal generated by a single element, i.e. the set of all multiples of that element, is said to be a *principal ideal*.

Lemma 9 *Any ideal J in the ring $\mathbb{Z}$ is a principal ideal.*

Proof If 0 is the only element of J, then 0 generates J. Otherwise there is a nonzero $a \in J$ with minimum absolute value. For any $b \in J$, we can write $b = qa + r$, for some $q, r \in \mathbb{Z}$ with $|r| < |a|$. By the definition of an ideal, $r \in J$ and so, by the definition of a, $r = 0$. Thus a generates J. $\qquad\square$

Proposition 10 *Any $a, b \in \mathbb{Z}$ have a greatest common divisor $d = (a, b)$. Moreover, for any $c \in \mathbb{Z}$, there exist $x, y \in \mathbb{Z}$ such that*

$$ax + by = c$$

if and only if d divides c.

Proof Let J be the ideal generated by a and b. By Lemma 9, J is generated by a single element d. Since $a, b \in J$, d is a common divisor of a and b. On the other hand, since $d \in J$, there exist $u, v \in \mathbb{Z}$ such that $d = au + bv$. Hence any common divisor of a and b also divides d. Thus $d = (a, b)$. The final statement of the proposition follows immediately since, by definition, $c \in J$ if and only if there exist $x, y \in \mathbb{Z}$ such that $ax + by = c$. $\qquad\square$

It is readily shown that if the 'linear Diophantine' equation $ax + by = c$ has a solution $x_0, y_0 \in \mathbb{Z}$, then all solutions $x, y \in \mathbb{Z}$ are given by the formula

$$x = x_0 + kb/d, \quad y = y_0 - ka/d,$$

where $d = (a, b)$ and k is an arbitrary integer.

Proposition 10 provides a new proof for the existence of greatest common divisors and, in addition, it shows that the greatest common divisor of two integers can be represented as a linear combination of them. This representation is usually referred to as the *Bézout identity*, although it was already known to Bachet (1624) and even earlier to the Hindu mathematicians Aryabhata (499) and Brahmagupta (628).

In exactly the same way that we proved Proposition 10 – or, alternatively, by induction from Proposition 10 – we can prove

Proposition 11 *Any finite set $a_1, \ldots, a_n$ of elements of $\mathbb{Z}$ has a greatest common divisor $d = (a_1, \ldots, a_n)$. Moreover, for any $c \in \mathbb{Z}$, there exist $x_1, \ldots, x_n \in \mathbb{Z}$ such that*

$$a_1 x_1 + \cdots + a_n x_n = c$$

if and only if d divides c.

The proof which we gave for Proposition 10 is a pure existence proof – it does not help us to find the greatest common divisor. The following constructive proof was already given in Euclid's *Elements* (Book VII, Proposition 2). Let a, b be arbitrary integers. Since $(0, b) = b$, we may assume $a \neq 0$. Then there exist integers q, r such that

$$b = qa + r, \quad |r| < |a|.$$

Put $a_0 = b$, $a_1 = a$ and repeatedly apply this procedure:

$$a_0 = q_1 a_1 + a_2, \quad |a_2| < |a_1|,$$
$$a_1 = q_2 a_2 + a_3, \quad |a_3| < |a_2|,$$
$$\cdots$$
$$a_{N-2} = q_{N-1} a_{N-1} + a_N, \quad |a_N| < |a_{N-1}|,$$
$$a_{N-1} = q_N a_N.$$

The process must eventually terminate as shown, because otherwise we would obtain an infinite sequence of positive integers with no least element. We claim that a_N is a greatest common divisor of a and b. In fact, working forwards from the first equation we see that any common divisor c of a and b divides each a_k and so, in particular, a_N. On the other hand, working backwards from the last equation we see that a_N divides each a_k and so, in particular, a and b.

The Bézout identity can also be obtained in this way, although Euclid himself lacked the necessary algebraic notation. Define sequences $(x_k), (y_k)$ by the recurrence relations

$$x_{k+1} = x_{k-1} - q_k x_k, \quad y_{k+1} = y_{k-1} - q_k y_k \quad (1 \leq k < N),$$

with the starting values

$$x_0 = 0, \quad x_1 = 1, \quad \text{resp. } y_0 = 1, \quad y_1 = 0.$$

It is easily shown by induction that $a_k = a x_k + b y_k$ and so, in particular, $a_N = a x_N + b y_N$.

The Euclidean algorithm is quite practical. For example, the reader may use it to verify that 13 is the greatest common divisor of 2171 and 5317, and that

$$49 \times 5317 - 120 \times 2171 = 13.$$

However, the first proof given for Proposition 10 also has its uses: there is some advantage in separating the conceptual from the computational and the proof actually rests on more general principles, since there are quadratic number fields whose ring of integers is a 'principal ideal domain' that does not possess any Euclidean algorithm.

It is not visibly obvious that the binomial coefficients

$$^{m+n}C_n = (m + 1) \cdots (m + n)/1 \cdot 2 \cdots \cdots n$$

are integers for all positive integers m, n, although it is apparent from their combinatorial interpretation. However, the property is readily proved by induction, using the relation

$$^{m+n}C_n = {}^{m+n-1}C_n + {}^{m+n-1}C_{n-1}.$$

Binomial coefficients have other arithmetic properties. Hermite observed that $^{m+n}C_n$ is divisible by the integers $(m+n)/(m,n)$ and $(m+1)/(m+1,n)$. In particular, the *Catalan numbers* $(n+1)^{-1}\, ^{2n}C_n$ are integers. The following proposition is a substantial generalization of these results and illustrates the application of Proposition 10.

Proposition 12 *Let (a_n) be a sequence of nonzero integers such that, for all $m, n \geq 1$, every common divisor of a_m and a_n divides a_{m+n}, and every common divisor of a_m and a_{m+n} divides a_n. Then, for all $m, n \geq 1$,*

(i) $(a_m, a_n) = a_{(m,n)}$;
(ii) $A_{m,n} := a_{m+1} \cdots a_{m+n}/a_1 \cdots a_n \in \mathbb{Z}$;
(iii) $A_{m,n}$ *is divisible by* $a_{m+n}/(a_m, a_n)$, *by* $a_{m+1}/(a_{m+1}, a_n)$ *and by* $a_{n+1}/(a_m, a_{n+1})$;
(iv) $(A_{m,n-1}, A_{m+1,n}, A_{m-1,n+1}) = (A_{m-1,n}, A_{m+1,n-1}, A_{m,n+1})$.

Proof The hypotheses imply that

$$(a_m, a_n) = (a_m, a_{m+n}) \quad \text{for all } m, n \geq 1.$$

Since $a_m = (a_m, a_m)$, it follows by induction that $a_m | a_{km}$ for all $k \geq 1$. Moreover,

$$(a_{km}, a_{(k+1)m}) = a_m,$$

since every common divisor of a_{km} and $a_{(k+1)m}$ divides a_m.

Put $d = (m, n)$. Then $m = dm'$, $n = dn'$, where $(m', n') = 1$. Thus there exist integers u, v such that $m'u - n'v = 1$. By replacing u, v by $u + tn', v + tm'$ with any $t > \max\{|u|, |v|\}$, we may assume that u and v are both positive. Then

$$(a_{mu}, a_{nv}) = (a_{(n'v+1)d}, a_{n'vd}) = a_d.$$

Since a_d divides (a_m, a_n) and (a_m, a_n) divides (a_{mu}, a_{nv}), this implies $(a_m, a_n) = a_d$. This proves (i).

Since $a_1 | a_{m+1}$, it is evident that $A_{m,1} \in \mathbb{Z}$ for all $m \geq 1$. We assume that $n > 1$ and $A_{m,n} \in \mathbb{Z}$ for all smaller values of n and all $m \geq 1$. Since it is trivial that $A_{0,n} \in \mathbb{Z}$, we assume also that $m \geq 1$ and $A_{m,n} \in \mathbb{Z}$ for all smaller values of m. By Proposition 10, there exist $x, y \in \mathbb{Z}$ such that

$$a_m x + a_n y = a_{m+n},$$

since (a_m, a_n) divides a_{m+n}. Since

$$A_{m,n} = \frac{a_{m+1} \cdots a_{m+n}}{a_1 \cdots a_n} = \frac{a_m a_{m+1} \cdots a_{m+n-1}}{a_1 \cdots a_n} x + \frac{a_{m+1} \cdots a_{m+n-1}}{a_1 \cdots a_{n-1}} y,$$

our induction hypotheses imply that $A_{m,n} \in \mathbb{Z}$. This proves (ii).

Since

$$a_{m+n} A_{m,n-1} = a_n A_{m,n},$$

a_{m+n} divides $(a_n, a_{m+n}) A_{m,n}$ and, since $(a_n, a_{m+n}) = (a_m, a_n)$, this in turn implies that $a_{m+n}/(a_m, a_n)$ divides $A_{m,n}$.

Similarly, since

$$a_{m+1}A_{m+1,n} = a_{m+n+1}A_{m,n}, \quad a_{m+1}A_{m+1,n-1} = a_n A_{m,n},$$

a_{m+1} divides $(a_n, a_{m+n+1})A_{m,n}$ and, since $(a_n, a_{m+n+1}) = (a_{m+1}, a_n)$, it follows that $a_{m+1}/(a_{m+1}, a_n)$ divides $A_{m,n}$. In the same way, since

$$a_{n+1}A_{m,n+1} = a_{m+n+1}A_{m,n}, \quad a_{n+1}A_{m-1,n+1} = a_m A_{m,n},$$

a_{n+1} divides $(a_m, a_{m+n+1})A_{m,n}$ and hence $a_{n+1}/(a_m, a_{n+1})$ divides $A_{m,n}$. This proves (iii).

By multiplying by $a_1 \cdots a_{n+1}/a_{m+2} \cdots a_{m+n-1}$, we see that (iv) is equivalent to

$$(a_n a_{n+1}a_{m+1}, a_{n+1}a_{m+n}a_{m+n+1}, a_m a_{m+1}a_{m+n})$$
$$= (a_{n+1}a_m a_{m+1}, a_n a_{n+1}a_{m+n}, a_{m+1}a_{m+n}a_{m+n+1}).$$

Since here the two sides are interchanged when m and n are interchanged, it is sufficient to show that any common divisor e of the three terms on the right is also a common divisor of the three terms on the left. We have

$$(a_{n+1}a_m a_{m+1}, a_n a_{n+1}a_{m+1}) = a_{n+1}a_{m+1}(a_m, a_n) = a_{n+1}a_{m+1}(a_m, a_{m+n})$$
$$= (a_{n+1}a_m a_{m+1}, a_{m+1}a_{n+1}a_{m+n}),$$

and similarly

$$(a_n a_{n+1}a_{m+n}, a_{n+1}a_{m+n}a_{m+n+1}) = (a_n a_{n+1}a_{m+n}, a_{m+1}a_{n+1}a_{m+n}),$$
$$(a_{m+1}a_{m+n}a_{m+n+1}, a_m a_{m+1}a_{m+n}) = (a_{m+1}a_{m+n}a_{m+n+1}, a_{m+1}a_{n+1}a_{m+n}).$$

Hence if we put $g = a_{m+1}a_{n+1}a_{m+n}$, then

$$(e, g) = (e, a_n a_{n+1}a_{m+1}) = (e, a_{n+1}a_{m+n}a_{m+n+1}) = (e, a_m a_{m+1}a_{m+n})$$

and if we put $f = (e, g)$, then

$$1 = (e/f, a_n a_{n+1}a_{m+1}/f) = (e/f, a_{n+1}a_{m+n}a_{m+n+1}/f) = (e/f, a_m a_{m+1}a_{m+n}/f).$$

Hence $(e/f, P/f^3) = 1$, where

$$P = a_n a_{n+1}a_{m+1} \cdot a_{n+1}a_{m+n}a_{m+n+1} \cdot a_m a_{m+1}a_{m+n}.$$

But P is divisible by e^3, since we can also write

$$P = a_{n+1}a_m a_{m+1} \cdot a_n a_{n+1}a_{m+n} \cdot a_{m+1}a_{m+n}a_{m+n+1}.$$

Hence the previous relation implies $e/f = 1$. Thus $e = f$ is a common divisor of $a_n a_{n+1}a_{m+1}, a_{n+1}a_{m+n}a_{m+n+1}$ and $a_m a_{m+1}a_{m+n}$, as we wished to show. □

For the binomial coefficient case, i.e. $a_n = n$, the property (iv) of Proposition 12 was discovered empirically by Gould (1972) and then proved by Hillman and Hoggatt (1972). It states that if in the *Pascal triangle* one picks out the hexagon surrounding a particular element, then the greatest common divisor of three alternately chosen vertices is equal to the greatest common divisor of the remaining three vertices. Hillman and Hoggatt also gave generalizations along the lines of Proposition 12.

The hypotheses of Proposition 12 are also satisfied if $a_n = q^n - 1$, for some integer $q > 1$, since in this case $a_{m+n} = a_m a_n + a_m + a_n$. The corresponding q-binomial coefficients were studied by Gauss and, as mentioned in Chapter XIII, they play a role in the theory of partitions.

We may also take (a_n) to be the sequence defined recurrently by

$$a_1 = 1, \quad a_2 = c, \quad a_{n+2} = c a_{n+1} + b a_n (n \geq 1),$$

where b and c are coprime positive integers. Indeed it is easily shown by induction that

$$(a_n, a_{n+1}) = (b, a_{n+1}) = 1 \quad \text{for all } n \geq 1.$$

By induction on m one may also show that

$$a_{m+n} = a_{m+1} a_n + b a_m a_{n-1} \quad \text{for all } m \geq 1, \ n > 1.$$

It follows that the hypotheses of Proposition 12 are satisfied. In particular, for $b = c = 1$, they are satisfied by the sequence of *Fibonacci numbers*.

We consider finally extensions of our results to more general algebraic structures. An integral domain R is said to be a *Bézout domain* if any $a, b \in R$ have a common divisor of the form $au + bv$ for some $u, v \in R$. Since such a common divisor is necessarily a greatest common divisor, any Bézout domain is a GCD domain. It is easily seen, by induction on the number of generators, that an integral domain is a Bézout domain if and only if every finitely generated ideal is a principal ideal. Thus Propositions 10 and 11 continue to hold if $\mathbb{Z}$ is replaced by any Bézout domain.

An integral domain R is said to be a *principal ideal domain* if every ideal is a principal ideal.

Lemma 13 *An integral domain R is a principal ideal domain if and only if it is a Bézout domain satisfying the chain condition*

(#) there exists no infinite sequence (a_n) of elements of R such that a_{n+1} is a proper divisor of a_n for every n.

Proof It is obvious that any principal ideal domain is a Bézout domain. Suppose R is a Bézout domain, but not a principal ideal domain. Then R contains an ideal J which is not finitely generated. Hence there exists a sequence (b_n) of elements of J such that b_{n+1} is not in the ideal J_n generated by $b_1, \ldots, b_n$. But J_n is a principal ideal. If a_n generates J_n, then a_{n+1} is a proper divisor of a_n for every n. Thus the chain condition is violated.

Suppose now that R is a Bézout domain containing a sequence (a_n) such that a_{n+1} is a proper divisor of a_n for every n. Let J denote the set of all elements of R which are divisible by at least one term of this sequence. Then J is an ideal. For if $a_j | b$ and $a_k | c$, where $j \leq k$, then also $a_k | b$ and hence $a_k | bx + cy$ for all $x, y \in R$. If J were generated by a single element a, we would have $a | a_n$ for every n. On the other hand, since $a \in J$, $a_N | a$ for some N. Hence $a_N | a_{N+1}$. Since a_{N+1} is a proper divisor of a_N, this is a contradiction. Thus R is not a principal ideal domain. $\square$

It follows from the remarks at the end of Section 1 that a principal ideal domain is factorial, i.e. any element which is neither zero nor a unit can be represented as a product of finitely many irreducibles and the representation is essentially unique.

In the next section we will show that the ring $K[t]$ of all polynomials in one indeterminate t with coefficients from an arbitrary field K is a principal ideal domain.

It may be shown that the ring of all algebraic integers is a Bézout domain, and likewise the ring of all functions which are holomorphic in a nonempty connected open subset G of the complex plane $\mathbb{C}$. However, neither is a principal ideal domain. In the former case there are no irreducibles, since any algebraic integer a has the factorization $a = \sqrt{a} \cdot \sqrt{a}$. In the latter case $z - \zeta$ is an irreducible for any $\zeta \in G$, but the chain condition is violated. For example, take

$$a_n(z) = f(z)/(z - \zeta_1) \cdots (z - \zeta_n),$$

where $f(z)$ is a non-identically vanishing function which is holomorphic in G and has infinitely many zeros $\zeta_1, \zeta_2, \ldots$ in G.

3 Polynomials

In this section we study the most important example of a principal ideal domain other than $\mathbb{Z}$, namely the ring $K[t]$ of all polynomials in t with coefficients from an arbitrary field K (e.g., $K = \mathbb{Q}$ or $\mathbb{C}$).

The attitude adopted towards polynomials in algebra is different from that adopted in analysis. In analysis we regard 't' as a variable which can take different values; in algebra we regard 't' simply as a symbol, an 'indeterminate', on which we can perform various algebraic operations. Since the concept of function is so pervasive, the algebraic approach often seems mysterious at first sight and it seems worthwhile taking the time to give a precise meaning to an 'indeterminate'.

Let R be an integral domain (e.g., $R = \mathbb{Z}$ or $\mathbb{Q}$). A *polynomial* with coefficients from R is defined to be a sequence $f = (a_0, a_1, a_2, \ldots)$ of elements of R in which at most finitely many terms are nonzero. The sum and product of two polynomials

$$f = (a_0, a_1, a_2, \ldots), \quad g = (b_0, b_1, b_2, \ldots)$$

are defined by

$$f + g = (a_0 + b_0, a_1 + b_1, a_2 + b_2, \ldots),$$
$$fg = (a_0 b_0, a_0 b_1 + a_1 b_0, a_0 b_2 + a_1 b_1 + a_2 b_0, \ldots).$$

It is easily verified that these are again polynomials and that the set $R[t]$ of all polynomials with coefficients from R is a commutative ring with $O = (0, 0, 0, \ldots)$ as zero element. (By dropping the requirement that at most finitely many terms are nonzero, we obtain the ring $R[[t]]$ of all *formal power series* with coefficients from R.)

We define the *degree* $\partial(f)$ of a polynomial $f = (a_0, a_1, a_2, \ldots) \neq O$ to be the greatest integer n for which $a_n \neq 0$ and we put

$$|f| = 2^{\partial(f)}, \quad |O| = 0.$$

It is easily verified that, for all polynomials f, g,

$$|f + g| \leq \max\{|f|, |g|\}, \quad |fg| = |f||g|.$$

Since $|f| \geq 0$, with equality if and only if $f = O$, the last property implies that $R[t]$ *is an integral domain*. Thus we can define divisibility in $R[t]$, as explained in Section 1.

The set of all polynomials of the form $(a_0, 0, 0, \ldots)$ is a subdomain isomorphic to R. By identifying this set with R, we may regard R as embedded in $R[t]$. The only units in $R[t]$ are the units in R, since $1 = ef$ implies $1 = |e||f|$ and hence $|e| = 1$.

If we put $t = (0, 1, 0, 0, \ldots)$, then

$$t^2 = tt = (0, 0, 1, 0, \ldots), \quad t^3 = tt^2 = (0, 0, 0, 1, \ldots), \ldots.$$

Hence if the polynomial $f = (a_0, a_1, a_2, \ldots)$ has degree n, then it can be uniquely expressed in the form

$$f = a_0 + a_1 t + \cdots + a_n t^n \quad (a_n \neq 0).$$

We refer to the elements $a_0, a_1, \ldots, a_n$ of R as the *coefficients* of f. In particular, a_0 is the *constant* coefficient and a_n the *highest* coefficient. We say that f is *monic* if its highest coefficient $a_n = 1$.

If also

$$g = b_0 + b_1 t + \cdots + b_m t^m \quad (b_m \neq 0),$$

then the sum and product assume their familiar forms:

$$f + g = (a_0 + b_0) + (a_1 + b_1)t + (a_2 + b_2)t^2 + \cdots,$$
$$fg = a_0 b_0 + (a_0 b_1 + a_1 b_0)t + (a_0 b_2 + a_1 b_1 + a_2 b_0)t^2 + \cdots.$$

Suppose now that $R = K$ is a field, and let

$$f = a_0 + a_1 t + \cdots + a_n t^n \quad (a_n \neq 0),$$
$$g = b_0 + b_1 t + \cdots + b_m t^m \quad (b_m \neq 0)$$

be any two nonzero elements of $K[t]$. If $|g| < |f|$, i.e. if $m < n$, then $g = qf + r$, with $q = O$ and $r = g$. Suppose on the other hand that $|f| \leq |g|$. Then

$$g = a_n^{-1} b_m t^{m-n} f + g^{\dagger},$$

where $g^{\dagger} \in K[t]$ and $|g^{\dagger}| < |g|$. If $|f| \leq |g^{\dagger}|$, the process can be repeated with $g^{\dagger}$ in place of g. Continuing in this way, we obtain $q, r \in K[t]$ such that

$$g = qf + r, \quad |r| < |f|.$$

Moreover, q and r are uniquely determined, since if also

$$g = q_1 f + r_1, \quad |r_1| < |f|,$$

then

$$(q - q_1)f = r_1 - r, \quad |r_1 - r| < |f|,$$

which is only possible if $q = q_1$.

Ideals in $K[t]$ can be defined in the same way as for $\mathbb{Z}$ and the proof of Lemma 9 remains valid. Thus $K[t]$ is a principal ideal domain and, *a fortiori*, a GCD domain.

The Euclidean algorithm can also be applied in $K[t]$ in the same way as for $\mathbb{Z}$ and again, from the sequence of polynomials $f_0, f_1, \ldots, f_N$ which it provides to determine the greatest common divisor f_N of f_0 and f_1 we can obtain polynomials u_k, v_k such that

$$f_k = f_1 u_k + f_0 v_k \quad (0 \le k \le N).$$

We can actually say more for polynomials than for integers, since if

$$f_{k-1} = q_k f_k + f_{k+1}, \quad |f_{k+1}| < |f_k|,$$

then $|f_{k-1}| = |q_k||f_k|$ and hence, by induction,

$$|f_{k-1}||u_k| = |f_0|, |f_{k-1}||v_k| = |f_1| \quad (1 < k \le N).$$

It may be noted in passing that the Euclidean algorithm can also be applied in the ring $K[t, t^{-1}]$ of *Laurent polynomials*. A Laurent polynomial $f \ne O$, with coefficients from the field K, has the form

$$f = a_m t^m + a_{m+1} t^{m+1} + \cdots + a_n t^n,$$

where $m, n \in \mathbb{Z}$ with $m \le n$ and $a_j \in K$ with $a_m a_n \ne 0$. Thus we can write $f = t^m f_0$, where $f_0 \in K[t]$. Put

$$|f| = 2^{n-m}, \quad |O| = 0;$$

then the division algorithm for ordinary polynomials implies one for Laurent polynomials: for any $f, g \in K[t, t^{-1}]$ with $f \ne O$, there exist $q, r \in K[t, t^{-1}]$ such that $g = qf + r, |r| < |f|$.

We return now to ordinary polynomials. The general definition for integral domains in Section 1 means, in the present case, that a polynomial $p \in K[t]$ is *irreducible* if it has positive degree and if every proper divisor has degree zero.

It follows that any polynomial of degree 1 is irreducible. However, there may exist also irreducible polynomials of higher degree. For example, we will show shortly that the polynomial $t^2 - 2$ is irreducible in $\mathbb{Q}[t]$. For $K = \mathbb{C}$, however, every irreducible polynomial has degree 1, by the fundamental theorem of algebra (Theorem I.30) and Proposition 14 below. It follows that, for $K = \mathbb{R}$, every irreducible polynomial has degree 1 or 2. (For if a real polynomial $f(t)$ has a root $\alpha \in \mathbb{C} \backslash \mathbb{R}$, its conjugate $\bar{\alpha}$ is also a root and $f(t)$ has the real irreducible factor $(t - \alpha)(t - \bar{\alpha})$.)

It is obvious that the chain condition (#) of Section 1 holds in the integral domain $K[t]$, since if g is a proper divisor of f, then $|g| < |f|$. It follows that any polynomial of positive degree can be represented as a product of finitely many irreducible polynomials and that the representation is essentially unique.

We now consider the connection between polynomials in the sense of algebra (polynomial forms) and polynomials in the sense of analysis (polynomial functions). Let K be a field and $f \in K[t]$:

$$f = a_0 + a_1 t + \cdots + a_n t^n.$$

If we replace 't' by $c \in K$ we obtain an element of K, which we denote by $f(c)$:

$$f(c) = a_0 + a_1 c + \cdots + a_n c^n.$$

A rapid procedure ('Horner's rule') for calculating $f(c)$ is to use the recurrence relations

$$f_0 = a_n, \quad f_j = f_{j-1}c + a_{n-j} \quad (j = 1, \ldots, n).$$

It is readily shown by induction that

$$f_j = a_n c^j + a_{n-1} c^{j-1} + \cdots + a_{n-j},$$

and hence $f(c) = f_n$ is obtained with just n multiplications and n additions.

It is easily seen that $f = g + h$ implies $f(c) = g(c) + h(c)$, and $f = gh$ implies $f(c) = g(c)h(c)$. Thus the mapping $f \to f(c)$ is a 'homomorphism' of $K[t]$ into K. A simple consequence is the so-called *remainder theorem*:

Proposition 14 *Let K be a field and $c \in K$. If $f \in K[t]$, then*

$$f = (t - c)g + f(c),$$

for some $g \in K[t]$.

In particular, f is divisible by $t - c$ if and only if $f(c) = 0$.

Proof We already know that there exist $q, r \in K[t]$ such that

$$f = (t - c)q + r, \quad |r| \le 1.$$

Thus $r \in K$ and the homomorphism properties imply that $f(c) = r$. □

We say that $c \in K$ is a *root* of the polynomial $f \in K[t]$ if $f(c) = 0$.

Proposition 15 *Let K be a field. If $f \in K[t]$ is a polynomial of degree $n \ge 0$, then f has at most n distinct roots in K.*

Proof If f is of degree 0, then $f = c$ is a nonzero element of K and f has no roots. Suppose now that $n \ge 1$ and the result holds for polynomials of degree less than n. If c is a root of f then, by Proposition 14, $f = (t - c)g$ for some $g \in K[t]$. Since g has degree $n - 1$, it has at most $n - 1$ roots. But every root of f distinct from c is a root of g. Hence f has at most n roots. □

We consider next properties of the integral domain $R[t]$, when R is an integral domain rather than a field (e.g., $R = \mathbb{Z}$). The famous Pythagorean proof that $\sqrt{2}$ is irrational is considerably generalized by the following result:

Proposition 16 *Let R be a GCD domain and K its field of fractions. Let*

$$f = a_0 + a_1 t + \cdots + a_n t^n$$

be a polynomial of degree $n > 0$ with coefficients $a_j \in R$ ($0 \le j \le n$). If $c \in K$ is a root of f and $c = ab^{-1}$, where $a, b \in R$ and $(a, b) = 1$, then $b | a_n$ and $a | a_0$.

In particular, if f is monic, then $c \in R$.

Proof We have

$$a_0 b^n + a_1 a b^{n-1} + \cdots + a_{n-1} a^{n-1} b + a_n a^n = 0.$$

Hence $b | a_n a^n$ and $a | a_0 b^n$. Since $(a^n, b) = (a, b^n) = 1$, by Proposition 3(v), the result follows from Proposition 3(ii).   □

The polynomial $t^2 - 2$ has no integer roots, since 0, 1, −1 are not roots and if $c \in \mathbb{Z}$ and $c \ne 0, 1, -1$, then $c^2 \ge 4$. Consequently, by Proposition 16, the polynomial $t^2 - 2$ also has no rational roots. It now follows from Proposition 14 that $t^2 - 2$ is irreducible in $\mathbb{Q}[t]$, since it has no divisors of degree 1.

Proposition 16 was known to Euler (1774) for the case $R = \mathbb{Z}$. In this case it shows that to obtain all rational roots of a polynomial with rational coefficients we need test only a finite number of possibilities, which can be explicitly enumerated. For example, if $z \in \mathbb{Z}$, the cubic polynomial $t^3 + zt + 1$ has no rational roots unless $z = 0$ or $z = -2$.

It was shown by Gauss (1801), again for the case $R = \mathbb{Z}$, that Proposition 16 may itself be considerably generalized. His result may be formulated in the following way:

Proposition 17 *Let $f, g \in R[t]$, where R is a GCD domain with field of fractions K. Then g divides f in $R[t]$ if and only if g divides f in $K[t]$ and the greatest common divisor of the coefficients of g divides the greatest common divisor of the coefficients of f.*

Proof For any polynomial $f \in R[t]$, let $c(f)$ denote the greatest common divisor of its coefficients. We say that f is *primitive* if $c(f) = 1$. We show first that the product $f = gh$ of two primitive polynomials g, h is again primitive.

Let

$$g = b_0 + b_1 t + \cdots , \quad h = c_0 + c_1 t + \cdots , \quad f = a_0 + a_1 t + \cdots ,$$

and assume on the contrary that the coefficients a_i have a common divisor d which is not a unit. Then d does not divide all the coefficients b_j, nor all the coefficients c_k. Let b_m, c_n be the first coefficients of g, h which are not divisible by d. Then

$$a_{m+n} = \sum_{j+k=m+n} b_j c_k$$

and d divides every term on the right, except possibly $b_m c_n$. In fact, since $d | a_{m+n}$, d must also divide $b_m c_n$. Hence we cannot have both $(d, b_m) = 1$ and $(d, c_n) = 1$.

Consequently we can replace d by a proper divisor d', again not a unit, for which $m' + n' > m + n$. Since there exists a divisor d for which $m + n$ is a maximum, this yields a contradiction.

Now let f, g be polynomials in $R[t]$ such that g divides f in $K[t]$. Thus $f = gH$, where $H \in K[t]$. We can write $H = ab^{-1}h_0$, where a, b are coprime elements of R and h_0 is a primitive polynomial in $R[t]$. Also

$$f = c(f)f_0, \quad g = c(g)g_0,$$

where f_0, g_0 are primitive polynomials in $R[t]$. Hence

$$bc(f)f_0 = ac(g)g_0h_0.$$

Since g_0h_0 is primitive, it follows that

$$bc(f) = ac(g).$$

If $H \in R[t]$, then $b = 1$ and so $c(g)|c(f)$. On the other hand, if $c(g)|c(f)$, then $bc(f)/c(g) = a$. Since $(a, b) = 1$, this implies that $b = 1$ and $H \in R[t]$. $\square$

Corollary 18 *If R is a GCD domain, then $R[t]$ is also a GCD domain. If, moreover, R is a factorial domain, then $R[t]$ is also a factorial domain.*

proof Let K denote the field of fractions of R. Since $K[t]$ is a GCD domain and $R[t] \subseteq K[t]$, $R[t]$ is certainly an integral domain. If $f, g \in R[t]$, then there exists a primitive polynomial $h_0 \in R[t]$ which is a greatest common divisor of f and g in $K[t]$. It follows from Proposition 17 that

$$h = (c(f), c(g))h_0$$

is a greatest common divisor of f and g in $R[t]$.

This proves the first statement of the corollary. It remains to show that if R also satisfies the chain condition (#), then $R[t]$ does likewise. But if $f_n \in R[t]$ and $f_{n+1}|f_n$ for every n, then f_n must be of constant degree for all large n. The second statement of the corollary now also follows from Proposition 17 and the chain condition in R. $\square$

It follows by induction that in the statement of Corollary 18 we may replace $R[t]$ by the ring $R[t_1, \ldots, t_m]$ of all polynomials in finitely many indeterminates $t_1, \ldots, t_m$ with coefficients from R. In particular, if K is a field, then any polynomial $f \in K[t_1, \ldots, t_m]$ such that $f \notin K$ can be represented as a product of finitely many irreducible polynomials and the representation is essentially unique.

It is now easy to give examples of GCD domains which are not Bézout domains. Let R be a GCD domain which is not a field (e.g., $R = \mathbb{Z}$). Then some $a_0 \in R$ is neither zero nor a unit. By Corollary 18, $R[t]$ is a GCD domain and, by Proposition 17, the greatest common divisor in $R[t]$ of the polynomials a_0 and t is 1. If there existed $g, h \in R[t]$ such that

$$a_0g + th = 1,$$

where $g = b_0 + b_1 t + \cdots$, then by equating constant coefficients we would obtain $a_0 b_0 = 1$, which is a contradiction. Thus $R[t]$ is not a Bézout domain.

As an application of the preceding results we show that if $a_1, \ldots, a_n$ are distinct integers, then the polynomial

$$f = \prod_{j=1}^{n} (t - a_j) - 1$$

is irreducible in $\mathbb{Q}[t]$. Assume, on the contrary, that $f = gh$, where $g, h \in \mathbb{Q}[t]$ and have positive degree. We may suppose without loss of generality that $g \in \mathbb{Z}[t]$ and that the greatest common divisor of the coefficients of g is 1. Since $f \in \mathbb{Z}[t]$, it then follows from Proposition 17 that also $h \in \mathbb{Z}[t]$. Thus $g(a_j)$ and $h(a_j)$ are integers for every j. Since $g(a_j)h(a_j) = -1$, it follows that $g(a_j) = -h(a_j)$. Thus the polynomial $g + h$ has the distinct roots $a_1, \ldots, a_n$. Since $g + h$ has degree less than n, this is possible only if $g + h = O$. Hence $f = -g^2$. But, since the highest coefficient of f is 1, this is a contradiction.

In general, it is not an easy matter to determine if a polynomial with rational coefficients is irreducible in $\mathbb{Q}[t]$. However, the following *irreducibility criterion*, due to Eisenstein (1850), is sometimes useful:

Proposition 19 *If*

$$f(t) = a_0 + a_1 t + \cdots + a_{n-1} t^{n-1} + t^n$$

is a monic polynomial of degree n with integer coefficients such that $a_0, a_1, \ldots, a_{n-1}$ are all divisible by some prime p, but a_0 is not divisible by p^2, then f is irreducible in $\mathbb{Q}[t]$.

Proof Assume on the contrary that f is reducible. Then there exist polynomials $g(t), h(t)$ of positive degrees l, m with *integer* coefficients such that $f = gh$. If

$$g(t) = b_0 + b_1 t + \cdots + b_l t^l,$$
$$h(t) = c_0 + c_1 t + \cdots + c_m t^m,$$

then $a_0 = b_0 c_0$. The hypotheses imply that exactly one of b_0, c_0 is divisible by p. Without loss of generality, assume it to be b_0. Since p divides $a_1 = b_0 c_1 + b_1 c_0$, it follows that $p | b_1$. Since p divides $a_2 = b_0 c_2 + b_1 c_1 + b_2 c_0$, it now follows that $p | b_2$. Proceeding in this way, we see that p divides b_j for every $j \leq l$. But, since $b_l c_m = 1$, this yields a contradiction.   □

It follows from Proposition 19 that, for any prime p, the p-th *cyclotomic polynomial*

$$\Phi_p(x) = x^{p-1} + x^{p-2} + \cdots + 1$$

is irreducible in $\mathbb{Q}[x]$. For $\Phi_p(x) = (x^p - 1)/(x - 1)$ and, if we put $x = 1 + t$, the transformed polynomial

$$\{(1+t)^p - 1\}/t = t^{p-1} + {}^pC_{p-1}t^{p-2} + \cdots + {}^pC_2 t + p$$

satisfies the hypotheses of Proposition 19.

For any field K, we define the *formal derivative* of a polynomial $f \in K[t]$,

$$f = a_0 + a_1 t + \cdots + a_n t^n,$$

to be the polynomial

$$f' = a_1 + 2a_2 t + \cdots + na_n t^{n-1}.$$

If the field K is of *characteristic* 0 (see Chapter I, §8), then $\partial(f') = \partial(f) - 1$.

Formal derivatives share the following properties with the derivatives of real analysis:

(i) $(f + g)' = f' + g'$;
(ii) $(cf)' = cf'$ for any $c \in K$;
(iii) $(fg)' = f'g + fg'$;
(iv) $(f^k)' = kf^{k-1}f'$ for any $k \in \mathbb{N}$.

The first two properties are easily established and the last two properties then need only be verified for monomials $f = t^m$, $g = t^n$.

We can use formal derivatives to determine when a polynomial is *square-free*:

Proposition 20 Let f be a polynomial of positive degree with coefficients from a field K. If f is relatively prime to its formal derivative f', then f is a product of irreducible polynomials, no two of which differ by a constant factor. Conversely, if f is such a product and if K has characteristic 0, then f is relatively prime to f'.

Proof If $f = g^2 h$ for some polynomials $g, h \in K[t]$ with $\partial(g) > 0$ then, by the rules above,

$$f' = 2gg'h + g^2 h'.$$

Hence $g | f'$ and f, f' are not relatively prime.

On the other hand, if $f = p_1 \cdots p_m$ is a product of essentially distinct irreducible polynomials p_j, then

$$f' = p_1' p_2 \cdots p_m + p_1 p_2' p_3 \cdots p_m + \cdots + p_1 \cdots p_{m-1} p_m'.$$

If the field K has characteristic 0, then p_1' is of lower degree than p_1 and is not the zero polynomial. Thus the first term on the right is not divisible by p_1, but all the other terms are. Therefore $p_1 \nmid f'$, and hence $(f', p_1) = 1$. Similarly, $(f', p_j) = 1$ for $1 < j \leq m$. Since essentially distinct irreducible polynomials are relatively prime, it follows that $(f', f) = 1$. $\square$

For example, it follows from Proposition 20 that the polynomial $t^n - 1 \in K[t]$ is square-free if the characteristic of the field K does not divide the positive integer n.

4 Euclidean Domains

An integral domain R is said to be *Euclidean* if it possesses a Euclidean algorithm, i.e. if there exists a map $\delta: R \to \mathbb{N} \cup \{0\}$ such that, for any $a, b \in R$ with $a \neq 0$, there exist $q, r \in R$ with the properties

$$b = qa + r, \quad \delta(r) < \delta(a).$$

It follows that $\delta(a) > \delta(0)$ for any $a \neq 0$. For there exist $q_1, a_1 \in R$ such that

$$0 = q_1 a + a_1, \quad \delta(a_1) < \delta(a),$$

and if $a_n \neq 0$ there exist $q_{n+1}, a_{n+1} \in R$ such that

$$0 = q_{n+1} a_n + a_{n+1}, \quad \delta(a_{n+1}) < \delta(a_n).$$

Repeatedly applying this process, we must arrive at $a_N = 0$ for some N, since the sequence $\{\delta(a_n)\}$ cannot decrease forever, and we then have $\delta(0) = \delta(a_N) < \cdots < \delta(a_1) < \delta(a)$.

By replacing δ by $\delta - \delta(0)$ *we may, and will, assume that* $\delta(0) = 0, \delta(a) > 0$ if $a \neq 0$.

Since the proof of Lemma 9 remains valid if $\mathbb{Z}$ is replaced by R and $|a|$ by $\delta(a)$, *any Euclidean domain is a principal ideal domain.*

The *polynomial ring* $K[t]$ is a Euclidean domain with $\delta(a) = |a| = 2^{\partial(a)}$. Polynomial rings are characterized among all Euclidean domains by the following result:

Proposition 21 *For a Euclidean domain R, the following conditions are equivalent:*

(i) *for any $a, b \in R$ with $a \neq 0$, there exist unique $q, r \in R$ such that $b = qa + r$, $\delta(r) < \delta(a)$;*

(ii) *for any $a, b, c \in R$ with $c \neq 0$,*

$$\delta(a + b) \leq \max\{\delta(a), \delta(b)\}, \quad \delta(a) \leq \delta(ac).$$

Moreover, if one or other of these two conditions holds, then either R is a field and $\delta(a) = \delta(1)$ for every $a \neq 0$, or $R = K[t]$ for some field K and δ is an increasing function of $| \, |$.

Proof Suppose first that (i) holds. If $a \neq 0, c \neq 0$, then from $0 = 0a - 0 = ca - ac$, we obtain $\delta(ac) \geq \delta(a)$, and this holds also if $a = 0$. If we take $c = -1$ and replace a by $-a$, we get $\delta(-a) = \delta(a)$. Since $b = 0(a + b) + b = 1(a + b) + (-a)$, it follows that either $\delta(b) \geq \delta(a + b)$ or $\delta(a) \geq \delta(a + b)$. Thus (i) $\Rightarrow$ (ii).

Suppose next that (ii) holds. Assume that, for some $a, b \in R$ with $a \neq 0$, there exist pairs q, r and q', r' such that

$$b = qa + r = q'a + r', \quad \max\{\delta(r), \delta(r')\} < \delta(a).$$

From (ii) we obtain first $\delta(-r) = \delta(r)$ and then $\delta(r' - r) \leq \max\{\delta(r), \delta(r')\} < \delta(a)$. Since $r' - r = a(q - q')$, this implies $q - q' = 0$ and hence $r' - r = 0$. Thus (ii) $\Rightarrow$ (i).

Suppose now that (i) and (ii) both hold. Then $\delta(1) \leq \delta(a)$ for any $a \neq 0$, since $a = 1a$. Furthermore, $\delta(a) = \delta(ae)$ for any unit e, since

$$\delta(a) \leq \delta(ae) \leq \delta(aee^{-1}) = \delta(a).$$

On the other hand, $\delta(a) = \delta(ae)$ for some $a \neq 0$ implies that e is a unit. For from

$$a = qae + r, \quad \delta(r) < \delta(ae),$$

we obtain $r = (1-qe)a$, $\delta(r) < \delta(a)$, and hence $1 - qe = 0$. In particular, $\delta(e) = \delta(1)$ if and only if e is a unit.

The set K of all $a \in R$ such that $\delta(a) \leq \delta(1)$ thus consists of 0 and all units of R. Since $a, b \in K$ implies $a - b \in K$, it follows that K is a field. We assume that $K \neq R$, since otherwise we have the first alternative of the proposition.

Choose $x \in R \backslash K$ so that

$$\delta(x) = \min_{a \in R \backslash K} \delta(a).$$

For any $a \in R \backslash K$, there exist $q_0, r_0 \in R$ such that

$$a = q_0 x + r_0, \quad \delta(r_0) < \delta(x),$$

i.e. $r_0 \in K$. Then $\delta(q_0) < \delta(q_0 x) = \delta(a - r_0) \leq \delta(a)$. If $\delta(q_0) \geq \delta(x)$, i.e. if $q_0 \in R \backslash K$, then in the same way there exist $q_1, r_1 \in R$ such that

$$q_0 = q_1 x + r_1, r_1 \in K, \quad \delta(q_1) < \delta(q_0).$$

After finitely many repetitions of this process we must arrive at some $q_{n-1} \in K$. Putting $r_n = q_{n-1}$, we obtain

$$a = r_n x^n + r_{n-1} x^{n-1} + \cdots + r_0,$$

where $r_0, \ldots, r_n \in K$ and $r_n \neq 0$. Since $\delta(r_j x^j) = \delta(x^j)$ if $r_j \neq 0$ and $\delta(x^j) < \delta(x^{j+1})$ for every j, it follows that $\delta(a) = \delta(x^n)$. Since the representation $a = qx^n + r$ with $\delta(r) < \delta(x^n)$ is unique, it follows that $r_0, \ldots, r_n$ are uniquely determined by a. Define a map $\psi : R \to K[t]$ by

$$\psi(r_n x^n + r_{n-1} x^{n-1} + \cdots + r_0) = r_n t^n + r_{n-1} t^{n-1} + \cdots + r_0.$$

Then ψ is a bijection and actually an isomorphism, since it preserves sums and products. Furthermore $\delta(a) >, =,$ or $< \delta(b)$ according as $|\psi(a)| >, =,$ or $< |\psi(b)|$. $\square$

Some significant examples of principal ideal domains are provided by quadratic fields, which will be studied in Chapter III. Any quadratic number field has the form $\mathbb{Q}(\sqrt{d})$, where $d \in \mathbb{Z}$ is square-free and $d \neq 1$. The set $\mathcal{O}_d$ of all algebraic integers in $\mathbb{Q}(\sqrt{d})$ is an integral domain. In the equivalent language of binary quadratic forms, it was known to Gauss that $\mathcal{O}_d$ is a principal ideal domain for nine negative values of d, namely

$$d = -1, -2, -3, -7, -11, -19, -43, -67, -163.$$

Heilbronn and Linfoot (1934) showed that there was at most one additional negative value of d for which $\mathcal{O}_d$ is a principal ideal domain. Stark (1967) proved that this additional value does not in fact exist, and soon afterwards it was observed that a gap in a previous proof by Heegner (1952) could be filled without difficulty. It is conjectured that $\mathcal{O}_d$ is a principal ideal domain for infinitely many positive values of d, but this remains unproved.

Much work has been done on determining for which quadratic number fields $\mathbb{Q}(\sqrt{d})$ the ring of integers $\mathcal{O}_d$ is a Euclidean domain. Although we regard being Euclidean more as a useful property than as an important concept, we report here the results which have been obtained for their intrinsic interest.

The ring $\mathcal{O}_d$ is said to be *norm-Euclidean* if it is Euclidean when one takes $\delta(a)$ to be the absolute value of the *norm* of a. It has been shown that $\mathcal{O}_d$ is norm-Euclidean for precisely the following values of d:

$$d = -11, -7, -3, -2, -1, 2, 3, 5, 6, 7, 11, 13, 17, 19, 21, 29, 33, 37, 41, 57, 73.$$

It is known that, for $d < 0$, $\mathcal{O}_d$ is Euclidean only if it is norm-Euclidean. Comparing the two lists, we see that for $d = -19, -43, -67, -163$, $\mathcal{O}_d$ is a principal ideal domain, but not a Euclidean domain. On the other hand it is also known that, for $d = 69$, $\mathcal{O}_d$ is Euclidean but not norm-Euclidean.

5 Congruences

The invention of a new notation often enables one to replace a long, involved argument by simple and mechanical algebraic operations. This is well illustrated by the congruence notation.

Two integers a and b are said to be *congruent modulo* a third integer m if m divides $a - b$, and this is denoted by $a \equiv b \bmod m$. For example,

$$13 \equiv 4 \bmod 3, \quad 13 \equiv -7 \bmod 5, \quad 19 \equiv 7 \bmod 4.$$

The notation is a modification by Gauss of the notation $a = b \bmod m$ used by Legendre, as Gauss explicitly acknowledged ($D.A.$, §2). (If a and b are not congruent modulo m, we write $a \not\equiv b \bmod m$.) Congruence has, in fact, many properties in common with equality:

(C1) $a \equiv a \bmod m$ *for all* a, m;   (reflexive law)
(C2) *if* $a \equiv b \bmod m$, *then* $b \equiv a \bmod m$;   (symmetric law)
(C3) *if* $a \equiv b$ *and* $b \equiv c \bmod m$, *then* $a \equiv c \bmod m$;   (transitive law)
(C4) *if* $a \equiv a'$ *and* $b \equiv b' \bmod m$, *then* $a + b \equiv a' + b'$ *and*
      $ab \equiv a'b' \bmod m$.   (replacement laws)

The proofs of these properties are very simple. For any a, m we have $a - a = 0 = m \cdot 0$. If m divides $a - b$, then it also divides $b - a = -(a - b)$. If m divides both $a - b$ and $b - c$, then it also divides $(a - b) + (b - c) = a - c$. Finally, if m divides both $a - a'$ and $b - b'$, then it also divides $(a - a') + (b - b') = (a + b) - (a' + b')$ and $(a - a')b + a'(b - b') = ab - a'b'$.

The properties **(C1)**–**(C3)** state that congruence mod m is an *equivalence relation*. Since $a = b$ implies $a \equiv b \bmod m$, it is a coarsening of the equivalence relation of equality (but coincides with it if $m = 0$). The corresponding equivalence classes are called *residue classes*. The set $\mathbb{Z}$ with equality replaced by congruence mod m will be denoted by $\mathbb{Z}_{(m)}$. If $m > 0$, $\mathbb{Z}_{(m)}$ has cardinality m, since an arbitrary integer a can be uniquely represented in the form $a = qm + r$, where $r \in \{0, 1, \ldots, m - 1\}$ and $q \in \mathbb{Z}$. The particular r which represents a given $a \in \mathbb{Z}$ is referred to as the *least non-negative residue* of $a \bmod m$.

The replacement laws imply that the associative, commutative and distributive laws for addition and multiplication are inherited from $\mathbb{Z}$ by $\mathbb{Z}_{(m)}$. Hence $\mathbb{Z}_{(m)}$ is a commutative ring, with 0 as an identity element for addition and 1 as an identity element for multiplication. However, $\mathbb{Z}_{(m)}$ is not an integral domain if m is composite, since if $m = m'm''$ with $1 < m' < m$, then

$$m'm'' \equiv 0, \text{ but } m' \not\equiv 0, m'' \not\equiv 0 \bmod m.$$

On the other hand, if $ab \equiv ac \bmod m$ and $(a, m) = 1$, then $b \equiv c \bmod m$, by Proposition 3(ii). Thus factors which are relatively prime to the modulus can be cancelled.

In algebraic terms, $\mathbb{Z}_{(m)}$ is the *quotient ring* $\mathbb{Z}/m\mathbb{Z}$ of $\mathbb{Z}$ with respect to the ideal $m\mathbb{Z}$ generated by m, and the elements of $\mathbb{Z}_{(m)}$ are the *cosets* of this ideal. For convenience, rather than necessity, we suppose from now on that $m > 1$.

Congruences enter implicitly into many everyday problems. For example, the ring $\mathbb{Z}_{(2)}$ contains two distinct elements, 0 and 1, with the addition and multiplication tables

$$0 + 0 = 1 + 1 = 0, 0 + 1 = 1 + 0 = 1,$$
$$0 \cdot 0 = 0 \cdot 1 = 1 \cdot 0 = 0, 1 \cdot 1 = 1.$$

This is the arithmetic of *odds* (1) and *evens* (0), which is used by electronic computers.

Again, to determine the day of the week on which one was born, from the date and day of the week today, is an easy calculation in the arithmetic of $\mathbb{Z}_{(7)}$ (remembering that $366 \equiv 2 \bmod 7$).

The well-known tests for divisibility of an integer by 3 or 9 are easily derived by means of congruences. Let the positive integer a have the decimal representation

$$a = a_0 + a_1 10 + \cdots + a_n 10^n,$$

where $a_0, a_1, \ldots, a_n \in \{0, 1, \ldots, 9\}$. Since $10 \equiv 1 \bmod m$, where $m = 3$ or 9, the replacement laws imply that $10^k \equiv 1 \bmod m$ for any positive integer k and hence

$$a \equiv a_0 + a_1 + \cdots + a_n \bmod m.$$

Thus a is divisible by 3 or 9 if and only if the sum of its digits is so divisible.

This can be used to check the accuracy of arithmetical calculations. Any equation involving only additions and multiplications must remain valid when equality is replaced by congruence mod m. For example, suppose we wish to check if

$$7714 \times 3036 = 23,419,804.$$

Taking congruences mod 9, we have on the left side $19 \times 12 \equiv 1 \times 3 \equiv 3$ and on the right side $5 + 14 + 12 \equiv 5 + 5 + 3 \equiv 4$. Since $4 \not\equiv 3 \bmod 9$, the original equation is incorrect (the 8 should be a 7).

Since the distinct squares in $\mathbb{Z}_{(4)}$ are 0 and 1, it follows that an integer $a \equiv 3 \bmod 4$ cannot be represented as the sum of two squares of integers. Similarly, since the distinct squares in $\mathbb{Z}_{(8)}$ are 0,1,4, an integer $a \equiv 7 \bmod 8$ cannot be represented as the sum of three squares of integers.

The oldest known work on number theory is a Babylonian cuneiform text, from at least as early as 1600 B.C., which contains a list of right-angled triangles whose side lengths are all exact multiples of the unit length. By Pythagoras' theorem, the problem is to find positive integers x, y, z such that

$$x^2 + y^2 = z^2.$$

For example, $3, 4, 5$ and $5, 12, 13$ are solutions. The number of solutions listed suggests that the Babylonians not only knew the theorem of Pythagoras, but also had some rule for finding such *Pythagorean triples*. There are in fact infinitely many, and a rule for finding them all is given by Euclid in his *Elements* (Book X, Lemma 1 following Proposition 28). This rule will now be derived.

We may assume that x and y are relatively prime since, if x, y, z is a Pythagorean triple for which x and y have greatest common divisor d, then $d^2|z^2$ and hence $d|z$, so that $x/d, y/d, z/d$ is also a Pythagorean triple. If x and y are relatively prime, then they are not both even and without loss of generality we may assume that x is odd. If y were also odd, we would have

$$z^2 = x^2 + y^2 \equiv 1 + 1 \equiv 2 \bmod 4,$$

which is impossible. Hence y is even and z is odd. Then 2 is a common divisor of $z + x$ and $z - x$, and is actually their greatest common divisor, since $(x, y) = 1$ implies $(x, z) = 1$. Since

$$(y/2)^2 = (z + x)/2 \cdot (z - x)/2$$

and the two factors on the right are relatively prime, they are also squares:

$$(z + x)/2 = a^2, \quad (z - x)/2 = b^2,$$

where $a > b > 0$ and $(a, b) = 1$. Then

$$x = a^2 - b^2, \quad y = 2ab, \quad z = a^2 + b^2.$$

Moreover a and b cannot both be odd, since z is odd.

Conversely, if x, y, z are defined by these formulas, where a and b are relatively prime positive integers with $a > b$ and either a or b even, then x, y, z is a Pythagorean triple. Moreover x is odd, since z is odd and y even, and it is easily verified that $(x, y) = 1$. For given x and z, a^2 and b^2 are uniquely determined, and hence a and b are also. Thus different couples a, b give different solutions x, y, z.

To return to congruences, we now consider the structure of the ring $\mathbb{Z}_{(m)}$. If $a \equiv a' \bmod m$ and $(a, m) = 1$, then also $(a', m) = 1$. Hence we may speak of an element of $\mathbb{Z}_{(m)}$ as being relatively prime to m. The set of all elements of $\mathbb{Z}_{(m)}$ which are relatively prime to m will be denoted by $\mathbb{Z}_{(m)}^{\times}$. If a is a *unit* of the ring $\mathbb{Z}_{(m)}$, then clearly $a \in \mathbb{Z}_{(m)}^{\times}$. The following proposition shows that, conversely, if $a \in \mathbb{Z}_{(m)}^{\times}$, then a is a unit of the ring $\mathbb{Z}_{(m)}$.

Proposition 22 *The set $\mathbb{Z}_{(m)}^{\times}$ is a commutative group under multiplication.*

Proof By Proposition 3(iv), $\mathbb{Z}_{(m)}^{\times}$ is closed under multiplication. Since multiplication is associative and commutative, it only remains to show that any $a \in \mathbb{Z}_{(m)}^{\times}$ has an inverse $a^{-1} \in \mathbb{Z}_{(m)}^{\times}$.

The elements of $\mathbb{Z}_{(m)}^{\times}$ may be taken to be the positive integers $c_1, \ldots, c_h$ which are less than m and relatively prime to m, and we may choose the notation so that $c_1 = 1$. Since $ac_j \equiv ac_k \bmod m$ implies $c_j \equiv c_k \bmod m$, the elements $ac_1, \ldots, ac_h$ are distinct elements of $\mathbb{Z}_{(m)}^{\times}$ and hence are a permutation of $c_1, \ldots, c_h$. In particular, $ac_i \equiv c_1 \bmod m$ for one and only one value of i. (The existence of inverses also follows from the Bézout identity $au + mv = 1$, since this implies $au \equiv 1 \bmod m$. Hence the Euclidean algorithm provides a way of calculating a^{-1}.) □

Corollary 23 *If p is a prime, then $\mathbb{Z}_{(p)}$ is a finite field with p elements.*

Proof We already know that $\mathbb{Z}_{(p)}$ is a commutative ring, whose distinct elements are represented by the integers $0, 1, \ldots, p - 1$. Since p is a prime, $\mathbb{Z}_{(p)}^{\times}$ consists of all nonzero elements of $\mathbb{Z}_{(p)}$. Since $\mathbb{Z}_{(p)}^{\times}$ is a multiplicative group, by Proposition 22, it follows that $\mathbb{Z}_{(p)}$ is a field. □

The finite field $\mathbb{Z}_{(p)}$ will be denoted from now on by the more usual notation $\mathbb{F}_p$. Corollary 23, in conjunction with Proposition 15, implies that if p is a prime and f a polynomial of degree $n \geq 1$, then the congruence

$$f(x) \equiv 0 \bmod p$$

has at most n mutually incongruent solutions mod p. This is no longer true if the modulus is not a prime. For example, the congruence $x^2 - 1 \equiv 0 \bmod 8$ has the distinct solutions $x \equiv 1, 3, 5, 7 \bmod 8$.

The *order* of the group $\mathbb{Z}_{(m)}^{\times}$, i.e. the number of positive integers less than m and relatively prime to m, is traditionally denoted by $\varphi(m)$, with the convention that $\varphi(1) = 1$. For example, if p is a prime, then $\varphi(p) = p - 1$. More generally, for any positive integer k,

$$\varphi(p^k) = p^k - p^{k-1},$$

since the elements of $\mathbb{Z}_{(p^k)}$ which are not in $\mathbb{Z}_{(p^k)}^{\times}$ are the multiples jp with $0 \leq j < p^{k-1}$. By Proposition 4, if $m = m'm''$, where $(m', m'') = 1$, then $\varphi(m) = \varphi(m')\varphi(m'')$. Together with what we have just proved, this implies that if an arbitrary positive integer m has the factorization

$$m = p_1^{k_1} \cdots p_s^{k_s}$$

as a product of positive powers of distinct primes, then

$$\varphi(m) = p_1^{k_1-1}(p_1 - 1) \cdots p_s^{k_s-1}(p_s - 1).$$

In other words,

$$\varphi(m) = m \prod_{p|m}(1 - 1/p).$$

The function $\varphi(m)$ was first studied by Euler and is known as Euler's *phi*-function (or 'totient' function), although it was Gauss who decided on the letter φ. Gauss (*D.A.*, §39) also established the following property:

Proposition 24 *For any positive integer n,*

$$\sum_{d|n}\varphi(d) = n,$$

where the summation is over all positive divisors d of n.

Proof Let d be a positive divisor of n and let S_d denote the set of all positive integers $m \le n$ such that $(m,n) = d$. Since $(m,n) = d$ if and only if $(m/d, n/d) = 1$, the cardinality of S_d is $\varphi(n/d)$. Moreover every positive integer $m \le n$ belongs to exactly one such set S_d. Hence

$$n = \sum_{d|n}\varphi(n/d) = \sum_{d|n}\varphi(d),$$

since n/d runs through the positive divisors of n at the same time as d.   □

Much of the significance of Euler's function stems from the following property:

Proposition 25 *If m is a positive integer and a an integer relatively prime to m, then*

$$a^{\varphi(m)} \equiv 1 \bmod m.$$

Proof Let $c_1, \ldots, c_h$, where $h = \varphi(m)$, be the distinct elements of $\mathbb{Z}_{(m)}^{\times}$. As we saw in the proof of Proposition 22, the elements $ac_1, \ldots, ac_h$ of $\mathbb{Z}_{(m)}^{\times}$ are just a permutation of $c_1, \ldots, c_h$. Forming their product, we obtain $a^h c_1 \cdots c_h \equiv c_1 \cdots c_h \bmod m$. Since the c's are relatively prime to m, they can be cancelled and we are left with $a^h \equiv 1 \bmod m$.   □

Corollary 26 *If p is a prime and a an integer not divisible by p, then $a^{p-1} \equiv 1 \bmod p$.*

Corollary 26 was stated without proof by Fermat (1640) and is commonly known as 'Fermat's little theorem'. The first published proof was given by Euler (1736), who later (1760) proved the general Proposition 25.

Proposition 25 is actually a very special case of Lagrange's theorem that the order of a subgroup of a finite group divides the order of the whole group. In the present case the whole group is $\mathbb{Z}_{(m)}^{\times}$ and the subgroup is the cyclic group generated by a.

Euler gave also another proof of Corollary 26, which has its own interest. For any two integers a, b and any prime p we have, by the binomial theorem,

$$(a+b)^p = \sum_{k=0}^{p} {}^pC_k a^k b^{p-k},$$

where the binomial coefficients

$${}^pC_k = (p-k+1)\cdots p / 1 \cdot 2 \cdots \cdot k$$

are integers. Moreover p divides ${}^pC_k$ for $0 < k < p$, since p divides ${}^pC_k \cdot k!$ and is relatively prime to $k!$ It follows that

$$(a+b)^p \equiv a^p + b^p \bmod p.$$

In particular, $(a+1)^p \equiv a^p + 1 \bmod p$, from which we obtain by induction $a^p \equiv a \bmod p$ for every integer a. If p does not divide a, the factor a can be cancelled to give $a^{p-1} \equiv 1 \bmod p$.

The first part of the second proof actually shows that *in any commutative ring R, of prime characteristic p, the map $a \to a^p$ is a homomorphism*:

$$(a+b)^p = a^p + b^p, \quad (ab)^p = a^p b^p.$$

(As defined in §8 of Chapter I, R has *characteristic k* if k is the least positive integer such that the sum of k 1's is 0, and has *characteristic zero* if there is no such positive integer.) By way of illustration, we give one important application of this result.

We showed in §3 that, for any prime p, the polynomial

$$\Phi_p(x) = x^{p-1} + x^{p-2} + \cdots + 1$$

is irreducible in $\mathbb{Q}[x]$. The roots in $\mathbb{C}$ of $\Phi_p(x)$ are the p-th roots of unity, other than 1. By a quite different argument we now show that, for any positive integer n, the 'primitive' n-th roots of unity are the roots of a monic polynomial $\Phi_n(x)$ with integer coefficients which is irreducible in $\mathbb{Q}[x]$. The uniquely determined polynomial $\Phi_n(x)$ is called the n-th *cyclotomic polynomial*.

Let ζ be a *primitive* n-th root of unity, i.e. $\zeta^n = 1$ but $\zeta^k \neq 1$ for $0 < k < n$. It follows from Corollary 18 that ζ is a root of some monic irreducible polynomial $f(x) \in \mathbb{Z}[x]$ which divides $x^n - 1$. If p is a prime which does not divide n, then ζ^p is also a primitive n-th root of unity and, for the same reason, ζ^p is a root of some monic irreducible polynomial $g(x) \in \mathbb{Z}[x]$ which divides $x^n - 1$.

We show first that $g(x) = f(x)$. Assume on the contrary that $g(x) \neq f(x)$. Then

$$x^n - 1 = f(x)g(x)h(x)$$

for some $h(x) \in \mathbb{Z}[x]$. Since ζ is a root of $g(x^p)$, we also have

$$g(x^p) = f(x)k(x)$$

for some $k(x) \in \mathbb{Z}[x]$. If $\bar{f}(x), \ldots$ denotes the polynomial in $\mathbb{F}_p[x]$ obtained from $f(x), \ldots$ by reducing the coefficients mod p,

then

$$x^n - 1 = \bar{f}(x)\bar{g}(x)\bar{h}(x), \quad \bar{g}(x^p) = \bar{f}(x)\bar{k}(x).$$

But $\bar{g}(x^p) = \bar{g}(x)^p$, since $\mathbb{F}_p[x]$ is a ring of characteristic p and $a^p = a$ for every $a \in \mathbb{F}_p$. Hence any irreducible factor $\bar{e}(x)$ of $\bar{f}(x)$ in $\mathbb{F}_p[x]$ also divides $\bar{g}(x)$. Consequently $\bar{e}(x)^2$ divides $x^n - 1$ in $\mathbb{F}_p[x]$. But $x^n - 1$ is relatively prime to its formal derivative nx^{n-1}, since $p\nmid n$, and so is square-free. This is the desired contradiction.

By applying this repeatedly for the same or different primes p, we see that ζ^m is a root of $f(x)$ for any positive integer m less than n and relatively prime to n. If ω is any n-th root of unity, then $\omega = \zeta^k$ for a unique k such that $0 \le k < n$. If $(k, n) \ne 1$, then $\omega^d = 1$ for some proper divisor d of n (cf. Lemma 31 below). If such an ω were a root of $f(x)$, then $f(x)$ would divide $x^d - 1$, which is impossible since ζ is not a root of $x^d - 1$. Hence $f(x)$ does not depend on the original choice of primitive n-th root of unity, its roots being all the primitive n-th roots of unity. The polynomial $f(x)$ will now be denoted by $\Phi_n(x)$. Since $x^n - 1$ is square-free, we have

$$x^n - 1 = \prod_{d \mid n} \Phi_d(x).$$

This yields a new proof of Proposition 24, since $\Phi_d(x)$ has degree $\varphi(d)$.

As an application of Fermat's little theorem (Corollary 26) we now prove

Proposition 27 *If p is a prime, then $(p - 1)! + 1$ is divisible by p.*

Proof Since $1! + 1 = 2$, we may suppose that the prime p is odd. By Corollary 26, the polynomial $f(t) = t^{p-1} - 1$ has the distinct roots $1, 2, \ldots, p - 1$ in the field $\mathbb{F}_p$. But the polynomial $g(t) = (t - 1)(t - 2) \cdots (t - p + 1)$ has the same roots. Since $f(t) - g(t)$ is a polynomial of degree less than $p - 1$, it follows from Proposition 15 that $f(t) - g(t)$ is the zero polynomial. In particular, $f(t)$ and $g(t)$ have the same constant coefficient. Since $(-1)^{p-1} = 1$, this yields the result. $\square$

Proposition 27 is known as *Wilson's theorem*, although the first published proof was given by Lagrange (1773). Lagrange observed also that $(n - 1)! + 1$ is divisible by n *only* if n is prime. For suppose $n = n'n''$, where $1 < n', n'' < n$. If $n' \ne n''$, then both n' and n'' occur as factors in $(n - 1)!$ and hence n divides $(n - 1)!$ If $n' = n'' > 2$ then, since $n > 2n'$, both n' and $2n'$ occur as factors in $(n - 1)!$ and again n divides $(n - 1)!$ Finally, if $n = 4$, then n divides $(n - 1)! + 2$.

As another application of Fermat's little theorem, we prove *Euler's criterion for quadratic residues.* If p is a prime and a an integer not divisible by p, we say that a is a *quadratic residue*, or *quadratic nonresidue*, of p according as there exists, or does not exist, an integer c such that $c^2 \equiv a \bmod p$. Thus a is a quadratic residue of p if and only if it is a square in $\mathbb{F}_p^\times$. Euler's criterion is the first statement of the following proposition:

Proposition 28 *If p is an odd prime and a an integer not divisible by p, then*

$$a^{(p-1)/2} \equiv 1 \text{ or } -1 \bmod p,$$

according as a is a quadratic residue or nonresidue of p.

Moreover, exactly half of the integers $1, 2, \ldots, p - 1$ are quadratic residues of p.

Proof If a is a quadratic residue of p, then $a \equiv c^2 \bmod p$ for some integer c and hence, by Fermat's little theorem,

$$a^{(p-1)/2} \equiv c^{p-1} \equiv 1 \bmod p.$$

Since the polynomial $t^{(p-1)/2} - 1$ has at most $(p-1)/2$ roots in the field $\mathbb{F}_p$, it follows that there are at most $r := (p-1)/2$ distinct quadratic residues of p. On the other hand, no two of the integers $1^2, 2^2, \ldots, r^2$ are congruent mod p, since $u^2 \equiv v^2 \bmod p$ implies $u \equiv v$ or $u \equiv -v \bmod p$. Hence there are exactly $(p-1)/2$ distinct quadratic residues of p and, if b is a quadratic nonresidue of p, then $b^{(p-1)/2} \not\equiv 1 \bmod p$. Since $b^{p-1} \equiv 1 \bmod p$, and

$$b^{p-1} - 1 = (b^{(p-1)/2} - 1)(b^{(p-1)/2} + 1),$$

we must have $b^{(p-1)/2} \equiv -1 \bmod p$. $\qquad\square$

Corollary 29 *If p is an odd prime, then -1 is a quadratic residue of p if $p \equiv 1 \bmod 4$ and a quadratic nonresidue of p if $p \equiv 3 \bmod 4$.*

Euler's criterion may also be used to determine for what primes 2 is a quadratic residue:

Proposition 30 *For any odd prime p, 2 is a quadratic residue of p if $p \equiv \pm 1 \bmod 8$ and a quadratic nonresidue if $p \equiv \pm 3 \bmod 8$.*

Proof Let A denote the set of all even integers a such that $p/2 < a < p$, and let B denote the set of all even integers b such that $0 < b < p/2$. Since $A \cup B$ is the set of all positive even integers less than p, it has cardinality $r := (p-1)/2$. Evidently $a \in A$ if and only if $p - a$ is odd and $0 < p - a < p/2$. Hence the integers $1, 2, \ldots, r$ are just the elements of B, together with the integers $p - a(a \in A)$. If we denote the cardinality of A by $\#A$, it follows that

$$\begin{aligned} r! &= \prod_{a \in A}(p-a) \prod_{b \in B} b \\ &\equiv (-1)^{\#A} \prod_{a \in A} a \prod_{b \in B} b \bmod p \\ &= (-1)^{\#A} 2^r r! \end{aligned}$$

Thus $2^r \equiv (-1)^{\#A} \bmod p$ and hence, by Proposition 28, 2 is a quadratic residue or nonresidue of p according as $\#A$ is even or odd. But $\#A = k$ if $p = 4k + 1$ and $\#A = k + 1$ if $p = 4k + 3$. The result follows. $\qquad\square$

We now introduce some simple group-theoretical concepts. Let G be a finite group and $a \in G$. Then there exist $j, k \in \mathbb{N}$ with $j < k$ such that $a^j = a^k$. Thus $a^{k-j} = 1$, where 1 is the identity element of G. The *order* of a is the least positive integer d such that $a^d = 1$.

Lemma 31 *Let G be a finite group of order n and a an element of G of order d. Then*

(i) *for any $k \in \mathbb{N}$, $a^k = 1$ if and only if d divides k;*

(ii) *for any $k \in \mathbb{N}$, a^k has order $d/(k, d)$;*
(iii) $H = \{1, a, \ldots, a^{d-1}\}$ *is a subgroup of G and d divides n.*

Proof Any $k \in \mathbb{N}$ can be written in the form $k = qd + r$, where $q \geq 0$ and $0 \leq r < d$. Since $a^{qd} = (a^d)^q = 1$, we have $a^k = 1$ if and only if $a^r = 1$, i.e. if and only if $r = 0$, by the definition of d.

It follows that if a^k has order e, then $ke = [k, d]$. Since $[k, d] = kd/(k, d)$, this implies $e = d/(k, d)$. In particular, a^k again has order d if and only if $(k, d) = 1$.

If $0 \leq j, k < d$, put $i = j + k$ if $j + k < d$ and $i = j + k - d$ if $j + k \geq d$. Then $a^j a^k = a^i$, and so H contains the product of any two of its elements. If $0 < k < d$, then $a^k a^{d-k} = 1$, and so H contains also the inverse of any one of its elements. Finally d divides n, by Lagrange's theorem that the order of a subgroup divides the order of the whole group. $\qquad\square$

The subgroup H in Lemma 31 is the *cyclic subgroup generated by a*. For $G = \mathbb{Z}_{(m)}^\times$, the case which we will be interested in, there is no need to appeal to Lagrange's theorem, since $\mathbb{Z}_{(m)}^\times$ has order $\varphi(m)$ and d divides $\varphi(m)$, by Proposition 25 and Lemma 31(i).

A group G is *cyclic* if it coincides with the cyclic subgroup generated by one of its elements. For example, the n-th roots of unity in $\mathbb{C}$ form a cyclic group generated by $e^{2\pi i/n}$. In fact the generators of this group are just the primitive n-th roots of unity.

Our next result provides a sufficient condition for a finite group to be cyclic.

Lemma 32 *A finite group G of order n is cyclic if, for each positive divisor d of n, there are at most d elements of G whose order divides d.*

Proof If H is a cyclic subgroup of G, then its order d divides n. Since all its elements are of order dividing d, the hypothesis of the lemma implies that any element of G whose order divides d must be in H. Furthermore, H contains exactly $\varphi(d)$ elements of order d since, if a generates H, a^k has order d if and only if $(k, d) = 1$.

For each divisor d of n, let $\psi(d)$ denote the number of elements of G of order d. Then, by what we have just proved, either $\psi(d) = 0$ or $\psi(d) = \varphi(d)$. But $\sum_{d|n} \psi(d) = n$, since the order of each element is a divisor of n, and $\sum_{d|n} \varphi(d) = n$, by Proposition 24. Hence we must have $\psi(d) = \varphi(d)$ for every $d|n$. In particular, the group G has $\psi(n) = \varphi(n)$ elements of order n. $\qquad\square$

The condition of Lemma 32 is also necessary. For let G be a finite cyclic group of order n, generated by the element a, and let d be a divisor of n. An element $x \in G$ has order dividing d if and only if $x^d = 1$. Thus the elements a^k of G of order dividing d are given by $k = jn/d$, with $j = 0, 1, \ldots, d - 1$.

We now return from group theory to number theory.

Proposition 33 *For any prime p, the multiplicative group $\mathbb{F}_p^\times$ of the field $\mathbb{F}_p$ is cyclic.*

Proof Put $G = \mathbb{F}_p^\times$ and denote the order of G by n. For any divisor d of n, the polynomial $t^d - 1$ has at most d roots in $\mathbb{F}_p$. Hence there are at most d elements of G whose order divides d. The result now follows from Lemma 32. $\qquad\square$

The same argument shows that, for an arbitrary field K, any finite subgroup of the multiplicative group of K is cyclic.

In the terminology of number theory, an integer which generates $\mathbb{Z}_{(m)}^{\times}$ is said to be a *primitive root* of m. Primitive roots may be used to replace multiplications mod m by additions mod $\varphi(m)$ in the same way that logarithms were once used in analysis. If g is a primitive root of m, then the elements of $\mathbb{Z}_{(m)}^{\times}$ are precisely $1, g, g^2, \ldots, g^{n-1}$, where $n = \varphi(m)$. Thus for each $a \in \mathbb{Z}_{(m)}^{\times}$ we have $a \equiv g^\alpha \bmod m$ for a unique index α $(0 \le \alpha < n)$. We can construct a table of these indices once and for all. If $a \equiv g^\alpha$ and $b \equiv g^\beta$, then $ab \equiv g^{\alpha+\beta}$. By replacing $\alpha + \beta$ by its least non-negative residue γ mod n and going backwards in our table we can determine c such that $ab \equiv c \bmod m$.

For any prime p, an essentially complete proof for the existence of primitive roots of p was given by Euler (1774). Jacobi (1839) constructed tables of indices for all primes less than 1000.

We now use primitive roots to prove a general property of polynomials with coefficients from a finite field:

Proposition 34 *If $f(x_1, \ldots, x_n)$ is a polynomial of degree less than n in n variables with coefficients from the finite field $\mathbb{F}_p$, then the number of zeros of f in $\mathbb{F}_p^n$ is divisible by the characteristic p. In particular, $(0, \ldots, 0)$ is not the only zero of f if f has no constant term.*

Proof Put $K = \mathbb{F}_p$ and $g = 1 - f^{p-1}$. If $\alpha = (a_1, \ldots, a_n)$ is a zero of f, then $g(\alpha) = 1$. If α is not a zero of f, then $f(\alpha)^{p-1} = 1$ and $g(\alpha) = 0$. Hence the number N of zeros of f satisfies

$$N \equiv \sum_{\alpha \in K^n} g(\alpha) \bmod p.$$

We will complete the proof by showing that

$$\sum_{\alpha \in K^n} g(\alpha) = 0.$$

Since g has degree less than $n(p-1)$, it is a constant linear combination of polynomials of the form $x_1^{k_1} \cdots x_n^{k_n}$, where $k_1 + \cdots + k_n < n(p-1)$. Thus $k_j < p-1$ for at least one j. Since

$$\sum_{\alpha \in K^n} a_1^{k_1} \cdots a_n^{k_n} = \left(\sum_{a_1 \in K} a_1^{k_1} \right) \cdots \left(\sum_{a_n \in K} a_n^{k_n} \right),$$

it is enough to show that $S_k := \sum_{a \in K} a^k$ is zero for $0 \le k < p-1$. If $k = 0$, then $a^k = 1$ and $S_0 = p \cdot 1 = 0$. Suppose $1 \le k < p-1$ and let b be a generator for the multiplicative group $K^{\times}$ of K. Then $c := b^k \neq 1$ and

$$S_k = \sum_{j=1}^{p-1} c^j = c(c^{p-1} - 1)/(c-1) = 0. \qquad \square$$

The general case of Proposition 34 was first proved by Warning (1936), after the particular case had been proved by Chevalley (1936). As an illustration, the particular case implies that, for any integers a, b, c and any prime p, the congruence $ax^2 + by^2 + cz^2 \equiv 0 \bmod p$ has a solution in integers x, y, z not all divisible by p.

If m is not a prime, then $\mathbb{Z}_{(m)}$ is not a field. However, we now show that the group $\mathbb{Z}_{(m)}^{\times}$ is cyclic also if $m = p^2$ is the square of a prime.

Let g be a primitive root of p. It follows from the binomial theorem that

$$(g + p)^p \equiv g^p \bmod p^2.$$

Hence, if $g^p \equiv g \bmod p^2$, then $(g + p)^p \not\equiv g + p \bmod p^2$. Thus, by replacing g by $g + p$ if necessary, we may assume that $g^{p-1} \not\equiv 1 \bmod p^2$. If the order of g in $\mathbb{Z}_{(p^2)}^{\times}$ is d, then d divides $\varphi(p^2) = p(p-1)$. But $\varphi(p) = p-1$ divides d, since $g^d \equiv 1 \bmod p^2$ implies $g^d \equiv 1 \bmod p$ and g is a primitive root of p. Since p is prime and $d \neq p - 1$, it follows that $d = p(p - 1)$, i.e. $\mathbb{Z}_{(p^2)}^{\times}$ is cyclic with g as generator.

We briefly state some further results about primitive roots, although we will not use them. Gauss (*D.A.*, §89–92) showed that *the group $\mathbb{Z}_{(m)}^{\times}$ is cyclic if and only if* $m \in \{2, 4, p^k, 2p^k\}$, where p is an odd prime and $k \in \mathbb{N}$. Evidently 1 is a primitive root of 2 and 3 is a primitive root of 4. *If g is a primitive root of p^2, where p is an odd prime, then g is a primitive root of p^k for every $k \in \mathbb{N}$; and if $g' = g$ or $g + p^k$, according as g is odd or even, then g' is a primitive root of $2p^k$.*

By Fermat's little theorem, if p is prime, then $a^{p-1} \equiv 1 \bmod p$ for every $a \in \mathbb{Z}$ such that $(a, p) = 1$. With the aid of primitive roots we will now show that there exist also composite integers n such that $a^{n-1} \equiv 1 \bmod n$ for every $a \in \mathbb{Z}$ such that $(a, n) = 1$.

Proposition 35 *For any integer $n > 1$, the following two statements are equivalent:*

(i) $a^{n-1} \equiv 1 \bmod n$ *for every integer a such that $(a, n) = 1$;*
(ii) *n is a product of distinct primes and, for each prime $p | n$, $p - 1$ divides $n - 1$.*

Proof Suppose first that (i) holds and assume that, for some prime p, $p^2 | n$. As we have just proved, there exists a primitive root g of p^2. Evidently $p \nmid g$. It is easily seen that there exists $c \in \mathbb{N}$ such that $a = g + cp^2$ is relatively prime to n; in fact we can take c to be the product of the distinct prime factors of n, other than p, which do not divide g. Since n divides $a^{n-1} - 1$, also p^2 divides $a^{n-1} - 1$. But a, like g, is a primitive root of p^2, and so its order in $\mathbb{Z}_{(p^2)}^{\times}$ is $\varphi(p^2) = p(p - 1)$. Hence $p(p - 1)$ divides $n - 1$. But this contradicts $p | n$.

Now let p be any prime divisor of n and let g be a primitive root of p. In the same way as before, there exists $c \in \mathbb{N}$ such that $a = g + cp$ is relatively prime to n. Arguing as before, we see that $\varphi(p) = p - 1$ divides $n - 1$. This proves that (i) implies (ii).

Suppose next that (ii) holds and let a be any integer relatively prime to n. If p is a prime factor of n, then $p \nmid a$ and hence $a^{p-1} \equiv 1 \bmod p$. Since $p - 1$ divides $n - 1$, it follows that $a^{n-1} \equiv 1 \bmod p$. Thus $a^{n-1} - 1$ is divisible by each prime factor of n and hence, since n is squarefree, also by n itself.   □

Proposition 35 was proved by Carmichael (1910), and a composite integer n with the equivalent properties stated in the proposition is said to be a *Carmichael number*.

Any Carmichael number n must be odd, since it has an odd prime factor p such that $p-1$ divides $n-1$. Furthermore a Carmichael number must have more than two prime factors. For assume $n = pq$, where $1 < p < q < n$ and $q-1$ divides $n-1$. Since $q \equiv 1 \bmod (q-1)$, it follows that

$$0 \equiv pq - 1 \equiv p - 1 \bmod (q - 1),$$

which contradicts $p < q$.

The composite integer $561 = 3 \times 11 \times 17$ is a Carmichael number, since 560 is divisible by 2,10 and 16, and it is in fact the smallest Carmichael number. The taxi-cab number 1729, which Hardy reckoned to Ramanujan was uninteresting, is also a Carmichael number, since $1729 = 7 \times 13 \times 19$. Indeed it is not difficult to show that if $p, 2p - 1$ and $3p - 2$ are all primes, with $p > 3$, then their product is a Carmichael number. Recently Alford, Granville and Pomerance (1994) confirmed a long-standing conjecture by proving that there are infinitely many Carmichael numbers.

Our next topic is of greater importance. Many arithmetical problems require for their solution the determination of an integer which is congruent to several given integers according to various given moduli. We consider first a simple, but important, special case.

Proposition 36 *Let* $m = m'm''$, *where* m' *and* m'' *are relatively prime integers. Then, for any integers* a', a'', *there exists an integer* a, *which is uniquely determined* $\bmod m$, *such that*

$$a \equiv a' \bmod m', \quad a \equiv a'' \bmod m''.$$

Moreover, a *is relatively prime to* m *if and only if* a' *is relatively prime to* m' *and* a'' *is relatively prime to* m''.

Proof By Proposition 22, there exist integers c', c'' such that

$$c'm'' \equiv 1 \bmod m', \quad c''m' \equiv 1 \bmod m''.$$

Thus $e' := c'm''$ is congruent to $1 \bmod m'$ and congruent to $0 \bmod m''$. Similarly $e'' := c''m'$ is congruent to $0 \bmod m'$ and congruent to $1 \bmod m''$. It follows that $a = a'e' + a''e''$ is congruent to $a' \bmod m'$ and congruent to $a'' \bmod m''$.

It is evident that if $b \equiv a \bmod m$, then also $b \equiv a' \bmod m'$ and $b \equiv a'' \bmod m''$. Conversely, if b satisfies these two congruences, then $b - a \equiv 0 \bmod m'$ and $b - a \equiv 0 \bmod m''$. Hence $b - a \equiv 0 \bmod m$, by Proposition 3(i).

Since m' and m'' are relatively prime, it follows from Proposition 3(iv) that $(a, m) = 1$ if and only if $(a, m') = (a, m'') = 1$. Since $a \equiv a' \bmod m'$ implies $(a, m') = (a', m')$, and $a \equiv a'' \bmod m''$ implies $(a, m'') = (a'', m'')$, this proves the last statement of the proposition. □

In algebraic terms, Proposition 36 says that if $m = m'm''$, where m' and m'' are relatively prime integers, then the ring $\mathbb{Z}_{(m)}$ is (isomorphic to) the direct sum of the rings $\mathbb{Z}_{(m')}$ and $\mathbb{Z}_{(m'')}$. Furthermore, the group $\mathbb{Z}_{(m)}^{\times}$ is (isomorphic to) the direct product of the groups $\mathbb{Z}_{(m')}^{\times}$ and $\mathbb{Z}_{(m'')}^{\times}$.

Proposition 36 can be considerably generalized:

Proposition 37 *For any integers $m_1, \ldots, m_n$ and $a_1, \ldots, a_n$, the simultaneous congruences*

$$x \equiv a_1 \bmod m_1, \ldots, x \equiv a_n \bmod m_n$$

have a solution x if and only if

$$a_j \equiv a_k \bmod(m_j, m_k) \quad for\ 1 \le j < k \le n.$$

Moreover, y is also a solution if and only if

$$y \equiv x \bmod[m_1, \ldots, m_n].$$

proof The necessity of the conditions is trivial. For if x is a solution and if $d_{jk} = (m_j, m_k)$ is the greatest common divisor of m_j and m_k, then $a_j \equiv x \equiv a_k \bmod d_{jk}$. Also, if y is another solution, then $y - x$ is divisible by $m_1, \ldots, m_n$ and hence also by their least common multiple $[m_1, \ldots, m_n]$.

We prove the sufficiency of the conditions by induction on n. Suppose first that $n = 2$ and $a_1 \equiv a_2 \bmod d$, where $d = (m_1, m_2)$. By the Bézout identity,

$$d = x_1 m_1 - x_2 m_2$$

for some $x_1, x_2 \in \mathbb{Z}$. Since $a_1 - a_2 = kd$ for some $k \in \mathbb{Z}$, it follows that

$$x := a_1 - kx_1 m_1 = a_2 - kx_2 m_2$$

is a solution.

Suppose next that $n > 2$ and the result holds for all smaller values of n. Then there exists $x' \in \mathbb{Z}$ such that

$$x' \equiv a_i \bmod m_i \quad for\ 1 \le i < n,$$

and x' is uniquely determined mod m', where $m' = [m_1, \ldots, m_{n-1}]$. Since any solution of the two congruences

$$x \equiv x' \bmod m', x \equiv a_n \bmod m_n$$

is a solution of the given congruences, we need only show that $x' \equiv a_n \bmod(m', m_n)$. But, by the distributive law connecting greatest common divisors and least common multiples,

$$(m', m_n) = [(m_1, m_n), \ldots, (m_{n-1}, m_n)].$$

Since $x' \equiv a_i \equiv a_n \bmod(m_i, m_n)$ for $1 \le i < n$, it follows that $x' \equiv a_n \bmod(m', m_n)$.
$\square$

Corollary 38 *Let $m_1, \ldots, m_n$ be integers, any two of which are relatively prime, and let $m = m_1 \cdots m_n$ be their product. Then, for any given integers $a_1, \ldots, a_n$, there is a unique integer $x \bmod m$ such that*

$$x \equiv a_1 \bmod m_1, \ldots, x \equiv a_n \bmod m_n.$$

Moreover, x is relatively prime to m if and only if a_i is relatively prime to m_i for $1 \le i \le n$.

Corollary 38 can also be proved by an extension of the argument used to prove Proposition 36. Both Proposition 37 and Corollary 38 are referred to as the *Chinese remainder theorem*. Sunzi (4th century A.D.) gave a procedure for obtaining the solution $x = 23$ of the simultaneous congruences

$$x \equiv 2 \bmod 3, \quad x \equiv 3 \bmod 5, \quad x \equiv 2 \bmod 7.$$

Qin Jiushao (1247) gave a general procedure for solving simultaneous congruences, the moduli of which need not be pairwise relatively prime, although he did not state the necessary condition for the existence of a solution. The problem appears to have its origin in the construction of calendars.

6 Sums of Squares

Which positive integers n can be represented as a sum of two squares of integers? The question is answered completely by the following proposition, which was stated by Girard (1625). Fermat (1645) claimed to have a proof, but the first published proof was given by Euler (1754).

Proposition 39 *A positive integer n can be represented as a sum of two squares if and only if for each prime $p \equiv 3 \bmod 4$ that divides n, the highest power of p dividing n is even.*

Proof We observe first that, since

$$(x^2 + y^2)(u^2 + v^2) = (xu + yv)^2 + (xv - yu)^2,$$

any product of sums of two squares is again a sum of two squares.

Suppose $n = x^2 + y^2$ for some integers x, y and that n is divisible by a prime $p \equiv 3 \bmod 4$. Then $x^2 \equiv -y^2 \bmod p$. But -1 is not a square in the field $\mathbb{F}_p$, by Corollary 29. Consequently we must have $y^2 \equiv x^2 \equiv 0 \bmod p$. Thus p divides both x and y. Hence p^2 divides n and $(n/p)^2 = (x/p)^2 + (y/p)^2$. It follows by induction that the highest power of p which divides n is even.

Thus the condition in the statement of the proposition is necessary. Suppose now that this condition is satisfied. Then $n = qm^2$, where q is square-free and the only possible prime divisors of q are 2 and primes $p \equiv 1 \bmod 4$. Since $m^2 = m^2 + 0^2$ and $2 = 1^2 + 1^2$, it follows from our initial observation that n is a sum of two squares if every prime $p \equiv 1 \bmod 4$ is a sum of two squares. Following Gauss (1832), we will prove this with the aid of complex numbers.

A complex number $\gamma = a + bi$ is said to be a *Gaussian integer* if $a, b \in \mathbb{Z}$. The set of all Gaussian integers will be denoted by $\mathscr{G}$. Evidently $\gamma \in \mathscr{G}$ implies $\bar{\gamma} \in \mathscr{G}$, where $\bar{\gamma} = a - bi$ is the complex conjugate of γ. Moreover $\alpha, \beta \in \mathscr{G}$ implies $\alpha \pm \beta \in \mathscr{G}$ and $\alpha\beta \in \mathscr{G}$. Thus $\mathscr{G}$ is a commutative ring. In fact $\mathscr{G}$ is an integral domain, since it is a subset of the field $\mathbb{C}$. We are going to show that $\mathscr{G}$ can be given the structure of a Euclidean domain.

Define the *norm* of a complex number $\gamma = a + bi$ to be

$$N(\gamma) = \gamma\bar{\gamma} = a^2 + b^2.$$

Then $N(\gamma) \geq 0$, with equality if and only if $\gamma = 0$, and $N(\gamma_1 \gamma_2) = N(\gamma_1)N(\gamma_2)$. If $\gamma \in \mathscr{G}$, then $N(\gamma)$ is an ordinary integer. Furthermore, γ is a unit in $\mathscr{G}$, i.e. γ divides 1 in $\mathscr{G}$, if and only if $N(\gamma) = 1$.

We wish to show that if $\alpha, \beta \in \mathscr{G}$ and $\alpha \neq 0$, then there exist $\kappa, \rho \in \mathscr{G}$ such that

$$\beta = \kappa\alpha + \rho, \quad N(\rho) < N(\alpha).$$

We have $\beta\alpha^{-1} = r + si$, where $r, s \in \mathbb{Q}$. Choose $a, b \in \mathbb{Z}$ so that

$$|r - a| \leq 1/2, \quad |s - b| \leq 1/2.$$

If $\kappa = a + bi$, then $\kappa \in \mathscr{G}$ and

$$N(\beta\alpha^{-1} - \kappa) \leq 1/4 + 1/4 = 1/2 < 1.$$

Hence if $\rho = \beta - \kappa\alpha$, then $\rho \in \mathscr{G}$ and $N(\rho) < N(\alpha)$.

It follows that we can apply to $\mathscr{G}$ the whole theory of divisibility in a Euclidean domain. Now let p be a prime such that $p \equiv 1 \bmod 4$. We will show that p is a sum of two squares by constructing $\beta \in \mathscr{G}$ for which $N(\beta) = p$.

By Corollary 29, there exists an integer a such that $a^2 \equiv -1 \bmod p$. Put $\alpha = a + i$. Then $N(\alpha) = a\bar{\alpha} = a^2 + 1$ is divisible by p in $\mathbb{Z}$ and hence also in $\mathscr{G}$. However, neither α nor $\bar{\alpha}$ is divisible by p in $\mathscr{G}$, since αp^{-1} and $\bar{\alpha} p^{-1}$ are not in $\mathscr{G}$. Thus p is not a prime in $\mathscr{G}$ and consequently, since $\mathscr{G}$ is a Euclidean domain, it has a factorization $p = \beta\gamma$, where neither β nor γ is a unit. Hence $N(\beta) > 1, N(\gamma) > 1$. Since

$$N(\beta)N(\gamma) = N(p) = p^2,$$

it follows that $N(\beta) = N(\gamma) = p$. $\qquad\qquad\square$

Proposition 39 solves the problem of representing a positive integer as a sum of two squares. What if we allow more than two squares? When congruences were first introduced in §5, it was observed that a positive integer $a \equiv 7 \bmod 8$ could not be represented as a sum of three squares. It was first completely proved by Gauss (1801) that a positive integer can be represented as a sum of three squares if and only if it is not of the form $4^n a$, where $n \geq 0$ and $a \equiv 7 \bmod 8$. The proof of this result is more difficult, and will be given in Chapter VII.

It was conjectured by Bachet (1621) that *every* positive integer can be represented as a sum of four squares. Fermat claimed to have a proof, but the first published proof was given by Lagrange (1770), using earlier ideas of Euler (1751). The proof of the four-squares theorem we will give is similar to that just given for the two-squares theorem, with complex numbers replaced by quaternions.

Proposition 40 *Every positive integer n can be represented as a sum of four squares.*

Proof A quaternion $\gamma = a + bi + cj + dk$ will be said to be a *Hurwitz integer* if a, b, c, d are either all integers or all halves of odd integers. The set of all Hurwitz integers will be denoted by $\mathscr{H}$. Evidently $\gamma \in \mathscr{H}$ implies $\bar{\gamma} \in \mathscr{H}$, where $\bar{\gamma} = a - bi - cj - dk$. Moreover $\alpha, \beta \in \mathscr{H}$ implies $\alpha \pm \beta \in \mathscr{H}$. We will show that $\alpha, \beta \in \mathscr{H}$ also implies $\alpha\beta \in \mathscr{H}$.

Evidently $\gamma \in \mathcal{H}$ if and only if it can be written in the form $\gamma = a_0 h + a_1 i + a_2 j + a_3 k$, where $a_0, a_1, a_2, a_3 \in \mathbb{Z}$ and $h = (1 + i + j + k)/2$. It is obvious that the product of h with i, j or k is again in $\mathcal{H}$ and it is easily verified that $h^2 = h - 1$. It follows that $\mathcal{H}$ is closed under multiplication and hence is a ring.

Define the *norm* of a quaternion $\gamma = a + bi + cj + dk$ to be

$$N(\gamma) = \gamma \bar{\gamma} = a^2 + b^2 + c^2 + d^2.$$

Then $N(\gamma) \geq 0$, with equality if and only if $\gamma = 0$. Moreover, since $\overline{\gamma_1 \gamma_2} = \bar{\gamma}_2 \bar{\gamma}_1$,

$$N(\gamma_1 \gamma_2) = \gamma_1 \gamma_2 \bar{\gamma}_2 \bar{\gamma}_1 = \gamma_1 \bar{\gamma}_1 \gamma_2 \bar{\gamma}_2 = N(\gamma_1) N(\gamma_2).$$

If $\gamma \in \mathcal{H}$, then $N(\gamma) = \gamma \bar{\gamma} \in \mathcal{H}$ and hence $N(\gamma)$ is an ordinary integer. Furthermore, γ is a unit in $\mathcal{H}$, i.e. γ divides 1 in $\mathcal{H}$, if and only if $N(\gamma) = 1$.

We now show that a Euclidean algorithm may be defined on $\mathcal{H}$. Suppose $\alpha, \beta \in \mathcal{H}$ and $\alpha \neq 0$. Then

$$\beta \alpha^{-1} = r_0 + r_1 i + r_2 j + r_3 k,$$

where $r_0, r_1, r_2, r_3 \in \mathbb{Q}$. If $\kappa = a_0 h + a_1 i + a_2 j + a_3 k$, then

$$\beta \alpha^{-1} - \kappa = (r_0 - a_0/2) + (r_1 - a_0/2 - a_1)i + (r_2 - a_0/2 - a_2)j$$
$$+ (r_3 - a_0/2 - a_3)k.$$

We can choose $a_0 \in \mathbb{Z}$ so that $|2r_0 - a_0| \leq 1/2$ and then choose $a_v \in \mathbb{Z}$ so that $|r_v - a_0/2 - a_v| \leq 1/2$ ($v = 1, 2, 3$). Then $\kappa \in \mathcal{H}$ and

$$N(\beta \alpha^{-1} - \kappa) \leq 1/16 + 3/4 = 13/16 < 1.$$

Thus if we set $\rho = \beta - \kappa \alpha$, then $\rho \in \mathcal{H}$ and

$$N(\rho) = N(\beta \alpha^{-1} - \kappa)N(\alpha) < N(\alpha).$$

By repeating this division process finitely many times we see that any $\alpha, \beta \in \mathcal{H}$ have a *greatest common right divisor* $\delta = (\alpha, \beta)_r$. Furthermore, there is a *left Bézout identity*: $\delta = \xi \alpha + \eta \beta$ for some $\xi, \eta \in \mathcal{H}$.

If a positive integer n is a sum of four squares, say $n = a^2 + b^2 + c^2 + d^2$, then $n = \gamma \bar{\gamma}$, where $\gamma = a + bi + cj + dk \in \mathcal{H}$. Since the norm of a product is the product of the norms, it follows that any product of sums of four squares is again a sum of four squares. Hence to prove the proposition we need only show that any prime p is a sum of four squares.

We show first that there exist integers a, b such that $a^2 + b^2 \equiv -1 \bmod p$. This follows from the illustration given for Proposition 34, but we will give a direct proof.

If $p = 2$, we can take $a = 1, b = 0$. If $p \equiv 1 \bmod 4$ then, by Corollary 29, there exists an integer a such that $a^2 \equiv -1 \bmod p$ and we can take $b = 0$. Suppose now that $p \equiv 3 \bmod 4$. Let c be the least positive quadratic non-residue of p. Then $c \geq 2$ and $c - 1$ is a quadratic residue of p. On the other hand, -1 is a quadratic non-residue of p, by Corollary 29. Hence, by Proposition 28, $-c$ is a quadratic residue. Thus there exist integers a, b such that

$$a^2 \equiv -c, b^2 \equiv c - 1 \bmod p,$$

and then $a^2 + b^2 \equiv -1 \bmod p$.

Put $\alpha = 1 + ai + bj$. Then p divides $N(\alpha) = \alpha\bar{\alpha} = 1 + a^2 + b^2$ in $\mathbb{Z}$ and hence also in $\mathscr{H}$. However, p does not divide either α or $\bar{\alpha}$ in $\mathscr{H}$, since αp^{-1} and $\bar{\alpha} p^{-1}$ are not in $\mathscr{H}$.

Let $\gamma = (p, \alpha)_r$. Then $p = \beta\gamma$ for some $\beta \in \mathscr{H}$. If β were a unit, p would be a right divisor of γ and hence also of α, which is a contradiction. Therefore $N(\beta) > 1$. Evidently $\gamma\bar{\alpha}$ is a common right divisor of $p\bar{\alpha}$ and $\alpha\bar{\alpha}$, and the Bézout representation for γ implies that $\gamma\bar{\alpha} = (p\bar{\alpha}, \alpha\bar{\alpha})_r$. Since $p\bar{\alpha} = \bar{\alpha}p$ and p divides $\alpha\bar{\alpha}$, it follows that p is a right divisor of $\gamma\bar{\alpha}$. Since p does not divide $\bar{\alpha}$, γ is not a unit and hence $N(\gamma) > 1$. Since

$$N(\beta)N(\gamma) = N(p) = p^2,$$

we must have $N(\beta) = N(\gamma) = p$.

Thus if $\gamma = c_0 + c_1 i + c_2 j + c_3 k$, then $c_0^2 + c_1^2 + c_2^2 + c_3^2 = p$. If $c_0, \ldots, c_3$ are all integers, we are finished. Otherwise $c_0, \ldots, c_3$ are all halves of odd integers. Hence we can write $c_v = 2d_v + e_v$, where $d_v \in \mathbb{Z}$ and $e_v = \pm 1/2$. If we put

$$\delta = d_0 + d_1 i + d_2 j + d_3 k, \quad \varepsilon = e_0 + e_1 i + e_2 j + e_3 k,$$

then $\gamma = 2\delta + \varepsilon$ and $N(\varepsilon) = 1$. Hence $\theta := \gamma\bar{\varepsilon} = 2\delta\bar{\varepsilon} + 1$ has all its coordinates integers and $N(\theta) = N(\gamma) = p$. $\square$

In his *Meditationes Algebraicae*, which also contains the first statement in print of Wilson's theorem, Waring (1770) stated that every positive integer is a sum of at most 4 positive integral squares, of at most 9 positive integral cubes and of at most 19 positive integral fourth powers. The statement concerning squares was proved by Lagrange in the same year, as we have seen. The statement concerning cubes was first proved by Wieferich (1909), with a gap filled by Kempner (1912), and the statement concerning fourth powers was first proved by Balasubramanian, Deshouillers and Dress (1986).

In a later edition of his book, Waring (1782) raised the same question for higher powers. *Waring's problem* was first solved by Hilbert (1909), who showed that, for each $k \in \mathbb{N}$, there exists $\gamma_k \in \mathbb{N}$ such that every positive integer is a sum of at most γ_k k-th powers. The least possible value of γ_k is traditionally denoted by $g(k)$. For example, $g(2) = 4$, since $7 = 2^2 + 3 \cdot 1^2$ is not a sum of less than 4 squares.

A lower bound for $g(k)$ was already derived by Euler (c. 1772). Let $m = \lfloor (3/2)^k \rfloor$ denote the greatest integer $\le (3/2)^k$ and take

$$n = 2^k m - 1.$$

Since $1 \le n < 3^k$, the only k-th powers of which n can be the sum are 0^k, 1^k and 2^k. Since the number of powers 2^k must be less than m, and since $n = (m - 1)2^k + (2^k - 1)1^k$, the least number of k-th powers with sum n is $m + 2^k - 2$. Hence $g(k) \ge w(k)$, where

$$w(k) = \lfloor (3/2)^k \rfloor + 2^k - 2.$$

In particular,

$$w(2) = 4, \quad w(3) = 9, \quad w(4) = 19, \quad w(5) = 37, \quad w(6) = 73.$$

By the results stated above, $g(k) = w(k)$ for $k = 2, 3, 4$ and this has been shown to hold also for $k = 5$ by Chen (1964) and for $k = 6$ by Pillai (1940).

Hilbert's method of proof yielded rather large upper bounds for $g(k)$. A completely new approach was developed in the 1920's by Hardy and Littlewood, using their analytic 'circle' method. They showed that, for each $k \in \mathbb{N}$, there exists $\Gamma_k \in \mathbb{N}$ such that every sufficiently large positive integer is a sum of at most Γ_k k-th powers. The least possible value of Γ_k is traditionally denoted by $G(k)$. For example, $G(2) = 4$, since no positive integer $n \equiv 7 \bmod 8$ is a sum of less than four squares. Davenport (1939) showed that $G(4) = 16$, but these are the only two values of k for which today $G(k)$ is known exactly.

It is obvious that $G(k) \leq g(k)$, and in fact $G(k) < g(k)$ for all $k > 2$. In particular, Dickson (1939) showed that 23 and 239 are the only positive integers which require the maximum 9 cubes. Hardy and Littlewood obtained the upper bound $G(k) \leq (k-2)2^{k-1}+5$, but this has been repeatedly improved by Hardy and Littlewood themselves, Vinogradov and others. For example, Wooley (1992) has shown that $G(k) \leq k(\log k + \log\log k + O(1))$.

By using the upper bound for $G(k)$ of Vinogradov (1935), it was shown by Dickson, Pillai and Niven (1936–1944) that $g(k) = w(k)$ for any given $k > 6$, *provided that*

$$(3/2)^k - \lfloor(3/2)^k\rfloor \leq 1 - \lfloor(3/2)^k\rfloor/2^k.$$

It is possible that this inequality holds for every $k \in \mathbb{N}$. For a given k, it may be checked by direct calculation, and Kubina and Wunderlich (1990) have verified in this way that the inequality holds if $k \leq 471600000$. Furthermore, using a p-adic extension by Ridout (1957) of the theorem of Roth (1955) on the approximation of algebraic numbers by rationals, Mahler (1957) proved that there exists $k_0 \in \mathbb{N}$ such that the inequality holds for all $k \geq k_0$. However, the proof does not provide a means of estimating k_0.

Thus we have the bizarre situation that $G(k)$ is known for only two values of k, that $g(k)$ is known for a vast number of values of k and is given by a simple formula, probably for all k, but the information about $g(k)$ is at present derived from information about $G(k)$. Is it too much to hope that an examination of the numerical data will reveal some pattern in the fractional parts of $(3/2)^k$?

7 Further Remarks

There are many good introductory books on the theory of numbers, e.g. Davenport [4], LeVeque [28] and Scholz [41]. More extensive accounts are given in Hardy and Wright [15], Hua [18], Narkiewicz [33] and Niven *et al.* [34].

Historical information is provided by Dickson [5], Smith [42] and Weil [46], as well as the classics Euclid [11], Gauss [13] and Dirichlet [6]. Gauss's masterpiece is quoted here and in the text as '*D.A.*'

The reader is warned that, besides its use in §1, the word 'lattice' also has quite a different mathematical meaning, which will be encountered in Chapter VIII.

The basic theory of divisibility is discussed more thoroughly than in the usual texts by Stieltjes [43]. For Proposition 6, see Prüfer [35]. In the theory of groups, Schreier's refinement theorem and the Jordan–Hölder theorem may be viewed as generalizations of Propositions 6 and 7. These theorems are stated and proved in Chapter I, §3 of Lang [23]. The fundamental theorem of arithmetic (Proposition 7) is usually attributed to Gauss (*D.A.*, §16). However, it is really contained in Euclid's *Elements* (Book VII, Proposition 31 and Book IX, Proposition 14), except for the appropriate terminology. Perhaps this is why Euler and his contemporaries simply assumed it without proof.

Generalizations of the fundamental theorem of arithmetic to other algebraic structures are discussed in Chap. 2 of Jacobson [21]. For factorial domains, see Samuel [39].

Our discussion of the fundamental theorem did not deal with the practical problems of deciding if a given integer is prime or composite and, in the latter case, of obtaining its factorization into primes. Evidently if the integer a is composite, its least prime factor p satisfies $p^2 \le a$. In former days one used this observation in conjunction with tables, such as [24], [25], [26]. With new methods and supercomputers, the primality of integers with hundreds of digits can now be determined without difficulty. The progress in this area may be traced through the survey articles [48], [7] and [27]. Factorization remains a more difficult problem, and this difficulty has found an important application in *public-key cryptography*; see Rivest *et al.* [37].

For Proposition 12, cf. Hillman and Hoggatt [17]. A proof that the ring of all algebraic integers is a Bézout domain is given on p. 86 of Mann [31]. The ring of all functions which are holomorphic in a given region was shown to be a Bézout domain by Wedderburn (1915); see Narasimhan [32].

For Gauss's version of Proposition 17, see *D.A.*, §42. It is natural to ask if Corollary 18 remains valid if the polynomial ring $R[t]$ is replaced by the ring $R[[t]]$ of formal power series. The ring $K[[t_1, \ldots, t_m]]$ of all formal power series in finitely many indeterminates with coefficients from an arbitrary field K is indeed a factorial domain. However, if R is a factorial domain, the integral domain $R[[t]]$ of all formal power series in t with coefficients from R need not be factorial. For an example in which R is actually a complete local ring, see Salmon [38].

For generalizations of Eisenstein's irreducibility criterion (Proposition 19), see Gao [12]. Proposition 21 is proved in Rhai [36]. Euclidean domains are studied further in Samuel [40]. Quadratic fields $\mathbb{Q}(\sqrt{d})$ whose ring of integers $\mathcal{O}_d$ is Euclidean are discussed in Clark [3], Dubois and Steger [8] and Eggleton *et al.* [9].

Congruences are discussed in all the books on number theory cited above. In connection with Lemma 32 we mention a result of Frobenius (1895). Frobenius proved that if G is a finite group of order n and if d is a positive divisor of n, then the number of elements of G whose order divides d is a multiple of d. He conjectured that if the number is exactly d, then these elements form a (normal) subgroup of G. The conjecture can be reduced to the case where G is simple, since a counterexample of minimal order must be a noncyclic simple group. By appealing to the recent classification of all finite simple groups (see Chapter V, §7), the proof of the conjecture was completed by Iiyori and Yamaki [20].

There is a table of primitive roots on pp. 52–56 of Hua [18]. For more extensive tables, see Western and Miller [47].

It is easily seen that an even square is never a primitive root, that an odd square (including 1) is a primitive root only for the prime $p = 2$, and that -1 is a primitive root only for the primes $p = 2, 3$. Artin (1927) conjectured that if the integer a is not a square or -1, then it is a primitive root for infinitely many primes p. (A quantitative form of the conjecture is considered in Chapter IX.) If the conjecture is not true, then it is almost true, since it has been shown by Heath-Brown [16] that there are at most 3 square-free positive integers a for which it fails.

A finite subgroup of the multiplicative group of a division ring need not be cyclic. For example, if $\mathbb{H}$ is the division ring of Hamilton's quaternions, $\mathbb{H}^\times$ contains the non-cyclic subgroup $\{\pm 1, \pm i, \pm j, \pm k\}$ of order 8. All possible finite subgroups of the multiplicative group of a division ring have been determined (with the aid of *class field theory*) by Amitsur [2].

For Carmichael numbers, see Alford *et al.* [1].

Galois (1830) showed that there were other finite fields besides $\mathbb{F}_p$ and indeed, as Moore (1893) later proved, he found them all. Finite fields have the following basic properties:

(i) The number of elements in a finite field is a prime power p^n, where $n \in \mathbb{N}$ and the prime p is the characteristic of the field.

(ii) For any prime power $q = p^n$, there is a finite field $\mathbb{F}_q$ containing exactly q elements. Moreover the field $\mathbb{F}_q$ is unique, up to isomorphism, and is the splitting field of the polynomial $t^q - t$ over $\mathbb{F}_p$.

(iii) For any finite field $\mathbb{F}_q$, the multiplicative group $\mathbb{F}_q^\times$ of nonzero elements is cyclic.

(iv) If $q = p^n$, the map $\sigma: a \to a^p$ is an automorphism of $\mathbb{F}_q$ and the distinct automorphisms of $\mathbb{F}_q$ are the powers $\sigma^k (k = 0, 1, \ldots, n - 1\}$.

The theorem of Chevalley and Warning (Proposition 34) extends immediately to arbitrary finite fields. Proofs and more detailed information on finite fields may be found in Lidl and Niederreiter [30] and in Joly [22].

A celebrated theorem of Wedderburn (1905) states that any finite division ring is a field, i.e. the commutative law of multiplication is a consequence of the other field axioms if the number of elements is finite. Here is a purely algebraic proof.

Assume there exists a finite division ring which is not a field and let D be one of minimum cardinality. Let C be the centre of D and $a \in D \backslash C$. The set M of all elements of D which commute with a is a field, since it is a division ring but not the whole of D. Evidently M is a maximal subfield of D which contains a. If $[D : C] = n$ and $[M : C] = m$ then, by Proposition I.32, $[D : M] = m$ and $n = m^2$. Thus m is independent of a.

If C has cardinality q, then D has cardinality q^n, M has cardinality q^m and the number of conjugates of a in D is $(q^n - 1)/(q^m - 1)$. Since this holds for every $a \in D \backslash C$, the partition of the multiplicative group of D into conjugacy classes shows that

$$q^n - 1 = q - 1 + r(q^n - 1)/(q^m - 1)$$

for some positive integer r. Hence $q - 1$ is divisible by

$$(q^n - 1)/(q^m - 1) = 1 + q^m + \cdots + q^{m(m-1)}.$$

Since $n > m > 1$, this is a contradiction.

For the history of the Chinese remainder theorem (not only in China), see Libbrecht [29].

We have developed the arithmetic of quaternions only as far as is needed to prove the four-squares theorem. A fuller account was given in the original (1896) paper of Hurwitz [19]. For more information about sums of squares, see Grosswald [14] and also Chapter XIII. For Waring's problem, see Waring [45], Ellison [10] and Vaughan [44].

8 Selected References

[1] W.R. Alford, A. Granville and C. Pomerance, There are infinitely many Carmichael numbers, *Ann. Math.* **139** (1994), 703–722.

[2] S.A. Amitsur, Finite subgroups of division rings, *Trans. Amer. Math. Soc.* **80** (1955), 361–386.

[3] D.A. Clark, A quadratic field which is Euclidean but not norm-Euclidean, *Manuscripta Math.* **83** (1994), 327–330.

[4] H. Davenport, *The higher arithmetic*, 7th ed., Cambridge University Press, 1999.

[5] L.E. Dickson, *History of the theory of numbers*, 3 vols., Carnegie Institute, Washington, D.C., 1919–1923. [Reprinted, Chelsea, New York, 1992.]

[6] P.G.L. Dirichlet, *Lectures on number theory,* with supplements by R. Dedekind, English transl. by J. Stillwell, American Mathematical Society, Providence, R.I., 1999. [German original, 1894.]

[7] J.D. Dixon, Factorization and primality tests, *Amer. Math. Monthly* **91** (1984), 333–352.

[8] D.W. Dubois and A. Steger, A note on division algorithms in imaginary quadratic fields, *Canad. J. Math.* **10** (1958), 285–286.

[9] R.B. Eggleton, C.B. Lacampagne and J.L. Selfridge, Euclidean quadratic fields, *Amer. Math. Monthly* **99** (1992), 829–837.

[10] W.J. Ellison, Waring's problem, *Amer. Math. Monthly* **78** (1971), 10–36.

[11] Euclid, *The thirteen books of Euclid's elements*, English translation by T.L. Heath, 2nd ed., reprinted in 3 vols., Dover, New York, 1956.

[12] S. Gao, Absolute irreducibility of polynomials via Newton polytopes, *J. Algebra* **237** (2001), 501–520.

[13] C.F. Gauss, *Disquisitiones arithmeticae*, English translation by A.A. Clarke, revised by W.C. Waterhouse, Springer, New York, 1986. [Latin original, 1801.]

[14] E. Grosswald, *Representations of integers as sums of squares*, Springer-Verlag, New York, 1985.

[15] G.H. Hardy and E.M. Wright, *An introduction to the theory of numbers*, 6th ed., Oxford University Press, 2008.

[16] D.R. Heath-Brown, Artin's conjecture for primitive roots, *Quart. J. Math. Oxford Ser.* (2) **37** (1986), 27–38.

[17] A.P. Hillman and V.E. Hoggatt, Exponents of primes in generalized binomial coefficients, *J. Reine Angew. Math.* **262/3** (1973), 375–380.

[18] L.K. Hua, *Introduction to number theory*, English translation by P. Shiu, Springer-Verlag, Berlin, 1982.

[19] A. Hurwitz, Über die Zahlentheorie der Quaternionen, *Mathematische Werke, Band II*, pp. 303–330, Birkhäuser, Basel, 1933.

[20] N. Iiyori and H. Yamaki, On a conjecture of Frobenius, *Bull. Amer. Math. Soc. (N.S)* **25** (1991), 413–416.

[21] N. Jacobson, *Basic Algebra I*, 2nd ed., W.H. Freeman, New York, 1985.

[22] J.-R. Joly, Équations et variétés algébriques sur un corps fini, *Enseign. Math.* (2) **19** (1973), 1–117.

[23] S. Lang, *Algebra*, corrected reprint of 3rd ed., Addison-Wesley, Reading, Mass., 1994.

[24] D.H. Lehmer, *Guide to tables in the theory of numbers*, National Academy of Sciences, Washington, D.C., reprinted 1961.

[25] D.N. Lehmer, *List of prime numbers from 1 to 10,006,721*, reprinted, Hafner, New York, 1956.

[26] D.N. Lehmer, *Factor table for the first ten millions*, reprinted, Hafner, New York, 1956.

[27] A.K. Lenstra, Primality testing, *Proc. Symp. Appl. Math.* **42** (1990), 13–25.

[28] W.J. LeVeque, *Fundamentals of number theory*, reprinted Dover, Mineola, N.Y., 1996.

[29] U. Libbrecht, *Chinese mathematics in the thirteenth century*, MIT Press, Cambridge, Mass., 1973.

[30] R. Lidl and H. Niederreiter, *Finite fields*, 2nd ed., Cambridge University Press, 1997.

[31] H.B. Mann, *Introduction to algebraic number theory*, Ohio State University, Columbus, Ohio, 1955.

[32] R. Narasimhan, *Complex analysis in one variable*, Birkhäuser, Boston, Mass., 1985.

[33] W. Narkiewicz, *Number theory*, English translation by S. Kanemitsu, World Scientific, Singapore, 1983.

[34] I. Niven, H.S. Zuckerman and H.L. Montgomery, *An introduction to the theory of numbers*, 5th ed., Wiley, New York, 1991.

[35] H. Prüfer, Untersuchungen über Teilbarkeitseigenschaften, *J. Reine Angew. Math.* **168** (1932), 1–36.

[36] T.-S. Rhai, A characterization of polynomial domains over a field, *Amer. Math. Monthly* **69** (1962), 984–986.

[37] R.L. Rivest, A. Shamir and L. Adleman, A method for obtaining digital signatures and public-key cryptosystems, *Comm. ACM* **21** (1978), 120–126.

[38] P. Salmon, Sulla fattorialità delle algebre graduate e degli anelli locali, *Rend. Sem. Mat. Univ. Padova* **41** (1968), 119–138.

[39] P. Samuel, Unique factorization, *Amer. Math. Monthly* **75** (1968), 945–952.

[40] P. Samuel, About Euclidean rings, *J. Algebra* **19** (1971), 282–301.

[41] A. Scholz, *Einführung in die Zahlentheorie*, revised and edited by B. Schoeneberg, 5th ed., de Gruyter, Berlin, 1973.

[42] H.J.S. Smith, Report on the theory of numbers, *Collected mathematical papers, Vol.1*, pp. 38–364, reprinted, Chelsea, New York, 1965. [Original, 1859–1865.]

[43] T.J. Stieltjes, Sur la théorie des nombres, *Ann. Fac. Sci. Toulouse* **4** (1890), 1–103. [Reprinted in Tome 2, pp. 265–377 of T.J. Stieltjes, *Oeuvres complètes*, 2 vols., Noordhoff, Groningen, 1914–1918.]

[44] R.C. Vaughan, *The Hardy–Littlewood method*, 2nd ed., Cambridge Tracts in Mathematics **125**, Cambridge University Press, 1997.

[45] E. Waring, *Meditationes algebraicae*, English transl. of 1782 edition by D. Weeks, Amer. Math. Soc., Providence, R.I., 1991.

[46] A. Weil, *Number theory: an approach through history*, Birkhäuser, Boston, Mass., 1984.

[47] A.E. Western and J.C.P. Miller, *Tables of indices and primitive roots*, Royal Soc. Math. Tables, Vol. 9, Cambridge University Press, London, 1968.

[48] H.C. Williams, Primality testing on a computer, *Ars Combin.* **5** (1978), 127–185.

Additional References

M. Agarwal, N. Kayal and N. Saxena, PRIMES is in P, *Ann. of Math.* **160** (2004), 781–793. [An unconditional deterministic polynomial-time algorithm for determining if an integer > 1 is prime or composite.]

A. Granville, It is easy to determine whether a given integer is prime, *Bull. Amer. Math. Soc. (N.S.)* **42** (2005), 3–38.

III

More on Divisibility

In this chapter the theory of divisibility is developed further. The various sections of the chapter are to a large extent independent. We consider in turn the law of quadratic reciprocity, quadratic fields, multiplicative functions, and linear Diophantine equations.

1 The Law of Quadratic Reciprocity

Let p be an odd prime. An integer a which is not divisible by p is said to be a *quadratic residue*, or *quadratic nonresidue*, of p according as the congruence

$$x^2 \equiv a \bmod p$$

has, or has not, a solution x. We will speak of the *quadratic nature* of a mod p, meaning whether a is a quadratic residue or nonresidue of p.

Let q be an odd prime different from p. The *law of quadratic reciprocity* connects the quadratic nature of q mod p with the quadratic nature of p mod q. It states that if either p or q is congruent to 1 mod 4, then the quadratic nature of q mod p is the same as the quadratic nature of p mod q, but if both p and q are congruent to 3 mod 4 then the quadratic nature of q mod p is different from the quadratic nature of p mod q.

This remarkable result plays a key role in the arithmetic theory of quadratic forms. It was discovered empirically by Euler (1783). Legendre (1785) gave a partial proof and later (1798) introduced the convenient 'Legendre symbol'. The first complete proofs were given by Gauss (1801) in his *Disquisitiones Arithmeticae*. Indeed the result so fascinated Gauss that during the course of his lifetime he gave eight proofs, four of them resting on completely different principles: an induction argument, the theory of binary quadratic forms, properties of sums of roots of unity, and a combinatorial lemma. The proof we are now going to give is also of a combinatorial nature. Its idea originated with Zolotareff (1872), but our treatment is based on Rousseau (1994).

Let n be a positive integer and let X be the set $\{0, 1, \ldots, n - 1\}$. As in §7 of Chapter I, a permutation α of X is said to be *even* or *odd* according as the total number of inversions of order it induces is even or odd. If a is an integer relatively prime to n, then the map $\pi_a : X \to X$ defined by

$$\pi_a(x) = ax \bmod n$$

W.A. Coppel, *Number Theory: An Introduction to Mathematics*, Universitext, DOI: 10.1007/978-0-387-89486-7_3, © Springer Science + Business Media, LLC 2009

is a permutation of X. We define the *Jacobi symbol* (a/n) to be $\mathrm{sgn}(\pi_a)$, i.e.

$$(a/n) = 1 \text{ or } -1$$

according as the permutation π_a is even or odd. Thus $(a/1) = 1$, for every integer a. (The definition is sometimes extended by putting $(a/n) = 0$ if a and n are not relatively prime.)

Proposition 1 *For any positive integer n and any integers a,b relatively prime to n, the Jacobi symbol has the following properties:*

(i) $(1/n) = 1$,
(ii) $(a/n) = (b/n)$ *if* $a \equiv b \bmod n$,
(iii) $(ab/n) = (a/n)(b/n)$,
(iv) $(-1/n) = 1$ *if* $n \equiv 1$ *or* $2 \bmod 4$ *and* $= -1$ *if* $n \equiv 3$ *or* $0 \bmod 4$.

Proof The first two properties follow at once from the definition of the Jacobi symbol. If a and b are both relatively prime to n, then so also is their product ab. Since $\pi_{ab} = \pi_a \pi_b$, we have $\mathrm{sgn}(\pi_{ab}) = \mathrm{sgn}(\pi_a)\mathrm{sgn}(\pi_b)$, which implies (iii). We now evaluate $(-1/n)$. Since the map $\pi_{-1}: x \to -x \bmod n$ fixes 0 and reverses the order of $1, \ldots, n-1$, the total number of inversions of order is $(n-2) + (n-3) + \cdots + 1 = (n-1)(n-2)/2$. It follows that $(-1/n) = (-1)^{(n-1)/2}$ or $(-1)^{(n-2)/2}$ according as n is odd or even. This proves (iv). $\qquad\square$

Proposition 2 *For any relatively prime positive integers m, n,*

(i) *if m and n are both odd, then* $(m/n)(n/m) = (-1)^{(m-1)(n-1)/4}$;
(ii) *if m is odd and n even, then* $(m/n) = 1$ *or* $(-1)^{(m-1)/2}$ *according as* $n \equiv 2$ *or* $0 \bmod 4$.

Proof The cyclic permutation $\tau: x \to x + 1 \bmod n$ of the set $X = \{0, 1, \ldots, n-1\}$ has sign $(-1)^{n-1}$, since the number of inversions of order is $n - 1$. Hence, for any integer $b \geq 0$ and any integer a relatively prime to n, the linear permutation

$$\tau^b \pi_a: x \to ax + b \bmod n$$

of X has $\mathrm{sign}(-1)^{b(n-1)}(a/n)$.
    Put $Y = \{0, 1, \ldots, m-1\}$ and $P = X \times Y$. We consider two transformations μ and v of P, defined by

$$\mu(x, y) = (mx + y \bmod n, y), \quad v(x, y) = (x, x + ny \bmod m).$$

For each fixed y, μ defines a permutation of the set (X, y) with sign $(-1)^{y(n-1)}(m/n)$. Since $\sum_{y=0}^{m-1} y = m(m-1)/2$, it follows that the permutation μ of P has sign

$$\mathrm{sgn}(\mu) = (-1)^{m(m-1)(n-1)/2}(m/n)^m.$$

Similarly the permutation v of P has sign

$$\mathrm{sgn}(v) = (-1)^{n(m-1)(n-1)/2}(n/m)^n,$$

and hence $\alpha := \nu\mu^{-1}$ has sign

$$\text{sgn}(\alpha) = (-1)^{(m+n)(m-1)(n-1)/2}(m/n)^m(n/m)^n.$$

But α is the permutation $(mx + y \bmod n, y) \to (x, x + ny \bmod m)$ and its sign can be determined directly in the following way.

Put $Z = \{0, 1, \ldots, mn - 1\}$. By Proposition II.36, for any $(x, y) \in P$ there is a unique $z \in Z$ such that

$$z \equiv x \bmod n, \quad z \equiv y \bmod m.$$

Moreover, any $z \in Z$ is obtained in this way from a unique $(x, y) \in P$. For any $z \in Z$, we will denote by $\rho(z)$ the corresponding element of P. Then the permutation α can be written in the form $\rho(mx + y) \to \rho(x + ny)$. Since ρ is a bijective map, the sign of the permutation α of P will be the same as the sign of the permutation $\beta = \rho^{-1}\alpha\rho: mx + y \to x + ny$ of Z. An inversion of order for β occurs when both $mx + y > mx' + y'$ and $x + ny < x' + ny'$, i.e. when both $m(x - x') > y' - y$ and $x - x' < n(y' - y)$. But these inequalities imply $mn(x - x') > x - x'$ and hence $x > x', y' > y$. Conversely, if $x > x', y' > y$, then

$$m(x - x') \geq m > y' - y, \quad n(y' - y) \geq n > x - x'.$$

Since the number of $(x, y), (x', y') \in P$ with $x > x', y < y'$ is $m(m-1)/2 \cdot n(n-1)/2$, it follows that the sign of the permutation α is $(-1)^{mn(m-1)(n-1)/4}$. Comparing this expression with the expression previously found, we obtain

$$(m/n)^m(n/m)^n = (-1)^{(mn+2m+2n)(m-1)(n-1)/4}.$$

This simplifies to the first statement of the proposition if m and n are both odd, and to the second statement if m is odd and n even. $\square$

Corollary 3 *For any odd positive integer n, $(2/n) = 1$ or -1 according as $n \equiv \pm 1$ or $\pm 5 \bmod 8$.*

Proof Since the result is already known for $n = 1$, we suppose $n > 1$. Then either n or $n - 2$ is congruent to $1 \bmod 4$ and so, by Proposition 1 and Proposition 2(i),

$$(2/n) = (-1/n)((n - 2)/n) = (-1/n)(n/(n - 2)) = (-1)^{(n-1)/2}(2/(n - 2)).$$

Iterating, we obtain $(2/n) = (-1)^h$, where $h = (n - 1)/2 + (n - 3)/2 + \cdots + 1 = (n^2 - 1)/8$. The result follows. $\square$

The value of (a/n) when n is even is completely determined by Propositions 1 and 2. The evaluation of (a/n) when n is odd reduces by these propositions and Corollary 3 to the evaluation of (m/n) for odd $m > 1$. Although Proposition 2 does not provide a formula for the Jacobi symbol in this case, it does provide a method for its rapid evaluation, as we now show.

If m and n are relatively prime odd positive integers, we can write $m = 2hn + \varepsilon_1 n_1$, where $h \in \mathbb{Z}$, $\varepsilon_1 = \pm 1$ and n_1 is an odd positive integer less than n. Then n and n_1 are also relatively prime and

$$(m/n) = (\varepsilon_1/n)(n_1/n).$$

If $n_1 = 1$, we are finished. Otherwise, using Proposition 2(i), we obtain

$$(m/n) = (-1)^{(n_1-1)(n-1)/4}(\varepsilon_1/n)(n/n_1) = \pm(n/n_1),$$

where the minus sign holds if and only if n and $\varepsilon_1 n_1$ are both congruent to 3 mod 4. The process can now be repeated with m, n replaced by n, n_1. After finitely many steps the process must terminate with $n_s = 1$.

As an example,

$$
\left(\frac{2985}{1951}\right) = \left(\frac{-1}{1951}\right)\left(\frac{917}{1951}\right) = -\left(\frac{1951}{917}\right)
$$
$$
= -\left(\frac{117}{917}\right) = -\left(\frac{917}{117}\right)
$$
$$
= -\left(\frac{-1}{117}\right)\left(\frac{19}{117}\right) = -\left(\frac{117}{19}\right)
$$
$$
= -\left(\frac{3}{19}\right) = \left(\frac{19}{3}\right) = \left(\frac{1}{3}\right) = 1.
$$

Further properties of the Jacobi symbol can be derived from those already established.

Proposition 4 *If n, n' are positive integers and if a is an integer relatively prime to n such that $n' \equiv n \bmod 4a$, then $(a/n') = (a/n)$.*

Proof If $a = -1$ then, since $n' \equiv n \bmod 4$, $(a/n') = (a/n)$, by Proposition 1(iv). If $a = 2$ then, since n and n' are odd and $n' \equiv n \bmod 8$, $(a/n') = (a/n)$, by Corollary 3. Consequently, by Proposition 1(iii), it is sufficient to prove the result for odd $a > 1$.

If n is even, the result now follows from Proposition 2(ii). If n is odd, it follows from Proposition 2(i) and Proposition 1.   □

Proposition 5 *If the integer a is relatively prime to the odd positive integers n and n', then $(a/nn') = (a/n)(a/n')$.*

Proof We have $a \equiv a' \bmod nn'$ for some $a' \in \{1, 2, \ldots, nn'\}$. Since nn' is odd, we can choose $j \in \{0, 1, 2, 3\}$ so that $a'' = a' + jnn'$ satisfies $a'' \equiv 1 \bmod 4$. Then, by Propositions 1 and 2,

$$
(a/nn') = (a''/nn') = (nn'/a'') = (n/a'')(n'/a'') = (a''/n)(a''/n') = (a/n)(a/n').
$$
                                                                                    □

Proposition 5 reduces the evaluation of (a/n) for odd positive n to the evaluation of (a/p), where p is an odd prime. This is where we make the connection with quadratic residues:

Proposition 6 *If p is an odd prime and a an integer not divisible by p, then $(a/p) = 1$ or -1 according as a is a quadratic residue or nonresidue of p. Moreover, exactly half of the integers $1, \ldots, p - 1$ are quadratic residues of p.*

Proof If a is a quadratic residue of p, there exists an integer x such that $x^2 \equiv a \bmod p$ and hence

$$
(a/p) = (x^2/p) = (x/p)(x/p) = 1.
$$

Let g be a primitive root mod p. Then the integers $1, g, \ldots, g^{p-2}$ mod p are just a rearrangement of the integers $1, 2, \ldots, p - 1$. The permutation

$$\pi_g : x \to gx \ (\mathrm{mod}\ p)$$

fixes 0 and cyclically permutes the remaining elements $1, g, \ldots, g^{p-2}$. Since the number of inversions of order is $p - 2$, it follows that $(g/p) = -1$. For any integer a not divisible by p there is a unique $k \in \{0, 1, \ldots, p - 2\}$ such that $a \equiv g^k$ mod p. Hence

$$(a/p) = (g^k/p) = (g/p)^k = (-1)^k.$$

Thus $(a/p) = 1$ if and only if k is even and then $a \equiv x^2$ mod p with $x = g^{k/2}$.

This proves the first statement of the proposition. Since exactly half the integers in the set $\{0, 1, \ldots, p-2\}$ are even, it also proves again (cf. Proposition II.28) the second statement.   □

The law of quadratic reciprocity can now be established without difficulty:

Theorem 7 *Let p and q be distinct odd primes. Then the quadratic natures of p mod q and q mod p are the same if $p \equiv 1$ or $q \equiv 1$ mod 4, but different if $p \equiv q \equiv 3$ mod 4.*

Proof The result follows at once from Proposition 6 since, by Proposition 2(i), if either $p \equiv 1$ or $q \equiv 1$ mod 4 then $(p/q) = (q/p)$, but if $p \equiv q \equiv 3$ mod 4 then $(p/q) = -(q/p)$.   □

Legendre (1798) *defined* $(a/p) = 1$ or -1 according as a was a quadratic residue or nonresidue of p, and Jacobi (1837) extended this definition to (a/n) for any odd positive integer n relatively prime to a by setting

$$(a/n) = \prod_p (a/p),$$

where p runs through the prime divisors of n, each occurring as often as its multiplicity. Propositions 5 and 6 show that these definitions of Legendre and Jacobi are equivalent to the definition adopted here. The relations $(-1/p) = (-1)^{(p-1)/2}$ and $(2/p) = (-1)^{(p^2-1)/8}$ for odd primes p are often called the *first and second supplements* to the law of quadratic reciprocity.

It should be noted that, if the congruence $x^2 \equiv a$ mod n is soluble then $(a/n) = 1$, but the converse need not hold when n is not prime. For example, if $n = 21$ and $a = 5$ then the congruence $x^2 \equiv 5$ mod 21 is insoluble, since both the congruences $x^2 \equiv 5$ mod 3 and $x^2 \equiv 5$ mod 7 are insoluble, but

$$\left(\frac{5}{21}\right) = \left(\frac{5}{3}\right)\left(\frac{5}{7}\right) = (-1)^2 = 1.$$

The Jacobi symbol finds an interesting application in the proof of the following result:

Proposition 8 *If a is an integer which is not a perfect square, then there exist infinitely many primes p not dividing a for which $(a/p) = -1$.*

Proof Suppose first that $a = -1$. Since $(-1/p) = (-1)^{(p-1)/2}$, we wish to show that there are infinitely many primes $p \equiv 3 \bmod 4$. Clearly 7 is such a prime. Let $\{p_1, \ldots, p_m\}$ be any finite set of such primes greater than 3. Adapting Euclid's proof of the infinity of primes (which is reproduced at the beginning of Chapter IX), we put

$$b = 4p_1 \cdots p_m + 3.$$

Then b is odd, but not divisible by 3 or by any of the primes $p_1, \ldots, p_m$. Since $b \equiv 3 \bmod 4$, at least one prime divisor q of b must satisfy $q \equiv 3 \bmod 4$. Thus the set $\{3, p_1, \ldots, p_m\}$ does not contain all primes $p \equiv 3 \bmod 4$.

Suppose next that $a = \pm 2$. Then $(a/5) = -1$. Let $\{p_1, \ldots, p_m\}$ be any finite set of primes greater than 3 such that $(a/p_i) = -1$ $(i = 1, \ldots, m)$ and put

$$b = 8p_1 \cdots p_m \pm 3,$$

where the $\pm$ sign is chosen according as $a = \pm 2$. Then b is not divisible by 3 or by any of the primes $p_1, \ldots, p_m$. Since $b \equiv \pm 3 \bmod 8$, we have $(2/b) = -1$ and $(a/b) = -1$ in both cases. If $b = q_1 \cdots q_n$ is the representation of b as a product of primes (repetitions allowed), then

$$(a/b) = (a/q_1) \cdots (a/q_n)$$

and hence $(a/q_j) = -1$ for at least one j. Consequently the result holds also in this case.

Consider now the general case. We may assume that a is square-free, since if $a = a'b^2$, where a' is square-free, then $(a/p) = (a'/p)$ for every prime p not dividing a. Thus we can write

$$a = \varepsilon 2^e r_1 \cdots r_h,$$

where $\varepsilon = \pm 1, e = 0$ or 1, and $r_1, \ldots, r_h$ are distinct odd primes. By what we have already proved, we may assume $h \geq 1$.

Let $\{p_1, \ldots, p_m\}$ be any finite set of odd primes not containing any of the primes $r_1, \ldots, r_h$. By Proposition 6, there exists an integer c such that $(c/r_1) = -1$. Since the moduli are relatively prime in pairs, by Corollary II.38 the simultaneous congruences

$$x \equiv 1 \bmod 8, \quad x \equiv 1 \bmod p_i (i = 1, \ldots, m),$$
$$x \equiv c \bmod r_1, \quad x \equiv 1 \bmod r_j (j = 2, \ldots, h),$$

have a positive solution $x = b$. Then b is not divisible by any of the odd primes $p_1, \ldots, p_m$ or $r_1, \ldots, r_h$. Moreover $(-1/b) = (2/b) = 1$, since $b \equiv 1 \bmod 8$. Since $(r_j/b) = (b/r_j)$ for $1 \leq j \leq h$, it follows that

$$(a/b) = (\varepsilon/b)(2/b)^e (r_1/b) \cdots (r_h/b)$$
$$= (b/r_1)(b/r_2) \cdots (b/r_h) = (c/r_1)(1/r_2) \cdots (1/r_h) = -1.$$

As in the special case previously considered, this implies that $(a/q) = -1$ for some prime q dividing b, and the result follows.   □

A second proof of the law of quadratic reciprocity will now be given. Let p be an odd prime and, for any integer a not divisible by p, with Legendre *define*

$$(a/p) = 1 \quad \text{or} \quad -1$$

according as a is a quadratic residue or quadratic nonresidue of p. It follows from Euler's criterion (Proposition II.28) that

$$(ab/p) = (a/p)(b/p)$$

for any integers a, b not divisible by p. Also, by Corollary II.29,

$$(-1/p) = (-1)^{(p-1)/2}.$$

Now let q be an odd prime distinct from p and let $K = \mathbb{F}_q$ be the finite field containing q elements. Since $p \neq q$, the polynomial $t^p - 1$ has no repeated factors in K and thus has p distinct roots in some field $L \supseteq K$. If ζ is any root other than 1, then the (cyclotomic) polynomial

$$f(t) = t^{p-1} + t^{p-2} + \cdots + 1$$

has the roots $\zeta^k (k = 1, \ldots, p - 1)$.

Consider the *Gauss sum*

$$\tau = \sum_{x=1}^{p-1} (x/p)\zeta^x.$$

Instead of summing from 1 to $p - 1$, we can just as well sum over any set of representatives of $\mathbb{F}_p^\times$:

$$\tau = \sum_{x \not\equiv 0 \bmod p} (x/p)\zeta^x.$$

Since q is odd, $(x/p)^q = (x/p)$ and hence, since L has characteristic q,

$$\tau^q = \sum_{x \not\equiv 0 \bmod p} (x/p)\zeta^{xq}.$$

If we put $y = xq$ then, since

$$(x/p) = (q^2x/p) = (qy/p) = (q/p)(y/p),$$

we obtain

$$\tau^q = \sum_{y \not\equiv 0 \bmod p} (q/p)(y/p)\zeta^y = (q/p)\tau.$$

Furthermore,

$$\tau^2 = \sum_{u,v \not\equiv 0 \bmod p} (u/p)(v/p)\zeta^u\zeta^v = \sum_{u,v \not\equiv 0 \bmod p} (uv/p)\zeta^{u+v}$$

or, putting $v = uw$,

$$\tau^2 = \sum_{w \not\equiv 0 \bmod p} (w/p) \sum_{u \not\equiv 0 \bmod p} \zeta^{u(1+w)}.$$

Since the coefficients of t^{p-1} and t^{p-2} in $f(t)$ are 1, the sum of the roots is -1 and thus

$$\sum_{u \not\equiv 0 \bmod p} \zeta^{au} = -1 \quad \text{if } a \not\equiv 0 \bmod p.$$

On the other hand, if $a \equiv 0 \bmod p$, then $\zeta^{au} = 1$ and

$$\sum_{u \not\equiv 0 \bmod p} \zeta^{au} = p - 1.$$

Hence

$$\tau^2 = (-1/p)(p-1) - \sum_{w \not\equiv 0, -1 \bmod p} (w/p) = (-1/p)p - \sum_{w \not\equiv 0 \bmod p} (w/p).$$

Since there are equally many quadratic residues and quadratic nonresidues, the last sum vanishes and we obtain finally

$$\tau^2 = (-1)^{(p-1)/2} p.$$

Thus $\tau \neq 0$ and from the previous expression for τ^q we now obtain

$$\tau^{q-1} = (q/p).$$

But

$$\tau^{q-1} = (\tau^2)^{(q-1)/2} = \{(-1)^{(p-1)/2} p\}^{(q-1)/2}$$

and $p^{(q-1)/2} = (p/q)$, by Proposition II.28 again. Hence

$$(q/p) = (-1)^{(p-1)(q-1)/4}(p/q),$$

which is the law of quadratic reciprocity.

The preceding proof is a variant of the sixth proof of Gauss (1818). Already in 1801 Gauss had shown that if p is an odd prime, then

$$\sum_{k=0}^{p-1} e^{2\pi i k^2/p} = \pm\sqrt{p} \text{ or } \pm i\sqrt{p} \text{ according as } p \equiv 1 \text{ or } p \equiv 3 \bmod 4.$$

After four more years of labour he managed to show that in fact the $+$ signs must be taken. From this result he obtained his fourth proof of the law of quadratic reciprocity. The sixth proof avoided this sign determination, but Gauss's result is of interest in itself. Dirichlet (1835) derived it by a powerful analytic method, which is readily generalized. Although we will make no later use of it, we now present Dirichlet's argument.

For any positive integers m, n, we define the *Gauss sum* $G(m, n)$ by

$$G(m, n) = \sum_{v=0}^{n-1} e^{2\pi i v^2 m/n}.$$

Instead of summing from 0 to $n - 1$ we can just as well sum over any complete set of representatives of the integers mod n:

$$G(m, n) = \sum_{v \bmod n} e^{2\pi i v^2 m/n}.$$

Gauss sums have a useful multiplicative property:

Proposition 9 *If m, n, n' are positive integers, with n and n' relatively prime, then*

$$G(mn', n)G(mn, n') = G(m, nn').$$

Proof When v and v' run through complete sets of representatives of the integers mod n and mod n' respectively, $\mu = vn' + v'n$ runs through a complete set of representatives of the integers mod nn'. Moreover

$$\mu^2 m = (vn' + v'n)^2 m \equiv (v^2 n'^2 + v'^2 n^2)m \bmod nn'.$$

It follows that

$$G(mn', n)G(mn, n') = \sum_{v \bmod n} \sum_{v' \bmod n'} e^{2\pi i (mn'^2 v^2 + mn^2 v'^2)/nn'}$$

$$= \sum_{\mu \bmod nn'} e^{2\pi i \mu^2 m/nn'} = G(m, nn'). \qquad \square$$

A deeper result is the following reciprocity formula, due to Schaar (1848):

Proposition 10 *For any positive integers m, n,*

$$G(m, n) = \sqrt{\frac{n}{m}} C \sum_{\mu=0}^{2m-1} e^{-\pi i \mu^2 n/2m},$$

where $C = (1 + i)/2$.

Proof Let $f : \mathbb{R} \to \mathbb{C}$ be a function which is continuously differentiable when restricted to the interval $[0, n]$ and which vanishes outside this interval. Since the sum

$$F(t) = \sum_{k=-\infty}^{\infty} f(t + k)$$

has only finitely many nonzero terms, the function F has period 1 and is continuously differentiable, except possibly for jump discontinuities when t is an integer. Therefore, by Dirichlet's convergence criterion in the theory of Fourier series,

$$\{F(+0) + F(-0)\}/2 = \lim_{N \to \infty} \sum_{h=-N}^{N} \int_0^1 e^{-2\pi i h t} F(t)dt.$$

But

$$\int_0^1 e^{-2\pi i h t} F(t)dt = \sum_{k=-\infty}^{\infty} \int_0^1 e^{-2\pi i h t} f(t+k)dt$$

$$= \sum_{k=-\infty}^{\infty} \int_k^{k+1} e^{-2\pi i h t} f(t)dt = \int_0^n e^{-2\pi i h t} f(t)dt.$$

Thus we obtain

$$f(0)/2 + f(1) + \cdots + f(n-1) + f(n)/2 = \lim_{N \to \infty} \sum_{h=-N}^{N} \int_0^n e^{-2\pi i h t} f(t)dt. \quad (*)$$

This is a simple form of *Poisson's summation formula* (which makes an appearance also in Chapters IX and X).

In particular, if we take $f(t) = e^{2\pi i t^2 m/n}$ $(0 \le t \le n)$, where m is also a positive integer, then the left side of $(*)$ is just the Gauss sum $G(m, n)$. We will now evaluate the right side of $(*)$ for this case. Put $h = 2mq + \mu$, where q and μ are integers and $0 \le \mu < 2m$. Then

$$e^{-2\pi i h t} f(t) = e^{2\pi i m(t-nq)^2/n} e^{-2\pi i \mu t}.$$

As h runs through all the integers, q does also and μ runs independently through the integers $0, \ldots, 2m - 1$. Hence

$$\lim_{N \to \infty} \sum_{h=-N}^{N} \int_0^n e^{-2\pi i h t} f(t)dt = \sum_{\mu=0}^{2m-1} \lim_{Q \to \infty} \sum_{q=-Q}^{Q} \int_0^n e^{2\pi i m(t-nq)^2/n} e^{-2\pi i \mu t} dt$$

$$= \sum_{\mu=0}^{2m-1} \lim_{Q \to \infty} \sum_{q=-Q}^{Q} \int_{-qn}^{-(q-1)n} e^{2\pi i t^2 m/n} e^{-2\pi i \mu t} dt$$

$$= \sum_{\mu=0}^{2m-1} \int_{-\infty}^{\infty} e^{2\pi i t^2 m/n} e^{-2\pi i \mu t} dt$$

$$= \sum_{\mu=0}^{2m-1} \int_{-\infty}^{\infty} e^{2\pi i m(t-\mu n/2m)^2/n} e^{-\pi i \mu^2 n/2m} dt$$

$$= \sum_{\mu=0}^{2m-1} e^{-\pi i \mu^2 n/2m} \int_{-\infty}^{\infty} e^{2\pi i t^2 m/n} dt$$

$$= \sqrt{\frac{n}{m}} C \sum_{\mu=0}^{2m-1} e^{-\pi i \mu^2 n/2m},$$

where C is the *Fresnel integral*

$$C = \int_{-\infty}^{\infty} e^{2\pi it^2} dt.$$

(This is an important example of an infinite integral which converges, although the integrand does not tend to zero.) From $(*)$ we now obtain the formula for $G(m, n)$ in the statement of the proposition. To determine the value of the constant C, take $m = 1$, $n = 3$. We obtain $i\sqrt{3} = \sqrt{3}C(1 + i)$, which simplifies to $C = (1 + i)/2$. □

From Proposition 10 with $m = 1$ we obtain

$$G(1, n) = \sum_{v=0}^{n-1} e^{2\pi iv^2/n} = \begin{cases} (1 + i)\sqrt{n} & \text{if } n \equiv 0 \ (\text{mod } 4), \\ \sqrt{n} & \text{if } n \equiv 1 \ (\text{mod } 4), \\ 0 & \text{if } n \equiv 2 \ (\text{mod } 4), \\ i\sqrt{n} & \text{if } n \equiv 3 \ (\text{mod } 4). \end{cases}$$

If m and n are both odd, it follows that

$$G(1, mn) = G(1, m)G(1, n) \quad \text{if either } m \equiv 1 \text{ or } n \equiv 1 \bmod 4,$$
$$= -G(1, m)G(1, n) \quad \text{if } m \equiv n \equiv 3 \bmod 4;$$

i.e.

$$G(1, mn) = (-1)^{(m-1)(n-1)/4}G(1, m)G(1, n).$$

If, in addition, m and n are relatively prime, then $G(m, n) G(n, m) = G(1, mn)$, by Proposition 9. Hence, if the integers m, n are odd, positive and relatively prime, then

$$G(m, n)G(n, m) = (-1)^{(m-1)(n-1)/4}G(1, m)G(1, n).$$

For any odd, positive relatively prime integers m, n, put

$$\rho(m, n) = G(m, n)/G(1, n).$$

Then

$$\rho(1, n) = 1,$$
$$\rho(m, n) = \rho(m', n) \quad \text{if } m \equiv m' \bmod n,$$
$$\rho(m, n)\rho(n, m) = (-1)^{(m-1)(n-1)/4}.$$

We claim that $\rho(m, n)$ is just the Jacobi symbol (m/n). This is evident if $m = 1$ and, by Proposition 2(i), if $\rho(m, n) = (m/n)$, then also $\rho(n, m) = (n/m)$.

Hence if the claim is not true for all m, n, there is a pair m, n with $1 < m < n$ such that

$$\rho(m, n) \neq (m/n),$$

but $\rho(\mu, \nu) = (\mu/\nu)$ for all odd, positive relatively prime integers μ, ν with $\mu < m$. We can write $n = km + r$ for some positive integers k, r with $r < m$.

Then

$$\rho(n, m) = \rho(r, m) = (r/m) = (n/m).$$

Since $\rho(m, n) \neq (m/n)$, this yields a contradiction. Thus, *if* n *is an odd positive integer,*

$$G(m, n) = (m/n)G(1, n)$$

for any odd positive integer m *relatively prime to* n.

In fact this relation holds also if m is negative, since

$$\overline{G(1, n)} = (-1)^{(n-1)/2}G(1, n) \quad \text{and} \quad G(-m, n) = \overline{G(m, n)}.$$

(It may be shown that the relation holds also if m is even.) As we have already obtained an explicit formula for $G(1, n)$, we now have also an explicit evaluation of $G(m, n)$.

2 Quadratic Fields

Let ζ be a complex number which is not rational, but whose square is rational. Since $\zeta \notin \mathbb{Q}$, a complex number α has at most one representation of the form $\alpha = r + s\zeta$, where $r, s \in \mathbb{Q}$. Let $\mathbb{Q}(\zeta)$ denote the set of all complex numbers α which have a representation of this form. Then $\mathbb{Q}(\zeta)$ is a *field*, since it is closed under subtraction and multiplication and since, if r and s are not both zero,

$$(r + s\zeta)^{-1} = (r - s\zeta)/(r^2 - s^2\zeta^2).$$

Evidently $\mathbb{Q}(\zeta) = \mathbb{Q}(t\zeta)$ for any nonzero rational number t. Conversely, if $\mathbb{Q}(\zeta) = \mathbb{Q}(\zeta^*)$, then $\zeta^* = t\zeta$ for some nonzero rational number t. For $\zeta^* = r + s\zeta$, where $r, s \in \mathbb{Q}$ and $s \neq 0$, and hence

$$r^2 = \zeta^{*2} - 2s\zeta\zeta^* + s^2\zeta^2.$$

Thus $\zeta\zeta^*$ is rational, and so is $\zeta\zeta^*/\zeta^2 = \zeta^*/\zeta$.

It follows that without loss of generality we may assume that $\zeta^2 = d$ is a square-free integer. Then $dt^2 \in \mathbb{Z}$ for some $t \in \mathbb{Q}$ implies $t \in \mathbb{Z}$. If $\zeta^{*2} = d^*$ is also a square-free integer, then $\mathbb{Q}(\zeta) = \mathbb{Q}(\zeta^*)$ if and only if $d = d^*$ and $\zeta^* = \pm\zeta$.

The *quadratic field* $\mathbb{Q}(\sqrt{d})$ is said to be *real* if $d > 0$ and *imaginary* if $d < 0$. We define the *conjugate* of an element $\alpha = r + s\sqrt{d}$ of the quadratic field $\mathbb{Q}(\sqrt{d})$ to be the element $\alpha' = r - s\sqrt{d}$. It is easily verified that

$$(\alpha + \beta)' = \alpha' + \beta', \quad (\alpha\beta)' = \alpha'\beta'.$$

Since the map $\sigma : \alpha \to \alpha'$ is also bijective, it is an *automorphism* of the field $\mathbb{Q}(\sqrt{d})$. Since $\alpha' = \alpha$ if and only if $s = 0$, the rational field $\mathbb{Q}$ is the fixed point set of σ. Since $(\alpha')' = \alpha$, the automorphism σ is an 'involution'.

We define the *norm* of an element $\alpha = r + s\sqrt{d}$ of the quadratic field $\mathbb{Q}(\sqrt{d})$ to be the rational number

$$N(\alpha) = \alpha\alpha' = r^2 - ds^2.$$

Evidently $N(\alpha) = N(\alpha')$, and $N(\alpha) = 0$ if and only if $\alpha = 0$. From the relation $(\alpha\beta)' = \alpha'\beta'$ we obtain

$$N(\alpha\beta) = N(\alpha)N(\beta).$$

An element α of the quadratic field $\mathbb{Q}(\sqrt{d})$ is said to be an *integer* of this field if it is a root of a quadratic polynomial $t^2 + at + b$ with coefficients $a, b \in \mathbb{Z}$. (Equivalently, the integers of $\mathbb{Q}(\sqrt{d})$ are the elements which are *algebraic integers*.)

It follows from Proposition II.16 that $\alpha \in \mathbb{Q}$ is an integer of the field $\mathbb{Q}(\sqrt{d})$ if and only if $\alpha \in \mathbb{Z}$. Suppose now that $\alpha = r + s\sqrt{d}$, where $r, s \in \mathbb{Q}$ and $s \neq 0$. Then α is a root of the quadratic polynomial

$$f(x) = (x - \alpha)(x - \alpha') = x^2 - 2rx + r^2 - ds^2.$$

Moreover, this is the unique monic quadratic polynomial with rational coefficients which has α as a root.

Consequently, if α is an integer of $\mathbb{Q}(\sqrt{d})$, then so also is its conjugate α' and its norm $N(\alpha) = r^2 - ds^2$ is an ordinary integer.

Proposition 11 *Let d be a square-free integer and define ω by*

$$\omega = \sqrt{d} \qquad \text{if } d \equiv 2 \text{ or } 3 \bmod 4,$$
$$= (\sqrt{d} - 1)/2 \quad \text{if } d \equiv 1 \bmod 4.$$

Then α is an integer of the quadratic field $\mathbb{Q}(\sqrt{d})$ if and only if $\alpha = a + b\omega$ for some $a, b \in \mathbb{Z}$.

Proof Suppose $\alpha = r + s\sqrt{d}$, where $r, s \in \mathbb{Q}$. As we have seen, if $s = 0$ then α is an integer of $\mathbb{Q}(\sqrt{d})$ if and only if $r \in \mathbb{Z}$. If $s \neq 0$, then α is an integer of $\mathbb{Q}(\sqrt{d})$ if and only if $a = 2r$ and $b = r^2 - ds^2$ are ordinary integers. If a is even, i.e. if $r \in \mathbb{Z}$, then $b \in \mathbb{Z}$ if and only if $ds^2 \in \mathbb{Z}$ and hence, since d is square-free, if and only if $s \in \mathbb{Z}$. If a is odd, then $a^2 \equiv 1 \bmod 4$ and hence $b \in \mathbb{Z}$ if and only if $4ds^2 \equiv 1 \bmod 4$. Since d is square-free, this implies that $2s \in \mathbb{Z}$, $s \notin \mathbb{Z}$. Hence $2s$ is odd and $d \equiv 1 \bmod 4$. Conversely, if $2r$ and $2s$ are odd integers and $d \equiv 1 \bmod 4$, then $r^2 - ds^2 \in \mathbb{Z}$. The result follows. □

Since $\omega^2 = -\omega + (d - 1)/4$ in the case $d \equiv 1 \bmod 4$, it follows directly from Proposition 11 that the set $\mathcal{O}_d$ of all integers of the field $\mathbb{Q}(\sqrt{d})$ is closed under subtraction and multiplication and consequently is a ring. In fact $\mathcal{O}_d$ is an integral domain, since $\mathcal{O}_d \subseteq \mathbb{Q}(\sqrt{d})$.

For example, $\mathcal{O}_{-1} = \mathcal{G}$ is the ring of Gaussian integers $a + bi$, where $a, b \in \mathbb{Z}$. They form a square 'lattice' in the complex plane. Similarly $\mathcal{O}_{-3} = \mathcal{E}$ is the ring of all complex numbers $a + b\rho$, where $a, b \in \mathbb{Z}$ and $\rho = (i\sqrt{3} - 1)/2$ is a cube root of unity. These *Eisenstein integers* were studied by Eisenstein (1844). They form a hexagonal 'lattice' in the complex plane.

We have already seen in §6 of Chapter II that the ring $\mathscr{G}$ of Gaussian integers is a Euclidean domain, with $\delta(\alpha) = N(\alpha)$. We now show that the ring $\mathscr{E}$ of Eisenstein integers is also a Euclidean domain, with $\delta(\alpha) = N(\alpha)$. If $\alpha, \beta \in \mathscr{E}$ and $\alpha \neq 0$, then

$$\beta \alpha^{-1} = \beta \alpha' / \alpha \alpha' = r + s\rho,$$

where $r, s \in \mathbb{Q}$. Choose $a, b \in \mathbb{Z}$ so that

$$|r - a| \leq 1/2, \quad |s - b| \leq 1/2.$$

If $\kappa = a + b\rho$, then $\kappa \in \mathscr{E}$ and

$$N(\beta \alpha^{-1} - \kappa) = \{r - a - (s - b)/2\}^2 + 3\{(s - b)/2\}^2$$
$$\leq (3/4)^2 + 3(1/4)^2 = 3/4 < 1.$$

Thus $\beta - \kappa\alpha \in \mathscr{E}$ and $N(\beta - \kappa\alpha) < N(\alpha)$.

Since $\mathscr{G}$ and $\mathscr{E}$ are Euclidean domains, the divisibility theory of Chapter II is valid for them. As an application, we prove

Proposition 12 *The equation $x^3 + y^3 = z^3$ has no solutions in nonzero integers.*

Proof Assume on the contrary that such a solution exists and choose one for which $|xyz|$ is a minimum. Then $(x, y) = (x, z) = (y, z) = 1$. If 3 did not divide xyz, then x^3, y^3 and z^3 would be congruent to $\pm 1 \bmod 9$, which contradicts $x^3 + y^3 = z^3$. So, without loss of generality, we may assume that $3|z$. Then $x^3 + y^3 \equiv 0 \bmod 3$ and, again without loss of generality, we may assume that $x \equiv 1 \bmod 3$, $y \equiv -1 \bmod 3$. This implies that

$$x^2 - xy + y^2 \equiv 3 \bmod 9.$$

If $x + y$ and $x^2 - xy + y^2$ have a common prime divisor p, then p divides $3xy$, since $3xy = (x + y)^2 - (x^2 - xy + y^2)$, and this implies $p = 3$, since $(x, y) = 1$. Since

$$(x + y)(x^2 - xy + y^2) = x^3 + y^3 = z^3 \equiv 0 \bmod 27,$$

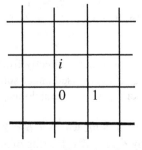

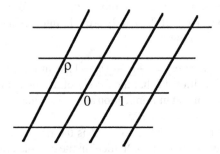

$\mathscr{G} = \mathscr{O}_{-1}$: *Gaussian integers* $\mathscr{E} = \mathscr{O}_{-3}$: *Eisenstein integers*

Fig. 1. Gaussian and Eisenstein integers.

it follows that

$$x + y = 9a^3,$$
$$x^2 - xy + y^2 = 3b^3,$$

where $a, b \in \mathbb{Z}$ and $3 \nmid b$.

We now shift operations to the Euclidean domain $\mathscr{E}$ of Eisenstein integers. We have

$$x^2 - xy + y^2 = (x + y\rho)(x + y\rho^2),$$

where $\rho = (i\sqrt{3} - 1)/2$ is a cube root of unity. Put $\lambda = 1 - \rho$, so that $(1 + \rho)\lambda^2 = 3$. Then λ is a common divisor of $x + y\rho$ and $x + y\rho^2$, since

$$x + y\rho = x + y - y\lambda,$$
$$x + y\rho^2 = x - 2y + y\lambda$$

and $x + y \equiv 0 \equiv x - 2y \bmod 3$. In fact λ is the greatest common divisor of $x + y\rho$ and $x + y\rho^2$ since, for all $m, n \in \mathbb{Z}$,

$$(m + n + n\rho)(x + y\rho^2) - (n + m\rho + n\rho)(x + y\rho) = (mx + ny)\lambda$$

and we can choose m, n so that $mx + ny = 1$. Since $\lambda^2 = -3\rho$ and since ρ is a unit, from $(x + y\rho)(x + y\rho^2) = 3b^3$ and the unique factorization of b in $\mathscr{E}$, we now obtain

$$x + y\rho = \varepsilon\lambda(c + d\rho)^3,$$

where $c, d \in \mathbb{Z}$ and ε is a unit. From

$$(x + y\rho)/\lambda = x - \lambda(x + y)/3 = x - 3a^3\lambda$$

and

$$(c + d\rho)^3 = c^3 - 3cd^2 + d^3 + 3cd(c - d)\rho,$$

by reducing mod 3 we get

$$\varepsilon(c^3 + d^3) \equiv 1 \bmod 3.$$

Since the units in $\mathscr{E}$ are $\pm 1, \pm\rho, \pm\rho^2$ (by the following Proposition 13), this implies $\varepsilon = \pm 1$. In fact we may suppose $\varepsilon = 1$, by changing the signs of c and d. Equating coefficients of ρ, we now get

$$a^3 = cd(c - d).$$

But $(c, d) = 1$, since $(x, y) = 1$, and hence also $(c, c - d) = (d, c - d) = 1$. It follows that $c = z_1^3, d = y_1^3, c - d = x_1^3$ for some $x_1, y_1, z_1 \in \mathbb{Z}$. Thus $x_1^3 + y_1^3 = z_1^3$ and

$$|x_1 y_1 z_1| = |a| = |z/3b| < |xyz|.$$

But this contradicts the definition of x, y, z. $\square$

The proof of Proposition 12 illustrates how problems involving ordinary integers may be better understood by viewing them as integers in a larger field of algebraic numbers.

We now return to the study of an arbitrary quadratic field $\mathbb{Q}(\sqrt{d})$, where d is a square-free integer. For convenience of writing we put $J = \mathcal{O}_d$. As in Chapter II, we say that $\varepsilon \in J$ is a *unit* if there exists $\eta \in J$ such that $\varepsilon\eta = 1$. For example, 1 and -1 are units. The set U of all units is evidently an abelian group under multiplication. Moreover, if $\varepsilon \in U$, then also $\varepsilon' \in U$.

If ε is a unit, then $N(\varepsilon) = \pm 1$, since $\varepsilon\eta = 1$ implies $N(\varepsilon)N(\eta) = 1$. Conversely, if $\varepsilon \in J$ and $N(\varepsilon) = \pm 1$, then ε is a unit, since $N(\varepsilon) = \varepsilon\varepsilon'$ and $\varepsilon' \in J$. (Note, however, that $N(\alpha) = \pm 1$ does not imply $\alpha \in J$. For example, in $\mathbb{Q}(\sqrt{-1})$, $\alpha = (3+4i)/5 \notin \mathcal{G}$, although $N(\alpha) = 1$.) It follows that, when $d \equiv 2$ or $3 \bmod 4$, $\alpha = a + b\sqrt{d}$ is a unit if and only if $a, b \in \mathbb{Z}$ and

$$a^2 - db^2 = \pm 1.$$

On the other hand, when $d \equiv 1 \bmod 4$, $\alpha = a + b(\sqrt{d} - 1)/2$ is a unit if and only if $a, b \in \mathbb{Z}$ and

$$(b - 2a)^2 - db^2 = \pm 4.$$

But if $b, c \in \mathbb{Z}$ and $c^2 - db^2 = \pm 4$, then $c^2 \equiv b^2 \bmod 4$ and hence $c \equiv b \bmod 2$.

Consequently, the units of J are determined by the solutions of the Diophantine equations $x^2 - dy^2 = \pm 4$ or $x^2 - dy^2 = \pm 1$, according as $d \equiv 1$ or $d \not\equiv 1 \bmod 4$. This makes it possible to determine all units, as we now show.

Proposition 13 *The units of $\mathcal{O}_{-1}$ are $\pm 1, \pm i$ and the units of $\mathcal{O}_{-3}$ are ± 1, $(\pm 1 \pm i\sqrt{3})/2$. For every other square-free integer $d < 0$, the only units of $\mathcal{O}_d$ are ± 1.*

For each square-free integer $d > 0$, there exists a unit $\varepsilon_0 > 1$ such that all units of $\mathcal{O}_d$ are given by $\pm\varepsilon_0^n (n \in \mathbb{Z})$.

Proof Suppose first that $d < 0$. Then only the Diophantine equations with the $+$ signs need to be considered. If $d < -4$, the only solutions of $x^2 - dy^2 = 4$ are $y = 0, x = \pm 2$. If $d < -4$ or if $d = -2$, the only solutions of $x^2 - dy^2 = 1$ are $y = 0, x = \pm 1$. In these cases the only units are ± 1. (The group U is a cyclic group of order 2, with -1 as generator.) If $d = -3$, the only solutions of $x^2 - dy^2 = 4$ are $y = 0, x = \pm 2$ and $y = \pm 1, x = \pm 1$. Hence the units are $\pm 1, \pm\rho, \pm\rho^2$, where $\rho = (i\sqrt{3} - 1)/2$. (The group U is a cyclic group of order 6, with $-\rho$ as generator.) If $d = -1$, the only solutions of $x^2 + y^2 = 1$ are $y = 0, x = \pm 1$ and $y = \pm 1, x = 0$. Hence the units are $\pm 1, \pm i$. (The group U is a cyclic group of order 4, with i as generator.)

Suppose next that $d > 0$. With the aid of continued fractions it will be shown in §4 of Chapter IV that the equation $x^2 - dy^2 = 1$ always has a solution in positive integers and, by doubling them, so also does the equation $x^2 - dy^2 = 4$. Hence there always exists a unit $\varepsilon > 1$. For any unit $\varepsilon > 1$ we have $\varepsilon > \pm\varepsilon'$, since $\varepsilon' = \varepsilon^{-1}$ or $-\varepsilon^{-1}$. If $\varepsilon = a + b\omega$, where ω is defined as in Proposition 11 and $a, b \in \mathbb{Z}$, then $\varepsilon' = a - b\omega$ or $a - b - b\omega$, according as $d \not\equiv 1$ or $d \equiv 1 \bmod 4$. Since ω is positive, $\varepsilon > \varepsilon'$ yields $b > 0$ and $\varepsilon > -\varepsilon'$ then yields $a > 0$. Thus every unit $\varepsilon > 1$ has the form $a + b\omega$, where

$a, b \in \mathbb{N}$. Consequently there is a least unit $\varepsilon_0 > 1$. Then, for any unit $\varepsilon > 1$, there is a positive integer n such that $\varepsilon_0^n \le \varepsilon < \varepsilon_0^{n+1}$. Since $\varepsilon \varepsilon_0^{-n}$ is a unit and $1 \le \varepsilon \varepsilon_0^{-n} < \varepsilon_0$, we must actually have $\varepsilon = \varepsilon_0^n$. (The group U is the direct product of the cyclic group of order 2 generated by -1 and the infinite cyclic group generated by ε_0.) □

As an example, take $d = 2$. Then $\varepsilon_0 = 1 + \sqrt{2}$ is a unit. Since $\varepsilon_0 > 1$ and all units greater than 1 have the form $a + b\sqrt{2}$ with $a, b \in \mathbb{N}$, it follows that all units are given by $\pm\varepsilon_0^n (n \in \mathbb{Z})$.

Having determined the units, we now consider more generally the theory of divisibility in the integral domain J. If $\alpha, \beta \in J$ and β is a proper divisor of α, then $N(\beta)$ is a proper divisor in $\mathbb{Z}$ of $N(\alpha)$ and hence $|N(\beta)| < |N(\alpha)|$. Consequently the chain condition (#) of Chapter II is satisfied. It follows that any element of J which is neither zero nor a unit is a product of finitely many irreducibles. Thus it only remains to determine the irreducibles. However, this is not such a simple matter, as the following examples indicate.

The ring $\mathscr{G}$ of Gaussian integers is a Euclidean domain. However, an ordinary prime p may or may not be irreducible in $\mathscr{G}$. For example, $2 = (1 + i)(1 - i)$ and neither factor is one of the units $\pm 1, \pm i$. On the other hand, 3 has no proper divisor $\alpha = a + bi$ which is not a unit, since $N(3) = 9$ and $N(\alpha) = a^2 + b^2 = \pm 3$ has no solutions in integers a, b.

Again, consider the ring $\mathscr{O}_{-5}$ of integers of the field $\mathbb{Q}(\sqrt{-5})$. An element $\alpha = a + b\sqrt{-5}$ of $\mathscr{O}_{-5}$ cannot have norm $N(\alpha) = a^2 + 5b^2$ equal to ± 2 or ± 3, since the square of any ordinary integer is congruent to $0, 1$ or $4 \bmod 5$. It follows that, in the factorizations

$$6 = 2 \cdot 3 = (1 + \sqrt{-5})(1 - \sqrt{-5}),$$

all four factors are irreducible and the factorizations are essentially distinct, since $N(2) = 4$, $N(3) = 9$ and $N(1 \pm \sqrt{-5}) = 6$. Thus 2 is not a 'prime' in $\mathscr{O}_{-5}$ and the 'fundamental theorem of arithmetic' does not hold.

It was shown by Kummer and Dedekind in the 19th century that uniqueness of factorization could be restored by considering ideals instead of elements. Any nonzero proper ideal of $\mathscr{O}_d$ can be represented as a product of finitely many prime ideals and the representation is unique except for the order of the factors. This result will now be established.

A nonempty subset A of a commutative ring R is an *ideal* if $a, b \in A$ and $x, y \in R$ imply $ax + by \in A$. For example, R and $\{0\}$ are ideals. If $a_1, \ldots, a_m \in R$, then the set $(a_1, \ldots, a_m)$ of all elements $a_1 x_1 + \cdots + a_m x_m$ with $x_j \in R$ $(1 \le j \le m)$ is an ideal, the ideal *generated* by $a_1, \ldots, a_m$. An ideal generated by a single element is a *principal ideal*.

If A and B are ideals in R, then the set AB of all finite sums $a_1 b_1 + \cdots + a_n b_n$ with $a_j \in A$ and $b_j \in B$ $(1 \le j \le n; n \in \mathbb{N})$ is also an ideal, the *product* of A and B. For any ideals A, B, C we have

$$AB = BA, \quad (AB)C = A(BC),$$

since multiplication in R is commutative and associative.

An ideal $A \neq \{0\}$ is said to be *divisible* by an ideal B, and B is said to be a *factor* of A, if there exists an ideal C such that $A = BC$. For example, A is divisible by itself and by R, since $A = AR$. Thus R is an identity element for multiplication of ideals.

Now take $R = \mathcal{O}_d$ to be the ring of all integers of the quadratic field $\mathbb{Q}(\sqrt{d})$. We will show that in this case much more can be said.

Proposition 14 *Let $A \neq \{0\}$ be an ideal in $\mathcal{O}_d$. Then there exist $\beta, \gamma \in A$ such that every $\alpha \in A$ can be uniquely represented in the form*

$$\alpha = m\beta + n\gamma \quad (m, n \in \mathbb{Z}).$$

Furthermore, if ω is defined as in Proposition 11, we may take $\beta = a$, $\gamma = b + c\omega$, where $a, b, c \in \mathbb{Z}$, $a > 0$, $c > 0$, c divides both a and b, and ac divides $\gamma \gamma'$, i.e.

$$b^2 - dc^2 \equiv 0 \bmod ac \quad \text{if } d \equiv 2 \text{ or } 3 \bmod 4,$$
$$b(b - c) - (d - 1)c^2/4 \equiv 0 \bmod ac \quad \text{if } d \equiv 1 \bmod 4.$$

Proof Since A is an ideal, the set J of all $z \in \mathbb{Z}$ such that $y + z\omega \in A$ for some $y \in \mathbb{Z}$ is an ideal in $\mathbb{Z}$. Moreover $J \neq \{0\}$, since $A \neq \{0\}$ and $\alpha \in A$ implies $\alpha\omega \in A$. Since $\mathbb{Z}$ is a principal ideal domain, it follows that there exists $c > 0$ such that $J = \{nc : n \in \mathbb{Z}\}$. Since $c \in J$, there exists $b \in \mathbb{Z}$ such that $\gamma := b + c\omega \in A$.

Moreover A contains some nonzero $x \in \mathbb{Z}$, since $\alpha \in A$ implies $\alpha\alpha' \in A$. Since the set I of all $x \in \mathbb{Z} \cap A$ is an ideal in $\mathbb{Z}$, there exists $a > 0$ such that $I = \{ma : m \in \mathbb{Z}\}$. For any $\alpha = y + z\omega \in A$ we have $z = nc$ for some $n \in \mathbb{Z}$ and $\alpha - n\gamma = y - nb = ma$ for some $m \in \mathbb{Z}$. Thus $\alpha = m\beta + n\gamma$ with $\beta = a$. The representation is unique, since γ is irrational.

Since $\beta\omega \in A$, we have

$$a\omega = ra + s(b + c\omega) \quad \text{for unique } r, s \in \mathbb{Z}.$$

Thus $a = sc$ and $ra + sb = 0$, which together imply $b = -rc$. Since $\gamma\omega \in A$, we have also

$$(b + c\omega)\omega = ma + n(b + c\omega) \quad \text{for unique } m, n \in \mathbb{Z}.$$

If $d \equiv 2$ or $3 \bmod 4$, then $\omega^2 = d$. In this case $n = -r$, $cd = ma - rb$ and hence $dc^2 = mac + b^2$. If $d \equiv 1 \bmod 4$, then $\omega^2 = -\omega + (d-1)/4$. Hence $n = -(r+1)$, $c(d-1)/4 = ma - rb - b$ and $(d-1)c^2/4 = mac + b(b - c)$. $\qquad\square$

If A is an ideal in $\mathcal{O}_d$, then the set $A' = \{\alpha' : \alpha \in A\}$ of all conjugates of elements of A is also an ideal in $\mathcal{O}_d$. We call A' the *conjugate* of A.

Proposition 15 *If $A \neq \{0\}$ is an ideal in $\mathcal{O}_d$, then $AA' = l\mathcal{O}_d$ for some $l \in \mathbb{N}$.*

Proof Choose β, γ so that $A = \{m\beta + n\gamma : m, n \in \mathbb{Z}\}$. Then AA' consists of all integral linear combinations of $\beta\beta', \beta\gamma', \beta'\gamma$ and $\gamma\gamma'$. Furthermore $r = \beta\beta'$, $s = \beta\gamma' + \beta'\gamma$ and $t = \gamma\gamma'$ are all in $\mathbb{Z}$. If l is the greatest common divisor of r, s and t, then $l \in AA'$, by the Bézout identity, and hence $l\mathcal{O}_d \subseteq AA'$.

On the other hand, $\beta\gamma'$ and $\beta'\gamma$ are roots of the quadratic equation

$$x^2 - sx + rt = 0$$

with integer coefficients $s = \beta\gamma' + \beta'\gamma$ and $rt = \beta\beta'\gamma\gamma'$. It follows that $\beta\gamma'/l$ and $\beta'\gamma/l$ are roots of the quadratic equation

$$y^2 - (s/l)y + rt/l^2 = 0,$$

which also has integer coefficients. Since $\beta\gamma'/l$ and $\beta'\gamma/l$ are in $\mathbb{Q}(\sqrt{d})$, this means that they are in $\mathcal{O}_d$. Thus $\beta\gamma'$ and $\beta'\gamma$ are in $l\mathcal{O}_d$. Since also $\beta\beta'$ and $\gamma\gamma'$ are in $l\mathcal{O}_d$, it follows that $AA' \subseteq l\mathcal{O}_d$.   □

If in the proof of Proposition 15 we choose $\beta = a$ and $\gamma = b + c\omega$ as in the statement of Proposition 14, then in the statement of Proposition 15 we will have $l = ac$. Since the proof of this when $d \equiv 1 \bmod 4$ is similar, we give the proof only for $d \equiv 2$ or $3 \bmod 4$. In this case $\omega = \sqrt{d}$ and hence $r = a^2$, $s = 2ab$, $t = b^2 - dc^2$. We wish to show that ac is the greatest common divisor of r, s and t. Thus if we put

$$a = cu, b = cv, t = acw,$$

then we wish to show that $u, 2v$ and w have greatest common divisor 1. Since $uw = v^2 - d$ and d is square-free, a common divisor greater than 1 can only be 2. But if 2 were a common divisor, we would have $v^2 \equiv d \bmod 4$, which is impossible, because $d \equiv 2$ or $3 \bmod 4$.

We can now show that multiplication of ideals satisfies the cancellation law:

Proposition 16 *If A, B, C are ideals in $\mathcal{O}_d$ with $A \neq \{0\}$, then $AB = AC$ implies $B = C$.*

Proof By multiplying by the conjugate A' of A we obtain $AA'B = AA'C$ and hence, by Proposition 15, $lB = lC$ for some positive integer l. But this implies $B = C$.   □

Proposition 17 *Let A and B be nonzero ideals in $\mathcal{O}_d$. Then A is divisible by B if and only if $A \subseteq B$.*

Proof If $A = BC$ for some ideal C, then $A \subseteq B$, by the definition of the product of two ideals.

Conversely, suppose $A \subseteq B$. By Proposition 15, $BB' = l\mathcal{O}_d$ for some positive integer l. Hence $AB' \subseteq l\mathcal{O}_d$. It follows that $AB' = lC$ for some ideal C. Thus $AB' = BB'C$ and so, by Proposition 16, $A = BC$.   □

Corollary 18 *Let A and B be nonzero ideals in $\mathcal{O}_d$. If D is the set of all elements $a + b$, with $a \in A$ and $b \in B$, then D is an ideal and is a factor of both A and B. Moreover, every common factor of A and B is also a factor of D.*

Proof It follows at once from its definition that D is an ideal. Moreover D contains both A and B, since 0 is an element of any ideal. Evidently also any ideal C which contains both A and B also contains D. The result now follows from Proposition 17.   □

In the terminology of Chapter II, §1, this shows that *any two nonzero ideals in $\mathcal{O}_d$ have a greatest common divisor.*

In a commutative ring R, an ideal $A \neq R, \{0\}$ is said to be *irreducible* if its only factors are A and R. It is said to be *maximal* if the only ideals containing A are A and R. It is said to be *prime* if, whenever A divides the product of two ideals, it also divides at least one of the factors.

By Proposition 17, an ideal in $\mathcal{O}_d$ is irreducible if and only if it is maximal. As we saw in §1 of Chapter II, the existence of greatest common divisors implies that an ideal in $\mathcal{O}_d$ is irreducible if and only if it is prime. (These equivalences do not hold in all commutative rings, but they do hold for the ring of all algebraic integers in any given algebraic number field, and also for the rings associated with algebraic curves.)

Proposition 19 *A nonzero ideal A in $\mathcal{O}_d$ has only finitely many factors.*

Proof Since $AA' = l\mathcal{O}_d$ for some positive integer l, any factor B of A is also a factor of $l\mathcal{O}_d$ and so contains l. Proposition 14 implies, in particular, that B is generated by two elements, say $B = (\beta_1, \beta_2)$. *A fortiori*, $B = (\beta_1, \beta_2, l)$ and hence, for any $\gamma_1, \gamma_2 \in \mathcal{O}_d$, also

$$B = (\beta_1 - l\gamma_1, \beta_2 - l\gamma_2, l).$$

We can choose $\gamma_1 \in \mathcal{O}_d$ so that in the representation

$$\beta_1 - l\gamma_1 = a_1 + b_1\omega \quad (a_1, b_1 \in \mathbb{Z})$$

we have $0 \le a_1, b_1 < l$. Similarly we can choose $\gamma_2 \in \mathcal{O}_d$ so that in the representation

$$\beta_2 - l\gamma_2 = a_2 + b_2\omega \quad (a_2, b_2 \in \mathbb{Z})$$

we have $0 \le a_2, b_2 < l$. It follows that there are at most l^4 different possibilities for the ideal B. $\square$

Corollary 20 *There exists no infinite sequence $\{A_n\}$ of nonzero ideals in $\mathcal{O}_d$ such that, for every n, A_{n+1} divides A_n and $A_{n+1} \neq A_n$.*

In the terminology of Chapter II, this shows that *the set of all nonzero ideals in $\mathcal{O}_d$ satisfies the chain condition* (#). Since also the conclusion of Proposition II.1 holds, the argument in §1 of Chapter II now shows that any nonzero proper ideal in $\mathcal{O}_d$ is a product of finitely many prime ideals and the representation is unique apart from the order of the factors.

It remains to determine the prime ideals. This is accomplished by the following three propositions.

Proposition 21 *For each prime ideal P in $\mathcal{O}_d$ there is a unique prime number p such that P divides $p\mathcal{O}_d$. Furthermore, for any prime number p there is a prime ideal P in $\mathcal{O}_d$ such that exactly one of the following alternatives holds:*

(i) $p\mathcal{O}_d = PP'$ and $P \neq P'$;
(ii) $p\mathcal{O}_d = P = P'$;
(iii) $p\mathcal{O}_d = P^2$ and $P = P'$.

Proof If P is a prime ideal in $\mathcal{O}_d$, then $PP' = l\mathcal{O}_d$ for some positive integer l. Moreover $l > 1$, since $l \in P$. If $l = mn$, where m and n are positive integers greater than 1, then P divides either $m\mathcal{O}_d$ or $n\mathcal{O}_d$. By repeating the argument it follows that P divides $p\mathcal{O}_d$ for some prime divisor p of l. The prime number p is uniquely determined by the prime ideal P since, by the Bézout identity, if P contained distinct primes it would also contain their greatest common divisor 1.

Now let p be any prime number and let the factorisation of $p\mathcal{O}_d$ into a product of positive powers of distinct prime ideals be

$$p\mathcal{O}_d = P_1^{e_1} \cdots P_s^{e_s}.$$

If we put $Q_j = P_j' (1 \leq j \leq s)$, then also

$$p\mathcal{O}_d = Q_1^{e_1} \cdots Q_s^{e_s}.$$

But $P_j Q_j = n_j \mathcal{O}_d$ for some integer $n_j > 1$ and hence

$$p^2 = n_1^{e_1} \cdots n_s^{e_s}.$$

Evidently the only possibilities are

(i)′ $s = 2, n_1 = n_2 = p, e_1 = e_2 = 1$;
(ii)′ $s = 1, n_1 = p^2, e_1 = 1$;
(iii)′ $s = 1, n_1 = p, e_1 = 2$.

Since the factorisation is unique apart from order, this yields the three possibilities in the statement of the proposition. $\square$

Proposition 21 does not tell us which of the three possibilities holds for a given prime p. For $p \neq 2$, the next result gives an answer in terms of the Legendre symbol.

Proposition 22 *Let p be an odd prime. Then, in the statement of Proposition 21, (i), (ii), or (iii) holds according as*

$$p \nmid d \text{ and } (d/p) = 1, \quad p \nmid d \text{ and } (d/p) = -1, \quad \text{or } p \mid d.$$

Proof Suppose first that $p \nmid d$ and that there exists $a \in \mathbb{Z}$ such that $a^2 \equiv d \bmod p$. Then $p \nmid a$ and $a^2 - d = pb$ for some $b \in \mathbb{Z}$. If $P = (p, a + \sqrt{d})$, then $P' = (p, a - \sqrt{d})$ and

$$PP' = p(p, a + \sqrt{d}, a - \sqrt{d}, b).$$

Since $(p, a + \sqrt{d}, a - \sqrt{d}, b)$ contains $2a$, which is relatively prime to p, it also contains 1. Hence $PP' = p\mathcal{O}_d$. Furthermore $P \neq P'$, since $P = P'$ would imply $2a \in P$ and hence $1 \in P$. We do not need to prove that P is a prime ideal, since what we have already established is incompatible with cases (ii) and (iii) of Proposition 21.

Suppose next that $p|d$. Then $d = pe$ for some $e \in \mathbb{Z}$ and $p \nmid e$, since d is square-free. If $P = (p, \sqrt{d})$, then

$$P^2 = p(p, \sqrt{d}, e) = p\mathcal{O}_d,$$

since $(p, e) = 1$. Since we cannot be in cases (i) or (ii) of Proposition 21, we must be in case (iii).

Suppose conversely that either (i) or (iii) of Proposition 21 holds. Then the corresponding prime ideal P contains p. Choose $\beta = a$ and $\gamma = b + c\omega$ as in Proposition 14, so that

$$P = \{m\beta + n\gamma : m, n \in \mathbb{Z}\}.$$

In the present case we must have $a = p$, since $p \in P$ and $1 \notin P$. We must also have $c = 1$, since $PP' = p\mathcal{O}_d$ implies $ac = p$. With these values of a and c the final condition of Proposition 14 takes the form

$$b^2 \equiv d \bmod p \qquad \text{if } d \equiv 2 \text{ or } 3 \bmod 4,$$
$$b(b - 1) \equiv (d - 1)/4 \bmod p \quad \text{if } d \equiv 1 \bmod 4.$$

Thus in the latter case $(2b - 1)^2 \equiv d \bmod p$. In either case if $p \nmid d$, then $(d/p) = 1$.

This proves that if $p \nmid d$ and $(d/p) = -1$, then we must be in case (ii) of Proposition 21. $\square$

Proposition 23 *Let $p = 2$. Then, in the statement of Proposition 21, (i),(ii), or (iii) holds according as*

$$d \equiv 1 \bmod 8, d \equiv 5 \bmod 8, \text{ or } d \equiv 2, 3 \bmod 4.$$

Proof Since the proof is similar to that of the previous proposition, we will omit some of the detail. Suppose first that $d \equiv 1 \bmod 8$. If $P = (2, (1 - \sqrt{d})/2)$, then $P' = (2, (1 + \sqrt{d})/2)$ and

$$PP' = 2(2, (1 - \sqrt{d})/2, (1 + \sqrt{d})/2, (1 - d)/8).$$

It follows that $PP' = 2\mathcal{O}_d$ and $P \neq P'$.

Suppose next that $d \equiv 2 \bmod 4$. Then $d = 2e$, where e is odd. If $P = (2, \sqrt{d})$, then

$$P^2 = 2(2, \sqrt{d}, e) = 2\mathcal{O}_d.$$

Similarly, if $d \equiv 3 \bmod 4$ and $P = (2, 1 + \sqrt{d})$, then

$$P^2 = 2(2, 1 + \sqrt{d}, (1 + d)/2 + \sqrt{d}) = 2\mathcal{O}_d.$$

Suppose conversely that either (i) or (iii) of Proposition 21 holds. Then the corresponding prime ideal P contains 2. Choose $\beta = a$ and $\gamma = b + c\omega$ as in Proposition 14, so that

$$P = \{m\beta + n\gamma : m, n \in \mathbb{Z}\}.$$

In the present case we must have $a = 2$, $c = 1$ and

$$b(b - 1) \equiv (d - 1)/4 \bmod 2 \quad \text{if } d \equiv 1 \bmod 4.$$

Since $b(b - 1)$ is even, it follows that $d \not\equiv 5 \bmod 8$.

This proves that if $d \equiv 5 \bmod 8$, then we must be in case (ii) of Proposition 21. $\square$

Proposition 22 uses only Legendre's definition of the Legendre symbol. What does the law of quadratic reciprocity tell us? By Proposition 4, if p and q are distinct odd primes and d an integer not divisible by p such that $q \equiv p \bmod 4d$, then $(d/p) = (d/q)$. Consequently, by Proposition 22, whether (i) or (ii) holds in Proposition 21 depends only on the residue class of $p \bmod 4d$. Thus, for given d, we need determine the behaviour of only finitely many primes p.

We mention without proof some further properties of the ring $\mathcal{O}_d$. We say that two nonzero ideals A, $\tilde{A}$ in $\mathcal{O}_d$ are *equivalent*, and we write $A \sim \tilde{A}$, if there exist nonzero principal ideals (a), $(\tilde{a})$ such that $(a)A = (\tilde{a})\tilde{A}$. It is easily verified that this is indeed an equivalence relation. Moreover, if $A \sim \tilde{A}$ and $B \sim \tilde{B}$, then $AB \sim \tilde{A}\tilde{B}$. Consequently, if we call an equivalence class of ideals an *ideal class*, we can without ambiguity define the product of two ideal classes. The set of ideal classes acquires in this way the structure of a commutative group, the ideal class containing the conjugate A' of A being the inverse of the ideal class containing A. It may be shown that this *ideal class group* is finite. The order of the group, i.e. the number of different ideal classes, is called the *class number* of the quadratic field $\mathbb{Q}(\sqrt{d})$ and is traditionally denoted by $h(d)$. The ring $\mathcal{O}_d$ is a principal ideal domain if and only if $h(d) = 1$. (It may be shown that $\mathcal{O}_d$ is a factorial domain only if it is a principal ideal domain.)

The theory of quadratic fields has been extensively generalized. An *algebraic number field* K is a field containing the field $\mathbb{Q}$ of rational numbers and of finite dimension as a vector space over $\mathbb{Q}$. An *algebraic integer* is a root of a monic polynomial $x^n + a_1 x^{n-1} + \cdots + a_n$ with coefficients $a_1, \ldots, a_n \in \mathbb{Z}$. The set of all algebraic integers in a given algebraic number field K is a ring $\mathcal{O}(K)$. It may be shown that, also in $\mathcal{O}(K)$, any nonzero proper ideal can be represented as a product of prime ideals and the representation is unique except for the order of the factors. One may also construct the *ideal class group* of K and show that it is finite, its order being the *class number* of K.

Some of the motivation for these generalizations came from 'Fermat's last theorem'. Fermat (c. 1640) asserted that the equation $x^n + y^n = z^n$ has no solutions in positive integers x, y, z if $n > 2$. In Proposition 12 we proved Fermat's assertion for $n = 3$. To prove the assertion in general it is sufficient to prove it when $n = 4$ and when $n = p$ is an odd prime, since if $x^{km} + y^{km} = z^{km}$, then $(x^k)^m + (y^k)^m = (z^k)^m$. Fermat himself gave a proof for $n = 4$, which is reproduced in Chapter XIII. Proofs for $n = 3, 5$ and 7 were given by Euler (1760–1770), Legendre (1825) and Lamé (1839) respectively.

Kummer (1850) made a remarkable advance beyond this by proving that the assertion holds whenever $n = p$ is a 'regular' prime. Here a prime p is said to be *regular* if it does not divide the class number of the *cyclotomic field* $\mathbb{Q}(\zeta_p)$, obtained by adjoining to $\mathbb{Q}$ a p-th root of unity ζ_p. Kummer converted his result into a practical test by further proving that a prime $p > 3$ is regular if and only if it does not divide the numerator of any of the *Bernoulli numbers* $B_2, B_4, \ldots, B_{p-3}$.

The only irregular primes less than 100 are $37, 59$ and 67. Other methods for dealing with irregular primes were devised by Kummer (1857) and Vandiver (1929). By

1993 Fermat's assertion had been established in this way for all n less than four million. However, these methods did not lead to a complete proof of 'Fermat's last theorem'. As will be seen in Chapter XIII, a complete solution was first found by Wiles (1995), using quite different methods.

3 Multiplicative Functions

We define a function $f : \mathbb{N} \to \mathbb{C}$ to be an *arithmetical function*. The set of all arithmetical functions can be given the structure of a commutative ring in the following way.

For any two functions $f, g : \mathbb{N} \to \mathbb{C}$, we define their *convolution* or *Dirichlet product* $f * g : \mathbb{N} \to \mathbb{C}$ by

$$f * g(n) = \sum_{d|n} f(d)g(n/d).$$

Dirichlet multiplication satisfies the usual commutative and associative laws:

Lemma 24 *For any three functions* $f, g, h : \mathbb{N} \to \mathbb{C}$,

$$f * g = g * f, \quad f * (g * h) = (f * g) * h.$$

Proof Since n/d runs through the positive divisors of n at the same time as d,

$$f * g(n) = \sum_{d|n} f(d)g(n/d)$$
$$= \sum_{d|n} f(n/d)g(d) = g * f(n).$$

To prove the associative law, put $G = g * h$. Then

$$f * G(n) = \sum_{de=n} f(d)G(e) = \sum_{de=n} f(d) \sum_{d'd''=e} g(d')h(d'')$$
$$= \sum_{dd'd''=n} f(d)g(d')h(d'').$$

Similarly, if we put $F = f * g$, we obtain

$$F * h(n) = \sum_{de=n} F(e)h(d) = \sum_{de=n} \sum_{d'd''=e} f(d')g(d'')h(d)$$
$$= \sum_{dd'd''=n} f(d')g(d'')h(d).$$

Hence $F * h(n) = f * G(n)$. $\square$

For any two functions $f, g : \mathbb{N} \to \mathbb{C}$, we define their *sum* $f + g : \mathbb{N} \to \mathbb{C}$ in the natural way:

$$(f + g)(n) = f(n) + g(n).$$

It is obvious that addition is commutative and associative, and that the distributive law holds:

$$f * (g + h) = f * g + f * h.$$

The function $\delta : \mathbb{N} \to \mathbb{C}$, defined by

$$\delta(n) = 1 \text{ or } 0 \quad \text{according as } n = 1 \text{ or } n > 1,$$

acts as an identity element for Dirichlet multiplication:

$$\delta * f = f \quad \text{for every } f : \mathbb{N} \to \mathbb{C},$$

since

$$\delta * f(n) = \sum_{d|n} \delta(d) f(n/d) = f(n).$$

Thus the set $\mathscr{A}$ of all arithmetical functions is indeed a commutative ring.

For any function $f : \mathbb{N} \to \mathbb{C}$ which is not identically zero, put $|f| = v(f)^{-1}$, where $v(f)$ is the least positive integer n such that $f(n) \neq 0$, and put $|O| = 0$. Then

$$|f * g| = |f||g|, |f + g| \leq \max(|f|, |g|) \quad \text{for all } f, g \in \mathscr{A}.$$

Hence the ring $\mathscr{A}$ of all arithmetical functions is actually an integral domain. It is readily shown that the set of all $f \in \mathscr{A}$ such that $|f| < 1$ is an ideal, but not a principal ideal. (Although $\mathscr{A}$ is not a principal ideal domain, it may be shown that it is a *factorial* domain.)

The next result shows that the functions $f \in \mathscr{A}$ such that $|f| = 1$ are the *units* in the ring $\mathscr{A}$:

Lemma 25 *For any function $f : \mathbb{N} \to \mathbb{C}$, there is a function $f^{-1} : \mathbb{N} \to \mathbb{C}$ such that $f^{-1} * f = \delta$ if and only if $f(1) \neq 0$. The inverse f^{-1} is uniquely determined and $f^{-1}(1) f(1) = 1$.*

Proof Suppose $g : \mathbb{N} \to \mathbb{C}$ has the property that $g * f = \delta$. Then $g(1) f(1) = 1$. Thus $g(1)$ is non-zero and uniquely determined. If $n > 1$, then

$$\sum_{d|n} g(d) f(n/d) = 0.$$

Hence

$$g(n) f(1) = - \sum_{d|n, d < n} g(d) f(n/d).$$

It follows by induction that $g(n)$ is uniquely determined for every $n \in \mathbb{N}$. Conversely, if g is defined inductively in this way, then $g * f = \delta$. □

It follows from Lemma 25 that the set of all arithmetical functions $f : \mathbb{N} \to \mathbb{C}$ such that $f(1) \neq 0$ is an abelian group under Dirichlet multiplication.

A function $f : \mathbb{N} \to \mathbb{C}$ is said to be *multiplicative* if it is not identically zero and if

$$f(mn) = f(m)f(n) \quad \text{for all } m, n \text{ with } (m, n) = 1.$$

It follows that $f(1) = 1$, since $f(n) \neq 0$ for some n and $f(n) = f(n)f(1)$. Any multiplicative function $f : \mathbb{N} \to \mathbb{C}$ is uniquely determined by its values at the prime powers, since if

$$m = p_1^{\alpha_1} \cdots p_s^{\alpha_s},$$

where $p_1, \ldots, p_s$ are distinct primes and $\alpha_1, \ldots, \alpha_s \in \mathbb{N}$, then

$$f(m) = f(p_1^{\alpha_1}) \cdots f(p_s^{\alpha_s}).$$

If

$$m = \prod_p p^{\alpha_p}, \quad n = \prod_p p^{\beta_p},$$

where $\alpha_p, \beta_p \geq 0$, then

$$(m, n) = \prod_p p^{\gamma_p}, \quad [m, n] = \prod_p p^{\delta_p},$$

where $\gamma_p = \min\{\alpha_p, \beta_p\}$ and $\delta_p = \max\{\alpha_p, \beta_p\}$. Since either $\gamma_p = \alpha_p$ and $\delta_p = \beta_p$, or $\gamma_p = \beta_p$ and $\delta_p = \alpha_p$, it follows that, for any multiplicative function f and all $m, n \in \mathbb{N}$,

$$f((m, n))f([m, n]) = \prod_p f(p^{\gamma_p})f(p^{\delta_p}) = \prod_p f(p^{\alpha_p})f(p^{\beta_p}) = f(m)f(n).$$

As we saw in §5 of Chapter II, it follows from Proposition II.4 that Euler's φ-function is multiplicative. Also, the functions $i : \mathbb{N} \to \mathbb{C}$ and $j : \mathbb{N} \to \mathbb{C}$, defined by

$$i(n) = 1, \; j(n) = n \quad \text{for every } n \in \mathbb{N},$$

are obviously multiplicative. Further examples of multiplicative functions can be constructed with the aid of the next two propositions.

Proposition 26 *If $f, g : \mathbb{N} \to \mathbb{C}$ are multiplicative functions, then their Dirichlet product $h = f * g$ is also multiplicative.*

Proof We have

$$h(n) = \sum_{d \mid n} f(d)g(n/d).$$

Suppose $n = n'n''$, where n' and n'' are relatively prime. Then, by Proposition II.4,

$$h(n) = \sum_{d'|n', d''|n''} f(d'd'')g(n'n''/d'd'')$$

$$= \sum_{d'|n', d''|n''} f(d')f(d'')g(n'/d')g(n''/d'')$$

$$= \sum_{d'|n'} f(d')g(n'/d') \sum_{d''|n''} f(d'')g(n''/d'') = h(n')h(n''). \qquad \square$$

Proposition 27 *If* $f : \mathbb{N} \to \mathbb{C}$ *is a multiplicative function, then its Dirichlet inverse* $f^{-1} : \mathbb{N} \to \mathbb{C}$ *is also multiplicative.*

Proof Assume on the contrary that $g := f^{-1}$ is not multiplicative and let n', n'' be relatively prime positive integers such that $g(n'n'') \neq g(n')g(n'')$. We suppose n', n'' chosen so that the product $n = n'n''$ is minimal. Since f is multiplicative, $f(1) = 1$ and hence $g(1) = 1$. Consequently $n' > 1, n'' > 1$ and

$$0 = \sum_{d'|n'} g(d')f(n'/d') = \sum_{d''|n''} g(d'')f(n''/d'') = \sum_{d|n} g(d)f(n/d).$$

But

$$\sum_{d|n} g(d)f(n/d) = g(n)f(1) + \sum_{d'|n', d''|n'', d'd''<n} g(d'd'')f(n'n''/d'd'')$$

$$= g(n) + \sum_{d'|n', d''|n'', d'd''<n} g(d')g(d'')f(n'/d')f(n''/d'')$$

$$= g(n) - g(n')g(n'') + \sum_{d'|n'} g(d')f(n'/d') \cdot \sum_{d''|n''} g(d'')f(n''/d'')$$

$$= g(n) - g(n')g(n'').$$

Thus we have a contradiction. $\qquad \square$

It follows from Propositions 26 and 27 that under Dirichlet multiplication the multiplicative functions form a subgroup of the group of all functions $f : \mathbb{N} \to \mathbb{C}$ with $f(1) \neq 0$. The further subgroup generated by i and j contains some interesting functions. Let $\tau(n)$ denote the number of positive divisors of n, and let $\sigma(n)$ denote the sum of the positive divisors of n:

$$\tau(n) = \sum_{d|n} 1, \quad \sigma(n) = \sum_{d|n} d.$$

In other words,

$$\tau = i * i, \quad \sigma = i * j,$$

and hence, by Proposition 26, τ and σ are multiplicative functions. Thus they are uniquely determined by their values at the prime powers. But if p is prime and $\alpha \in \mathbb{N}$, the divisors of p^α are $1, p, \ldots, p^\alpha$ and hence

$$\tau(p^\alpha) = \alpha + 1, \quad \sigma(p^\alpha) = (p^{\alpha+1} - 1)/(p - 1).$$

By Proposition II.24, Euler's φ-function satisfies $i * \varphi = j$. Thus $\varphi = i^{-1} * j$, and Propositions 26 and 27 provide a new proof that Euler's φ-function is multiplicative. Since

$$\tau * \varphi = i * i * \varphi = i * j = \sigma,$$

we also obtain the new relation

$$\sigma(n) = \sum_{d\mid n} \tau(n/d)\varphi(d).$$

The *Möbius function* $\mu : \mathbb{N} \to \mathbb{C}$ is defined to be the Dirichlet inverse i^{-1}. Thus $\mu * i = \delta$ or, in other words,

$$\sum_{d\mid n} \mu(d) = 1 \text{ or } 0 \quad \text{according as } n = 1 \text{ or } n > 1.$$

Instead of this inductive definition, we may explicitly characterize the Möbius function in the following way:

Proposition 28 *For any $n \in \mathbb{N}$,*

$$\mu(n) = \begin{cases} 1 & \textit{if } n = 1, \\ (-1)^s & \textit{if } n \textit{ is a product of } s \textit{ distinct primes,} \\ 0 & \textit{if } n \textit{ is divisible by the square of a prime.} \end{cases}$$

Proof It is trivial that $\mu(1) = 1$. Suppose p is prime and $\alpha \in \mathbb{N}$. Since the divisors of p^α are $1, p, \ldots, p^\alpha$, we have

$$\mu(1) + \mu(p) + \cdots + \mu(p^\alpha) = 0.$$

Since this holds for all $\alpha \in \mathbb{N}$, it follows that $\mu(p) = -\mu(1) = -1$, whereas $\mu(p^\alpha) = 0$ if $\alpha > 1$. Since the Möbius function is multiplicative, by Proposition 27, the general formula follows.   □

The function defined by the statement of Proposition 28 had already appeared in work of Euler (1748), but Möbius (1832) discovered the basic property which we have adopted as a definition. From this property we can easily derive the *Möbius inversion formula*:

Proposition 29 *For any function $f : \mathbb{N} \to \mathbb{C}$, if $\hat{f} : \mathbb{N} \to \mathbb{C}$ is defined by*

$$\hat{f}(n) = \sum_{d\mid n} f(d),$$

then

$$f(n) = \sum_{d\mid n} \hat{f}(d)\mu(n/d) = \sum_{d\mid n} \hat{f}(n/d)\mu(d).$$

Furthermore, for any function $\hat{f} : \mathbb{N} \to \mathbb{C}$, there is a unique function $f : \mathbb{N} \to \mathbb{C}$ such that $\hat{f}(n) = \sum_{d\mid n} f(d)$ for every $n \in \mathbb{N}$.

Proof Let $f : \mathbb{N} \to \mathbb{C}$ be given and put $\hat{f} = f * i$. Then

$$\hat{f} * \mu = f * i * \mu = f * \delta = f.$$

Conversely, let $\hat{f} : \mathbb{N} \to \mathbb{C}$ be given and put $f = \hat{f} * \mu$. Then $f * i = \hat{f} * \delta = \hat{f}$. Moreover, by the first part of the proof, this is the only possible choice for f. □

If we apply Proposition 29 to Euler's φ-function then, by Proposition II.24, we obtain the formula

$$\varphi(n) = n \sum_{d|n} \mu(d)/d.$$

In particular, if $n = p^\alpha$, where p is prime and $\alpha \in \mathbb{N}$, then

$$\varphi(p^\alpha) = \mu(1)p^\alpha + \mu(p)p^{\alpha-1} = p^\alpha(1 - 1/p).$$

Since φ is multiplicative, we recover in this way the formula

$$\varphi(n) = n \prod_{p|n}(1 - 1/p) \quad \text{for every } n \in \mathbb{N}.$$

The σ-function arises in the study of perfect numbers, to which the Pythagoreans attached much significance. A positive integer n is said to be *perfect* if it is the sum of its (positive) divisors other than itself, i.e. if $\sigma(n) = 2n$.

For example, 6 and 28 are perfect, since

$$6 = 1 + 2 + 3, \quad 28 = 1 + 2 + 4 + 7 + 14.$$

It is an age-old conjecture that there are no odd perfect numbers. However, the even perfect numbers are characterized by the following result:

Proposition 30 *An even positive integer is perfect if and only if it has the form* $2^t(2^{t+1} - 1)$, *where $t \in \mathbb{N}$ and $2^{t+1} - 1$ is prime.*

Proof Let n be any even positive integer and write $n = 2^t m$, where $t \geq 1$ and m is odd. Then, since σ is multiplicative, $\sigma(n) = d\sigma(m)$, where

$$d := \sigma(2^t) = 2^{t+1} - 1.$$

If $m = d$ and d is prime, then $\sigma(m) = 1 + d = 2^{t+1}$ and consequently $\sigma(n) = 2^{t+1}m = 2n$.

On the other hand, if $\sigma(n) = 2n$, then $d\sigma(m) = 2^{t+1}m$. Since d is odd, it follows that $m = dq$ for some $q \in \mathbb{N}$. Hence

$$\sigma(m) = 2^{t+1}q = (1 + d)q = q + m.$$

Thus q is the only proper divisor of m. Hence $q = 1$ and $m = d$ is prime. □

The sufficiency of the condition in Proposition 30 was proved in Euclid's *Elements* (Book IX, Proposition 36). The necessity of the condition was proved over two thousand years later by Euler. The condition is quite restrictive. In the first place, if $2^m - 1$ is prime for some $m \in \mathbb{N}$, then m must itself be prime. For, if $m = rs$, where $1 < r < m$, then with $a = 2^s$ we have

$$2^m - 1 = a^r - 1 = (a - 1)(a^{r-1} + a^{r-2} + \cdots + 1).$$

A prime of the form $M_p := 2^p - 1$ is said to be a *Mersenne prime* in honour of Mersenne (1644), who gave a list of all primes $p \leq 257$ for which, he claimed, M_p was prime. However, he included two values of p for which M_p is composite and omitted three values of p for which M_p is prime. The correct list is now known to be

$$p = 2, 3, 5, 7, 13, 17, 19, 31, 61, 89, 107, 127.$$

The first four even perfect numbers, namely 6, 28, 496 and 8128, which correspond to the values $p = 2, 3, 5$ and 7, were known to the ancient Greeks.

That M_{11} is not prime follows from $2^{11} - 1 = 2047 = 23 \times 89$. The factor 23 is not found simply by guesswork. It was already known to Fermat (1640) that if p is an odd prime, then any divisor of M_p is congruent to $1 \bmod 2p$. It is sufficient to establish this for prime divisors. But if q is a prime divisor of M_p, then $2^p \equiv 1 \bmod q$. Hence the order of 2 in $\mathbb{F}_q^\times$ divides p and, since it is not 1, it must be exactly p. Hence, by Lemma II.31 with $G = \mathbb{F}_q^\times$, p divides $q - 1$. Thus $q \equiv 1 \bmod p$ and actually $q \equiv 1 \bmod 2p$, since q is necessarily odd.

The least 39 Mersenne primes are now known. The hunt for more uses thousands of linked personal computers and the following test, which was stated by Lucas (1878), but first completely proved by D.H. Lehmer (1930):

Proposition 31 *Define the sequence* (S_n) *recurrently by*

$$S_1 = 4, \quad S_{n+1} = S_n^2 - 2 \ (n \geq 1).$$

Then, for any odd prime p, $M_p := 2^p - 1$ *is prime if and only if it divides* S_{p-1}.

Proof Put

$$\omega = 2 + \sqrt{3}, \quad \omega' = 2 - \sqrt{3}.$$

Since $\omega\omega' = 1$, it is easily shown by induction that

$$S_n = \omega^{2^{n-1}} + \omega'^{2^{n-1}} \quad (n \geq 1).$$

Let q be a prime and let J denote the set of all real numbers of the form $a + b\sqrt{3}$, where $a, b \in \mathbb{Z}$. Evidently J is a commutative ring. By identifying two elements $a + b\sqrt{3}$ and $\tilde{a} + \tilde{b}\sqrt{3}$ of J when $a \equiv \tilde{a}$ and $b \equiv \tilde{b} \bmod q$, we obtain a finite commutative ring J_q containing q^2 elements. The set $J_q^\times$ of all invertible elements of J_q is a commutative group containing at most $q^2 - 1$ elements, since $0 \notin J_q^\times$.

Suppose first that M_p divides S_{p-1} and assume that M_p is composite. If q is the least prime divisor of M_p, then $q^2 \le M_p$ and $q \ne 2$. By hypothesis,

$$\omega^{2^{p-2}} + \omega'^{2^{p-2}} \equiv 0 \bmod q.$$

Now consider ω and ω' as elements of J_q. By multiplying by $\omega^{2^{p-2}}$, we obtain $\omega^{2^{p-1}} = -1$ and hence $\omega^{2^p} = 1$. Thus $\omega \in J_q^\times$ and the order of ω in $J_q^\times$ is exactly 2^p. Hence

$$2^p \le q^2 - 1 \le M_p - 1 = 2^p - 2,$$

which is a contradiction.

Suppose next that $M_p = q$ is prime. Then $q \equiv -1 \bmod 8$, since $p \ge 3$. Since $(2/q) = (-1)^{(q^2-1)/8}$, it follows that 2 is a quadratic residue of q. Thus there exists an integer a such that

$$a^2 \equiv 2 \bmod q.$$

Furthermore $q \equiv 1 \bmod 3$, since $2^2 \equiv 1$ and hence $2^{p-1} \equiv 1 \bmod 3$. Thus q is a quadratic residue of 3. Since $q \equiv -1 \bmod 4$, it follows from the law of quadratic reciprocity that 3 is a quadratic nonresidue of q. Hence, by Euler's criterion (Proposition II.28),

$$3^{(q-1)/2} \equiv -1 \bmod q.$$

Consider the element $\tau = a^{q-2}(1 + \sqrt{3})$ of J_q. We have

$$\tau^2 = 2^{q-2} \cdot 2\omega = \omega,$$

since $2^{q-1} \equiv 1 \bmod q$. On the other hand,

$$(1 + \sqrt{3})^q = 1 + 3^{(q-1)/2}\sqrt{3} = 1 - \sqrt{3}$$

and hence

$$\tau^q = a^{q-2}(1 - \sqrt{3}).$$

Consequently,

$$\omega^{(q+1)/2} = \tau^{q+1} = a^{q-2}(1 - \sqrt{3}) \cdot a^{q-2}(1 + \sqrt{3}) = 2^{q-2}(-2) = -1.$$

Multiplying by $\omega'^{(q+1)/4}$, we obtain $\omega^{(q+1)/4} = -\omega'^{(q+1)/4}$. In other words, since $(q + 1)/4 = 2^{p-2}$,

$$S_{p-1} = \omega^{2^{p-2}} + \omega'^{2^{p-2}} \equiv 0 \bmod q. \qquad \square$$

It is conjectured that there are infinitely many Mersenne primes, and hence infinitely many even perfect numbers. A heuristic argument of Gillies (1964), as modified by Wagstaff (1983), suggests that the number of primes $p \le x$ for which M_p is prime is asymptotic to $(e^\gamma / \log 2) \log x$, where γ is Euler's constant (Chapter IX, §4) and thus $e^\gamma / \log 2 = 2.570 \ldots$.

We turn now from the primality of $2^m - 1$ to the primality of $2^m + 1$. It is easily seen that if $2^m + 1$ is prime for some $m \in \mathbb{N}$, then m must be a power of 2. For, if $m = rs$, where $r > 1$ is odd, then with $a = 2^s$ we have

$$2^m + 1 = a^r + 1 = (a + 1)(a^{r-1} - a^{r-2} + \cdots + 1).$$

Put $F_n := 2^{2^n} + 1$. Thus, in particular,

$$F_0 = 3, \quad F_1 = 5, \quad F_2 = 17, \quad F_3 = 257, \quad F_4 = 65537.$$

Evidently $F_{n+1} - 2 = (F_n - 2)F_n$, from which it follows by induction that

$$F_n - 2 = F_0 F_1 \cdots F_{n-1} \quad (n \geq 1).$$

Since F_n is odd, this implies that $(F_m, F_n) = 1$ if $m \neq n$. As a byproduct, we have a proof that there are infinitely many primes.

It is easily verified that F_n itself is prime for $n \leq 4$. It was conjectured by Fermat that the 'Fermat numbers' F_n are all prime. However, this was disproved by Euler, who showed that 641 divides F_5. In fact

$$641 = 5 \cdot 2^7 + 1 = 5^4 + 2^4.$$

Thus $5 \cdot 2^7 \equiv -1 \bmod 641$ and hence $2^{32} \equiv -5^4 \cdot 2^{28} \equiv -(-1)^4 \equiv -1 \bmod 641$.

Fermat may have been as wrong as possible, since no F_n with $n > 4$ is known to be prime, although many have been proved to be composite. The Fermat numbers which *are* prime found an unexpected application to the construction of regular polygons by ruler and compass, the only instruments which Euclid allowed himself. It was shown by Gauss, at the age of 19, that a regular polygon of m sides can be constructed by ruler and compass if the order $\varphi(m)$ of $\mathbb{Z}_{(m)}^{\times}$ is a power of 2. It follows from the formula $\varphi(p^\alpha) = p^{\alpha-1}(p - 1)$, and the multiplicative nature of Euler's function, that $\varphi(m)$ is a power of 2 if and only if m has the form $2^k \cdot p_1 \cdots p_s$, where $k \geq 0$ and $p_1, \ldots, p_s$ are distinct Fermat primes. (Wantzel (1837) showed that a regular polygon of m sides cannot be constructed by ruler and compass unless m has this form.) Gauss's result, in which he took particular pride, was a forerunner of Galois theory and is today usually established as an application of that theory.

The factor 641 of F_5 is not found simply by guesswork. Indeed, we now show that any divisor of F_n must be congruent to $1 \bmod 2^{n+1}$. It is sufficient to establish this for prime divisors. But if p is a prime divisor of F_n, then $2^{2^n} \equiv -1 \bmod p$ and hence $2^{2^{n+1}} \equiv 1 \bmod p$. Thus the order of 2 in $\mathbb{F}_p^{\times}$ is exactly 2^{n+1}. Hence 2^{n+1} divides $p - 1$ and $p \equiv 1 \bmod 2^{n+1}$.

With a little more effort we can show that any divisor of F_n must be congruent to $1 \bmod 2^{n+2}$ if $n > 1$. For, if p is a prime divisor of F_n and $n > 1$, then $p \equiv 1 \bmod 8$ by what we have already proved. Hence, by Proposition II.30, 2 is a quadratic residue of p. Thus there exists an integer a such that $a^2 \equiv 2 \bmod p$. Since $a^{2^{n+1}} \equiv -1 \bmod p$ and $a^{2^{n+2}} \equiv 1 \bmod p$, the order of a in $\mathbb{F}_p^{\times}$ is exactly 2^{n+2} and hence 2^{n+2} divides $p-1$.

It follows from the preceding result that 641 is the first possible candidate for a prime divisor of F_5, since $128k + 1$ is not prime for $k = 1, 3, 4$ and $257 = F_3$ is relatively prime to F_5.

The hunt for Fermat primes today uses supercomputers and the following test due to Pépin (1877):

Proposition 32 *If $m > 1$, then $N := 2^m + 1$ is prime if and only if $3^{(N-1)/2} + 1$ is divisible by N.*

Proof Suppose first that N divides $a^{(N-1)/2} + 1$ for some integer a. If p is any prime divisor of N, then $a^{(N-1)/2} \equiv -1 \bmod p$ and hence $a^{N-1} \equiv 1 \bmod p$. Thus, since p is necessarily odd, the order of a in $\mathbb{F}_p^\times$ divides $N - 1 = 2^m$, but does not divide $(N - 1)/2 = 2^{m-1}$. Hence the order of a must be exactly 2^m. Consequently, by Lemma II.31 with $G = \mathbb{F}_p^\times$, 2^m divides $p - 1$. Thus

$$2^m \leq p - 1 \leq N - 1 = 2^m,$$

which implies that $N = p$ is prime.

To prove the converse we use the law of quadratic reciprocity. Suppose $N = p$ is prime. Then $p > 3$, since $m > 1$. From $2 \equiv -1 \bmod 3$ we obtain $p \equiv (-1)^m + 1 \bmod 3$. Since $3 \nmid p$, it follows that $p \equiv -1 \bmod 3$. Thus p is a quadratic non-residue of 3. But $p \equiv 1 \bmod 4$, since $m > 1$. Consequently, by the law of quadratic reciprocity, 3 is a quadratic non-residue of p. Hence, by Euler's criterion, $3^{(p-1)/2} \equiv -1 \bmod p$. $\square$

By means of Proposition 32 it has been shown that F_{14} is composite, even though no nontrivial factors are known!

4 Linear Diophantine Equations

A *Diophantine equation* is an algebraic equation with integer coefficients of which the integer solutions are required. The name honours Diophantus of Alexandria (3rd century A.D.), who solved many problems of this type, although the surviving books of his *Arithmetica* do not treat the linear problems with which we will be concerned.

We wish to determine integers $x_1, \ldots, x_n$ such that

$$a_{11}x_1 + \cdots + a_{1n}x_n = c_1$$
$$a_{21}x_1 + \cdots + a_{2n}x_n = c_2$$
$$\cdots$$
$$a_{m1}x_1 + \cdots + a_{mn}x_n = c_m,$$

where the coefficients a_{jk} and the right sides c_j are given integers ($1 \leq j \leq m$, $1 \leq k \leq n$). We may write the system, in matrix notation, as

$$Ax = c.$$

The problem may also be put geometrically. A nonempty set $M \subseteq \mathbb{Z}^m$ is said to be a $\mathbb{Z}$–*module*, or simply a *module*, if $a, b \in M$ and $x, y \in \mathbb{Z}$ imply $xa + yb \in M$.

For example, if $a_1, \ldots, a_n$ is a finite subset of $\mathbb{Z}^m$, then the set M of all linear combinations $x_1a_1 + \cdots + x_na_n$ with $x_1, \ldots, x_n \in \mathbb{Z}$ is a module, the module *generated* by $a_1, \ldots, a_n$. If we take $a_1, \ldots, a_n$ to be the columns of the matrix A, then M is the set of all vectors Ax with $x \in \mathbb{Z}^n$ and the system $Ax = c$ is soluble if and only if $c \in M$.

If a module M is generated by the elements $a_1, \ldots, a_n$, then it is also generated by the elements $b_1, \ldots, b_n$, where

$$b_k = u_{1k}a_1 + \cdots + u_{nk}a_n \quad (u_{jk} \in \mathbb{Z} : 1 \le j, \ k \le n),$$

if the matrix $U = (u_{jk})$ is invertible. Here an $n \times n$ matrix U of integers is said to be *invertible* if there exists an $n \times n$ matrix U^{-1} of integers such that $U^{-1}U = I_n$ or, equivalently, $UU^{-1} = I_n$.

For example, if $ax + by = 1$, then the matrix

$$U = \begin{pmatrix} a & b \\ -y & x \end{pmatrix}$$

is invertible, with inverse

$$U^{-1} = \begin{pmatrix} x & -b \\ y & a \end{pmatrix}.$$

It may be shown, although we will not use it, that an $n \times n$ matrix U is invertible if and only if its determinant $\det U$ is a *unit*, i.e. $\det U = \pm 1$. Under matrix multiplication, the set of all invertible $n \times n$ matrices of integers is a group, usually denoted by $GL_n(\mathbb{Z})$.

To solve the linear Diophantine system $Ax = c$ we replace it by a system $By = c$, where $B = AU$ for some invertible matrix U. The idea is to choose U so that B has such a simple form that y can be determined immediately, and then $x = Uy$.

We will use the elementary fact that interchanging two columns of a matrix A, or adding an integral multiple of one column to another column, is equivalent to postmultiplying A by a suitable invertible matrix U. In fact U is obtained by performing the same column operation on the identity matrix I_n. In the following discussion 'matrix' will mean 'matrix with entries from $\mathbb{Z}$'.

Proposition 33 *If $A = (a_1 \cdots a_n)$ is a $1 \times n$ matrix, then there exists an invertible $n \times n$ matrix U such that*

$$AU = (d \ 0 \cdots 0)$$

if and only if d is a greatest common divisor of $a_1, \ldots, a_n$.

Proof Suppose first that there exists such a matrix U. Since

$$A = (d \ 0 \cdots 0)U^{-1},$$

d is a common divisor of $a_1, \ldots, a_n$. On the other hand,

$$d = a_1 b_1 + \cdots + a_n b_n,$$

where $b_1, \ldots, b_n$ is the first column of U. Hence any common divisor of $a_1, \ldots, a_n$ divides d. Thus d is a greatest common divisor of $a_1, \ldots, a_n$.

Suppose next that $a_1, \ldots, a_n$ have greatest common divisor d. Since there is nothing to do if $n = 1$, we assume $n > 1$ and use induction on n. Then if d' is the greatest common divisor of $a_2, \ldots, a_n$, there exists an invertible $(n-1) \times (n-1)$ matrix V' such that

$$(a_2 \cdots a_n)V' = (d'\, 0 \cdots 0).$$

Since d is the greatest common divisor of a_1 and d', there exist integers u, v such that

$$a_1 u + d' v = d.$$

Put $V = I_1 \oplus V'$ and $W = W' \oplus I_{n-2}$, where

$$W' = \begin{pmatrix} u & -d'/d \\ v & a_1/d \end{pmatrix}.$$

Then V and W are invertible, and

$$(a_1\, a_2 \cdots a_n)VW = (a_1\, d'\, 0 \cdots 0)W = (d\, 0 \cdots 0).$$

Thus we can take $U = VW$. $\qquad\square$

Corollary 34 *For any given integers $a_1, \ldots, a_n$, there exists an invertible $n \times n$ matrix U with $a_1, \ldots, a_n$ as its first row if and only if the greatest common divisor of $a_1, \ldots, a_n$ is 1.*

Proof An invertible matrix U has $a_1, \ldots, a_n$ as its first row if and only if

$$(a_1\, a_2 \cdots a_n) = (1\, 0 \cdots 0)U. \qquad\square$$

If U is invertible, then its transpose is also invertible. It follows that there exists an invertible $n \times n$ matrix with $a_1, \ldots, a_n$ as its first column also if and only if the greatest common divisor of $a_1, \ldots, a_n$ is 1.

Proposition 35 *For any $m \times n$ matrix A, there exists an invertible $n \times n$ matrix U such that $B = AU$ has the form*

$$B = (B_1 O),$$

where B_1 is an $m \times r$ submatrix of rank r.

Proof Let A have rank r. If $r = n$, there is nothing to do. If $r < n$ and we denote the columns of A by $a_1, \ldots, a_n$, then there exist $x_1, \ldots, x_n \in \mathbb{Z}$, not all zero, such that

$$x_1 a_1 + \cdots + x_n a_n = O.$$

Moreover, we may assume that $x_1, \ldots, x_n$ have greatest common divisor 1. Then, by Corollary 34, there exists an invertible $n \times n$ matrix U' with $x_1, \ldots, x_n$ as its last column. Hence $A' := AU'$ has its last column zero. If $r < n - 1$, we can apply the same argument to the submatrix formed by the first $n - 1$ columns of A', and so on until we arrive at a matrix B of the required form. $\qquad\square$

The elements $b_1, \ldots, b_r$ of a module M are said to be a *basis* for M if they generate M and are linearly independent, i.e. $x_1 b_1 + \cdots + x_r b_r = O$ for some $x_1, \ldots, x_r \in \mathbb{Z}$ implies that $x_1 = \cdots = x_r = 0$. If O is the only element of M, we say also that O is a basis for M.

In geometric terms, Proposition 35 says that any finitely generated module $M \subseteq \mathbb{Z}^m$ has a finite basis, and that a finite set of generators is a basis if and only if its elements are linearly independent over $\mathbb{Q}$. Hence any two bases have the same cardinality.

Proposition 36 *For any $m \times n$ matrix A, the set N of all $x \in \mathbb{Z}^n$ such that $Ax = O$ is a module with a finite basis.*

Proof It is evident that N is a module. By Proposition 35, there exists an invertible $n \times n$ matrix U such that $AU = B = (B_1 O)$, where B_1 is an $m \times r$ submatrix of rank r. Hence $By = O$ if and only if the first r coordinates of y vanish. By taking y to be the vector with k-th coordinate 1 and all other coordinates 0, for each k such that $r < k \leq n$, we see that the equation $By = O$ has $n - r$ linearly independent solutions $y^{(1)}, \ldots, y^{(n-r)}$ such that all solutions are given by

$$y = z_1 y^{(1)} + \cdots + z_{n-r} y^{(n-r)},$$

where $z_1, \ldots, z_{n-r}$ are arbitrary integers. If we put $x^{(j)} = U y^{(j)}$, it follows that $x^{(1)}, \ldots, x^{(n-r)}$ are a basis for the module N. $\square$

An $m \times n$ matrix $B = (b_{jk})$ will be said to be in *echelon form* if the following two conditions are satisfied:

(i) $b_{jk} = 0$ for all j if $k > r$;
(ii) $b_{jk} \neq 0$ for some j if $k \leq r$ and, if m_k is the least such j, then $1 \leq m_1 < m_2 < \cdots < m_r \leq m$.

Evidently $r = \text{rank} B$.

Proposition 37 *For any $m \times n$ matrix A, there exists an invertible $n \times n$ matrix U such that $B = AU$ is in echelon form.*

Proof By Proposition 35, we may suppose that A has the form $(A_1 O)$, where A_1 is an $m \times r$ submatrix of rank r, and by replacing A_1 by A we may suppose that A itself has rank n. We are going to show that there exists an invertible $n \times n$ matrix U such that, if $AU = B = (b_{jk})$, then $b_{jk} = 0$ for all $j < k$.

If $m = 1$, this follows from Proposition 33. We assume $m > 1$ and use induction on m. Then the first $m - 1$ rows of A may be assumed to have already the required triangular form. If $n \leq m$, there is nothing more to do. If $n > m$, we can take $U = I_{m-1} \oplus U'$, where U' is an invertible $(n - m + 1) \times (n - m + 1)$ matrix such that

$$(a_{m,m}\, a_{m,m+1} \cdots a_{m,n}) U' = (a'\, 0 \cdots 0).$$

Replacing B by A, we now suppose that for A itself we have $a_{jk} = 0$ for all $j < k$. Since A still has rank n, each column of A contains a nonzero entry. If the first nonzero entry in the k-th column appears in the m_k-th row, then $m_k \geq k$. By permuting the columns, if necessary, we may suppose in addition that $m_1 \leq m_2 \leq \cdots \leq m_n$.

Suppose $m_1 = m_2$. Let a and b be the entries in the m_1-th row of the first and second columns, and let d be their greatest common divisor. Then $d \neq 0$ and there exist integers u, v such that $au + bv = d$. If we put $U = V \oplus I_{n-2}$, where

$$V = \begin{pmatrix} u & -b/d \\ v & a/d \end{pmatrix},$$

then U is invertible. Moreover, the last $n - 2$ columns of $B = AU$ are the same as in A and the first $m_1 - 1$ entries of the first two columns are still zero. However, $b_{m_1 1} = d$ and $b_{m_1 2} = 0$. By permuting the last $n - 1$ columns, if necessary, we obtain a matrix A', of the same form as A, with $m'_1 \leq m'_2 \leq \cdots \leq m'_n$, where $m'_1 = m_1$ and $m'_2 + \cdots + m'_n > m_2 + \cdots + m_n$.

By repeating this process finitely many times, we will obtain a matrix in echelon form. $\qquad\square$

Corollary 38 *If A is an $m \times n$ matrix of rank m, then there exists an invertible $n \times n$ matrix U such that $AU = B = (b_{jk})$, where*

$$b_{jj} \neq 0, \ b_{jk} = 0 \ \text{if} \ j < k \quad (1 \leq j \leq m, \ 1 \leq k \leq n).$$

Before proceeding further we consider the uniqueness of the echelon form. Let $T = (t_{jk})$ be any $r \times r$ matrix which is lower triangular and invertible, i.e. $t_{jk} = 0$ if $j < k$ and the diagonal elements t_{jj} are units. It is easily seen that if $U = T \oplus I_{n-r}$, and if B is an echelon form for a matrix A with rank r, then BU is also an echelon form for A. We will show that all possible echelon forms for A are obtained in this way.

Suppose $B' = BU$ is in echelon form, for some invertible $n \times n$ matrix U, and write

$$B = (B_1 \ O),$$

where B_1 is an $m \times r$ submatrix. If

$$U = \begin{pmatrix} U_1 & U_2 \\ U_3 & U_4 \end{pmatrix},$$

then from $(B_1 \ O)U = (B'_1 \ O)$ we obtain $U_2 = O$, since $B_1 U_2 = O$ and B_1 has rank r. Consequently U_1 is invertible and we can equally well take $U_3 = O, \ U_4 = I$. Let $b_1, \ldots, b_r$ be the columns of B_1 and $b'_1, \ldots, b'_r$ the columns of B'_1. If $U_1 = (t_{jk})$, then

$$b'_k = t_{1k}b_1 + \cdots + t_{rk}b_r \quad (1 \leq k \leq r).$$

Taking $k = 1$, we obtain $m'_1 \geq m_1$ and so, by symmetry, $m'_1 = m_1$. Since $m'_k > m'_1$ for $k > 1$, it follows that $t_{1k} = 0$ for $k > 1$. Taking $k = 2$, we now obtain in the same way $m'_2 = m_2$. Proceeding in this manner, we see that U_1 is a lower triangular matrix.

We return now to the linear Diophantine equation

$$Ax = c.$$

The set of all $c \in \mathbb{Z}^m$ for which there exists a solution $x \in \mathbb{Z}^n$ is evidently a module $L \subseteq \mathbb{Z}^m$. If U is an invertible matrix such that $B = AU$ is in echelon form, then x is a solution of the given system if and only if $y = U^{-1}x$ is a solution of the transformed system

$$By = c.$$

But the latter system is soluble if and only if c is an integral linear combination of the first r columns $b_1, \ldots, b_r$ of B. Since $b_1, \ldots, b_r$ are linearly independent, they form a basis for L.

To determine if a given system $Ax = c$ is soluble, we may use the customary methods of linear algebra over the field $\mathbb{Q}$ of rational numbers to test if c is linearly dependent on $b_1, \ldots, b_r$; then express it as a linear combination of $b_1, \ldots, b_r$, and finally check that the coefficients $y_1, \ldots, y_r$ are all integers. The solutions of the original system are given by $x = Uy$, where y is any vector in $\mathbb{Z}^n$ whose first r coordinates are $y_1, \ldots, y_r$.

If M_1 and M_2 are modules in $\mathbb{Z}^m$, their *intersection* $M_1 \cap M_2$ is again a module. The set of all $a \in \mathbb{Z}^m$ such that $a = a_1 + a_2$ for some $a_1 \in M_1$ and $a_2 \in M_2$ is also a module, which will be denoted by $M_1 + M_2$ and called the *sum* of M_1 and M_2. If M_1 and M_2 are finitely generated, then $M_1 + M_2$ is evidently finitely generated. We will show that $M_1 \cap M_2$ is also finitely generated.

Since $M_1 + M_2$ is a finitely generated module in $\mathbb{Z}^m$, it has a basis $a_1, \ldots, a_n$. Since M_1 and M_2 are contained in $M_1 + M_2$, their generators $b_1, \ldots, b_p$ and $c_1, \ldots, c_q$ have the form

$$b_i = \sum_{k=1}^{n} u_{ki} a_k,$$

$$c_j = \sum_{k=1}^{n} v_{kj} a_k,$$

for some $u_{ki}, v_{kj} \in \mathbb{Z}$. Then $a \in M_1 \cap M_2$ if and only if

$$a = \sum_{i=1}^{p} y_i b_i = \sum_{j=1}^{q} z_j c_j$$

for some $y_i, z_j \in \mathbb{Z}$. Since $a_1, \ldots, a_n$ is a basis for $M_1 + M_2$, this is equivalent to

$$\sum_{i=1}^{p} u_{ki} y_i = \sum_{j=1}^{q} v_{kj} z_j$$

or, in matrix notation, $By = Cz$. But this is equivalent to the homogeneous system $Ax = O$, where

$$A = (B - C), \quad x = \begin{pmatrix} y \\ z \end{pmatrix},$$

and by Proposition 36 the module of solutions of this system has a finite basis.

Suppose the modules $M_1, M_2 \subseteq \mathbb{Z}^m$ are generated by the columns of the $m \times n_1$, $m \times n_2$ matrices A_1, A_2. Evidently M_2 is a submodule of M_1 if and only if each column of A_2 can be expressed as a linear combination of the columns of A_1, i.e. if and only if there exists an $n_1 \times n_2$ matrix X such that

$$A_1 X = A_2.$$

We say in this case that A_1 is a *left divisor* of A_2, or that A_2 is a *right multiple* of A_1.

We may also define greatest common divisors and least common multiples for matrices. An $m \times p$ matrix D is a *greatest common left divisor* of A_1 and A_2 if it is a left divisor of both A_1 and A_2, and if every left divisor C of both A_1 and A_2 is also a left divisor of D. An $m \times q$ matrix H is a *least common right multiple* of A_1 and A_2 if it is a right multiple of both A_1 and A_2, and if every right multiple G of both A_1 and A_2 is also a right multiple of H. It will now be shown that these objects exist and have simple geometrical interpretations.

Let M_1, M_2 be the modules defined by the matrices A_1, A_2. We will show that if the sum $M_1 + M_2$ is defined by the matrix D, then D is a greatest common left divisor of A_1 and A_2. In fact D is a common left divisor of A_1 and A_2, since M_1 and M_2 are contained in $M_1 + M_2$. On the other hand, any common left divisor C of A_1 and A_2 defines a module which contains $M_1 + M_2$, since it contains both M_1 and M_2, and so C is a left divisor of D.

A similar argument shows that if the intersection $M_1 \cap M_2$ is defined by the matrix H, then H is a least common right multiple of A_1 and A_2.

The sum $M_1 + M_2$ is defined, in particular, by the block matrix $(A_1 \ A_2)$. There exists an invertible $(n_1 + n_2) \times (n_1 + n_2)$ matrix U such that

$$(A_1 \ A_2)U = (D' \ O),$$

where D' is an $m \times r$ submatrix of rank r. If

$$U = \begin{pmatrix} U_1 & U_2 \\ U_3 & U_4 \end{pmatrix},$$

is the corresponding partition of U, then

$$A_1 U_1 + A_2 U_3 = D'.$$

On the other hand,

$$(A_1 \ A_2) = (D' \ O)U^{-1}.$$

If

$$U^{-1} = \begin{pmatrix} V_1 & V_2 \\ V_3 & V_4 \end{pmatrix}$$

is the corresponding partition of U^{-1}, then

$$A_1 = D'V_1, \quad A_2 = D'V_2.$$

Thus D' is a common left divisor of A_1 and A_2, and the previous relation implies that it is a greatest common left divisor. It follows that *any* greatest common left divisor D of A_1 and A_2 has a right 'Bézout' representation $D = A_1 X_1 + A_2 X_2$.

We may also define coprimeness for matrices. Two matrices A_1, A_2 of size $m \times n_1, m \times n_2$ are *left coprime* if I_m is a greatest common left divisor. If M_1, M_2 are the modules defined by A_1, A_2, this means that $M_1 + M_2 = \mathbb{Z}^m$. The definition may also be reformulated in several other ways:

Proposition 39 *For any $m \times n$ matrix A, the following conditions are equivalent:*

(i) *for some, and hence every, partition $A = (A_1 \ A_2)$, the submatrices A_1 and A_2 are left coprime;*

(ii) *there exists an $n \times m$ matrix $A^\dagger$ such that $AA^\dagger = I_m$;*

(iii) *there exists an $(n - m) \times n$ matrix A^c such that*

$$\begin{pmatrix} A \\ A^c \end{pmatrix}$$

is invertible;

(iv) *there exists an invertible $n \times n$ matrix V such that $AV = (I_m \ O)$.*

Proof If $A = (A_1 \ A_2)$ for some left coprime matrices A_1, A_2, then there exist X_1, X_2 such that $A_1 X_1 + A_2 X_2 = I_m$ and hence (ii) holds. On the other hand, if (ii) holds then, for any partition $A = (A_1 \ A_2)$, there exist X_1, X_2 such that $A_1 X_1 + A_2 X_2 = I_m$ and hence A_1, A_2 are left coprime.

Thus (i) $\Leftrightarrow$ (ii). Suppose now that (ii) holds. Then A has rank m and hence there exists an invertible $n \times n$ matrix U such that $A = (D \ O)U$, where the $m \times m$ matrix D is non-singular. In fact D is invertible, since $AA^\dagger = I_m$ implies that D is a left divisor of I_m. Consequently, by changing U, we may assume $D = I_m$. If we now take $A^c = (O \ I_{n-m})U$, we see that (ii) $\Rightarrow$ (iii).

It is obvious that (iii) $\Rightarrow$ (iv) and that (iv) $\Rightarrow$ (ii).   □

We now consider the extension of these results to other rings besides $\mathbb{Z}$. Let R be an arbitrary ring. A nonempty set $M \subseteq R^m$ is said to be an R-*module* if $a, b \in M$ and $x, y \in R$ imply $xa + yb \in M$. The module M is *finitely generated* if it contains elements $a_1, \ldots, a_n$ such that every element of M has the form $x_1 a_1 + \cdots + x_n a_n$ for some $x_1, \ldots, x_n \in R$.

It is easily seen that if R is a *Bézout domain*, then the whole of the preceding discussion in this section remains valid if 'module' is interpreted to mean 'R-module' and 'matrix' to mean 'matrix with entries from R'. In particular, we may take $R = K[t]$ to be the ring of all polynomials in one indeterminate with coefficients from an arbitrary field K. However, both $\mathbb{Z}$ and $K[t]$ are principal ideal domains. In this case further results may be obtained.

Proposition 40 *If R is a principal ideal domain and M a finitely generated R-module, then any submodule L of M is also finitely generated. Moreover, if M is generated by n elements, so also is L.*

Proof Suppose M is generated by $a_1, \ldots, a_n$. If $n = 1$, then any $b \in L$ has the form $b = xa_1$ for some $x \in R$ and the set of all x which appear in this way is an ideal of R. Since R is a principal ideal domain, it follows that L is generated by a single element b_1, where $b_1 = x'a_1$ for some $x' \in R$.

Suppose now that $n > 1$ and that, for each $m < n$, any submodule of a module generated by m elements is also generated by m elements. Any $b \in L$ has the form

$$b = x_1 a_1 + \cdots + x_n a_n$$

for some $x_1, \ldots, x_n \in R$ and the set of all x_1 which appear in this way is an ideal of R. Since R is a principal ideal domain, it follows that there is a fixed $b_1 \in L$ such that $b = y_1 b_1 + b'$ for some $y_1 \in R$ and some b' in the module M' generated by $a_2, \ldots, a_n$. The set of all b' which appear in this way is a submodule L' of M'. By the induction hypothesis, L' is generated by $n - 1$ elements and hence L is generated by n elements.   □

Just as it is useful to define vector spaces abstractly over an arbitrary field K, so it is useful to define modules abstractly over an arbitrary ring R. An abelian group M, with the group operation denoted by $+$, is said to be an R-*module* if, with any $a \in M$ and any $x \in R$, there is associated an element $xa \in M$ so that the following properties hold, for all $a, b \in M$ and all $x, y \in R$:

(i) $x(a + b) = xa + xb$,
(ii) $(x + y)a = xa + ya$,
(iii) $(xy)a = x(ya)$,
(iv) $1a = a$.

The proof of Proposition 40 remains valid for modules in this abstract sense. However, a finitely generated module need not now have a basis. For, even if it is generated by a single element a, we may have $xa = O$ for some nonzero $x \in R$. Nevertheless, we are going to show that, if R is a principal ideal domain, all finitely generated R-modules can be completely characterized.

Let R be a principal ideal domain and M a finitely generated R-module, with generators $a_1, \ldots, a_n$, say. The set N of all $x = (x_1, \ldots, x_n) \in R^n$ such that

$$x_1 a_1 + \cdots + x_n a_n = O$$

is evidently a module in R^n. Hence N is finitely generated, by Proposition 40. The given module M is isomorphic to the quotient module R^n / N.

Let $f_1, \ldots, f_m$ be a set of generators for N and let $e_1, \ldots, e_n$ be a basis for R^n. Then

$$f_j = a_{j1} e_1 + \cdots + a_{jn} e_n \quad (1 \le j \le m),$$

for some $a_{jk} \in R$. The module M is completely determined by the matrix $A = (a_{jk})$. However, we can change generators and change bases.

If we put

$$f_i' = v_{i1} f_1 + \cdots + v_{im} f_m \quad (1 \le i \le m),$$

where $V = (v_{ij})$ is an invertible $m \times m$ matrix, then $f_1', \ldots, f_m'$ is also a set of generators for N. If we put

$$e_k = u_{k1} e_1' + \cdots + u_{kn} e_n' \quad (1 \le k \le n),$$

where $U = (u_{k\ell})$ is an invertible $n \times n$ matrix, then $e_1', \ldots, e_n'$ is also a basis for R^n. Moreover

$$f_i' = b_{i1} e_1' + \cdots + b_{in} e_n' \quad (1 \le i \le m),$$

where the $m \times n$ matrix $B = (b_{i\ell})$ is given by $B = VAU$.

The idea is to choose V and U so that B is as simple as possible. This is made precise in the next proposition, first proved by H.J.S. Smith (1861) for $R = \mathbb{Z}$. The corresponding matrix S is known as the *Smith normal form* of A.

Proposition 41 *Let R be a principal ideal domain and let A be an m × n matrix with entries from R. If A has rank r, then there exist invertible m × m, n × n matrices V, U with entries from R such that S = VAU has the form*

$$S = \begin{pmatrix} D & O \\ O & O \end{pmatrix},$$

where D = diag[d₁, ..., dᵣ] is a diagonal matrix with nonzero entries d_i and $d_i | d_j$ for $1 \le i \le j \le r$.

Proof We show first that it is enough to obtain a matrix which satisfies all the requirements except the divisibility conditions for the d's.

If a, b are nonzero elements of R with greatest common divisor d, then there exist $u, v \in R$ such that $au + bv = d$. It is easily verified that

$$\begin{pmatrix} 1 & 1 \\ -bv/d & au/d \end{pmatrix} \begin{pmatrix} a & 0 \\ 0 & b \end{pmatrix} \begin{pmatrix} u & -b/d \\ v & a/d \end{pmatrix} = \begin{pmatrix} d & 0 \\ 0 & ab/d \end{pmatrix},$$

and the outside matrices on the left-hand side are both invertible. By applying this process finitely many times, a non-singular diagonal matrix $D' = \mathrm{diag}[d'_1, \ldots, d'_r]$ may be transformed into a non-singular diagonal matrix $D = \mathrm{diag}[d_1, \ldots, d_r]$ which satisfies $d_i | d_j$ for $1 \le i \le j \le r$.

Consider now an arbitrary matrix A. By applying Proposition 35 to the transpose of A, we may reduce the problem to the case where A has rank m and then, by Corollary 38, we may suppose further that $a_{jj} \ne 0$, $a_{jk} = 0$ for all $j < k$.

It is now sufficient to show that, for any 2×2 matrix

$$A = \begin{pmatrix} a & 0 \\ b & c \end{pmatrix},$$

with nonzero entries a, b, c, there exist invertible 2×2 matrices U, V such that VAU is a diagonal matrix. Moreover, we need only prove this when the greatest common divisor $(a, b, c) = 1$. But then there exists $q \in R$ such that $(a, b+qc) = 1$. In fact, take q to be the product of the distinct primes which divide a but not b. For any prime divisor p of a, if $p|b$, then $p \nmid c$, $p \nmid q$ and hence $p \nmid (b+qc)$; if $p \nmid b$, then $p|q$ and again $p \nmid (b+qc)$.

If we put $b' = b + qc$, then there exist $x, y \in R$ such that $ax + b'y = 1$, and hence $ax + by = 1 - qcy$. It is easily verified that

$$\begin{pmatrix} x & y \\ -b' & a \end{pmatrix} \begin{pmatrix} a & 0 \\ b & c \end{pmatrix} \begin{pmatrix} 1 & -cy \\ q & 1 - qcy \end{pmatrix} = \begin{pmatrix} 1 & 0 \\ 0 & ac \end{pmatrix},$$

and the outside matrices on the left-hand side are both invertible.   □

In the important special case $R = \mathbb{Z}$, there is a more constructive proof of Proposition 41. Obviously we may suppose $A \ne O$. By interchanges of rows and columns we can arrange that a_{11} is the nonzero entry of A with minimum absolute value. If there is an entry a_{1k} ($k > 1$) in the first row which is not divisible by a_{11}, then we can write $a_{1k} = za_{11} + a'_{1k}$, where $z, a'_{1k} \in \mathbb{Z}$ and $|a'_{1k}| < |a_{11}|$. By subtracting z times the first column from the k-th column we replace a_{1k} by a'_{1k}. Thus we obtain a new matrix A in which the minimum absolute value of the nonzero entries has been reduced.

On the other hand, if a_{11} divides a_{1k} for all $k > 1$ then, by subtracting multiples of the first column from the remaining columns, we can arrange that $a_{1k} = 0$ for all $k > 1$. If there is now an entry $a_{j1}(j > 1)$ in the first column which is not divisible by a_{11} then, by subtracting a multiple of the first row from the j-th row, the minimum absolute value of the nonzero entries can again be reduced. Otherwise, by subtracting multiples of the first row from the remaining rows, we can bring A to the form

$$\begin{pmatrix} a_{11} & O \\ O & A' \end{pmatrix}.$$

Since $A \neq O$ and the minimum absolute value of the nonzero entries cannot be reduced indefinitely, we must in any event arrive at a matrix of this form after a finite number of steps. The same procedure can now be applied to the submatrix A', and so on until we obtain a matrix

$$\begin{pmatrix} D' & O \\ O & O \end{pmatrix},$$

where D' is a diagonal matrix with the same rank as A. As in the first part of the proof of Proposition 41, we can now replace D' by a diagonal matrix D which satisfies the divisibility conditions.

Clearly this constructive proof is also valid for any Euclidean domain R and, in particular, for the polynomial ring $R = K[t]$, where K is an arbitrary field.

It will now be shown that the Smith normal form of a matrix A is uniquely determined, apart from replacing each d_k by an arbitrary unit multiple. For, if S' is another Smith normal form, then $S' = V'SU'$ for some invertible $m \times m$, $n \times n$ matrices V', U'. Since d_1 divides all entries of S, it also divides all entries of S'. In particular, $d_1 | d_1'$. In the same way $d_1' | d_1$ and hence d_1' is a unit multiple of d_1. To show that d_k' is a unit multiple of d_k, also for $k > 1$, it is quickest to use determinants (Chapter V, §1). Since $d_1 \cdots d_k$ divides all $k \times k$ subdeterminants or *minors* of S, it also divides all $k \times k$ minors of S'. In particular, $d_1 \cdots d_k | d_1' \cdots d_k'$. Similarly, $d_1' \cdots d_k' | d_1 \cdots d_k$ and hence $d_1' \cdots d_k'$ is a unit multiple of $d_1 \cdots d_k$. The conclusion now follows by induction on k.

The products $\Delta_k := d_1 \cdots d_k$ $(1 \leq k \leq r)$ are known as the *invariant factors* of the matrix A. A similar argument to that in the preceding paragraph shows that Δ_k is the greatest common divisor of all $k \times k$ minors of A.

Two $m \times n$ matrices A, B are said to be *equivalent* if there exist invertible $m \times m$, $n \times n$ matrices V, U such that $B = VAU$. Since equivalence is indeed an 'equivalence relation', the uniqueness of the Smith normal form implies that two $m \times n$ matrices A, B are equivalent if and only if they have the same rank and the same invariant factors.

We return now from matrices to modules. Let R be a principal ideal domain and M a finitely generated R-module, with generators $a_1, \ldots, a_n$. The Smith normal form tells us that M has generators $a_1', \ldots, a_n'$, where

$$a_k = u_{k1}a_1' + \cdots + u_{kn}a_n' \quad (1 \leq k \leq n)$$

for some invertible matrix $U = (u_{k\ell})$, such that $d_k a_k' = O$ $(1 \leq k \leq r)$. Moreover,

$$x_1 a_1' + \cdots + x_n a_n' = O$$

implies $x_k = y_k d_k$ for some $y_k \in R$ if $1 \leq k \leq r$ and $x_k = 0$ if $r < k \leq n$. In particular, $x_k a'_k = O$ for $1 \leq k \leq n$, and thus the module M is the direct sum of the submodules $M'_1, \ldots, M'_n$ generated by $a'_1, \ldots, a'_n$ respectively.

If N_k denotes the set of all $x \in R$ such that $x a'_k = O$, then N_k is the principal ideal of R generated by d_k for $1 \leq k \leq r$ and $N_k = \{0\}$ for $r < k \leq n$. The divisibility conditions on the d's imply that $N_{k+1} \subseteq N_k$ ($1 \leq k < r$). If $N_k = R$ for some k, then a'_k contributes nothing as a generator and may be omitted.

Evidently the submodule M' generated by $a'_1, \ldots, a'_r$ consists of all $a \in M$ such that $xa = O$ for some nonzero $x \in R$, and the submodule M'' generated by $a'_{r+1}, \ldots, a'_n$ has $a'_{r+1}, \ldots, a'_n$ as a basis. Thus we have proved the *structure theorem for finitely generated modules over a principal ideal domain*:

Proposition 42 *Let R be a principal ideal domain and M a finitely generated R-module. Then M is the direct sum of two submodules M' and M'', where M' consists of all $a \in M$ such that $xa = O$ for some nonzero $x \in R$ and M'' has a finite basis.*

Moreover, M' is the direct sum of s submodules $Ra_1, \ldots, Ra_s$, such that

$$0 \subset N_s \subseteq \cdots \subseteq N_1 \subset R,$$

where N_k is the ideal consisting of all $x \in R$ such that $xa_k = O$ ($1 \leq k \leq s$).

The uniquely determined submodule M' is called the *torsion submodule* of M. The *free submodule* M'' is not uniquely determined, although the number of elements in a basis is uniquely determined. Of course, for a particular M one may have $M' = \{O\}$ or $M'' = \{O\}$.

Any abelian group A, with the group operation denoted by $+$, may be regarded as a $\mathbb{Z}$-module by defining na to be the sum $a + \cdots + a$ with n summands if $n \in \mathbb{N}$, to be O if $n = 0$, and to be $-(a + \cdots + a)$ with $-n$ summands if $-n \in \mathbb{N}$. The structure theorem in this case becomes the *structure theorem for finitely generated abelian groups*: any finitely generated abelian group A is the direct product of finitely many finite or infinite cyclic subgroups. The finite cyclic subgroups have orders $d_1, \ldots, d_s$, where $d_1 > 1$ if $s > 0$ and $d_i | d_j$ if $i \leq j$. In particular, A is the direct product of a finite subgroup A' (of order $d_1 \cdots d_r$), its *torsion subgroup*, and a *free* subgroup A''.

The fundamental structure theorem also has an important application to linear algebra. Let V be a vector space over a field K and $T : V \to V$ a linear transformation. We can give V the structure of a $K[t]$-module by defining, for any $v \in V$ and any $f = a_0 + a_1 t + \cdots + a_n t^n \in K[t]$,

$$fv = a_0 v + a_1 T v + \cdots + a_n T^n v.$$

If V is finite-dimensional, then for any $v \in V$ there is a nonzero polynomial f such that $fv = O$. In this case the fundamental structure theorem says that V is the direct sum of finitely many subspaces $V_1, \ldots, V_s$ which are invariant under T. If V_i has dimension $n_i \geq 1$, then there exists a vector $w_i \in V_i$ such that $w_i, T w_i, \ldots, T^{n_i - 1} w_i$ are a vector space basis for V_i ($1 \leq i \leq s$). There is a uniquely determined monic polynomial m_i of degree n_i such that $m_i(T) w_i = O$ and, finally, $m_i | m_j$ if $i \leq j$.

The Smith normal form can be used to solve systems of linear ordinary differential equations with constant coefficients. Such a system has the form

$$a_{11}(D)x_1 + \cdots + a_{1n}(D)x_n = c_1(t)$$
$$a_{21}(D)x_1 + \cdots + a_{2n}(D)x_n = c_2(t)$$

$$\cdots$$

$$a_{m1}(D)x_1 + \cdots + a_{mn}(D)x_n = c_m(t),$$

where the coefficients $a_{jk}(D)$ are polynomials in $D = d/dt$ with complex coefficients and the right sides $c_j(t)$ are, say, infinitely differentiable functions of the time t. Since $\mathbb{C}[s]$ is a Euclidean domain, we can bring the coefficient matrix $A = (a_{jk}(D))$ to Smith normal form and thus replace the given system by an equivalent system in which the variables are 'uncoupled'.

For the polynomial ring $R = K[t]$ it is possible to say more about R-modules than for an arbitrary Euclidean domain, since the absolute value

$$|f| = 2^{\partial(f)} \quad \text{if } f \neq O, |O| = 0,$$

has not only the Euclidean property, but also the properties

$$|f + g| \leq \max\{|f|, |g|\}, \quad |fg| = |f||g| \quad \text{for any } f, g \in R.$$

For any $a \in R^m$, where $R = K[t]$, define $|a|$ to be the maximum absolute value of any of its coordinates. Then a basis for a module $M \subseteq R^m$ can be obtained in the following way. Suppose $M \neq O$ and choose a nonzero element a_1 of M for which $|a_1|$ is a minimum. If there is an element of M which is not of the form p_1a_1 with $p_1 \in R$, choose one, a_2, for which $|a_2|$ is a minimum. If there is an element of M which is not of the form $p_1a_1 + p_2a_2$ with $p_1, p_2 \in R$, choose one, a_3, for which $|a_3|$ is a minimum. And so on.

Evidently $|a_1| \leq |a_2| \leq \cdots$. We will show that $a_1, a_2, \ldots$ are linearly independent for as long as the the process can be continued, and thus ultimately a basis is obtained.

If this is not the case, then there exists a positive integer $k \leq m$ such that $a_1, \ldots, a_k$ are linearly independent, but $a_1, \ldots, a_{k+1}$ are not. Hence there exist $s_1, \ldots, s_{k+1} \in R$ with $s_{k+1} \neq 0$ such that $s_1a_1 + \cdots + s_{k+1}a_{k+1} = O$. For each $j \leq k$, there exist q_j, $r_j \in R$ such that

$$s_j = q_js_{k+1} + r_j, \quad |r_j| < |s_{k+1}|.$$

Put

$$a'_{k+1} = a_{k+1} + q_1a_1 + \cdots + q_ka_k, \quad b_k = r_1a_1 + \cdots + r_ka_k.$$

Since a_{k+1} is not of the form $p_1a_1 + \cdots + p_ka_k$, neither is a'_{k+1} and hence $|a'_{k+1}| \geq |a_{k+1}|$. Furthermore, $|b_k| \leq \max_{1 \leq j \leq k} |r_j||a_j| < |s_{k+1}||a_{k+1}|$. Since $b_k = -s_{k+1}a'_{k+1}$, by construction, this is a contradiction.

A basis for M which is obtained in this way will be called a *minimal basis*. It is not difficult to show that a basis $a_1, \ldots, a_n$ is a minimal basis if and only if $|a_1| \leq \cdots \leq |a_n|$ and the sum $|a_1| + \cdots + |a_n|$ is minimal. Although a minimal basis is not uniquely determined, the values $|a_1|, \ldots, |a_n|$ are uniquely determined.

5 Further Remarks

For the history of the law of quadratic reciprocity, see Frei [16]. The first two proofs by Gauss of the law of quadratic reciprocity appeared in §§125–145 and §262 of [17]. A simplified account of Gauss's inductive proof has been given by Brown [7]. The proofs most commonly given use 'Gauss's lemma' and are variants of Gauss's third proof. The first proof given here, due to Rousseau [46], is of this general type, but it does not use Gauss's lemma and is based on a natural definition of the Jacobi symbol. For an extension of this definition of Zolotareff to algebraic number fields, see Cartier [9].

For Dirichlet's evaluation of Gauss sums, see [33]. A survey of Gauss sums is given in Berndt and Evans [6].

The extension of the law of quadratic reciprocity to arbitrary algebraic number fields was the subject of Hilbert's 9th Paris problem. Although such generalizations lie outside the scope of the present work, it may be worthwhile to give a brief glimpse. Let $K = \mathbb{Q}$ be the field of rational numbers and let $L = \mathbb{Q}(\sqrt{d})$ be a quadratic extension of K. If p is a prime in K, the law of quadratic reciprocity may be interpreted as describing how the ideal generated by p in L factors into prime ideals. Now let K be an arbitrary algebraic number field and let L be any finite extension of K. Quite generally, we may ask how the arithmetic of the extension L is determined by the arithmetic of K. The general reciprocity law, conjectured by Artin in 1923 and proved by him in 1927, gives an answer in the form of an isomorphism between two groups, provided the Galois group of L over K is abelian. For an introduction, see Wyman [54] and, for more detail, Tate [51]. The outstanding problem is to find a meaningful extension to the case when the Galois group is non-abelian. Some intriguing conjectures are provided by the Langlands program, for which see also Gelbart [18].

The law of quadratic reciprocity has an analogue for polynomials with coefficients from a finite field. Let $\mathbb{F}_q$ be a finite field containing q elements, where q is a power of an *odd* prime. If $g \in \mathbb{F}_q[x]$ is a monic irreducible polynomial of positive degree, then for any $f \in \mathbb{F}_q[x]$ not divisible by g we define (f/g) to be 1 if f is congruent to a square mod g, and -1 otherwise. The law of quadratic reciprocity, which in the case of prime q was stated by Dedekind (1857) and proved by Artin (1924), says that

$$(f/g)(g/f) = (-1)^{mn(q-1)/2}$$

for any distinct monic irreducible polynomials $f, g \in \mathbb{F}_q[x]$ of positive degrees m, n. Artin also developed a theory of ideals, analogous to that for quadratic number fields, for the field obtained by adjoining to $\mathbb{F}_q[x]$ an element ω with $\omega^2 = D(x)$, where $D(x) \in \mathbb{F}_q[x]$ is square-free; see [3].

Quadratic fields are treated in the third volume of Landau [30]. There is also a useful resumé accompanying the tables in Ince [23].

A complex number is said to be *algebraic* if it is a root of a monic polynomial with rational coefficients and *transcendental* otherwise. Hence a complex number is algebraic if and only if it is an element of some algebraic number field.

For an introduction to the theory of algebraic number fields, see Samuel [47]. This vast theory may be approached in a variety of ways. For a more detailed treatment the student may choose from Hecke [22], Hasse [20], Lang [32], Narkiewicz [38] and Neukirch [39]. There are useful articles in Cassels and Fröhlich [10], and Artin [2] treats also algebraic functions.

For the early history of Fermat's last theorem, see Vandiver [52], Ribenboim [41] and Kummer [28]. Further references will be given in Chapter XIII.

Arithmetical functions are discussed in Apostol [1], McCarthy [35] and Sivaramakrishnan [48]. The term 'Dirichlet product' comes from the connection with Dirichlet series, which will be considered in Chapter IX, §6. The ring of all arithmetical functions was shown to be a factorial domain by Cashwell and Everett (1959); the result is proved in [48].

In the form $f(a \wedge b) f(a \vee b) = f(a) f(b)$, the concept of multiplicative function can be extended to any map $f : L \to \mathbb{C}$, where L is a lattice. Möbius inversion can be extended to any locally finite partially ordered set and plays a significant role in modern combinatorics; see Bender and Goldman [5], Rota [45] and Barnabei et al. [4].

The early history of perfect numbers and Fermat numbers is described in Dickson [13]. It has been proved that any odd perfect number, if such a thing exists, must be greater than 10^{300} and have at least 8 distinct prime factors. On the other hand, if an odd perfect number N has at most k distinct prime factors, then $N < 4^{4^k}$ and thus all such N can be found by a finite amount of computation. See te Riele [42] and Heath-Brown [21].

The proof of the Lucas–Lehmer test for Mersenne primes follows Rosen [43] and Bruce [8]. For the conjectured distribution of Mersenne primes, see Wagstaff [53]. The construction of regular polygons by ruler and compass is discussed in Hadlock [19], Jacobson [24] and Morandi [36].

Much of the material in §4 is also discussed in Macduffee [34] and Newman [40]. Corollary 34 was proved by Hermite (1849), who later (1851) also proved Corollary 38. Indeed the latter result is the essential content of *Hermite's normal form*, which will be encountered in Chapter VIII, §2.

It is clear that Corollary 34 remains valid if the underlying ring $\mathbb{Z}$ is replaced by any principal ideal domain. There have recently been some noteworthy extensions to more general rings. It may be asked, for an arbitrary commutative ring R and any $a_1, \ldots, a_n \in R$, does there exist an invertible $n \times n$ matrix U with entries from R which has $a_1, \ldots, a_n$ as its first row? It is obviously necessary that there exist $x_1, \ldots, x_n \in R$ such that

$$a_1 x_1 + \cdots + a_n x_n = 1,$$

i.e. that the ideal generated by $a_1, \ldots, a_n$ be the whole ring R. If $n = 2$, this necessary condition is also sufficient, by the same observation as when invertibility of matrices was first considered for $R = \mathbb{Z}$. However, if $n > 2$ there exist even factorial domains R for which the condition is not sufficient. In 1976 Quillen and Suslin independently proved the twenty-year-old conjecture that it *is* sufficient if $R = K[t_1, \ldots, t_d]$ is the ring of polynomials in finitely many indeterminates with coefficients from an arbitrary field K.

By pursuing an analogy between projective modules in algebra and vector bundles in topology, Serre (1955) had been led to conjecture that, for $R = K[t_1, \ldots, t_d]$, if an R-module has a finite basis and is the direct sum of two submodules, then each of these submodules has a finite basis. Seshadri (1958) proved the conjecture for $d = 2$ and in the same year Serre showed that, for arbitrary d, it would follow from the result which Quillen and Suslin subsequently proved.

For proofs of these results and for later developments, see Lam [29], Fitchas and Galligo [14], and Swan [50]. There is a short proof of the Quillen–Suslin theorem in Lang [31].

For Smith's normal form, see Smith [49] and Kaplansky [27]. It was shown by Wedderburn (1915) that Smith's normal form also holds for matrices of holomorphic functions, even though the latter do not form a principal ideal domain; see Narasimhan [37].

Finitely generated commutative groups are important, not only because more can be said about them, but also because they arise in practice. *Dirichlet's unit theorem* says that the units of an algebraic number field form a finitely generated commutative group. As will be seen in Chapter XIII, §4, *Mordell's theorem* says that the rational points of an elliptic curve also form a finitely generated commutative group.

Modules over a polynomial ring $K[s]$ play an important role in what electrical engineers call *linear systems theory*. Connected accounts are given in Kalman [26], Rosenbrock [44] and Kailath [25]. For some further mathematical developments, see Forney [15], Coppel [11], and Coppel and Cullen [12].

6 Selected References

[1] T.M. Apostol, *Introduction to analytic number theory*, Springer-Verlag, New York, 1976.

[2] E. Artin, *Algebraic numbers and algebraic functions*, Nelson, London, 1968.

[3] E. Artin, Quadratische Körper im Gebiet der höheren Kongruenzen I, II, *Collected Papers* (ed. S. Lang and J.T. Tate), pp. 1–94, reprinted, Springer-Verlag, New York, 1986.

[4] M. Barnabei, A. Brini and G.-C. Rota, The theory of Möbius functions, *Russian Math. Surveys* **41** (1986), no. 3, 135–188.

[5] E.A. Bender and J.R. Goldman, On the application of Möbius inversion in combinatorial analysis, *Amer. Math. Monthly* **82** (1975), 789–803.

[6] B.C. Berndt and R.J. Evans, The determination of Gauss sums, *Bull. Amer. Math. Soc. (N.S.)* **5** (1981), 107–129.

[7] E. Brown, The first proof of the quadratic reciprocity law, revisited, *Amer. Math. Monthly* **88** (1981), 257–264.

[8] J.W. Bruce, A really trivial proof of the Lucas–Lehmer test, *Amer. Math. Monthly* **100** (1993), 370–371.

[9] P. Cartier, Sur une généralisation des symboles de Legendre–Jacobi, *Enseign. Math.* **16** (1970), 31–48.

[10] J.W.S. Cassels and A. Fröhlich (ed.), *Algebraic number theory*, Academic Press, London, 1967.

[11] W.A. Coppel, Matrices of rational functions, *Bull. Austral. Math. Soc.* **11** (1974), 89–113.

[12] W.A. Coppel and D.J. Cullen, Strong system equivalence (II), *J. Austral. Math. Soc. B* **27** (1985), 223–237.

[13] L.E. Dickson, *History of the theory of numbers, Vol. I*, reprinted, Chelsea, New York, 1992.

[14] N. Fitchas and A. Galligo, Nullstellensatz effectif et conjecture de Serre (théorème de Quillen–Suslin) pour le calcul formel, *Math. Nachr.* **149** (1990), 231–253.

[15] G.D. Forney, Minimal bases of rational vector spaces, with applications to multivariable linear systems, *SIAM J. Control* **13** (1975), 493–520.

[16] G. Frei, The reciprocity law from Euler to Eisenstein, *The intersection of history and mathematics* (ed. C. Sasaki, M. Sugiura and J.W. Dauben), pp. 67–90, Birkhäuser, Basel, 1994.

[17] C.F. Gauss, *Disquisitiones arithmeticae*, English translation by A.A. Clarke, revised by W.C. Waterhouse, Springer, New York, 1986. [Latin original, 1801]

[18] S. Gelbart, An elementary introduction to the Langlands program, *Bull. Amer. Math. Soc.* (*N.S.*) **10** (1984), 177–219.

[19] C.R. Hadlock, *Field theory and its classical problems*, Carus Mathematical Monographs no. 19, Mathematical Association of America, Washington, D.C., 1978. [Reprinted in paperback, 2000]

[20] H. Hasse, *Number theory*, English transl. by H.G. Zimmer, Springer-Verlag, Berlin, 1980.

[21] D.R. Heath-Brown, Odd perfect numbers, *Math. Proc. Cambridge Philos. Soc.* **115** (1994), 191–196.

[22] E. Hecke, *Lectures on the theory of algebraic numbers*, English transl. by G.U. Brauer, J.R. Goldman and R. Kotzen, Springer-Verlag, New York, 1981. [German original, 1923]

[23] E.L. Ince, *Cycles of reduced ideals in quadratic fields*, Mathematical Tables Vol. IV, British Association, London, 1934.

[24] N. Jacobson, *Basic Algebra I*, 2nd ed., W.H. Freeman, New York, 1985.

[25] T. Kailath, *Linear systems*, Prentice–Hall, Englewood Cliffs, N.J., 1980.

[26] R.E. Kalman, Algebraic theory of linear systems, *Topics in mathematical system theory* (R.E. Kalman, P.L. Falb and M.A. Arbib), pp. 237–339, McGraw-Hill, New York, 1969.

[27] I. Kaplansky, Elementary divisors and modules, *Trans. Amer. Math. Soc.* **66** (1949), 464–491.

[28] E. Kummer, *Collected Papers, Vol. I* (ed. A. Weil), Springer-Verlag, Berlin, 1975.

[29] T.Y. Lam, *Serre's conjecture*, Lecture Notes in Mathematics **635**, Springer-Verlag, Berlin, 1978.

[30] E. Landau, *Vorlesungen über Zahlentheorie*, 3 vols., Hirzel, Leipzig, 1927. [Reprinted, Chelsea, New York, 1969]

[31] S. Lang, *Algebra*, corrected reprint of 3rd ed., Addison-Wesley, Reading, Mass., 1994.

[32] S. Lang, *Algebraic number theory*, 2nd ed., Springer-Verlag, New York, 1994.

[33] G. Lejeune-Dirichlet, *Werke*, Band I, pp. 237–256, reprinted Chelsea, New York, 1969.

[34] C.C. Macduffee, *The theory of matrices*, corrected reprint, Chelsea, New York, 1956.

[35] P.J. McCarthy, *Introduction to arithmetical functions*, Springer-Verlag, New York, 1986.

[36] P. Morandi, *Field and Galois theory*, Springer-Verlag, New York, 1996.

[37] R. Narasimhan, *Complex analysis in one variable*, Birkhäuser, Boston, Mass., 1985.

[38] W. Narkiewicz, *Elementary and analytic theory of algebraic numbers*, 2nd ed., Springer-Verlag, Berlin, 1990.

[39] J. Neukirch, *Algebraic number theory*, English transl. by N. Schappacher, Springer, Berlin, 1999.

[40] M. Newman, *Integral matrices*, Academic Press, New York, 1972.

[41] P. Ribenboim, *13 Lectures on Fermat's last theorem*, Springer-Verlag, New York, 1979.

[42] H.J.J. te Riele, Perfect numbers and aliquot sequences, *Computational methods in number theory* (ed. H.W. Lenstra Jr. and R. Tijdeman), Part I, pp. 141–157, Mathematical Centre Tracts **154**, Amsterdam, 1982.

[43] M.L. Rosen, A proof of the Lucas–Lehmer test, *Amer. Math. Monthly* **95** (1988), 855–856.

[44] H.H. Rosenbrock, *State-space and multivariable theory*, Nelson, London, 1970.

[45] G.-C. Rota, On the foundations of combinatorial theory I. Theory of Möbius functions, *Z. Wahrsch. Verw. Gebiete* **2** (1964), 340–368.

[46] G. Rousseau, On the Jacobi symbol, *J. Number Theory* **48** (1994), 109–111.

[47] P. Samuel, *Algebraic theory of numbers*, English transl. by A.J. Silberger, Houghton Mifflin, Boston, Mass., 1970.

[48] R. Sivaramakrishnan, *Classical theory of arithmetic functions*, M. Dekker, New York, 1989.

[49] H.J.S. Smith, *Collected mathematical papers, Vol. 1*, pp. 367–409, reprinted, Chelsea, New York, 1965.

[50] R.G. Swan, Gubeladze's proof of Anderson's conjecture, *Azumaya algebras, actions and modules* (ed. D. Haile and J. Osterburg), pp. 215–250, Contemporary Mathematics **124**, Amer. Math. Soc., Providence, R.I., 1992.

[51] J. Tate, Problem 9: The general reciprocity law, *Mathematical developments arising from Hilbert problems* (ed. F.E. Browder), pp. 311–322, Proc. Symp. Pure Math. **28**, Part 2, Amer. Math. Soc., Providence, Rhode Island, 1976.

[52] H.S. Vandiver, Fermat's last theorem: its history and the nature of the known results concerning it, *Amer. Math. Monthly* **53** (1946), 555–578.

[53] S.S. Wagstaff Jnr., Divisors of Mersenne numbers, *Math. Comp.* **40** (1983), 385–397.

[54] B.F. Wyman, What is a reciprocity law?, *Amer. Math. Monthly* **79** (1972), 571–586.

Additional References

J. Bernstein and S. Gelbart (ed.), *Introduction to the Langlands program*, Birkhäuser, Boston, 2003.

E. Frenkel, Recent advances in the Langlands program, *Bull. Amer. Math. Soc. (N.S.)* **41** (2004), 151–184.

T. Metsänkylä, Catalan's conjecture: another old Diophantine problem solved, *Bull. Amer. Math. Soc. (N.S.)* **41** (2004), 43–57.

IV

Continued Fractions and Their Uses

1 The Continued Fraction Algorithm

Let $\xi = \xi_0$ be an irrational real number. Then we can write

$$\xi_0 = a_0 + \xi_1^{-1},$$

where $a_0 = \lfloor \xi_0 \rfloor$ is the greatest integer $\leq \xi_0$ and where $\xi_1 > 1$ is again an irrational number. Hence the process can be repeated indefinitely:

$$\xi_1 = a_1 + \xi_2^{-1}, \quad (a_1 = \lfloor \xi_1 \rfloor, \xi_2 > 1),$$
$$\xi_2 = a_2 + \xi_3^{-1}, \quad (a_2 = \lfloor \xi_2 \rfloor, \xi_3 > 1),$$
$$\cdots$$

By construction, $a_n \in \mathbb{Z}$ for all $n \geq 0$ and $a_n \geq 1$ if $n \geq 1$. The uniquely determined infinite sequence $[a_0, a_1, a_2, \ldots]$ is called the *continued fraction expansion* of ξ. The continued fraction expansion of ξ_n is $[a_n, a_{n+1}, a_{n+2}, \ldots]$.

For example, the 'golden ratio' $\tau = (1 + \sqrt{5})/2$ has the continued fraction expansion $[1, 1, 1, \ldots]$, since $\tau = 1 + \tau^{-1}$. Similarly, $\sqrt{2}$ has the continued fraction expansion $[1, 2, 2, \ldots]$, since $\sqrt{2} + 1 = 2 + 1/(\sqrt{2} + 1)$.

The relation between ξ_n and ξ_{n+1} may be written as a linear fractional transformation:

$$\xi_n = (a_n \xi_{n+1} + 1)/(1\xi_{n+1} + 0).$$

An arbitrary linear fractional transformation

$$\xi = (\alpha \xi' + \beta)/(\gamma \xi' + \delta)$$

is completely determined by its matrix

$$T = \begin{pmatrix} \alpha & \beta \\ \gamma & \delta \end{pmatrix}.$$

This description is convenient, because if we make a further linear fractional transformation

$$\xi' = (\alpha' \xi'' + \beta')/(\gamma' \xi'' + \delta')$$

W.A. Coppel, *Number Theory: An Introduction to Mathematics*, Universitext, DOI: 10.1007/978-0-387-89486-7_4, © Springer Science + Business Media, LLC 2009

with matrix

$$T' = \begin{pmatrix} \alpha' & \beta' \\ \gamma' & \delta' \end{pmatrix},$$

then, as is easily verified, the matrix

$$T'' = \begin{pmatrix} \alpha'' & \beta'' \\ \gamma'' & \delta'' \end{pmatrix}$$

of the composite transformation

$$\xi = (\alpha'' \xi'' + \beta'')/(\gamma'' \xi'' + \delta'')$$

is given by the matrix product $T'' = TT'$.

It follows that, if we set

$$A_k = \begin{pmatrix} a_k & 1 \\ 1 & 0 \end{pmatrix},$$

then the matrix of the linear fractional transformation which expresses ξ in terms of ξ_{n+1} is

$$T_n = A_0 \cdots A_n.$$

It is readily verified by induction that

$$T_n = \begin{pmatrix} p_n & p_{n-1} \\ q_n & q_{n-1} \end{pmatrix},$$

i.e.,

$$\xi = (p_n \xi_{n+1} + p_{n-1})/(q_n \xi_{n+1} + q_{n-1}),$$

where the elements p_n, q_n satisfy the recurrence relations

$$p_n = a_n p_{n-1} + p_{n-2}, \quad q_n = a_n q_{n-1} + q_{n-2} \ (n \geq 0), \tag{1}$$

with the conventional starting values

$$p_{-2} = 0, \quad p_{-1} = 1, \quad \text{resp.} \ q_{-2} = 1, \quad q_{-1} = 0. \tag{2}$$

In particular,

$$p_0 = a_0, \quad p_1 = a_1 a_0 + 1, \quad q_0 = 1, \quad q_1 = a_1.$$

Since $\det A_k = -1$, by taking determinants we obtain

$$p_n q_{n-1} - p_{n-1} q_n = (-1)^{n+1} \quad (n \geq 0). \tag{3}$$

By (1) also,

$$\begin{pmatrix} p_n & p_{n-1} \\ q_n & q_{n-1} \end{pmatrix} \begin{pmatrix} 1 & 1 \\ -a_n & 0 \end{pmatrix} = \begin{pmatrix} p_{n-2} & p_n \\ q_{n-2} & q_n \end{pmatrix},$$

from which, by taking determinants again, we get

$$p_n q_{n-2} - p_{n-2} q_n = (-1)^n a_n \quad (n \geq 0). \tag{4}$$

It follows from (1)–(2) that p_n and q_n are integers, and from (3) that they are coprime. Since $a_n \geq 1$ for $n \geq 1$, we have

$$1 = q_0 \leq q_1 < q_2 < \cdots .$$

Thus $q_n \geq n$ for $n \geq 1$. (In fact, since $q_n \geq q_{n-1} + q_{n-2}$ for $n \geq 1$, it is readily shown by induction that $q_n > \tau^{n-1}$ for $n > 1$, where $\tau = (1 + \sqrt{5})/2$.)

Since $q_n > 0$ for $n \geq 0$, we can rewrite (3), (4) in the forms

$$p_n/q_n - p_{n-1}/q_{n-1} = (-1)^{n+1}/q_{n-1} q_n \quad (n \geq 1), \tag{3'}$$

$$p_n/q_n - p_{n-2}/q_{n-2} = (-1)^n a_n / q_{n-2} q_n \quad (n \geq 2). \tag{4'}$$

It follows that the sequence $\{p_{2n}/q_{2n}\}$ is increasing, the sequence $\{p_{2n+1}/q_{2n+1}\}$ is decreasing, and every member of the first sequence is less than every member of the second sequence. Hence both sequences have limits and actually, since $q_n \to \infty$, the limits of the two sequences are the same.

From

$$\xi = (p_n \xi_{n+1} + p_{n-1})/(q_n \xi_{n+1} + q_{n-1})$$

we obtain

$$q_n \xi - p_n = (p_{n-1} q_n - p_n q_{n-1})/(q_n \xi_{n+1} + q_{n-1}) = (-1)^n/(q_n \xi_{n+1} + q_{n-1}).$$

Hence $\xi > p_n/q_n$ if n is even and $\xi < p_n/q_n$ if n is odd. It follows that $p_n/q_n \to \xi$ as $n \to \infty$. Consequently different irrational numbers have different continued fraction expansions.

Since ξ lies between p_n/q_n and p_{n+1}/q_{n+1}, we have

$$|p_{n+2}/q_{n+2} - p_n/q_n| < |\xi - p_n/q_n| < |p_{n+1}/q_{n+1} - p_n/q_n|.$$

By (3)' and (4)' we can rewrite this in the form

$$a_{n+2}/q_n q_{n+2} < |\xi - p_n/q_n| < 1/q_n q_{n+1} \quad (n \geq 0). \tag{5}$$

Hence

$$q_{n+2}^{-1} < |q_n \xi - p_n| < q_{n+1}^{-1},$$

which shows that $|q_n \xi - p_n|$ decreases as n increases. It follows that $|\xi - p_n/q_n|$ also decreases as n increases.

The rational number p_n/q_n is called the n-th *convergent* of ξ. The integers a_n are called the *partial quotients* and the real numbers ξ_n the *complete quotients* in the continued fraction expansion of ξ.

The continued fraction algorithm can be applied also when $\xi = \xi_0$ is rational, but in this case it is really the same as the Euclidean algorithm. For suppose $\xi_n = b_n/c_n$, where b_n and c_n are integers and $c_n > 0$. We can write

$$b_n = a_n c_n + c_{n+1},$$

where $a_n = \lfloor \xi_n \rfloor$ and c_{n+1} is an integer such that $0 \leq c_{n+1} < c_n$. Thus ξ_{n+1} is defined if and only if $c_{n+1} \neq 0$, and then $\xi_{n+1} = c_n/c_{n+1}$. Since the sequence of positive integers $\{c_n\}$ cannot decrease for ever, the continued fraction algorithm for a rational number ξ always terminates. At the last stage of the algorithm we have simply

$$\xi_N = a_N,$$

where $a_N > 1$ if $N > 0$. The uniquely determined finite sequence $[a_0, a_1, \ldots, a_N]$ is called the continued fraction expansion of ξ.

Convergents and *complete quotients* can be defined as before; all the properties derived for ξ irrational continue to hold for ξ rational, provided we do not go past $n = N$. The relation

$$\xi = (p_{N-1}\xi_N + p_{N-2})/(q_{N-1}\xi_N + q_{N-2})$$

now shows that

$$\xi = p_N/q_N.$$

Consequently different rational numbers have different continued fraction expansions.

Now let $a_0, a_1, a_2, \ldots$ be any infinite sequence of integers with $a_n \geq 1$ for $n \geq 1$. If we define integers p_n, q_n by the recurrence relations (1)–(2), our previous argument shows that the sequence $\{p_{2n}/q_{2n}\}$ is increasing, the sequence $\{p_{2n+1}/q_{2n+1}\}$ is decreasing, and the two sequences have a common limit ξ. If we put $\xi_0 = \xi$ and

$$\xi_{n+1} = -(q_{n-1}\xi - p_{n-1})/(q_n\xi - p_n) \quad (n \geq 0),$$

our previous argument shows also that $\xi_{n+1} > 1$ $(n \geq 0)$. Since

$$\xi_n = a_n + \xi_{n+1}^{-1},$$

it follows that $a_n = \lfloor \xi_n \rfloor$. Hence ξ is irrational and $[a_0, a_1, a_2, \ldots]$ is its continued fraction expansion.

Similarly it may be seen that, for any finite sequence of integers $a_0, a_1, \ldots, a_N$, with $a_n \geq 1$ for $1 \leq n < N$ and $a_N > 1$ if $N > 0$, there is a rational number ξ with $[a_0, a_1, \ldots, a_N]$ as its continued fraction expansion.

We will write simply $\xi = [a_0, a_1, \ldots, a_N]$ if ξ is rational and $\xi = [a_0, a_1, a_2, \ldots]$ if ξ is irrational.

We will later have need of the following result:

Lemma 0 *Let ξ be an irrational number with complete quotients ξ_n and convergents p_n/q_n. If η is any irrational number different from ξ, and if we define η_{n+1} by*

$$\eta = (p_n\eta_{n+1} + p_{n-1})/(q_n\eta_{n+1} + q_{n-1}),$$

then $-1 < \eta_n < 0$ for all large n.

Moreover, if $\xi > 1$ and $\eta < 0$, then $-1 < \eta_n < 0$ for all $n > 0$.

Proof We have

$$\eta_{n+1} = (q_{n-1}\eta - p_{n-1})/(p_n - q_n\eta).$$

Hence

$$\begin{aligned}
\theta_{n+1} &:= q_n\eta_{n+1} + q_{n-1} \\
&= (p_nq_{n-1} - p_{n-1}q_n)/(p_n - q_n\eta) \\
&= (-1)^{n+1}/(p_n - q_n\eta) \\
&= (-1)^n/q_n(\eta - p_n/q_n).
\end{aligned}$$

Since $p_n/q_n \to \xi \neq \eta$ and $q_n \to \infty$, it follows that $\theta_n \to 0$. Since

$$\eta_{n+1} = -(q_{n-1} - \theta_{n+1})/q_n,$$

we conclude that $-1 < \eta_{n+1} < 0$ for all large n.

Suppose now that $\xi > 1$ and $\eta < 0$. It is readily verified that $\eta_n = a_n + 1/\eta_{n+1}$. But $a_n = \lfloor \xi_n \rfloor \geq 1$ for all $n \geq 0$. Consequently $\eta_n < 0$ implies $1/\eta_{n+1} < -1$ and thus $-1 < \eta_{n+1} < 0$. Since $\eta_0 < 0$, it follows by induction that $-1 < \eta_n < 0$ for all $n > 0$. $\qquad\square$

The complete quotients of a real number may be characterized in the following way:

Proposition 1 *If $\eta > 1$ and*

$$\xi = (p\eta + p')/(q\eta + q'),$$

where p, q, p', q' are integers such that $pq' - p'q = \pm 1$ and $q > q' > 0$, then η is a complete quotient of ξ and p'/q', p/q are corresponding consecutive convergents of ξ.

Proof The relation $pq' - p'q = \pm 1$ implies that p and q are relatively prime. Since $q > 0$, p/q has a finite continued fraction expansion

$$p/q = [a_0, a_1, \ldots, a_{n-1}] = p_{n-1}/q_{n-1}$$

and $q = q_{n-1}$, $p = p_{n-1}$. In fact, since $q > 1$, we have $n > 1$, $a_{n-1} \geq 2$ and $q_{n-1} > q_{n-2}$. From

$$p_{n-1}q_{n-2} - p_{n-2}q_{n-1} = (-1)^n = \varepsilon(pq' - p'q),$$

where $\varepsilon = \pm 1$, we obtain

$$p_{n-1}(q_{n-2} - \varepsilon q') = q_{n-1}(p_{n-2} - \varepsilon p').$$

Hence q_{n-1} divides $q_{n-2} - \varepsilon q'$. Since $0 < q_{n-2} < q_{n-1}$ and $0 < q' < q_{n-1}$, it follows that $q' = q_{n-2}$ if $\varepsilon = 1$ and $q' = q_{n-1} - q_{n-2}$ if $\varepsilon = -1$. Hence $p' = p_{n-2}$ if $\varepsilon = 1$ and $p' = p_{n-1} - p_{n-2}$ if $\varepsilon = -1$. Thus

$$\xi = (p_{n-1}\eta + p_{n-2})/(q_{n-1}\eta + q_{n-2}),$$
$$\text{resp. } (p_{n-1}\eta + p_{n-1} - p_{n-2})/(q_{n-1}\eta + q_{n-1} - q_{n-2}).$$

Since $\eta > 1$, its continued fraction expansion has the form $[a_n, a_{n+1}, \ldots]$, where $a_n \geq 1$. It follows that ξ has the continued fraction expansion

$$[a_0, a_1, \ldots, a_{n-1}, a_n, \ldots], \quad \text{resp.} \quad [a_0, a_1, \ldots, a_{n-1} - 1, 1, a_n, \ldots].$$

In either case p'/q' and p/q are consecutive convergents of ξ and η is the corresponding complete quotient.   □

A complex number ζ is said to be *equivalent* to a complex number ω if there exist integers a, b, c, d with $ad - bc = \pm 1$ such that

$$\zeta = (a\omega + b)/(c\omega + d),$$

and *properly equivalent* if actually $ad - bc = 1$. Then ω is also equivalent, resp. properly equivalent, to ζ, since

$$\omega = (d\zeta - b)/(-c\zeta + a).$$

By taking $a = d = 1$ and $b = c = 0$, we see that any complex number ζ is properly equivalent to itself. It is not difficult to verify also that if ζ is equivalent to ω and ω equivalent to χ, then ζ is equivalent to χ, and the same holds with 'equivalence' replaced by 'proper equivalence'. Thus equivalence and proper equivalence are indeed 'equivalence relations'.

For any coprime integers b, d, there exist integers a, c such that $ad - bc = 1$. Since

$$b/d = (a \cdot 0 + b)/(c \cdot 0 + d),$$

it follows that any rational number is properly equivalent to 0, and hence any two rational numbers are properly equivalent. The situation is more interesting for irrational numbers:

Proposition 2 *Two irrational numbers ξ, η are equivalent if and only if their continued fraction expansions $[a_0, a_1, a_2, \ldots], [b_0, b_1, b_2, \ldots]$ have the same 'tails', i.e. there exist integers $m \geq 0$ and $n \geq 0$ such that*

$$a_{m+k} = b_{n+k} \quad \text{for all } k \geq 0.$$

Proof If the continued fraction expansions of ξ and η have the same tails, then some complete quotient ξ_m of ξ coincides with some complete quotient η_n of η. But ξ is equivalent to ξ_m, since $\xi = (p_{m-1}\xi_m + p_{m-2})/(q_{m-1}\xi_m + q_{m-2})$ and $p_{m-1}q_{m-2} - p_{m-2}q_{m-1} = (-1)^m$, and similarly η is equivalent to η_n. Hence ξ and η are equivalent.

Suppose on the other hand that ξ and η are equivalent. Then

$$\eta = (a\xi + b)/(c\xi + d)$$

for some integers a, b, c, d such that $ad - bc = \pm 1$. By changing the signs of all four we may suppose that $c\xi + d > 0$. From the relation

$$\xi = (p_{n-1}\xi_n + p_{n-2})/(q_{n-1}\xi_n + q_{n-2})$$

between ζ and its complete quotient ξ_n it follows that

$$\eta = (a_n\xi_n + b_n)/(c_n\xi_n + d_n),$$

where

$$a_n = ap_{n-1} + bq_{n-1}, \quad b_n = ap_{n-2} + bq_{n-2},$$
$$c_n = cp_{n-1} + dq_{n-1}, \quad d_n = cp_{n-2} + dq_{n-2},$$

and hence

$$a_nd_n - b_nc_n = (ad - bc)(p_{n-1}q_{n-2} - p_{n-2}q_{n-1}) = \pm 1.$$

The inequalities

$$|q_{n-1}\zeta - p_{n-1}| < 1/q_n, \quad |q_{n-2}\zeta - p_{n-2}| < 1/q_{n-1}$$

imply that

$$|c_n - (c\zeta + d)q_{n-1}| < |c|/q_n, \quad |d_n - (c\zeta + d)q_{n-2}| < |c|/q_{n-1}.$$

Since $c\zeta + d > 0$, $q_{n-1} > q_{n-2}$ and $q_n \to \infty$ as $n \to \infty$, it follows that $c_n > d_n > 0$ for sufficiently large n. Then, by Proposition 1, ξ_n is a complete quotient also of η. Thus the continued fraction expansions of ζ and η have a common tail. $\square$

2 Diophantine Approximation

The subject of *Diophantine approximation* is concerned with finding integer or rational solutions for systems of inequalities. For problems in one dimension the continued fraction algorithm is a most helpful tool, as we will now see.

Proposition 3 *Let $p_n/q_n(n \geq 1)$ be a convergent of the real number ζ. If p, q are integers such that $0 < q \leq q_n$ and $p \neq p_n$ if $q = q_n$, then*

$$|q\zeta - p| \geq |q_{n-1}\zeta - p_{n-1}| > |q_n\zeta - p_n|$$

and

$$|\zeta - p/q| > |\zeta - p_n/q_n|.$$

Proof It follows from (3) that the simultaneous linear equations

$$\lambda p_{n-1} + \mu p_n = p, \quad \lambda q_{n-1} + \mu q_n = q,$$

have integer solutions, namely

$$\lambda = (-1)^{n-1}(p_nq - q_np), \quad \mu = (-1)^n(p_{n-1}q - q_{n-1}p).$$

The hypotheses on p, q imply that $\lambda \neq 0$. If $\mu = 0$, then

$$|q\zeta - p| = |\lambda(q_{n-1}\zeta - p_{n-1})| \geq |q_{n-1}\zeta - p_{n-1}|.$$

Thus we now assume $\mu \neq 0$. Since $q \leq q_n$, λ and μ cannot both be positive and hence, since $q > 0$, $\lambda\mu < 0$. Then

$$q\xi - p = \lambda(q_{n-1}\xi - p_{n-1}) + \mu(q_n\xi - p_n)$$

and both terms on the right have the same sign. Hence

$$|q\xi - p| = |\lambda(q_{n-1}\xi - p_{n-1})| + |\mu(q_n\xi - p_n)|$$
$$\geq |q_{n-1}\xi - p_{n-1}|.$$

This proves the first statement of the proposition. The second statement follows, since

$$|\xi - p/q| = q^{-1}|q\xi - p| > q^{-1}|q_n\xi - p_n|$$
$$= (q_n/q)|\xi - p_n/q_n|$$
$$\geq |\xi - p_n/q_n|. \qquad \square$$

To illustrate the application of Proposition 3, consider the continued fraction expansion of $\pi = 3.14159265358\ldots$. We easily find that it begins $[3, 7, 15, 1, 292, \ldots]$. It follows that the first five convergents of π are

$$3/1, \quad 22/7, \quad 333/106, \quad 355/113, \quad 103993/33102.$$

Using the inequality $|\xi - p_n/q_n| < 1/q_nq_{n+1}$ and choosing $n = 3$ so that a_{n+1} is large, we obtain

$$0 < 355/113 - \pi < 0.000000267\cdots.$$

The approximation $355/113$ to π was first given by the Chinese mathematician Zu Chongzhi in the 5th century A.D. Proposition 3 shows that it is a better approximation to π than any other rational number with denominator ≤ 113.

In general, a rational number p'/q', where p', q' are integers and $q' > 0$, may be said to be a *best approximation* to a real number ξ if

$$|\xi - p/q| > |\xi - p'/q'|$$

for all different rational numbers p/q whose denominator q satisfies $0 < q \leq q'$. Thus Proposition 3 says that any convergent p_n/q_n $(n \geq 1)$ of ξ is a best approximation of ξ. However, these are not the only best approximations. It may be shown that, if p_{n-2}/q_{n-2} and p_{n-1}/q_{n-1} are consecutive convergents of ξ, then any rational number of the form

$$(cp_{n-1} + p_{n-2})/(cq_{n-1} + q_{n-2}),$$

where c is an integer such that $a_n/2 < c \leq a_n$ is a best approximation of ξ. Furthermore, every best approximation of ξ has this form if, when a_n is even, one allows also $c = a_n/2$.

It follows that $355/113$ is a better approximation to π than any other rational number with denominator less than 16604, since $292/2 = 146$ and $146 \times 113 + 106 = 16604$.

The complete continued fraction expansion of π is not known. However, it was discovered by Cotes (1714) and then proved by Euler (1737) that the complete continued fraction expansion of $e = 2.71828182459\ldots$ is given by $e - 1 = [1, 1, 2, 1, 1, 4, 1, 1, 6, \ldots]$.

The preceding results may also be applied to the construction of calendars. The solar year has a length of about 365.24219 mean solar days. The continued fraction expansion of $\lambda = (0.24219)^{-1}$ begins $[4, 7, 1, 3, 24, \ldots]$. Hence the first five convergents of λ are

$$4/1, \quad 29/7, \quad 33/8, \quad 128/31, \quad 3105/752.$$

It follows that

$$0 < 128/31 - \lambda < 0.0000428$$

and 128/31 is a better approximation to λ than any other rational number with denominator less than 380. The Julian calendar, by adding a day every 4 years, estimated the year at 365.25 days. The Gregorian calendar, by adding 97 days every 400 years, estimates the year at 365.2425 days. Our analysis shows that, if we added instead 31 days every 128 years, we would obtain the much more precise estimate of 365.2421875 days.

Best approximations also find an application in the selection of gear ratios, and continued fractions were already used for this purpose by Huygens (1682) in constructing his planetarium (a mechanical model for the solar system).

The next proposition describes another way in which the continued fraction expansion provides good rational approximations.

Proposition 4 *If p, q are coprime integers with $q > 0$ such that, for some real number ξ,*

$$|\xi - p/q| < 1/2q^2,$$

then p/q is a convergent of ξ.

Proof Let p_n/q_n be the convergents of ξ and assume that p/q is not a convergent. We show first that $q < q_N$ for some $N > 0$. This is obvious if ξ is irrational. If $\xi = p_N/q_N$ is rational, then

$$1/q_N \leq |qp_N - pq_N|/q_N = |q\xi - p| < 1/2q.$$

Hence $q < q_N$ and $N > 0$.

It follows that $q_{n-1} \leq q < q_n$ for some $n > 0$. By Proposition 3,

$$|q_{n-1}\xi - p_{n-1}| \leq |q\xi - p| < 1/2q.$$

Hence

$$\begin{aligned}
1/qq_{n-1} &\leq |qp_{n-1} - pq_{n-1}|/qq_{n-1} \\
&= |p_{n-1}/q_{n-1} - p/q| \\
&\leq |p_{n-1}/q_{n-1} - \xi| + |\xi - p/q| \\
&< 1/2qq_{n-1} + 1/2q^2.
\end{aligned}$$

But this implies $q < q_{n-1}$, which is a contradiction. $\qquad\qquad \square$

As an application of Proposition 4 we prove

Proposition 5 *Let d be a positive integer which is not a square and m an integer such that $0 < m^2 < d$. If x, y are positive integers such that*

$$x^2 - dy^2 = m,$$

then x/y is a convergent of the irrational number $\sqrt{d}$.

Proof Suppose first that $m > 0$. Then $x/y > \sqrt{d}$ and

$$0 < x/y - \sqrt{d} = m/(xy + y^2\sqrt{d}) < \sqrt{d}/2y^2\sqrt{d} = 1/2y^2.$$

Hence x/y is a convergent of $\sqrt{d}$, by Proposition 4.

Suppose next that $m < 0$. Then $y/x > 1/\sqrt{d}$ and

$$0 < y/x - 1/\sqrt{d} = -m/d(xy + x^2/\sqrt{d}) < 1/\sqrt{d}(xy + x^2/\sqrt{d}) < 1/2x^2.$$

Hence y/x is a convergent of $1/\sqrt{d}$. But, since $1/\sqrt{d} = 0 + 1/\sqrt{d}$, the convergents of $1/\sqrt{d}$ are $0/1$ and the reciprocals of the convergents of $\sqrt{d}$.   □

In the next section we will show that the continued fraction expansion of $\sqrt{d}$ has a particularly simple form.

It was shown by Vahlen (1895) that at least one of any two consecutive convergents of ξ satisfies the inequality of Proposition 4. Indeed, since consecutive convergents lie on opposite sides of ξ,

$$|p_n/q_n - \xi| + |p_{n-1}/q_{n-1} - \xi| = |p_n/q_n - p_{n-1}/q_{n-1}|$$
$$= 1/q_n q_{n-1} \leq 1/2q_n^2 + 1/2q_{n-1}^2,$$

with equality only if $q_n = q_{n-1}$. This proves the assertion, except when $n = 1$ and $q_1 = q_0 = 1$. But in this case $a_1 = 1, 1 \leq \xi_1 < 2$ and hence

$$|\xi - p_1/q_1| = |\xi - a_0 - 1| = 1 - \xi_1^{-1} < 1/2.$$

It was shown by Borel (1903) that at least one of any three consecutive convergents of ξ satisfies the sharper inequality

$$|\xi - p/q| < 1/\sqrt{5}q^2.$$

In fact this is obtained by taking $r = 1$ in the following more general result, due to Forder (1963) and Wright (1964).

Proposition 6 *Let ξ be an irrational number with the continued fraction expansion $[a_0, a_1, \ldots]$ and the convergents p_n/q_n. If, for some positive integer r,*

$$|\xi - p_n/q_n| \geq 1/(r^2 + 4)^{1/2}q_n^2 \quad for \ n = m - 1, m, m + 1,$$

then $a_{m+1} < r$.

Proof If we put $s = (r^2 + 4)^{1/2}/2$, then s is irrational. For otherwise $2s$ would be an integer and from $(2s + r)(2s - r) = 4$ we would obtain $2s + r = 4, 2s - r = 1$ and hence $r = 3/2$, which is a contradiction.

By the hypotheses of the proposition,

$$1/q_{m-1}q_m = |p_{m-1}/q_{m-1} - p_m/q_m| = |\xi - p_{m-1}/q_{m-1}| + |\xi - p_m/q_m|$$
$$\geq (q_{m-1}^{-2} + q_m^{-2})/2s$$

and hence

$$q_m^2 - 2s q_{m-1} q_m + q_{m-1}^2 \leq 0.$$

Furthermore, this inequality also holds when q_{m-1}, q_m are replaced by q_m, q_{m+1}. Consequently q_{m-1}/q_m and q_{m+1}/q_m both satisfy the inequality $t^2 - 2st + 1 \leq 0$. Since

$$t^2 - 2st + 1 = (t - s + r/2)(t - s - r/2),$$

it follows that

$$s - r/2 < q_{m-1}/q_m < q_{m+1}/q_m < s + r/2,$$

the first and last inequalities being strict because s is irrational. Hence

$$a_{m+1} = q_{m+1}/q_m - q_{m-1}/q_m < s + r/2 - (s - r/2) = r. \qquad \square$$

It follows from Proposition 6 with $r = 1$ that, for any irrational number ξ, there exist infinitely many rational numbers $p/q = p_n/q_n$ such that

$$|\xi - p/q| < 1/\sqrt{5}q^2.$$

Here the constant $\sqrt{5}$ is best possible. For take any $c > \sqrt{5}$. If there exists a rational number p/q, with $q > 0$ and $(p, q) = 1$, such that

$$|\xi - p/q| < 1/cq^2,$$

then p/q is a convergent of ξ, by Proposition 4. But for any convergent p_n/q_n we have

$$|\xi - p_n/q_n| = 1/q_n(q_n \xi_{n+1} + q_{n-1}).$$

If we take $\xi = \tau := (1 + \sqrt{5})/2$, then also $\xi_{n+1} = \tau$ and $p_n = q_{n+1}$. Hence

$$|\tau - q_{n+1}/q_n| = 1/q_n^2(\tau + q_{n-1}/q_n),$$

where $\tau + q_{n-1}/q_n \to \tau + \tau^{-1} = \sqrt{5}$, since $q_n/q_{n-1} \to \tau$. Thus, for any $c > \sqrt{5}$, there exist at most finitely many rational numbers p/q such that

$$|\tau - p/q| < 1/cq^2.$$

It follows from Proposition 6 with $r = 2$ that if

$$|\xi - p_n/q_n| \geq 1/\sqrt{8}q_n^2 \quad \text{for all large } n,$$

then $a_n = 1$ for all large n. The constant $\sqrt{8}$ is again best possible, since a similar argument to that just given shows that if $\sigma := 1 + \sqrt{2} = [2, 2, \ldots]$ then, for any $c > \sqrt{8}$, there exist at most finitely many rational numbers p/q such that

$$|\sigma - p/q| < 1/cq^2.$$

It follows from Proposition 6 with $r = 3$ that if

$$|\xi - p_n/q_n| \geq 1/\sqrt{13}q_n^2 \quad \text{for all large } n,$$

then $a_n \in \{1, 2\}$ for all large n.

For any irrational ξ, with continued fraction expansion $[a_0, a_1, \ldots]$ and convergents p_n/q_n, put

$$M(\xi) = \varlimsup_{n \to \infty} q_n^{-1}|q_n\xi - p_n|^{-1}.$$

It follows from Proposition 2 that $M(\xi) = M(\eta)$ if ξ and η are equivalent. The results just established show that $M(\xi) \geq \sqrt{5}$ for every ξ. If $M(\xi) < \sqrt{8}$, then $a_n = 1$ for all large n; hence ξ is equivalent to τ and $M(\xi) = M(\tau) = \sqrt{5}$. If $M(\xi) < \sqrt{13}$, then $a_n \in \{1, 2\}$ for all large n.

An irrational number ξ is said to be *badly approximable* if $M(\xi) < \infty$. The inequalities

$$a_{n+2}/q_nq_{n+2} < |\xi - p_n/q_n| < 1/q_nq_{n+1}$$

imply

$$a_{n+1} \leq q_{n+1}/q_n < q_n^{-1}|q_n\xi - p_n|^{-1}$$

and

$$q_n^{-1}|q_n\xi - p_n|^{-1} < q_{n+2}/a_{n+2}q_n \leq q_{n+1}/q_n + 1 \leq a_{n+1} + 2.$$

Hence ξ is badly approximable if and only if its partial quotients a_n are bounded.

It is obvious that ξ is badly approximable if there exists a constant $c > 0$ such that

$$|\xi - p/q| > c/q^2$$

for every rational number p/q. Conversely, if ξ is badly approximable, then there exists such a constant $c > 0$. This is clear when p and q are coprime integers, since if p/q is *not* a convergent of ξ then, by Proposition 4,

$$|\xi - p/q| \geq 1/2q^2.$$

On the other hand, if $p = \lambda p'$, $q = \lambda q'$, where p', q' are coprime, then

$$|\xi - p/q| = |\xi - p'/q'| \geq c/q'^2 = \lambda^2 c/q^2 \geq c/q^2.$$

Some of the applications of badly approximable numbers stem from the following characterization: a real number θ is badly approximable if and only if there exists a constant $c' > 0$ such that

$$|e^{2\pi iq\theta} - 1| \geq c'/q \quad \text{for all } q \in \mathbb{N}.$$

To establish this, put $q\theta = p + \delta$, where $p \in \mathbb{Z}$ and $|\delta| \leq 1/2$. Then

$$|e^{2\pi i q\theta} - 1| = 2|\sin \pi q\theta| = 2|\sin \pi \delta|$$

and the result follows from the previous characterization, since $(\sin x)/x$ decreases from 1 to $2/\pi$ as x increases from 0 to $\pi/2$.

3 Periodic Continued Fractions

A complex number ζ is said to be a *quadratic irrational* if it is a root of a monic quadratic polynomial $t^2 + rt + s$ with rational coefficients r, s, but is not itself rational. Since $\zeta \notin \mathbb{Q}$, the rational numbers r, s are uniquely determined by ζ.

Equivalently, ζ is a quadratic irrational if it is a root of a quadratic polynomial

$$f(t) = At^2 + Bt + C$$

with integer coefficients A, B, C such that $B^2 - 4AC$ is not the square of an integer. The integers A, B, C are uniquely determined up to a common factor and are uniquely determined up to sign if we require that they have greatest common divisor 1. The corresponding integer $D = B^2 - 4AC$ is then uniquely determined and is called the *discriminant* of ζ. A quadratic irrational is real if and only if its discriminant is positive.

It is readily verified that if a quadratic irrational ζ is equivalent to a complex number ω, i.e. if

$$\zeta = (\alpha\omega + \beta)/(\gamma\omega + \delta),$$

where $\alpha, \beta, \gamma, \delta \in \mathbb{Z}$ and $\alpha\delta - \beta\gamma = \pm 1$, then ω is also a quadratic irrational. Moreover, if ζ is a root of the quadratic polynomial $f(t) = At^2 + Bt + C$, where A, B, C are integers with greatest common divisor 1, then ω is a root of the quadratic polynomial

$$g(t) = A't^2 + B't + C',$$

where

$$A' = \alpha^2 A + \alpha\gamma B + \gamma^2 C,$$
$$B' = 2\alpha\beta A + (\alpha\delta + \beta\gamma)B + 2\gamma\delta C,$$
$$C' = \beta^2 A + \beta\delta B + \delta^2 C,$$

and hence

$$B'^2 - 4A'C' = B^2 - 4AC = D.$$

Since

$$A = \delta^2 A' - \gamma\delta B' + \gamma^2 C',$$
$$B = -2\beta\delta A' + (\alpha\delta + \beta\gamma)B' - 2\alpha\gamma C',$$
$$C = \beta^2 A' - \alpha\beta B' + \alpha^2 C',$$

A', B', C' also have greatest common divisor 1.

If ζ is a quadratic irrational, we define the *conjugate* ζ' of ζ to be the other root of the quadratic polynomial $f(t)$ which has ζ as a root. If

$$\zeta = (\alpha\omega + \beta)/(\gamma\omega + \delta),$$

where $\alpha, \beta, \gamma, \delta \in \mathbb{Z}$ and $\alpha\delta - \beta\gamma = \pm 1$, then evidently

$$\zeta' = (\alpha\omega' + \beta)/(\gamma\omega' + \delta).$$

Suppose now that $\zeta = \xi$ is real and that the integers A, B, C are uniquely determined by requiring not only $(A, B, C) = 1$ but also $A > 0$. The real quadratic irrational ξ is said to be *reduced* if $\xi > 1$ and $-1 < \xi' < 0$. If ξ is reduced then, since $\xi > \xi'$, we must have

$$\xi = (-B + \sqrt{D})/2A, \quad \xi' = (-B - \sqrt{D})/2A.$$

Thus the inequalities $\xi > 1$ and $-1 < \xi' < 0$ imply

$$0 < \sqrt{D} + B < 2A < \sqrt{D} - B.$$

Conversely, if the coefficients A, B, C of $f(t)$ satisfy these inequalities, where $D = B^2 - 4AC > 0$, then one of the roots of $f(t)$ is reduced. For $B < 0 < A$ and so the roots ξ, ξ' of $f(t)$ have opposite signs. If ξ is the positive root, then ξ and ξ' are given by the preceding formulas and hence $\xi > 1, -1 < \xi' < 0$. It should be noted also that if ξ is reduced, then $B^2 < D$ and hence $C < 0$.

We return now to continued fractions. If ξ is a real quadratic irrational, then its complete quotients ξ_n are all quadratic irrationals and, conversely, if some complete quotient ξ_n is a quadratic irrational, then ξ is also a quadratic irrational.

The continued fraction expansion $[a_0, a_1, a_2, \ldots]$ of a real number ξ is said to be *eventually periodic* if there exist integers $m \geq 0$ and $h > 0$ such that

$$a_n = a_{n+h} \quad \text{for all } n \geq m.$$

The continued fraction expansion is then conveniently denoted by

$$[a_0, a_1, \ldots, a_{m-1}, \overline{a_m, \ldots, a_{m+h-1}}].$$

The continued fraction expansion is said to be *periodic* if it is eventually periodic with $m = 0$.

Equivalently, the continued fraction expansion of ξ is eventually periodic if $\xi_m = \xi_{m+h}$ for some $m \geq 0$ and $h > 0$, and periodic if this holds with $m = 0$. The *period* of the continued fraction expansion, in either case, is the least positive integer h with this property.

We are going to show that there is a close connection between real quadratic irrationals and eventually periodic continued fractions.

Proposition 7 *A real number ξ is a reduced quadratic irrational if and only if its continued fraction expansion is periodic.*

Moreover, if $\xi = [\overline{a_0, \ldots, a_{h-1}}]$, then $-1/\xi' = [\overline{a_{h-1}, \ldots, a_0}]$.

Proof Suppose first that $\xi = [\overline{a_0, \ldots, a_{h-1}}]$ has a periodic continued fraction expansion. Then $a_0 = a_h \geq 1$ and hence $\xi > 1$. Furthermore, since

$$\xi = (p_{h-1}\xi_h + p_{h-2})/(q_{h-1}\xi_h + q_{h-2})$$

and $\xi_h = \xi$, ξ is an irrational root of the quadratic polynomial

$$f(t) = q_{h-1}t^2 + (q_{h-2} - p_{h-1})t - p_{h-2}.$$

Thus ξ is a quadratic irrational. Since $f(0) = -p_{h-2} < 0$ and

$$f(-1) = q_{h-1} - q_{h-2} + p_{h-1} - p_{h-2} > 0$$

(even for $h = 1$), it follows that $-1 < \xi' < 0$. Thus ξ is reduced.

If ξ is a reduced quadratic irrational, then its complete quotients ξ_n, which are all quadratic irrationals, are also reduced, by Lemma 0 with $\eta = \xi'$. Since $\xi'_n = a_n + 1/\xi'_{n+1}$ and $-1 < \xi'_n < 0$, we have

$$a_n = \lfloor -1/\xi'_{n+1} \rfloor.$$

Thus ξ_n, ξ'_n are the roots of a uniquely determined polynomial

$$f_n(t) = A_n t^2 + B_n t + C_n,$$

where A_n, B_n, C_n are integers with greatest common divisor 1 and $A_n > 0$. Furthermore, $D = B_n^2 - 4A_n C_n$ is independent of n and positive. Since ξ_n is reduced, we have

$$\xi_n = (-B_n + \sqrt{D})/2A_n, \quad \xi'_n = (-B_n - \sqrt{D})/2A_n,$$

where

$$0 < \sqrt{D} + B_n < 2A_n < \sqrt{D} - B_n.$$

If we put $g = \lfloor \sqrt{D} \rfloor$, then $-B_n \in \{1, \ldots, g\}$ and, for a given value of B_n, there are at most $-B_n$ possible values for A_n. Consequently the number of distinct pairs A_n, B_n does not exceed $1 + \cdots + g = g(g+1)/2$. Hence we must have

$$\xi_j = \xi_k, \quad \xi'_j = \xi'_k$$

for some j, k such that $0 \leq j < k \leq g(g+1)/2$. If $j = 0$, this already proves that the continued fraction expansion of ξ is periodic. If $j > 0$, then

$$a_{j-1} = \lfloor -1/\xi'_j \rfloor = \lfloor -1/\xi'_k \rfloor = a_{k-1}$$

and hence

$$\xi_{j-1} = a_{j-1} + 1/\xi_j = a_{k-1} + 1/\xi_k = \xi_{k-1}.$$

Repeating this argument j times, we obtain $\xi_0 = \xi_{k-j}$. Thus ξ has a periodic continued fraction expansion in any case.

If the period is h, so that $\xi = [\overline{a_0, \ldots, a_{h-1}}]$, then $\xi'_0 = \xi'_h$ and the relation $a_n = \lfloor -1/\xi'_{n+1} \rfloor$ implies that $-1/\xi' = [\overline{a_{h-1}, \ldots, a_0}]$. $\qquad \square$

The proof of Proposition 7 shows that the period is at most $g(g+1)/2$ and thus is certainly less than D. By counting the pairs of integers A, B for which not only

$$0 < \sqrt{D} + B < 2A < \sqrt{D} - B,$$

but also $D \equiv B^2 \bmod 4A$, it may be shown that the period is at most $O(\sqrt{D} \log D)$. (The *Landau order symbol* used here is defined under 'Notations'.)

Proposition 8 *A real number ξ is a quadratic irrational if and only if its continued fraction expansion is eventually periodic.*

Proof Suppose first that the continued fraction expansion of ξ is eventually periodic. Then some complete quotient ξ_m has a periodic continued fraction expansion and hence is a quadratic irrational, by Proposition 7. But this implies that ξ also is a quadratic irrational.

Suppose next that ξ is a quadratic irrational. We will prove that the continued fraction expansion of ξ is eventually periodic by showing that some complete quotient ξ_{n+1} is reduced. Since we certainly have $\xi_{n+1} > 1$, we need only show that $-1 < \xi'_{n+1} < 0$. But $\xi' \neq \xi$ and $\xi' = (p_n \xi'_{n+1} + p_{n-1})/(q_n \xi'_{n+1} + q_{n-1})$. Hence, by Lemma 0, $-1 < \xi'_{n+1} < 0$ for all large n. $\square$

It follows from Proposition 8 that any real quadratic irrational is badly approximable, since its partial quotients are bounded. It follows from Propositions 7 and 8 that there are only finitely many inequivalent quadratic irrationals with a given discriminant $D > 0$, since any real quadratic irrational is equivalent to a reduced one and only finitely many pairs of integers A, B satisfy the inequalities

$$0 < \sqrt{D} + B < 2A < \sqrt{D} - B.$$

Proposition 8 is due to Euler and Lagrange. It was first shown by Euler (1737) that a real number is a quadratic irrational if its continued fraction expansion is eventually periodic, and the converse was proved by Lagrange (1770). Proposition 7 was first stated and proved by Galois (1829), although it was implicit in the work of Lagrange (1773) on the reduction of binary quadratic forms. Proposition 7 provides a simple proof of the following result due to Legendre:

Proposition 9 *For any real number ξ, the following two conditions are equivalent:*

(i) $\xi > 1$, ξ is irrational and ξ^2 is rational;
(ii) *the continued fraction expansion of ξ has the form* $[a_0, \overline{a_1, \ldots, a_h}]$, *where* $a_h = 2a_0$ *and* $a_i = a_{h-i}$ *for* $i = 1, \ldots, h-1$.

Proof Suppose first that (i) holds. Then ξ is a quadratic irrational, since it is a root of the polynomial $t^2 - \xi^2$. The continued fraction expansion of ξ cannot be periodic, by Proposition 7, since $\xi' = -\xi < -1$. However, the continued fraction expansion of ξ_1 is periodic, since $\xi_1 > 1$ and $1/\xi'_1 = \xi' - a_0 < -1$. Thus $\xi_1 = [\overline{a_1, \ldots, a_h}]$ for some $h \geq 1$. By Proposition 7 also,

$$-1/\xi'_1 = [\overline{a_h, \ldots, a_1}].$$

But

$$-1/\xi_1' = \xi + a_0 = [2a_0, \overline{a_1, \ldots, a_h}].$$

Comparing this with the previous expression, we see that (ii) holds.

Suppose, conversely, that (ii) holds. Then ξ is irrational, $a_0 > 0$ and hence $\xi > 1$. Moreover $\xi_1 = [\overline{a_1, \ldots, a_h}]$ is a reduced quadratic irrational and

$$-1/\xi_1' = [\overline{a_h, \ldots, a_1}] = [2a_0, \overline{a_1, \ldots, a_h}] = a_0 + \xi.$$

Hence $\xi' = a_0 + 1/\xi_1' = -\xi$ and $\xi^2 = -\xi\xi'$ is rational. $\square$

4 Quadratic Diophantine Equations

We are interested in finding all integers x, y such that

$$ax^2 + bxy + cy^2 + dx + ey + f = 0, \tag{6}$$

where $a, \ldots, f$ are given integers. Writing (6) as a quadratic equation for x,

$$ax^2 + (by + d)x + cy^2 + ey + f = 0,$$

we see that if a solution exists for some y, then the discriminant

$$(by + d)^2 - 4a(cy^2 + ey + f)$$

must be a perfect square. Thus

$$(b^2 - 4ac)y^2 + 2(bd - 2ae)y + d^2 - 4af = z^2$$

for some integer z. If we put

$$p := b^2 - 4ac, \quad q := bd - 2ae, \quad r := d^2 - 4af,$$

we have a quadratic equation for y,

$$py^2 + 2qy + r - z^2 = 0,$$

whose discriminant must also be a perfect square. Thus

$$q^2 - p(r - z^2) = w^2$$

for some integer w. Thus if (6) has a solution in integers, so also does the equation

$$w^2 - pz^2 = q^2 - pr.$$

Moreover, from all solutions in integers of the latter equation we may obtain, by retracing our steps, all solutions in integers of the original equation (6).

Thus we now restrict our attention to finding all integers x, y such that

$$x^2 - dy^2 = m, \tag{7}$$

where d and m are given integers.

The equation (7) has the remarkable property, which was known to Brahmagupta (628) and later rediscovered by Euler (1758), that if we have solutions for two values m_1, m_2 of m, then we can derive a solution for their product $m_1 m_2$. This follows from the identity

$$(x_1^2 - dy_1^2)(x_2^2 - dy_2^2) = x^2 - dy^2,$$

where

$$x = x_1 x_2 + d y_1 y_2, \quad y = x_1 y_2 + y_1 x_2.$$

(In fact, Brahmagupta's identity is just a restatement of the norm relation $N(\alpha\beta) = N(\alpha)N(\beta)$ for elements α, β of a quadratic field.) In particular, from two solutions of the equation

$$x^2 - dy^2 = 1, \tag{8}$$

a third solution can be obtained by *composition* in this way.

Composition of solutions is evidently commutative and associative. In fact the solutions of (8) form an abelian group under composition, with the trivial solution $1, 0$ as identity element and the solution $x, -y$ as the inverse of the solution x, y. Also, by composing an arbitrary solution x, y of (8) with the trivial solution $-1, 0$ we obtain the solution $-x, -y$.

Suppose first that $d < 0$. Evidently (7) is insoluble if $m < 0$ and $x = y = 0$ is the only solution if $m = 0$. If $m > 0$, there are at most finitely many solutions and we may find them all by testing, for each non-negative integer $y \leq (-m/d)^{1/2}$, whether $m + dy^2$ is a perfect square.

Suppose now that $d > 0$. If $d = e^2$ is a perfect square, then (7) is equivalent to the finite set of simultaneous linear Diophantine equations

$$x - ey = m', \quad x + ey = m'',$$

where m', m'' are any integers such that $m'm'' = m$. Thus *we now suppose also that d is not a perfect square.* Then $\xi = \sqrt{d}$ is irrational.

If $0 < m^2 < d$ then, by Proposition 5, any positive solution x, y of (7) has the form $x = p_n, y = q_n$, where p_n/q_n is a convergent of ξ. In particular, all positive solutions of $x^2 - dy^2 = \pm 1$ are obtained in this way.

On the other hand, as we now show, if p_n/q_n is any convergent of ξ then

$$|p_n^2 - dq_n^2| < 2\sqrt{d}.$$

If $n = 0$, then $|p_0^2 - dq_0^2| = |a_0^2 - d|$, where $a_0 < \sqrt{d} < a_0 + 1$ and so $0 < d - a_0^2 < \sqrt{d} + a_0 < 2\sqrt{d}$. Now suppose $n > 0$. Then $|p_n - q_n\xi| < q_{n+1}^{-1}$ and hence

$$|p_n^2 - dq_n^2| = |p_n - q_n\xi||p_n - q_n\xi + 2q_n\xi|$$
$$< q_{n+1}^{-1}(q_{n+1}^{-1} + 2q_n\xi) < 2\xi.$$

An easy congruence argument shows that the equation

$$x^2 - dy^2 = -1 \tag{9}$$

has no solutions in integers unless $d \equiv 1 \bmod 4$ or $d \equiv 2 \bmod 8$. It will now be shown that the equation (8), on the other hand, always has solutions in positive integers.

Proposition 10 *Let d be a positive integer which is not a perfect square. Suppose* $\xi = \sqrt{d}$ *has complete quotients* ξ_n, *convergents* p_n/q_n, *and continued fraction expansion* $[a_0, \overline{a_1, \ldots, a_h}]$ *of period h.*

Then $p_n^2 - dq_n^2 = \pm 1$ *if and only if* $n = kh - 1$ *for some integer* $k > 0$ *and in this case*

$$p_{kh-1}^2 - dq_{kh-1}^2 = (-1)^{kh}.$$

Proof From $\xi = (p_n\xi_{n+1} + p_{n-1})/(q_n\xi_{n+1} + q_{n-1})$ we obtain

$$(p_n - q_n\xi)\xi_{n+1} = q_{n-1}\xi - p_{n-1}.$$

Multiplying by $(-1)^{n+1}(p_n + q_n\xi)$, we get

$$s_n\xi_{n+1} = \xi + r_n,$$

where

$$s_n = (-1)^{n+1}(p_n^2 - dq_n^2), \quad r_n = (-1)^n(p_{n-1}p_n - dq_{n-1}q_n).$$

Thus s_n and r_n are integers. Moreover, since $\xi_{n+1+kh} = \xi_{n+1}$ and ξ is irrational, $s_{n+kh} = s_n$ and $r_{n+kh} = r_n$ for all positive integers k.

If $p_n^2 - dq_n^2 = \pm 1$, then actually $p_n^2 - dq_n^2 = (-1)^{n+1}$, since p_n/q_n is less than or greater than ξ according as n is even or odd. Hence $s_n = 1$ and $\xi_{n+1} = \xi + r_n$. Taking integral parts, we get $a_{n+1} = a_0 + r_n$. Consequently

$$\xi_{n+2}^{-1} = \xi_{n+1} - a_{n+1} = \xi - a_0 = \xi_1^{-1}.$$

Thus $\xi_{n+2} = \xi_1$, which implies that $n = kh - 1$ for some positive integer k.

On the other hand, if $n = kh - 1$ for some positive integer k, then $\xi_{n+2} = \xi_1$ and hence

$$\xi_{n+1} - a_{n+1} = \xi - a_0.$$

Thus $\xi_{n+1} = \xi + a_{n+1} - a_0$, which implies that $s_n = 1$, since ξ is irrational. $\quad\square$

It follows from Proposition 10 that, if d is a positive integer which is not a perfect square, then the equation (8) always has a solution in positive integers and all such solutions are given by

$$x = p_{kh-1}, \quad y = q_{kh-1} \quad (k = 1, 2, \ldots) \text{ if } h \text{ is even,}$$
$$x = p_{2kh-1}, \quad y = q_{2kh-1} \quad (k = 1, 2, \ldots) \text{ if } h \text{ is odd.}$$

The least solution in positive integers, obtained by taking $k = 1$, is called the *fundamental solution* of (8).

On the other hand, the equation (9) has a solution in positive integers if and only if h is odd and all such solutions are then given by

$$x = p_{kh-1}, \quad y = q_{kh-1} \quad (k = 1, 3, 5, \ldots).$$

The least solution in positive integers, obtained by taking $k = 1$, is called the *fundamental solution* of (9).

To illustrate these results, suppose $d = a^2 + 1$ for some $a \in \mathbb{N}$. Since $\sqrt{d} = [a, \overline{2a}]$, the equation $x^2 - dy^2 = -1$ has the fundamental solution $x = a, y = 1$ and the equation $x^2 - dy^2 = 1$ has the fundamental solution $x = 2a^2 + 1, y = 2a$. Again, suppose $d = a^2 + a$ for some $a \in \mathbb{N}$. Since $\sqrt{d} = [a, \overline{2, 2a}]$, the equation $x^2 - dy^2 = -1$ is insoluble, but the equation $x^2 - dy^2 = 1$ has the fundamental solution $x = 2a + 1, y = 2$.

It is not difficult to obtain upper bounds for the fundamental solutions. Since $\xi = \sqrt{d}$ is a root of the polynomial $t^2 - d$ and since its complete quotients ξ_n are reduced for $n \geq 1$, they have the form

$$\xi_n = (-B_n + \sqrt{D})/2A_n,$$

where $D = 4d, 0 < -B_n < \sqrt{D}$ and $A_n \geq 1$. Therefore $a_0 = \lfloor \xi \rfloor < \sqrt{d}$ and $a_n = \lfloor \xi_n \rfloor < 2\sqrt{d}$ for $n \geq 1$. If we put $\alpha = \lfloor \sqrt{d} \rfloor$, it is easily shown by induction that

$$p_n \leq (\alpha + \alpha^{-1})^{n+1}/2, \quad q_n \leq (\alpha + \alpha^{-1})^n \quad (n \geq 0).$$

These inequalities may now be combined with any upper bound for the period h (cf. §3).

Under composition, the fundamental solution of (8) generates an infinite cyclic group $\mathscr{C}$ of solutions of (8). Furthermore, by composing the fundamental solution of (9) with any element of $\mathscr{C}$ we obtain infinitely many solutions of (9). We are going to show that, by composing also with the trivial solution $-1, 0$ of (8), all integral solutions of (8) and (9) are obtained in this way. This can be proved by means of continued fractions, but the following argument due to Nagell (1950) provides additional information.

Proposition 11 *Let d be a positive integer which is not a perfect square, let m be a positive integer, and let x_0, y_0 be the fundamental solution of the equation (8).*
If the equation

$$u^2 - dv^2 = m \tag{10}$$

has an integral solution, then it actually has one for which $u^2 \leq m(x_0 + 1)/2$, $dv^2 \leq m(x_0 - 1)/2$.
Similarly, if the equation

$$u^2 - dv^2 = -m \tag{11}$$

has an integral solution, then it actually has one for which $u^2 \leq m(x_0 - 1)/2$, $dv^2 \leq m(x_0 + 1)/2$.

Proof By composing a given solution of (10) with any solution in the subgroup $\mathscr{C}$ of solutions of (8) which is generated by the solution x_0, y_0 we obtain again a solution of (10). Let u_0, v_0 be the solution of (10) obtained in this way for which v_0 has its least non-negative value. Then $u_0^2 = m + dv_0^2$ also has its least value and by changing the sign of u_0 we may suppose $u_0 > 0$. By composing the solution u_0, v_0 of (10) with the inverse of the fundamental solution x_0, y_0 of (8) we obtain the solution

$$u = x_0 u_0 - d y_0 v_0, \quad v = x_0 v_0 - y_0 u_0$$

of (10). Since

$$u = x_0 u_0 - d y_0 v_0 = x_0 u_0 - [(x_0^2 - 1)(u_0^2 - m)]^{1/2} > 0,$$

we must have

$$x_0 u_0 - d y_0 v_0 \geq u_0.$$

Hence

$$(x_0 - 1)^2 u_0^2 \geq d^2 y_0^2 v_0^2 = (x_0^2 - 1)(u_0^2 - m).$$

Thus

$$(x_0 - 1)/(x_0 + 1) \geq 1 - m/u_0^2,$$

which implies $u_0^2 \leq m(x_0 + 1)/2$ and hence $d v_0^2 \leq m(x_0 - 1)/2$.

For the equation (11) we begin in the same way. Then from

$$(x_0 v_0)^2 = (y_0^2 + 1/d)(u_0^2 + m) > y_0^2 u_0^2$$

we obtain $v = x_0 v_0 - y_0 u_0 > 0$ and hence $x_0 v_0 - y_0 u_0 \geq v_0$. Thus

$$d(x_0 - 1)^2 v_0^2 \geq d y_0^2 u_0^2$$

and hence

$$(x_0 - 1)^2 (u_0^2 + m) \geq (x_0^2 - 1) u_0^2.$$

The argument can now be completed in the same way as before. $\qquad \square$

The proof of Proposition 11 shows that if (10), or (11), has an integral solution, then we obtain all solutions by finding the *finitely many* solutions u, v which satisfy the inequalities in the statement of Proposition 11 and composing them with all solutions in $\mathscr{C}$ of (8).

The only solutions x, y of (8) for which $x^2 \leq (x_0 + 1)/2$ are the trivial ones $x = \pm 1, y = 0$. Hence any solution of (8) is in $\mathscr{C}$ or is obtained by reversing the signs of a solution in $\mathscr{C}$.

If u, v is a positive solution of (9) such that $u^2 \leq (x_0 - 1)/2, dv^2 \leq (x_0 + 1)/2$, then $x = u^2 + dv^2$, $y = 2uv$ is a positive solution of (8) such that $x \leq x_0$. Hence $(x, y) = (x_0, y_0)$ is the fundamental solution of (8) and $u^2 = (x_0 - 1)/2$, $dv^2 = (x_0 + 1)/2$. Thus (u, v) is uniquely determined and is the fundamental solution of (9). Hence, if (9) has a solution, any solution is obtained by composing the fundamental solution of (9) with an element of $\mathscr{C}$ or by reversing the signs of such a solution.

A necessary condition for the solubility in integers of the equation (9) is that d may be represented as a sum of two squares. For the period h of the continued fraction expansion $\xi = \sqrt{d} = [a_0, \overline{a_1, \ldots, a_h}]$ must be odd, say $h = 2m + 1$. It follows from Proposition 9 that

$$\xi_{m+1} = [\overline{a_m, \ldots, a_1, 2a_0, a_1 \ldots, a_m}],$$

and then from Proposition 7 that $\zeta_{m+1} = -1/\zeta'_{m+1}$. But, by the proof of Proposition 10,

$$s_m \zeta_{m+1} = \zeta + r_m,$$

where s_m and r_m are integers. Hence

$$-1 = \zeta_{m+1} \zeta'_{m+1} = (\zeta + r_m)(-\zeta + r_m)/s_m^2 = (r_m^2 - d)/s_m^2,$$

and thus $d = r_m^2 + s_m^2$. The formulas for s_m and r_m show that, if p_n/q_n are the convergents of $\sqrt{d}$, then $d = x^2 + y^2$ with

$$x = p_{m-1} p_m - d q_{m-1} q_m, \quad y = p_m^2 - d q_m^2.$$

Unfortunately, the equation (9) may be insoluble, even though d is a sum of two squares. As an example, take $d = 34 = 5^2 + 3^2$. It is easily verified that the fundamental solution of the equation $x^2 - 34y^2 = 1$ is $x_0 = 35$, $y_0 = 6$. If the equation $u^2 - 34v^2 = -1$ were soluble in integers, then, by Proposition 11, it would have a solution u, v such that $34v^2 \leq 18$, which is clearly impossible.

As already observed, the equation (9) has no integral solutions if $d \equiv 3 \bmod 4$. It will now be shown that (9) does have integral solutions if $d = p$ is prime and $p \equiv 1 \bmod 4$. For let x, y be the fundamental solution of the equation (8). Since any square is congruent to 0 or 1 mod 4, we must have $y^2 \equiv 0$ and $x^2 \equiv 1$. Thus $y = 2z$ for some positive integer z and

$$(x - 1)(x + 1) = 4pz^2.$$

Since x is odd, $x - 1$ and $x + 1$ have greatest common divisor 2. It follows that there exist positive integers u, v such that

$$either \; x - 1 = 2pu^2, \; x + 1 = 2v^2 \quad or \quad x - 1 = 2u^2, \; x + 1 = 2pv^2.$$

In the first case $v^2 - pu^2 = 1$, which contradicts the choice of x, y as the fundamental solution of (8), since $v < x$. Thus only the second case is possible and then $u^2 - pv^2 = -1$. (In fact, u, v is the fundamental solution of (9).)

This proves again that *any prime $p \equiv 1 \bmod 4$ may be represented as a sum of two squares,* and moreover shows that an explicit construction for this representation is provided by the continued fraction expansion of $\sqrt{p}$.

The representation of a prime $p \equiv 1 \bmod 4$ in the form $x^2 + y^2$ is actually unique, apart from interchanging x and y and changing their signs. For suppose

$$x^2 + y^2 = p = u^2 + v^2,$$

where x, y, u, v are all positive integers. Then

$$y^2 u^2 - x^2 v^2 = (p - x^2)u^2 - x^2(p - u^2) = p(u^2 - x^2).$$

Hence $yu \equiv \varepsilon xv \bmod p$, where $\varepsilon = \pm 1$. On the other hand,

$$p^2 = (x^2 + y^2)(u^2 + v^2) = (xu + \varepsilon yv)^2 + (xv - \varepsilon yu)^2.$$

Since the second term on the right is divisible by p^2, we must have $xv = \varepsilon yu$ or $xu = -\varepsilon yv$. Evidently $\varepsilon = 1$ in the first case and $\varepsilon = -1$ in the second case. Since $(x, y) = (u, v) = 1$, it follows that either $x = u, y = v$ or $x = v, y = u$.

The equation $x^2 - dy^2 = 1$, where d is a positive integer which is not a perfect square, is generally known as *Pell's equation*, following an erroneous attribution of Euler. The problem of finding its integral solutions was issued as a challenge by Fermat (1657). In the same year Brouncker and Wallis gave a method of solution which is essentially the same as the solution by continued fractions. The first complete proof that a nontrivial solution always exists was given by Lagrange (1768).

Unknown to them all, the problem had been considered centuries earlier by Hindu mathematicians. Special cases of Pell's equation were solved by Brahmagupta (628) and a general method of solution, which was described by Bhascara II (1150), was known to Jayadeva at least a century earlier. No proofs were given, but their method is a modification of the solution by continued fractions and is often faster in practice. Bhascara found the fundamental solution of the equation $x^2 - 61y^2 = 1$, namely

$$x = 1766319049, \quad y = 226153980,$$

a remarkable achievement for the era.

5 The Modular Group

We recall that a complex number w is said to be *equivalent* to a complex number z if there exist integers a, b, c, d with $ad - bc = \pm 1$ such that

$$w = (az + b)/(cz + d).$$

Since we can write

$$w = (az + b)(c\bar{z} + d)/|cz + d|^2,$$

the imaginary parts are related by

$$\mathscr{I}w = (ad - bc)\mathscr{I}z/|cz + d|^2.$$

Consequently $\mathscr{I}w$ and $\mathscr{I}z$ have the same sign if $ad - bc = 1$ and opposite signs if $ad - bc = -1$. Since the map $z \to -z$ interchanges the upper and lower half-planes, we may restrict attention to z's in the *upper half-plane* $\mathscr{H} = \{z \in \mathbb{C} : \mathscr{I}z > 0\}$ and to w's which are *properly equivalent* to them, i.e. with $ad - bc = 1$.

A *modular transformation* is a map $f : \mathscr{H} \to \mathscr{H}$ of the form

$$f(z) = (az + b)/(cz + d),$$

where $a, b, c, d \in \mathbb{Z}$ and $ad - bc = 1$. Such a map is bijective and its inverse is again a modular transformation:

$$f^{-1}(z) = (dz - b)/(-cz + a).$$

Furthermore, if

$$g(z) = (a'z + b')/(c'z + d')$$

is another modular transformation, then the composite map $h = g \circ f$ is again a modular transformation:

$$h(z) = (a''z + b'')/(c''z + d''),$$

where

$$a'' = a'a + b'c, \quad b'' = a'b + b'd,$$
$$c'' = c'a + d'c, \quad d'' = c'b + d'd,$$

and hence

$$a''d'' - b''c'' = (a'd' - b'c')(ad - bc) = 1.$$

It follows that the set Γ of all modular transformations is a group. Moreover, composition of modular transformations corresponds to multiplication of the corresponding matrices:

$$\begin{pmatrix} a'' & b'' \\ c'' & d'' \end{pmatrix} = \begin{pmatrix} a' & b' \\ c' & d' \end{pmatrix} \begin{pmatrix} a & b \\ c & d \end{pmatrix}.$$

However, the same modular transformation is obtained if the signs of a, b, c, d are all changed (and in no other way). It follows that the *modular group* Γ is isomorphic to the factor group $SL_2(\mathbb{Z})/\{\pm I\}$ of the special linear group $SL_2(\mathbb{Z})$ of all 2×2 integer matrices with determinant 1 by its centre $\{\pm I\}$.

Proposition 12 *The modular group Γ is generated by the transformations*

$$T(z) = z + 1, \quad S(z) = -1/z.$$

Proof It is evident that $S, T \in \Gamma$ and $S^2 = I$ is the identity transformation. Any $g \in \Gamma$ has the form

$$g(z) = (az + b)/(cz + d),$$

where $a, b, c, d \in \mathbb{Z}$ and $ad - bc = 1$. If $c = 0$, then $a = d = \pm 1$ and $g = T^m$, where $m = b/d \in \mathbb{Z}$. Similarly if $a = 0$, then $b = -c = \pm 1$ and $g = ST^m$, where $m = d/c \in \mathbb{Z}$. Suppose now that $ac \neq 0$. For any $n \in \mathbb{Z}$ we have

$$ST^{-n}g(z) = (a'z + b')/(c'z + d'),$$

where $a' = -c, b' = -d, c' = a - nc$ and $d' = b - nd$. We can choose $n = m_1$ so that for $g_1 = ST^{-m_1}g$ we have $|c'| < |a|$ and hence $|a'| + |c'| < |a| + |c|$. If $a'c' \neq 0$, the argument can be repeated with g_1 in place of g. After finitely many repetitions we must obtain

$$ST^{-m_k} \cdots ST^{-m_1}g = T^m \text{ or } ST^m.$$

Since $S^{-1} = S$ and $(T^n)^{-1} = T^{-n}$, it follows that

$$g = T^{m_1} S \cdots T^{m_k} S T^m \text{ or } g = T^{m_1} S \cdots T^{m_k} T^m. \qquad \square$$

The proof of Proposition 12 may be regarded as an analogue of the continued fraction algorithm, since

$$T^{m_1} S \cdots T^{m_k} S T^m z = m_1 - \cfrac{1}{m_2 - \cfrac{\ddots}{\; - \cfrac{1}{m_k - \cfrac{1}{m + z}}}}.$$

Obviously Γ is also generated by S and $R := ST$. The transformation R has order 3, since

$$R(z) = -1/(z+1), \quad R^2(z) = -(z+1)/z, \quad R^3(z) = z.$$

We are going to show that all other relations between the generators S and R are consequences of the relations $S^2 = R^3 = I$, so that Γ is the *free product* of a cyclic group of order 2 and a cyclic group of order 3.

Partition the upper half-plane $\mathscr{H}$ by putting

$$A = \{z \in \mathscr{H} : \mathscr{R}z < 0\}, \quad B = \{z \in \mathscr{H} : \mathscr{R}z \geq 0\}.$$

It is easily verified that

$$SA \subset B, \quad RB \subset A, \quad R^2 B \subset A$$

(where the inclusions are strict). If $g' = SR^{\varepsilon_1} SR^{\varepsilon_2} \cdots SR^{\varepsilon_n}$ for some $n \geq 1$, where $\varepsilon_j \in \{1, 2\}$, it follows that $g'B \subset B$ and $g'SA \subset B$. Similarly, if $g'' = R^{\varepsilon_1} S \cdots R^{\varepsilon_n}$, then $g''B \subset A$ and $g''SA \subset A$. By taking account of the relations $S^2 = R^3 = I$, every $g \in \Gamma$ can be written in one of the forms

$$I, \ S, \ g', \ g'', \ g'S, \ g''S.$$

But, by what has just been said, no element except the first is the identity transformation.

The modular group is *discrete*, since there exists a neighbourhood of the identity transformation which contains no other element of Γ.

Proposition 13 *The open set*

$$F = \{z \in \mathscr{H} : -1/2 < \mathscr{R}z < 1/2, |z| > 1\}$$

(see Figure 1) is a fundamental domain for the modular group Γ*, i.e. distinct points of* F *are not equivalent and each point of* $\mathscr{H}$ *is equivalent to some point of* F *or its boundary* ∂F.

Proof For any $z \in \mathbb{C}$ we write $z = x + iy$, where $x, y \in \mathbb{R}$. We show first that no two points of F are equivalent. Assume on the contrary that there exist distinct points $z, z' \in F$ with $y' \geq y$ such that

$$z' = (az + b)/(cz + d)$$

for some $a, b, c, d \in \mathbb{Z}$ with $ad - bc = 1$. If $c = 0$, then $a = d = \pm 1$, $b \neq 0$ and $z' = z + b/d$, which is impossible for $z, z' \in F$. Hence $c \neq 0$. Since

$$y' = y/|cz + d|^2,$$

we have $|cz + d| \leq 1$. Thus $|z + d/c| \leq 1/|c|$, which is impossible not only if $|c| \geq 2$ but also if $c = \pm 1$.

We now show that any $z_0 \in \mathcal{H}$ is equivalent to a point of the *closure* $\bar{F} = F \cup \partial F$. We can choose $m_0 \in \mathbb{Z}$ so that $z_1 = z_0 + m_0$ satisfies $|x_1| \leq 1/2$. If $|z_1| \geq 1$, there is nothing more to do. Thus we now suppose $|z_1| < 1$. Put $z_2 = -1/z_1$. Then

$$y_2 = y_1/|z_1|^2 > y_1$$

and actually $y_2 \geq 2y_1$ if $y_1 \leq 1/2$, since then $|z_1|^2 \leq 1/4 + 1/4 = 1/2$. We now repeat the process, with z_2 in place of z_0, and choose $m_2 \in \mathbb{Z}$ so that $z_3 = z_2 + m_2$ satisfies $|x_3| \leq 1/2$. From $z_3 = (m_2 z_1 - 1)/z_1$ we obtain

$$|z_3|^2 = \{(m_2 x_1 - 1)^2 + (m_2 y_1)^2\}/(x_1^2 + y_1^2).$$

Assume $|z_3| < 1$. Then $m_2 \neq 0$ and also $m_2 \neq \pm 1$, since $|1 \pm x_1| \geq 1/2 \geq |x_1|$. If $|m_2| \geq 2$, then $|z_3|^2 \geq 4|y_1|^2$ and hence $y_1 < 1/2$. Thus in passing from z_1 to z_3 we obtain either $z_3 \in \bar{F}$ or $y_3 = y_2 \geq 2y_1$. Hence, after repeating the process finitely many times we must obtain a point $z_{2k+1} \in \bar{F}$. $\square$

Proposition 13 implies that the sets $\{g(\bar{F}) : g \in \Gamma\}$ form a *tiling* of $\mathcal{H}$, since

$$\mathcal{H} = \bigcup_{g \in \Gamma} g(\bar{F}), \quad g(F) \cap g'(F) = \emptyset \text{ if } g, g' \in \Gamma \text{ and } g \neq g'.$$

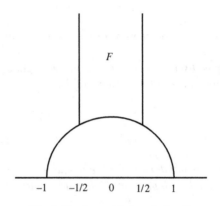

Fig. 1. Fundamental domain for Γ.

This is illustrated in Figure 2, where the domain $g(F)$ is represented simply by the group element g.

There is an interesting connection between the modular group and binary quadratic forms. The *discriminant* of a binary quadratic form

$$f = ax^2 + bxy + cy^2$$

with coefficients $a, b, c \in \mathbb{R}$ is $D := b^2 - 4ac$. The quadratic form is *indefinite* (i.e. assumes both positive and negative values) if and only if $D > 0$, and *positive definite* (i.e. assumes only positive values unless $x = y = 0$) if and only if $D < 0$, $a > 0$, which implies also $c > 0$. (If $D = 0$, the quadratic form is proportional to the square of a linear form.)

If we make a linear change of variables

$$x = \alpha x' + \beta y', \quad y = \gamma x' + \delta y',$$

where $\alpha, \beta, \gamma, \delta \in \mathbb{Z}$ and $\alpha\delta - \beta\gamma = 1$, the quadratic form f is transformed into the quadratic form

$$f' = a'x'^2 + b'x'y' + c'y'^2,$$

where

$$a' = a\alpha^2 + b\alpha\gamma + c\gamma^2,$$
$$b' = 2a\alpha\beta + b(\alpha\delta + \beta\gamma) + 2c\gamma\delta,$$
$$c' = a\beta^2 + b\beta\delta + c\delta^2,$$

and hence

$$b'^2 - 4a'c' = b^2 - 4ac = D.$$

The quadratic forms f and f' are said to be *properly equivalent*.

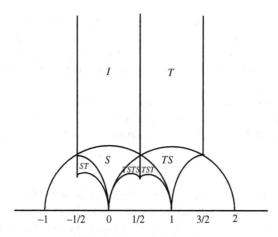

Fig. 2. Tiling of $\mathscr{H}$ by Γ.

Thus properly equivalent forms have the same discriminant. As the name implies, proper equivalence is indeed an equivalence relation. Moreover, any form properly equivalent to an indefinite form is again indefinite, and any form properly equivalent to a positive definite form is again positive definite.

We will now show that any binary quadratic form is properly equivalent to one which is in some sense canonical. The indefinite and positive definite cases will be treated separately.

Suppose first that f is positive definite, so that $D < 0$, $a > 0$ and $c > 0$. With the quadratic form f we associate a point $\tau(f)$ of the upper half-plane $\mathcal{H}$, namely

$$\tau(f) = (-b + i\sqrt{-D})/2a.$$

Thus $\tau(f)$ is the root with positive imaginary part of the polynomial $at^2 + bt + c$. Conversely, for any given $D < 0$ and $\tau \in \mathcal{H}$, there is a unique positive definite quadratic form f with discriminant D such that $\tau(f) = \tau$. In fact, if $\tau = \xi + i\eta$, where $\xi, \eta \in \mathbb{R}$ and $\eta > 0$, we must take

$$a = \sqrt{(-D)}/2\eta, \quad b = -2a\xi, \quad c = (b^2 - D)/4a.$$

Let f', as above, be a form properly equivalent to f. If $t = (\alpha t' + \beta)/(\gamma t' + \delta)$, then

$$at^2 + bt + c = (a't'^2 + b't' + c')/(\gamma t' + \delta)^2.$$

It follows that if $\tau = \tau(f)$ and $\tau' = \tau(f')$, then $\tau = (\alpha \tau' + \beta)/(\gamma \tau' + \delta)$. Thus τ' is properly equivalent to τ, in the terminology introduced in Section 1.

By Proposition 13 we may choose the change of variables so that $\tau' \in \bar{F}$, i.e.

$$-1/2 \leq \mathcal{R}\tau' \leq 1/2, \quad |\tau'| \geq 1.$$

It is easily verified that this is the case if and only if for f' we have

$$|b'| \leq a', \quad 0 < a' \leq c'.$$

Such a quadratic form f' is said to be *reduced*. Thus every positive definite binary quadratic form is properly equivalent to a reduced form. (It is possible to ensure that every positive definite binary quadratic form is properly equivalent to a unique reduced form by slightly restricting the definition of 'reduced', but we will have no need of this.)

If the coefficients of f are integers, then so also are the coefficients of f' and τ, τ' are complex quadratic irrationals. There are only finitely many reduced forms f with integer coefficients and with a given discriminant $D < 0$. For, if f is reduced, then

$$4b^2 \leq 4a^2 \leq 4ac = b^2 - D$$

and hence $b^2 \leq -D/3$. Since $4ac = b^2 - D$, for each of the finitely many possible values of b there are only finitely many possible values for a and c.

A quadratic form $f = ax^2 + bxy + cy^2$ is said to be *primitive* if the coefficients a, b, c are integers with greatest common divisor 1. For any integer $D < 0$, let $h^{\dagger}(D)$

denote the number of primitive positive definite quadratic forms with discriminant D which are properly inequivalent. By what has been said, $h^\dagger(D)$ is finite.

Consider next the indefinite case:

$$f = ax^2 + bxy + cy^2$$

where $a, b, c \in \mathbb{R}$ and $D > 0$. If $a \neq 0$, we can write

$$f = a(x - \xi y)(x - \eta y),$$

where ξ, η are the distinct real roots of the polynomial $at^2 + bt + c$. It follows from Lemma 0 that, if ξ and η are irrational, then f is properly equivalent to a form f' for which $\xi' > 1$ and $-1 < \eta' < 0$. Such a quadratic form f' is said to be *reduced*. Evidently f' is reduced if and only if $-f'$ is reduced. Thus we may suppose $a' > 0$, and then f' is reduced if and only if

$$0 < \sqrt{D} + b' < 2a' < \sqrt{D} - b'.$$

If the coefficients of f are integers and the positive integer D is not a square, then $a \neq 0$ and ξ, η are conjugate real quadratic irrationals. In this case, as we already saw in Section 3, there are only finitely many reduced forms with discriminant D. For any integer $D > 0$ which is not a square, let $h^\dagger(D)$ denote the number of primitive quadratic forms with discriminant D which are properly inequivalent. By what has been said, $h^\dagger(D)$ is finite.

It should be noted that, for any quadratic form f with integer coefficients, the discriminant $D \equiv 0$ or $1 \bmod 4$. Moreover, for any $D \equiv 0$ or $1 \bmod 4$, there is a quadratic form f with integer coefficients and with discriminant D; for example,

$$f = x^2 - Dy^2/4 \qquad \text{if } D \equiv 0 \bmod 4,$$
$$f = x^2 + xy + (1 - D)y^2/4 \qquad \text{if } D \equiv 1 \bmod 4.$$

The preceding results for quadratic forms can also be restated in terms of quadratic fields. By making correspond to the ideal with basis $\beta = a, \gamma = b + c\omega$ in the quadratic field $\mathbb{Q}(\sqrt{d})$ the binary quadratic form

$$\{\beta\beta' x^2 + (\beta\gamma' + \beta'\gamma)xy + \gamma\gamma' y^2\}/ac,$$

one can establish a bijective map between 'strict' equivalence classes of ideals in $\mathbb{Q}(\sqrt{d})$ and proper equivalence classes of binary quadratic forms with discriminant D, where

$$D = 4d \quad \text{if } d \equiv 2 \text{ or } 3 \bmod 4,$$
$$D = d \quad \text{if } d \equiv 1 \bmod 4.$$

(The middle coefficient b of $f = ax^2 + bxy + cy^2$ was not required to be even in order to obtain this one-to-one correspondence.) Since any ideal class is either a strict ideal class or the union of two strict ideal classes, the finiteness of the class number $h(d)$ of the quadratic field $\mathbb{Q}(\sqrt{d})$ thus follows from the finiteness of $h^\dagger(D)$.

6 Non-Euclidean Geometry

There is an important connection between the modular group and the non-Euclidean geometry of Bolyai (1832) and Lobachevski (1829). It was first pointed out by Beltrami (1868) that their *hyperbolic geometry* is the geometry on a manifold of constant curvature. In the model of Poincaré (1882) for two-dimensional hyperbolic geometry the underlying space is taken to be the upper half-plane $\mathcal{H}$. A 'line' is either a semi-circle with centre on the real axis or a half-line perpendicular to the real axis. It follows that through any two distinct points there passes exactly one 'line'. However, through a given point not on a given 'line' there passes more than one 'line' having no point in common with the given 'line'.

Although Euclid's parallel axiom fails to hold, all the other axioms of Euclidean geometry are satisfied. Poincaré's model shows that if Euclidean geometry is free from contradiction, then so also is hyperbolic geometry. Before the advent of non-Euclidean geometry there had been absolute faith in Euclidean geometry. It is realized today that it is a matter for experiment to determine what kind of geometry best describes our physical world.

Poincaré's model will now be examined in more detail (with the constant curvature normalized to have the value -1). A curve γ in $\mathcal{H}$ is specified by a continuously differentiable function $z(t) = x(t) + iy(t)$ $(a \leq t \leq b)$. The (hyperbolic) *length* of γ is defined to be

$$\ell(\gamma) = \int_a^b y(t)^{-1}|dz/dt|dt.$$

It follows from this definition that the 'line' segment joining two points z, w of $\mathcal{H}$ has length

$$d(z, w) = \ln \frac{|z - \bar{w}| + |z - w|}{|z - \bar{w}| - |z - w|}.$$

It may be shown that any other curve joining z and w has greater length. Thus the 'lines' are *geodesics*.

For any $z_0 \in \mathcal{H}$, there is a unique geodesic through z_0 in any specified direction. Also, for any distinct real numbers ξ, η, there is a unique geodesic which intersects the real axis at ξ, η, namely the semicircle with centre at $(\xi + \eta)/2$. (By abuse of language we say 'ξ', for example, when we mean the point $(\xi, 0)$.)

A linear fractional transformation

$$z' = f(z) = (az + b)/(cz + d),$$

where $a, b, c, d \in \mathbb{R}$ and $ad - bc = 1$, maps the upper half-plane $\mathcal{H}$ onto itself and maps 'lines' onto 'lines'. Moreover, if the curve γ is mapped onto the curve γ', then $\ell(\gamma) = \ell(\gamma')$, since $\mathscr{I}f(z) = \mathscr{I}z/|cz + d|^2$ and $df/dz = 1/|cz + d|^2$. In particular,

$$d(z, w) = d(z', w').$$

Thus a linear fractional transformation of the above form is an *isometry*. It may be shown that any isometry is either a linear fractional transformation of this form or is obtained by composing such a transformation with the (orientation-reversing) transformation $x + iy \to -x + iy$. For any two 'lines' L and L', there is an isometry which maps L onto L'.

We may define *angles* to be the same as in Euclidean geometry, since any linear fractional transformation is conformal. The (hyperbolic) *area* of a domain $D \subset \mathcal{H}$, defined by

$$\mu(D) = \iint_D y^{-2} dx dy,$$

is invariant under any isometry. In particular, this gives $\pi - (\alpha + \beta + \gamma)$ for the area of a 'triangle' with angles α, β, γ. Since the angles are non-negative, the area of a 'triangle' is at most π and, since the area is necessarily positive, the sum of the angles of a 'triangle' is less than π.

For example, if F is the fundamental domain of the modular group Γ, then $\bar{F}$ is a 'triangle' with angles $\pi/3, \pi/3, 0$ and hence the area of $\bar{F}$ is $\pi - 2\pi/3 = \pi/3$. For any fixed $z_0 \in F$ on the imaginary axis, we may characterize F as the set of all $z \in \mathcal{H}$ such that, for every $g \in \Gamma$ with $g \neq I$,

$$d(z, z_0) < d(z, g(z_0)) = d(g^{-1}(z), z_0).$$

By identifying two points z, z' of $\mathcal{H}$ if $z' = g(z)$ for some $g \in \Gamma$ we obtain the *quotient space* $\mathcal{M} = \mathcal{H}/\Gamma$. Equivalently, we may regard $\mathcal{M}$ as the closure $\bar{F}$ of the fundamental domain F with the boundary point $-1/2 + iy$ identified with the boundary point $1/2 + iy$ ($1 \leq y < \infty$) and the boundary point $-e^{-i\theta}$ identified with the boundary point $e^{i\theta}$ ($0 < \theta < \pi/2$).

Since the elements of Γ are isometries of $\mathcal{H}$, the metric on $\mathcal{H}$ induces a metric on $\mathcal{M}$ in which the geodesics are the projections onto $\mathcal{M}$ of the geodesics in $\mathcal{H}$. Thus if we regard $\mathcal{M}$ as $\bar{F}$ with appropriate boundary points identified, then a geodesic in $\mathcal{M}$ will be a sequence of geodesic arcs in F, each with initial point and endpoint on the boundary of F, so that the initial point of one arc is the point identified to the endpoint of the preceding arc.

Let L be a geodesic in $\mathcal{H}$ which intersects the real axis in irrational points ξ, η such that $\xi > 1, -1 < \eta < 0$ and let

$$\xi = [a_0, a_1, a_2, \ldots], \quad -1/\eta = [a_{-1}, a_{-2}, \ldots]$$

be the continued fraction expansions of ξ and $-1/\eta$. If we choose ξ and $\eta = \xi'$ to be conjugate quadratic irrationals then, by Proposition 7, the doubly-infinite sequence

$$[\ldots, a_{-2}, a_{-1}, a_0, a_1, a_2, \ldots]$$

is periodic and it is not difficult to see that the geodesic in $\mathcal{M}$ obtained by projection from L is closed. Artin (1924) showed that there are other geodesics which behave very differently. Let the convergents of ξ be p_n/q_n and put

$$\xi = (p_{n-1}\xi_n + p_{n-2})/(q_{n-1}\xi_n + q_{n-2}), \quad \eta = (p_{n-1}\eta_n + p_{n-2})/(q_{n-1}\eta_n + q_{n-2}).$$

Then

$$\xi_n = [a_n, a_{n+1}, \ldots], \quad -1/\eta_n = [a_{n-1}, a_{n-2}, \ldots],$$

and $\xi_n > 1$, $-1 < \eta_n < 0$. Moreover, if n is even, then ξ and η are properly equivalent to ξ_n and η_n respectively. If we choose ξ so that the sequence $a_0, a_1, a_2, \ldots$ contains each finite sequence of positive integers (and hence contains it infinitely often), then the corresponding geodesic in $\mathcal{M}$ passes arbitrarily close to every point of $\mathcal{M}$ and to every direction at that point.

Some much-studied subgroups of the modular group are the *congruence subgroups* $\Gamma(n)$, consisting of all linear fractional transformations $z \to (az + b)/(cz + d)$ in Γ congruent to the identity transformation, i.e.

$$a \equiv d \equiv \pm 1, \quad b \equiv c \equiv 0 \mod n.$$

We may in the same way investigate the geodesics in the *quotient space* $\mathcal{H}/\Gamma(n)$. In the case $n = 3$ it has been shown by Lehner and Sheingorn (1984) that there is an interesting connection with the *Markov spectrum*.

In Section 2 we defined, for any irrational number ξ with convergents p_n/q_n,

$$M(\xi) = \overline{\lim_{n \to \infty}} \, q_n^{-1} |q_n \xi - p_n|^{-1},$$

and we noted that $M(\xi) = M(\eta)$ if ξ and η are equivalent. It is not difficult to show that there are uncountably many inequivalent ξ for which $M(\xi) = 3$. However, it was shown by Markov (1879/80) that there is a sequence of real quadratic irrationals $\xi^{(k)}$ such that $M(\xi) < 3$ if and only if ξ is equivalent to $\xi^{(k)}$ for some k. If $\mu_k = M(\xi^{(k)})$, then $\mu_1 < \mu_2 < \mu_3 < \cdots$ and $\mu_k \to 3$ as $k \to \infty$. Although μ_k is irrational, μ_k^2 is rational. The first few values are

$$\mu_1 = 5^{1/2} = 2.236\ldots, \quad \mu_2 = 8^{1/2} = 2.828\ldots,$$
$$\mu_3 = (221)^{1/2}/5 = 2.973\ldots, \quad \mu_4 = (1517)^{1/2}/13 = 2.996\ldots.$$

As we already showed in Section 2, we can take $\xi^{(1)} = (1 + \sqrt{5})/2$ and $\xi^{(2)} = 1 + \sqrt{2}$.

Lehner and Sheingorn showed that the simple closed geodesics in $\mathcal{H}/\Gamma(3)$ are just the projections of the geodesics in $\mathcal{H}$ whose endpoints ξ, η on the real axis are conjugate quadratic irrationals equivalent to $\xi^{(k)}$ for some k.

There is a recursive procedure for calculating the quantities μ_k and $\xi^{(k)}$. A *Markov triple* is a triple (u, v, w) of positive integers such that

$$u^2 + v^2 + w^2 = 3uvw.$$

If (u, v, w) is a Markov triple, then so also are $(3uw - v, u, w)$ and $(3uv - w, u, v)$. They are distinct from the original triple if $u = \max(u, v, w)$, since then $u < 3uw - v$ and $u < 3uv - w$. They are also distinct from one another if $w < v$. Starting from the trivial triple $(1, 1, 1)$, all Markov triples can be obtained by repeated applications of this process. The successive values of $u = \max(u, v, w)$ are $1, 2, 5, 13, 29, \ldots$. The numbers μ_k and $\xi^{(k)}$ are the corresponding successive values of $(9 - 4/u^2)^{1/2}$ and $(9 - 4/u^2)^{1/2}/2 + 1/2 + v/uw$.

It was conjectured by Frobenius (1913) that a Markov triple is uniquely determined by its greatest element. This has been verified whenever the greatest element does not exceed 10^{140}. It has also been proved when the greatest element is a prime (and in some other cases) by Baragar (1996), using the theory of quadratic fields.

7 Complements

There is an important analogue of the continued fraction algorithm for infinite series. Let K be an arbitrary field and let F denote the set of all formal Laurent series

$$f = \sum_{n \in \mathbb{Z}} \alpha_n t^n$$

with coefficients $\alpha_n \in K$ such that $\alpha_n \neq 0$ for at most finitely many $n > 0$. If

$$g = \sum_{n \in \mathbb{Z}} \beta_n t^n$$

is also an element of F, and if we define addition and multiplication by

$$f + g = \sum_{n \in \mathbb{Z}} (\alpha_n + \beta_n) t^n, \quad fg = \sum_{n \in \mathbb{Z}} \gamma_n t^n,$$

where $\gamma_n = \sum_{j+k=n} \alpha_j \beta_k$, then F acquires the structure of a commutative ring. In fact, F is a field. For, if $f = \sum_{n \leq v} \alpha_n t^n$, where $\alpha_v \neq 0$, we obtain $g = \sum_{n \leq -v} \beta_n t^n$ such that $fg = 1$ by solving successively the equations

$$\alpha_v \beta_{-v} = 1$$
$$\alpha_v \beta_{-v-1} + \alpha_{v-1} \beta_{-v} = 0$$
$$\alpha_v \beta_{-v-2} + \alpha_{v-1} \beta_{-v-1} + \alpha_{v-2} \beta_{-v} = 0$$
$$\ldots \ldots$$

Define the absolute value of an element $f = \sum_{n \in \mathbb{Z}} \alpha_n t^n$ of F by putting

$$|0| = 0, \quad |f| = 2^{v(f)} \quad \text{if } f \neq 0,$$

where $v(f)$ is the greatest integer n such that $\alpha_n \neq 0$. It is easily verified that

$$|fg| = |f||g|, |f + g| \leq \max(|f|, |g|),$$

and $|f + g| = \max(|f|, |g|)$ if $|f| \neq |g|$.
   For any $f = \sum_{n \in \mathbb{Z}} \alpha_n t^n \in F$, let

$$\lfloor f \rfloor = \sum_{n \geq 0} \alpha_n t^n, \quad \{f\} = \sum_{n < 0} \alpha_n t^n$$

denote respectively its polynomial and strictly proper parts. Then $|\{f\}| < 1$, and $|\lfloor f \rfloor| = |f|$ if $|f| \geq 1$, i.e. if $\lfloor f \rfloor \neq 0$.

If $f_0 := f$ is not the formal Laurent series of a rational function, we can write

$$f_0 = a_0 + 1/f_1,$$

where $a_0 = \lfloor f_0 \rfloor$ and $|f_1| > 1$. In the same way,

$$f_1 = a_1 + 1/f_2,$$

where $a_1 = \lfloor f_1 \rfloor$ and $|f_2| > 1$. Continuing in this way, we obtain the *continued fraction expansion* $[a_0, a_1, a_2, \ldots]$ of f. In the same way as for real numbers, if we define polynomials p_n, q_n by the recurrence relations

$$p_n = a_n p_{n-1} + p_{n-2}, \quad q_n = a_n q_{n-1} + q_{n-2} \quad (n \geq 0),$$

with $p_{-2} = q_{-1} = 0$, $p_{-1} = q_{-2} = 1$, then

$$p_n q_{n-1} - p_{n-1} q_n = (-1)^{n+1} \quad (n \geq 0),$$
$$f = (p_n f_{n+1} + p_{n-1})/(q_n f_{n+1} + q_{n-1}) \quad (n \geq 0),$$

and so on. In addition, however, we now have

$$|a_n| = |f_n| > 1 \quad (n \geq 1),$$

from which we obtain by induction

$$|p_n| = |a_n||p_{n-1}| > |p_{n-1}|, |q_n| = |a_n||q_{n-1}| > |q_{n-1}| \quad (n \geq 1).$$

Hence

$$|p_n| = |a_0 a_1 \cdots a_n|, |q_n| = |a_1 \cdots a_n| \quad (n \geq 1).$$

From the relation $q_n f - p_n = (-1)^n/(q_n f_{n+1} + q_{n-1})$ we further obtain

$$|q_n f - p_n| = |q_{n+1}|^{-1},$$

since

$$|q_n f_{n+1} + q_{n-1}| = |q_n f_{n+1}| = |q_n||a_{n+1}| = |q_{n+1}|.$$

In particular, $|q_n f - p_n| < 1$ and hence

$$p_n = \lfloor q_n f \rfloor, |\{q_n f\}| = |q_{n+1}|^{-1} \quad (n \geq 1).$$

Thus p_n is readily determined from q_n. Furthermore,

$$|f - p_n/q_n| = |q_n|^{-1}|q_{n+1}|^{-1} \to 0 \quad \text{as } n \to \infty.$$

The rational function p_n/q_n is called the *n-th convergent* of f. The polynomials a_n are called the *partial quotients*, and the Laurent series f_n the *complete quotients*, in the continued fraction expansion of f.

The continued fraction algorithm can also be applied when f is the formal Laurent expansion of a rational function, but in this case the process terminates after a finite number of steps. If $a_0, a_1, a_2, \ldots$ is any finite or infinite sequence of polynomials with $|a_n| > 1$ for $n \geq 1$, there is a unique formal Laurent series f with $[a_0, a_1, a_2, \ldots]$ as its continued fraction expansion.

For formal Laurent series there are sharper Diophantine properties than for real numbers:

Proposition 14 *Let f be a formal Laurent series with convergents p_n/q_n and let p, q be polynomials with $q \neq O$.*

(i) *If $|q| < |q_{n+1}|$ and $p/q \neq p_n/q_n$, then*

$$|qf - p| \geq |q_{n-1}f - p_{n-1}| = |q_n|^{-1}.$$

(ii) *If $|qf - p| < |q|^{-1}$, then p/q is a convergent of f.*

Proof (i) Assume on the contrary that $|qf - p| < |q_n|^{-1}$. Since

$$q_n(qf - p) - q(q_nf - p_n) = qp_n - pq_n \neq O$$

and $|q_n||qf - p| < 1$, we must have

$$|q||q_{n+1}|^{-1} = |q||q_nf - p_n| = |qp_n - pq_n| \geq 1,$$

which is contrary to hypothesis.

(ii) Assume that p/q is not a convergent of f. If $f = p_N/q_N$ is a rational function then $|q| < |q_N|$, since

$$1 \leq |qp_N - pq_N| = |qf - p||q_N| < |q|^{-1}|q_N|.$$

Thus, whether or not f is rational, we can choose n so that $|q_n| \leq |q| < |q_{n+1}|$. Hence, by (i),

$$|qf - p| \geq |q_n|^{-1} \geq |q|^{-1},$$

which is a contradiction. □

It was shown by Abel (1826) that, for any complex polynomial $D(t)$ which is not a square, the 'Pell' equation $X^2 - D(t)Y^2 = 1$ has a solution in polynomials $X(t), Y(t)$ of positive degree if and only if $\sqrt{D(t)}$ may be represented as a periodic continued fraction: $\sqrt{D(t)} = [a_0, \overline{a_1, \ldots, a_h}]$, where $a_h = 2a_0$ and $a_i = a_{h-i}(i = 1, \ldots, h - 1)$ are polynomials of positive degree. By differentiation one obtains

$$XX'/Y = Y'D + (1/2)YD'.$$

It follows that Y divides X', since X and Y are relatively prime, and

$$(X + Y\sqrt{D})' = (X + Y\sqrt{D})X'/Y\sqrt{D}.$$

Thus the 'abelian' integral

$$\int X'(t)dt/Y(t)\sqrt{D(t)}$$

is actually the elementary function $\log\{X(t) + Y(t)\sqrt{D(t)}\}$.

Some remarkable results have recently been obtained on the approximation of algebraic numbers by rational numbers, which deserve to be mentioned here, even though the proofs are beyond our scope.

A complex number ζ is said to be an *algebraic number*, or simply *algebraic, of degree d* if it is a root of a polynomial of degree d with rational coefficients which is irreducible over the rational field $\mathbb{Q}$. Thus an algebraic number of degree 2 is just a quadratic irrational.

For any irrational number ξ, there exist infinitely many rational numbers p/q such that

$$|\xi - p/q| < 1/q^2,$$

since the inequality is satisfied by any convergent of ξ. It was shown by Roth (1955) that if ξ is a real algebraic number of degree $d \geq 2$ then, for any given $\varepsilon > 0$, there exist only finitely many rational numbers p/q with $q > 0$ such that

$$|\xi - p/q| < 1/q^{2+\varepsilon}.$$

The proof does not provide a bound for the magnitude of the rational numbers which satisfy the inequality, but it does provide a bound for their number. Roth's result was the culmination of a line of research that was begun by Thue (1909), and further developed by Siegel (1921) and Dyson (1947).

A sharpening of Roth's result has been *conjectured* by Lang (1965): if ξ is a real algebraic number of degree $d \geq 2$ then, for any given $\varepsilon > 0$, there exist only finitely many rational numbers p/q with $q > 1$ such that

$$|\xi - p/q| < 1/q^2(\log q)^{1+\varepsilon}.$$

An even stronger sharpening has been conjectured by P.M. Wong (1989) in which $(\log q)^{1+\varepsilon}$ is replaced by $(\log q)(\log \log q)^{1+\varepsilon}$ with $q > 2$.

For real algebraic numbers of degree 2 we already know more than this. For, if ξ is a real quadratic irrational, its partial quotients are bounded and so there exists a constant $c = c(\xi) > 0$ such that $|\xi - p/q| > c/q^2$ for every rational number p/q. It is a long-standing conjecture that this is false for any real algebraic number ξ of degree $d > 2$.

It is not difficult to show that Roth's theorem may be restated in the following homogeneous form: if

$$L_1(u, v) = \alpha u + \beta v, \quad L_2(u, v) = \gamma u + \delta v,$$

are linearly independent linear forms with algebraic coefficients $\alpha, \beta, \gamma, \delta$, then, for any given $\varepsilon > 0$, there exist at most finitely many integers x, y, not both zero, such that

$$|L_1(x, y)L_2(x, y)| < \max(|x|, |y|)^{-\varepsilon}.$$

The *subspace theorem* of W. Schmidt (1972) generalizes Roth's theorem in this form to higher dimensions. In the stronger form given it by Vojta (1989) it says: if $L_1(u), \ldots, L_n(u)$ are linearly independent linear forms in n variables $u = (u_1, \ldots, u_n)$ with (real or complex) algebraic coefficients, then there exist finitely many proper linear subspaces $V_1, \ldots, V_h$ of $\mathbb{Q}^n$ such that every nonzero $x = (x_1, \ldots, x_n) \in \mathbb{Z}^n$ for which

$$|L_1(x) \cdots L_n(x)| < \|x\|^{-\varepsilon},$$

where $\|x\| = \max(|x_1|, \ldots, |x_n|)$, is contained in some subspace V_i, except for finitely many points whose number may depend on ε. A new proof of Schmidt's subspace theorem has been given by Faltings and Wüstholz (1994). The subspace theorem has also been given a more quantitative form by Schmidt (1989) and Evertse (1996). These results have immediate applications to the simultaneous approximation of several algebraic numbers.

Vojta (1987) has developed a remarkable analogy between the approximation of algebraic numbers by rationals and the theory of Nevanlinna (1925) on the value distribution of meromorphic functions, in which Roth's theorem corresponds to Nevanlinna's second main theorem. Although the analogy is largely formal, it is suggestive in both directions. It has already led to new proofs for the theorems of Roth and Schmidt, and to a proof of the Mordell conjecture (discussed below) which is quite different from the original proof by Faltings.

Roth's theorem has an interesting application to Diophantine equations. Let

$$f(z) = a_0 z^n + a_1 z^{n-1} + \cdots + a_n$$

be a polynomial of degree $n \geq 3$ with integer coefficients whose roots are distinct and not rational. Let

$$f(u, v) = a_0 u^n + a_1 u^{n-1} v + \cdots + a_n v^n$$

be the corresponding homogeneous polynomial and let $g(u, v)$ be a polynomial of degree $m \geq 0$ with integer coefficients. We will deduce from Roth's theorem that the equation

$$f(x, y) = g(x, y)$$

has at most finitely many solutions in integers if $m \leq n - 3$. This was already proved by Thue for $m = 0$.

Assume on the contrary that there exist infinitely many solutions in integers. Without loss of generality we may assume that there exist infinitely many integer solutions x, y for which $|x| \leq |y|$. Then there exists a constant $c_1 > 0$ such that

$$|g(x, y)| \leq c_1 |y|^m.$$

Over the complex field $\mathbb{C}$ the homogeneous polynomial $f(u, v)$ has a factorization

$$f(u, v) = a_0 \prod_{j=1}^{n} (u - \zeta_j v),$$

where $\zeta_1, \ldots, \zeta_n$ are distinct algebraic numbers which are not rational. For at least one j we must have, for infinitely many x, y,

$$|a_0||x - \zeta_j y|^n \leq c_1 |y|^m$$

and hence

$$|x - \zeta_j y| \leq c_2 |y|^{m/n},$$

where $c_2 = (c_1/|a_0|)^{1/n}$. If $k \neq j$, then

$$|x - \zeta_k y| \geq |(\zeta_j - \zeta_k)y| - |x - \zeta_j y|$$
$$\geq c_3|y| - c_2|y|^{m/n} \geq c_4|y|,$$

where c_3, c_4 are positive constants. It follows that

$$|a_0||x - \zeta_j y|c_4^{n-1}|y|^{n-1} \leq |f(x, y)| = |g(x, y)| \leq c_1|y|^m$$

and hence

$$|\zeta_j - x/y| \leq c_5/|y|^{n-m},$$

where the positive constant c_5 depends only on the coefficients of f and g. Evidently this implies that ζ_j is real. Since ζ_j is not rational and $m \leq n - 3$, we now obtain a contradiction to Roth's theorem.

It is actually possible to characterize all polynomial Diophantine equations with infinitely many solutions. Let $F(x, y)$ be a polynomial with rational coefficients which is irreducible over $\mathbb{C}$. It was shown by Siegel (1929), by combining his own results on the approximation of algebraic numbers with results of Mordell and Weil concerning the rational points on elliptic curves and Jacobian varieties, that if the equation

$$F(x, y) = 0 \qquad\qquad (*)$$

has infinitely many integer solutions, then there exist polynomials or Laurent polynomials $\phi(t), \psi(t)$ (not both constant) with coefficients from either the rational field $\mathbb{Q}$ or a real quadratic field $\mathbb{Q}(\sqrt{d})$, where $d > 0$ is a square-free integer, such that $F(\phi(t), \psi(t))$ is identically zero. If $\phi(t), \psi(t)$ are Laurent polynomials with coefficients from $\mathbb{Q}(\sqrt{d})$, they may be chosen to be invariant when t is replaced by t^{-1} and the coefficients are replaced by their conjugates in $\mathbb{Q}(\sqrt{d})$.

This implies, in particular, that the algebraic curve defined by $(*)$ may be transformed by a birational transformation with rational coefficients into either a linear equation $ax + by + c = 0$ or a Pellian equation $x^2 - dy^2 - m = 0$. It is not significant that the birational transformation has rational, rather than integral, coefficients since, by combining a result of Mahler (1934) with the *Mordell conjecture*, it may be seen that the same conclusions hold if the equation $(*)$ has infinitely many solutions in rational numbers whose denominators involve only finitely many primes.

The conjecture of Mordell (1922) says that the equation $(*)$ has at most finitely many *rational* solutions if the algebraic curve defined by $(*)$ has genus $g > 1$. (The concept of *genus* will not be formally defined here, but we mention that the genus of an irreducible plane algebraic curve may be calculated by a procedure due to M. Noether.) The conjecture has now been proved by Faltings (1983), as will be mentioned in Chapter XIII. As mentioned also at the end of Chapter XIII, if the algebraic curve defined by $(*)$ has genus 1, then explicit bounds may be obtained for the number of integral points. It was already shown by Hilbert and Hurwitz (1890) that the algebraic curve defined by $(*)$ has genus 0 if and only if it is birationally equivalent over $\mathbb{Q}$ either to a line or to a conic. There then exist rational functions $\phi(t), \psi(t)$ (not both constant) with coefficients either from $\mathbb{Q}$ or from a quadratic extension of $\mathbb{Q}$ such that

$F(\phi(t), \psi(t))$ is identically zero. The coefficients may be taken from $\mathbb{Q}$ if the curve has at least one non-singular rational point.

Thus in retrospect, and quite unfairly, Siegel's remarkable result may be seen as simply picking out those curves of genus 0 which have infinitely many integral points, a problem which had already been treated by Maillet (1919).

In this connection it may be mentioned that the formula for Pythagorean triples given in §5 of Chapter II may be derived from the parametrization of the unit circle $x^2 + y^2 = 1$ by the rational functions

$$x(t) = (1 - t^2)/(1 + t^2), \quad y(t) = 2t/(1 + t^2).$$

8 Further Remarks

More extensive accounts of the theory of continued fractions are given in the books of Rockett and Szusz [45] and Perron [41]. Many historical references are given in Brezinski [12]. The first systematic account of the subject, which it is still a delight to read, was given in 1774 by Lagrange [32] in his additions to the French translation of Euler's *Algebra*.

The continued fraction algorithm is such a useful tool that there have been many attempts to generalize it to higher dimensions. Jacobi, in a paper published posthumously (1868), defined a continued fraction algorithm in $\mathbb{R}^2$. Perron (1907) extended his definition to $\mathbb{R}^n$ and proved that convergence holds in the following weak sense: for a given nonzero $x \in \mathbb{R}^n$, the Jacobi-Perron algorithm constructs recursively a sequence of bases $\mathscr{B}^k = \{b_1^k, \ldots, b_n^k\}$ of $\mathbb{Z}^n$ such that, for each $j \in \{1, \ldots, n\}$, the angle between the line Ob_j^k and the line Ox tends to zero as $k \to \infty$. More recently, other algorithms have been proposed for which convergence holds in the strong sense that, for each $j \in \{1, \ldots, n\}$, the distance of b_j^k from the line Ox tends to zero as $k \to \infty$. See Brentjes [11], Ferguson [22], Just [28] and Lagarias [31].

Proposition 2 was first proved by Serret [51]. Proposition 3 was proved by Lagrange. The complete characterization of best approximations is proved in the book of Perron.

Lambert (1766) proved that π was irrational by using a continued fraction expansion for $\tan x$. For the continued fraction expansion of π, see Choong *et al.* [15]. Badly approximable numbers are thoroughly surveyed by Shallit [52].

The theory of Diophantine approximation is treated more comprehensively in the books of Koksma [30], Cassels [13] and Schmidt [47].

The estimate $O(\sqrt{D} \log D)$ for the period of the continued fraction expansion of a quadratic irrational with discriminant D is proved by elementary means in the book of Rockett and Szusz. Further references are given in Podsypanin [42].

The ancient Hindu method of solving Pell's equation is discussed in Selenius [49]. Tables for solving the Diophantine equation $x^2 - dy^2 = m$, where $m^2 < d$, are given in Patz [39]. Pell's equation plays a role in the negative solution of Hilbert's tenth problem, which asks for an algorithm to determine whether an arbitrary polynomial Diophantine equation is solvable in integers. See Davis *et al.* [18] and Jones and Matijasevic [26].

The continued fraction construction for the representation of a prime $p \equiv 1 \bmod 4$ as a sum of two squares is due to Legendre. Some other constructions are given in Chapter V of Davenport [17] and in Wagon [61]. A construction for the representation of any positive integer as a sum of four squares is given by Rousseau [46].

The modular group is the basic example of a *Fuchsian group*, i.e. a discrete subgroup of the group $PSL_2(\mathbb{R})$ of all linear fractional transformations $z \to (az + b)/(cz + d)$, where $a, b, c, d \in \mathbb{R}$ and $ad - bc = 1$. Fuchsian groups are studied from different points of view in the books of Katok [29], Beardon [7], Lehner [36], and Vinberg and Shvartsman [58].

The significance of Fuchsian groups stems in part from the uniformization theorem, which characterizes Riemann surfaces. A *Riemann surface* is a 1-dimensional complex manifold. Two Riemann surfaces are *conformally equivalent* if there is a bijective holomorphic map from one to the other. The *uniformization theorem*, first proved by Koebe and Poincaré independently in 1907, says that any Riemann surface is conformally equivalent to exactly one of the following:

  (i)   the complex plane $\mathbb{C}$,
 (ii)   the Riemann sphere $\mathbb{C} \cup \{\infty\}$,
(iii)   the cylinder $\mathbb{C}/G$, where G is the cyclic group generated by the translation $z \to z + 1$,
 (iv)   a torus $\mathbb{C}/G$, where G is the abelian group generated by the translations $z \to z+1$ and $z \to z + \tau$ for some $\tau \in \mathscr{H}$ (the upper half-plane),
  (v)   a quotient space $\mathscr{H}/G$, where G is a Fuchsian group which acts *freely* on $\mathscr{H}$, i.e. if $z \in \mathscr{H}$, $g \in G$ and $g \neq I$, then $g(z) \neq z$.

(It should be noted that, since the modular group does not act freely on $\mathscr{H}$, the corresponding 'Riemann surface' is *ramified*.) For more information on the uniformization theorem, see Abikoff [1], Bers [9], Farkas and Kra [21], Jost [27], Beardon and Stephenson [8], and He and Schramm [24].

For the equivalence between quadratic fields and binary quadratic forms, see Zagier [63]. The class number $h(d)$ of the quadratic field $\mathbb{Q}(\sqrt{d})$ has been deeply investigated, originally by exploiting this equivalence. Dirichlet (1839) obtained an analytic formula for $h(d)$ with the aid of his theorem on primes in an arithmetic progression (which will be proved in Chapter X). A clearly motivated proof of Dirichlet's formula is given in Hasse [23], and there are some interesting observations on the formula in Stark [56].

It was conjectured by Gauss (1801), in the language of quadratic forms, that $h(d) \to \infty$ as $d \to -\infty$. This was first proved by Heilbronn (1934). Siegel (1935) showed that actually

$$\log h(d)/\log|d| \to 1/2 \quad \text{as } d \to -\infty.$$

Generalizations of these results to arbitrary algebraic number fields are given in books on algebraic number theory, e.g. Narkiewicz [38].

Siegel (1943) has given a natural generalization of the modular group to higher dimensions. Instead of the upper half-plane $\mathscr{H}$, we consider the space $\mathscr{H}_n$ of all complex $n \times n$ matrices $Z = X + iY$, where X, Y are real symmetric matrices and Y is positive definite. If the real $2n \times 2n$ matrix

$$M = \begin{pmatrix} A & B \\ C & D \end{pmatrix}$$

is *symplectic*, i.e. if $M^t J M = J$, where

$$J = \begin{pmatrix} O & I \\ -I & O \end{pmatrix},$$

then the linear fractional transformation $Z \to (AZ + B)(CZ + D)^{-1}$, maps $\mathscr{H}_n$ onto itself. Siegel's modular group Γ_n is the group of all such transformations. The generalized upper half-plane $\mathscr{H}_n$ is itself just a special case of the vast theory of symmetric Riemannian spaces initiated by E. Cartan (1926/7). See Siegel [54] and Helgason [25].

The development of non-Euclidean geometry is traced in Bonola [10]. (This edition also contains translations of works by Bolyai and Lobachevski.) The basic properties of Poincaré's model, here only stated, are proved in the books of Katok [29] and Beardon [7].

For the connection between continued fractions and geodesics, see Artin [5] and Sheingorn [53]. For the Markov spectrum see not only the books of Cassels [13] and Rockett and Szusz [45], but also Cusick and Flahive [16] and Baragar [6].

The theory of continued fractions for formal Laurent series is developed further in de Mathan [37]. The corresponding theory of Diophantine approximation is surveyed in Lasjaunias [35]. The polynomial Pell equation is discussed by Schmidt [48]. For formal Laurent series there is a multidimensional generalization which is quite different from those for real numbers; see Antoulas [4].

Roth's theorem and Schmidt's subspace theorem are proved in Schmidt [47]. See also Faltings and Wüstholz [20] and Evertse [19]. Nevanlinna's theory of the value distribution of meromorphic functions is treated in the recent book of Cherry and Ye [14]. For Vojta's work see, for example, [59] and [60]. It should be noted, though, that this area is still in a state of flux, besides using techniques beyond our scope. For an overview, see Lang [34].

Siegel's theorem on Diophantine equations with infinitely many solutions is proved with the aid of non-standard analysis by Robinson and Roquette [44]; the proof is reproduced in Stepanov [57]. The theorem is discussed from the standpoint of *Diophantine geometry* in Serre [50]. Any algebraic curve over $\mathbb{Q}$ of genus zero which has a nonsingular rational point can be parametrized by rational functions *effectively*; see Poulakis [43].

It is worth noting that if $F(x, y)$ is a polynomial with rational coefficients which is irreducible over $\mathbb{Q}$, but not over $\mathbb{C}$, then the curve $F(x, y) = 0$ has at most finitely many rational points. For any rational point is a common root of at least two distinct complex-irreducible factors of F and any two such factors have at most finitely many common complex roots.

In conclusion we mention some further applications of continued fractions. A procedure, due to Vincent (1836), for separating the roots of a polynomial with integer coefficients has acquired some practical value with the advent of modern computers. See Alesina and Galuzzi [3].

Continued fractions play a role in the small divisor problems of classical mechanics. As an example, suppose the function f is holomorphic in some neighbourhood of the origin and $f(z) = \lambda z + O(z^2)$, where $\lambda = e^{2\pi i\theta}$ for some irrational θ. It is readily shown that there exists a formal power series h which linearizes f, i.e. $f(h(z)) = h(\lambda z)$. Brjuno (1971) proved that this formal power series converges in a neighbourhood of the origin if $\sum_{n\geq 0}(\log q_{n+1})/q_n < \infty$, where q_n is the denominator of the n-th convergent of θ. It was shown by Yoccoz (1995) that this condition is also necessary. In fact, if $\sum_{n\geq 0}(\log q_{n+1})/q_n = \infty$, the conclusion fails even for $f(z) = \lambda z(1 - z)$. See Yoccoz [62] and Pérez-Marco [40].

Our discussion of continued fractions has neglected their analytic theory. The outstanding work of Stieltjes (1894) on the *problem of moments*, which was extended by Hamburger (1920) and R. Nevanlinna (1922) from the half-line to the whole line, not only gave birth to the Stieltjes integral but also contributed to the development of functional analysis. For modern accounts, see Akhiezer [2], Landau [33] and Simon [55].

9 Selected References

[1] W. Abikoff, The uniformization theorem, *Amer. Math. Monthly* **88** (1981), 574–592.

[2] N.I. Akhiezer, *The classical moment problem*, Hafner, New York, 1965.

[3] A. Alesina and M. Galuzzi, A new proof of Vincent's theorem, *Enseign. Math.* **44** (1998), 219–256.

[4] A.C. Antoulas, On recursiveness and related topics in linear systems, *IEEE Trans. Automat. Control* **31** (1986), 1121–1135.

[5] E. Artin, Ein mechanisches System mit quasiergodischen Bahnen, *Abh. Math. Sem. Univ. Hamburg* **3** (1924), 170–175. [*Collected Papers*, pp. 499–504, Addison-Wesley, Reading, Mass., 1965.]

[6] A. Baragar, On the unicity conjecture for Markoff numbers, *Canad. Math. Bull.* **39** (1996), 3–9.

[7] A.F. Beardon, *The geometry of discrete groups*, Springer-Verlag, New York, 1983.

[8] A.F. Beardon and K. Stephenson, The uniformization theorem for circle packings, *Indiana Univ. Math. J.* **39** (1990), 1383–1425.

[9] L. Bers, On Hilbert's 22nd problem, *Mathematical developments arising from Hilbert problems* (ed. F.E. Browder), pp. 559–609, Proc. Symp. Pure Math. **28**, Part 2, Amer. Math. Soc., Providence, R.I., 1976.

[10] R. Bonola, *Non-Euclidean geometry*, English transl. by H.S. Carslaw, reprinted Dover, New York, 1955.

[11] A.J. Brentjes, *Multi-dimensional continued fraction algorithms*, Mathematics Centre Tracts **145**, Amsterdam, 1981.

[12] C. Brezinski, *History of continued fractions and Padé approximants*, Springer-Verlag, Berlin, 1991.

[13] J.W.S. Cassels, *An introduction to Diophantine approximation*, Cambridge University Press, 1957.

[14] W. Cherry and Z. Ye, *Nevanlinna's theory of value distribution*, Springer-Verlag, New York, 2000.

[15] K.Y. Choong, D.E. Daykin and C.R. Rathbone, Rational approximations to π, *Math. Comp.* **25** (1971), 387–392.

[16] T.W. Cusick and M.E. Flahive, *The Markoff and Lagrange spectra*, Mathematical Surveys and Monographs **30**, Amer. Math. Soc., Providence, R.I., 1989.

[17] H. Davenport, *The higher arithmetic*, 7th ed., Cambridge University Press, 1999.

[18] M. Davis, Y. Matijasevic and J. Robinson, Hilbert's tenth problem. Diophantine equations: positive aspects of a negative solution, *Mathematical developments arising from Hilbert problems* (ed. F.E. Browder), pp. 323–378, Proc. Symp. Pure Math. **28**, Part 2, Amer. Math. Soc., Providence, R.I., 1976.

[19] J.H. Evertse, An improvement of the quantitative subspace theorem, *Compositio Math.* **101** (1996), 225–311.

[20] G. Faltings and G. Wüstholz, Diophantine approximation on projective spaces, *Invent. Math.* **116** (1994), 109–138.

[21] H.M. Farkas and I. Kra, *Riemann surfaces*, Springer-Verlag, New York, 1980.

[22] H. Ferguson, A short proof of the existence of vector Euclidean algorithms, *Proc. Amer. Math. Soc.* **97** (1986), 8–10.

[23] H. Hasse, *Vorlesungen über Zahlentheorie*, Zweite Auflage, Springer-Verlag, Berlin, 1964.

[24] Z.-H. He and O. Schramm, On the convergence of circle packings to the Riemann map, *Invent. Math.* **125** (1996), 285–305.

[25] S. Helgason, *Differential geometry, Lie groups, and symmetric spaces*, Academic Press, New York, 1978. [Corrected reprint, Amer. Math. Soc., Providence, R.I., 2001]

[26] J.P. Jones and Y.V. Matijasevic, Proof of recursive insolvability of Hilbert's tenth problem, *Amer. Math. Monthly* **98** (1991) 689–709.

[27] J. Jost, *Compact Riemann surfaces*, transl. by R.R. Simha, Springer-Verlag, Berlin, 1997.

[28] B. Just, Generalizing the continued fraction algorithm to arbitrary dimensions, *SIAM J. Comput.* **21** (1992), 909–926.

[29] S. Katok, *Fuchsian groups*, University of Chicago Press, 1992.

[30] J.F. Koksma, *Diophantische Approximationen*, Springer-Verlag, Berlin, 1936.

[31] J.C. Lagarias, Geodesic multidimensional continued fractions, *Proc. London Math. Soc.* (3) **69** (1994), 464–488.

[32] J.L. Lagrange, *Oeuvres*, t. VII, pp. 5–180, reprinted Olms Verlag, Hildesheim, 1973.

[33] H.J. Landau, The classical moment problem: Hilbertian proofs, *J. Funct. Anal.* **38** (1980), 255–272.

[34] S. Lang, *Number Theory III: Diophantine geometry*, Encyclopaedia of Mathematical Sciences Vol. 60, Springer-Verlag, Berlin, 1991.

[35] A. Lasjaunias, A survey of Diophantine approximation in fields of power series, *Monatsh. Math.* **130** (2000), 211–229.

[36] J. Lehner, *Discontinuous groups and automorphic functions*, Mathematical Surveys VIII, Amer. Math. Soc., Providence, R.I., 1964.

[37] B. de Mathan, Approximations diophantiennes dans un corps local, *Bull. Soc. Math. France Suppl. Mém.* **21** (1970), Chapitre IV.

[38] W. Narkiewicz, *Elementary and analytic theory of algebraic numbers*, 2nd ed., Springer-Verlag, Berlin, 1990.

[39] W. Patz, *Tafel der regelmässigen Kettenbrüche und ihrer vollständigen Quotienten für die Quadratwurzeln aus den natürlichen Zahlen von 1-10000*, Akademie-Verlag, Berlin, 1955.

[40] R. Pérez-Marco, Fixed points and circle maps, *Acta Math.* **179** (1997), 243–294.

[41] O. Perron, *Die Lehre von den Kettenbrüchen*, Dritte Auflage, Teubner, Stuttgart, Band I, 1954; Band II, 1957. (Band II treats the analytic theory of continued fractions.)

[42] E.V. Podsypanin, Length of the period of a quadratic irrational, *J. Soviet Math.* **18** (1982), 919–923.

[43] D. Poulakis, Bounds for the minimal solution of genus zero Diophantine equations, *Acta Arith.* **86** (1998), 51–90.

[44] A. Robinson and P. Roquette, On the finiteness theorem of Siegel and Mahler concerning Diophantine equations, *J. Number Theory* **7** (1975), 121–176.

[45] A.M Rockett and P. Szusz, *Continued fractions*, World Scientific, River Edge, N.J., 1992.

[46] G. Rousseau, On a construction for the representation of a positive integer as the sum of four squares, *Enseign. Math.* (2) **33** (1987), 301–306.

[47] W.M. Schmidt, *Diophantine approximation*, Lecture Notes in Mathematics **785**, Springer-Verlag, Berlin, 1980.

[48] W.M. Schmidt, On continued fractions and diophantine approximation in power series fields, *Acta Arith.* **95** (2000), 139–166.

[49] C.-O. Selenius, Rationale of the chakravala process of Jayadeva and Bhaskara II, *Historia Math.* **2** (1975), 167–184.

[50] J.-P. Serre, *Lectures on the Mordell–Weil theorem*, English transl. by M. Brown from notes by M. Waldschmidt, Vieweg & Sohn, Braunschweig, 1989.

[51] J.A. Serret, Developpements sur une classe d'équations, *J. Math. Pures Appl.* **15** (1850), 152–168.

[52] J. Shallit, Real numbers with bounded partial quotients, *Enseign. Math.* **38** (1992), 151–187.

[53] M. Sheingorn, Continued fractions and congruence subgroup geodesics, *Number theory with an emphasis on the Markoff spectrum* (ed. A.D. Pollington and W. Moran), pp. 239–254, Lecture Notes in Pure and Applied Mathematics **147**, Dekker, New York, 1993.

[54] C.L. Siegel, Symplectic geometry, *Amer. J. Math.* **65** (1943), 1–86. [*Gesammelte Abhandlungen, Band II,* pp. 274–359, Springer-Verlag, Berlin, 1966.]

[55] B. Simon, The classical moment problem as a self-adjoint finite difference operator, *Adv. in Math.* **137** (1998), 82–203.

[56] H.M. Stark, Dirichlet's class-number formula revisited, *A tribute to Emil Grosswald*: *Number theory and related analysis* (ed. M. Knopp and M. Sheingorn), pp. 571–577, Contemporary Mathematics **143**, Amer. Math. Soc., Providence, R.I., 1993.

[57] S.A. Stepanov, *Arithmetic of algebraic curves*, English transl. by I. Aleksanova, Consultants Bureau, New York, 1994.

[58] E.B. Vinberg and O.V. Shvartsman, *Discrete groups of motions of spaces of constant curvature*, Geometry II, pp. 139–248, Encyclopaedia of Mathematical Sciences Vol. 29, Springer-Verlag, Berlin, 1993.

[59] P. Vojta, *Diophantine approximations and value distribution theory*, Lecture Notes in Mathematics **1239**, Springer-Verlag, Berlin, 1987.

[60] P. Vojta, A generalization of theorems of Faltings and Thue–Siegel–Roth–Wirsing, *J. Amer. Math. Soc.* **5** (1992), 763–804.

[61] S. Wagon, The Euclidean algorithm strikes again, *Amer. Math. Monthly* **97** (1990), 125–129.

[62] J.-C. Yoccoz, Théorème de Siegel, nombres de Bruno et polynômes quadratiques, *Astérisque* **231** (1995), 3–88.

[63] D.B. Zagier, *Zetafunktionen und quadratische Körper*, Springer-Verlag, Berlin, 1981.

Additional References

M. Laczkovich, On Lambert's proof of the irrationality of π, *Amer. Math. Monthly* **104** (1997), 439–443.

Anitha Srinivasan, A really simple proof of the Markoff conjecture for prime powers, *Preprint*.

V

Hadamard's Determinant Problem

It was shown by Hadamard (1893) that, if all elements of an $n \times n$ matrix of complex numbers have absolute value at most μ, then the determinant of the matrix has absolute value at most $\mu^n n^{n/2}$. For each positive integer n there exist complex $n \times n$ matrices for which this upper bound is attained. For example, the upper bound is attained for $\mu = 1$ by the matrix $(\omega^{jk})(1 \leq j, k \leq n)$, where ω is a primitive n-th root of unity. This matrix is real for $n = 1, 2$. However, Hadamard also showed that if the upper bound is attained for a real $n \times n$ matrix, where $n > 2$, then n is divisible by 4.

Without loss of generality one may suppose $\mu = 1$. A real $n \times n$ matrix for which the upper bound $n^{n/2}$ is attained in this case is today called a *Hadamard matrix*. It is still an open question whether an $n \times n$ Hadamard matrix exists for every positive integer n divisible by 4.

Hadamard's inequality played an important role in the theory of linear integral equations created by Fredholm (1900), and partly for this reason many proofs and generalizations were soon given. Fredholm's approach to linear integral equations has been superseded, but Hadamard's inequality has found connections with several other branches of mathematics, such as number theory, combinatorics and group theory. Hadamard matrices have been used to enhance the precision of spectrometers, to design agricultural experiments and to correct errors in messages transmitted by spacecraft.

The moral is that a good mathematical problem will in time find applications. Although the case where n is divisible by 4 has a richer theory, we will also treat other cases of Hadamard's determinant problem, since progress with them might lead to progress also for Hadamard matrices.

1 What is a Determinant?

The system of two simultaneous linear equations

$$\alpha_{11}\xi_1 + \alpha_{12}\xi_2 = \beta_1$$
$$\alpha_{21}\xi_1 + \alpha_{22}\xi_2 = \beta_2$$

W.A. Coppel, *Number Theory: An Introduction to Mathematics*, Universitext, DOI: 10.1007/978-0-387-89486-7_5, © Springer Science + Business Media, LLC 2009

has, if $\delta_2 = \alpha_{11}\alpha_{22} - \alpha_{12}\alpha_{21}$ is nonzero, the unique solution

$$\xi_1 = (\beta_1\alpha_{22} - \beta_2\alpha_{12})/\delta_2, \quad \xi_2 = -(\beta_1\alpha_{21} - \beta_2\alpha_{11})/\delta_2.$$

If $\delta_2 = 0$, then either there is no solution or there is more than one solution.

Similarly the system of three simultaneous linear equations

$$\alpha_{11}\xi_1 + \alpha_{12}\xi_2 + \alpha_{13}\xi_3 = \beta_1$$
$$\alpha_{21}\xi_1 + \alpha_{22}\xi_2 + \alpha_{23}\xi_3 = \beta_2$$
$$\alpha_{31}\xi_1 + \alpha_{32}\xi_2 + \alpha_{33}\xi_3 = \beta_3$$

has a unique solution if and only if $\delta_3 \neq 0$, where

$$\delta_3 = \alpha_{11}\alpha_{22}\alpha_{33} + \alpha_{12}\alpha_{23}\alpha_{31} + \alpha_{13}\alpha_{21}\alpha_{32}$$
$$- \alpha_{11}\alpha_{23}\alpha_{32} - \alpha_{12}\alpha_{21}\alpha_{33} - \alpha_{13}\alpha_{22}\alpha_{31}.$$

These considerations may be extended to any finite number of simultaneous linear equations. The system

$$\alpha_{11}\xi_1 + \alpha_{12}\xi_2 + \cdots + \alpha_{1n}\xi_n = \beta_1$$
$$\alpha_{21}\xi_1 + \alpha_{22}\xi_2 + \cdots + \alpha_{2n}\xi_n = \beta_2$$
$$\cdots$$
$$\alpha_{n1}\xi_1 + \alpha_{n2}\xi_2 + \cdots + \alpha_{nn}\xi_n = \beta_n$$

has a unique solution if and only if $\delta_n \neq 0$, where

$$\delta_n = \sum \pm \alpha_{1k_1}\alpha_{2k_2} \cdots \alpha_{nk_n},$$

the sum being taken over all $n!$ permutations $k_1, k_2, \ldots, k_n$ of $1, 2, \ldots, n$ and the sign chosen being $+$ or $-$ according as the permutation is even or odd, as defined in Chapter I, §7.

It has been tacitly assumed that the given quantities $\alpha_{jk}, \beta_j (j, k = 1, \ldots, n)$ are real numbers, in which case the solution $\xi_k (k = 1, \ldots, n)$ also consists of real numbers. However, everything that has been said remains valid if the given quantities are elements of an arbitrary field F, in which case the solution also consists of elements of F. Since δ_n is an element of F which is uniquely determined by the matrix

$$A = \begin{bmatrix} \alpha_{11} & \cdots & \alpha_{1n} \\ & \cdots & \\ \alpha_{n1} & \cdots & \alpha_{nn} \end{bmatrix},$$

it will be called the *determinant* of the matrix A and denoted by det A.

Determinants appear in the work of the Japanese mathematician Seki (1683) and in a letter of Leibniz (1693) to l'Hospital, but neither had any influence on later developments. The rule which expresses the solution of a system of linear equations by quotients of determinants was stated by Cramer (1750), but the study of determinants for their own sake began with Vandermonde (1771). The word 'determinant' was first used in the present sense by Cauchy (1812), who gave a systematic account of their theory. The diffusion of this theory throughout the mathematical world owes much to the clear exposition of Jacobi (1841).

For the practical solution of linear equations Cramer's rule is certainly inferior to the age-old method of elimination of variables. Even many of the theoretical uses to which determinants were once put have been replaced by simpler arguments from linear algebra, to the extent that some have advocated banning determinants from the curriculum. However, determinants have a geometrical interpretation which makes their survival desirable.

Let $M_n(\mathbb{R})$ denote the set of all $n \times n$ matrices with entries from the real field $\mathbb{R}$. If $A \in M_n(\mathbb{R})$, then the linear map $x \to Ax$ of $\mathbb{R}^n$ into itself multiplies the volume of any parallelotope by a fixed factor $\mu(A) \geq 0$. Evidently

(i)'' $\mu(AB) = \mu(A)\mu(B)$ for all $A, B \in M_n(\mathbb{R})$,
(ii)'' $\mu(D) = |\alpha|$ for any diagonal matrix $D = \operatorname{diag}[1, \ldots, 1, \alpha] \in M_n(\mathbb{R})$.

(A matrix $A = (\alpha_{jk})$ is denoted by $\operatorname{diag}[\alpha_{11}, \alpha_{22}, \ldots, \alpha_{nn}]$ if $\alpha_{jk} = 0$ whenever $j \neq k$ and is then said to be *diagonal*.) It may be shown (e.g., by representing A as a product of elementary matrices in the manner described below) that $\mu(A) = |\det A|$. The sign of the determinant also has a geometrical interpretation: $\det A \gtrless 0$ according as the linear map $x \to Ax$ preserves or reverses orientation.

Now let F be an arbitrary field and let $M_n = M_n(F)$ denote the set of all $n \times n$ matrices with entries from F. We intend to show that determinants, as defined above, have the properties:

(i)' $\det(AB) = \det A \cdot \det B$ for all $A, B \in M_n$,
(ii)' $\det D = \alpha$ for any diagonal matrix $D = \operatorname{diag}[1, \ldots, 1, \alpha] \in M_n$,

and, moreover, that these two properties actually characterize determinants. To avoid notational complexity, we consider first the case $n = 2$.

Let $\mathscr{E}$ denote the set of all matrices $A \in M_2$ which are products of finitely many matrices of the form U_λ, V_μ, where

$$U_\lambda = \begin{bmatrix} 1 & \lambda \\ 0 & 1 \end{bmatrix}, \quad V_\mu = \begin{bmatrix} 1 & 0 \\ \mu & 1 \end{bmatrix},$$

and $\lambda, \mu \in F$. The set $\mathscr{E}$ is a group under matrix multiplication, since multiplication is associative, $I \in \mathscr{E}$, $\mathscr{E}$ is obviously closed under multiplication and U_λ, V_μ have inverses $U_{-\lambda}, V_{-\mu}$ respectively.

We are going to show that, *if $A \in M_2$ and $A \neq O$, then there exist $S, T \in \mathscr{E}$ and $\delta \in F$ such that $SAT = \operatorname{diag}[1, \delta]$.*

For any $\rho \neq 0$, put

$$W = \begin{bmatrix} 0 & -1 \\ 1 & 0 \end{bmatrix}, \quad R_\rho = \begin{bmatrix} \rho^{-1} & 0 \\ 0 & \rho \end{bmatrix}.$$

Then $W = U_{-1}V_1U_{-1} \in \mathscr{E}$ and also $R_\rho \in \mathscr{E}$ since, if $\sigma = 1 - \rho$, $\rho' = \rho^{-1}$ and $\tau = \rho^2 - \rho$, then

$$R_\rho = V_{-1}U_\sigma V_{\rho'}U_\tau.$$

Let

$$A = \begin{bmatrix} \alpha & \beta \\ \gamma & \delta \end{bmatrix},$$

where at least one of $\alpha, \beta, \gamma, \delta$ is nonzero. By multiplying A on the left, or on the right, or both by W we may suppose that $\alpha \neq 0$. Now, by multiplying A on the right or left by R_α, we may suppose that $\alpha = 1$. Next, by multiplying A on the right by $U_{-\beta}$, we may further suppose that $\beta = 0$. Finally, by multiplying A on the left by $V_{-\gamma}$, we may also suppose that $\gamma = 0$.

The preceding argument is valid even if F is a division ring. In what follows we will use the commutativity of multiplication in F.

We are now going to show that *if* $d : \mathcal{E} \to F$ *is a map such that* $d(ST) = d(S)d(T)$ *for all* $S, T \in \mathcal{E}$, *then either* $d(S) = 0$ *for every* $S \in \mathcal{E}$ *or* $d(S) = 1$ *for every* $S \in \mathcal{E}$.

If $d(T) = 0$ for some $T \in \mathcal{E}$, then $d(I) = d(T)d(T^{-1}) = 0$ and $d(S) = d(I)d(S) = 0$ for every $S \in \mathcal{E}$. Thus we now suppose $d(S) \neq 0$ for every $S \in \mathcal{E}$. Then, in the same way, $d(I) = 1$ and $d(S^{-1}) = d(S)^{-1}$ for every $S \in \mathcal{E}$.

It is easily verified that

$$U_\lambda U_\mu = U_{\lambda+\mu}, \qquad V_\lambda V_\mu = V_{\lambda+\mu},$$
$$W^{-1} = -W, \qquad W^{-1}V_\mu W = U_{-\mu}.$$

It follows that

$$d(V_\mu) = d(U_{-\mu}) = d(U_\mu)^{-1}.$$

Also, for any $\rho \neq 0$,

$$R_\rho^{-1} U_\lambda R_\rho = U_{\lambda\rho^2}.$$

Hence $d(U_{\lambda\rho^2}) = d(U_\lambda)$ and $d(U_{\lambda(\rho^2-1)}) = 1$.

If the field F contains more than three elements, then $\rho^2 - 1 \neq 0$ for some nonzero $\rho \in F$. Since $\lambda(\rho^2 - 1)$ runs through the nonzero elements of F at the same time as λ, it follows that $d(U_\lambda) = 1$ for every $\lambda \in F$. Hence also $d(V_\mu) = 1$ for every $\mu \in F$ and $d(S) = 1$ for all $S \in \mathcal{E}$.

If F contains 2 elements, then $d(S) = 1$ for every $S \in \mathcal{E}$ is the only possibility. If F contains 3 elements, then $d(S) = \pm 1$ for every $S \in \mathcal{E}$. Hence $d(S^{-1}) = d(S)$ and $d(S^2) = 1$. Since $U_2 = U_1^2$ and $U_1 = U_2^{-1}$, this implies $d(U_\lambda) = 1$ for every $\lambda \in F$, and the rest follows as before.

The preceding discussion is easily extended to higher dimensions. Put

$$U_{ij}(\lambda) = I_n + \lambda E_{ij},$$

for any $i, j \in \{1, \dots, n\}$ with $i \neq j$, where E_{ij} is the $n \times n$ matrix with all entries 0 except the (i, j)-th, which is 1, and let $SL_n(F)$ denote the set of all $A \in M_n$ which are products of finitely many matrices $U_{ij}(\lambda)$. Then $SL_n(F)$ is a group under matrix multiplication.

If $A \in M_n$ and $A \neq O$, then there exist $S, T \in SL_n(F)$ and a positive integer $r \leq n$ such that

$$SAT = \mathrm{diag}[1_{r-1}, \delta, 0_{n-r}]$$

for some nonzero $\delta \in F$. The matrix A is *singular* if $r < n$ and *nonsingular* if $r = n$. Hence $A = (\alpha_{jk})$ is nonsingular if and only if its transpose $A^t = (\alpha_{kj})$ is nonsingular. In the nonsingular case we need multiply A on only one side by a matrix from $SL_n(F)$ to bring it to the form

$$D_\delta = \mathrm{diag}[1_{n-1}, \delta].$$

For if $SAT = D_\delta$, then $SA = D_\delta T^{-1}$ and this implies $SA = S'D_\delta$ for some $S' \in SL_n(F)$, since

$$
\begin{aligned}
D_\delta U_{ij}(\lambda) &= U_{ij}(\lambda \delta^{-1}) D_\delta && \text{if } i < j = n, \\
D_\delta U_{ij}(\lambda) &= U_{ij}(\delta \lambda) D_\delta && \text{if } j < i = n, \\
D_\delta U_{ij}(\lambda) &= U_{ij}(\lambda) D_\delta && \text{if } i, j \neq n \text{ and } i \neq j.
\end{aligned}
$$

In the same way as for $n = 2$ it may be shown that, if $d : SL_n(F) \to F$ is a map such that $d(ST) = d(S)d(T)$ for all $S, T \in SL_n(F)$, then either $d(S) = 0$ for every S or $d(S) = 1$ for every S.

Theorem 1 *There exists a unique map* $d : M_n \to F$ *such that*

(i)' $d(AB) = d(A)d(B)$ *for all* $A, B \in M_n$,
(ii)' *for any* $\alpha \in F$, *if* $D_\alpha = \mathrm{diag}[1_{n-1}, \alpha]$, *then* $d(D_\alpha) = \alpha$.

Proof We consider first uniqueness. Since $d(I) = d(D_1) = 1$, we must have $d(S) = 1$ for every $S \in SL_n(F)$, by what we have just said. Also, if

$$H = \mathrm{diag}[\eta_1, \ldots, \eta_{n-1}, 0],$$

then $d(H) = 0$, since $H = D_0 H$. In particular, $d(O) = 0$. If $A \in M_n$ and $A \neq O$, there exist $S, T \in SL_n(F)$ such that

$$SAT = \mathrm{diag}[1_{r-1}, \delta, 0_{n-r}],$$

where $1 \le r \le n$ and $\delta \neq 0$. It follows that $d(A) = 0$ if $r < n$, i.e. if A is singular. On the other hand if $r = n$, i.e. if A is nonsingular, then $SAT = D_\delta$ and hence $d(A) = \delta$. This proves uniqueness.

We consider next existence. For any $A = (\alpha_{jk}) \in M_n$, define

$$\det A = \sum_{\sigma \in \mathscr{S}_n} (\mathrm{sgn}\,\sigma) \alpha_{1\sigma 1} \alpha_{2\sigma 2} \cdots \alpha_{n\sigma n},$$

where σ is a permutation of $1, 2, \ldots, n$, $\mathrm{sgn}\,\sigma = 1$ or -1 according as the permutation σ is even or odd, and the summation is over the symmetric group $\mathscr{S}_n$ of all permutations. Several consequences of this definition will now be derived.

(i) *if every entry in some row of A is 0, then* $\det A = 0$.

Proof Every summand vanishes in the expression for $\det A$.   □

(ii) *if the matrix B is obtained from the matrix A by multiplying all entries in one row by λ, then* $\det B = \lambda \det A$.

Proof This is also clear, since in the expression for $\det A$ each summand contains exactly one factor from any given row. □

(iii) *if two rows of A are the same, then* $\det A = 0$.

Proof Suppose for definiteness that the first and second rows are the same, and let τ be the permutation which interchanges 1 and 2 and leaves fixed every $k > 2$. Then τ is odd and we can write

$$\det A = \sum_{\sigma \in \mathscr{A}_n} \alpha_{1\sigma 1} \alpha_{2\sigma 2} \cdots \alpha_{n\sigma n} - \sum_{\sigma \in \mathscr{A}_n} \alpha_{1\sigma \tau 1} \alpha_{2\sigma \tau 2} \cdots \alpha_{n\sigma \tau n},$$

where $\mathscr{A}_n$ is the alternating group of all even permutations. In the second sum

$$\alpha_{1\sigma \tau 1} \alpha_{2\sigma \tau 2} \cdots \alpha_{n\sigma \tau n} = \alpha_{1\sigma 2} \alpha_{2\sigma 1} \alpha_{3\sigma 3} \cdots \alpha_{n\sigma n} = \alpha_{2\sigma 2} \alpha_{1\sigma 1} \alpha_{3\sigma 3} \cdots \alpha_{n\sigma n},$$

because the first and second rows are the same. Hence the two sums cancel. □

(iv) *if the matrix B is obtained from the matrix A by adding a scalar multiple of one row to a different row, then* $\det B = \det A$.

Proof Suppose for definiteness that B is obtained from A by adding λ times the second row to the first. Then

$$\det B = \sum_{\sigma \in \mathscr{S}_n} (\operatorname{sgn} \sigma) \alpha_{1\sigma 1} \alpha_{2\sigma 2} \cdots \alpha_{n\sigma n} + \lambda \sum_{\sigma \in \mathscr{S}_n} (\operatorname{sgn} \sigma) \alpha_{2\sigma 1} \alpha_{2\sigma 2} \cdots \alpha_{n\sigma n}.$$

The first sum is $\det A$ and the second sum is 0, by (iii), since it is the determinant of the matrix obtained from A by replacing the first row by the second. □

(v) *if A is singular, then* $\det A = 0$.

Proof If A is singular, then some row of A is a linear combination of the remaining rows. Thus by subtracting from this row scalar multiples of the remaining rows we can replace it by a row of 0's. For the new matrix B we have $\det B = 0$, by (i). On the other hand, $\det B = \det A$, by (iv). □

(vi) *if* $A = \operatorname{diag}[\delta_1, \ldots, \delta_n]$, *then* $\det A = \delta_1 \cdots \delta_n$. *In particular,* $\det D_\alpha = \alpha$.

Proof In the expression for $\det A$ the only possible nonzero summand is that for which σ is the identity permutation, and the identity permutation is even. □

(vii) $\det(AB) = \det A \cdot \det B$ *for all* $A, B \in M_n$.

Proof If A is singular, then AB is also and so, by (v), $\det(AB) = 0 = \det A \cdot \det B$. Thus we now suppose that A is nonsingular. Then there exists $S \in SL_n(F)$ such that $SA = D_\delta$ for some nonzero $\delta \in F$. Since, by the definition of $SL_n(F)$, left multiplication by S corresponds to a finite number of operations of the type considered in (iv) we have

$$\det A = \det(SA) = \det D_\delta$$

and

$$\det(AB) = \det(SAB) = \det(D_\delta B).$$

But $\det D_\delta = \delta$, by (vi), and $\det(D_\delta B) = \delta \det B$, by (ii). Therefore $\det(AB) = \det A \cdot \det B$.

This completes the proof of existence.   □

Corollary 2 *If $A \in M_n$ and if A^t is the transpose of A, then $\det A^t = \det A$.*

Proof The map $d : M_n \to F$ defined by $d(A) = \det A^t$ also has the properties (i)′, (ii)′.   □

The proof of Theorem 1 shows further that $SL_n(F)$ *is the special linear group, consisting of all $A \in M_n$ with $\det A = 1$.*

We do not propose to establish here all the properties of determinants which we may later require. However, we note that if

$$A = \begin{bmatrix} B & 0 \\ C & D \end{bmatrix}$$

is a partitioned matrix, where B and D are square matrices of smaller size, then

$$\det A = \det B \cdot \det D.$$

It follows that if $A = (\alpha_{jk})$ is *lower triangular* (i.e. $\alpha_{jk} = 0$ for all j, k with $j < k$) or *upper triangular* (i.e. $\alpha_{jk} = 0$ for all j, k with $j > k$), then

$$\det A = \alpha_{11}\alpha_{22}\cdots\alpha_{nn}.$$

2 Hadamard Matrices

We begin by obtaining an upper bound for $\det(A^t A)$, where A is an $n \times m$ real matrix. If $m = n$, then $\det(A^t A) = (\det A)^2$ and bounding $\det(A^t A)$ is the same as Hadamard's problem of bounding $|\det A|$. However, as we will see in §3, the problem is of interest also for $m < n$.

In the statement of the following result we denote by $\|v\|$ the Euclidean norm of a vector $v = (\alpha_1, \dots, \alpha_n) \in \mathbb{R}^n$. Thus $\|v\| \geq 0$ and $\|v\|^2 = \alpha_1^2 + \cdots + \alpha_n^2$. The geometrical interpretation of the result is that a parallelotope with given side lengths has maximum volume when the sides are orthogonal.

Proposition 3 *Let A be an $n \times m$ real matrix with linearly independent columns $v_1, \dots, v_m$. Then*

$$\det(A^t A) \leq \prod_{k=1}^{m} \|v_k\|^2,$$

with equality if and only if $A^t A$ is a diagonal matrix.

Proof We are going to construct inductively mutually orthogonal vectors $w_1, \ldots, w_m$ such that w_k is a linear combination of $v_1, \ldots, v_k$ in which the coefficient of v_k is 1 ($1 \leq k \leq m$). Take $w_1 = v_1$ and suppose $w_1, \ldots, w_{k-1}$ have been determined. If we take

$$w_k = v_k - \alpha_1 w_1 - \cdots - \alpha_{k-1} w_{k-1},$$

where $\alpha_j = \langle v_k, w_j \rangle$, then $\langle w_k, w_j \rangle = 0$ ($1 \leq j < k$). Moreover, $w_k \neq 0$, since $v_1, \ldots, v_k$ are linearly independent. (This is the same process as in §10 of Chapter I, but without the normalization.)

If B is the matrix with columns $w_1, \ldots, w_m$ then, by construction,

$$B^t B = \operatorname{diag}[\delta_1, \ldots, \delta_m]$$

is a diagonal matrix with diagonal entries $\delta_k = \|w_k\|^2$ and $AT = B$ for some upper triangular matrix T with 1's in the main diagonal. Since $\det T = 1$, we have

$$\det(A^t A) = \det(B^t B) = \prod_{k=1}^{m} \|w_k\|^2.$$

But

$$\|v_k\|^2 = \|w_k\|^2 + |\alpha_1|^2 \|w_1\|^2 + \cdots + |\alpha_{k-1}|^2 \|w_{k-1}\|^2$$

and hence $\|w_k\|^2 \leq \|v_k\|^2$, with equality only if $w_k = v_k$. The result follows. $\square$

Corollary 4 *Let $A = (\alpha_{jk})$ be an $n \times m$ real matrix such that $|\alpha_{jk}| \leq 1$ for all j, k. Then*

$$\det(A^t A) \leq n^m,$$

with equality if and only if $\alpha_{jk} = \pm 1$ for all j, k and $A^t A = n I_m$.

Proof We may assume that the columns of A are linearly independent, since otherwise $\det(A^t A) = 0$. If v_k is the k-th column of A, then $\|v_k\|^2 \leq n$, with equality if and only if $|\alpha_{jk}| = 1$ for $1 \leq j \leq n$. The result now follows from Proposition 3. $\square$

An $n \times m$ matrix $A = (\alpha_{jk})$ will be said to be an *H-matrix* if $\alpha_{jk} = \pm 1$ for all j, k and $A^t A = n I_m$. If, in addition, $m = n$ then A will be said to be a *Hadamard matrix of order n*.

If A is an $n \times m$ H-matrix, then $m \leq n$. Furthermore, if A is a Hadamard matrix of order n then, for any $m < n$, the submatrix formed by the first m columns of A is an H-matrix. (This distinction between H-matrices and Hadamard matrices is convenient, but not standard. It is an unproven conjecture that any H-matrix can be completed to a Hadamard matrix.)

The transpose A^t of a Hadamard matrix A is again a Hadamard matrix, since $A^t = n A^{-1}$ commutes with A. The 1×1 unit matrix is a Hadamard matrix, and so is the 2×2 matrix

$$\begin{bmatrix} 1 & 1 \\ 1 & -1 \end{bmatrix}.$$

There is one rather simple procedure for constructing H-matrices. If $A = (\alpha_{jk})$ is an $n \times m$ matrix and $B = (\beta_{i\ell})$ a $q \times p$ matrix, then the $nq \times mp$ matrix

$$\begin{bmatrix} \alpha_{11}B & \alpha_{12}B & \cdots & \alpha_{1m}B \\ \alpha_{21}B & \alpha_{22}B & \cdots & \alpha_{2m}B \\ & \cdots & \cdots & \\ \alpha_{n1}B & \alpha_{n2}B & \cdots & \alpha_{nm}B \end{bmatrix},$$

with entries $\alpha_{jk}\beta_{i\ell}$, is called the *Kronecker product* of A and B and is denoted by $A \otimes B$. It is easily verified that

$$(A \otimes B)(C \otimes D) = AC \otimes BD$$

and

$$(A \otimes B)^t = A^t \otimes B^t.$$

It follows directly from these rules of calculation that if A_1 is an $n_1 \times m_1$ H-matrix and A_2 an $n_2 \times m_2$ H-matrix, then $A_1 \otimes A_2$ is an $n_1 n_2 \times m_1 m_2$ H-matrix. Consequently, since there exist Hadamard matrices of orders 1 and 2, there also exist Hadamard matrices of order any power of 2. This was already known to Sylvester (1867).

Proposition 5 *Let $A = (\alpha_{jk})$ be an $n \times m$ H-matrix. If $n > 1$, then n is even and any two distinct columns of A have the same entries in exactly $n/2$ rows. If $n > 2$, then n is divisible by 4 and any three distinct columns of A have the same entries in exactly $n/4$ rows.*

Proof If $j \neq k$, then

$$\alpha_{1j}\alpha_{1k} + \cdots + \alpha_{nj}\alpha_{nk} = 0.$$

Since $\alpha_{ij}\alpha_{ik} = 1$ if the j-th and k-th columns have the same entry in the i-th row and $= -1$ otherwise, the number of rows in which the j-th and k-th columns have the same entry is $n/2$.

If j, k, ℓ are all different, then

$$\sum_{i=1}^{n}(\alpha_{ij} + \alpha_{ik})(\alpha_{ij} + \alpha_{i\ell}) = \sum_{i=1}^{n}\alpha_{ij}^2 = n.$$

But $(\alpha_{ij} + \alpha_{ik})(\alpha_{ij} + \alpha_{i\ell}) = 4$ if the j-th, k-th and ℓ-th columns all have the same entry in the i-th row and $= 0$ otherwise. Hence the number of rows in which the j-th, k-th and ℓ-th columns all have the same entry is exactly $n/4$. $\square$

Thus the order n of a Hadamard matrix must be divisible by 4 if $n > 2$. It is unknown if a Hadamard matrix of order n exists for every n divisible by 4. However, it is known for $n \leq 424$ and for several infinite families of n. We restrict attention here to the family of Hadamard matrices constructed by Paley (1933).

The following lemma may be immediately verified by matrix multiplication.

Lemma 6 *Let C be an n × n matrix, with 0's on the main diagonal and all other entries 1 or −1, such that*

$$C^t C = (n-1)I_n.$$

If C is skew-symmetric (i.e. $C^t = -C$), then $C + I$ is a Hadamard matrix of order n, whereas if C is symmetric (i.e. $C^t = C$), then

$$\begin{bmatrix} C+I & C-I \\ C-I & -C-I \end{bmatrix}$$

is a Hadamard matrix of order 2n.

Proposition 7 *If q is a power of an odd prime, there exists a $(q+1) \times (q+1)$ matrix C with 0's on the main diagonal and all other entries 1 or −1, such that*

(i) $C^t C = q I_{q+1}$,
(ii) *C is skew-symmetric if $q \equiv 3 \bmod 4$ and symmetric if $q \equiv 1 \bmod 4$.*

Proof Let F be a finite field containing q elements. Since q is odd, not all elements of F are squares. For any $a \in F$, put

$$\chi(a) = \begin{cases} 0 & \text{if } a = 0, \\ 1 & \text{if } a \neq 0 \text{ and } a = c^2 \text{ for some } c \in F, \\ -1 & \text{if } a \text{ is not a square.} \end{cases}$$

If $q = p$ is a prime, then F is the field of integers modulo p and $\chi(a) = (a/p)$ is the Legendre symbol studied in Chapter III. The following argument may be restricted to this case, if desired.

Since the multiplicative group of F is cyclic, we have

$$\chi(ab) = \chi(a)\chi(b) \quad \text{for all } a, b \in F.$$

Since the number of nonzero elements which are squares is equal to the number which are non-squares, we also have

$$\sum_{a \in F} \chi(a) = 0.$$

It follows that, for any $c \neq 0$,

$$\sum_{b \in F} \chi(b)\chi(b+c) = \sum_{b \neq 0} \chi(b)^2 \chi(1 + cb^{-1}) = \sum_{x \neq 1} \chi(x) = -1.$$

Let $0 = a_0, a_1, \ldots, a_{q-1}$ be an enumeration of the elements of F and define a $q \times q$ matrix $Q = (q_{jk})$ by

$$q_{jk} = \chi(a_j - a_k) \quad (0 \leq j, k < q).$$

Thus Q has 0's on the main diagonal and ±1's elsewhere. Also, by what has been said in the previous paragraph, if J_m denotes the $m \times m$ matrix with all entries 1, then

$$QJ_q = 0, \quad Q^tQ = qI_q - J_q.$$

Furthermore, since $\chi(-1) = (-1)^{(q-1)/2}$, Q is symmetric if $q \equiv 1 \bmod 4$ and skew-symmetric if $q \equiv 3 \bmod 4$. If e_m denotes the $1 \times m$ matrix with all entries 1, it follows that the matrix

$$C = \begin{bmatrix} 0 & e_q \\ \pm e_q^t & Q \end{bmatrix},$$

where the $\pm$ sign is chosen according as $q \equiv \pm 1 \bmod 4$, satisfies the various requirements.   □

By combining Lemma 6 with Proposition 7 we obtain Paley's result that, for any odd prime power q, there exists a Hadamard matrix of order $q + 1$ if $q \equiv 3 \bmod 4$ and of order $2(q + 1)$ if $q \equiv 1 \bmod 4$. Together with the Kronecker product construction, this establishes the existence of Hadamard matrices for all orders $n \equiv 0 \bmod 4$ with $n \leq 100$, except $n = 92$.

A Hadamard matrix of order 92 was found by Baumert, Golomb and Hall (1962), using a computer search and the following method proposed by Williamson (1944). Let A, B, C, D be $d \times d$ matrices with entries ± 1 and let

$$H = \begin{bmatrix} A & D & B & C \\ -D & A & -C & B \\ -B & C & A & -D \\ -C & -B & D & A \end{bmatrix},$$

i.e. $H = A \otimes I + B \otimes i + C \otimes j + D \otimes k$, where the 4×4 matrices I, i, j, k are matrix representations of the unit quaternions. It may be immediately verified that H is a Hadamard matrix of order $n = 4d$ if

$$A^tA + B^tB + C^tC + D^tD = 4dI_d$$

and

$$X^tY = Y^tX$$

for every two distinct matrices X, Y from the set $\{A, B, C, D\}$. The first infinite class of Hadamard matrices of Williamson type was found by Turyn (1972), who showed that they exist for all orders $n = 2(q + 1)$, where q is a prime power and $q \equiv 1 \bmod 4$. Lagrange's theorem that any positive integer is a sum of four squares suggests that Hadamard matrices of Williamson type may exist for all orders $n \equiv 0 \bmod 4$.

The Hadamard matrices constructed by Paley are either symmetric or of the form $I + S$, where S is skew-symmetric. It has been conjectured that in fact Hadamard matrices of both these types exist for all orders $n \equiv 0 \bmod 4$.

3 The Art of Weighing

It was observed by Yates (1935) that, if several quantities are to be measured, more accurate results may be obtained by measuring suitable combinations of them than by measuring each separately. Suppose, for definiteness, that we have m objects whose weights are to be determined and we perform $n \geq m$ weighings. The whole experiment may be represented by an $n \times m$ matrix $A = (\alpha_{jk})$. If the k-th object is not involved in the j-th weighing, then $\alpha_{jk} = 0$; if it is involved, then $\alpha_{jk} = +1$ or -1 according as it is placed in the left-hand or right-hand pan of the balance. The individual weights $\xi_1, \ldots, \xi_m$ are connected with the observed results $\eta_1, \ldots, \eta_n$ of the weighings by the system of linear equations

$$y = Ax, \tag{1}$$

where $x = (\xi_1, \ldots, \xi_m)^t \in \mathbb{R}^m$ and $y = (\eta_1, \ldots, \eta_n)^t \in \mathbb{R}^n$.

We will again denote by $\|y\|$ the Euclidean norm $(|\eta_1|^2 + \cdots + |\eta_n|^2)^{1/2}$ of the vector y. Let $\bar{x} \in \mathbb{R}^m$ have as its coordinates the correct weights and let $\bar{y} = A\bar{x}$. If, because of errors of measurement, y ranges over the ball $\|y - \bar{y}\| \leq \rho$ in $\mathbb{R}^n$, then x ranges over the ellipsoid $(x - \bar{x})^t A^t A(x - \bar{x}) \leq \rho^2$ in $\mathbb{R}^m$. Since the volume of the ellipsoid is $[\det(A^t A)]^{-1/2}$ times the volume of the ball, we may regard the best choice of the design matrix A to be that for which the ellipsoid has minimum volume. Thus we are led to the problem of maximizing $\det(A^t A)$ among all $n \times m$ matrices $A = (\alpha_{jk})$ with $\alpha_{jk} \in \{0, -1, 1\}$.

A different approach to the best choice of design matrix leads (by §2) to a similar result. If $n > m$ the linear system (1) is overdetermined. However, the least squares estimate for the solution of (1) is

$$x = Cy,$$

where $C = (A^t A)^{-1} A^t$. Let $a_k \in \mathbb{R}^n$ be the k-th column of A and let $c_k \in \mathbb{R}^n$ be the k-th row of C. Since $CA = I_m$, we have $c_k a_k = 1$. If y ranges over the ball $\|y - \bar{y}\| \leq \rho$ in $\mathbb{R}^n$, then ξ_k ranges over the real interval $|\xi_k - \bar{\xi}_k| \leq \rho \|c_k\|$. Thus we may regard the optimal choice of the design matrix A for measuring ξ_k to be that for which $\|c_k\|$ is a minimum.

By Schwarz's inequality (Chapter I, §4),

$$\|c_k\| \|a_k\| \geq 1,$$

with equality only if c_k^t is a scalar multiple of a_k. Also $\|a_k\| \leq n^{1/2}$, since all elements of A have absolute value at most 1. Hence $\|c_k\| \geq n^{-1/2}$, with equality if and only if all elements of a_k have absolute value 1 and $c_k^t = a_k/n$. It follows that the design matrix A is optimal for measuring *each* of $\xi_1, \ldots, \xi_m$ if all elements of A have absolute value 1 and $A^t A = n I_m$. Moreover, in this case the least squares estimate for the solution of (1) is simply $x = A^t y/n$. Thus the individual weights are easily determined from the observed measurements by additions and subtractions, followed by a division by n.

Suppose, for example, that $m = 3$ and $n = 4$. If we take

$$A = \begin{bmatrix} + & + & + \\ + & + & - \\ - & + & + \\ + & - & + \end{bmatrix},$$

where $+$ and $-$ stand for 1 and -1 respectively, then $A^t A = 4I_3$. With this experimental design the individual weights may all be determined with twice the accuracy of the weighing procedure.

The next result shows, in particular, that if we wish to maximize $\det(A^t A)$ among the $n \times m$ matrices A with all entries 0, 1 or -1, then we may restrict attention to those with all entries 1 or -1.

Proposition 8 *Let α, β be real numbers with $\alpha < \beta$ and let $\mathscr{S}$ be the set of all $n \times m$ matrices $A = (\alpha_{jk})$ such that $\alpha \le \alpha_{jk} \le \beta$ for all j, k. Then there exists an $n \times m$ matrix $M = (\mu_{jk})$ such that $\mu_{jk} \in \{\alpha, \beta\}$ for all j, k and*

$$\det(M^t M) = \max_{A \in \mathscr{S}} \det(A^t A).$$

Proof For any $n \times m$ real matrix A, either the symmetric matrix $A^t A$ is positive definite and $\det(A^t A) > 0$, or $A^t A$ is positive semidefinite and $\det(A^t A) = 0$. Since the result is obvious if $\det(A^t A) = 0$ for every $A \in \mathscr{S}$, we assume that $\det(A^t A) > 0$ for some $A \in \mathscr{S}$. This implies $m \le n$. Partition such an A in the form

$$A = (v B),$$

where v is the first column of A and B is the remainder. Then

$$A^t A = \begin{bmatrix} v^t v & v^t B \\ B^t v & B^t B \end{bmatrix}$$

and $B^t B$ is also a positive definite symmetric matrix. By multiplying $A^t A$ on the left by

$$\begin{bmatrix} I & -v^t B (B^t B)^{-1} \\ O & I \end{bmatrix}$$

and taking determinants, we see that

$$\det(A^t A) = f(v) \det(B^t B),$$

where

$$f(v) = v^t v - v^t B (B^t B)^{-1} B^t v.$$

We can write $f(v) = v^t Q v$, where

$$Q = I - P, \quad P = B (B^t B)^{-1} B^t.$$

From $P^t = P = P^2$ we obtain $Q^t = Q = Q^2$. Hence $Q = Q^t Q$ is a positive semidefinite symmetric matrix.

If $v = \theta v_1 + (1 - \theta) v_2$, where v_1 and v_2 are fixed vectors and $\theta \in \mathbb{R}$, then $f(v)$ is a quadratic polynomial $q(\theta)$ in θ whose leading coefficient

$$v_1^t Q v_1 - v_2^t Q v_1 - v_1^t Q v_2 + v_2^t Q v_2$$

is nonnegative, since Q is positive semidefinite. It follows that $q(\theta)$ attains its maximum value in the interval $0 \le \theta \le 1$ at an endpoint.

Put

$$\mu = \sup_{A \in \mathscr{S}} \det(A^t A).$$

Since $\det(A^t A)$ is a continuous function of the mn variables α_{jk} and $\mathscr{S}$ may be regarded as a compact set in $\mathbb{R}^{mn}$, μ is finite and there exists a matrix $A \in \mathscr{S}$ for which
$\det(A^t A) = \mu$. By repeatedly applying the argument of the preceding paragraph to
this A we may replace it by one for which every entry in the first column is either α or
β and for which also $\det(A^t A) = \mu$. These operations do not affect the submatrix B
formed by the last $m - 1$ columns of A. By interchanging the k-th column of A with
the first, which does not alter the value of $\det(A^t A)$, we may apply the same argument
to every other column of A.   □

The proof of Proposition 8 actually shows that if C is a compact subset of $\mathbb{R}^n$ and
if $\mathscr{S}$ is the set of all $n \times m$ matrices A whose columns are in C, then there exists an
$n \times m$ matrix M whose columns are extreme points of C such that

$$\det(M^t M) = \sup_{A \in \mathscr{S}} \det(A^t A).$$

Here $e \in C$ is said to be an *extreme point* of C if there do not exist distinct $v_1, v_2 \in C$
and $\theta \in (0, 1)$ such that $e = \theta v_1 + (1 - \theta) v_2$.

The preceding discussion concerns weighings by a chemical balance. If instead
we use a spring balance, then we are similarly led to the problem of maximizing
$\det(B^t B)$ among all $n \times m$ matrices $B = (\beta_{jk})$ with $\beta_{jk} = 1$ or 0 according as the k-th
object is or is not involved in the j-th weighing. Moreover other types of measurement
lead to the same problem. A spectrometer sorts electromagnetic radiation into bundles
of rays, each bundle having a characteristic wavelength. Instead of measuring the
intensity of each bundle separately, we can measure the intensity of various combinations of bundles by using masks with open or closed slots.

It will now be shown that in the case $m = n$ the chemical and spring balance
problems are essentially equivalent.

Lemma 9 *If B is an $(n - 1) \times (n - 1)$ matrix of 0's and 1's, and if J_n is the $n \times n$
matrix whose entries are all 1, then*

$$A = J_n - \begin{bmatrix} O & O \\ O & 2B \end{bmatrix},$$

*is an $n \times n$ matrix of 1's and -1's, whose first row and column contain only 1's, such
that*

$$\det A = (-2)^{n-1} \det B.$$

*Moreover, every $n \times n$ matrix of 1's and -1's, whose first row and column contain only
1's, is obtained in this way.*

Proof Since

$$A = \begin{bmatrix} 1 & O \\ e_{n-1}^t & I \end{bmatrix} \begin{bmatrix} 1 & e_{n-1} \\ O & -2B \end{bmatrix},$$

where e_m denotes a row of m 1's, the matrix A has determinant $(-2)^{n-1} \det B$. The rest of the lemma is obvious. □

Let A be an $n \times n$ matrix with entries ± 1. By multiplying rows and columns of A by -1 we can make all elements in the first row and first column equal to 1 without altering the value of $\det(A^t A)$. It follows from Lemma 9 that if α_n is the maximum of $\det(A^t A)$ among all $n \times n$ matrices $A = (\alpha_{jk})$ with $\alpha_{jk} \in \{-1, 1\}$, and if β_{n-1} is the maximum of $\det(B^t B)$ among all $(n-1) \times (n-1)$ matrices $B = (\beta_{jk})$ with $\beta_{jk} \in \{0, 1\}$, then

$$\alpha_n = 2^{2n-2} \beta_{n-1}.$$

4 Some Matrix Theory

In rectangular coordinates the equation of an ellipse with centre at the origin has the form

$$Q := ax^2 + 2bxy + cy^2 = \text{const.} \tag{*}$$

This is not the form in which the equation of an ellipse is often written, because of the 'cross product' term $2bxy$. However, we can bring it to that form by rotating the axes, so that the major axis of the ellipse lies along one coordinate axis and the minor axis along the other. This is possible because the major and minor axes are perpendicular to one another. These assertions will now be verified analytically.

In matrix notation, $Q = z^t A z$, where

$$A = \begin{bmatrix} a & b \\ b & c \end{bmatrix}, \quad z = \begin{bmatrix} x \\ y \end{bmatrix}.$$

A rotation of coordinates has the form $z = Tw$, where

$$T = \begin{bmatrix} \cos\theta & -\sin\theta \\ \sin\theta & \cos\theta \end{bmatrix}, \quad w = \begin{bmatrix} u \\ v \end{bmatrix}.$$

Then $Q = w^t B w$, where $B = T^t A T$. Multiplying out, we obtain

$$B = \begin{bmatrix} a' & b' \\ b' & c' \end{bmatrix},$$

where

$$b' = b(\cos^2\theta - \sin^2\theta) - (a - c)\sin\theta\cos\theta.$$

To eliminate the cross product term we choose θ so that $b(\cos^2\theta - \sin^2\theta) = (a - c)\sin\theta\cos\theta$; i.e., $2b\cos 2\theta = (a - c)\sin 2\theta$, or

$$\tan 2\theta = 2b/(a - c).$$

The preceding argument applies equally well to a hyperbola, since it is also described by an equation of the form $(*)$. We now wish to extend this result to higher dimensions. An n-dimensional conic with centre at the origin has the form

$$Q := x^t A x = \text{const.},$$

where $x \in \mathbb{R}^n$ and A is an $n \times n$ real symmetric matrix. The analogue of a rotation is a linear transformation $x = Ty$ which preserves Euclidean lengths, i.e. $x^t x = y^t y$. This holds for all $y \in \mathbb{R}^n$ if and only if

$$T^t T = I.$$

A matrix T which satisfies this condition is said to be *orthogonal*. Then $T^t = T^{-1}$ and hence also $TT^t = I$.

The single most important fact about real symmetric matrices is the *principal axes transformation*:

Theorem 10 *If H is an $n \times n$ real symmetric matrix, then there exists an $n \times n$ real orthogonal matrix U such that $U^t H U$ is a diagonal matrix:*

$$U^t H U = \text{diag}[\lambda_1, \ldots, \lambda_n].$$

Proof Let $f : \mathbb{R}^n \to \mathbb{R}$ be the map defined by

$$f(x) = x^t H x.$$

Since f is continuous and the unit sphere $S = \{x \in \mathbb{R}^n : x^t x = 1\}$ is compact,

$$\lambda_1 := \sup_{x \in S} f(x)$$

is finite and there exists an $x_1 \in S$ such that $f(x_1) = \lambda_1$. We are going to show that, if $x \in S$ and $x^t x_1 = 0$, then also $x^t H x_1 = 0$.

For any real ε, put

$$y = (x_1 + \varepsilon x)/(1 + \varepsilon^2)^{1/2}.$$

Then also $y \in S$, since x and x_1 are orthogonal vectors of unit length. Hence $f(y) \le f(x_1)$, by the definition of x_1. But $x_1^t H x = x^t H x_1$, since H is symmetric, and hence

$$f(y) = \{f(x_1) + 2\varepsilon x^t H x_1 + \varepsilon^2 f(x)\}/(1 + \varepsilon^2).$$

For small $|\varepsilon|$ it follows that

$$f(y) = f(x_1) + 2\varepsilon x^t H x_1 + O(\varepsilon^2).$$

If $x^t H x_1$ were different from zero, we could choose ε to have the same sign as it and obtain the contradiction $f(y) > f(x_1)$.

On the intersection of the unit sphere S with the hyperplane $x^t x_1 = 0$, the function f attains its maximum value λ_2 at some point x_2. Similarly, on the intersection of the unit sphere S with the $(n-2)$-dimensional subspace of all x such that $x^t x_1 = x^t x_2 = 0$, the function f attains its maximum value λ_3 at some point x_3. Proceeding in this way we obtain n mutually orthogonal unit vectors $x_1, \ldots, x_n$. Moreover $x_j^t H x_j = \lambda_j$ and, by the argument of the previous paragraph, $x_j^t H x_k = 0$ if $j > k$. It follows that the matrix U with columns $x_1, \ldots, x_n$ satisfies all the requirements. $\square$

It should be noted that, if U is any orthogonal matrix such that $U^t H U = \text{diag}[\lambda_1, \ldots, \lambda_n]$ then, since $U U^t = I$, the columns $x_1, \ldots, x_n$ of U satisfy

$$Hx_j = \lambda_j x_j \ (1 \leq j \leq n).$$

That is, λ_j is an *eigenvalue* of H and x_j a corresponding *eigenvector* $(1 \leq j \leq n)$.

A real symmetric matrix A is *positive definite* if $x^t A x > 0$ for every real vector $x \neq 0$ (and *positive semi-definite* if $x^t A x \geq 0$ for every real vector x with equality for some $x \neq 0$). It follows from Theorem 10 that two real symmetric matrices can be simultaneously diagonalized, if one of them is positive definite, although the transforming matrix may not be orthogonal:

Proposition 11 *If A and B are $n \times n$ real symmetric matrices, with A positive definite, then there exists an $n \times n$ nonsingular real matrix T such that $T^t A T$ and $T^t B T$ are both diagonal matrices.*

Proof By Theorem 10, there exists a real orthogonal matrix U such that $U^t A U$ is a diagonal matrix:

$$U^t A U = \text{diag}[\lambda_1, \ldots, \lambda_n].$$

Moreover, $\lambda_j > 0 \ (1 \leq j \leq n)$, since A is positive definite. Hence there exists $\delta_j > 0$ such that $\delta_j^2 = 1/\lambda_j$. If $D = \text{diag}[\delta_1, \ldots, \delta_n]$, then $D^t U^t A U D = I$. By Theorem 10 again, there exists a real orthogonal matrix V such that

$$V^t(D^t U^t B U D)V = \text{diag}[\mu_1, \ldots, \mu_n]$$

is a diagonal matrix. Hence we can take $T = U D V$.   □

Proposition 11 will now be used to obtain an inequality due to Fischer (1908):

Proposition 12 *If G is a positive definite real symmetric matrix, and if*

$$G = \begin{bmatrix} G_1 & G_2 \\ G_2^t & G_3 \end{bmatrix}$$

is any partition of G, then

$$\det G \leq \det G_1 \cdot \det G_3,$$

with equality if and only if $G_2 = 0$.

Proof Since G_3 is also positive definite, we can write $G = Q^t H Q$, where

$$Q = \begin{bmatrix} I & 0 \\ G_3^{-1}G_2^t & I \end{bmatrix}, \quad H = \begin{bmatrix} H_1 & 0 \\ 0 & G_3 \end{bmatrix},$$

and $H_1 = G_1 - G_2 G_3^{-1} G_2^t$. Since $\det G = \det H_1 \cdot \det G_3$, we need only show that $\det H_1 \leq \det G_1$, with equality only if $G_2 = 0$.

Since G_1 and H_1 are both positive definite, they can be simultaneously diagonalized. Thus, if G_1 and H_1 are $p \times p$ matrices, there exists a nonsingular real matrix T

such that

$$T^t G_1 T = \text{diag}[\gamma_1, \ldots, \gamma_p], \quad T^t H_1 T = \text{diag}[\delta_1, \ldots, \delta_p].$$

Since G_3^{-1} is positive definite, $u^t(G_1 - H_1)u \geq 0$ for any $u \in \mathbb{R}^p$. Hence $\gamma_i \geq \delta_i > 0$ for $i = 1, \ldots, p$ and $\det G_1 \geq \det H_1$. Moreover $\det G_1 = \det H_1$ only if $\gamma_i = \delta_i$ for $i = 1, \ldots, p$.

Hence if $\det G_1 = \det H_1$, then $G_1 = H_1$, i.e. $G_2 G_3^{-1} G_2^t = 0$. Thus $w^t G_3^{-1} w = 0$ for any vector $w = G_2^t v$. Since $w^t G_3^{-1} w = 0$ implies $w = 0$, it follows that $G_2 = 0$.   □

From Proposition 12 we obtain by induction

Proposition 13 *If* $G = (\gamma_{jk})$ *is an* $m \times m$ *positive definite real symmetric matrix, then*

$$\det G \leq \gamma_{11} \gamma_{22} \cdots \gamma_{mm},$$

with equality if and only if G *is a diagonal matrix.*

By applying Proposition 13 to the matrix $G = A^t A$, we obtain again Proposition 3. Proposition 13 may be sharpened in the following way:

Proposition 14 *If* $G = (\gamma_{jk})$ *is an* $m \times m$ *positive definite real symmetric matrix, then*

$$\det G \leq \gamma_{11} \prod_{j=2}^{m} (\gamma_{jj} - \gamma_{1j}^2/\gamma_{11}),$$

with equality if and only if $\gamma_{jk} = \gamma_{1j}\gamma_{1k}/\gamma_{11}$ *for* $2 \leq j < k \leq m$.

Proof If

$$T = \begin{bmatrix} 1 & g \\ 0 & I_{m-1} \end{bmatrix},$$

where $g = (-\gamma_{12}/\gamma_{11}, \ldots, -\gamma_{1m}/\gamma_{11})$, then

$$T^t G T = \begin{bmatrix} \gamma_{11} & 0 \\ 0 & H \end{bmatrix},$$

where $H = (\eta_{jk})$ is an $(m-1) \times (m-1)$ positive definite real symmetric matrix with entries

$$\eta_{jk} = \gamma_{jk} - \gamma_{1j}\gamma_{1k}/\gamma_{11} \quad (2 \leq j \leq k \leq m).$$

Since $\det G = \gamma_{11} \det H$, the result now follows from Proposition 13.   □

Some further inequalities for the determinants of positive definite matrices will now be derived, which will be applied to Hadamard's determinant problem in the next section. We again denote by J_m the $m \times m$ matrix whose entries are all 1.

Lemma 15 *If $C = \alpha I_m + \beta J_m$ for some real α, β, then*

$$\det C = \alpha^{m-1}(\alpha + m\beta).$$

Moreover, if $\det C \neq 0$, then $C^{-1} = \gamma I_m + \delta J_m$, where $\delta = -\beta \alpha^{-1}(\alpha + m\beta)^{-1}$ and $\gamma = \alpha^{-1}$.

Proof Subtract the first row of C from each of the remaining rows, and then add to the first column of the resulting matrix each of the remaining columns. These operations do not alter the determinant and replace C by an upper triangular matrix with main diagonal entries $\alpha + m\beta$ (once) and α ($m - 1$ times). Hence $\det C = \alpha^{m-1}(\alpha + m\beta)$.

If $\det C \neq 0$ and if γ, δ are defined as in the statement of the lemma, then from $J_m^2 = m J_m$ it follows directly that

$$(\alpha I_m + \beta J_m)(\gamma I_m + \delta J_m) = I_m. \qquad \square$$

Proposition 16 *Let $G = (\gamma_{jk})$ be an $m \times m$ positive definite real symmetric matrix such that $|\gamma_{jk}| \geq \beta$ for all j, k and $\gamma_{jj} \leq \alpha + \beta$ for all j, where $\alpha, \beta > 0$. Then*

$$\det G \leq \alpha^{m-1}(\alpha + m\beta). \qquad (2)$$

Moreover, equality holds if and only if there exists a diagonal matrix D, with main diagonal elements ± 1, such that

$$DGD = \alpha I_m + \beta J_m.$$

Proof The result is trivial if $m = 1$ and is easily verified if $m = 2$. We assume $m > 2$ and use induction on m. By replacing G by DGD, where D is a diagonal matrix whose main diagonal elements have absolute value 1, we may suppose that $\gamma_{1k} \geq 0$ for $2 \leq k \leq m$. Since the determinant is a linear function of its rows, we have

$$\det G = (\gamma_{11} - \beta)\delta + \eta,$$

where δ is the determinant of the matrix obtained from G by omitting the first row and column and η is the determinant of the matrix H obtained from G by replacing γ_{11} by β. By the induction hypothesis,

$$\delta \leq \alpha^{m-2}(\alpha + m\beta - \beta).$$

If $\eta \leq 0$, it follows that

$$\det G \leq \alpha^{m-1}(\alpha + m\beta - \beta) < \alpha^{m-1}(\alpha + m\beta).$$

Thus we now suppose $\eta > 0$. Then H is positive definite, since the submatrix obtained by omitting the first row and column is positive definite. By Proposition 14,

$$\eta \leq \beta \prod_{j=2}^{m} (\gamma_{jj} - \gamma_{1j}^2/\beta),$$

with equality only if $\gamma_{jk} = \gamma_{1j}\gamma_{1k}/\beta$ for $2 \le j < k \le m$. Hence $\eta \le \alpha^{m-1}\beta$, with equality only if $\gamma_{jj} = \alpha + \beta$ for $2 \le j \le m$ and $\gamma_{jk} = \beta$ for $1 \le j < k \le m$. Consequently

$$\det G \le \alpha^{m-1}(\alpha + m\beta - \beta) + \alpha^{m-1}\beta = \alpha^{m-1}(\alpha + m\beta),$$

with equality only if $G = \alpha I_m + \beta J_m$. $\square$

A square matrix will be called a *signed permutation matrix* if each row and column contains only one nonzero entry and this entry is 1 or -1.

Proposition 17 *Let* $G = (\gamma_{jk})$ *be an* $m \times m$ *positive definite real symmetric matrix such that* $\gamma_{jj} \le \alpha + \beta$ *for all* j *and either* $\gamma_{jk} = 0$ *or* $|\gamma_{jk}| \ge \beta$ *for all* j, k*, where* $\alpha, \beta > 0$.

Suppose in addition that $\gamma_{ik} = \gamma_{jk} = 0$ *implies* $\gamma_{ij} \ne 0$. *Then*

$$\det G \le \alpha^{m-2}(\alpha + m\beta/2)^2 \qquad\qquad \text{if } m \text{ is even,}$$
$$\det G \le \alpha^{m-2}(\alpha + (m+1)\beta/2)(\alpha + (m-1)\beta/2) \quad \text{if } m \text{ is odd.}$$
$$\tag{3}$$

Moreover, equality holds if and only if there is a signed permutation matrix U *such that*

$$U^t G U = \begin{bmatrix} L & 0 \\ 0 & M \end{bmatrix},$$

where

$$L = M = \alpha I_{m/2} + \beta J_{m/2} \qquad\qquad \text{if } m \text{ is even,}$$
$$L = \alpha I_{(m+1)/2} + \beta J_{(m+1)/2}, \ M = \alpha I_{(m-1)/2} + \beta J_{(m-1)/2} \quad \text{if } m \text{ is odd.}$$

Proof We are going to establish the inequality

$$\det G \le \alpha^{m-2}(\alpha + s\beta)(\alpha + m\beta - s\beta), \tag{4}$$

where s is the maximum number of zero elements in any row of G. Since, as a function of the real variable s, the quadratic on the right of (4) attains its maximum value for $s = m/2$, and has the same value for $s = (m + 1)/2$ as for $s = (m - 1)/2$, this will imply (3). It will also imply that if equality holds in (3), then $s = m/2$ if m is even and $s = (m + 1)/2$ or $(m - 1)/2$ if m is odd.

For $m = 2$ it is easily verified that (4) holds. We assume $m > 2$ and use induction. By performing the same signed permutation on rows and columns, we may suppose that the second row of G has the maximum number s of zero elements, and that all nonzero elements of the first row are positive and precede the zero elements. All the hypotheses of the proposition remain satisfied by the matrix G after this operation.

Let s' be the number of zero elements in the first row and put $r' = m - s'$. As in the proof of Proposition 16, we have

$$\det G = (\gamma_{11} - \beta)\delta + \eta,$$

where δ is the determinant of the matrix obtained from G by omitting the first row and column and η is the determinant of the matrix H obtained from G by replacing γ_{11} by β. We partition H in the form

$$H = \begin{bmatrix} L & N \\ N^t & M \end{bmatrix},$$

where L, M are square matrices of orders r', s' respectively. By construction all elements in the first row of L are positive and all elements in the first row of N are zero. Furthermore, by the hypotheses of the proposition, all elements of M have absolute value $\geq \beta$.

By the induction hypothesis,

$$\delta \leq \alpha^{m-3}(\alpha + s\beta)(\alpha + m\beta - \beta - s\beta).$$

If $\eta \leq 0$, it follows immediately that (4) holds with strict inequality. Thus we now suppose $\eta > 0$. Then H is positive definite and hence, by Fischer's inequality (Proposition 12), $\eta \leq \det L \cdot \det M$, with equality only if $N = 0$. But, by Proposition 14,

$$\det L \leq \beta \prod_{j=2}^{r'} (\gamma_{jj} - \gamma_{1j}^2/\beta) \leq \alpha^{r'-1}\beta$$

and, by Proposition 16,

$$\det M \leq \alpha^{s'-1}(\alpha + s'\beta).$$

Hence

$$\det G \leq \alpha^{m-2}(\alpha + s\beta)(\alpha + m\beta - \beta - s\beta) + \alpha^{m-2}\beta(\alpha + s'\beta),$$

Since $s' \leq s$, it follows that (4) holds and actually with strict inequality if $s' \neq s$.

If equality holds in (4) then, by Proposition 14, we must have $L = \alpha I_{r'} + \beta J_{r'}$, and by Proposition 16 after normalization we must also have $M = \alpha I_{s'} + \beta J_{s'}$. □

5 Application to Hadamard's Determinant Problem

We have seen that, if A is an $n \times m$ real matrix with all entries ± 1, then $\det(A^t A) \leq n^m$, with strict inequality if $n > 2$ and n is not divisible by 4. The question arises, what is the maximum value of $\det(A^t A)$ in such a case? In the present section we use the results of the previous section to obtain some answers to this question. We consider first the case where n is odd.

Proposition 18 *Let $A = (\alpha_{jk})$ be an $n \times m$ matrix with $\alpha_{jk} = \pm 1$ for all j, k. If n is odd, then*

$$\det(A^t A) \leq (n - 1)^{m-1}(n - 1 + m).$$

Moreover, equality holds if and only if $n \equiv 1 \bmod 4$ and, after changing the signs of some columns of A,

$$A^t A = (n - 1)I_m + J_m.$$

Proof We may assume $\det(A^t A) \neq 0$ and thus $m \leq n$. Then $A^t A = G = (\gamma_{jk})$ is a positive definite real symmetric matrix. For all j, k,

$$\gamma_{jk} = \alpha_{1j}\alpha_{1k} + \cdots + \alpha_{nj}\alpha_{nk}$$

is an integer and $\gamma_{jj} = n$. Moreover γ_{jk} is odd for all j, k, being the sum of an odd number of ± 1's. Hence the matrix G satisfies the hypotheses of Proposition 16 with $\alpha = n - 1$ and $\beta = 1$. Everything now follows from Proposition 16, except for the remark that if equality holds we must have $n \equiv 1 \bmod 4$.

But if $G = (n - 1)I_m + J_m$, then $\gamma_{jk} = 1$ for $j \neq k$. It now follows, by the argument used in the proof of Proposition 5, that any two distinct columns of A have the same entries in exactly $(n + 1)/2$ rows, and any three distinct columns of A have the same entries in exactly $(n + 3)/4$ rows. Thus $n \equiv 1 \bmod 4$.   □

Even if $n \equiv 1 \bmod 4$ there is no guarantee that that the upper bound in Proposition 18 is attained. However the question may be reduced to the existence of H-matrices if $m \neq n$. For suppose $m \leq n - 1$ and there exists an $(n - 1) \times m$ H-matrix B. If we put

$$A = \begin{bmatrix} B \\ e_m \end{bmatrix},$$

where e_m again denotes a row of m 1's, then $A^t A = (n - 1)I_m + J_m$.

On the other hand if $m = n$, then equality in Proposition 18 can hold only under very restrictive conditions. For in this case

$$(\det A)^2 = \det A^t A = (n - 1)^{n-1}(2n - 1)$$

and, since n is odd, it follows that $2n - 1$ is the square of an integer. It is an open question whether the upper bound in Proposition 18 is always attained when $m = n$ and $2n - 1$ is a square. However the nature of an extremal matrix, if one exists, can be specified rather precisely:

Proposition 19 *If $A = (\alpha_{jk})$ is an $n \times n$ matrix with $n > 1$ odd and $\alpha_{jk} = \pm 1$ for all j, k, then*

$$\det(A^t A) \leq (n - 1)^{n-1}(2n - 1).$$

Moreover if equality holds, then $n \equiv 1 \bmod 4$, $2n - 1 = s^2$ for some integer s and, after changing the signs of some rows and columns of A, the matrix A must satisfy

$$A^t A = (n - 1)I_n + J_n, \qquad AJ_n = sJ_n.$$

Proof By Proposition 18 and the preceding remarks, it only remains to show that if there exists an A such that $A^t A = (n - 1)I_n + J_n$ then, by changing the signs of some rows, we can ensure that also $AJ_n = sJ_n$.

Since $\det(AA^t) = \det(A^t A)$, it follows from Proposition 18 that there exists a diagonal matrix D with $D^2 = I_n$ such that

$$DAA^t D = (n - 1)I_n + J_n = A^t A.$$

Replacing A by DA, we obtain $AA^t = A^tA$. Then A commutes with A^tA and hence also with J_n. Thus the rows and columns of A all have the same sum s and $AJ_n = sJ_n = A^tJ_n$. Moreover $s^2 = 2n - 1$, since

$$s^2 J_n = s A^t J_n = A^t A J_n = (2n - 1) J_n.$$ $\square$

The maximum value of $\det(A^tA)$ when $n \equiv 3 \bmod 4$ is still a bit of a mystery. We now consider the remaining case when n is even, but not divisible by 4.

Proposition 20 *Let $A = (\alpha_{jk})$ be an $n \times m$ matrix with $2 \le m \le n$ and $\alpha_{jk} = \pm 1$ for all j, k. If $n \equiv 2 \bmod 4$ and $n > 2$, then*

$$\det(A^tA) \le (n-2)^{m-2}(n-2+m)^2 \qquad \text{if } m \text{ is even,}$$
$$\det(A^tA) \le (n-2)^{m-2}(n-1+m)(n-3+m) \quad \text{if } m \text{ is odd.}$$

Moreover, equality holds if and only if there is a signed permutation matrix U such that

$$U^t A^t A U = \begin{bmatrix} L & 0 \\ 0 & M \end{bmatrix},$$

where

$$L = M = (n-2)I_{m/2} + 2J_{m/2} \qquad \text{if } m \text{ is even,}$$
$$L = (n-2)I_{(m+1)/2} + 2J_{(m+1)/2}, \ M = (n-2)I_{(m-1)/2} + 2J_{(m-1)/2} \quad \text{if } m \text{ is odd.}$$

Proof We need only show that $G = A^tA$ satisfies the hypotheses of Proposition 17 with $\alpha = n - 2$ and $\beta = 2$. We certainly have $\gamma_{jj} = n$. Moreover all γ_{jk} are even, since n is even and

$$\gamma_{jk} = \alpha_{1j}\alpha_{1k} + \cdots + \alpha_{nj}\alpha_{nk}.$$

Hence $|\gamma_{jk}| \ge 2$ if $\gamma_{jk} \ne 0$. Finally, if j, k, ℓ, are all different and $\gamma_{j\ell} = \gamma_{k\ell} = 0$, then

$$\sum_{i=1}^n (\alpha_{ij} + \alpha_{ik})(\alpha_{ij} + \alpha_{i\ell}) = n + \gamma_{jk}.$$

Since $n \equiv 2 \bmod 4$, it follows that also $\gamma_{jk} \equiv 2 \bmod 4$ and thus $\gamma_{jk} \ne 0$. $\square$

Again there is no guarantee that the upper bound in Proposition 20 is attained. However the question may be reduced to the existence of H-matrices if $m \ne n, n - 1$. For suppose $m \le n - 2$ and there exists an $(n - 2) \times m$ H-matrix B. If we put

$$A = \begin{bmatrix} B \\ C \end{bmatrix},$$

where

$$C = \begin{bmatrix} e_r & e_s \\ e_r & -e_s \end{bmatrix},$$

and $r + s = m$, then

$$A^t A = \begin{bmatrix} (n-2)I_r + 2J_r & 0 \\ 0 & (n-2)I_s + 2J_s \end{bmatrix}.$$

Thus the upper bound in Proposition 20 is attained by taking $r = s = m/2$ when m is even and $r = (m+1)/2$, $s = (m-1)/2$ when m is odd.

Suppose now that $m = n$ and

$$A^t A = \begin{bmatrix} L & 0 \\ 0 & L \end{bmatrix},$$

where $L = (n-2)I_{n/2} + 2J_{n/2}$. If B is the $n \times (n-1)$ submatrix of A obtained by omitting the last column, then

$$B^t B = \begin{bmatrix} L & 0 \\ 0 & M \end{bmatrix},$$

where $M = (n-2)I_{n/2-1} + 2J_{n/2-1}$. Thus if the upper bound in Proposition 20 is attained for $m = n$, then it is also attained for $m = n - 1$. Furthermore, since

$$\det(AA^t) = \det(A^t A),$$

it follows from Proposition 20 that there exists a signed permutation matrix U such that

$$U A A^t U^t = A^t A.$$

Replacing A by UA, we obtain $AA^t = A^t A$. Then A commutes with $A^t A$. If

$$A = \begin{bmatrix} X & Y \\ Z & W \end{bmatrix},$$

is the partition of A into square submatrices of order $n/2$, it follows that X, Y, Z, W all commute with L and hence with $J_{n/2}$. This means that the entries in any row or any column of X have the same sum, which we will denote by x. Similarly the entries in any row or any column of Y, Z, W have the same sum, which will be denoted by y, z, w respectively. We may assume $x, y, w \ge 0$ by replacing A by

$$\begin{bmatrix} I_{n/2} & 0 \\ 0 & \pm I_{n/2} \end{bmatrix} A \begin{bmatrix} \pm I_{n/2} & 0 \\ 0 & \pm I_{n/2} \end{bmatrix},$$

We have

$$X^t X + Z^t Z = Y^t Y + W^t W = L, \quad X^t Y + Z^t W = 0,$$

and

$$XX^t + YY^t = ZZ^t + WW^t = L, \quad XZ^t + YW^t = 0.$$

Postmultiplying by J, we obtain

$$x^2 + z^2 = y^2 + w^2 = 2n - 2, \quad xy + zw = 0,$$

and

$$x^2 + y^2 = z^2 + w^2 = 2n - 2, \quad xz + yw = 0.$$

Adding, we obtain $x^2 = w^2$ and hence $x = w$. Thus $z^2 = y^2$ and actually $z = -y$, since $xy + zw = 0$.

This shows, in particular, that if the upper bound in Proposition 20 is attained for $m = n \equiv 2 \bmod 4$, then $2n - 2 = x^2 + y^2$, where x and y are integers. By Proposition II.39, such a representation is possible if and only if, for every prime $p \equiv 3 \bmod 4$, the highest power of p which divides $n - 1$ is even. Hence the upper bound in Proposition 20 is never attained if $m = n = 22$. On the other hand if $m = n = 6$, then $2n - 2 = 10 = 9 + 1$ and an extremal matrix A is obtained by taking $W = X = J_3$ and $Z = -Y = 2I_3 - J_3$.

It is an open question whether the upper bound in Proposition 20 is always attained when $m = n$ and $2n - 2$ is a sum of two squares. It is also unknown if, when an extremal matrix exists, one can always take $W = X$ and $Z = -Y$.

6 Designs

A *design* (in the most general sense) is a pair $(P, \mathcal{B})$, where P is a finite set of elements, called *points*, and $\mathcal{B}$ is a collection of subsets of P, called *blocks*. If $p_1, \ldots, p_v$ are the points of the design and $B_1, \ldots, B_b$ the blocks, then the *incidence matrix* of the design is the $v \times b$ matrix $A = (\alpha_{ij})$ of 0's and 1's defined by

$$\alpha_{ij} = \begin{cases} 1 & \text{if } p_i \in B_j, \\ 0 & \text{if } p_i \notin B_j. \end{cases}$$

Conversely, any $v \times b$ matrix $A = (\alpha_{ij})$ of 0's and 1's defines in this way a design. However, two such matrices define the same design if one can be obtained from the other by permutations of the rows and columns.

We will be interested in designs with rather more structure. A 2-*design* or, especially in older literature, a 'balanced incomplete block design' (*BIBD*) is a design, with more than one point and more than one block, in which each block contains the same number k of points, each point belongs to the same number r of blocks, and every pair of distinct points occurs in the same number λ of blocks.

Thus each column of the incidence matrix contains k 1's and each row contains r 1's. Counting the total number of 1's in two ways, by columns and by rows, we obtain

$$bk = vr.$$

Similarly, by counting in two ways the 1's which lie below the 1's in the first row, we obtain

$$r(k - 1) = \lambda(v - 1).$$

Thus if v, k, λ are given, then r and b are determined and we may speak of a 2-(v, k, λ) design. Since $v > 1$ and $b > 1$, we have

$$1 < k < v, \quad 1 \le \lambda < r.$$

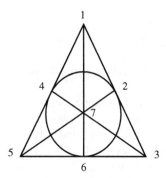

Fig. 1. The Fano plane.

A $v \times b$ matrix $A = (\alpha_{ij})$ of 0's and 1's is the incidence matrix of a 2-design if and only if, for some positive integers k, r, λ,

$$\sum_{i=1}^{v} \alpha_{ij} = k, \quad \sum_{k=1}^{b} \alpha_{ik}^2 = r, \quad \sum_{k=1}^{b} \alpha_{ik}\alpha_{jk} = \lambda \quad \text{if } i \neq j \; (1 \leq i, j \leq v),$$

or in other words,

$$e_v A = k e_b, \quad AA^t = (r - \lambda)I_v + \lambda J_v, \tag{5}$$

where e_n is the $1 \times n$ matrix with all entries 1, I_n is the $n \times n$ unit matrix and J_n is the $n \times n$ matrix with all entries 1.

Designs have been used extensively in the design of agricultural and other experiments. To compare the yield of v varieties of a crop on b blocks of land, it would be expensive to test each variety separately on each block. Instead we can divide each block into k plots and use a 2-(v, k, λ) design, where $\lambda = bk(k-1)/v(v-1)$. Then each variety is used exactly $r = bk/v$ times, no variety is used more than once in any block, and any two varieties are used together in exactly λ blocks. As an example, take $v = 4, b = 6, k = 2$ and hence $\lambda = 1, r = 3$.

Some examples of 2-designs are the finite projective planes. In fact a *projective plane* of *order n* may be defined as a 2-(v, k, λ) design with

$$v = n^2 + n + 1, \quad k = n + 1, \quad \lambda = 1.$$

It follows that $b = v$ and $r = k$. The blocks in this case are called 'lines'. The projective plane of order 2, or *Fano plane*, is illustrated in Figure 1. There are seven points and seven blocks, the blocks being the six triples of collinear points and the triple of points on the circle.

Consider now an arbitrary 2-(v, k, λ) design. By (5) and Lemma 15,

$$\det(AA^t) = (r - \lambda)^{v-1}(r - \lambda + \lambda v) > 0,$$

since $r > \lambda$. This implies the inequality $b \geq v$, due to Fisher (1940), since AA^t would be singular if $b < v$.

A 2-design is said to be *square* or (more commonly, but misleadingly) 'symmetric' if $b = v$, i.e. if the number of blocks is the same as the number of points. Thus any projective plane is a square 2-design.

For a square 2-(v, k, λ) design, $k = r$ and the incidence matrix A is itself nonsingular. The first relation (5) is now equivalent to $J_v A = k J_v$. Since $k = r$, the sum of the entries in any row of A is also k and thus $J_v A^t = k J_v$. By multiplying the second relation (5) on the left by A^{-1} and on the right by A, we further obtain

$$A^t A = (r - \lambda) I_v + \lambda J_v.$$

Thus A^t is also the incidence matrix of a square 2-(v, k, λ) design, the *dual* of the given design.

This partly combinatorial argument may be replaced by a more general matrix one:

Lemma 21 *Let a, b, k be real numbers and $n > 1$ an integer. There exists a nonsingular real $n \times n$ matrix A such that*

$$AA^t = aI + bJ, \quad JA = kJ, \tag{6}$$

if and only if $a > 0$, $a + bn > 0$ and $k^2 = a + bn$. Moreover any such matrix A also satisfies

$$A^t A = aI + bJ, \quad JA^t = kJ. \tag{7}$$

Proof We show first that if A is any real $n \times n$ matrix satisfying (6), then $a + bn = k^2$. In fact, since $J^2 = nJ$, the first relation in (6) implies $J A A^t J = (a + bn) n J$, whereas the second implies $J A A^t J = k^2 n J$.

We show next that the symmetric matrix $G := aI + bJ$ is positive definite if and only if $a > 0$ and $a + bn > 0$. By Lemma 15, $\det G = a^{n-1} (a + bn)$. If G is positive definite, its determinant is positive. Since all principal submatrices are also positive definite, we must have $a^{i-1}(a + bi) > 0$ for $1 \leq i \leq n$. In particular, $a + b > 0$, $a(a + 2b) > 0$, which is only possible if $a > 0$. It now follows that also $a + bn > 0$.

Conversely, suppose $a > 0$ and $a + bn > 0$. Then $\det G > 0$ and there exist nonzero real numbers h, k such that $a = h^2$, $a + bn = k^2$. If we put $C = hI + (k - h)n^{-1}J$, then $JC = kJ$ and

$$C^2 = h^2 I + \{2h(k - h) + (k - h)^2\} n^{-1} J = aI + bJ = G.$$

Since $\det G > 0$, this shows that $G = CC^t$ is positive definite and C is nonsingular.

Finally, let A be any nonsingular real $n \times n$ matrix satisfying (6). Since A is nonsingular, AA^t is a positive definite symmetric matrix and hence $a > 0$, $a + bn > 0$. Since $AA^t = C^2$ and $C^t = C$, we have $A = CU$, where U is orthogonal. Hence $A^t = U^t C$ and $C = UA^t$. From $JC = kJ$ we obtain $kJ = JA = JCU = kJU$. Thus $J = JU$ and $JA^t = JUA^t = JC = kJ$. Moreover $U^t JU = J$, since $J^t = J$, and hence

$$A^t A = U^t C^2 U = U^t (aI + bJ) U = aI + bJ. \qquad \square$$

In Chapter VII we will derive necessary and sufficient conditions for the existence of a nonsingular *rational* $n \times n$ matrix A such that $AA^t = aI + bJ$, and thus in particular obtain some basic restrictions on the parameters v, k, λ for the existence of a square 2-(v, k, λ) design. These were first obtained by Bruck, Ryser and Chowla (1949/50).

We now consider the relationship between designs and Hadamard's determinant problem. By passing from A to $B = (J_n - A^t)/2$, it may be seen immediately that equality holds in Proposition 19 if and only if there exists a 2-(n, k, λ) design, where $k = (n - s)/2$, $\lambda = (n + 1 - 2s)/4$ and $s^2 = 2n - 1$.

We now show that with any Hadamard matrix $A = (\alpha_{jk})$ of order $n = 4d$ there is associated a 2-$(4d - 1, 2d - 1, d - 1)$ design. Assume without loss of generality that all elements in the first row and column of A are 1. We take $P = \{2, \ldots, n\}$ as the set of points and $\mathcal{B} = \{B_2, \ldots, B_n\}$ as the set of blocks, where $B_k = \{j \in P : \alpha_{jk} = 1\}$. Then B_k has cardinality $|B_k| = n/2 - 1$ for $k = 2, \ldots, n$. Moreover, if T is any subset of P with $|T| = 2$, then the number of blocks containing T is $n/4 - 1$. The argument may also be reversed to show that any 2-$(4d - 1, 2d - 1, d - 1)$ design is associated in this way with a Hadamard matrix of order $4d$.

In particular, for $d = 2$, the 2-$(7, 3, 1)$ design associated with the Hadamard matrix $H_2 \otimes H_2 \otimes H_2$, where

$$H_2 = \begin{bmatrix} 1 & 1 \\ 1 & -1 \end{bmatrix},$$

is the projective plane of order 2 (Fano plane) illustrated in Figure 1.

The connection between Hadamard matrices and designs may also be derived by a matrix argument. If

$$A = \begin{bmatrix} 1 & e_{n-1} \\ e^t_{n-1} & \tilde{A} \end{bmatrix},$$

is a Hadamard matrix of order $n = 4d$, normalized so that its first row and column contain only 1's, then $B = (J_{n-1} + \tilde{A})/2$ is a matrix of 0's and 1's such that

$$J_{4d-1}B = (2d - 1)J_{4d-1}, \quad BB^t = dI_{4d-1} + (d - 1)J_{4d-1}.$$

The optimal spring balance design of order $4d - 1$, which is obtained by taking $C = (J_{n-1} - \tilde{A})/2$, is a 2-$(4d - 1, 2d, d)$ design, since

$$J_{4d-1}C = 2dJ_{4d-1}, \quad CC^t = dI_{4d-1} + dJ_{4d-1}.$$

The notion of 2-design will now be generalized. Let t, v, k, λ be positive integers with $v \geq k \geq t$. A t-(v, k, λ) design, or simply a t-design, is a pair $(P, \mathcal{B})$, where P is a set of cardinality v and $\mathcal{B}$ is a collection of subsets of P, each of cardinality k, such that any subset of P of cardinality t is contained in exactly λ elements of $\mathcal{B}$. The elements of P will be called *points* and the elements of $\mathcal{B}$ will be called *blocks*. A t-(v, k, λ) design with $\lambda = 1$ is known as a *Steiner system*. The *automorphism group* of a t-design is the group of all permutations of the points which map blocks onto blocks.

If $t = 1$, then each point is contained in exactly λ blocks and so the number of blocks is $\lambda v/k$. Suppose now that $t > 1$. Let S be a fixed subset of P of cardinality $t - 1$ and let λ' be the number of blocks which contain S. Consider the number of pairs (T, B), where $B \in \mathcal{B}$, $S \subseteq T \subseteq B$ and $|T| = t$. By first fixing B and varying T we see that this number is $\lambda'(k - t + 1)$. On the other hand, by first fixing T and varying B we see that this number is $\lambda(v - t + 1)$. Hence

$$\lambda' = \lambda(v - t + 1)/(k - t + 1)$$

does not depend on the choice of S and a t-(v, k, λ) design $(P, \mathscr{B})$ is also a $(t - 1)$-(v, k, λ') design. By repeating this argument, we see that each point is contained in exactly r blocks, where

$$r = \lambda(v - t + 1) \cdots (v - 1)/(k - t + 1) \cdots (k - 1),$$

and the total number of blocks is $b = rv/k$. In particular, any t-design with $t > 2$ is also a 2-design.

With any Hadamard matrix $A = (\alpha_{jk})$ of order $n = 4d$ there is, in addition, associated a 3-$(4d, 2d, d - 1)$ design. For assume without loss of generality that all elements in the first column of A are 1. We take $P = \{1, 2, \ldots, n\}$ as the set of points and $\{B_2, \ldots, B_n, B_2', \ldots, B_n'\}$ as the set of blocks, where $B_k = \{j \in P : \alpha_{jk} = 1\}$ and $B_k' = \{j \in P : \alpha_{jk} = -1\}$. Then, by Proposition 5, $|B_k| = |B_k'| = n/2$ for $k = 2, \ldots, n$. If T is any subset of P with $|T| = 3$, say $T = \{i, j, \ell\}$, then the number of blocks containing T is the number of $k > 1$ such that $\alpha_{ik} = \alpha_{jk} = \alpha_{\ell k}$. But, by Proposition 5 again, the number of columns of A which have the same entries in rows i, j, ℓ is $n/4$ and this includes the first column. Hence T is contained in exactly $n/4 - 1$ blocks. Again the argument may be reversed to show that any 3-$(4d, 2d, d - 1)$ design is associated in this way with a Hadamard matrix of order $4d$.

7 Groups and Codes

A group is said to be *simple* if it contains more than one element and has no *normal* subgroups besides itself and the subgroup containing only the identity element. The finite simple groups are in some sense the building blocks from which all finite groups are constructed. There are several infinite families of them: the cyclic groups C_p of prime order p, the alternating groups $\mathscr{A}_n$ of all even permutations of n objects ($n \geq 5$), the groups $PSL_n(q)$ derived from the general linear groups of all invertible linear transformations of an n-dimensional vector space over a finite field of $q = p^m$ elements ($n \geq 2$ and $q > 3$ if $n = 2$), and some other families similar to the last which are analogues for a finite field of the simple Lie groups.

In addition to these infinite families there are 26 *sporadic* finite simple groups. (The *classification theorem* states that there are no other finite simple groups besides those already mentioned. The proof of the classification theorem at present occupies thousands of pages, scattered over a variety of journals, and some parts are actually still unpublished.) All except five of the sporadic groups were found in the years 1965–1981. However, the first five were found by Mathieu (1861,1873): M_{12} is a 5-fold transitive group of permutations of 12 objects of order $12 \cdot 11 \cdot 10 \cdot 9 \cdot 8$ and M_{11} the subgroup of all permutations in M_{12} which fix one of the objects; M_{24} is a 5-fold transitive group of permutations of 24 objects of order $24 \cdot 23 \cdot 22 \cdot 21 \cdot 20 \cdot 48$, M_{23} the subgroup of all permutations in M_{24} which fix one of the objects and M_{22} the subgroup of all permutations which fix two of the objects. The Mathieu groups may be defined in several ways, but the definitions by means of Hadamard matrices that we are going to give are certainly competitive with the others.

Two $n \times n$ Hadamard matrices H_1, H_2 are said to be *equivalent* if one may be obtained from the other by interchanging two rows or two columns, or by changing the sign of a row or a column, or by any finite number of such operations. Otherwise expressed, $H_2 = PH_1Q$, where P and Q are signed permutation matrices. An *automorphism* of a Hadamard matrix H is an equivalence of H with itself: $H = PHQ$. Since $P = HQ^{-1}H^{-1}$, the automorphism is uniquely determined by Q. Under matrix multiplication all admissible Q form a group $\mathscr{G}$, the *automorphism group* of the Hadamard matrix H. Evidently $-I \in \mathscr{G}$ and $-I$ commutes with all elements of $\mathscr{G}$. The factor group $\mathscr{G}/\{\pm I\}$, obtained by identifying Q and $-Q$, may be called the *reduced automorphism group* of H.

To illustrate these concepts we will show that all Hadamard matrices of order 12 are equivalent. In fact rather more is true:

Proposition 22 *Any Hadamard matrix of order 12 may be brought to the form*

$$
\begin{array}{ccc|ccc|ccc|ccc}
+ & + & + & + & + & + & + & + & + & + & + & + \\
+ & + & + & + & + & + & - & - & - & - & - & - \\
+ & + & + & - & - & - & + & + & + & - & - & - \\
+ & - & + & - & + & - & - & + & - & + & - & + \\
+ & + & - & - & - & + & - & - & + & + & - & + \\
- & + & + & + & - & - & + & - & - & + & - & + \\
+ & - & + & - & - & + & + & - & - & + & + & - \\
+ & - & + & + & - & - & - & - & + & - & + & + \\
+ & + & - & - & + & - & + & - & - & - & + & + \\
- & + & + & - & + & - & - & - & + & + & + & - \\
+ & + & - & + & - & - & - & + & - & + & + & - \\
- & + & + & - & - & + & - & + & - & - & + & + \\
\end{array}
\qquad (*)
$$

(where $+$ *stands for* 1 *and* $-$ *for* -1*) by changing the signs of some rows and columns, by permuting the columns, and by permuting the first three rows and the last seven rows.*

Proof Let $A = (\alpha_{jk})$ be a Hadamard matrix of order 12. By changing the signs of some columns we may assume that all elements of the first row are $+1$. Then, by the orthogonality relations, half the elements of any other row are $+1$. By permuting the columns we may assume that all elements in the first half of the second row are $+1$. It now follows from the orthogonality relations that in any row after the second the sum of all elements in each half is zero. Hence, by permuting the columns within each half we may assume that the third row is the same as the third row of the array $(*)$ displayed above. In the r-th row, where $r > 3$, let ρ_k be the sum of the entries in the k-th block of three columns ($k = 1, 2, 3, 4$). The orthogonality relations now imply that

$$\rho_1 = \rho_4 = -\rho_2 = -\rho_3.$$

In the s-th row, where $s > 3$ and $s \neq r$, let σ_k be the sum of the entries in the k-th block of three columns. Then also

$$\sigma_1 = \sigma_4 = -\sigma_2 = -\sigma_3.$$

If $\rho_1 = \pm 3$, then all elements of the same triple of columns in the r-th row have the same sign and orthogonality to the s-th row implies $\sigma_1 = 0$, which is impossible because σ_1 is odd. Hence $\rho_1 = \pm 1$. By changing the signs of some rows we may assume that $\rho_1 = 1$ for every $r > 3$. By permuting columns within each block of three we may also normalize the 4-th row, so that the first four rows are now the same as the first four rows of the array (*).

In any row after the third, within a given block of three columns two elements have the same sign and the third element the opposite sign. Moreover, these signs depend only on the block and not on the row, since $\rho_1 = 1$. The scalar product of the triples from two different rows belonging to the same block of columns is 3 if the exceptional elements have the same position in the triple and is -1 otherwise. Since the two rows are orthogonal, the exceptional elements must have the same position in exactly one of the four blocks of columns. Thus if two rows after the 4-th have the same triple of elements in the k-th block as the 4-th row, then they have no other triple in common with the 4-th row or with one another. But this implies that if one of the two rows is given, then the other is uniquely determined. Hence no other row besides these two has the same triple of elements in the k-th block as the 4-th row. Since there are eight rows after the 4-th, and since each has exactly one triple in common with the 4-th row, it follows that, for each $k \in \{1, 2, 3, 4\}$, exactly two of them have the same triple in the k-th block as the 4-th row.

The first four rows are unaltered by the following operations:

(i) interchange of the first and last columns of any triple of columns,
(ii) interchange of the second and third triple of columns, and then interchange of the second and third rows,
(iii) interchange of the first and fourth triple of columns, then interchange of the second and third rows and change of sign of these two rows,
(iv) interchange of the second and fourth triple of columns and change of their signs, then interchange of the first and third rows.

If we denote the elements of the r-th row $(r > 4)$ by $\xi_1, \ldots, \xi_{12}$, then we have

$$\xi_1 + \xi_2 + \xi_3 = 1 = \xi_{10} + \xi_{11} + \xi_{12},$$
$$\xi_4 + \xi_5 + \xi_6 = -1 = \xi_7 + \xi_8 + \xi_9,$$
$$\xi_2 - \xi_5 - \xi_8 + \xi_{11} = 2.$$

In particular in the 5-th row we have $\alpha_{52} - \alpha_{55} - \alpha_{58} + \alpha_{5,11} = 2$. Thus α_{52} and $\alpha_{5,11}$ cannot both be -1 and by an operation (iii) we may assume that $\alpha_{52} = 1$. Similarly α_{55} and α_{58} cannot both be 1 and by an operation (ii) we may assume that $\alpha_{58} = -1$. Then $\alpha_{55} = \alpha_{5,11}$ and by an operation (iv) we may assume that $\alpha_{55} = \alpha_{5,11} = -1$. By operations (i) we may finally assume that the 5-th row is the same as the 5-th row of the array (*).

As we have already shown, exactly one row after the 5-th row has the same triple $+ - +$ in the last block of columns as the 4-th and 5-th rows and this row must be the same as the 6-th row of the array (*). By permuting the last seven rows we may assume that this row is also the 6-th row of the given matrix, that the 7-th and 8-th rows have the same first triple of elements as the 4-th row, that the 9-th and 10-th rows have the same second triple of elements as the 4-th row, and that the 11-th and 12-th rows have the same third triple of elements as the 4-th row.

In any row after the 6-th we have, in addition to the relations displayed above, $\xi_{11} = 1, \xi_{10} + \xi_{12} = 0$ and

$$\xi_1 - \xi_4 - \xi_7 = \xi_2 - \xi_5 - \xi_8 = \xi_3 - \xi_6 - \xi_9 = 1.$$

In the 7-th and 8-th rows we have $\xi_1 = \xi_3 = 1, \xi_2 = -1$, and hence $\xi_5 = \xi_8 = -1$, $\xi_4 = -\xi_6 = -\xi_7 = \xi_9$. Since the first six rows are still unaltered by an operation (ii), and also by interchanging the first and third columns of the last block, we may assume that $\alpha_{74} = -1, \alpha_{7,10} = 1$. The 7-th and 8-th rows are now uniquely determined and are the same as the 7-th and 8-th rows of the array (∗).

In any row after the 8-th we have

$$\xi_2 - \xi_6 - \xi_7 + \xi_{12} = 2 = \xi_2 - \xi_4 - \xi_9 + \xi_{10}.$$

In the 9-th and 10-th rows we have $\xi_5 = \xi_{11} = 1$ and $\xi_4 = \xi_6 = -1$. Hence $\xi_2 = -\xi_8 = 1, \xi_1 = \xi_7 = -\xi_3 = -\xi_9$, and finally $\xi_9 = \xi_{10} = -\xi_{12}$. Thus the 9-th and 10-th rows are together uniquely determined and may be ordered so as to coincide with the corresponding rows of the array (∗). Similarly the 11-th and 12-th rows are together uniquely determined and may be ordered so as to coincide with the corresponding rows of the displayed array.   □

It follows from Proposition 22 that, for any five distinct rows of a Hadamard matrix of order 12, there exists exactly one pair of columns which either agree in all these rows or disagree in all these rows. Indeed, by permuting the rows we may arrange that the five given rows are the first five rows. Now, by Proposition 22, we may assume that the matrix has the form (∗). But it is evident that in this case there is exactly one pair of columns which either agree or disagree in all the first five rows, namely the 10-th and 12-th columns.

Hence a 5-(12, 6, 1) design is obtained by taking the points to be elements of the set $P = \{1, \ldots, 12\}$ and the blocks to be the $12 \cdot 11$ subsets B_{jk}, B'_{jk} with $j, k \in P$ and $j \neq k$, where

$$B_{jk} = \{i \in P : \alpha_{ij} = \alpha_{ik}\}, \quad B'_{jk} = \{i \in P : \alpha_{ij} \neq \alpha_{ik}\}.$$

The Mathieu group M_{12} may be defined as the automorphism group of this design or as the reduced automorphism group of any Hadamard matrix of order 12.

It is certainly not true in general that all Hadamard matrices of the same order n are equivalent. For example, there are 60 equivalence classes of Hadamard matrices of order 24. The Mathieu group M_{24} is connected with the Hadamard matrix of order 24 which is constructed by Paley's method, described in §2. The connection is not as immediate as for M_{12}, but the ideas involved are of general significance, as we now explain.

A sequence $x = (\xi_1, \ldots, \xi_n)$ of n 0's and 1's may be regarded as a vector in the n-dimensional vector space $V = \mathbb{F}_2^n$ over the field of two elements. If we define the *weight* $|x|$ of the vector x to be the number of nonzero coordinates ξ_k, then

(i) $|x| \geq 0$ with equality if and only if $x = 0$,

(ii) $|x + y| \leq |x| + |y|$.

The vector space V acquires the structure of a metric space if we define the *(Hamming)* *distance* between the vectors x and y to be $d(x, y) = |x - y|$.

A *binary linear code* is a subspace U of the vector space V. If U has dimension k, then a *generator matrix* for the code is a $k \times n$ matrix G whose rows form a basis for U. The *automorphism group* of the code is the group of all permutations of the n coordinates which map U onto itself. An $[n, k, d]$-*binary code* is one for which V has dimension n, U has dimension k and d is the least weight of any nonzero vector in U.

There are useful connections between codes and designs. Corresponding to any design with incidence matrix A there is the binary linear code generated over $\mathbb{F}_2$ by the rows of A. Given a binary linear code U, on the other hand, a theorem of Assmus and Mattson (1969) provides conditions under which the nonzero vectors in U with minimum weight form the rows of the incidence matrix of a t-design.

Suppose now that H is a Hadamard matrix of order n, normalized so that all elements in the first row are 1. Then $A = (H + J_n)/2$ is a matrix of 0's and 1's with all elements in the first row 1. The code $C(H)$ *defined* by the Hadamard matrix H is the subspace generated by the rows of A, considered as vectors in the n-dimensional vector space $V = \mathbb{F}_2^n$.

In particular, take $H = H_{24}$ to be the Hadamard matrix of order 24 formed by Paley's construction:

$$H_{24} = I_{24} + \begin{bmatrix} 0 & e_{23} \\ -e_{23}^t & Q \end{bmatrix},$$

where $Q = (q_{jk})$ with $q_{jk} = 0$ if $j = k$ and otherwise $= 1$ or -1 according as $j - k$ is or is not a square mod 23 ($0 \leq j, k \leq 22$). It may be shown that the *extended binary Golay code* $G_{24} = C(H_{24})$ is a 12-dimensional subspace of $\mathbb{F}_2^{24}$, that the minimum weight of any nonzero vector in G_{24} is 8, and that the sets of nonzero coordinates of the vectors $x \in G_{24}$ with $|x| = 8$ form the blocks of a 5-(24, 8, 1) design. The Mathieu group M_{24} may be defined as the automorphism group of this design or as the automorphism group of the code G_{24}.

Again, suppose that $H^{(m)}$ is the Hadamard matrix of order $n = 2^m$ defined by

$$H^{(m)} = H_2 \otimes \cdots \otimes H_2 \quad (m \text{ factors}),$$

where

$$H_2 = \begin{bmatrix} 1 & 1 \\ 1 & -1 \end{bmatrix}.$$

The *first-order Reed–Muller code* $R(1, m) = C(H^{(m)})$ is an $(m+1)$-dimensional subspace of $\mathbb{F}_2^n$ and the minimum weight of any nonzero vector in $R(1, m)$ is 2^{m-1}. It may be mentioned that the 3-$(2^m, 2^{m-1}, 2^{m-2} - 1)$ design associated with the Hadamard matrix $H^{(m)}$ has a simple geometrical interpretation. Its points are the points of m-dimensional affine space over the field of two elements, and its blocks are the hyperplanes of this space (not necessarily containing the origin).

In electronic communication a message is sent as a sequence of 'bits' (an abbreviation for *binary digits*), which may be realised physically by *off* or *on* and which may be denoted mathematically by 0 or 1. On account of noise the message received may differ slightly from that transmitted, and in some situations it is extremely important to detect and correct the errors. One way of doing so would be to send the same message many times, but it is an inefficient way. Instead suppose the message is composed of *codewords* of length n, taken from a subspace U of the vector space $V = \mathbb{F}_2^n$. There are 2^k different codewords, where k is the dimension of U. If the minimum weight of any nonzero vector in U is d, then any two distinct codewords differ in at least d places. Hence if a codeword $u \in U$ is transmitted and the received vector $v \in V$ contains less than $d/2$ errors, then v will be closer to u than to any other codeword. Thus if we are confident that any transmitted codeword will contain less than $d/2$ errors, we can correct them all by replacing each received vector by the codeword nearest to it.

The Golay code and the first-order Reed–Muller codes are of considerable practical importance in this connection. For the first-order Reed–Muller codes there is a fast algorithm for finding the nearest codeword to any received vector. Photographs of Mars taken by the Mariner 9 spacecraft were transmitted to Earth, using the code $R(1, 5)$.

Other *error-correcting codes* are used with compact discs to ensure high quality sound reproduction by eliminating imperfections due, for example, to dust particles.

8 Further Remarks

Kowalewski [22] gives a useful traditional account of determinants. Muir [28] is a storehouse of information on special types of determinants; the early Japanese work is described in Mikami [27].

Another approach to determinants, based on the work of Grassmann (1844), should be mentioned here, as it provides easy access to their formal properties and is used in the theory of differential forms. If V is an n-dimensional vector space over a field F, then there exists an associative algebra E, of dimension 2^n as a vector space over F, such that

(a) $V \subseteq E$,
(b) $v^2 = 0$ for every $v \in V$,
(c) V generates E, i.e. each element of E can be expressed as a sum of a scalar multiple of the unit element 1 and of a finite number of products of elements of V.

The associative algebra E, which is uniquely determined by these properties, is called the *Grassmann algebra* or *exterior algebra* of the vector space V. It is easily seen that any two products of n elements of V differ only by a scalar factor. Hence, for any linear transformation $A: V \rightarrow V$, there exists $d(A) \in F$ such that

$$(Av_1) \cdots (Av_n) = d(A)v_1 \cdots v_n \quad \text{for all } v_1, \ldots, v_n \in V.$$

Evidently $d(AB) = d(A)d(B)$ and in fact $d(A) = \det A$, if we identify A with its matrix with respect to some fixed basis of V. This approach to determinants is developed in Bourbaki [6]; see also Barnabei *et al.* [4].

Dieudonné (1943) has extended the notion of determinant to matrices with entries from a division ring; see Artin [1] and Cohn [9]. For a very different method, see Gelfand and Retakh [13].

Hadamard's original paper of 1893 is reproduced in [16]. Surveys on Hadamard matrices have been given by Hedayat and Wallis [19], Seberry and Yamada [34], and Craigen and Wallis [11]. Weighing designs are treated in Raghavarao [31]. For applications of Hadamard matrices to spectrometry, see Harwit and Sloane [18]. The proof of Proposition 8 is due to Shahriari [35].

Our proof of Theorem 10 is a pure existence proof. A more constructive approach was proposed by Jacobi (1846). If one applies to $n \times n$ matrices the method which we used for 2×2 matrices, one can annihilate a symmetric pair of off-diagonal entries. By choosing at each step an off-diagonal pair with maximum absolute value, one obtains a sequence of orthogonal transforms of the given symmetric matrix which converges to a diagonal matrix.

Calculating the eigenvalues of a real symmetric matrix has important practical applications, e.g. to problems of small oscillations in dynamical systems. Householder [21] and Golub and van Loan [14] give accounts of the various computational methods available.

Gantmacher [12] and Horn and Johnson [20] give general treatments of matrix theory, including the inequalities of Hadamard and Fischer. Our discussion of the Hadamard determinant problem for matrices of order not divisible by 4 is mainly based on Wojtas [37]. Further references are given in Neubauer and Ratcliffe [29].

Results of Brouwer (1983) are used in [29] to show that the upper bound in Proposition 19 is attained for infinitely many values of n. It follows that the upper bound in Proposition 20, with $m = n$, is also attained for infinitely many values of n. For if the $n \times n$ matrix A satisfies

$$A^t A = (n-1)I_n + J_n,$$

then the $2n \times 2n$ matrix

$$\bar{A} = \begin{bmatrix} A & A \\ A & -A \end{bmatrix}$$

satisfies

$$\tilde{A}^t \tilde{A} = \begin{bmatrix} L & O \\ O & L \end{bmatrix},$$

where $L = 2A^t A = (2n-2)I_n + 2J_n$.

There are introductions to design theory in Ryser [33], Hall [17], and van Lint and Wilson [25]. For more detailed information, see Brouwer [7], Lander [23] and Beth et al. [5]. Applications of design theory are treated in Chapter XIII of [5].

We mention two interesting results which are proved in Chapter 16 of Hall [17]. Given positive integers v, k, λ with $\lambda < k < v$:

(i) If $k(k-1) = \lambda(v-1)$ and if there exists a $v \times v$ matrix A of rational numbers such that

$$AA^t = (k-\lambda)I + \lambda J,$$

then A may be chosen so that in addition $JA = kJ$.

(ii) If there exists a $v \times v$ matrix A of integers such that

$$AA^t = (k - \lambda)I + \lambda J, \, JA = kJ,$$

then every entry of A is either 0 or 1, and thus A is the incidence matrix of a square 2-design.

For introductions to the classification theorem for finite simple groups, see Aschbacher [2] and Gorenstein [15]. Detailed information about the finite simple groups is given in Conway *et al.* [10]. There is a remarkable connection between the largest sporadic simple group, nicknamed the 'Monster', and modular forms; see Ray [32].

Good introductions to coding theory are given by van Lint [24] and Pless [30]. MacWilliams and Sloane [26] is more comprehensive, but less up-to-date. Assmus and Mattson [3] is a useful survey article. Connections between codes, designs and graphs are treated in Cameron and van Lint [8]. The historical account in Thompson [36] recaptures the excitement of scientific discovery.

9 Selected References

[1] E. Artin, *Geometric algebra*, reprinted, Wiley, New York, 1988. [Original edition, 1957]

[2] M. Aschbacher, The classification of the finite simple groups, *Math. Intelligencer* **3** (1980/81), 59–65.

[3] E.F. Assmus Jr. and H.F. Mattson Jr., Coding and combinatorics, *SIAM Rev.* **16** (1974), 349–388.

[4] M. Barnabei, A. Brini and G.-C. Rota, On the exterior calculus of invariant theory, *J. Algebra* **96** (1985), 120–160.

[5] T. Beth, D. Jungnickel and H. Lenz, *Design theory*, 2nd ed., 2 vols., Cambridge University Press, 1999.

[6] N. Bourbaki, *Algebra I, Chapters 1–3*, Hermann, Paris, 1974. [French original, 1948]

[7] A.E. Brouwer, Block designs, *Handbook of combinatorics* (ed. R.L. Graham, M. Grötschel and L. Lóvasz), Vol. I, pp. 693–745, Elsevier, Amsterdam, 1995.

[8] P.J. Cameron and J.H. van Lint, *Designs, graphs, codes and their links*, Cambridge University Press, 1991.

[9] P.M. Cohn, *Algebra*, 2nd ed., Vol. 3, Wiley, Chichester, 1991.

[10] J.H. Conway *et al.*, *Atlas of finite groups*, Clarendon Press, Oxford, 1985.

[11] R. Craigen and W.D. Wallis, Hadamard matrices: 1893–1993, *Congr. Numer.* **97** (1993), 99–129.

[12] F.R. Gantmacher, *The theory of matrices*, English transl. by K. Hirsch, 2 vols., Chelsea, New York, 1960.

[13] I.M. Gelfand and V.S. Retakh, A theory of noncommutative determinants and characteristic functions of graphs, *Functional Anal. Appl.* **26** (1992), 231–246.

[14] G.H. Golub and C.F. van Loan, *Matrix computations*, 3rd ed., Johns Hopkins University Press, Baltimore, MD, 1996.

[15] D. Gorenstein, Classifying the finite simple groups, *Bull. Amer. Math. Soc. (N.S.)* **14** (1986), 1–98.

[16] J. Hadamard, Résolution d'une question relative aux determinants, *Selecta*, pp. 136–142, Gauthier-Villars, Paris, 1935 and *Oeuvres*, Tome I, pp. 239–245, CNRS, Paris, 1968.

[17] M. Hall, *Combinatorial theory*, 2nd ed., Wiley, New York, 1986.

[18] M. Harwit and N.J.A. Sloane, *Hadamard transform optics*, Academic Press, New York, 1979.

[19] A. Hedayat and W.D. Wallis, Hadamard matrices and their applications, *Ann. Statist.* **6** (1978), 1184–1238.

[20] R.A. Horn and C.A. Johnson, *Matrix analysis*, Cambridge University Press, 1985.

[21] A.S. Householder, *The theory of matrices in numerical analysis*, Blaisdell, New York, 1964.

[22] G. Kowalewski, *Einführung in die Determinantentheorie*, 4th ed., de Gruyter, Berlin, 1954.

[23] E.S. Lander, *Symmetric designs: an algebraic approach*, London Mathematical Society Lecture Note Series **74**, Cambridge University Press, 1983.

[24] J.H. van Lint, *Introduction to coding theory*, 3rd ed., Springer, Berlin, 2000.

[25] J.H. van Lint and R.M. Wilson, *A course in combinatorics*, Cambridge University Press, 1992.

[26] F.J. MacWilliams and N.J.A. Sloane, *The theory of error-correcting codes*, 2 vols., North-Holland, Amsterdam, 1977.

[27] Y. Mikami, *The development of mathematics in China and Japan*, 2nd ed., Chelsea, New York, 1974.

[28] T. Muir, *The theory of determinants in the historical order of development*, reprinted in 2 vols, Dover, New York, 1960.

[29] M.G. Neubauer and A.J. Ratcliffe, The maximum determinant of ± 1 matrices, *Linear Algebra Appl.* **257** (1997), 289–306.

[30] V. Pless, *Introduction to the theory of error-correcting codes*, 3rd ed., Wiley, New York, 1998.

[31] D. Raghavarao, *Constructions and combinatorial problems in design of experiments*, Wiley, New York, 1971.

[32] U. Ray, Generalized Kac–Moody algebras and some related topics, *Bull. Amer. Math. Soc. (N.S.)* **38** (2001), 1–42.

[33] H.J. Ryser, *Combinatorial mathematics*, Mathematical Association of America, 1963.

[34] J. Seberry and M. Yamada, Hadamard matrices, sequences and block designs, *Contemporary design theory* (ed. J.H. Dinitz and D.R. Stinson), pp. 431–560, Wiley, New York, 1992.

[35] S. Shahriari, On maximizing det $X^t X$, *Linear and multilinear algebra* **36** (1994), 275–278.

[36] T.M. Thompson, *From error-correcting codes through sphere packings to simple groups*, Mathematical Association of America, 1983.

[37] M. Wojtas, On Hadamard's inequality for the determinants of order non-divisible by 4, *Colloq. Math.* **12** (1964), 73–83.

VI

Hensel's p-adic Numbers

The ring $\mathbb{Z}$ of all integers has a very similar algebraic structure to the ring $\mathbb{C}[z]$ of all polynomials in one variable with complex coefficients. This similarity extends to their fields of fractions: the field $\mathbb{Q}$ of rational numbers and the field $\mathbb{C}(z)$ of rational functions in one variable with complex coefficients. Hensel (1899) had the bold idea of pushing this analogy even further. For any $\zeta \in \mathbb{C}$, the ring $\mathbb{C}[z]$ may be embedded in the ring $\mathbb{C}_\zeta[[z]]$ of all functions $f(z) = \sum_{n\geq 0} \alpha_n (z-\zeta)^n$ with complex coefficients α_n which are holomorphic at ζ, and the field $\mathbb{C}(z)$ may be embedded in the field $\mathbb{C}_\zeta((z))$ of all functions $f(z) = \sum_{n\in\mathbb{Z}} \alpha_n (z - \zeta)^n$ with complex coefficients α_n which are meromorphic at ζ, i.e. $\alpha_n \neq 0$ for at most finitely many $n < 0$. Hensel constructed, for each prime p, a ring $\mathbb{Z}_p$ of all 'p-adic integers' $\sum_{n\geq 0} \alpha_n p^n$, where $\alpha_n \in \{0, 1, \ldots, p-1\}$, and a field $\mathbb{Q}_p$ of all 'p-adic numbers' $\sum_{n\in\mathbb{Z}} \alpha_n p^n$, where $\alpha_n \in \{0, 1, \ldots, p-1\}$ and $\alpha_n \neq 0$ for at most finitely many $n < 0$. This led him to arithmetic analogues of various analytic results and even to analytic methods of proving them. Hensel's idea of concentrating attention on one prime at a time has proved very fruitful for algebraic number theory. Furthermore, his methods enable the theory of algebraic numbers and the theory of algebraic functions of one variable to be developed completely in parallel.

Hensel simply defined p-adic integers by their power series expansions. We will adopt a more general approach, due to Kürschák (1913), which is based on absolute values.

1 Valued Fields

Let F be an arbitrary field. An *absolute value* on F is a map $|\ | : F \to \mathbb{R}$ with the following properties:

(V1) $|0| = 0, |a| > 0$ *for all* $a \in F$ *with* $a \neq 0$;
(V2) $|ab| = |a||b|$ *for all* $a, b \in F$;
(V3) $|a + b| \leq |a| + |b|$ *for all* $a, b \in F$.

A field with an absolute value will be called simply a *valued field*.

A *non-archimedean absolute value* on F is a map $|\ | : F \to \mathbb{R}$ with the properties **(V1)**, **(V2)** and

(V3)' $|a + b| \leq \max(|a|, |b|)$ *for all* $a, b \in F$.

W.A. Coppel, *Number Theory: An Introduction to Mathematics*, Universitext,
DOI: 10.1007/978-0-387-89486-7_6, © Springer Science + Business Media, LLC 2009

A non-archimedean absolute value is indeed an absolute value, since **(V1)** implies that **(V3)**′ is a strengthening of **(V3)**. An absolute value is said to be *archimedean* if it is not non-archimedean.

The inequality **(V3)** is usually referred to as the *triangle inequality* and **(V3)**′ as the 'strong triangle', or *ultrametric*, inequality.

If F is a field with an absolute value $| \, |$, then the set of real numbers $|a|$ for all nonzero $a \in F$ is clearly a subgroup of the multiplicative group of positive real numbers. This subgroup will be called the *value group* of the valued field.

Here are some examples to illustrate these definitions:

(i) An arbitrary field F has a *trivial* non-archimedean absolute value defined by

$$|0| = 0, \quad |a| = 1 \quad \text{if } a \neq 0.$$

(ii) The ordinary absolute value

$$|a| = a \quad \text{if } a \geq 0, \quad |a| = -a \quad \text{if } a < 0,$$

defines an archimedean absolute value on the field $\mathbb{Q}$ of rational numbers. We will denote this absolute value by $| \, |_\infty$ to avoid confusion with other absolute values on $\mathbb{Q}$ which will now be defined.

If p is a fixed prime, any rational number $a \neq 0$ can be uniquely expressed in the form $a = ep^v m/n$, where $e = \pm 1$, $v = v_p(a)$ is an integer and m, n are relatively prime positive integers which are not divisible by p. It is easily verified that a non-archimedean absolute value is defined on $\mathbb{Q}$ by putting

$$|0|_p = 0, \quad |a|_p = p^{-v_p(a)} \quad \text{if } a \neq 0.$$

We call this the *p-adic absolute value*.

(iii) Let $F = K(t)$ be the field of all rational functions in one indeterminate with coefficients from some field K. Any rational function $f \neq 0$ can be uniquely expressed in the form $f = g/h$, where g and h are relatively prime polynomials with coefficients from K and h is *monic* (i.e., has leading coefficient 1). If we denote the degrees of g and h by $\partial(g)$ and $\partial(h)$, then a non-archimedean absolute value is defined on F by putting, for a fixed $q > 1$,

$$|0|_\infty = 0, \quad |f|_\infty = q^{\partial(g) - \partial(h)} \quad \text{if } f \neq 0.$$

Other absolute values on F can be defined in the following way. If $p \in K[t]$ is a fixed irreducible polynomial, then any rational function $f \neq 0$ can be uniquely expressed in the form $f = p^v g/h$, where $v = v_p(f)$ is an integer, g and h are relatively prime polynomials with coefficients from K which are not divisible by p, and h is monic. It is easily verified that a non-archimedean absolute value is defined on F by putting, for a fixed $q > 1$,

$$|0|_p = 0, \quad |f|_p = q^{-\partial(p) v_p(f)} \quad \text{if } f \neq 0.$$

(iv) Let $F = K((t))$ be the field of all formal Laurent series $f(t) = \sum_{n \in \mathbb{Z}} a_n t^n$ with coefficients $a_n \in K$ such that $a_n \neq 0$ for at most finitely many $n < 0$. A non-archimedean absolute value is defined on F by putting, for a fixed $q > 1$,

$$|0| = 0, \quad |f| = q^{-v(f)} \quad \text{if } f \neq 0,$$

where $v(f)$ is the least integer n such that $a_n \neq 0$.

(v) Let $F = C_\zeta((z))$ denote the field of all complex-valued functions $f(z) = \sum_{n \in \mathbb{Z}} a_n (z - \zeta)^n$ which are meromorphic at $\zeta \in \mathbb{C}$. Any $f \in F$ which is not identically zero can be uniquely expressed in the form $f(z) = (z - \zeta)^v g(z)$, where $v = v_\zeta(f)$ is an integer, g is holomorphic at ζ and $g(\zeta) \neq 0$. A non-archimedean absolute value is defined on F by putting, for a fixed $q > 1$,

$$|0|_\zeta = 0, \quad |f|_\zeta = q^{-v_\zeta(f)} \quad \text{if } f \neq 0.$$

It should be noted that in examples (iii) and (iv) the restriction of the absolute value to the ground field K is the trivial absolute value, and the same holds in example (v) for the restriction of the absolute value to $\mathbb{C}$. For all the absolute values considered in examples (iii)–(v) the value group is an infinite cyclic group.

We now derive some simple properties common to all absolute values. The notation in the statement of the following lemma is a bit sloppy, since we use the same symbol to denote the unit elements of both F and $\mathbb{R}$ (as we have already done for the zero elements).

Lemma 1 *In any field F with an absolute value $|\ |$ the following properties hold:*

  (i) $|1| = 1$, $|-1| = 1$ *and, more generally,* $|a| = 1$ *for every $a \in F$ which is a root of unity;*
 (ii) $|-a| = |a|$ *for every $a \in F$;*
(iii) $||a| - |b||_\infty \leq |a - b|$ *for all $a, b \in F$, where $|\ |_\infty$ is the ordinary absolute value on $\mathbb{R}$;*
(iv) $|a^{-1}| = |a|^{-1}$ *for every $a \in F$ with $a \neq 0$.*

Proof By taking $a = b = 1$ in **(V2)** and using **(V1)**, we obtain $|1| = 1$. If $a^n = 1$ for some positive integer n, it now follows from **(V2)** that $\alpha = |a|$ satisfies $\alpha^n = 1$. Since $\alpha > 0$, this implies $\alpha = 1$. In particular, $|-1| = 1$. Taking $b = -1$ in **(V2)**, we now obtain (ii).

Replacing a by $a - b$ in **(V3)**, we obtain

$$|a| - |b| \leq |a - b|.$$

Since a and b may be interchanged, by (ii), this implies (iii). Finally, if we take $b = a^{-1}$ in **(V2)** and use (i), we obtain (iv). $\qquad \square$

It follows from Lemma 1(i) that a finite field admits only the trivial absolute value.

We show next how non-archimedean and archimedean absolute values may be distinguished from one another. The notation in the statement of the following proposition is very sloppy, since we use the same symbol to denote both the positive integer n and the sum $1 + 1 + \cdots + 1$ (n summands), although the latter may be 0 if the field has prime characteristic.

Proposition 2 *Let F be a field with an absolute value $|\ |$. Then the following properties are equivalent:*

(i) $|2| \leq 1$;
(ii) $|n| \leq 1$ *for every positive integer n;*
(iii) *the absolute value $|\ |$ is non-archimedean.*

Proof It is trivial that (iii) $\Rightarrow$ (i). Suppose now that (i) holds. Then $|2^k| = |2|^k \leq 1$ for any positive integer k. An arbitrary positive integer n can be written to the base 2 in the form

$$n = a_0 + a_1 2 + \cdots + a_g 2^g,$$

where $a_i \in \{0, 1\}$ for all $i < g$ and $a_g = 1$. Then

$$|n| \leq |a_0| + |a_1| + \cdots + |a_g| \leq g + 1.$$

Now consider the powers n^k. Since $n < 2^{g+1}$, we have $n^k < 2^{k(g+1)}$ and hence

$$n^k = b_0 + b_1 2 + \cdots + b_h 2^h,$$

where $b_j \in \{0, 1\}$ for all $j < h$, $b_h = 1$ and $h < k(g + 1)$. Thus

$$|n|^k = |n^k| \leq h + 1 \leq k(g + 1).$$

Taking k-th roots and letting $k \to \infty$, we obtain $|n| \leq 1$, since $k^{1/k} = e^{(\log k)/k} \to 1$ and likewise $(g + 1)^{1/k} = e^{(\log(g+1))/k} \to 1$. Thus (i) $\Rightarrow$ (ii).

Suppose next that (ii) holds. Then, since the binomial coefficients are positive integers,

$$|x + y|^n = |(x + y)^n| = \left| \sum_{k=0}^{n} \binom{n}{k} x^k y^{n-k} \right|$$
$$\leq \sum_{k=0}^{n} |x|^k |y|^{n-k}$$
$$\leq (n + 1)\rho^n,$$

where $\rho = \max(|x|, |y|)$. Taking n-th roots and letting $n \to \infty$, we obtain $|x + y| \leq \rho$. Thus (ii) $\Rightarrow$ (iii). $\qquad \square$

It follows from Proposition 2 that for an archimedean absolute value the sequence $(|n|)$ is unbounded, since $|2^k| \to \infty$ as $k \to \infty$. Consequently, for any $a, b \in F$ with $a \neq 0$, there is a positive integer n such that $|na| > |b|$. The name 'archimedean' is used because of the analogy with the archimedean axiom of geometry. It follows also from Proposition 2 that any absolute value on a field of prime characteristic is non-archimedean, since there are only finitely many distinct values of $|n|$.

2 Equivalence

If λ, μ, α are positive real numbers with $\alpha < 1$, then

$$\left(\frac{\lambda}{\lambda + \mu}\right)^{\alpha} + \left(\frac{\mu}{\lambda + \mu}\right)^{\alpha} > \frac{\lambda}{\lambda + \mu} + \frac{\mu}{\lambda + \mu} = 1$$

and hence

$$\lambda^{\alpha} + \mu^{\alpha} > (\lambda + \mu)^{\alpha}.$$

It follows that if $|\ |$ is an absolute value on a field F and if $0 < \alpha < 1$, then $|\ |^{\alpha}$ is also an absolute value, since

$$|a + b|^{\alpha} \leq (|a| + |b|)^{\alpha} \leq |a|^{\alpha} + |b|^{\alpha}.$$

Actually, if $|\ |$ is a *non-archimedean* absolute value on a field F, then it follows directly from the definition that, for any $\alpha > 0$, $|\ |^{\alpha}$ is also a non-archimedean absolute value on F. However, if $|\ |$ is an *archimedean* absolute value on F then, for all large $\alpha > 0$, $|\ |^{\alpha}$ is not an absolute value on F. For $|2| > 1$ and hence, if $\alpha > \log 2/\log |2|$,

$$|1 + 1|^{\alpha} > 2 = |1|^{\alpha} + |1|^{\alpha}.$$

Proposition 3 *Let $|\ |_1$ and $|\ |_2$ be absolute values on a field F such that $|a|_2 < 1$ for any $a \in F$ with $|a|_1 < 1$. If $|\ |_1$ is nontrivial, then there exists a real number $\rho > 0$ such that*

$$|a|_2 = |a|_1^{\rho} \quad \text{for every } a \in F.$$

Proof By taking inverses we see that also $|a|_2 > 1$ for any $a \in F$ with $|a|_1 > 1$. Choose $b \in F$ with $|b|_1 > 1$. For any nonzero $a \in F$ we have $|a|_1 = |b|_1^{\gamma}$, where

$$\gamma = \log |a|_1/\log |b|_1.$$

Let m, n be integers with $n > 0$ such that $m/n > \gamma$. Then $|a|_1^n < |b|_1^m$ and hence $|a^n/b^m|_1 < 1$. Therefore also $|a^n/b^m|_2 < 1$ and by reversing the argument we obtain

$$m/n > \log |a|_2/\log |b|_2.$$

Similarly if m', n' are integers with $n' > 0$ such that $m'/n' < \gamma$, then

$$m'/n' < \log |a|_2/\log |b|_2.$$

It follows that

$$\log |a|_2/\log |b|_2 = \gamma = \log |a|_1/\log |b|_1.$$

Thus if we put $\rho = \log |b|_2/\log |b|_1$, then $\rho > 0$ and $|a|_2 = |a|_1^{\rho}$. This holds trivially also for $a = 0$. $\qquad\square$

Two absolute values, $| \ |_1$ and $| \ |_2$, on a field F are said to be *equivalent* when, for any $a \in F$,

$$|a|_1 < 1 \quad \text{if and only if } |a|_2 < 1.$$

This implies that $|a|_1 > 1$ if and only if $|a|_2 > 1$ and hence also that $|a|_1 = 1$ if and only if $|a|_2 = 1$. Thus if one absolute value is trivial, so also is the other. It now follows from Proposition 3 that *two absolute values, $| \ |_1$ and $| \ |_2$, on a field F are equivalent if and only if there exists a real number $\rho > 0$ such that $|a|_2 = |a|_1^\rho$ for every $a \in F$.*

We have seen that the field $\mathbb{Q}$ of rational numbers admits the p-adic absolute values $| \ |_p$ in addition to the ordinary absolute value $| \ |_\infty$. These absolute values are all inequivalent since, if p and q are distinct primes,

$$|p|_p < 1, \quad |p|_q = 1, \quad |p|_\infty = p > 1.$$

It was first shown by Ostrowski (1918) that these are essentially the only absolute values on $\mathbb{Q}$:

Proposition 4 *Every nontrivial absolute value $| \ |$ of the rational field $\mathbb{Q}$ is equivalent either to the ordinary absolute value $| \ |_\infty$ or to a p-adic absolute value $| \ |_p$ for some prime p.*

Proof Let b, c be integers > 1. By writing c to the base b, we obtain

$$c = c_m b^m + c_{m-1} b^{m-1} + \cdots + c_0,$$

where $0 \le c_j < b \ (j = 0, \ldots, m)$ and $c_m \ne 0$. Then $m \le \log c / \log b$, since $c_m \ge 1$. If we put $\mu = \max_{1 \le d < b} |d|$, it follows from the triangle inequality that

$$|c| \le \mu(1 + \log c / \log b)\{\max(1, |b|)\}^{\log c / \log b}.$$

Taking $c = a^n$ we obtain, for any $a > 1$,

$$|a| \le \mu^{1/n}(1 + n \log a / \log b)^{1/n}\{\max(1, |b|)\}^{\log a / \log b}$$

and hence, letting $n \to \infty$,

$$|a| \le \{\max(1, |b|)\}^{\log a / \log b}.$$

Suppose first that $|a| > 1$ for some $a > 1$. It follows that $|b| > 1$ for every $b > 1$ and

$$|b|^{1/\log b} \ge |a|^{1/\log a}.$$

In fact, since a and b may now be interchanged,

$$|b|^{1/\log b} = |a|^{1/\log a}.$$

Thus $\rho = \log |a| / \log a$ is a positive real number independent of $a > 1$ and $|a| = a^\rho$. It follows that $|a| = |a|_\infty^\rho$ for every rational number a. Thus the absolute value is equivalent to the ordinary absolute value.

Suppose next that $|a| \leq 1$ for every $a > 1$ and so for every $a \in \mathbb{Z}$. Since the absolute value on $\mathbb{Q}$ is nontrivial, we must have $|a| < 1$ for some integer $a \neq 0$. The set M of all $a \in \mathbb{Z}$ such that $|a| < 1$ is a proper ideal in $\mathbb{Z}$ and hence is generated by an integer $p > 1$. We will show that p must be a prime. Suppose $p = bc$, where b and c are positive integers. Since $|b||c| = |p| < 1$, we may assume without loss of generality that $|b| < 1$. Then $b \in M$ and thus $b = pd$ for some $d \in \mathbb{Z}$. Hence $cd = 1$ and so $c = 1$. Thus p has no nontrivial factorization.

Every rational number $a \neq 0$ can be expressed in the form $a = p^{v}b/c$, where v is an integer and b, c are integers not divisible by p. Hence $|b| = |c| = 1$ and $|a| = |p|^{v}$. We can write $|p| = p^{-\rho}$, for some real number $\rho > 0$. Then $|a| = p^{-v\rho} = |a|_{p}^{\rho}$, and thus the absolute value is equivalent to the p-adic absolute value. $\qquad\square$

Similarly, the absolute values on the field $F = K(t)$ considered in example (iii) of §1 are all inequivalent and it may be shown that any nontrivial absolute value on F whose restriction to K is trivial is equivalent to one of these absolute values.

In example (ii) of §1 we have made a specific choice in each class of equivalent absolute values. The choice which has been made ensures the validity of the *product formula*: for any nonzero $a \in \mathbb{Q}$,

$$|a|_{\infty} \prod_{p} |a|_{p} = 1,$$

where $|a|_{p} \neq 1$ for at most finitely many p.

Similarly, in example (iii) of §1 the absolute values have been chosen so that, for any nonzero $f \in K(t)$, $|f|_{\infty} \prod_{p} |f|_{p} = 1$, where $|f|_{p} \neq 1$ for at most finitely many p.

The following *approximation theorem*, due to Artin and Whaples (1945), treats several absolute values simultaneously. For p-adic absolute values of the rational field $\mathbb{Q}$ the result also follows from the Chinese remainder theorem (Corollary II.38).

Proposition 5 *Let* $|\;|_{1}, \ldots, |\;|_{m}$ *be nontrivial pairwise inequivalent absolute values of an arbitrary field F and let $x_{1}, \ldots, x_{m}$ be any elements of F. Then for each real $\varepsilon > 0$ there exists an $x \in F$ such that*

$$|x - x_{k}|_{k} < \varepsilon \quad \text{for } 1 \leq k \leq m.$$

Proof During the proof we will more than once use the fact that if $f_{n}(x) = x^{n}(1 + x^{n})^{-1}$, then $|f_{n}(a)| \to 0$ or 1 as $n \to \infty$ according as $|a| < 1$ or $|a| > 1$.

We show first that *there exists an $a \in F$ such that*

$$|a|_{1} > 1, \quad |a|_{k} < 1 \quad \text{for } 2 \leq k \leq m.$$

Since $|\;|_{1}$ and $|\;|_{2}$ are nontrivial and inequivalent, there exist $b, c \in F$ such that

$$|b|_{1} < 1, \quad |b|_{2} \geq 1,$$
$$|c|_{1} \geq 1, \quad |c|_{2} < 1.$$

If we put $a = b^{-1}c$, then $|a|_1 > 1$, $|a|_2 < 1$. This proves the assertion for $m = 2$. We now assume $m > 2$ and use induction. Then there exist $b, c \in F$ such that

$$|b|_1 > 1, \quad |b|_k < 1 \quad \text{for } 1 < k < m,$$
$$|c|_1 > 1, \quad |c|_m < 1.$$

If $|b|_m < 1$ we can take $a = b$. If $|b|_m = 1$ we can take $a = b^n c$ for sufficiently large n. If $|b|_m > 1$ we can take $a = f_n(b)c$ for sufficiently large n.

Thus for each $i \in \{1, \ldots, m\}$ we can choose $a_i \in F$ so that

$$|a_i|_i > 1, \quad |a_i|_k < 1 \quad \text{for all } k \neq i.$$

Then

$$x = x_1 f_n(a_1) + \cdots + x_m f_n(a_m)$$

satisfies the requirements of the proposition for sufficiently large n.   □

It follows from Proposition 5, that if $|\ |_1, \ldots, |\ |_m$ are nontrivial pairwise inequivalent absolute values of a field F, then there exists an $a \in F$ such that $|a|_k > 1$ ($k = 1, \ldots, m$). Consequently the absolute values are *multiplicatively independent*, i.e. if $\rho_1, \ldots, \rho_m$ are nonnegative real numbers, not all zero, then for some nonzero $a \in F$,

$$|a|_1^{\rho_1} \cdots |a|_m^{\rho_m} \neq 1.$$

3 Completions

Any field F with an absolute value $|\ |$ has the structure of a metric space, with the metric

$$d(a, b) = |a - b|,$$

and thus has an associated topology. Since $|a| < 1$ if and only if $a^n \to 0$ as $n \to \infty$, it follows that two absolute values are equivalent if and only if the induced topologies are the same.

When we use topological concepts in connection with valued fields we will always refer to the topology induced by the metric space structure. In this sense addition and multiplication are continuous operations, since

$$|(a + b) - (a_0 + b_0)| \leq |a - a_0| + |b - b_0|,$$
$$|ab - a_0 b_0| \leq |a - a_0||b| + |a_0||b - b_0|.$$

Inversion is also continuous at any point $a_0 \neq 0$, since if $|a - a_0| < |a_0|/2$ then $|a_0| < 2|a|$ and

$$|a^{-1} - a_0^{-1}| = |a - a_0||a|^{-1}|a_0|^{-1} < 2|a_0|^{-2}|a - a_0|.$$

Thus a valued field is a *topological field*.

It will now be shown that the procedure by which Cantor extended the field of rational numbers to the field of real numbers can be generalized to any valued field.

Let F be a field with an absolute value $|\ |$. A sequence (a_n) of elements of F is said to *converge* to an element a of F, and a is said to be the *limit* of the sequence (a_n), if for each real $\varepsilon > 0$ there is a corresponding positive integer $N = N(\varepsilon)$ such that

$$|a_n - a| < \varepsilon \quad \text{for all } n \geq N.$$

It is easily seen that the limit of a convergent sequence is uniquely determined.

A sequence (a_n) of elements of F is said to be a *fundamental sequence* if for each $\varepsilon > 0$ there is a corresponding positive integer $N = N(\varepsilon)$ such that

$$|a_m - a_n| < \varepsilon \quad \text{for all } m, n \geq N.$$

Any convergent sequence is a fundamental sequence, since

$$|a_m - a_n| \leq |a_m - a| + |a_n - a|,$$

but the converse need not hold. However, any fundamental sequence is bounded since, if $m = N(1)$, then for $n \geq m$ we have

$$|a_n| \leq |a_m - a_n| + |a_m| < 1 + |a_m|.$$

Thus $|a_n| \leq \mu$ for all n, where $\mu = \max\{|a_1|, \ldots, |a_{m-1}|, 1 + |a_m|\}$.

The preceding definitions are specializations of the definitions for an arbitrary metric space (cf. Chapter I, §4). We now take advantage of the algebraic structure of F. Let $A = (a_n)$ and $B = (b_n)$ be two fundamental sequences. We write $A = B$ if $a_n = b_n$ for all n, and we define the sum and product of A and B to be the sequences

$$A + B = (a_n + b_n), \quad AB = (a_n b_n).$$

These are again fundamental sequences. For we can choose $\mu \geq 1$ so that $|a_n| \leq \mu$, $|b_n| \leq \mu$ for all n and then choose a positive integer N so that

$$|a_m - a_n| < \varepsilon/2\mu, \quad |b_m - b_n| < \varepsilon/2\mu \quad \text{for all } m, n \geq N.$$

It follows that, for all $m, n \geq N$,

$$|(a_m + b_m) - (a_n + b_n)| \leq |a_m - a_n| + |b_m - b_n| < \varepsilon/2\mu + \varepsilon/2\mu \leq \varepsilon,$$

and similarly

$$|a_m b_m - a_n b_n| \leq |a_m - a_n||b_m| + |a_n||b_m - b_n| < (\varepsilon/2\mu)\mu + (\varepsilon/2\mu)\mu = \varepsilon.$$

It is easily seen that the set $\mathscr{F}$ of all fundamental sequences is a commutative ring with respect to these operations. The subset of all constant sequences (a), i.e. $a_n = a$ for all n, forms a field isomorphic to F. Thus we may regard F as embedded in $\mathscr{F}$.

Let $\mathscr{N}$ denote the subset of $\mathscr{F}$ consisting of all sequences (a_n) which converge to 0. Evidently $\mathscr{N}$ is a subring of $\mathscr{F}$ and actually an ideal, since any fundamental sequence is bounded. We will show that $\mathscr{N}$ is even a maximal ideal.

Let (a_n) be a fundamental sequence which is not in $\mathcal{N}$. Then there exists $\mu > 0$ such that $|a_v| \geq \mu$ for infinitely many v. Since $|a_m - a_n| < \mu/2$ for all $m, n \geq N$, it follows that $|a_n| > \mu/2$ for all $n \geq N$. Put $b_n = a_n^{-1}$ if $a_n \neq 0$, $b_n = 0$ if $a_n = 0$. Then (b_n) is a fundamental sequence since, for $m, n \geq N$,

$$|b_m - b_n| = |(a_n - a_m)/a_m a_n| \leq 4\mu^{-2}|a_n - a_m|.$$

Since $(1) - (b_n a_n) \in \mathcal{N}$, the ideal generated by (a_n) and $\mathcal{N}$ contains the constant sequence (1) and hence every sequence in $\mathcal{F}$. Since this holds for each sequence $(a_n) \in \mathcal{F} \backslash \mathcal{N}$, the ideal $\mathcal{N}$ is maximal.

Consequently (see Chapter I, §8) the quotient $\bar{F} = \mathcal{F}/\mathcal{N}$ is a field. Since (0) is the only constant sequence in $\mathcal{N}$, by mapping each constant sequence into the coset of $\mathcal{N}$ which contains it we obtain a field in $\bar{F}$ isomorphic to F. Thus we may regard F as embedded in $\bar{F}$.

It follows from Lemma 1(iii), and from the completeness of the field of real numbers, that $|A| = \lim_{n \to \infty} |a_n|$ exists for any fundamental sequence $A = (a_n)$. Moreover,

$$|A| \geq 0, \quad |AB| = |A||B|, \quad |A + B| \leq |A| + |B|.$$

Furthermore $|A| = 0$ if and only if $A \in \mathcal{N}$. It follows that $|B| = |C|$ if $B - C \in \mathcal{N}$, since

$$|B| \leq |B - C| + |C| = |C| \leq |C - B| + |B| = |B|.$$

Thus we may consider $|\ |$ as defined on $\bar{F} = \mathcal{F}/\mathcal{N}$, and it is then an absolute value on the field $\bar{F}$ which coincides with the original absolute value when restricted to the field F.

If $A = (a_n)$ is a fundamental sequence, and if A_m is the constant sequence (a_m), then $|A - A_m|$ can be made arbitrarily small by taking m sufficiently large. It follows that F is *dense* in $\bar{F}$, i.e. for any $\alpha \in \bar{F}$ and any $\varepsilon > 0$ there exists $a \in F$ such that $|\alpha - a| < \varepsilon$.

We show finally that $\bar{F}$ is *complete* as a metric space, i.e. every fundamental sequence of elements of $\bar{F}$ converges to an element of $\bar{F}$. For let (α_n) be a fundamental sequence in $\bar{F}$. Since F is dense in $\bar{F}$, for each n we can choose $a_n \in F$ so that $|a_n - \alpha_n| < 1/n$. Since

$$|a_m - a_n| \leq |a_m - \alpha_m| + |\alpha_m - \alpha_n| + |\alpha_n - a_n|,$$

it follows that (a_n) is also a fundamental sequence. Thus there exists $\alpha \in \bar{F}$ such that $\lim_{n \to \infty} |a_n - \alpha| = 0$. Since

$$|\alpha_n - \alpha| \leq |\alpha_n - a_n| + |a_n - \alpha|,$$

we have also $\lim_{n \to \infty} |\alpha_n - \alpha| = 0$. Thus the sequence (α_n) converges to α.

Summing up, we have proved

Proposition 6 *If F is a field with an absolute value $|\ |$, then there exists a field $\bar{F}$ containing F, with an absolute value $|\ |$ extending that of F, such that $\bar{F}$ is complete and F is dense in $\bar{F}$.*

It is easily seen that $\bar{F}$ is uniquely determined, up to an isomorphism which preserves the absolute value. The field $\bar{F}$ is called the *completion* of the valued field F. The density of F in $\bar{F}$ implies that the absolute value on the completion $\bar{F}$ is non-archimedean or archimedean according as the absolute value on F is non-archimedean or archimedean.

It is easy to see that in example (iv) of §1 the valued field $F = K((t))$ of all formal Laurent series is complete, i.e. it is its own completion. For let $\{f^{(k)}\}$ be a fundamental sequence in F. Given any positive integer N, there is a positive integer $M = M(N)$ such that $|f^{(k)} - f^{(j)}| < q^{-N}$ for $j, k \geq M$. Thus we can write

$$f^{(k)}(t) = \sum_{n \leq N} a_n t^n + \sum_{n > N} a_n^{(k)} t^n \quad \text{for all } k \geq M.$$

If $f(t) = \sum_{n \in \mathbb{Z}} a_n t^n$, then $\lim_{k \to \infty} |f^{(k)} - f| = 0$.

On the other hand, given any $f(t) = \sum_{n \in \mathbb{Z}} a_n t^n \in K((t))$, we have $|f^{(k)} - f| \to 0$ as $k \to \infty$, where $f^{(k)}(t) = \sum_{n \leq k} a_n t^n \in K(t)$. It follows that $K((t))$ is the completion of the field $K(t)$ of rational functions considered in example (iii) of §1, with the absolute value $| \ |_t$ corresponding to the irreducible polynomial $p(t) = t$ (for which $\partial(p) = 1$).

The completion of the rational field $\mathbb{Q}$ with respect to the p-adic absolute value $| \ |_p$ will be denoted by $\mathbb{Q}_p$, and the elements of $\mathbb{Q}_p$ will be called *p-adic numbers*.

The completion of the rational field $\mathbb{Q}$ with respect to the ordinary absolute value $| \ |_\infty$ is of course the real field $\mathbb{R}$. In §6 we will show that the only fields with a complete archimedean absolute value are the real field $\mathbb{R}$ and the complex field $\mathbb{C}$, and the absolute value has the form $| \ |_\infty^\rho$ for some $\rho > 0$. In fact $\rho \leq 1$, since $2^\rho \leq 1^\rho + 1^\rho = 2$. Thus an arbitrary archimedean valued field is equivalent to a subfield of $\mathbb{C}$ with the usual absolute value. (Hence, for a field with an archimedean absolute value $| \ |$, $|n| > 1$ for every integer $n > 1$ and $|n| \to \infty$ as $n \to \infty$.) Since this case may be considered well-known, we will in the following devote our attention primarily to the peculiarities of non-archimedean valued fields.

We will later be concerned with extending an absolute value on a field F to a field E which is a finite extension of F. Since all that matters for some purposes is that E is a vector space over F, it is useful to introduce the following definition.

Let F be a field with an absolute value $| \ |$ and let E be a vector space over F. A *norm* on E is a map $\| \ \| : E \to \mathbb{R}$ with the following properties:

(i) $\|a\| > 0$ for every $a \in E$ with $a \neq 0$;
(ii) $\|\alpha a\| = |\alpha| \|a\|$ for all $\alpha \in F$ and $a \in E$;
(iii) $\|a + b\| \leq \|a\| + \|b\|$ for all $a, b \in E$.

It follows from (ii) that $\|O\| = 0$. We will require only one result about normed vector spaces:

Lemma 7 *Let F be a complete valued field and let E be a finite-dimensional vector space over F. If $\| \ \|_1$ and $\| \ \|_2$ are both norms on E, then there exist positive constants σ, μ such that*

$$\sigma \|a\|_1 \leq \|a\|_2 \leq \mu \|a\|_1 \quad \text{for every } a \in E.$$

Proof Let $e_1, \ldots, e_n$ be a basis for the vector space E. Then any $a \in E$ can be uniquely represented in the form

$$a = \alpha_1 e_1 + \cdots + \alpha_n e_n,$$

where $\alpha_1, \ldots, \alpha_n \in F$. It is easily seen that

$$\|a\|_0 = \max_{1 \le i \le n} |\alpha_i|$$

is a norm on E, and it is sufficient to prove the proposition for $\| \ \|_2 = \| \ \|_0$. Since

$$\|a\|_1 \le \|a\|_0 (\|e_1\|_1 + \cdots + \|e_n\|_1),$$

we can take $\sigma = (\|e_1\|_1 + \cdots + \|e_n\|_1)^{-1}$. To establish the existence of μ we assume $n > 1$ and use induction, since the result is obviously true for $n = 1$.

Assume, contrary to the assertion, that there exists a sequence $a^{(k)} \in E$ such that

$$\|a^{(k)}\|_1 < \varepsilon_k \|a^{(k)}\|_0,$$

where $\varepsilon_k > 0$ and $\varepsilon_k \to 0$ as $k \to \infty$. We may suppose, without loss of generality, that

$$|\alpha_n^{(k)}| = \|a^{(k)}\|_0$$

and also, by replacing $a^{(k)}$ by $(\alpha_n^{(k)})^{-1} a^{(k)}$, that $\alpha_n^{(k)} = 1$. Thus $a^{(k)} = b^{(k)} + e_n$, where

$$b^{(k)} = \alpha_1^{(k)} e_1 + \cdots + \alpha_{n-1}^{(k)} e_{n-1},$$

and $\|a^{(k)}\|_1 \to 0$ as $k \to \infty$. The sequences $\alpha_i^{(k)}$ $(i = 1, \ldots, n-1)$ are fundamental sequences in F, since

$$\|b^{(j)} - b^{(k)}\|_1 \le \|b^{(j)} + e_n\|_1 + \|b^{(k)} + e_n\|_1 = \|a^{(j)}\|_1 + \|a^{(k)}\|_1$$

and, by the induction hypothesis,

$$|\alpha_i^{(j)} - \alpha_i^{(k)}| \le \mu_{n-1} \|b^{(j)} - b^{(k)}\|_1 \quad (i = 1, \ldots, n-1).$$

Hence, since F is complete, there exist $\alpha_i \in F$ such that $|\alpha_i^{(k)} - \alpha_i| \to 0$ $(i = 1, \ldots, n-1)$. Put

$$b = \alpha_1 e_1 + \cdots + \alpha_{n-1} e_{n-1}.$$

Since $\|b^{(k)} - b\|_1 \le \sigma_{n-1}^{-1} \|b^{(k)} - b\|_0$, it follows that $\|b^{(k)} - b\|_1 \to 0$. But if $a = b + e_n$, then

$$\|a\|_1 \le \|a - a^{(k)}\|_1 + \|a^{(k)}\|_1 = \|b - b^{(k)}\|_1 + \|a^{(k)}\|_1.$$

Letting $k \to \infty$, we obtain $a = 0$, which contradicts the definition of a. $\qquad \square$

4 Non-Archimedean Valued Fields

Throughout this section we denote by F a field with a non-archimedean absolute value $|\ |$. A basic property of such fields is the following simple lemma. It may be interpreted as saying that in ultrametric geometry every triangle is isosceles.

Lemma 8 *If $a, b \in F$ and $|a| < |b|$, then $|a + b| = |b|$.*

Proof We certainly have

$$|a + b| \le \max\{|a|, |b|\} = |b|.$$

On the other hand, since $b = (a + b) - a$, we have

$$|b| \le \max\{|a + b|, |-a|\}$$

and, since $|-a| = |a| < |b|$, this implies $|b| \le |a + b|$. □

It may be noted that if $a \ne 0$ and $b = -a$, then $|a| = |b|$ and $|a + b| < |b|$. From Lemma 8 it follows by induction that if $a_1, \ldots, a_n \in F$ and $|a_k| < |a_1|$ for $1 < k \le n$, then

$$|a_1 + \cdots + a_n| = |a_1|.$$

As an application we show that *if a field E is a finite extension of a field F, then the trivial absolute value on E is the only extension to E of the trivial absolute value on F.* By Proposition 2, any extension to E of the trivial absolute value on F must be non-archimedean. Suppose $\alpha \in E$ and $|\alpha| > 1$. Then α satisfies a polynomial equation

$$\alpha^n + c_{n-1}\alpha^{n-1} + \cdots + c_0 = 0$$

with coefficients $c_k \in F$. Since $|c_k| = 0$ or 1 and since $|\alpha^k| < |\alpha^n|$ if $k < n$, we obtain the contradiction $|\alpha^n| = |\alpha^n + c_{n-1}\alpha^{n-1} + \cdots + c_0| = 0$.

As another application we prove

Proposition 9 *If a field F has a non-archimedean absolute value $|\ |$, then the valuation on F can be extended to the polynomial ring $F[t]$ by defining the absolute value of $f(t) = a_0 + a_1 t + \cdots + a_n t^n$ to be $|f| = \max\{|a_0|, \ldots, |a_n|\}$.*

Proof We need only show that $|fg| = |f||g|$, since it is evident that $|f| = 0$ if and only if $f = 0$ and that $|f + g| \le |f| + |g|$. Let $g(t) = b_0 + b_1 t + \cdots + b_m t^m$. Then $f(t)g(t) = c_0 + c_1 t + \cdots + c_l t^l$, where

$$c_i = a_0 b_i + a_1 b_{i-1} + \cdots + a_i b_0.$$

If r is the least integer such that $|a_r| = |f|$ and s the least integer such that $|b_s| = |g|$, then $a_r b_s$ has strictly greatest absolute value among all products $a_j b_k$ with $j + k = r + s$. Hence $|c_{r+s}| = |a_r||b_s|$ and $|fg| \ge |f||g|$. On the other hand,

$$|fg| = \max_i |c_i| \le \max_{j,k} |a_j||b_k| = |f||g|.$$

Consequently $|fg| = |f||g|$. Clearly also $|f| = |a|$ if $f = a \in F$. (The absolute value on F can be further extended to the field $F(t)$ of rational functions by defining $|f(t)/g(t)|$ to be $|f|/|g|$.) □

It also follows at once from Lemma 8 that if a sequence (a_n) of elements of F converges to a limit $a \neq 0$, then $|a_n| = |a|$ for all large n. Hence the value group of the field F is the same as the value group of its completion $\bar{F}$. The next lemma has an especially appealing corollary.

Lemma 10 *Let F be a field with a non-archimedean absolute value $|\,|$. Then a sequence (a_n) of elements of F is a fundamental sequence if and only if $\lim_{n\to\infty} |a_{n+1} - a_n| = 0$.*

Proof If $|a_{n+1} - a_n| \to 0$, then for each $\varepsilon > 0$ there is a corresponding positive integer $N = N(\varepsilon)$ such that

$$|a_{n+1} - a_n| < \varepsilon \quad \text{for } n \geq N.$$

For any integer $k > 1$,

$$a_{n+k} - a_n = (a_{n+1} - a_n) + (a_{n+2} - a_{n+1}) + \cdots + (a_{n+k} - a_{n+k-1})$$

and hence

$$|a_{n+k} - a_n| \leq \max\{|a_{n+1} - a_n|, |a_{n+2} - a_{n+1}|, \ldots, |a_{n+k} - a_{n+k-1}|\} < \varepsilon \text{ for } n \geq N.$$

Thus (a_n) is a fundamental sequence. The converse follows at once from the definition of a fundamental sequence. $\qquad\square$

Corollary 11 *In a field F with a complete non-archimedean absolute value $|\,|$, an infinite series $\sum_{n=1}^{\infty} a_n$ of elements of F is convergent if and only if $|a_n| \to 0$.*

Let F be a field with a nontrivial non-archimedean absolute value $|\,|$ and put

$$R = \{a \in F : |a| \leq 1\},$$
$$M = \{a \in F : |a| < 1\},$$
$$U = \{a \in F : |a| = 1\}.$$

Then R is the union of the disjoint nonempty subsets M and U. It follows from the definition of a non-archimedean absolute value that R is a (commutative) ring containing the unit element of F and that, for any nonzero $a \in F$, either $a \in R$ or $a^{-1} \in R$ (or both). Moreover M is an ideal of R and U is a multiplicative group, consisting of all $a \in R$ such that also $a^{-1} \in R$. Thus a proper ideal of R cannot contain an element of U and hence M is the unique maximal ideal of R. Consequently (see again Chapter I, §8) the quotient R/M is a field.

We call R the *valuation ring*, M the *valuation ideal*, and R/M the *residue field* of the valued field F.

We draw attention to the fact that the 'closed unit ball' R is both open and closed in the topology induced by the absolute value. For if $a \in R$ and $|b - a| < 1$, then also $b \in R$. Furthermore, if $a_n \in R$ and $a_n \to a$ then $a \in R$, since $|a_n| = |a|$ for all large n. Similarly, the 'open unit ball' M is also both open and closed.

In particular, let $F = \mathbb{Q}$ be the field of rational numbers and $|\,| = |\,|_p$ the p-adic absolute value. In this case the valuation ring $R = R_p$ is the set of all rational numbers

m/n, where m and n are relatively prime integers, $n > 0$ and p does not divide n. The valuation ideal is $M = pR_p$ and the residue field $\mathbb{F}_p = R_p/pR_p$ is the finite field with p elements.

As another example, let $F = K(t)$ be the field of rational functions with coefficients from an arbitrary field K and let $|| = ||_t$ be the absolute value considered in example (iii) of §1 for the irreducible polynomial $p(t) = t$. In this case the valuation ring R is the set of all rational functions $f = g/h$, where g and h are relatively prime polynomials and h has nonzero constant term. The valuation ideal is $M = tR$ and the residue field R/M is isomorphic to K, since $f(t) \equiv f(0) \bmod M$ (i.e., $f(t) - f(0) \in M$).

Let $\bar{F}$ be the completion of F. If $\bar{R}$ and $\bar{M}$ are the valuation ring and valuation ideal of $\bar{F}$, then evidently

$$R = \bar{R} \cap F, \quad M = \bar{M} \cap F.$$

Moreover R is *dense* in $\bar{R}$ since, if $0 < \varepsilon \le 1$, for any $\alpha \in \bar{R}$ there exists $a \in F$ such that $|\alpha - a| < \varepsilon$ and then $a \in R$ (and $\alpha - a \in \bar{M}$). Furthermore the residue fields R/M and $\bar{R}/\bar{M}$ are isomorphic. For the map $a + M \to a + \bar{M} (a \in R)$ is an isomorphism of R/M onto a subfield of $\bar{R}/\bar{M}$ and this subfield is not proper (by the preceding bracketed remark).

The valuation ring of the field $\mathbb{Q}_p$ of p-adic numbers will be denoted by $\mathbb{Z}_p$ and its elements will be called *p-adic integers*. The ring $\mathbb{Z}$ of ordinary integers is dense in $\mathbb{Z}_p$, and the residue field of $\mathbb{Q}_p$ is the finite field $\mathbb{F}_p$ with p elements, since this is the residue field of $\mathbb{Q}$.

Similarly, the valuation ring of the field $K((t))$ of all formal Laurent series is the ring $K[[t]]$ of all formal power series $\sum_{n \ge 0} a_n t^n$. The polynomial ring $K[t]$ is dense in $K[[t]]$, and the residue field of $K((t))$ is K, since this is the residue field of $K(t)$ with the absolute value $||_t$.

A non-archimedean absolute value $||$ on a field F will be said to be *discrete* if there exists some $\delta \in (0, 1)$ such that $a \in F$ and $|a| \ne 1$ implies either $|a| < 1 - \delta$ or $|a| > 1 + \delta$. (This situation cannot arise for archimedean absolute values.)

A non-archimedean absolute value need not be discrete, but the examples of non-archimedean absolute values which we have given are all discrete.

Lemma 12 *Let F be a field with a nontrivial non-archimedean absolute value $||$, and let R and M be the corresponding valuation ring and valuation ideal. Then the absolute value is discrete if and only if M is a principal ideal. In this case the only nontrivial proper ideals of R are the powers $M^k (k = 1, 2, \ldots)$.*

Proof Suppose first that the absolute value $||$ is discrete and put $\mu = \sup_{a \in M} |a|$. Then $0 < \mu < 1$ and the supremum is attained, since $|a_n| \to \mu$ implies $|a_{n+1} a_n^{-1}| \to 1$. Thus $\mu = |\pi|$ for some $\pi \in M$. For any $a \in M$ we have $|a\pi^{-1}| \le 1$ and hence $a = \pi a'$, where $a' \in R$. Thus M is a principal ideal with generating element π.

Suppose next that M is a principal ideal with generating element π. If $|a| < 1$, then $a \in M$. Thus $a = \pi a'$, where $a' \in R$, and hence $|a| \le |\pi|$. Similarly if $|a| > 1$, then $a^{-1} \in M$. Thus $|a^{-1}| \le |\pi|$ and hence $|a| \ge |\pi|^{-1}$. This proves that the absolute value is discrete.

We now show that, for any nonzero $a \in M$, there is a positive integer k such that $|a| = |\pi|^k$. In fact we can choose k so that

$$|\pi|^{k+1} < |a| \leq |\pi|^k.$$

Then $|\pi| < |a\pi^{-k}| \leq 1$, which implies $|a\pi^{-k}| = 1$ and hence $|a| = |\pi|^k$. Thus the value group of the valued field F is the infinite cyclic group generated by $|\pi|$. The final statement of the lemma follows immediately. $\square$

It is clear that if an absolute value $|\,|$ on a field F is discrete, then its extension to the completion $\bar{F}$ of F is also discrete. Moreover, if π is a generating element for the valuation ideal of F, then it is also a generating element for the valuation ideal of $\bar{F}$.

Suppose now that not only is $M = (\pi)$ a principal ideal, but the residue field $k = R/M$ is finite. Then there exists a finite set $S \subseteq R$, with the same cardinality as k, such that for each $a \in R$ there is a unique $\alpha \in S$ for which $|\alpha - a| < 1$. Since the elements of k are the cosets $\alpha + M$, where $\alpha \in S$, we call S a *set of representatives* in R of the residue field. It is convenient to choose $\alpha = 0$ as the representative for M itself.

Under these hypotheses a rather explicit representation for the elements of the valued field can be derived:

Proposition 13 *Let F be a field with a non-archimedean absolute value $|\,|$, and let R and M be the corresponding valuation ring and valuation ideal. Suppose the absolute value is discrete, i.e. $M = (\pi)$ is a principal ideal. Suppose also that the residue field $k = R/M$ is finite, and let $S \subseteq R$ be a set of representatives of k with $0 \in S$.*

Then for each $a \in F$ there exists a unique bi-infinite sequence $(\alpha_n)_{n \in \mathbb{Z}}$, where $\alpha_n \in S$ for all $n \in \mathbb{Z}$ and $\alpha_n \neq 0$ for at most finitely many $n < 0$, such that

$$a = \sum_{n \in \mathbb{Z}} \alpha_n \pi^n.$$

If N is the least integer n such that $\alpha_n \neq 0$, then $|a| = |\pi|^N$. In particular, $a \in R$ if and only if $\alpha_n = 0$ for all $n < 0$.

If F is complete then, for any such bi-infinite sequence (α_n), the series $\sum_{n \in \mathbb{Z}} \alpha_n \pi^n$ is convergent with sum $a \in F$.

Proof Suppose $a \in F$ and $a \neq 0$. Then $|a| = |\pi|^N$ for some $N \in \mathbb{Z}$ and hence $|a\pi^{-N}| = 1$. There is a unique $\alpha_N \in S$ such that $|a\pi^{-N} - \alpha_N| < 1$. Then $|\alpha_N| = 1$, $|a\pi^{-N} - \alpha_N| \leq |\pi|$ and

$$a\pi^{-N} = \alpha_N + a_1 \pi,$$

where $a_1 \in R$. Similarly there is a unique $\alpha_{N+1} \in S$ such that

$$a_1 = \alpha_{N+1} + a_2 \pi,$$

where $a_2 \in R$. Continuing in this way we obtain, for any positive integer n,

$$a = \alpha_N \pi^N + \alpha_{N+1} \pi^{N+1} + \cdots + \alpha_{N+n} \pi^{N+n} + a_{n+1} \pi^{N+n+1},$$

where $\alpha_N, \alpha_{N+1}, \ldots, \alpha_{N+n} \in S$ and $a_{n+1} \in R$. Since $|a_{n+1}\pi^{N+n+1}| \to 0$ as $n \to \infty$, the series $\sum_{n \geq N} \alpha_n \pi^n$ converges with sum a.

On the other hand, it is clear that if $a = \sum_{n \geq N} \alpha_n \pi^n$, where $\alpha_n \in S$ and $\alpha_N \neq 0$, then the coefficients α_n must be determined in the above way.

If F is complete then, by Corollary 11, any series $\sum_{n \geq N} \alpha_n \pi^n$ is convergent, since $|\alpha_n \pi^n| \to 0$ as $n \to \infty$. $\qquad\qquad\qquad\qquad\qquad\qquad\qquad\qquad\qquad\qquad\qquad\qquad\qquad\square$

Corollary 14 *Every $a \in \mathbb{Q}_p$ can be uniquely expressed in the form*

$$a = \sum_{n \in \mathbb{Z}} \alpha_n p^n,$$

where $\alpha_n \in \{0, 1, \ldots, p-1\}$ and $\alpha_n \neq 0$ for at most finitely many $n < 0$. Conversely, any such series is convergent with sum $a \in \mathbb{Q}_p$. Furthermore $a \in \mathbb{Z}_p$ if and only if $\alpha_n = 0$ for all $n < 0$.

Thus we have now arrived at Hensel's starting-point. It is not difficult to show that if $a = \sum_{n \in \mathbb{Z}} \alpha_n p^n \in \mathbb{Q}_p$, then actually $a \in \mathbb{Q}$ if and only if the sequence of coefficients (α_n) is *eventually periodic*, i.e. there exist integers $h > 0$ and m such that $\alpha_{n+h} = \alpha_n$ for all $n \geq m$.

From Corollary 14 we can deduce again that the ring $\mathbb{Z}$ of ordinary integers is dense in the ring $\mathbb{Z}_p$ of p-adic integers. For, if

$$a = \sum_{n \geq 0} \alpha_n p^n \in \mathbb{Z}_p,$$

where $\alpha_n \in \{0, 1, \ldots, p-1\}$, then

$$a_k = \sum_{n=0}^{k} \alpha_n p^n \in \mathbb{Z}$$

and $|a - a_k| < p^{-k}$.

5 Hensel's Lemma

The analogy between p-adic absolute values and ordinary absolute values suggests that methods well-known in analysis may be applied also to arithmetic problems. We will illustrate this by showing how Newton's method for finding the real or complex roots of an equation can also be used to find p-adic roots. In fact the ultrametric inequality makes it possible to establish a stronger convergence criterion than in the classical case. The following proposition is modestly known as 'Hensel's lemma'.

Proposition 15 *Let F be a field with a complete non-archimedean absolute value $|\ |$ and let R be its valuation ring. Let*

$$f(x) = c_n x^n + c_{n-1} x^{n-1} + \cdots + c_0$$

be a polynomial with coefficients $c_0, \ldots, c_n \in R$ and let

$$f_1(x) = nc_n x^{n-1} + (n-1)c_{n-1}x^{n-2} + \cdots + c_1$$

be its formal derivative. If $|f(a_0)| < |f_1(a_0)|^2$ for some $a_0 \in R$, then the equation $f(a) = 0$ has a unique solution $a \in R$ such that $|a - a_0| < |f_1(a_0)|$.

Proof We consider first the existence of a and postpone discussion of its uniqueness. Put

$$\sigma := |f_1(a_0)| > 0, \quad \theta_0 := \sigma^{-2}|f(a_0)| < 1,$$

and let D_θ denote the set

$$\{a \in R : |f_1(a)| = \sigma, \ |f(a)| \le \theta\sigma^2\}.$$

Thus $a_0 \in D_{\theta_0}$ and $D_{\theta'} \subseteq D_\theta$ if $\theta' \le \theta$. We are going to show that, if $\theta \in (0,1)$, then the 'Newton' map

$$Ta = a^* := a - f(a)/f_1(a)$$

maps D_θ into D_{θ^2}.

We can write

$$f(x + y) = f(x) + f_1(x)y + \cdots + f_n(x)y^n,$$

where $f_1(x)$ has already been defined and $f_2(x), \ldots, f_n(x)$ are also polynomials with coefficients from R. We substitute

$$x = a, \ y = b := -f(a)/f_1(a),$$

where $a \in D_\theta$. Then $|f_j(a)| \le 1$, since $a \in R$ and $f_j(x) \in R[x]$ $(j = 1, \ldots, n)$. Furthermore

$$|b| = \sigma^{-1}|f(a)| \le \theta\sigma < \sigma.$$

Thus $b \in R$. Since $f(a) + f_1(a)b = 0$, it follows that $a^* = a + b$ satisfies

$$|f(a^*)| \le \max_{2 \le j \le n} |f_j(a)b^j| \le |b|^2 = \sigma^{-2}|f(a)|^2 \le \theta^2\sigma^2.$$

Similarly, since $f_1(a+b) - f_1(a)$ can be written as a polynomial in b with coefficients from R and with no constant term,

$$|f_1(a+b) - f_1(a)| \le |b| < \sigma = |f_1(a)|$$

and hence $|f_1(a^*)| = \sigma$. This completes the proof that $TD_\theta \subseteq D_{\theta^2}$.

Now put $a_k = T^k a_0$, so that

$$a_{k+1} - a_k = -f(a_k)/f_1(a_k).$$

It follows by induction from what we have proved that

$$|f(a_k)| \le \theta_0^{2^k} \sigma^2.$$

Since $\theta_0 < 1$ and $|a_{k+1} - a_k| = \sigma^{-1}|f(a_k)|$, this shows that $\{a_k\}$ is a fundamental sequence. Hence, since F is complete, $a_k \to a$ for some $a \in R$. Evidently $f(a) = 0$ and $|f_1(a)| = \sigma$. Since, for every $k \ge 1$,

$$|a_k - a_0| \le \max_{1 \le j \le k} |a_j - a_{j-1}| \le \theta_0 \sigma,$$

we also have $|a - a_0| \le \theta_0 \sigma < \sigma$.

To prove uniqueness, assume $f(\tilde{a}) = 0$ for some $\tilde{a} \ne a$ such that $|\tilde{a} - a_0| < \sigma$. If we put $b = \tilde{a} - a$, then

$$0 = f(\tilde{a}) - f(a) = f_1(a)b + \cdots + f_n(a)b^n.$$

From $b = \tilde{a} - a_0 - (a - a_0)$ we obtain $|b| < \sigma$. Since $b \ne 0$ and $|f_j(a)| \le 1$, it follows that, for $j \ge 2$,

$$|f_j(a)b^j| \le |b|^2 < \sigma|b| = |f_1(a)b|.$$

But this implies

$$|f(\tilde{a}) - f(a)| = |f_1(a)b| > 0,$$

which is a contradiction. $\qquad\square$

As an application of Proposition 15 we will determine which elements of the field $\mathbb{Q}_p$ of p-adic numbers are squares. Since $b = a^2$ implies $b = p^{2v}b'$, where $v \in \mathbb{Z}$ and $|b'|_p = 1$, we may restrict attention to the case $|b|_p = 1$.

Proposition 16 Suppose $b \in \mathbb{Q}_p$ and $|b|_p = 1$.
If $p \ne 2$, then $b = a^2$ for some $a \in \mathbb{Q}_p$ if and only if $|b - a_0^2|_p < 1$ for some $a_0 \in \mathbb{Z}$.
If $p = 2$, then $b = a^2$ for some $a \in \mathbb{Q}_2$ if and only if $|b - 1|_2 \le 2^{-3}$.

Proof Suppose first that $p \ne 2$. If $b = a^2$ for some $a \in \mathbb{Q}_p$, then $|a|_p = 1$ and $|a - a_0|_p < 1$ for some $a_0 \in \mathbb{Z}$, since $\mathbb{Z}$ is dense in $\mathbb{Z}_p$. Hence $|a_0|_p = 1$ and

$$|b - a_0^2|_p = |a - a_0|_p|a + a_0|_p \le |a - a_0|_p < 1.$$

Conversely, suppose $|b - a_0^2|_p < 1$ for some $a_0 \in \mathbb{Z}$. Then $|a_0^2|_p = 1$ and so $|a_0|_p = 1$. In Proposition 15 take $F = \mathbb{Q}_p$ and $f(x) = x^2 - b$. The hypotheses of the proposition are satisfied, since $|f(a_0)|_p < 1$ and $|f_1(a_0)|_p = |2a_0|_p = 1$, and hence $b = a^2$ for some $a \in \mathbb{Q}_p$.

Suppose next that $p = 2$. If $b = a^2$ for some $a \in \mathbb{Q}_2$, then $|a|_2 = 1$ and $|a - a_0|_2 \le 2^{-3}$ for some $a_0 \in \mathbb{Z}$, since $\mathbb{Z}$ is dense in $\mathbb{Z}_2$. Hence $|a_0|_2 = 1$ and

$$|b - a_0^2|_2 = |a - a_0|_2|a + a_0|_2 \le |a - a_0|_2 \le 2^{-3}.$$

Since a_0 is odd, we have $a_0 \equiv \pm 1 \bmod 4$ and $a_0^2 \equiv 1 \bmod 8$. Hence

$$|b - 1|_2 \le \max\{|b - a_0^2|_2, |a_0^2 - 1|_2\} \le 2^{-3}.$$

Conversely, suppose $|b - 1|_2 \leq 2^{-3}$. In Proposition 15 take $F = \mathbb{Q}_2$ and $f(x) = x^2 - b$. The hypotheses of the proposition are satisfied, since $|f(1)|_2 \leq 2^{-3}$ and $|f_1(1)|_2 = 2^{-1}$, and hence $b = a^2$ for some $a \in \mathbb{Q}_2$. $\qquad\qquad\square$

Corollary 17 *Let b be an integer not divisible by the prime p.*

If $p \neq 2$, then $b = a^2$ for some $a \in \mathbb{Q}_p$ if and only if b is a quadratic residue mod *p.*

If $p = 2$, then $b = a^2$ for some $a \in \mathbb{Q}_2$ if and only if $b \equiv 1 \bmod 8$.

It follows from Corollary 17 that $\mathbb{Q}_p$ cannot be given the structure of an *ordered* field. For, if p is odd, then $1 - p = a^2$ for some $a \in \mathbb{Q}_p$ and hence

$$a^2 + 1 + \cdots + 1 = 0,$$

where there are $p - 1$ 1's. Similarly, if $p = 2$, then $1 - 2^3 = a^2$ for some $a \in \mathbb{Q}_2$ and the same relation holds with 7 1's.

Suppose again that F is a field with a complete non-archimedean absolute value $|\ |$. Let R and M be the corresponding valuation ring and valuation ideal, and let $k = R/M$ be the residue field. For any $a \in R$ we will denote by $\bar{a}$ the corresponding element $a + M$ of k, and for any polynomial

$$f(x) = c_n x^n + c_{n-1} x^{n-1} + \cdots + c_0$$

with coefficients $c_0, \ldots, c_n \in R$, we will denote by

$$\bar{f}(x) = \bar{c}_n x^n + \bar{c}_{n-1} x^{n-1} + \cdots + \bar{c}_0$$

the polynomial whose coefficients are the corresponding elements of k.

The hypotheses of Proposition 15 are certainly satisfied if $|f(a_0)| < 1 = |f_1(a_0)|$. In this case Proposition 15 says that if

$$\bar{f}(x) = (x - \bar{a}_0)\bar{h}_0(x),$$

where $a_0 \in R, h_0(x) \in R[x]$ and $h_0(a_0) \notin M$, then

$$f(x) = (x - a)h(x),$$

where $a - a_0 \in M$, and $h(x) \in R[x]$. In other words, the factorization of $\bar{f}(x)$ in $k[x]$ can be 'lifted' to a factorization of $f(x)$ in $R[x]$. This form of Hensel's lemma can be generalized to factorizations where neither factor is linear, and the result is again known as Hensel's lemma!

Proposition 18 *Let F be a field with a complete non-archimedean absolute value $|\ |$. Let R and M be the valuation ring and valuation ideal of F, and $k = R/M$ the residue field.*

Let $f \in R[x]$ be a polynomial with coefficients in R and suppose there exist relatively prime polynomials $\phi, \psi \in k[x]$, with ϕ monic and $\partial(\phi) > 0$, such that $\bar{f} = \phi\psi$.

Then there exist polynomials $g, h \in R[x]$, with g monic and $\partial(g) = \partial(\phi)$, such that $\bar{g} = \phi, \bar{h} = \psi$ and $f = gh$.

Proof Put $n = \partial(f)$ and $m = \partial(\phi)$. Then $\partial(\psi) = \partial(\bar{f}) - \partial(\phi) \leq n - m$. There exist polynomials $g_1, h_1 \in R[x]$, with g_1 monic, $\partial(g_1) = m$ and $\partial(h_1) \leq n - m$, such that $\bar{g}_1 = \phi, \bar{h}_1 = \psi$. Since ϕ, ψ are relatively prime, there exist polynomials $\chi, \omega \in k[x]$ such that

$$\chi\phi + \omega\psi = 1,$$

and there exist polynomials $u, v \in R[x]$ such that $\bar{u} = \chi, \bar{v} = \omega$. Thus

$$f - g_1h_1 \in M[x], \quad ug_1 + vh_1 - 1 \in M[x].$$

If $f = g_1h_1$, there is nothing more to do. Otherwise, let π be the coefficient of $f - g_1h_1$ or of $ug_1 + vh_1 - 1$ which has maximum absolute value. Then

$$f - g_1h_1 \in \pi R[x], \quad ug_1 + vh_1 - 1 \in \pi R[x].$$

We are going to construct inductively polynomials $g_j, h_j \in R[x]$ such that

(i) $\bar{g}_j = \phi, \bar{h}_j = \psi$;
(ii) g_j is monic and $\partial(g_j) = m, \partial(h_j) \leq n - m$;
(iii) $g_j - g_{j-1} \in \pi^{j-1}R[x], h_j - h_{j-1} \in \pi^{j-1}R[x]$;
(iv) $f - g_jh_j \in \pi^j R[x]$.

This holds already for $j = 1$ with $g_0 = h_0 = 0$. Assume that, for some $k \geq 2$, it holds for all $j < k$ and put $f - g_jh_j = \pi^j\ell_j$, where $\ell_j \in R[x]$. Since g_1 is monic, the Euclidean algorithm provides polynomials $q_k, r_k \in R[x]$ such that

$$\ell_{k-1}v = q_kg_1 + r_k, \quad \partial(r_k) < \partial(g_1) = m.$$

Let $w_k \in R[x]$ be a polynomial of minimal degree such that all coefficients of $\ell_{k-1}u + q_kh_1 - w_k$ have absolute value at most $|\pi|$. Then

$$w_kg_1 + r_kh_1 - \ell_{k-1} = (ug_1 + vh_1 - 1)\ell_{k-1} - (\ell_{k-1}u + q_kh_1 - w_k)g_1 \in \pi R[x].$$

We will show that $\partial(w_k) \leq n - m$. Indeed otherwise

$$\partial(w_kg_1) > n \geq \partial(r_kh_1 - \ell_{k-1})$$

and hence, since g_1 is monic, $w_kg_1 + r_kh_1 - \ell_{k-1}$ has the same leading coefficient as w_k. Consequently the leading coefficient of w_k is in πR. Thus the polynomial obtained from w_k by omitting the term of highest degree satisfies the same requirements as w_k, which is a contradiction.

If we put

$$g_k = g_{k-1} + \pi^{k-1}r_k, \quad h_k = h_{k-1} + \pi^{k-1}w_k,$$

then (i)–(iii) are evidently satisfied for $j = k$. Moreover

$$f - g_kh_k = -\pi^{k-1}(w_kg_{k-1} + r_kh_{k-1} - \ell_{k-1}) - \pi^{2k-2}r_kw_k$$

and

$$w_kg_{k-1} + r_kh_{k-1} - \ell_{k-1}$$
$$= w_kg_1 + r_kh_1 - \ell_{k-1} + w_k(g_{k-1} - g_1) + r_k(h_{k-1} - h_1) \in \pi R[x].$$

Hence also (iv) is satisfied for $j = k$.

Put

$$g_j(x) = x^m + \sum_{i=0}^{m-1} a_i^{(j)} x^i, \quad h_j(x) = \sum_{i=0}^{n-m} \beta_i^{(j)} x^i.$$

By (iii), the sequences $(a_i^{(j)})$ and $(\beta_i^{(j)})$ are fundamental sequences for each i and hence, since F is complete, there exist $\alpha_i, \beta_i \in R$ such that

$$\alpha_i^{(j)} \to \alpha_i, \beta_i^{(j)} \to \beta_i \text{ as } j \to \infty.$$

If

$$g(x) = x^m + \sum_{i=0}^{m-1} \alpha_i x^i, \quad h_j(x) = \sum_{i=0}^{n-m} \beta_i x^i,$$

then, for each $j \geq 1$,

$$g - g_j \in \pi^j R[x], \quad h - h_j \in \pi^j R[x].$$

Since

$$f - gh = f - g_j h_j - (g - g_j)h - g_j(h - h_j),$$

it follows that $f - gh \in \pi^j R[x]$ for each $j \geq 1$. Hence $f = gh$. It is obvious that g and h have the other required properties.   □

As an application of this form of Hensel's lemma we prove

Proposition 19 *Let F be a field with a complete non-archimedean absolute value $|\ |$ and let*

$$f(t) = c_n t^n + c_{n-1} t^{n-1} + \cdots + c_0 \in F[t].$$

If $c_0 c_n \neq 0$ and, for some m such that $0 < m < n$,

$$|c_0| \leq |c_m|, \quad |c_n| \leq |c_m|,$$

with at least one of the two inequalities strict, then f is reducible over F.

Proof Suppose first that $|c_0| < |c_m|$ and $|c_n| \leq |c_m|$. Evidently we may choose m so that $|c_m| = \max_{0 < i < n} |c_i|$ and $|c_i| < |c_m|$ for $0 \leq i < m$. By multiplying f by c_m^{-1} we may further assume that, if R is the valuation ring of F, then $f(t) \in R[t]$, $c_m = 1$ and $|c_i| < 1$ for $0 \leq i < m$. Hence

$$\bar{f}(t) = t^m (\bar{c}_n t^{n-m} + \bar{c}_{n-1} t^{n-m-1} + \cdots + 1).$$

Since the two factors are relatively prime, it follows from Proposition 18 that f is reducible.

  If $|c_n| < |c_m|$ and $|c_0| \leq |c_m|$, then the same argument also applies to the polynomial $t^n f(t^{-1})$.   □

Proposition 19 shows that if a quadratic polynomial $at^2 + bt + c$ is irreducible, then $|b| \leq \max\{|a|, |c|\}$, with strict inequality if $|a| \neq |c|$. Proposition 19 will now be used to extend an absolute value on a given field to a finite extension of that field.

Proposition 20 *Let F be a field with a complete non-archimedean absolute value $|\ |$. If the field E is a finite extension of F, then the absolute value on F can be extended to an absolute value on E.*

Proof We will not only show that an extension of the absolute value exists, but we will provide an explicit expression for it.

Regard E as a vector space over F of finite dimension n, and with any $a \in E$ associate the linear transformation $L_a : E \rightarrow E$ defined by $L_a(x) = ax$. Then $\det L_a \in F$ and we claim that *an extended absolute value is given by the formula*

$$|a| = |\det L_a|^{1/n}.$$

Evidently $|a| \geq 0$, and equality holds only if $a = 0$, since $ax = 0$ for some $x \neq 0$ implies $a = 0$. Furthermore $|ab| = |a||b|$, since $L_{ab} = L_a L_b$ and hence $\det L_{ab} = (\det L_a)(\det L_b)$. If $a \in F$, then $L_a = aI_n$ and hence the proposed absolute value coincides with the original absolute value on F. It only remains to show that

$$|a - b| \leq \max(|a|, |b|) \quad \text{for all } a, b \in F.$$

In fact we may suppose $|a| \leq |b|$ and then, by dividing by b, we see that it is sufficient to show that $0 < |a| \leq 1$ implies $|1 - a| \leq 1$.

To simplify notation, write $A = L_a$ and let

$$f(t) = \det(tI - A) = t^n + c_{n-1}t^{n-1} + \cdots + c_0$$

be the *characteristic polynomial* of A. Then $c_i \in F$ for all i and $c_0 = (-1)^n \det A$. Let $g(t)$ be the monic polynomial in $F[t]$ of least positive degree such that $g(a) = 0$. Then $g(t)$ is irreducible, since the field E has no zero divisors. Evidently $g(t)$ is also the *minimal polynomial* of A. But, for an arbitrary linear transformation of an n-dimensional vector space, the characteristic polynomial divides the n-th power of the minimal polynomial (see M. Deuring, *Algebren*, p.4). It follows in the present case that $f(t) = g(t)^r$ for some positive integer r.

Suppose

$$g(t) = t^m + b_{m-1}t^{m-1} + \cdots + b_0$$

and let $a \in E$ satisfy $|a| \leq 1$ with respect to the proposed absolute value. Then $|c_0| = |\det A| \leq 1$ and hence, since $b_0^r = c_0$, $|b_0| \leq 1$. Since g is irreducible, it follows from Proposition 19 that $|b_j| \leq 1$ for all j. Since

$$g(1) = 1 + b_{m-1} + \cdots + b_0,$$

this implies $|g(1)| \leq 1$ and hence $|f(1)| \leq 1$. Since $f(1) = \det(I - A)$, this proves that $|1 - a| \leq 1$. $\qquad \square$

Finally we show that there is no other extension to E of the given absolute value on F besides the one constructed in the proof of Proposition 20.

Proposition 21 *Let F be a complete field with respect to the absolute value $|\ |$ and let the field E be a finite extension of F. Then there is at most one extension of the absolute value on F to an absolute value on E, and E is necessarily complete with respect to the extended absolute value.*

Proof Let $e_1, \ldots, e_n$ be a basis for E, regarded as a vector space over F. Then any $a \in E$ can be uniquely expressed in the form

$$a = \alpha_1 e_1 + \cdots + \alpha_n e_n,$$

where $\alpha_1, \ldots, \alpha_n \in F$. By Lemma 7, for any extended absolute value there exist positive real numbers σ, μ such that

$$\sigma |a| \leq \max_i |\alpha_i| \leq \mu |a| \quad \text{for every } a \in E.$$

It follows at once that E is complete. For if $a^{(k)}$ is a fundamental sequence, then $\alpha_i^{(k)}$ is a fundamental sequence in F for $i = 1, \ldots, n$. Since F is complete, there exist $\alpha_i \in F$ such that $\alpha_i^{(k)} \to \alpha_i (i = 1, \ldots, n)$ and then $a^{(k)} \to a$, where $a = \alpha_1 e_1 + \cdots + \alpha_n e_n$.

It will now be shown that there is at most one extension to E of the absolute value on F. Since we saw in §4 that the trivial absolute value on E is the only extension of the trivial absolute value on F, we may assume that the given absolute value on F is nontrivial. For a fixed $a \in E$, consider the powers $a, a^2, \ldots$. For each k we can write

$$a^k = \alpha_1^{(k)} e_1 + \cdots + \alpha_n^{(k)} e_n.$$

Since $|a| < 1$ if and only if $|a^k| \to 0$, it follows from the remarks at the beginning of the proof that $|a| < 1$ if and only if $|\alpha_i^{(k)}| \to 0$ $(i = 1, \ldots, n)$. This condition is independent of the absolute value on E. Thus if there exist two absolute values, $|\ |_1$ and $|\ |_2$, which extend the absolute value on F, then $|a|_1 < 1$ if and only if $|a|_2 < 1$. Hence, by Proposition 3, there exists a positive real number ρ such that

$$|a|_2 = |a|_1^\rho \quad \text{for every } a \in E.$$

In fact $\rho = 1$, since for some $a \in F$ we have $|a|_2 = |a|_1 > 1$. $\square$

6 Locally Compact Valued Fields

We prove first a theorem of Ostrowski (1918):

Theorem 22 *A complete archimedean valued field F is (isomorphic to) either the real field $\mathbb{R}$ or the complex field $\mathbb{C}$, and its absolute value is equivalent to the usual absolute value.*

Proof Since the valuation on it is archimedean, the field F has characteristic 0 and thus contains $\mathbb{Q}$. Since an archimedean absolute value on $\mathbb{Q}$ is equivalent to the usual absolute value, by replacing the given absolute value on F by an equivalent one we may assume that it reduces to the usual absolute value on $\mathbb{Q}$. Since the valuation on F

is complete, it now follows that F contains (a copy of) $\mathbb{R}$ and that the absolute value on F reduces to the usual absolute value on $\mathbb{R}$. If F contains an element i such that $i^2 = -1$, then F contains (a copy of) $\mathbb{C}$ and, by Proposition 21, the absolute value on F reduces to the usual absolute value on $\mathbb{C}$.

We now show that if $a \in F$ and $|a| < 1$, then $1 - a$ is a square in F. Let B be the set of all $x \in F$ such that $|x| \le |a|$ and, for any $x \in B$, put

$$Tx = (x^2 + a)/2.$$

Then also $Tx \in B$, since

$$|Tx| \le (|x|^2 + |a|)/2 \le (|a|^2 + |a|)/2 \le |a|.$$

Moreover, the map T is a contraction since, for all $x, y \in B$,

$$|Tx - Ty| = |x^2 - y^2|/2 = |x - y||x + y|/2 \le |a||x - y|.$$

Since F is complete and B is a closed subset of F, it follows from the contraction principle (Proposition I.26) that the map T has a fixed point $\bar{x} \in B$. Evidently $\bar{x} = (\bar{x}^2 + a)/2$ and

$$1 - a = 1 - 2\bar{x} + \bar{x}^2 = (1 - \bar{x})^2.$$

We show next that, if the polynomial $t^2 + 1$ does not have a root in F, then the valuation on F can be extended to the field $E = F(i)$, where $i^2 = -1$. Each $\gamma \in E$ has a unique representation $\gamma = a + ib$, where $a, b \in F$. We claim that $|\gamma| = \sqrt{|a^2 + b^2|}$ is an extension to E of the given valuation on F.

The only part of this claim which is not easily established is the triangle inequality. To prove it, we need only show that

$$|1 + \gamma| \le 1 + |\gamma| \quad \text{for every } \gamma \in E.$$

That is, we need only show that

$$|(1 + a)^2 + b^2| \le 1 + 2\sqrt{|a^2 + b^2|} + |a^2 + b^2| \quad \text{for all } a, b \in F.$$

Since, by the triangle inequality in F,

$$|(1 + a)^2 + b^2| \le 1 + 2|a| + |a^2 + b^2|,$$

it is enough to show that

$$|a| \le \sqrt{|a^2 + b^2|} \quad \text{for all } a, b \in F$$

or, since we may suppose $a \ne 0$,

$$1 \le |1 + c^2| \quad \text{for every } c \in F.$$

Assume, on the contrary, that $|1 + c^2| < 1$ for some $c \in F$. Then, by the previous part of the proof,

$$-c^2 = 1 - (1 + c^2) = x^2 \quad \text{for some } x \in F.$$

Since $c \ne 0$, this implies that $-1 = i^2$ for some $i \in F$, which is a contradiction.

Now $E = F(i)$ contains $\mathbb{C}$ and the absolute value on E reduces to the usual absolute value on $\mathbb{C}$. To prove the theorem it is enough to show that $E = \mathbb{C}$. For then $\mathbb{R} \subseteq F \subseteq \mathbb{C}$ and F has dimension 1 or 2 as a vector space over $\mathbb{R}$ according as $i \notin F$ or $i \in F$.

Assume on the contrary that there exists $\zeta \in E \backslash \mathbb{C}$. Consider the function $\varphi : \mathbb{C} \to \mathbb{R}$ defined by

$$\varphi(z) = |z - \zeta|$$

and put $r = \inf_{z \in \mathbb{C}} \varphi(z)$. Since $\varphi(0) = |\zeta|$ and $\varphi(z) > |\zeta|$ for $|z| > 2|\zeta|$, and since φ is continuous, the compact set $\{z \in \mathbb{C} : |z| \leq 2|\zeta|\}$ contains a point w such that $\varphi(w) = r$.

Thus if we put $\omega = \zeta - w$, then $\omega \neq 0$ and

$$0 < r = |\omega| \leq |\omega - z| \quad \text{for every } z \in \mathbb{C}.$$

We will show that $|\omega - z| = r$ for every $z \in \mathbb{C}$ such that $|z| < r$.

If $\varepsilon = e^{2\pi i/n}$, then

$$\omega^n - z^n = (\omega - z)(\omega - \varepsilon z) \cdots (\omega - \varepsilon^{n-1} z)$$

and hence

$$|\omega^n - z^n| \geq r^{n-1} |\omega - z|.$$

Thus $|\omega - z| \leq r|1 - z^n/\omega^n|$. Since $|z| < |\omega|$, by letting $n \to \infty$ we obtain $|\omega - z| \leq r$. But this is possible only if $|\omega - z| = r$.

Thus if $0 < |z| < r$, then ω may be replaced by $\omega - z$. It follows that $|\omega - nz| = r$ for every positive integer n. Hence $r \geq n|z| - r$, which yields a contradiction for sufficiently large n.   □

If a field F is locally compact with respect to an archimedean absolute value, then it is certainly complete and so, by Theorem 22, it is equivalent either to $\mathbb{R}$ or to $\mathbb{C}$ with the usual absolute value. It will now be shown that a field F is locally compact with respect to a non-archimedean absolute value if and only if it is a complete field of the type discussed in Proposition 13. It should be observed that a non-archimedean valued field F is locally compact if and only if its valuation ring R is compact, since then any closed ball in F is compact.

Proposition 23 *Let F be a field with a non-archimedean absolute value $|\ |$. Then F is locally compact with respect to the topology induced by the absolute value if and only if the following three conditions are satisfied:*

(i) *F is complete,*
(ii) *the absolute value $|\ |$ is discrete,*
(iii) *the residue field is finite.*

Proof As we have just observed, F is locally compact if and only if its valuation ring R is compact. Moreover, since R is a subset of the metric space F, it is compact if and only if any sequence of elements of R has a convergent subsequence.

The field F is certainly complete if it is locally compact, since any fundamental sequence is bounded. If the residue field is infinite, then there exists an infinite sequence (a_k) of elements of R such that $|a_k - a_j| = 1$ for $j \neq k$. Since the sequence (a_k) has no convergent subsequence, R is not compact. If the absolute value $| \ |$ is not discrete, then there exists an infinite sequence (a_k) of elements of R with

$$|a_1| < |a_2| < \cdots$$

and $|a_k| \to 1$ as $k \to \infty$. If $k > j$, then $|a_k - a_j| = |a_k|$ and again the sequence (a_k) has no convergent subsequence. Thus the conditions (i)–(iii) are all necessary for F to be locally compact.

Suppose now that the conditions (i)–(iii) are all satisfied and let $\sigma = (a_k)$ be a sequence of elements of R. In the notation of Proposition 13, let

$$a_k = \sum_{n \geq 0} \alpha_n^{(k)} \pi^n,$$

where $\alpha_n^{(k)} \in S$. Since S is finite, there exists $\alpha_0 \in S$ such that $\alpha_0^{(k)} = \alpha_0$ for infinitely many $a_k \in \sigma$. If σ_0 is the subsequence of σ containing those a_k for which $\alpha_0^{(k)} = \alpha_0$, then there exists $\alpha_1 \in S$ such that $\alpha_1^{(k)} = \alpha_1$ for infinitely many $a_k \in \sigma_0$. Similarly, if σ_1 is the subsequence of σ_0 containing those a_k for which $\alpha_1^{(k)} = \alpha_1$, then there exists $\alpha_2 \in S$ such that $\alpha_2^{(k)} = \alpha_2$ for infinitely many $a_k \in \sigma_1$. And so on. If $a^{(j)} \in \sigma_j$, then

$$a^{(j)} = \alpha_0 + \alpha_1 \pi + \cdots + \alpha_j \pi^j + \sum_{n \geq 0} \alpha_n(j) \pi^{j+1+n}.$$

But $a = \sum_{n \geq 0} \alpha_n \pi^n \in F$, since F is complete, and $|a^{(j)} - a| \leq |\pi|^{j+1}$. Thus the subsequence $(a^{(j)})$ of σ converges to a. $\qquad\square$

Corollary 24 *The field $\mathbb{Q}_p$ of p-adic numbers is locally compact, and the ring $\mathbb{Z}_p$ of p-adic integers is compact.*

Corollary 25 *If K is a finite field, then the field $K((t))$ of all formal Laurent series is locally compact, and the ring $K[[t]]$ of all formal power series is compact.*

We now show that all locally compact valued fields F with a non-archimedean absolute value can in fact be explicitly determined. It is convenient to treat the cases where F has prime characteristic and zero characteristic separately, since the arguments in the two cases are quite different.

Lemma 26 *Let F be a locally compact valued field with a nontrivial valuation. A normed vector space E over F is locally compact if and only if it is finite-dimensional.*

Proof Suppose first that E is finite-dimensional over F. If $e_1, \ldots, e_n$ is a basis for the vector space E, then any $a \in E$ can be uniquely represented in the form

$$a = \alpha_1 e_1 + \cdots + \alpha_n e_n,$$

where $\alpha_1, \ldots, \alpha_n \in F$, and

$$\|a\|_0 = \max_{1 \le i \le n} |\alpha_i|$$

is a norm on E. Since the field F is locally compact, it is also complete. Hence, by Lemma 7, there exist positive real constants σ, μ such that

$$\sigma \|a\|_0 \le \|a\| \le \mu \|a\|_0 \quad \text{for every } a \in E.$$

Consequently, if $\{a_k\}$ is a bounded sequence of elements of E then, for each $j \in \{1, \dots, n\}$, the corresponding coefficients $\{\alpha_{kj}\}$ form a bounded sequence of elements of F. Hence, since F is locally compact, there exists a subsequence $\{a_{k_v}\}$ such that each of the sequences $\{\alpha_{k_v j}\}$ converges in F, with limit β_j say $(j = 1, \dots, n)$. It follows that the subsequence $\{a_{k_v}\}$ converges in E with limit $b = \beta_1 e_1 + \dots + \beta_n e_n$. Thus E is locally compact.

Suppose next that E is infinite-dimensional over F. Since the valuation on F is nontrivial, there exists $\alpha \in F$ such that $r = |\alpha|$ satisfies $0 < r < 1$. Let V be any finite-dimensional subspace of E, let $u' \in E \backslash V$ and let

$$d = \inf_{v \in V} \|u' - v\|.$$

Since V is locally compact, $d > 0$ and $d = \|u' - v'\|$ for some $v' \in V$. Choose $k \in \mathbb{Z}$ so that $r^{k+1} < d \le r^k$ and put $w' = \alpha^{-k}(u' - v')$. For any $v \in V$,

$$\|\alpha^k v + v' - u'\| \ge d$$

and hence

$$\|w' - v\| \ge dr^{-k} > r.$$

On the other hand,

$$\|w'\| = dr^{-k} \le 1.$$

We now define a sequence $\{w_m\}$ of elements of E in the following way. Taking $V = \{O\}$ we obtain a vector w_1 with $r < \|w_1\| \le 1$. Suppose we have defined $w_1, \dots, w_m \in E$ so that, for $1 \le j \le m$, $\|w_j\| \le 1$ and $\|w_j - v_j\| > r$ for all v_j in the vector subspace V_{j-1} of E spanned by $w_1, \dots, w_{j-1}$. Then, taking $V = V_m$, we obtain a vector w_{m+1} such that $\|w_{m+1}\| \le 1$ and $\|w_{m+1} - v_{m+1}\| > r$ for all $v_{m+1} \in V_m$. Thus the process can be continued indefinitely. Since $\|w_m\| \le 1$ for all m and $\|w_m - w_j\| > r$ for $1 \le j < m$, the bounded sequence $\{w_m\}$ has no convergent subsequence. Thus E is not locally compact. $\qquad\square$

Proposition 27 *A non-archimedean valued field E with zero characteristic is locally compact if and only if, for some prime p, E is isomorphic to a finite extension of the field $\mathbb{Q}_p$ of p-adic numbers.*

Proof If E is a finite extension of the p-adic field $\mathbb{Q}_p$ then, since $\mathbb{Q}_p$ is locally compact, so also is E, by Lemma 26.

Suppose on the other hand that E is a locally compact valued field with zero characteristic. Then $\mathbb{Q} \subseteq E$. By Proposition 23, the residue field $k = R/M$ is finite and thus has prime characteristic p. It follows from Proposition 4 that the restriction to $\mathbb{Q}$ of the absolute value on E is (equivalent to) the p-adic absolute value. Hence, since E is necessarily complete, $\mathbb{Q}_p \subseteq E$. If E were infinite-dimensional as a vector space over $\mathbb{Q}_p$ then, by Lemma 26, it would not be locally compact. Hence E is a finite extension of $\mathbb{Q}_p$. $\qquad\square$

We consider next locally compact valued fields of prime characteristic.

Proposition 28 *A valued field F with prime characteristic p is locally compact if and only if F is isomorphic to the field $K((t))$ of formal Laurent series over a finite field K of characteristic p, with the absolute value defined in example (iv) of §1. The finite field K is the residue field of F.*

Proof We need only prove the necessity of the condition, since (Corollary 25) we have already established its sufficiency. Since F has prime characteristic, the absolute value on F is non-archimedean. Hence, by Proposition 23 and Lemma 12, the absolute value on F is discrete and the valuation ideal M is a principal ideal. Let π be a generating element for M. By Proposition 23 also, the residue field $k = R/M$ is finite. Evidently the characteristic of k must also be p. Let $q = p^f$ be the number of elements in k. Since F has characteristic p, for any $a, b \in F$,

$$(b - a)^p = b^p - a^p$$

and hence, by induction,

$$(b - a)^{p^n} = b^{p^n} - a^{p^n} \quad \text{for all } n \geq 1.$$

The multiplicative group of k is a cyclic group of order $q - 1$. Choose $a \in R$ so that $a + M$ generates this cyclic group. Then $|a^q - a| < 1$. By what we have just proved,

$$a^{q^{n+1}} - a^{q^n} = (a^q - a)^{q^n},$$

and hence (a^{q^n}) is a fundamental sequence, by Lemma 10. Since F is complete, by Proposition 23, it follows that $a^{q^n} \to \alpha \in R$. Moreover $\alpha^q = \alpha$, since

$$\lim_{n \to \infty} (a^{q^n})^q = \lim_{n \to \infty} a^{q^{n+1}},$$

and $\alpha - a \in M$, since $a^{q^{n+1}} - a^{q^n} \in M$ for every $n \geq 0$. Hence $\alpha \neq 0$ and $\alpha^{q-1} = 1$. Moreover $\alpha^j \neq 1$ for $1 \leq j < q - 1$, since $\alpha^j \equiv a^j \bmod M$. It follows that the set S consisting of 0 and the powers $1, \alpha, \ldots, \alpha^{q-1}$ is a set of representatives in R of the residue field k.

Since F has characteristic p, α generates a finite subring K of R. In fact K is a field, since $\beta^q = \beta$ for every $\beta \in K$ and so $\beta\beta^{q-2} = 1$ if $\beta \neq 0$. Since $S \subseteq K$ and the polynomial $x^q - x$ has at most q roots in K, we conclude that $S = K$. Thus K has q elements and is isomorphic to the residue field k.

Every element a of F has a unique representation

$$a = \sum_{n \in \mathbb{Z}} \alpha_n \pi^n,$$

where π is a generating element for the principal ideal M, $\alpha_n \in S$ and $\alpha_n \neq 0$ for at most finitely many $n < 0$. The map

$$a' = \sum_{n \in \mathbb{Z}} \alpha_n t^n \to a = \sum_{n \in \mathbb{Z}} \alpha_n \pi^n$$

is a bijection of the field $K((t))$ onto F. Since S is closed under addition this map preserves sums, and since S is also closed under multiplication it also preserves products. Finally, if N is the least integer such that $\alpha_N \neq 0$, then $|a| = |\pi|^N$ and $|a'| = \rho^{-N}$ for some fixed $\rho > 1$. Hence the map is an isomorphism of the valued field $K((t))$ onto F. $\qquad\square$

7 Further Remarks

Valued fields are discussed in more detail in the books of Cassels [1], Endler [3] and Ribenboim [5].

For still more forms of Hensel's lemma, see Ribenboim [6]. There are also generalizations to polynomials in several variables and to power series. The algorithmic implementation of Hensel's lemma is studied in von zur Gathen [4]. Newton's method for finding real or complex zeros is discussed in Stoer and Bulirsch [7], for example.

Proposition 20 continues to hold if the word 'complete' is omitted from its statement. However, the formula given in the proof of Proposition 20 defines an absolute value on E if and only if there is a *unique* extension of the absolute value on F to an absolute value on E; see Viswanathan [8].

Ostrowski's Theorem 22 has been generalized by weakening the requirement $|ab| = |a||b|$ to $|ab| \leq |a||b|$. Mazur (1938) proved that the only normed associative division algebras over $\mathbb{R}$ are $\mathbb{R}$, $\mathbb{C}$ and $\mathbb{H}$, and that the only normed associative division algebra over $\mathbb{C}$ is $\mathbb{C}$ itself. An elegant functional-analytic proof of the latter result was given by Gelfand (1941). See Chapter 8 (by Koecher and Remmert) of Ebbinghaus *et al.* [2].

8 Selected References

[1] J.W.S. Cassels, *Local fields*, Cambridge University Press, 1986.
[2] H.-D. Ebbinghaus *et al.*, *Numbers*, English transl. of 2nd German ed. by H.L.S. Orde, Springer-Verlag, New York, 1990.
[3] O. Endler, *Valuation theory*, Springer-Verlag, Berlin, 1972.
[4] J. von zur Gathen, Hensel and Newton methods in valuation rings, *Math. Comp.* **42** (1984), 637–661.
[5] P. Ribenboim, *The theory of classical valuations*, Springer-Verlag, New York, 1999.
[6] P. Ribenboim, Equivalent forms of Hensel's lemma, *Exposition. Math.* **3** (1985), 3–24.
[7] J. Stoer and R. Bulirsch, *Introduction to numerical analysis*, 3rd ed. (English transl.), Springer-Verlag, New York, 2002.
[8] T.M. Viswanathan, A characterisation of Henselian valuations via the norm, *Bol. Soc. Brasil. Mat.* **4** (1973), 51–53.

VII

The Arithmetic of Quadratic Forms

We have already determined the integers which can be represented as a sum of two squares. Similarly, one may ask which integers can be represented in the form $x^2 + 2y^2$ or, more generally, in the form $ax^2 + 2bxy + cy^2$, where a, b, c are given integers. The arithmetic theory of binary quadratic forms, which had its origins in the work of Fermat, was extensively developed during the 18th century by Euler, Lagrange, Legendre and Gauss. The extension to quadratic forms in more than two variables, which was begun by them and is exemplified by Lagrange's theorem that every positive integer is a sum of four squares, was continued during the 19th century by Dirichlet, Hermite, H.J.S. Smith, Minkowski and others. In the 20th century Hasse and Siegel made notable contributions. With Hasse's work especially it became apparent that the theory is more perspicuous if one allows the variables to be rational numbers, rather than integers. This opened the way to the study of quadratic forms over arbitrary fields, with pioneering contributions by Witt (1937) and Pfister (1965–67).

From this vast theory we focus attention on one central result, the *Hasse–Minkowski theorem*. However, we first study quadratic forms over an arbitrary field in the geometric formulation of Witt. Then, following an interesting approach due to Fröhlich (1967), we study quadratic forms over a *Hilbert field*.

1 Quadratic Spaces

The theory of quadratic spaces is simply another name for the theory of quadratic forms. The advantage of the change in terminology lies in its appeal to geometric intuition. It has in fact led to new results even at quite an elementary level. The new approach had its debut in a paper by Witt (1937) on the arithmetic theory of quadratic forms, but it is appropriate also if one is interested in quadratic forms over the real field or any other field.

For the remainder of this chapter *we will restrict attention to fields for which* $1 + 1 \neq 0$. Thus the phrase 'an arbitrary field' will mean 'an arbitrary field of characteristic $\neq 2$'. The proofs of many results make essential use of this restriction on the

W.A. Coppel, *Number Theory: An Introduction to Mathematics*, Universitext,
DOI: 10.1007/978-0-387-89486-7_7, © Springer Science + Business Media, LLC 2009

characteristic. For any field F, we will denote by $F^\times$ the multiplicative group of all nonzero elements of F. The squares in $F^\times$ form a subgroup $F^{\times 2}$ and any coset of this subgroup is called a *square class*.

Let V be a finite-dimensional vector space over such a field F. We say that V is a *quadratic space* if with each ordered pair u, v of elements of V there is associated an element (u, v) of F such that

(i) $(u_1 + u_2, v) = (u_1, v) + (u_2, v)$ for all $u_1, u_2, v \in V$;
(ii) $(\alpha u, v) = \alpha(u, v)$ for every $\alpha \in F$ and all $u, v \in V$;
(iii) $(u, v) = (v, u)$ for all $u, v \in V$.

It follows that

(i)' $(u, v_1 + v_2) = (u, v_1) + (u, v_2)$ for all $u, v_1, v_2 \in V$;
(ii)' $(u, \alpha v) = \alpha(u, v)$ for every $\alpha \in F$ and all $u, v \in V$.

Let $e_1, \ldots, e_n$ be a basis for the vector space V. Then any $u, v \in V$ can be uniquely expressed in the form

$$u = \sum_{j=1}^{n} \xi_j e_j, \quad v = \sum_{j=1}^{n} \eta_j e_j,$$

where $\xi_j, \eta_j \in F(j = 1, \ldots, n)$, and

$$(u, v) = \sum_{j,k=1}^{n} \alpha_{jk} \xi_j \eta_k,$$

where $\alpha_{jk} = (e_j, e_k) = \alpha_{kj}$. Thus

$$(u, u) = \sum_{j,k=1}^{n} \alpha_{jk} \xi_j \xi_k$$

is a *quadratic form* with coefficients in F. The quadratic space is completely determined by the quadratic form, since

$$(u, v) = \{(u + v, u + v) - (u, u) - (v, v)\}/2. \tag{1}$$

Conversely, for a given basis $e_1, \ldots, e_n$ of V, any $n \times n$ symmetric matrix $A = (\alpha_{jk})$ with elements from F, or the associated quadratic form $f(x) = x^t A x$, may be used in this way to give V the structure of a quadratic space.

Let $e_1', \ldots, e_n'$ be any other basis for V. Then

$$e_i = \sum_{j=1}^{n} \tau_{ji} e_j',$$

where $T = (\tau_{ij})$ is an invertible $n \times n$ matrix with elements from F. Conversely, any such matrix T defines in this way a new basis $e_1', \ldots, e_n'$. Since

$$(e_i, e_k) = \sum_{j,h=1}^{n} \tau_{ji} \beta_{jh} \tau_{hk},$$

where $\beta_{jh} = (e'_j, e'_h)$, the matrix $B = (\beta_{jh})$ is symmetric and

$$A = T^t B T. \tag{2}$$

Two symmetric matrices A, B with elements from F are said to be *congruent* if (2) holds for some invertible matrix T with elements from F. Thus congruence of symmetric matrices corresponds to a change of basis in the quadratic space. Evidently congruence is an equivalence relation, i.e. it is reflexive, symmetric and transitive. Two quadratic forms are said to be *equivalent over* F if their coefficient matrices are congruent. Equivalence over F of the quadratic forms f and g will be denoted by $f \sim_F g$ or simply $f \sim g$.

It follows from (2) that

$$\det A = (\det T)^2 \det B.$$

Thus, although $\det A$ is not uniquely determined by the quadratic space, if it is nonzero, its *square class* is uniquely determined. By abuse of language, we will call any representative of this square class the *determinant* of the quadratic space V and denote it by $\det V$.

Although quadratic spaces are better adapted for proving theorems, quadratic forms and symmetric matrices are useful for computational purposes. Thus a familiarity with both languages is desirable. However, we do not feel obliged to give two versions of each definition or result, and a version in one language may later be used in the other without explicit comment.

A vector v is said to be *orthogonal* to a vector u if $(u, v) = 0$. Then also u is orthogonal to v. The *orthogonal complement* $U^\perp$ of a subspace U of V is defined to be the set of all $v \in V$ such that $(u, v) = 0$ for every $u \in U$. Evidently $U^\perp$ is again a subspace. A subspace U will be said to be *non-singular* if $U \cap U^\perp = \{0\}$.

The whole space V is itself non-singular if and only if $V^\perp = \{0\}$. Thus V is non-singular if and only if some, and hence every, symmetric matrix describing it is non-singular, i.e. if and only if $\det V \neq 0$.

We say that a quadratic space V is the *orthogonal sum* of two subspaces V_1 and V_2, and we write $V = V_1 \perp V_2$, if $V = V_1 + V_2$, $V_1 \cap V_2 = \{0\}$ and $(v_1, v_2) = 0$ for all $v_1 \in V_1, v_2 \in V_2$.

If A_1 is a coefficient matrix for V_1 and A_2 a coefficient matrix for V_2, then

$$A = \begin{pmatrix} A_1 & 0 \\ 0 & A_2 \end{pmatrix}$$

is a coefficient matrix for $V = V_1 \perp V_2$. Thus $\det V = (\det V_1)(\det V_2)$. Evidently V is non-singular if and only if both V_1 and V_2 are non-singular.

If W is any subspace supplementary to the orthogonal complement $V^\perp$ of the whole space V, then $V = V^\perp \perp W$ and W is non-singular. Many problems for arbitrary quadratic spaces may be reduced in this way to non-singular quadratic spaces.

Proposition 1 *If a quadratic space V contains a vector u such that $(u, u) \neq 0$, then*

$$V = U \perp U^{\perp},$$

where $U = \langle u \rangle$ is the one-dimensional subspace spanned by u.

Proof For any vector $v \in V$, put $v' = v - \alpha u$, where $\alpha = (v, u)/(u, u)$. Then $(v', u) = 0$ and hence $v' \in U^{\perp}$. Since $U \cap U^{\perp} = \{0\}$, the result follows. □

A vector space basis $u_1, \ldots, u_n$ of a quadratic space V is said to be an *orthogonal basis* if $(u_j, u_k) = 0$ whenever $j \neq k$.

Proposition 2 *Any quadratic space V has an orthogonal basis.*

Proof If V has dimension 1, there is nothing to prove. Suppose V has dimension $n > 1$ and the result holds for quadratic spaces of lower dimension. If $(v, v) = 0$ for all $v \in V$, then any basis is an orthogonal basis, by (1). Hence we may assume that V contains a vector u_1 such that $(u_1, u_1) \neq 0$. If U_1 is the 1-dimensional subspace spanned by u_1 then, by Proposition 1,

$$V = U_1 \perp U_1^{\perp}.$$

By the induction hypothesis $U_1^{\perp}$ has an orthogonal basis $u_2, \ldots, u_n$, and $u_1, u_2, \ldots, u_n$ is then an orthogonal basis for V. □

Proposition 2 says that any symmetric matrix A is congruent to a diagonal matrix, or that the corresponding quadratic form f is equivalent over F to a diagonal form $\delta_1 \xi_1^2 + \cdots + \delta_n \xi_n^2$. Evidently $\det f = \delta_1 \cdots \delta_n$ and f is non-singular if and only if $\delta_j \neq 0$ ($1 \leq j \leq n$). If $A \neq 0$ then, by Propositions 1 and 2, we can take δ_1 to be any element of $F^{\times}$ which is represented by f.

Here $\gamma \in F^{\times}$ is said to be *represented* by a quadratic space V over the field F if there exists a vector $v \in V$ such that $(v, v) = \gamma$.

As an application of Proposition 2 we prove

Proposition 3 *If U is a non-singular subspace of the quadratic space V, then $V = U \perp U^{\perp}$.*

Proof Let $u_1, \ldots, u_m$ be an orthogonal basis for U. Then $(u_j, u_j) \neq 0$ ($1 \leq j \leq m$), since U is non-singular. For any vector $v \in V$, let $u = \alpha_1 u_1 + \cdots + \alpha_m u_m$, where $\alpha_j = (v, u_j)/(u_j, u_j)$ for each j. Then $u \in U$ and $(u, u_j) = (v, u_j)$ ($1 \leq j \leq m$). Hence $v - u \in U^{\perp}$. Since $U \cap U^{\perp} = \{0\}$, the result follows. □

It may be noted that if U is a non-singular subspace and $V = U \perp W$ for some subspace W, then necessarily $W = U^{\perp}$. For it is obvious that $W \subseteq U^{\perp}$ and $\dim W = \dim V - \dim U = \dim U^{\perp}$, by Proposition 3.

Proposition 4 *Let V be a non-singular quadratic space. If $v_1, \ldots, v_m$ are linearly independent vectors in V then, for any $\eta_1, \ldots, \eta_m \in F$, there exists a vector $v \in V$ such that $(v_j, v) = \eta_j (1 \leq j \leq m)$.*

Moreover, if U is any subspace of V, then

(i) $\dim U + \dim U^{\perp} = \dim V$;

(ii) $U^{\perp\perp} = U$;

(iii) $U^{\perp}$ is non-singular if and only if U is non-singular.

Proof There exist vectors $v_{m+1}, \ldots, v_n \in V$ such that $v_1, \ldots, v_n$ form a basis for V. If we put $\alpha_{jk} = (v_j, v_k)$ then, since V is non-singular, the $n \times n$ symmetric matrix $A = (\alpha_{jk})$ is non-singular. Hence, for any $\eta_1, \ldots, \eta_n \in F$, there exist *unique* $\xi_1, \ldots, \xi_n \in F$ such that $v = \xi_1 v_1 + \cdots + \xi_n v_n$ satisfies

$$(v_1, v) = \eta_1, \ldots, (v_n, v) = \eta_n.$$

This proves the first part of the proposition.

By taking $U = \langle v_1, \ldots, v_m \rangle$ and $\eta_1 = \cdots = \eta_m = 0$, we see that $\dim U^{\perp} = n - m$. Replacing U by $U^{\perp}$, we obtain $\dim U^{\perp\perp} = \dim U$. Since it is obvious that $U \subseteq U^{\perp\perp}$, this implies $U = U^{\perp\perp}$. Since U non-singular means $U \cap U^{\perp} = \{0\}$, (iii) follows at once from (ii). $\square$

We now introduce some further definitions. A vector u is said to be *isotropic* if $u \neq 0$ and $(u, u) = 0$. A subspace U of V is said to be *isotropic* if it contains an isotropic vector and *anisotropic* otherwise. A subspace U of V is said to be *totally isotropic* if every nonzero vector in U is isotropic, i.e. if $U \subseteq U^{\perp}$. According to these definitions, the trivial subspace $\{0\}$ is both anisotropic and totally isotropic.

A quadratic space V over a field F is said to be *universal* if it represents every $\gamma \in F^{\times}$, i.e. if for each $\gamma \in F^{\times}$ there is a vector $v \in V$ such that $(v, v) = \gamma$.

Proposition 5 *If a non-singular quadratic space V is isotropic, then it is universal.*

Proof Since V is isotropic, it contains a vector $u \neq 0$ such that $(u, u) = 0$. Since V is non-singular, it contains a vector w such that $(u, w) \neq 0$. Then w is linearly independent of u and by replacing w by a scalar multiple we may assume $(u, w) = 1$. If $v = \alpha u + w$, then $(v, v) = \gamma$ for $\alpha = \{\gamma - (w, w)\}/2$. $\square$

On the other hand, a non-singular universal quadratic space need not be isotropic. As an example, take F to be the finite field with three elements and V the 2-dimensional quadratic space corresponding to the quadratic form $\xi_1^2 + \xi_2^2$.

Proposition 6 *A non-singular quadratic form $f(\xi_1, \ldots, \xi_n)$ with coefficients from a field F represents $\gamma \in F^{\times}$ if and only if the quadratic form*

$$g(\xi_0, \xi_1, \ldots, \xi_n) = -\gamma \xi_0^2 + f(\xi_1, \ldots, \xi_n)$$

is isotropic.

Proof Obviously if $f(x_1, \ldots, x_n) = \gamma$ and $x_0 = 1$, then $g(x_0, x_1, \ldots, x_n) = 0$. Suppose on the other hand that $g(x_0, x_1, \ldots, x_n) = 0$ for some $x_j \in F$, not all zero. If $x_0 \neq 0$, then f certainly represents γ. If $x_0 = 0$, then f is isotropic and hence, by Proposition 5, it still represents γ. $\square$

Proposition 7 *Let V be a non-singular isotropic quadratic space. If $V = U \perp W$, then there exists $\gamma \in F^{\times}$ such that, for some $u \in U$ and $w \in W$,*

$$(u, u) = \gamma, \quad (w, w) = -\gamma.$$

Proof Since V is non-singular, so also are U and W, and since V contains an isotropic vector v', there exist $u' \in U$, $w' \in W$, not both zero, such that

$$(u', u') = -(w', w').$$

If this common value is nonzero, we are finished. Otherwise either U or W is isotropic. Without loss of generality, suppose U is isotropic. Since W is non-singular, it contains a vector w such that $(w, w) \neq 0$, and U contains a vector u such that $(u, u) = -(w, w)$, by Proposition 5.   □

We now show that the totally isotropic subspaces of a quadratic space are important for an understanding of its structure, even though they are themselves trivial as quadratic spaces.

Proposition 8 *All maximal totally isotropic subspaces of a quadratic space have the same dimension.*

Proof Let U_1 be a maximal totally isotropic subspace of the quadratic space V. Then $U_1 \subseteq U_1^\perp$ and $U_1^\perp \backslash U_1$ contains no isotropic vector. Since $V^\perp \subseteq U_1^\perp$, it follows that $V^\perp \subseteq U_1$. If V' is a subspace of V supplementary to $V^\perp$, then V' is non-singular and $U_1 = V^\perp + U_1'$, where $U_1' \subseteq V'$. Since U_1' is a maximal totally isotropic subspace of V', this shows that it is sufficient to establish the result when V itself is non-singular.

Let U_2 be another maximal totally isotropic subspace of V. Put $W = U_1 \cap U_2$ and let W_1, W_2 be subspaces supplementary to W in U_1, U_2 respectively. We are going to show that $W_2 \cap W_1^\perp = \{0\}$.

Let $v \in W_2 \cap W_1^\perp$. Since $W_2 \subseteq U_2$, v is isotropic and $v \in U_2^\perp \subseteq W^\perp$. Hence $v \in U_1^\perp$ and actually $v \in U_1$, since v is isotropic. Since $W_2 \subseteq U_2$ this implies $v \in W$, and since $W \cap W_2 = \{0\}$ this implies $v = 0$.

It follows that $\dim W_2 + \dim W_1^\perp \leq \dim V$. But, since V is now assumed non-singular, $\dim W_1 = \dim V - \dim W_1^\perp$, by Proposition 4. Hence $\dim W_2 \leq \dim W_1$ and, for the same reason, $\dim W_1 \leq \dim W_2$. Thus $\dim W_2 = \dim W_1$, and hence $\dim U_2 = \dim U_1$.   □

We define the *index*, ind V, of a quadratic space V to be the dimension of any maximal totally isotropic subspace. Thus V is anisotropic if and only if ind $V = 0$.

A field F is said to be *ordered* if it contains a subset P of *positive* elements, which is closed under addition and multiplication, such that F is the disjoint union of the sets $\{0\}$, P and $-P = \{-x : x \in P\}$. The rational field $\mathbb{Q}$ and the real field $\mathbb{R}$ are ordered fields, with the usual interpretation of 'positive'. For quadratic spaces over an ordered field there are other useful notions of index.

A subspace U of a quadratic space V over an ordered field F is said to be *positive definite* if $(u, u) > 0$ for all nonzero $u \in U$ and *negative definite* if $(u, u) < 0$ for all nonzero $u \in U$. Evidently positive definite and negative definite subspaces are anisotropic.

Proposition 9 *All maximal positive definite subspaces of a quadratic space V over an ordered field F have the same dimension.*

Proof Let U_+ be a maximal positive definite subspace of the quadratic space V. Since U_+ is certainly non-singular, we have $V = U_+ \perp W$, where $W = U_+^\perp$, and since U_+ is maximal, $(w, w) \le 0$ for all $w \in W$. Since $U_+ \subseteq V$, we have $V^\perp \subseteq W$. If U_- is a maximal negative definite subspace of W, then in the same way $W = U_- \perp U_0$, where $U_0 = U_-^\perp \cap W$. Evidently U_0 is totally isotropic and $U_0 \subseteq V^\perp$. In fact $U_0 = V^\perp$, since $U_- \cap V^\perp = \{0\}$. Since $(v, v) \ge 0$ for all $v \in U_+ \perp V^\perp$, it follows that U_- is a maximal negative definite subspace of V.

If U_+' is another maximal positive definite subspace of V, then $U_+' \cap W = \{0\}$ and hence

$$\dim U_+' + \dim W = \dim(U_+' + W) \le \dim V.$$

Thus $\dim U_+' \le \dim V - \dim W = \dim U_+$. But U_+ and U_+' can be interchanged. $\square$

If V is a quadratic space over an ordered field F, we define the *positive index* $\mathrm{ind}^+ V$ to be the dimension of any maximal positive definite subspace. Similarly all maximal negative definite subspaces have the same dimension, which we will call the *negative index* of V and denote by $\mathrm{ind}^- V$. The proof of Proposition 9 shows that

$$\mathrm{ind}^+ V + \mathrm{ind}^- V + \dim V^\perp = \dim V.$$

Proposition 10 *Let F denote the real field $\mathbb{R}$ or, more generally, an ordered field in which every positive element is a square. Then any non-singular quadratic form f in n variables with coefficients from F is equivalent over F to a quadratic form*

$$g = \xi_1^2 + \cdots + \xi_p^2 - \xi_{p+1}^2 - \cdots - \xi_n^2,$$

where $p \in \{0, 1, \ldots, n\}$ is uniquely determined by f. In fact,

$$\mathrm{ind}^+ f = p, \ \mathrm{ind}^- f = n - p, \ \mathrm{ind} f = \min(p, n - p).$$

Proof By Proposition 2, f is equivalent over F to a diagonal form $\delta_1 \eta_1^2 + \cdots + \delta_n \eta_n^2$, where $\delta_j \ne 0$ $(1 \le j \le n)$. We may choose the notation so that $\delta_j > 0$ for $j \le p$ and $\delta_j < 0$ for $j > p$. The change of variables $\xi_j = \delta_j^{1/2} \eta_j$ $(j \le p)$, $\xi_j = (-\delta_j)^{1/2} \eta_j$ $(j > p)$ now brings f to the form g. Since the corresponding quadratic space has a p-dimensional maximal positive definite subspace, $p = \mathrm{ind}^+ f$ is uniquely determined. Similarly $n - p = \mathrm{ind}^- f$, and the formula for $\mathrm{ind} f$ follows readily. $\square$

It follows that, for quadratic spaces over a field of the type considered in Proposition 10, a subspace is anisotropic if and only if it is either positive definite or negative definite.

Proposition 10 completely solves the problem of equivalence for real quadratic forms. (The uniqueness of p is known as *Sylvester's law of inertia*.) It will now be shown that the problem of equivalence for quadratic forms over a finite field can also be completely solved.

Lemma 11 *If V is a non-singular 2-dimensional quadratic space over a finite field $\mathbb{F}_q$, of (odd) cardinality q, then V is universal.*

Proof By choosing an orthogonal basis for V we are reduced to showing that if $\alpha, \beta,$
$\gamma \in \mathbb{F}_q^\times$, then there exist $\xi, \eta \in \mathbb{F}_q$ such that $\alpha\xi^2 + \beta\eta^2 = \gamma$. As ξ runs through $\mathbb{F}_q$,
$\alpha\xi^2$ takes $(q+1)/2 = 1 + (q-1)/2$ distinct values. Similarly, as η runs through $\mathbb{F}_q$,
$\gamma - \beta\eta^2$ takes $(q+1)/2$ distinct values. Since $(q+1)/2 + (q+1)/2 > q$, there exist
$\xi, \eta \in \mathbb{F}_q$ for which $\alpha\xi^2$ and $\gamma - \beta\eta^2$ take the same value. $\qquad\square$

Proposition 12 *Any non-singular quadratic form f in n variables over a finite field $\mathbb{F}_q$
is equivalent over $\mathbb{F}_q$ to the quadratic form*

$$\xi_1^2 + \cdots + \xi_{n-1}^2 + \delta\xi_n^2,$$

where $\delta = \det f$ is the determinant of f.

There are exactly two equivalence classes of non-singular quadratic forms in n
variables over $\mathbb{F}_q$, one consisting of those forms f whose determinant $\det f$ is a square
in $\mathbb{F}_q^\times$, and the other those for which $\det f$ is not a square in $\mathbb{F}_q^\times$.

Proof Since the first statement of the proposition is trivial for $n = 1$, we assume that
$n > 1$ and it holds for all smaller values of n. It follows from Lemma 11 that f represents 1 and hence, by the remark after the proof of Proposition 2, f is equivalent over
$\mathbb{F}_q$ to a quadratic form $\xi_1^2 + g(\xi_2, \ldots, \xi_n)$. Since f and g have the same determinant,
the first statement of the proposition now follows from the induction hypothesis.

Since $\mathbb{F}_q^\times$ contains $(q-1)/2$ distinct squares, every element of $\mathbb{F}_q^\times$ is either a square
or a square times a fixed non-square. The second statement of the proposition now fol-
lows from the first. $\qquad\square$

We now return to quadratic spaces over an arbitrary field. A 2-dimensional quadratic
space is said to be a *hyperbolic plane* if it is non-singular and isotropic.

Proposition 13 *For a 2-dimensional quadratic space V, the following statements are
equivalent:*

  (i) *V is a hyperbolic plane;*
 (ii) *V has a basis u_1, u_2 such that $(u_1, u_1) = (u_2, u_2) = 0$, $(u_1, u_2) = 1$;*
(iii) *V has a basis v_1, v_2 such that $(v_1, v_1) = 1$, $(v_2, v_2) = -1$, $(v_1, v_2) = 0$;*
 (iv) *$-\det V$ is a square in $F^\times$.*

Proof Suppose first that V is a hyperbolic plane and let u_1 be *any* isotropic
vector in V. If v is any linearly independent vector, then $(u_1, v) \neq 0$, since V is
non-singular. By replacing v by a scalar multiple we may assume that $(u_1, v) = 1$. If
we put $u_2 = v + \alpha u_1$, where $\alpha = -(v, v)/2$, then

$$(u_2, u_2) = (v, v) + 2\alpha = 0, \quad (u_1, u_2) = (u_1, v) = 1,$$

and u_1, u_2 is a basis for V.

If u_1, u_2 are isotropic vectors in V such that $(u_1, u_2) = 1$, then the vectors $v_1 = u_1 + u_2/2$ and $v_2 = u_1 - u_2/2$ satisfy (iii), and if v_1, v_2 satisfy (iii) then $\det V = -1$.

Finally, if (iv) holds then V is certainly non-singular. Let w_1, w_2 be an orthogonal
basis for V and put $\delta_j = (w_j, w_j)$ $(j = 1, 2)$. By hypothesis, $\delta_1\delta_2 = -\gamma^2$, where
$\gamma \in F^\times$. Since $\gamma w_1 + \delta_1 w_2$ is an isotropic vector, this proves that (iv) implies (i). $\quad\square$

Proposition 14 *Let V be a non-singular quadratic space. If U is a totally isotropic subspace with basis $u_1, \ldots, u_m$, then there exists a totally isotropic subspace U' with basis $u'_1, \ldots, u'_m$ such that*

$$(u_j, u'_k) = 1 \text{ or } 0 \text{ according as } j = k \text{ or } j \neq k.$$

Hence $U \cap U' = \{0\}$ and

$$U + U' = H_1 \perp \cdots \perp H_m,$$

where H_j is the hyperbolic plane with basis u_j, u'_j $(1 \leq j \leq m)$.

Proof Suppose first that $m = 1$. Since V is non-singular, there exists a vector $v \in V$ such that $(u_1, v) \neq 0$. The subspace H_1 spanned by u_1, v is a hyperbolic plane and hence, by Proposition 13, it contains a vector u'_1 such that $(u'_1, u'_1) = 0$, $(u_1, u'_1) = 1$. This proves the proposition for $m = 1$.

Suppose now that $m > 1$ and the result holds for all smaller values of m. Let W be the totally isotropic subspace with basis $u_2, \ldots, u_m$. By Proposition 4, there exists a vector $v \in W^\perp$ such that $(u_1, v) \neq 0$. The subspace H_1 spanned by u_1, v is a hyperbolic plane and hence it contains a vector u'_1 such that $(u'_1, u'_1) = 0$, $(u_1, u'_1) = 1$. Since H_1 is non-singular, $H_1^\perp$ is also non-singular and $V = H_1 \perp H_1^\perp$. Since $W \subseteq H_1^\perp$, the result now follows by applying the induction hypothesis to the subspace W of the quadratic space $H_1^\perp$. $\quad\square$

Proposition 15 *Any quadratic space V can be represented as an orthogonal sum*

$$V = V^\perp \perp H_1 \perp \cdots \perp H_m \perp V_0,$$

where $H_1, \ldots, H_m$ are hyperbolic planes and the subspace V_0 is anisotropic.

Proof Let V_1 be any subspace supplementary to $V^\perp$. Then V_1 is non-singular, by the definition of $V^\perp$. If V_1 is anisotropic, we can take $m = 0$ and $V_0 = V_1$. Otherwise V_1 contains an isotropic vector and hence also a hyperbolic plane H_1, by Proposition 14. By Proposition 3,

$$V_1 = H_1 \perp V_2,$$

where $V_2 = H_1^\perp \cap V_1$ is non-singular. If V_2 is anisotropic, we can take $V_0 = V_2$. Otherwise we repeat the process. After finitely many steps we must obtain a representation of the required form, possibly with $V_0 = \{0\}$. $\quad\square$

Let V and V' be quadratic spaces over the same field F. The quadratic spaces V, V' are said to be *isometric* if there exists a linear map $\varphi : V \to V'$ which is an *isometry*, i.e. it is bijective and

$$(\varphi v, \varphi v) = (v, v) \quad \text{for all } v \in V.$$

By (1), this implies

$$(\varphi u, \varphi v) = (u, v) \quad \text{for all } u, v \in V.$$

The concept of isometry is only another way of looking at equivalence. For if $\varphi : V \to V'$ is an isometry, then V and V' have the same dimension. If $u_1, \ldots, u_n$ is a basis for V and $u'_1, \ldots, u'_n$ a basis for V' then, since $(u_j, u_k) = (\varphi u_j, \varphi u_k)$, the isometry is completely determined by the change of basis in V' from $\varphi u_1, \ldots, \varphi u_n$ to $u'_1, \ldots, u'_n$.

A particularly simple type of isometry is defined in the following way. Let V be a quadratic space and w a vector such that $(w, w) \neq 0$. The map $\tau : V \to V$ defined by

$$\tau v = v - \{2(v, w)/(w, w)\}w$$

is obviously linear. If W is the non-singular one-dimensional subspace spanned by w, then $V = W \perp W^{\perp}$. Since $\tau v = v$ if $v \in W^{\perp}$ and $\tau v = -v$ if $v \in W$, it follows that τ is bijective. Writing $\alpha = -2(v, w)/(w, w)$, we have

$$(\tau v, \tau v) = (v, v) + 2\alpha(v, w) + \alpha^2(w, w) = (v, v).$$

Thus τ is an isometry. Geometrically, τ is a *reflection* in the hyperplane orthogonal to w. We will refer to $\tau = \tau_w$ as the reflection corresponding to the non-isotropic vector w.

Proposition 16 *If u, u' are vectors of a quadratic space V such that $(u, u) = (u', u') \neq 0$, then there exists an isometry $\varphi : V \to V$ such that $\varphi u = u'$.*

Proof Since

$$(u + u', u + u') + (u - u', u - u') = 2(u, u) + 2(u', u') = 4(u, u),$$

at least one of the vectors $u + u', u - u'$ is not isotropic. If $u - u'$ is not isotropic, the reflection τ corresponding to $w = u - u'$ has the property $\tau u = u'$, since $(u - u', u - u') = 2(u, u - u')$. If $u + u'$ is not isotropic, the reflection τ corresponding to $w = u + u'$ has the property $\tau u = -u'$. Since u' is not isotropic, the corresponding reflection σ maps u' onto $-u'$, and hence the isometry $\sigma \tau$ maps u onto u'. $\square$

The proof of Proposition 16 has the following interesting consequence:

Proposition 17 *Any isometry $\varphi : V \to V$ of a non-singular quadratic space V is a product of reflections.*

Proof Let $u_1, \ldots, u_n$ be an orthogonal basis for V. By Proposition 16 and its proof, there exists an isometry ψ, which is either a reflection or a product of two reflections, such that $\psi u_1 = \varphi u_1$. If U is the subspace with basis u_1 and W the subspace with basis $u_2, \ldots, u_n$, then $V = U \perp W$ and $W = U^{\perp}$ is non-singular. Since the isometry $\varphi_1 = \psi^{-1}\varphi$ fixes u_1, we have also $\varphi_1 W = W$. But if $\sigma : W \to W$ is a reflection, the extension $\tau : V \to V$ defined by $\tau u = u$ if $u \in U$, $\tau w = \sigma w$ if $w \in W$, is also a reflection. By using induction on the dimension n, it follows that φ_1 is a product of reflections, and hence so also is $\varphi = \psi \varphi_1$. $\square$

By a more elaborate argument E. Cartan (1938) showed that any isometry of an n-dimensional non-singular quadratic space is a product of at most n reflections.

Proposition 18 *Let V be a quadratic space with two orthogonal sum representations*

$$V = U \perp W = U' \perp W'.$$

If there exists an isometry $\varphi : U \to U'$, then there exists an isometry $\psi : V \to V$ such that $\psi u = \varphi u$ for all $u \in U$ and $\psi W = W'$. Thus if U is isometric to U', then W is isometric to W'.

Proof Let $u_1, \ldots, u_m$ and $u_{m+1}, \ldots, u_n$ be bases for U and W respectively. If $u'_j = \varphi u_j (1 \le j \le m)$, then $u'_1, \ldots, u'_m$ is a basis for U'. Let $u'_{m+1}, \ldots, u'_n$ be a basis for W'. The symmetric matrices associated with the bases $u_1, \ldots, u_n$ and $u'_1, \ldots, u'_n$ of V have the form

$$\begin{pmatrix} A & 0 \\ 0 & B \end{pmatrix}, \begin{pmatrix} A & 0 \\ 0 & C \end{pmatrix},$$

which we will write as $A \oplus B$, $A \oplus C$. Thus the two matrices $A \oplus B$, $A \oplus C$ are congruent. It is enough to show that this implies that B and C are congruent. For suppose $C = S^t B S$ for some invertible matrix $S = (\sigma_{ij})$. If we define $u''_{m+1}, \ldots, u''_n$ by

$$u'_i = \sum_{j=m+1}^{n} \sigma_{ji} u''_j \quad (m+1 \le i \le n),$$

then $(u''_j, u''_k) = (u_j, u_k)$ $(m+1 \le j, k \le n)$ and the linear map $\psi : V \to V$ defined by

$$\psi u_j = u'_j (1 \le j \le m), \quad \psi u_j = u''_j (m+1 \le j \le n),$$

is the required isometry.

By taking the bases for U, W, W' to be orthogonal bases we are reduced to the case in which A, B, C are diagonal matrices. We may choose the notation so that $A = \text{diag}[a_1, \ldots, a_m]$, where $a_j \ne 0$ for $j \le r$ and $a_j = 0$ for $j > r$. If $a_1 \ne 0$, i.e. if $r > 0$, and if we write $A' = \text{diag}[a_2, \ldots, a_m]$, then it follows from Propositions 1 and 16 that the matrices $A' \oplus B$ and $A' \oplus C$ are congruent. Proceeding in this way, we are reduced to the case $A = O$.

Thus we now suppose $A = O$. We may assume $B \ne O$, $C \ne O$, since otherwise the result is obvious. We may choose the notation also so that $B = O_s \oplus B'$ and $C = O_s \oplus C'$, where B' is non-singular and $0 \le s < n - m$. If $T^t(O_{m+s} \oplus C')T = O_{m+s} \oplus B'$, where

$$T = \begin{pmatrix} T_1 & T_2 \\ T_3 & T_4 \end{pmatrix},$$

then $T_4^t C' T_4 = B'$. Since B' is non-singular, so also is T_4 and thus B' and C' are congruent. It follows that B and C are also congruent. $\qquad\square$

Corollary 19 *If a non-singular subspace U of a quadratic space V is isometric to another subspace U', then $U^\perp$ is isometric to $U'^\perp$.*

Proof This follows at once from Proposition 18, since U' is also non-singular and

$$V = U \perp U^\perp = U' \perp U'^\perp.$$ □

The first statement of Proposition 18 is known as *Witt's extension theorem* and the second statement as *Witt's cancellation theorem*. It was Corollary 19 which was actually proved by Witt (1937).

There is also another version of the extension theorem, stating that if $\varphi : U \to U'$ is an isometry between two subspaces U, U' of a *non-singular* quadratic space V, then there exists an isometry $\psi : V \to V$ such that $\psi u = \varphi u$ for all $u \in U$. For non-singular U this has just been proved, and the singular case can be reduced to the non-singular by applying (several times, if necessary) the following lemma.

Lemma 20 *Let V be a non-singular quadratic space. If U, U' are singular subspaces of V and if there exists an isometry $\varphi : U \to U'$, then there exist subspaces $\bar{U}, \bar{U}'$, properly containing U, U' respectively and an isometry $\bar{\varphi} : \bar{U} \to \bar{U}'$ such that $\bar{\varphi}u = \varphi u$ for all $u \in U$.*

Proof By hypothesis there exists a nonzero vector $u_1 \in U \cap U^\perp$. Then U has a basis $u_1, \ldots, u_m$ with u_1 as first vector. By Proposition 4, there exists a vector $w \in V$ such that

$$(u_1, w) = 1, \quad (u_j, w) = 0 \quad \text{for } 1 < j \le m.$$

Moreover we may assume that $(w, w) = 0$, by replacing w by $w - \alpha u_1$, with $\alpha = (w, w)/2$. If W is the 1-dimensional subspace spanned by w, then $U \cap W = \{0\}$ and $\bar{U} = U + W$ contains U properly.

The same construction can be applied to U', with the basis $\varphi u_1, \ldots, \varphi u_m$, to obtain an isotropic vector w' and a subspace $\bar{U}' = U' + W'$. The linear map $\bar{\varphi} : \bar{U} \to \bar{U}'$ defined by

$$\bar{\varphi}u_j = \varphi u_j \ (1 \le j \le m), \quad \bar{\varphi}w = w',$$

is easily seen to have the required properties.   □

As an application of Proposition 18, we will consider the uniqueness of the representation obtained in Proposition 15.

Proposition 21 *Suppose the quadratic space V can be represented as an orthogonal sum*

$$V = U \perp H \perp V_0,$$

where U is totally isotropic, H is the orthogonal sum of m hyperbolic planes, and the subspace V_0 is anisotropic.
   Then $U = V^\perp$, $m = \text{ind } V - \dim V^\perp$, and V_0 is uniquely determined up to an isometry.

Proof Since H and V_0 are non-singular, so also is $W = H \perp V_0$. Hence, by the remark after the proof of Proposition 3, $U = W^\perp$. Since $U \subseteq U^\perp$, it follows that $U \subseteq V^\perp$. In fact $U = V^\perp$, since $W \cap V^\perp = \{0\}$.

The subspace H has two m-dimensional totally isotropic subspaces U_1, U_1' such that

$$H = U_1 + U_1', \quad U_1 \cap U_1' = \{0\}.$$

Evidently $V_1 := V^\perp + U_1$ is a totally isotropic subspace of V. In fact V_1 is maximal, since any isotropic vector in $U_1' \perp V_0$ is contained in U_1'. Thus $m = \text{ind } V - \dim V^\perp$ is uniquely determined and H is uniquely determined up to an isometry. If also

$$V = V^\perp \perp H' \perp V_0',$$

where H' is the orthogonal sum of m hyperbolic planes and V_0' is anisotropic then, by Proposition 18, V_0 is isometric to V_0'. $\quad\square$

Proposition 21 reduces the problem of equivalence for quadratic forms over an arbitrary field to the case of anisotropic forms. As we will see, this can still be a difficult problem, even for the field of rational numbers.

Two quadratic spaces V, V' over the same field F may be said to be *Witt-equivalent*, in symbols $V \approx V'$, if their anisotropic components V_0, V_0' are isometric. This is certainly an equivalence relation. The cancellation law makes it possible to define various algebraic operations on the set $\mathcal{W}(F)$ of all quadratic spaces over the field F, with equality replaced by Witt-equivalence. If we define $-V$ to be the quadratic space with the same underlying vector space as V but with (v_1, v_2) replaced by $-(v_1, v_2)$, then

$$V \perp (-V) \approx \{O\}.$$

If we define the *sum* of two quadratic spaces V and W to be $V \perp W$, then

$$V \approx V', \; W \approx W' \Rightarrow V \perp W \approx V' \perp W'.$$

Similarly, if we define the *product* of V and W to be the tensor product $V \otimes W$ of the underlying vector spaces with the quadratic space structure defined by

$$(\{v_1, w_1\}, \{v_2, w_2\}) = (v_1, v_2)(w_1, w_2),$$

then

$$V \approx V', \; W \approx W' \Rightarrow V \otimes W \approx V' \otimes W'.$$

It is readily seen that in this way $\mathcal{W}(F)$ acquires the structure of a commutative ring, the *Witt ring* of the field F.

2 The Hilbert Symbol

Again let F be any field of characteristic $\neq 2$ and $F^\times$ the multiplicative group of all nonzero elements of F. We define the *Hilbert symbol* $(a, b)_F$, where $a, b \in F^\times$, by

$$(a, b)_F = 1 \text{ if there exist } x, y \in F \text{ such that } ax^2 + by^2 = 1,$$
$$= -1 \text{ otherwise.}$$

By Proposition 6, $(a, b)_F = 1$ *if and only if the ternary quadratic form* $a\xi^2 + b\eta^2 - \zeta^2$ *is isotropic.*

The following lemma shows that the Hilbert symbol can also be defined in an asymmetric way:

Lemma 22 *For any field F and any $a, b \in F^{\times}$, $(a, b)_F = 1$ if and only if the binary quadratic form $f_a = \xi^2 - a\eta^2$ represents b. Moreover, for any $a \in F^{\times}$, the set G_a of all $b \in F^{\times}$ which are represented by f_a is a subgroup of $F^{\times}$.*

Proof Suppose first that $ax^2 + by^2 = 1$ for some $x, y \in F$. If a is a square, the quadratic form f_a is isotropic and hence universal. If a is not a square, then $y \neq 0$ and $(y^{-1})^2 - a(xy^{-1})^2 = b$.

Suppose next that $u^2 - av^2 = b$ for some $u, v \in F$. If $-ba^{-1}$ is a square, the quadratic form $a\xi^2 + b\eta^2$ is isotropic and hence universal. If $-ba^{-1}$ is not a square, then $u \neq 0$ and $a(vu^{-1})^2 + b(u^{-1})^2 = 1$.

It is obvious that if $b \in G_a$, then also $b^{-1} \in G_a$, and it is easily verified that if

$$\zeta_1 = \xi_1 \eta_1 + a\xi_2 \eta_2, \quad \zeta_2 = \xi_1 \eta_2 + \xi_2 \eta_1,$$

then

$$\zeta_1^2 - a\zeta_2^2 = (\xi_1^2 - a\xi_2^2)(\eta_1^2 - a\eta_2^2).$$

(In fact this is just Brahmagupta's identity, already encountered in §4 of Chapter IV.) It follows that G_a is a subgroup of $F^{\times}$.   □

Proposition 23 *For any field F, the Hilbert symbol has the following properties:*

  (i) $(a, b)_F = (b, a)_F$,
 (ii) $(a, bc^2)_F = (a, b)_F$ *for any* $c \in F^{\times}$,
(iii) $(a, 1)_F = 1$,
 (iv) $(a, -ab)_F = (a, b)_F$,
  (v) *if* $(a, b)_F = 1$, *then* $(a, bc)_F = (a, c)_F$ *for any* $c \in F^{\times}$.

Proof The first three properties follow immediately from the definition. The fourth property follows from Lemma 22. For, since G_a is a group and f_a represents $-a$, f_a represents $-ab$ if and only if it represents b. The proof of (v) is similar: if f_a represents b, then it represents bc if and only if it represents c.   □

The Hilbert symbol will now be evaluated for the real field $\mathbb{R} = \mathbb{Q}_{\infty}$ and the p-adic fields $\mathbb{Q}_p$ studied in Chapter VI. In these cases it will be denoted simply by $(a, b)_{\infty}$, resp. $(a, b)_p$. For the real field, we obtain at once from the definition of the Hilbert symbol

Proposition 24 *Let $a, b \in \mathbb{R}^{\times}$. Then $(a, b)_{\infty} = -1$ if and only if both $a < 0$ and $b < 0$.*

To evaluate $(a, b)_p$, we first note that we can write $a = p^{\alpha}a'$, $b = p^{\beta}b'$, where $\alpha, \beta \in \mathbb{Z}$ and $|a'|_p = |b'|_p = 1$. It follows from (i), (ii) of Proposition 23 that we may assume $\alpha, \beta \in \{0, 1\}$. Furthermore, by (ii), (iv) of Proposition 23 we may assume that α and β are not both 1. Thus we are reduced to the case where a is a p-adic unit and either b is a p-adic unit or $b = pb'$, where b' is a p-adic unit. To evaluate $(a, b)_p$ under these assumptions we will use the conditions for a p-adic unit to be a square which were derived in Chapter VI. It is convenient to treat the case $p = 2$ separately.

Proposition 25 *Let p be an odd prime and $a, b \in \mathbb{Q}_p$ with $|a|_p = |b|_p = 1$. Then*

(i) $(a, b)_p = 1$,
(ii) $(a, pb)_p = 1$ *if and only if $a = c^2$ for some $c \in \mathbb{Q}_p$.*

In particular, for any integers a,b not divisible by p, $(a, b)_p = 1$ and $(a, pb)_p = (a/p)$, where (a/p) is the Legendre symbol.

Proof Let $S \subseteq \mathbb{Z}_p$ be a set of representatives, with $0 \in S$, of the finite residue field $\mathbb{F}_p = \mathbb{Z}_p/p\mathbb{Z}_p$. There exist non-zero $a_0, b_0 \in S$ such that

$$|a - a_0|_p < 1, \quad |b - b_0|_p < 1.$$

But Lemma 11 implies that there exist $x_0, y_0 \in S$ such that

$$|a_0 x_0^2 + b_0 y_0^2 - 1|_p < 1.$$

Since $|x_0|_p \leq 1, |y_0|_p \leq 1$, it follows that

$$|a x_0^2 + b y_0^2 - 1|_p < 1.$$

Hence, by Proposition VI.16, $a x_0^2 + b y_0^2 = z^2$ for some $z \in \mathbb{Q}_p$. Since $z \neq 0$, this implies $(a, b)_p = 1$. This proves (i).

If $a = c^2$ for some $c \in \mathbb{Q}_p$, then $(a, pb)_p = 1$, by Proposition 23. Conversely, suppose there exist $x, y \in \mathbb{Q}_p$ such that $ax^2 + pby^2 = 1$. Then $|ax^2|_p \neq |pby^2|_p$, since $|a|_p = |b|_p = 1$. It follows that $|x|_p = 1, |y|_p \leq 1$. Thus $|ax^2 - 1|_p < 1$ and hence $ax^2 = z^2$ for some $z \in \mathbb{Q}_p^\times$. This proves (ii).

The special case where a and b are integers now follows from Corollary VI.17. $\square$

Corollary 26 *If p is an odd prime and if $a, b, c \in \mathbb{Q}_p$ are p-adic units, then the quadratic form $a\xi^2 + b\eta^2 + c\zeta^2$ is isotropic.*

Proof In fact, the quadratic form $-c^{-1}a\xi^2 - c^{-1}b\eta^2 - \zeta^2$ is isotropic, since $(-c^{-1}a, -c^{-1}b)_p = 1$, by Proposition 25. $\square$

Proposition 27 *Let $a, b \in \mathbb{Q}_2$ with $|a|_2 = |b|_2 = 1$. Then*

(i) $(a, b)_2 = 1$ *if and only if at least one of $a, b, a - 4, b - 4$ is a square in $\mathbb{Q}_2$;*
(ii) $(a, 2b)_2 = 1$ *if and only if either a or $a + 2b$ is a square in $\mathbb{Q}_2$.*

In particular, for any odd integers a, b, $(a, b)_2 = 1$ if and only if $a \equiv 1$ or $b \equiv 1 \bmod 4$, and $(a, 2b)_2 = 1$ if and only if $a \equiv 1$ or $a + 2b \equiv 1 \bmod 8$.

Proof Suppose there exist $x, y \in \mathbb{Q}_2$ such that $ax^2 + by^2 = 1$ and assume, for example, that $|x|_2 \geq |y|_2$. Then $|x|_2 \geq 1$ and $|x|_2 = 2^\alpha$, where $\alpha \geq 0$. By Corollary VI.14,

$$x = 2^\alpha(x_0 + 4x'), \quad y = 2^\alpha(y_0 + 4y'),$$

where $x_0 \in \{1, 3\}, y_0 \in \{0, 1, 2, 3\}$ and $x', y' \in \mathbb{Z}_2$. If a and b are not squares in $\mathbb{Q}_2$ then, by Proposition VI.16, $|a - 1|_2 > 2^{-3}$ and $|b - 1|_2 > 2^{-3}$. Thus

$$a = a_0 + 8a', \quad b = b_0 + 8b',$$

where $a_0, b_0 \in \{3, 5, 7\}$ and $a', b' \in \mathbb{Z}_2$. Hence

$$1 = ax^2 + by^2 = 2^{2a}(a_0 + b_0 y_0^2 + 8z'),$$

where $z' \in \mathbb{Z}_2$. Since a_0, b_0 are odd and $y_0^2 \equiv 0, 1$ or $4 \bmod 8$, we must have $\alpha = 0$, $y_0^2 \equiv 4 \bmod 8$ and $a_0 = 5$. Thus, by Proposition VI.16 again, $a - 4$ is a square in $\mathbb{Q}_2$. This proves that the condition in (i) is necessary.

If a is a square in $\mathbb{Q}_2$, then certainly $(a, b)_2 = 1$. If $a - 4$ is a square, then $a = 5 + 8a'$, where $a' \in \mathbb{Z}_2$, and $a + 4b = 1 + 8c'$, where $c' \in \mathbb{Z}_2$. Hence $a + 4b$ is a square in $\mathbb{Q}_2$ and the quadratic form $a\xi^2 + b\eta^2$ represents 1. This proves that the condition in (i) is sufficient.

Suppose next that there exist $x, y \in \mathbb{Q}_2$ such that $ax^2 + 2by^2 = 1$. By the same argument as for odd p in Proposition 25, we must have $|x|_2 = 1$, $|y|_2 \leq 1$. Thus $x = x_0 + 4x'$, $y = y_0 + 4y'$, where $x_0 \in \{1, 3\}$, $y_0 \in \{0, 1, 2, 3\}$ and $x', y' \in \mathbb{Z}_2$. Writing $a = a_0 + 8a'$, $b = b_0 + 8b'$, where $a_0, b_0 \in \{1, 3, 5, 7\}$ and $a', b' \in \mathbb{Z}_2$, we obtain $a_0 x_0^2 + 2b_0 y_0^2 \equiv 1 \bmod 8$. Since $2y_0^2 \equiv 0$ or $2 \bmod 8$, this implies either $a_0 \equiv 1$ or $a_0 + 2b_0 \equiv 1 \bmod 8$. Hence either a or $a + 2b$ is a square in $\mathbb{Q}_2$. It is obvious that, conversely, $(a, 2b)_2 = 1$ if either a or $a + 2b$ is a square in $\mathbb{Q}_2$.

The special case where a and b are integers again follows from Corollary VI.17. $\square$

For $F = \mathbb{R}$, the factor group $F^\times / F^{\times 2}$ is of order 2, with 1 and -1 as representatives of the two square classes. For $F = \mathbb{Q}_p$, with p odd, it follows from Corollary VI.17 that the factor group $F^\times / F^{\times 2}$ is of order 4. Moreover, if r is an integer such that $(r/p) = -1$, then $1, r, p, rp$ are representatives of the four square classes. Similarly for $F = \mathbb{Q}_2$, the factor group $F^\times / F^{\times 2}$ is of order 8 and $1, 3, 5, 7, 2, 6, 10, 14$ are representatives of the eight square classes. The Hilbert symbol $(a, b)_F$ for these representatives, and hence for all $a, b \in F^\times$, may be determined directly from Propositions 24, 25 and 27. The values obtained are listed in Table 1, where $\varepsilon = (-1/p)$ and thus $\varepsilon = \pm 1$ according as $p \equiv \pm 1 \bmod 4$.

It will be observed that each of the three symmetric matrices in Table 1 is a Hadamard matrix! In particular, in each row after the first row of $+$'s there are equally many $+$ and $-$ signs. This property turns out to be of basic importance and prompts the following definition:

A field F is a *Hilbert field* if some $a \in F^\times$ is not a square and if, for every such a, the subgroup G_a has index 2 in $F^\times$.

Thus the real field $\mathbb{R} = \mathbb{Q}_\infty$ and the p-adic fields $\mathbb{Q}_p$ are all Hilbert fields. We now show that in Hilbert fields further properties of the Hilbert symbol may be derived.

Proposition 28 *A field F is a Hilbert field if and only if some $a \in F^\times$ is not a square and the Hilbert symbol has the following additional properties:*

(i) *if $(a, b)_F = 1$ for every $b \in F^\times$, then a is a square in $F^\times$;*
(ii) *$(a, bc)_F = (a, b)_F (a, c)_F$ for all $a, b, c \in F^\times$.*

Proof Let F be a Hilbert field. Then (i) holds, since $G_a \neq F^\times$ if a is not a square. If $(a, b)_F = 1$ or $(a, c)_F = 1$, then (ii) follows from Proposition 23(v). Suppose now that $(a, b)_F = -1$ and $(a, c)_F = -1$. Then a is not a square and f_a does not represent b or c. Since F is a Hilbert field and $b, c \notin G_a$, it follows that $bc \in G_a$. Thus $(a, bc)_F = 1$. The converse is equally simple. $\square$

Table 1. Values of the Hilbert symbol $(a, b)_F$ for $F = \mathbb{Q}_v$

$\mathbb{Q}_\infty = \mathbb{R}$				$\mathbb{Q}_p : p$ odd			
$a\backslash b$	1	-1		$a\backslash b$	1	p	rp	r
1	$+$	$+$		1	$+$	$+$	$+$	$+$
-1	$+$	$-$		p	$+$	ε	$-\varepsilon$	$-$
				rp	$+$	$-\varepsilon$	ε	$-$
				r	$+$	$-$	$-$	$+$

where r is a primitive root mod p and
$$\varepsilon = (-1)^{(p-1)/2}$$

$\mathbb{Q}_2$

$a\backslash b$	1	3	6	2	14	10	5	7
1	$+$	$+$	$+$	$+$	$+$	$+$	$+$	$+$
3	$+$	$-$	$+$	$-$	$+$	$-$	$+$	$-$
6	$+$	$+$	$-$	$-$	$+$	$+$	$-$	$-$
2	$+$	$-$	$-$	$+$	$+$	$-$	$-$	$+$
14	$+$	$+$	$+$	$+$	$-$	$-$	$-$	$-$
10	$+$	$-$	$+$	$-$	$-$	$+$	$-$	$+$
5	$+$	$+$	$-$	$-$	$-$	$-$	$+$	$+$
7	$+$	$-$	$-$	$+$	$-$	$+$	$+$	$-$

The definition of a Hilbert field can be reformulated in terms of quadratic forms. If f is an anisotropic binary quadratic form with determinant d, then $-d$ is not a square and f is equivalent to a diagonal form $a(\xi^2 + d\eta^2)$. It follows that F is a Hilbert field if and only if there exists an anisotropic binary quadratic form and for each such form there is, apart from equivalent forms, exactly one other whose determinant is in the same square class. We are going to show that Hilbert fields can also be characterized by means of quadratic forms in 4 variables.

Lemma 29 *Let F be an arbitrary field and a, b elements of $F^\times$ with $(a, b)_F = -1$. Then the quadratic form*

$$f_{a,b} = \xi_1^2 - a\xi_2^2 - b(\xi_3^2 - a\xi_4^2)$$

is anisotropic. Morover, the set $G_{a,b}$ of all elements of $F^\times$ which are represented by $f_{a,b}$ is a subgroup of $F^\times$.

Proof Since $(a, b)_F = -1$, a is not a square and hence the binary form f_a is anisotropic. If $f_{a,b}$ were isotropic, some $c \in F^\times$ would be represented by both f_a and bf_a. But then $(a, c)_F = 1$ and $(a, bc)_F = 1$. Since $(a, b)_F = -1$, this contradicts Proposition 23.

Clearly if $c \in G_{a,b}$, then also $c^{-1} \in G_{a,b}$, and it is easily verified that if

$$\zeta_1 = \xi_1\eta_1 + a\xi_2\eta_2 + b\xi_3\eta_3 - ab\xi_4\eta_4, \quad \zeta_2 = \xi_1\eta_2 + \xi_2\eta_1 - b\xi_3\eta_4 + b\xi_4\eta_3,$$
$$\zeta_3 = \xi_1\eta_3 + \xi_3\eta_1 + a\xi_2\eta_4 - a\xi_4\eta_2, \quad \zeta_4 = \xi_1\eta_4 + \xi_4\eta_1 + \xi_2\eta_3 - \xi_3\eta_2,$$

then

$$\zeta_1^2 - a\zeta_2^2 - b\zeta_3^2 + ab\zeta_4^2 = (\xi_1^2 - a\xi_2^2 - b\xi_3^2 + ab\xi_4^2)(\eta_1^2 - a\eta_2^2 - b\eta_3^2 + ab\eta_4^2).$$

It follows that $G_{a,b}$ is a subgroup of $F^\times$.   □

Proposition 30 *A field F is a Hilbert field if and only if one of the following mutually exclusive conditions is satisfied:*

(A) *F is an ordered field and every positive element of F is a square;*
(B) *there exists, up to equivalence, one and only one anisotropic quaternary quadratic form over F.*

Proof Suppose first that the field F is of type (A). Then -1 is not a square, since $-1 + 1 = 0$ and any nonzero square is positive. By Proposition 10, any anisotropic binary quadratic form is equivalent over F to exactly one of the forms $\xi^2 + \eta^2, -\xi^2 - \eta^2$ and therefore F is a Hilbert field. Since the quadratic forms $\xi_1^2 + \xi_2^2 + \xi_3^2 + \xi_4^2$ and $-\xi_1^2 - \xi_2^2 - \xi_3^2 - \xi_4^2$ are anisotropic and inequivalent, the field F is not of type (B).

Suppose next that the field F is of type (B). The anisotropic quaternary quadratic form must be universal, since it is equivalent to any nonzero scalar multiple. Hence, for any $a \in F^\times$ there exists an anisotropic diagonal form

$$-a\xi_1^2 - b'\xi_2^2 - c'\xi_3^2 - d'\xi_4^2,$$

where $b', c', d' \in F^\times$. In particular, for $a = -1$, this shows that not every element of $F^\times$ is a square. The ternary quadratic form $h = -b'\xi_2^2 - c'\xi_3^2 - d'\xi_4^2$ is certainly anisotropic. If h does not represent 1, the quaternary quadratic form $-\xi_1^2 + h$ is also anisotropic and hence, by Witt's cancellation theorem, a must be a square. Consequently, if $a \in F^\times$ is not a square, then there exists an anisotropic form

$$-a\xi_1^2 + \xi_2^2 - b\xi_3^2 - c\xi_4^2.$$

Thus for any $a \in F^\times$ which is not a square, there exists $b \in F^\times$ such that $(a, b)_F = -1$. If $(a, b)_F = (a, b')_F = -1$ then, by Lemma 29, the forms

$$\xi_1^2 - a\xi_2^2 - b(\xi_3^2 - a\xi_4^2), \ \xi_1^2 - a\xi_2^2 - b'(\xi_3^2 - a\xi_4^2)$$

are anisotropic and thus equivalent. It follows from Witt's cancellation theorem that the binary forms $b(\xi_3^2 - a\xi_4^2)$ and $b'(\xi_3^2 - a\xi_4^2)$ are equivalent. Consequently $\xi_3^2 - a\xi_4^2$ represents bb' and $(a, bb')_F = 1$. Thus G_a has index 2 in $F^\times$ for any $a \in F^\times$ which is not a square, and F is a Hilbert field.

Suppose now that F is a Hilbert field. Then there exists $a \in F^\times$ which is not a square and, for any such a, there exists $b \in F^\times$ such that $(a, b)_F = -1$. Consequently, by Lemma 29, the quaternary quadratic form $f_{a,b}$ is anisotropic and represents 1. Conversely, any anisotropic quaternary quadratic form which represents 1 is equivalent to some form

$$g = \xi_1^2 - a\xi_2^2 - b(\xi_3^2 - c\xi_4^2)$$

with $a, b, c \in F^\times$. Evidently a and c are not squares, and if d is represented by $\xi_3^2 - c\xi_4^2$, then bd is not represented by $\xi_1^2 - a\xi_2^2$. Thus $(c, d)_F = 1$ implies $(a, bd)_F = -1$. In particular, $(a, b)_F = -1$ and hence $(c, d)_F = 1$ implies $(a, d)_F = 1$. By interchanging the roles of $\xi_1^2 - a\xi_2^2$ and $\xi_3^2 - c\xi_4^2$, we see that $(a, d)_F = 1$ also implies $(c, d)_F = 1$. Hence $(ac, d)_F = 1$ for all $d \in F^\times$. Thus ac is a square and g is equivalent to

$$f_{a,b} = \xi_1^2 - a\xi_2^2 - b(\xi_3^2 - a\xi_4^2).$$

We now show that $f_{a,b}$ and $f_{a',b'}$ are equivalent if $(a, b)_F = (a', b')_F = -1$. Suppose first that $(a, b')_F = -1$. Then $(a, bb')_F = 1$ and there exist $x_3, x_4 \in F$ such that $b' = b(x_3^2 - ax_4^2)$. Since

$$(x_3^2 - ax_4^2)(\xi_3^2 - a\xi_4^2) = \eta_3^2 - a\eta_4^2,$$

where $\eta_3 = x_3\xi_3 + ax_4\xi_4$, $\eta_4 = x_4\xi_3 + x_3\xi_4$, it follows that $f_{a,b'}$ is equivalent to $f_{a,b}$. For the same reason $f_{a,b'}$ is equivalent to $f_{a',b'}$ and thus $f_{a,b}$ is equivalent to $f_{a',b'}$. By symmetry, the same conclusion holds if $(a', b)_F = -1$. Thus we now suppose

$$(a, b')_F = (a', b)_F = 1.$$

But then $(a, bb')_F = (a', bb')_F = -1$ and so, by what we have already proved,

$$f_{a,b} \sim f_{a,bb'} \sim f_{a',bb'} \sim f_{a',b'}.$$

Together, the last two paragraphs show that if F is a Hilbert field, then all anisotropic quaternary quadratic forms which represent 1 are equivalent. Hence the Hilbert field F is of type (B) if every anisotropic quaternary quadratic form represents 1.

Suppose now that some anisotropic quaternary quadratic form does not represent 1. Then some scalar multiple of this form represents 1, but is not universal. Thus $f_{a,b}$ is not universal for some $a, b \in F^\times$ with $(a, b)_F = -1$. By Lemma 29, the set $G_{a,b}$ of all $c \in F^\times$ which are represented by $f_{a,b}$ is a subgroup of $F^\times$. In fact $G_{a,b} = G_a$, since $G_a \subseteq G_{a,b}$, $G_{a,b} \neq F^\times$ and G_a has index 2 in $F^\times$. Since $f_{a,b} \sim f_{b,a}$, we have also $G_{a,b} = G_b$. Thus $(a, c)_F = (b, c)_F$ for all $c \in F^\times$, and hence $(ab, c)_F = 1$ for all $c \in F^\times$. Thus ab is a square and $(a, a)_F = (a, b)_F = -1$. Since $(a, -a)_F = 1$, it follows that $(a, -1)_F = -1$. Hence $f_{a,b} \sim f_{a,a} \sim f_{a,-1}$. Replacing a, b by $-1, a$ we now obtain $(-1, -1)_F = -1$ and $f_{a,-1} \sim f_{-1,-1}$.

Thus the form

$$f = \xi_1^2 + \xi_2^2 + \xi_3^2 + \xi_4^2$$

is not universal and the subgroup P of all elements of $F^\times$ represented by f coincides with the set of all elements of $F^\times$ represented by $\xi^2 + \eta^2$. Hence $P + P \subseteq P$ and P is the set of all $c \in F^\times$ such that $(-1, c)_F = 1$. Consequently $-1 \notin P$ and F is the disjoint union of the sets $\{O\}$, P and $-P$. Thus F is an ordered field with P as the set of positive elements.

For any $c \in F^\times$, $c^2 \in P$. It follows that if $a, b \in P$ then $(-a, -b)_F = -1$, since $a\xi^2 + b\eta^2$ does not represent -1. Hence it follows that, if $a, b \in P$, then $(-a, -b)_F = -1 = (-1, -b)_F$ and $(-a, b)_F = 1 = (-1, b)_F$. Thus, for all $c \in F^\times$, $(-a, c)_F = (-1, c)_F$ and hence $(a, c)_F = 1$. Therefore a is a square and the Hilbert field F is of type (A). $\qquad \square$

Proposition 31 *If F is a Hilbert field of type (B), then any quadratic form f in more than 4 variables is isotropic.*

For any prime p, the field $\mathbb{Q}_p$ of p-adic numbers is a Hilbert field of type (B).

Proof The quadratic form f is equivalent to a diagonal form $a_1\xi_1^2 + \cdots + a_n\xi_n^2$, where $n > 4$. If $g = a_1\xi_1^2 + \cdots + a_4\xi_4^2$ is isotropic, then so also is f. If g is anisotropic then, since F is of type (B), it is universal and represents $-a_5$. This proves the first part of the proposition.

We already know that $\mathbb{Q}_p$ is a Hilbert field and we have already shown, after the proof of Corollary VI.17, that $\mathbb{Q}_p$ is not an ordered field. Hence $\mathbb{Q}_p$ is a Hilbert field of type (B). $\qquad \square$

Proposition 10 shows that two non-singular quadratic forms in n variables, with coefficients from a Hilbert field of type (A), are equivalent over F if and only if they have the same positive index. We consider next the equivalence of quadratic forms with coefficients from a Hilbert field of type (B). We will show that they are classified by their determinant and their Hasse invariant.

If a non-singular quadratic form f, with coefficients from a Hilbert field F, is equivalent to a diagonal form $a_1\xi_1^2 + \cdots + a_n\xi_n^2$, then its *Hasse invariant* is defined to be the product of Hilbert symbols

$$s_F(f) = \prod_{1 \le j < k \le n} (a_j, a_k)_F.$$

We write $s_p(f)$ for $s_F(f)$ when $F = \mathbb{Q}_p$. (It should be noted that some authors define the Hasse invariant with $\prod_{j \le k}$ in place of $\prod_{j < k}$). It must first be shown that this is indeed an invariant of f, and for this we make use of *Witt's chain equivalence theorem*:

Lemma 32 *Let V be a non-singular quadratic space over an arbitrary field F. If $\mathscr{B} = \{u_1, \ldots, u_n\}$ and $\mathscr{B}' = \{u'_1, \ldots, u'_n\}$ are both orthogonal bases of V, then there exists a chain of orthogonal bases $\mathscr{B}_0, \mathscr{B}_1, \ldots, \mathscr{B}_m$, with $\mathscr{B}_0 = \mathscr{B}$ and $\mathscr{B}_m = \mathscr{B}'$, such that $\mathscr{B}_{j-1}$ and $\mathscr{B}_j$ differ by at most 2 vectors for each $j \in \{1, \ldots, m\}$.*

Proof Since there is nothing to prove if $\dim V = n \le 2$, we assume that $n \ge 3$ and the result holds for all smaller values of n. Let $p = p(\mathscr{B})$ be the number of nonzero coefficients in the representation of u'_1 as a linear combination of $u_1, \ldots, u_n$. Without loss of generality we may suppose

$$u'_1 = \sum_{j=1}^{p} a_j u_j,$$

where $a_j \ne 0$ $(1 \le j \le p)$. If $p = 1$, we may replace u_1 by u'_1 and the result now follows by applying the induction hypothesis to the subspace of all vectors orthogonal to u'_1. Thus we now assume $p \ge 2$. We have

$$a_1^2(u_1, u_1) + \cdots + a_p^2(u_p, u_p) = (u_1', u_1') \neq 0,$$

and each summand on the left is nonzero. If the sum of the first two terms is zero, then $p > 2$ and either the sum of the first and third terms is nonzero or the sum of the second and third terms is nonzero. Hence we may suppose without loss of generality that

$$a_1^2(u_1, u_1) + a_2^2(u_2, u_2) \neq 0.$$

If we put

$$v_1 = a_1 u_1 + a_2 u_2, \quad v_2 = u_1 + b u_2, \quad v_j = u_j \quad \text{for } 3 \leq j \leq n,$$

where $b = -a_1(u_1, u_1)/a_2(u_2, u_2)$, then $\mathcal{B}_1 = \{v_1, \ldots, v_n\}$ is an orthogonal basis and $u_1' = v_1 + a_3 v_3 + \cdots + a_p v_p$. Thus $p(\mathcal{B}_1) < p(\mathcal{B})$. By replacing $\mathcal{B}$ by $\mathcal{B}_1$ and repeating the procedure, we must arrive after $s < n$ steps at an orthogonal basis $\mathcal{B}_s$ for which $p(\mathcal{B}_s) = 1$. The induction hypothesis can now be applied to $\mathcal{B}_s$ in the same way as for $\mathcal{B}$. $\qquad\square$

Proposition 33 *Let F be a Hilbert field. If the non-singular diagonal forms $a_1 \xi_1^2 + \cdots + a_n \xi_n^2$ and $b_1 \xi_1^2 + \cdots + b_n \xi_n^2$ are equivalent over F, then*

$$\prod_{1 \leq j < k \leq n} (a_j, a_k)_F = \prod_{1 \leq j < k \leq n} (b_j, b_k)_F.$$

Proof Suppose first that $n = 2$. Since $a_1 \xi_1^2 + a_2 \xi_2^2$ represents b_1, $\xi_1^2 + a_1^{-1} a_2 \xi_2^2$ represents $a_1^{-1} b_1$ and hence $(-a_1^{-1} a_2, a_1^{-1} b_1)_F = 1$. Thus $(a_1 b_1, -a_1 a_2 b_1^2)_F = 1$ and hence $(a_1 b_1, a_2 b_1)_F = 1$. But (Proposition 28 (ii)) the Hilbert symbol is multiplicative, since F is a Hilbert field. It follows that $(a_1, a_2)_F (b_1, a_1 a_2 b_1)_F = 1$. Since the determinants $a_1 a_2$ and $b_1 b_2$ are in the same square class, this implies $(a_1, a_2)_F = (b_1, b_2)_F$, as we wished to prove.

Suppose now that $n > 2$. Since the Hilbert symbol is symmetric, the product $\prod_{1 \leq j < k \leq n} (a_j, a_k)_F$ is independent of the ordering of $a_1, \ldots, a_n$. It follows from Lemma 32 that we may restrict attention to the case where $a_1 \xi_1^2 + a_2 \xi_2^2$ is equivalent to $b_1 \xi_1^2 + b_2 \xi_2^2$ and $a_j = b_j$ for all $j > 2$. Then $(a_1, a_2)_F = (b_1, b_2)_F$, by what we have already proved, and it is enough to show that

$$(a_1, c)_F (a_2, c)_F = (b_1, c)_F (b_2, c)_F \quad \text{for any } c \in F^\times.$$

But this follows from the multiplicativity of the Hilbert symbol and the fact that $a_1 a_2$ and $b_1 b_2$ are in the same square class. $\qquad\square$

Proposition 33 shows that the Hasse invariant is well-defined.

Proposition 34 *Two non-singular quadratic forms in n variables, with coefficients from a Hilbert field F of type (B), are equivalent over F if and only if they have the same Hasse invariant and their determinants are in the same square class.*

Proof Only the sufficiency of the conditions needs to be proved. Since this is trivial for $n = 1$, we suppose first that $n = 2$. It is enough to show that if

$$f = a(\xi_1^2 + d \xi_2^2), \quad g = b(\eta_1^2 + d \eta_2^2),$$

where $(a, ad)_F = (b, bd)_F$, then f is equivalent to g. The hypothesis implies $(-d, a)_F = (-d, b)_F$ and hence $(-d, ab)_F = 1$. Thus $\xi_1^2 + d\xi_2^2$ represents ab and f represents b. Since $\det f$ and $\det g$ are in the same square class, it follows that f is equivalent to g.

Suppose next that $n \geq 3$ and the result holds for all smaller values of n. Let $f(\xi_1, \ldots, \xi_n)$ and $g(\eta_1, \ldots, \eta_n)$ be non-singular quadratic forms with $\det f = \det g = d$ and $s_F(f) = s_F(g)$. By Proposition 31, the quadratic form

$$h(\xi_1, \ldots, \xi_n, \eta_1, \ldots, \eta_n) = f(\xi_1, \ldots, \xi_n) - g(\eta_1, \ldots, \eta_n)$$

is isotropic and hence, by Proposition 7, there exists some $a_1 \in F^\times$ which is represented by both f and g. Thus

$$f \sim a_1\xi_1^2 + f^*, \quad g \sim a_1\eta_1^2 + g^*,$$

where

$$f^* = a_2\xi_2^2 + \cdots + a_n\xi_n^2, \quad g^* = b_2\eta_2^2 + \cdots + b_n\eta_n^2.$$

Evidently $\det f^*$ and $\det g^*$ are in the same square class and $s_F(f) = cs_F(f^*)$, $s_F(g) = c's_F(g^*)$, where

$$c = (a_1, a_2 \cdots a_n)_F = (a_1, a_1)_F(a_1, d)_F = (a_1, b_2 \cdots b_n)_F = c'.$$

Hence $s_F(f^*) = s_F(g^*)$. It follows from the induction hypothesis that $f^* \sim g^*$, and so $f \sim g$. $\qquad\square$

3 The Hasse–Minkowski Theorem

Let a, b, c be nonzero squarefree integers which are relatively prime in pairs. It was proved by Legendre (1785) that the equation

$$ax^2 + by^2 + cz^2 = 0$$

has a nontrivial solution in integers x, y, z if and only if a, b, c are not all of the same sign and the congruences

$$u^2 \equiv -bc \bmod a, \quad v^2 \equiv -ca \bmod b, \quad w^2 \equiv -ab \bmod c$$

are all soluble.

It was first completely proved by Gauss (1801) that every positive integer which is not of the form $4^n(8k + 7)$ can be represented as a sum of three squares. Legendre had given a proof, based on the assumption that if a and m are relatively prime positive integers, then the arithmetic progression

$$a, a + m, \ a + 2m, \ldots$$

contains infinitely many primes. Although his proof of this assumption was faulty, his intuition that it had a role to play in the arithmetic theory of quadratic forms was inspired. The assumption was first proved by Dirichlet (1837) and will be referred to here as 'Dirichlet's theorem on primes in an arithmetic progression'. In the present chapter Dirichlet's theorem will simply be assumed, but it will be proved (in a quantitative form) in Chapter X.

It was shown by Meyer (1884), although the published proof was incomplete, that a quadratic form in five or more variables with integer coefficients is isotropic if it is neither positive definite nor negative definite.

The preceding results are all special cases of the *Hasse–Minkowski theorem*, which is the subject of this section. Let $\mathbb{Q}$ denote the field of rational numbers. By Ostrowski's theorem (Proposition VI.4), the completions $\mathbb{Q}_v$ of $\mathbb{Q}$ with respect to an arbitrary absolute value $|\,|_v$ are the field $\mathbb{Q}_\infty = \mathbb{R}$ of real numbers and the fields $\mathbb{Q}_p$ of p-adic numbers, where p is an arbitrary prime. The Hasse–Minkowski theorem has the following statement:

A non-singular quadratic form $f(\xi_1, \ldots, \xi_n)$ with coefficients from $\mathbb{Q}$ is isotropic in $\mathbb{Q}$ if and only if it is isotropic in every completion of $\mathbb{Q}$.

This concise statement contains, and to some extent conceals, a remarkable amount of information. (Its equivalence to Legendre's theorem when $n = 3$ may be established by elementary arguments.) The theorem was first stated and proved by Hasse (1923). Minkowski (1890) had derived necessary and sufficient conditions for the equivalence over $\mathbb{Q}$ of two non-singular quadratic forms with rational coefficients by using known results on quadratic forms with integer coefficients. The role of p-adic numbers was taken by congruences modulo prime powers. Hasse drew attention to the simplifications obtained by studying from the outset quadratic forms over the field $\mathbb{Q}$, rather than the ring $\mathbb{Z}$, and soon afterwards (1924) he showed that the theorem continues to hold if the rational field $\mathbb{Q}$ is replaced by an arbitrary algebraic number field (with its corresponding completions).

The condition in the statement of the theorem is obviously necessary and it is only its sufficiency which requires proof. Before embarking on this we establish one more property of the Hilbert symbol for the field $\mathbb{Q}$ of rational numbers.

Proposition 35 *For any $a, b \in \mathbb{Q}^\times$, the number of completions $\mathbb{Q}_v$ for which one has $(a, b)_v = -1$ (where v denotes either ∞ or an arbitrary prime p) is finite and even.*

Proof By Proposition 23, it is sufficient to establish the result when a and b are square-free integers such that ab is also square-free. Then $(a, b)_r = 1$ for any odd prime r which does not divide ab, by Proposition 25. We wish to show that $\prod_v (a, b)_v = 1$. Since the Hilbert symbol is multiplicative, it is sufficient to establish this in the following special cases: for $a = -1$ and $b = -1, 2, p$; for $a = 2$ and $b = p$; for $a = p$ and $b = q$, where p and q are distinct odd primes. But it follows from Propositions 24, 25 and 27 that

$$\prod_v (-1, -1)_v = (-1, -1)_\infty (-1, -1)_2 = (-1)(-1) = 1;$$

$$\prod_v (-1, 2)_v = (-1, 2)_\infty (-1, 2)_2 = 1 \cdot 1 = 1;$$

$$\prod_v (-1, p)_v = (-1, p)_p(-1, p)_2 = (-1/p)(-1)^{(p-1)/2};$$

$$\prod_v (2, p)_v = (2, p)_p(2, p)_2 = (2/p)(-1)^{(p^2-1)/8};$$

$$\prod_v (p, q)_v = (p, q)_p(p, q)_q(p, q)_2 = (q/p)(p/q)(-1)^{(p-1)(q-1)/4}.$$

Hence the proposition holds if and only if

$$(-1/p) = (-1)^{(p-1)/2}, (2/p) = (-1)^{(p^2-1)/8}, (q/p)(p/q) = (-1)^{(p-1)(q-1)/4}.$$

Thus it is actually equivalent to the law of quadratic reciprocity and its two 'supplements'. □

We are now ready to prove the Hasse–Minkowski theorem:

Theorem 36 *A non-singular quadratic form $f(\xi_1, \ldots, \xi_n)$ with rational coefficients is isotropic in $\mathbb{Q}$ if and only if it is isotropic in every completion $\mathbb{Q}_v$.*

Proof We may assume that the quadratic form is diagonal:

$$f = a_1\xi_1^2 + \cdots + a_n\xi_n^2,$$

where $a_k \in \mathbb{Q}^\times (k = 1, \ldots, n)$. Moreover, by replacing ξ_k by $r_k\xi_k$, we may assume that each coefficient a_k is a square-free integer.

The proof will be broken into three parts, according as $n = 2, n = 3$ or $n \geq 4$. The proofs for $n = 2$ and $n = 3$ are quite independent. The more difficult proof for $n \geq 4$ uses induction on n and Dirichlet's theorem on primes in an arithmetic progression.

(i) $n = 2$: We show first that if $a \in \mathbb{Q}^\times$ is a square in $\mathbb{Q}_v^\times$ for all v, then a is already a square in $\mathbb{Q}^\times$. Since a is a square in $\mathbb{Q}_\infty^\times$, we have $a > 0$. Let $a = \prod_p p^{\alpha_p}$ be the factorization of a into powers of distinct primes, where $\alpha_p \in \mathbb{Z}$ and $\alpha_p \neq 0$ for at most finitely many primes p. Since $|a|_p = p^{-\alpha_p}$ and a is a square in $\mathbb{Q}_p$, α_p must be even. But if $\alpha_p = 2\beta_p$ then $a = b^2$, where $b = \prod_p p^{\beta_p}$.

Suppose now that $f = a_1\xi_1^2 + a_2\xi_2^2$ is isotropic in $\mathbb{Q}_v$ for all v. Then $a := -a_1a_2$ is a square in $\mathbb{Q}_v$ for all v and hence, by what we have just proved, a is a square in $\mathbb{Q}$. But if $a = b^2$, then $a_1a_2^2 + a_2b^2 = 0$ and thus f is isotropic in $\mathbb{Q}$.

(ii) $n = 3$: By replacing f by $-a_3 f$ and ξ_3 by $a_3\xi_3$, we see that it is sufficient to prove the theorem for

$$f = a\xi^2 + b\eta^2 - \zeta^2,$$

where a and b are nonzero square-free integers. The quadratic form f is isotropic in $\mathbb{Q}_v$ if and only if $(a, b)_v = 1$. If $a = 1$ or $b = 1$, then f is certainly isotropic in $\mathbb{Q}$. Since f is not isotropic in $\mathbb{Q}_\infty$ if $a = b = -1$, this proves the result if $|ab| = 1$. We will assume that the result does not hold for some pair a, b and derive a contradiction. Choose a pair a, b for which the result does not hold and for which $|ab|$ has its minimum value. Then $a \neq 1, b \neq 1$ and $|ab| \geq 2$. Without loss of generality we may assume $|a| \leq |b|$, and then $|b| \geq 2$.

We are going to show that there exists an integer c such that $c^2 \equiv a \bmod b$. Since $\pm b$ is a product of distinct primes, it is enough to show that the congruence $x^2 \equiv a \bmod p$ is soluble for each prime p which divides b (by Corollary II.38). Since this is obvious if $a \equiv 0$ or $1 \bmod p$, we may assume that p is odd and a not divisible by p. Then, since f is isotropic in $\mathbb{Q}_p$, $(a, b)_p = 1$. Hence a is a square mod p by Proposition 25.

Consequently there exist integers c, d such that $a = c^2 - bd$. Moreover, by adding to c a suitable multiple of b we may assume that $|c| \leq |b|/2$. Then

$$|d| = |c^2 - a|/|b| \leq |b|/4 + 1 < |b|$$

and $d \neq 0$, since a is square-free and $a \neq 1$. We have

$$bd(a\xi^2 + b\eta^2 - \zeta^2) = aX^2 + dY^2 - Z^2,$$

where

$$X = c\xi + \zeta, \quad Y = b\eta, \quad Z = a\xi + c\zeta.$$

Moreover the linear transformation $\xi, \eta, \zeta \to X, Y, Z$ is invertible in any field of zero characteristic, since $c^2 - a \neq 0$. Hence, since f is isotropic in $\mathbb{Q}_v$ for all v, so also is $g = a\xi^2 + d\eta^2 - \zeta^2$. Since f is not isotropic in $\mathbb{Q}$, by hypothesis, neither is g. But this contradicts the original choice of f, since $|ad| < |ab|$.

It may be noted that for $n = 3$ it need only be assumed that f is isotropic in $\mathbb{Q}_p$ for all primes p. For the preceding proof used the fact that f is isotropic in $\mathbb{Q}_\infty$ only to exclude from consideration the quadratic form $-\xi^2 - \eta^2 - \zeta^2$ and this quadratic form is anisotropic also in $\mathbb{Q}_2$, by Proposition 27. In fact for $n = 3$ it need only be assumed that f is isotropic in $\mathbb{Q}_v$ for all v with at most one exception since, by Proposition 35, the number of exceptions must be even.

(iii) $n \geq 4$: We have

$$f = a_1\xi_1^2 + \cdots + a_n\xi_n^2,$$

where $a_1, \ldots, a_n$ are square-free integers. We write $f = g - h$, where

$$g = a_1\xi_1^2 + a_2\xi_2^2, h = -a_3\xi_3^2 - \cdots - a_n\xi_n^2.$$

Let S be the finite set consisting of ∞ and all primes p which divide $2a_1 \cdots a_n$. By Proposition 7, for each $v \in S$ there exists $c_v \in \mathbb{Q}_v^\times$ which is represented in $\mathbb{Q}_v$ by both g and h. We will show that we can take c_v to be the same nonzero integer c for every $v \in S$.

Let $v = p$ be a prime in S. By multiplying by a square in $\mathbb{Q}_p^\times$ we may assume that $c_p = p^{\varepsilon_p}c_p'$, where $\varepsilon_p = 0$ or 1 and $|c_p'|_p = 1$. If p is odd and if b_p is an integer such that $|c_p - b_p|_p \leq p^{-\varepsilon_p - 1}$, then $|b_p|_p = |c_p|_p$ and $|b_p c_p^{-1} - 1|_p \leq p^{-1}$. Hence $b_p c_p^{-1}$ is a square in $\mathbb{Q}_p^\times$, by Proposition VI.16, and we can replace c_p by b_p. Similarly if $p = 2$ and if b_2 is an integer such that $|c_2 - b_2|_2 \leq 2^{-\varepsilon_2 - 3}$, then $|b_2|_2 = |c_2|_2$ and $|b_2 c_2^{-1} - 1|_2 \leq 2^{-3}$. Hence $b_2 c_2^{-1}$ is a square in $\mathbb{Q}_2^\times$ and we can replace c_2 by b_2.

By the Chinese remainder theorem (Corollary II.38), the simultaneous congruences

$$c \equiv b_2 \bmod 2^{\varepsilon_2+3}, c \equiv b_p \bmod p^{\varepsilon_p+1} \quad \text{for every odd } p \in S,$$

have a solution $c \in \mathbb{Z}$, that is uniquely determined mod m, where $m = 4 \prod_{p \in S} p^{\varepsilon_p+1}$. In exactly the same way as before we can replace b_p by c for all primes $p \in S$. By choosing c to have the same sign as c_∞, we can take $c_v = c$ for all $v \in S$.

If $d = \prod_{p \in S} p^{\varepsilon_p}$ is the greatest common divisor of c and m then, by Dirichlet's theorem on primes in an arithmetic progression, there exists an integer k with the same sign as c such that

$$c/d + km/d = \pm q,$$

where q is a prime. If we put

$$a = c + km = \pm dq,$$

then q is the only prime divisor of a which is not in S and the quadratic forms

$$g^* = -a\xi_0^2 + a_1\xi_1^2 + a_2\xi_2^2, \quad h^* = a_3\xi_3^2 + \cdots + a_n\xi_n^2 + a\xi_{n+1}^2$$

are isotropic in $\mathbb{Q}_v$ for every $v \in S$, since $c^{-1}a$ is a square in $\mathbb{Q}_v^\times$.

For all primes p not in S, except $p = q$, a is not divisible by p. Hence, by the definition of S and Corollary 26, g^* is isotropic in $\mathbb{Q}_v$ for all v, except possibly $v = q$. Consequently, by the final remark of part (ii) of the proof, g^* is isotropic in $\mathbb{Q}$.

Suppose first that $n = 4$. In this case, in the same way, $h^* = a_3\xi_3^2 + a_4\xi_4^2 + a\xi_5^2$ is also isotropic in $\mathbb{Q}$. Hence, by Proposition 6, there exist $y_1, \ldots, y_4 \in \mathbb{Q}$ such that

$$a_1y_1^2 + a_2y_2^2 = a = -a_3y_3^2 - a_4y_4^2.$$

Thus f is isotropic in $\mathbb{Q}$.

Suppose next that $n \geq 5$ and the result holds for all smaller values of n. Then the quadratic form h^* is isotropic in $\mathbb{Q}_v$, not only for $v \in S$, but for all v. For if p is a prime which is not in S, then a_3, a_4, a_5 are not divisible by p. It follows from Corollary 26 that the quadratic form $a_3\xi_3^2 + a_4\xi_4^2 + a_5\xi_5^2$ is isotropic in $\mathbb{Q}_p$, and hence h^* is also. Since h^* is a non-singular quadratic form in $n - 1$ variables, it follows from the induction hypothesis that h^* is isotropic in $\mathbb{Q}$. The proof can now be completed in the same way as for $n = 4$.   □

Corollary 37 *A non-singular rational quadratic form in $n \geq 5$ variables is isotropic in $\mathbb{Q}$ if and only if it is neither positive definite nor negative definite.*

Proof This follows at once from Theorem 36, on account of Propositions 10 and 31.   □

Corollary 38 *A non-singular quadratic form over the rational field $\mathbb{Q}$ represents a nonzero rational number c in $\mathbb{Q}$ if and only if it represents c in every completion $\mathbb{Q}_v$.*

Proof Only the sufficiency of the condition requires proof. But if the rational quadratic form $f(\xi_1, \ldots, \xi_n)$ represents c in $\mathbb{Q}_v$ for all v then, by Theorem 36, the quadratic form

$$f^*(\xi_0, \xi_1, \ldots, \xi_n) = -c\xi_0^2 + f(\xi_1, \ldots, \xi_n)$$

is isotropic in $\mathbb{Q}$. Hence f represents c in $\mathbb{Q}$, by Proposition 6.   □

Proposition 39 *Two non-singular quadratic forms with rational coefficients are equivalent over $\mathbb{Q}$ if and only if they are equivalent over all completions $\mathbb{Q}_v$.*

Proof Again only the sufficiency of the condition requires proof. Let f and g be non-singular rational quadratic forms in n variables which are equivalent over $\mathbb{Q}_v$ for all v.

Suppose first that $n = 1$ and that $f = a\xi^2$, $g = b\eta^2$. By hypothesis, for every v there exists $t_v \in \mathbb{Q}_v^\times$ such that $b = at_v^2$. Thus ba^{-1} is a square in $\mathbb{Q}_v^\times$ for every v, and hence ba^{-1} is a square in $\mathbb{Q}^\times$, by part (i) of the proof of Theorem 36. Therefore f is equivalent to g over $\mathbb{Q}$.

Suppose now that $n > 1$ and the result holds for all smaller values of n. Choose some $c \in \mathbb{Q}^\times$ which is represented by f in $\mathbb{Q}$. Then f certainly represents c in $\mathbb{Q}_v$ and hence g represents c in $\mathbb{Q}_v$, since g is equivalent to f over $\mathbb{Q}_v$. Since this holds for all v, it follows from Corollary 38 that g represents c in $\mathbb{Q}$.

Thus, by the remark after the proof of Proposition 2, f is equivalent over $\mathbb{Q}$ to a quadratic form $c\xi_1^2 + f^*(\xi_2, \ldots, \xi_n)$ and g is equivalent over $\mathbb{Q}$ to a quadratic form $c\xi_1^2 + g^*(\xi_2, \ldots, \xi_n)$. Since f is equivalent to g over $\mathbb{Q}_v$, it follows from Witt's cancellation theorem that $f^*(\xi_2, \ldots, \xi_n)$ is equivalent to $g^*(\xi_2, \ldots, \xi_n)$ over $\mathbb{Q}_v$. Since this holds for every v, it follows from the induction hypothesis that f^* is equivalent to g^* over $\mathbb{Q}$, and so f is equivalent to g over $\mathbb{Q}$.   □

Corollary 40 *Two non-singular quadratic forms f and g in n variables with rational coefficients are equivalent over the rational field $\mathbb{Q}$ if and only if*

(i) *$(\det f)/(\det g)$ is a square in $\mathbb{Q}^\times$,*
(ii) *$\mathrm{ind}^+ f = \mathrm{ind}^+ g$,*
(iii) *$s_p(f) = s_p(g)$ for every prime p.*

Proof This follows at once from Proposition 39, on account of Propositions 10 and 34.   □

The *strong Hasse principle* (Theorem 36) says that a quadratic form is *isotropic* over the global field $\mathbb{Q}$ if (and only if) it is isotropic over all its local completions $\mathbb{Q}_v$. The so-named *weak Hasse principle* (Proposition 39) says that two quadratic forms are *equivalent* over $\mathbb{Q}$ if (and only if) they are equivalent over all $\mathbb{Q}_v$. These *local-global principles* have proved remarkably fruitful. They organize the subject, they can be extended to other situations and, even when they fail, they are still a useful guide. We describe some results which illustrate these remarks.

As mentioned at the beginning of this section, the strong Hasse principle continues to hold when the rational field is replaced by any algebraic number field. Waterhouse (1976) has established the weak Hasse principle for pairs of quadratic forms: if over every completion $\mathbb{Q}_v$ there is a change of variables taking both f_1 to g_1 and f_2 to g_2, then there is also such a change of variables over $\mathbb{Q}$. For quadratic forms over the field

$F = K(t)$ of rational functions in one variable with coefficients from a field K, the weak Hasse principle always holds, and the strong Hasse principle holds for $K = \mathbb{R}$, but not for all fields K.

The strong Hasse principle also fails for polynomial forms over $\mathbb{Q}$ of degree > 2. For example, Selmer (1951) has shown that the cubic equation $3x^3 + 4y^3 + 5z^3 = 0$ has no nontrivial solutions in $\mathbb{Q}$, although it has nontrivial solutions in every completion $\mathbb{Q}_v$. However, Gusić (1995) has proved the weak Hasse principle for non-singular ternary cubic forms.

Finally, we draw attention to a remarkable local-global principle of Rumely (1986) for algebraic integer solutions of arbitrary systems of polynomial equations

$$f_1(\xi_1, \ldots, \xi_n) = \cdots = f_r(\xi_1, \ldots, \xi_n) = 0$$

with rational coefficients.

We now give some applications of the results which have been established.

Proposition 41 *A positive integer can be represented as the sum of the squares of three integers if and only if it is not of the form $4^n b$, where $n \geq 0$ and $b \equiv 7 \bmod 8$.*

Proof The necessity of the condition is easily established. Since the square of any integer is congruent to $0, 1$ or $4 \bmod 8$, the sum of three squares cannot be congruent to 7. For the same reason, if there exist integers x, y, z such that $x^2 + y^2 + z^2 = 4^n b$, where $n \geq 1$ and b is odd, then x, y, z must all be even and thus $(x/2)^2 + (y/2)^2 + (z/2)^2 = 4^{n-1} b$. By repeating the argument n times, we see that there is no such representation if $b \equiv 7 \bmod 8$.

We show next that any positive integer which satisfies this necessary condition is the sum of three squares of *rational* numbers. We need only show that any positive integer $a \not\equiv 7 \bmod 8$, which is not divisible by 4, is represented in $\mathbb{Q}$ by the quadratic form

$$f = \xi_1^2 + \xi_2^2 + \xi_3^2.$$

For every odd prime p, f is isotropic in $\mathbb{Q}_p$, by Corollary 26, and hence any integer is represented in $\mathbb{Q}_p$ by f, by Proposition 5. By Corollary 38, it only remains to show that f represents a in $\mathbb{Q}_2$.

It is easily seen that if $a \equiv 1, 3$ or $5 \bmod 8$, then there exist integers $x_1, x_2, x_3 \in \{0, 1, 2\}$ such that

$$x_1^2 + x_2^2 + x_3^2 \equiv a \bmod 8.$$

Hence $a^{-1}(x_1^2 + x_2^2 + x_3^2)$ is a square in $\mathbb{Q}_2^\times$ and f represents a in $\mathbb{Q}_2$.

Again, if $a \equiv 2$ or $6 \bmod 8$, then $a \equiv 2, 6, 10$ or $14 \bmod 2^4$ and it is easily seen that there exist integers $x_1, x_2, x_3 \in \{0, 1, 2, 3\}$ such that

$$x_1^2 + x_2^2 + x_3^2 \equiv a \bmod 2^4.$$

Hence $a^{-1}(x_1^2 + x_2^2 + x_3^2)$ is a square in $\mathbb{Q}_2^\times$ and f represents a in $\mathbb{Q}_2$.

To complete the proof of the proposition we show, by an elegant argument due to Aubry (1912), that if f represents c in $\mathbb{Q}$ then it also represents c in $\mathbb{Z}$.

Let

$$(x, y) = \{f(x + y) - f(x) - f(y)\}/2$$

be the symmetric bilinear form associated with f, so that $f(x) = (x, x)$, and assume there exists a point $x \in \mathbb{Q}^3$ such that $(x, x) = c \in \mathbb{Z}$. If $x \notin \mathbb{Z}^3$, we can choose $z \in \mathbb{Z}^3$ so that each coordinate of z differs in absolute value by at most $1/2$ from the corresponding coordinate of x. Hence if we put $y = x - z$, then $y \neq 0$ and $0 < (y, y) \leq 3/4$.

If $x' = x - \lambda y$, where $\lambda = 2(x, y)/(y, y)$, then $x' \in \mathbb{Q}^3$ and $(x', x') = (x, x) = c$. Substituting $y = x - z$, we obtain

$$(y, y)x' = (y, y)x - 2(x, y)y = \{(z, z) - (x, x)\}x + 2\{(x, x) - (x, z)\}z.$$

If $m > 0$ is the least common denominator of the coordinates of x, so that $mx \in \mathbb{Z}^3$, it follows that

$$m(y, y)x' = \{(z, z) - c)\}mx + 2\{mc - (mx, z)\}z \in \mathbb{Z}^3.$$

But

$$m(y, y) = m\{(x, x) - 2(x, z) + (z, z)\} = mc - 2(mx, z) + m(z, z) \in \mathbb{Z}.$$

Thus if $m' > 0$ is the least common denominator of the coordinates of x', then m' divides $m(y, y)$. Hence $m' \leq (3/4)m$. If $x' \notin \mathbb{Z}^3$, we can repeat the argument with x replaced by x'. After performing the process finitely many times we must obtain a point $x^* \in \mathbb{Z}^3$ such that $(x^*, x^*) = c$. $\square$

As another application of the preceding results we now prove

Proposition 42 *Let n, a, b be integers with $n > 1$. Then there exists a nonsingular $n \times n$ rational matrix A such that*

$$A^t A = aI_n + bJ_n, \tag{3}$$

where J_n is the $n \times n$ matrix with all entries 1, if and only if $a > 0$, $a + bn > 0$ and

(i) *for n odd: $a + bn$ is a square and the quadratic form*

$$a\xi^2 + (-1)^{(n-1)/2}b\eta^2 - \zeta^2$$

is isotropic in $\mathbb{Q}$;

(ii) *for n even: $a(a + bn)$ is a square and either $n \equiv 0 \bmod 4$, or $n \equiv 2 \bmod 4$ and a is a sum of two squares.*

Proof If we put

$$B = \begin{bmatrix} 1 & 1 & \cdots & 1 & 1 \\ -1 & 1 & \cdots & 1 & 1 \\ 0 & -2 & \cdots & 1 & 1 \\ & \cdots & \cdots & \cdots & \\ 0 & 0 & \cdots & 1-n & 1 \end{bmatrix},$$

then $D := B^t B$ and $E := B^t J B$ are diagonal matrices:

$$D = \text{diag}[d_1, \ldots, d_{n-1}, n], \quad E = \text{diag}[0, \ldots, 0, n^2],$$

where $d_j = j(j+1)$ for $1 \le j < n$. Hence, if $C = D^{-1} B^t A B$, then

$$C^t D C = B^t A^t A B.$$

Thus the rational matrix A satisfies (3) if and only if the rational matrix C satisfies

$$C^t D C = aD + bE,$$

and consequently if and only if the diagonal quadratic forms

$$f = d_1 \xi_1^2 + \cdots + d_{n-1} \xi_{n-1}^2 + n\xi_n^2, \quad g = ad_1 \eta_1^2 + \cdots + ad_{n-1} \eta_{n-1}^2 + n(a + bn)\eta_n^2$$

are equivalent over $\mathbb{Q}$.

We now apply Corollary 40. Since $(\det g)/(\det f) = a^{n-1}(a + bn)$, the condition that $\det g / \det f$ be a square in $\mathbb{Q}^\times$ means that $a + bn$ is a nonzero square if n is odd and $a(a+bn)$ is a nonzero square if n is even. Since $\text{ind}^+ f = n$, the condition that $\text{ind}^+ g = \text{ind}^+ f$ means that $a > 0$ and $a + bn > 0$. The relation $s_p(g) = s_p(f)$ takes the form

$$\prod_{1 \le i < j < n} (ad_i, ad_j)_p \prod_{1 \le i < n} (ad_i, n(a + bn))_p = \prod_{1 \le i < j < n} (d_i, d_j)_p \prod_{1 \le i < n} (d_i, n)_p.$$

The multiplicativity and symmetry of the Hilbert symbol imply that

$$(ad_i, ad_j)_p = (a, a)_p (a, d_i d_j)_p (d_i, d_j)_p.$$

Since $(a, a)_p = (a, -1)_p$, it follows that $s_p(g) = s_p(f)$ if and only if

$$(a, -1)_p^{(n-1)(n-2)/2} (a, n)_p^{n-1} \prod_{1 \le i < n} (ad_i, a + bn)_p \prod_{1 \le i < j < n} (a, d_i d_j)_p = 1.$$

But

$$\prod_{1 \le i < j < n} d_i d_j = (d_1 \cdots d_{n-1})^{n-2}$$

and, by the definition of d_j, $d_1 \cdots d_{n-1}$ is in the same rational square class as n. Hence $s_p(g) = s_p(f)$ if and only if

$$(a, -1)_p^{(n-1)(n-2)/2} (a, n)_p (an, a + bn)_p = 1. \tag{4}$$

If n is odd, then $a + bn$ is a square and (4) reduces to $(a, (-1)^{(n-1)/2} n)_p = 1$. But, since $a + bn$ is a square, the quadratic form $a\xi^2 + bn\eta^2 - \zeta^2$ is isotropic in $\mathbb{Q}$ and thus $(a, bn)_p = 1$ for all p. Hence $(a, (-1)^{(n-1)/2} n)_p = 1$ for all p if and only if $(a, (-1)^{(n-1)/2} b)_p = 1$ for all p. Since $a > 0$, this is equivalent to (i).

If n is even, then $a(a + bn)$ is a square and (4) reduces to $(a, (-1)^{(n-2)/2} a)_p = 1$. Since $a > 0$, this holds for all p if and only if the ternary quadratic form

$$a\xi^2 + (-1)^{(n-2)/2}a\eta^2 - \zeta^2,$$

is isotropic in $\mathbb{Q}$. Thus it is certainly satisfied if $n \equiv 0 \bmod 4$. If $n \equiv 2 \bmod 4$ it is satisfied if and only if the quadratic form $\xi^2 + \eta^2 - a\zeta^2$ is isotropic. Thus it is satisfied if a is a sum of two squares. It is not satisfied if a is not a sum of two squares since then, by Proposition II.39, for some prime $p \equiv 3 \bmod 4$, the highest power of p which divides a is odd and

$$(a,a)_p = (a,-1)_p = (p,-1)_p = (-1)^{(p-1)/2} = -1. \qquad \square$$

It is worth noting that the last part of this proof shows that if a positive integer a is a sum of two rational squares, then it is also a sum of two squares of integers.

It follows at once from Proposition 42 that, for any positive integer n, there is an $n \times n$ *rational* matrix A such that $A^t A = nI_n$ if and only if either n is an odd square, or $n \equiv 2 \bmod 4$ and n is a sum of two squares, or $n \equiv 0 \bmod 4$ (the Hadamard matrix case).

In Chapter V we considered not only Hadamard matrices, but also designs. We now use Proposition 42 to derive the necessary conditions for the existence of square 2-designs which were obtained by Bruck, Ryser and Chowla (1949/50). Let v, k, λ be integers such that $0 < \lambda < k < v$ and $k(k-1) = \lambda(v-1)$. Since $k - \lambda + \lambda v = k^2$, it follows from Proposition 42 that there exists a $v \times v$ rational matrix A such that

$$A^t A = (k - \lambda)I_v + \lambda J_v$$

if and only if, *either v is even and $k - \lambda$ is a square, or v is odd and the quadratic form*

$$(k - \lambda)\xi^2 + (-1)^{(v-1)/2}\lambda\eta^2 - \zeta^2$$

is isotropic in $\mathbb{Q}$.

A projective plane of order d corresponds to a $(d^2 + d + 1, d + 1, 1)$ (square) 2-design. In this case Proposition 42 tells us that there is no projective plane of order d if d is not a sum of two squares and $d \equiv 1$ or $2 \bmod 4$. In particular, there is no projective plane of order 6.

The existence of projective planes of any prime power order follows from the existence of finite fields of any prime power order. (All known projective planes are of prime power order, but even for $d = 9$ there are projective planes of the same order d which are not isomorphic.) Since there is no projective plane of order 6, the least order in doubt is $d = 10$. The condition derived from Proposition 42 is obviously satisfied in this case, since

$$10\xi^2 - \eta^2 - \zeta^2 = 0$$

has the solution $\xi = \eta = 1, \zeta = 3$. However, Lam, Thiel and Swiercz (1989) have announced that, nevertheless, there is no projective plane of order 10. The result was obtained by a search involving thousands of hours time on a supercomputer and does not appear to have been independently verified.

4 Supplements

It was shown in the proof of Proposition 41 that if an integer can be represented as a sum of 3 squares of rational numbers, then it can be represented as a sum of 3 squares of integers. A similar argument was used by Cassels (1964) to show that if a polynomial can be represented as a sum of n squares of rational functions, then it can be represented as a sum of n squares of polynomials. This was immediately generalized by Pfister (1965) in the following way:

Proposition 43 *For any field F, if there exist scalars $\alpha_1, \ldots, \alpha_n \in F$ and rational functions $r_1(t), \ldots, r_n(t) \in F(t)$ such that*

$$p(t) = \alpha_1 r_1(t)^2 + \cdots + \alpha_n r_n(t)^2$$

is a polynomial, then there exist polynomials $p_1(t), \ldots, p_n(t) \in F[t]$ such that

$$p(t) = \alpha_1 p_1(t)^2 + \cdots + \alpha_n p_n(t)^2.$$

Proof Suppose first that $n = 1$. We can write $r_1(t) = p_1(t)/q_1(t)$, where $p_1(t)$ and $q_1(t)$ are relatively prime polynomials and $q_1(t)$ has leading coefficient 1. Since

$$p(t)q_1(t)^2 = \alpha_1 p_1(t)^2,$$

we must actually have $q_1(t) = 1$.

Suppose now that $n > 1$ and the result holds for all smaller values of n. We may assume that $\alpha_j \neq 0$ for all j, since otherwise the result follows from the induction hypothesis. Suppose first that the quadratic form

$$\phi = \alpha_1 \xi_1^2 + \cdots + \alpha_n \xi_n^2$$

is isotropic over F. In this case there exists an invertible linear transformation $\xi_j = \sum_{k=1}^{n} \tau_{jk} \eta_k$ with $\tau_{jk} \in F (1 \leq j, k \leq n)$ such that

$$\phi = \eta_1^2 - \eta_2^2 + \beta_3 \eta_3^2 + \cdots + \beta_n \eta_n^2,$$

where $\beta_j \in F$ for all $j > 2$. If we substitute

$$\eta_1 = \{p(t) + 1\}/2, \ \eta_2 = \{p(t) - 1\}/2, \ \eta_j = 0 \quad \text{for all } j > 2,$$

we obtain a representation for $p(t)$ of the required form.

Thus we now suppose that ϕ is anisotropic over F. This implies that ϕ is also anisotropic over $F(t)$, since otherwise there would exist a nontrivial representation

$$\alpha_1 q_1(t)^2 + \cdots + \alpha_n q_n(t)^2 = 0,$$

where $q_j(t) \in F[t] (1 \leq j \leq n)$, and by considering the terms of highest degree we would obtain a contradiction.

By hypothesis there exists a representation

$$p(t) = \alpha_1 \{f_1(t)/f_0(t)\}^2 + \cdots + \alpha_n \{f_n(t)/f_0(t)\}^2,$$

where $f_0(t), f_1(t), \ldots, f_n(t) \in F[t]$. Assume that f_0 does not divide f_j for some $j \in \{1, \ldots, n\}$. Then $d := \deg f_0 > 0$ and we can write

$$f_j(t) = g_j(t) f_0(t) + h_j(t),$$

where $g_j(t), h_j(t) \in F[t]$ and $\deg h_j < d$ $(1 \leq j \leq n)$.

Let

$$(x, y) = \{\phi(x + y) - \phi(x) - \phi(y)\}/2$$

be the symmetric bilinear form associated with the quadratic form ϕ and put

$$f = (f_1, \ldots, f_n), \quad g = (g_1, \ldots, g_n), \quad h = (h_1, \ldots, h_n).$$

If

$$f_0^* = \{(g, g) - p\} f_0 - 2\{(f, g) - p f_0\}, \quad f^* = \{(g, g) - p\} f - 2\{(f, g) - p f_0\} g,$$

and $f^* = (f_1^*, \ldots, f_n^*)$, then clearly $f_0^*, f_1^*, \ldots, f_n^* \in F[t]$. Since $(f, f) = p f_0^2$ and $g = (f - h)/f_0$, we can also write

$$f_0^* = (h, h)/f_0, \quad f^* = \{(h, h)f - 2(f, h)h\}/f_0^2.$$

It follows that $\deg f_0^* < d$ and $(f^*, f^*) = p f_0^{*2}$. Also $f_0^* \neq 0$, since $h \neq 0$ and ϕ is anisotropic. Thus

$$p(t) = \alpha_1 \{f_1^*(t)/f_0^*(t)\}^2 + \cdots + \alpha_n \{f_n^*(t)/f_0^*(t)\}^2.$$

If f_0^* does not divide f_j^* for some $j \in \{1, \ldots, n\}$, we can repeat the process. After at most d steps we must obtain a representation for $p(t)$ of the required form. $\quad\square$

It was already known to Hilbert (1888) that there is no analogue of Proposition 43 for polynomials in more than one variable. Motzkin (1967) gave the simple example

$$p(x, y) = 1 - 3x^2 y^2 + x^4 y^2 + x^2 y^4,$$

which is a sum of 4 squares in $\mathbb{R}(x, y)$, but is not a sum of any finite number of squares in $\mathbb{R}[x, y]$.

In the same paper in which he proved Proposition 43 Pfister introduced his *multiplicative forms*. The quadratic forms $f_a, f_{a,b}$ in §2 are examples of such forms. Pfister (1966) used his multiplicative forms to obtain several new results on the structure of the Witt ring and then (1967) to give a strong solution to Hilbert's 17th Paris problem. We restrict attention here to the latter application.

Let $g(x), h(x) \in \mathbb{R}[x]$ be polynomials in n variables $x = (\xi_1, \ldots, \xi_n)$ with real coefficients. The rational function $f(x) = g(x)/h(x)$ is said to be *positive definite* if $f(a) \geq 0$ for every $a \in \mathbb{R}^n$ such that $h(a) \neq 0$. Hilbert's 17th problem asks if every positive definite rational function can be represented as a sum of squares:

$$f(x) = f_1(x)^2 + \cdots + f_s(x)^2,$$

where $f_1(x), \ldots, f_s(x) \in \mathbb{R}(x)$. The question was answered affirmatively by Artin (1927). Artin's solution allowed the number s of squares to depend on the function f, and left open the possibility that there might be no uniform bound. Pfister showed that one can always take $s = 2^n$.

Finally we mention a conjecture of Oppenheim (1929–1953), that if $f(\xi_1, \ldots, \xi_n)$ is a non-singular isotropic real quadratic form in $n \geq 3$ variables, which is not a scalar multiple of a rational quadratic form, then $f(\mathbb{Z}^n)$ is dense in $\mathbb{R}$, i.e. for each $\alpha \in \mathbb{R}$ and $\varepsilon > 0$ there exist $z_1, \ldots, z_n \in \mathbb{Z}$ such that $|f(z_1, \ldots, z_n) - \alpha| < \varepsilon$. (It is not difficult to show that this is not always true for $n = 2$.) Raghunathan (1980) made a general conjecture about Lie groups, which he observed would imply Oppenheim's conjecture. Oppenheim's conjecture was then proved in this way by Margulis (1987), using deep results from the theory of Lie groups and ergodic theory. The full conjecture of Raghunathan has now also been proved by Ratner (1991).

5 Further Remarks

Lam [18] gives a good introduction to the arithmetic theory of quadratic spaces. The Hasse–Minkowski theorem is also proved in Serre [29]. Additional information is contained in the books of Cassels [4], Kitaoka [16], Milnor and Husemoller [20], O'Meara [22] and Scharlau [28].

Quadratic spaces were introduced (under the name 'metric spaces') by Witt [32]. This noteworthy paper also made several other contributions: Witt's cancellation theorem, the Witt ring, Witt's chain equivalence theorem and the Hasse invariant in its most general form (as described below). Quadratic spaces are treated not only in books on the arithmetic of quadratic forms, but also in works of a purely algebraic nature, such as Artin [1], Dieudonné [8] and Jacobson [15].

An important property of the Witt ring was established by Merkur'ev (1981). In one formulation it says that every element of order 2 in the *Brauer group* of a field F is represented by the Clifford algebra of some quadratic form over F. For a clear account, see Lewis [19].

Our discussion of Hilbert fields is based on Fröhlich [9]. It may be shown that any locally compact non-archimedean valued field is a Hilbert field. Fröhlich gives other examples, but rightly remarks that the notion of Hilbert field clarifies the structure of the theory, even if one is interested only in the p-adic case. (The name 'Hilbert field' is also given to fields for which Hilbert's irreducibility theorem is valid.)

In the study of quadratic forms over an arbitrary field F, the Hilbert symbol $(a, b/F)$ is a generalized quaternion algebra (more strictly, an equivalence class of such algebras) and the Hasse invariant is a tensor product of Hilbert symbols. See, for example, Lam [18].

Hasse's original proof of the Hasse–Minkowski theorem is reproduced in Hasse [13]. In principle it is the same as that given here, using a reduction argument due to Lagrange for $n = 3$ and Dirichlet's theorem on primes in an arithmetic progression for $n \geq 4$.

The book of Cassels contains a proof of Theorem 36 which does not use Dirichlet's theorem, but it uses intricate results on genera of quadratic forms and is not so 'clean'. However, Conway [6] has given an elementary approach to the *equivalence* of quadratic forms over $\mathbb{Q}$ (Proposition 39 and Corollary 40).

The book of O'Meara gives a proof of the Hasse–Minkowski theorem over any algebraic number field which avoids Dirichlet's theorem and is 'cleaner' than ours, but it uses deep results from *class field theory*. For the latter, see Cassels and Fröhlich [5], Garbanati [10] and Neukirch [21].

To determine if a rational quadratic form $f(\xi_1, \ldots, \xi_n) = \sum_{j,k=1}^{n} a_{jk}\xi_j\xi_k$ is isotropic by means of Theorem 36 one has to show that it is isotropic in infinitely many completions. Nevertheless, the problem is a finite one. Clearly one may assume that the coefficients a_{jk} are integers and, if the equation $f(x_1, \ldots, x_n) = 0$ has a nontrivial solution in rational numbers, then it also has a nontrivial solution in integers. But Cassels has shown by elementary arguments that if $f(x_1, \ldots, x_n) = 0$ for some $x_j \in \mathbb{Z}$, not all zero, then the x_j may be chosen so that

$$\max_{1 \le j \le n} |x_j| \le (3H)^{(n-1)/2},$$

where $H = \sum_{j,k=1}^{n} |a_{jk}|$. See Lemma 8.1 in Chapter 6 of [4].

Williams [31] gives a sharper result for the ternary quadratic form

$$g(\xi, \eta, \zeta) = a\xi^2 + b\eta^2 + c\zeta^2,$$

where a, b, c are integers with greatest common divisor $d > 0$. If $g(x, y, z) = 0$ for some integers x, y, z, not all zero, then these integers may be chosen so that

$$|x| \le |bc|^{1/2}/d, |y| \le |ca|^{1/2}/d, |z| \le |ab|^{1/2}/d.$$

The necessity of the Bruck–Ryser–Chowla conditions for the existence of symmetric block designs may also be established in a more elementary way, without also proving their sufficiency for rational equivalence. See, for example, Beth *et al.* [2]. For the non-existence of a projective plane of order 10, see C. Lam [17].

For various manifestations of the local-global principle, see Waterhouse [30], Hsia [14], Gusić [12] and Green *et al.* [11].

The work of Pfister instigated a flood of papers on the algebraic theory of quadratic forms. The books of Lam and Scharlau give an account of these developments. For Hilbert's 17th problem, see also Pfister [23], [24] and Rajwade [25].

Although a positive integer which is a sum of n rational squares is also a sum of n squares of integers, the same does not hold for higher powers. For example,

$$5906 = (149/17)^4 + (25/17)^4,$$

but there do not exist integers m, n such that $5906 = m^4 + n^4$, since $9^4 > 5906$, $2 \cdot 7^4 < 5906$ and $5906 - 8^4 = 1810$ is not a fourth power. For the representation of a polynomial as a sum of squares of polynomials, see Rudin [27].

For Oppenheim's conjecture, see Dani and Margulis [7], Borel [3] and Ratner [26].

6 Selected References

[1] E. Artin, *Geometric algebra*, reprinted, Wiley, New York, 1988. [Original edition, 1957]

[2]   T. Beth, D. Jungnickel and H. Lenz, *Design theory*, 2nd ed., 2 vols., Cambridge University Press, 1999.

[3]   A. Borel, Values of indefinite quadratic forms at integral points and flows on spaces of lattices, *Bull. Amer. Math. Soc. (N.S.)* **32** (1995), 184–204.

[4]   J.W.S. Cassels, *Rational quadratic forms*, Academic Press, London, 1978.

[5]   J.W.S. Cassels and A. Fröhlich (ed.), *Algebraic number theory*, Academic Press, London, 1967.

[6]   J.H. Conway, Invariants for quadratic forms, *J. Number Theory* **5** (1973), 390–404.

[7]   S.G. Dani and G.A. Margulis, Values of quadratic forms at integral points: an elementary approach, *Enseign. Math.* **36** (1990), 143–174.

[8]   J. Dieudonné, *La géométrie des groupes classiques*, 2nd ed., Springer-Verlag, Berlin, 1963.

[9]   A. Fröhlich, Quadratic forms 'à la' local theory, *Proc. Camb. Phil. Soc.* **63** (1967), 579–586.

[10] D. Garbanati, Class field theory summarized, *Rocky Mountain J. Math.* **11** (1981), 195–225.

[11] B. Green, F. Pop and P. Roquette, On Rumely's local-global principle, *Jahresber. Deutsch. Math.-Verein.* **97** (1995), 43–74.

[12] I. Gusić, Weak Hasse principle for cubic forms, *Glas. Mat. Ser. III* **30** (1995), 17–24.

[13] H. Hasse, *Mathematische Abhandlungen* (ed. H.W. Leopoldt and P. Roquette), Band I, de Gruyter, Berlin, 1975.

[14] J.S. Hsia, On the Hasse principle for quadratic forms, *Proc. Amer. Math. Soc.* **39** (1973), 468–470.

[15] N. Jacobson, *Basic Algebra I*, 2nd ed., Freeman, New York, 1985.

[16] Y. Kitaoka, *Arithmetic of quadratic forms*, Cambridge University Press, 1993.

[17] C.W.H. Lam, The search for a finite projective plane of order 10, *Amer. Math. Monthly* **98** (1991), 305–318.

[18] T.Y. Lam, *The algebraic theory of quadratic forms*, revised 2nd printing, Benjamin, Reading, Mass., 1980.

[19] D.W. Lewis, The Merkuryev–Suslin theorem, *Irish Math. Soc. Newsletter* **11** (1984), 29–37.

[20] J. Milnor and D. Husemoller, *Symmetric bilinear forms*, Springer-Verlag, Berlin, 1973.

[21] J. Neukirch, *Class field theory*, Springer-Verlag, Berlin, 1986.

[22] O.T. O'Meara, *Introduction to quadratic forms*, corrected reprint, Springer-Verlag, New York, 1999. [Original edition, 1963]

[23] A. Pfister, Hilbert's seventeenth problem and related problems on definite forms, *Mathematical developments arising from Hilbert problems* (ed. F.E. Browder), pp. 483–489, Proc. Symp. Pure Math. **28**, Part 2, Amer. Math. Soc., Providence, Rhode Island, 1976.

[24] A. Pfister, *Quadratic forms with applications to algebraic geometry and topology*, Cambridge University Press, 1995.

[25] A.R. Rajwade, *Squares,* Cambridge University Press, 1993.

[26] M. Ratner, Interactions between ergodic theory, Lie groups, and number theory, *Proceedings of the International Congress of Mathematicians: Zürich 1994*, pp. 157–182, Birkhäuser, Basel, 1995.

[27] W. Rudin, Sums of squares of polynomials, *Amer. Math. Monthly* **107** (2000), 813–821.

[28] W. Scharlau, *Quadratic and Hermitian forms,* Springer-Verlag, Berlin, 1985.

[29] J.-P. Serre, *A course in arithmetic*, Springer-Verlag, New York, 1973.

[30] W.C. Waterhouse, Pairs of quadratic forms, *Invent. Math.* **37** (1976), 157–164.

[31] K.S. Williams, On the size of a solution of Legendre's equation, *Utilitas Math.* **34** (1988), 65–72.

[32] E. Witt, Theorie der quadratischen Formen in beliebigen Körpern, *J. Reine Angew. Math.* **176** (1937), 31–44.

VIII

The Geometry of Numbers

It was shown by Hermite (1850) that if

$$f(x) = x^t A x$$

is a positive definite quadratic form in n real variables, then there exists a vector x with *integer* coordinates, not all zero, such that

$$f(x) \leq c_n (\det A)^{1/n},$$

where c_n is a positive constant depending only on n. Minkowski (1891) found a new and more geometric proof of Hermite's result, which gave a much smaller value for the constant c_n. Soon afterwards (1893) he noticed that his proof was valid not only for an n-dimensional ellipsoid $f(x) \leq$ const., but for any convex body which was symmetric about the origin. This led him to a large body of results, to which he gave the somewhat paradoxical name 'geometry of numbers'. It seems fair to say that Minkowski was the first to realize the importance of convexity for mathematics, and it was in his lattice point theorem that he first encountered it.

1 Minkowski's Lattice Point Theorem

A set $C \subseteq \mathbb{R}^n$ is said to be *convex* if $x_1, x_2 \in C$ implies $\theta x_1 + (1 - \theta) x_2 \in C$ for $0 < \theta < 1$. Geometrically, this means that whenever two points belong to the set the whole line segment joining them is also contained in the set.

The *indicator function* or 'characteristic function' of a set $S \subseteq \mathbb{R}^n$ is defined by $\chi(x) = 1$ or 0 according as $x \in S$ or $x \notin S$. If the indicator function is Lebesgue integrable, then the set S is said to have *volume*

$$\lambda(S) = \int_{\mathbb{R}^n} \chi(x) dx.$$

The indicator function of a convex set C is actually Riemann integrable. It is easily seen that if a convex set C is not contained in a hyperplane of $\mathbb{R}^n$, then its *interior* int C (see §4 of Chapter I) is not empty. It follows that $\lambda(C) = 0$ if and only if C is

W.A. Coppel, *Number Theory: An Introduction to Mathematics*, Universitext, DOI: 10.1007/978-0-387-89486-7_8, © Springer Science + Business Media, LLC 2009

contained in a hyperplane, and $0 < \lambda(C) < \infty$ if and only if C is bounded and is not contained in a hyperplane.

A set $S \subseteq \mathbb{R}^n$ is said to be *symmetric* (with respect to the origin) if $x \in S$ implies $-x \in S$. Evidently any (nonempty) symmetric convex set contains the origin.

A point $x = (\xi_1, \ldots, \xi_n) \in \mathbb{R}^n$ whose coordinates $\xi_1, \ldots, \xi_n$ are all integers will be called a *lattice point*. Thus the set of all lattice points in $\mathbb{R}^n$ is $\mathbb{Z}^n$.

These definitions are the ingredients for Minkowski's *lattice point theorem*:

Theorem 1 *Let C be a symmetric convex set in $\mathbb{R}^n$. If $\lambda(C) > 2^n$, or if C is compact and $\lambda(C) = 2^n$, then C contains a nonzero point of $\mathbb{Z}^n$.*

The proof of Theorem 1 will be deferred to §3. Here we illustrate the utility of the result by giving several applications, all of which go back to Minkowski himself.

Proposition 2 *If A is an $n \times n$ positive definite real symmetric matrix, then there exists a nonzero point $x \in \mathbb{Z}^n$ such that*

$$x^t A x \leq c_n (\det A)^{1/n},$$

where $c_n = (4/\pi)\{(n/2)!\}^{2/n}$.

Proof For any $\rho > 0$ the ellipsoid $x^t A x \leq \rho$ is a compact symmetric convex set. By putting $A = T^t T$, for some nonsingular matrix T, it may be seen that the volume of this set is $\kappa_n \rho^{n/2} (\det A)^{-1/2}$, where κ_n is the volume of the n-dimensional unit ball. It follows from Theorem 1 that the ellipsoid contains a nonzero lattice point if $\kappa_n \rho^{n/2} (\det A)^{-1/2} = 2^n$. But, as we will see in §4 of Chapter IX, $\kappa_n = \pi^{n/2}/(n/2)!$, where $x! = \Gamma(x + 1)$. This gives the value c_n for ρ. $\square$

It follows from Stirling's formula (Chapter IX, §4) that $c_n \sim 2n/\pi e$ for $n \to \infty$. Hermite had proved Proposition 2 with $c_n = (4/3)^{(n-1)/2}$. Hermite's value is smaller than Minkowski's for $n \leq 8$, but much larger for large n.

As a second application of Theorem 1 we prove Minkowski's *linear forms theorem*:

Proposition 3 *Let A be an $n \times n$ real matrix with determinant ± 1. Then there exists a nonzero point $x \in \mathbb{Z}^n$ such that $Ax = y = (\eta_k)$ satisfies*

$$|\eta_1| \leq 1, \quad |\eta_k| < 1 \quad for\ 1 < k \leq n.$$

Proof For any positive integer m, let C_m be the set of all $x \in \mathbb{R}^n$ such that $Ax \in D_m$, where

$$D_m = \{y = (\eta_k) \in \mathbb{R}^n : |\eta_1| \leq 1 + 1/m, \ |\eta_k| < 1 \quad for\ 2 \leq k \leq n\}.$$

Then C_m is a symmetric convex set, since A is linear and D_m is symmetric and convex. Moreover $\lambda(C_m) = 2^n(1 + 1/m)$, since $\lambda(D_m) = 2^n(1 + 1/m)$ and A is volume-preserving. Therefore, by Theorem 1, C_m contains a lattice point $x_m \neq O$. Since $C_m \subset C_1$ for all $m > 1$ and the number of lattice points in C_1 is finite, there exist only finitely many distinct points x_m. Thus there exists a lattice point $x \neq O$ which belongs to C_m for infinitely many m. Evidently x has the required properties. $\square$

The continued fraction algorithm enables one to find rational approximations to irrational numbers. The subject of *Diophantine approximation* is concerned with the more general problem of solving inequalities in integers. From Proposition 3 we can immediately obtain a result in this area due to Dirichlet (1842):

Proposition 4 *Let $A = (\alpha_{jk})$ be an $n \times m$ real matrix and let $t > 1$ be real. Then there exist integers $q_1, \ldots, q_m, p_1, \ldots, p_n$, with $0 < \max(|q_1|, \ldots, |q_m|) < t^{n/m}$, such that*

$$\left| \sum_{k=1}^{m} \alpha_{jk} q_k - p_j \right| \leq 1/t \quad (1 \leq j \leq n).$$

Proof Since the matrix

$$\begin{pmatrix} t^{-n/m} I_m & 0 \\ tA & tI_n \end{pmatrix}$$

has determinant 1, it follows from Proposition 3 that there exists a nonzero vector

$$x = \begin{pmatrix} q \\ -p \end{pmatrix} \in \mathbb{Z}^{n+m}$$

such that

$$|q_k| < t^{n/m} \quad (k = 1, \ldots, m),$$

$$\left| \sum_{k=1}^{m} \alpha_{jk} q_k - p_j \right| \leq 1/t \quad (j = 1, \ldots, n).$$

Since $q = O$ would imply $|p_j| < 1$ for all j and hence $p = O$, which contradicts $x \neq O$, we must have $\max_k |q_k| > 0$. $\qquad\square$

Corollary 5 *Let $A = (\alpha_{jk})$ be an $n \times m$ real matrix such that $Az \notin \mathbb{Z}^n$ for any nonzero vector $z \in \mathbb{Z}^m$. Then there exist infinitely many $(m+n)$-tuples $q_1, \ldots, q_m, p_1, \ldots, p_n$ of integers with greatest common divisor 1 and with arbitrarily large values of*

$$\|q\| = \max(|q_1|, \ldots, |q_m|)$$

such that

$$\left| \sum_{k=1}^{m} \alpha_{jk} q_k - p_j \right| < \|q\|^{-m/n} \quad (1 \leq j \leq n).$$

Proof Let $q_1, \ldots, q_m, p_1, \ldots, p_n$ be integers satisfying the conclusions of Proposition 4 for some $t > 1$. Evidently we may assume that $q_1, \ldots, q_m, p_1, \ldots, p_n$ have no common divisor greater than 1. For given $q_1, \ldots, q_m$, let δ_j be the distance of $\sum_{k=1}^{m} \alpha_{jk} q_k$ from the nearest integer and put $\delta = \max \delta_j$ $(1 \leq j \leq n)$. By hypothesis $0 < \delta < 1$, and by construction

$$\delta \leq 1/t < \|q\|^{-m/n}.$$

Choosing some $t' > 2/\delta$, we find a new set of integers $q_1', \ldots, q_m', p_1', \ldots, p_n'$ satisfying the same requirements with t replaced by t', and hence with $\delta' \leq 1/t' < \delta/2$. Proceeding in this way, we obtain a sequence of $(m + n)$-tuples of integers $q_1^{(v)}, \ldots, q_m^{(v)}, p_1^{(v)}, \ldots, p_n^{(v)}$ for which $\delta^{(v)} \to 0$ and hence $\|q^{(v)}\| \to \infty$, since we cannot have $q^{(v)} = q$ for infinitely many v. $\square$

The hypothesis of the corollary is certainly satisfied if $1, \alpha_{j1}, \ldots, \alpha_{jm}$ are linearly independent over the field $\mathbb{Q}$ of rational numbers for some $j \in \{1, \ldots, n\}$.

Minkowski also used his lattice point theorem to give the first proof that the discriminant of any algebraic number field, other than $\mathbb{Q}$, has absolute value greater than 1. The proof is given in most books on algebraic number theory.

2 Lattices

In the previous section we defined the set of lattice points to be $\mathbb{Z}^n$. However, this definition is tied to a particular coordinate system in $\mathbb{R}^n$. It is useful to consider lattices from a more intrinsic point of view. The key property is 'discreteness'.

With vector addition as the group operation, $\mathbb{R}^n$ is an abelian group. A subgroup Λ is said to be *discrete* if there exists a ball with centre O which contains no other point of Λ. (More generally, a subgroup H of a topological group G is said to be discrete if there exists an open set $U \subseteq G$ such that $H \cap U = \{e\}$, where e is the identity element of G.)

If Λ is a discrete subgroup of $\mathbb{R}^n$, then any bounded subset of $\mathbb{R}^n$ contains at most finitely many points of Λ since, if there were infinitely many, they would have an accumulation point and their differences would accumulate at O. In particular, Λ is a closed subset of $\mathbb{R}^n$.

Proposition 6 *If $x_1, \ldots, x_m$ are linearly independent vectors in $\mathbb{R}^n$, then the set*

$$\Lambda = \{\zeta_1 x_1 + \cdots + \zeta_m x_m : \zeta_1, \ldots, \zeta_m \in \mathbb{Z}\}$$

is a discrete subgroup of $\mathbb{R}^n$.

Proof It is clear that Λ is a subgroup of $\mathbb{R}^n$, since $x, y \in \Lambda$ implies $x - y \in \Lambda$. If Λ is not discrete, then there exist $y^{(v)} \in \Lambda$ with $|y^{(1)}| > |y^{(2)}| > \cdots$ and $|y^{(v)}| \to 0$ as $v \to \infty$. Let V be the vector subspace of $\mathbb{R}^n$ with basis $x_1, \ldots, x_m$ and for any vector

$$x = \alpha_1 x_1 + \cdots + \alpha_m x_m,$$

where $\alpha_k \in \mathbb{R}$ ($1 \leq k \leq m$), put

$$\|x\| = \max(|\alpha_1|, \ldots, |\alpha_m|).$$

This defines a norm on V. We have

$$y^{(v)} = \zeta_1^{(v)} x_1 + \cdots + \zeta_m^{(v)} x_m,$$

where $\zeta_k^{(v)} \in \mathbb{Z}$ ($1 \leq k \leq m$). Since any two norms on a finite-dimensional vector space are equivalent (Lemma VI.7), it follows that $\zeta_k^{(v)} \to 0$ as $v \to \infty$ ($1 \leq k \leq m$). Since $\zeta_k^{(v)}$ is an integer, this is only possible if $y^{(v)} = O$ for all large v, which is a contradiction. $\square$

The converse of Proposition 6 is also valid. In fact we will prove a sharper result:

Proposition 7 *If Λ is a discrete subgroup of $\mathbb{R}^n$, then there exist linearly independent vectors $x_1, \ldots, x_m$ in $\mathbb{R}^n$ such that*

$$\Lambda = \{\zeta_1 x_1 + \cdots + \zeta_m x_m : \zeta_1, \ldots, \zeta_m \in \mathbb{Z}\}.$$

Furthermore, if $y_1, \ldots, y_m$ is any maximal set of linearly independent vectors in Λ, we can choose $x_1, \ldots, x_m$ so that

$$\Lambda \cap \langle y_1, \ldots, y_k \rangle = \{\zeta_1 x_1 + \cdots + \zeta_k x_k : \zeta_1, \ldots, \zeta_k \in \mathbb{Z}\} \quad (1 \le k \le m),$$

where $\langle Y \rangle$ denotes the vector subspace generated by the set Y.

Proof Let S_1 denote the set of all $\alpha_1 > 0$ such that $\alpha_1 y_1 \in \Lambda$ and let μ_1 be the infimum of all $\alpha_1 \in S_1$. We are going to show that $\mu_1 \in S_1$. If this is not the case there exist $\alpha_1^{(v)} \in S_1$ with $\alpha_1^{(1)} > \alpha_1^{(2)} > \cdots$ and $\alpha_1^{(v)} \to \mu_1$ as $v \to \infty$. Since the ball $|x| \le (1 + \mu_1)|y_1|$ contains only finitely many points of Λ, this is a contradiction.

Any $\alpha_1 \in S_1$ can be written in the form $\alpha_1 = p\mu_1 + \theta$, where p is a positive integer and $0 \le \theta < \mu_1$. Since $\theta > 0$ would imply $\theta \in S_1$, contrary to the definition of μ_1, we must have $\theta = 0$. Hence if we put $x_1 = \mu_1 y_1$, then

$$\Lambda \cap \langle y_1 \rangle = \{\zeta_1 x_1 : \zeta_1 \in \mathbb{Z}\}.$$

Assume that, for some positive integer k $(1 \le k < m)$, we have found vectors $x_1, \ldots, x_k \in \Lambda$ such that

$$\Lambda \cap \langle y_1, \ldots, y_k \rangle = \{\zeta_1 x_1 + \cdots + \zeta_k x_k : \zeta_1, \ldots, \zeta_k \in \mathbb{Z}\}.$$

We will prove the proposition by showing that this assumption continues to hold when k is replaced by $k + 1$.

Any $x \in \Lambda \cap \langle y_1, \ldots, y_{k+1} \rangle$ has the form

$$x = \alpha_1 x_1 + \cdots + \alpha_k x_k + \alpha_{k+1} y_{k+1},$$

where $\alpha_1, \ldots, \alpha_{k+1} \in \mathbb{R}$. Let S_{k+1} denote the set of all $\alpha_{k+1} > 0$ which arise in such representations and let μ_{k+1} be the infimum of all $\alpha_{k+1} \in S_{k+1}$. We are going to show that $\mu_{k+1} \in S_{k+1}$. If $\mu_{k+1} \notin S_{k+1}$, there exist $\alpha_{k+1}^{(v)} \in S_{k+1}$ with $\alpha_{k+1}^{(1)} > \alpha_{k+1}^{(2)} > \cdots$ and $\alpha_{k+1}^{(v)} \to \mu_{k+1}$ as $v \to \infty$. Then Λ contains a point

$$x^{(v)} = \alpha_1^{(v)} x_1 + \cdots + \alpha_k^{(v)} x_k + \alpha_{k+1}^{(v)} y_{k+1},$$

where $\alpha_j^{(v)} \in \mathbb{R}$ $(1 \le j \le k)$. In fact, by subtracting an integral linear combination of $x_1, \ldots, x_k$ we may assume that $0 \le \alpha_j^{(v)} < 1$ $(1 \le j \le k)$. Since only finitely many points of Λ are contained in the ball $|x| \le |x_1| + \cdots + |x_k| + (1 + \mu_{k+1})|y_{k+1}|$, this is a contradiction.

Hence $\mu_{k+1} > 0$ and Λ contains a vector

$$x_{k+1} = \alpha_1 x_1 + \cdots + \alpha_k x_k + \mu_{k+1} y_{k+1}.$$

As for S_1, it may be seen that S_{k+1} consists of all positive integer multiples of μ_{k+1}. Hence any $x \in \Lambda \cap \langle y_1, \ldots, y_{k+1} \rangle$ has the form

$$x = \zeta_1 x_1 + \cdots + \zeta_k x_k + \zeta_{k+1} x_{k+1},$$

where $\zeta_1, \ldots, \zeta_k \in \mathbb{R}$ and $\zeta_{k+1} \in \mathbb{Z}$. Since

$$x - \zeta_{k+1} x_{k+1} \in \Lambda \cap \langle y_1, \ldots, y_k \rangle,$$

we must actually have $\zeta_1, \ldots, \zeta_k \in \mathbb{Z}$.   □

By being more specific in the proof of Proposition 7 it may be shown that there is a *unique* choice of $x_1, \ldots, x_m$ such that

$$y_1 = p_{11} x_1$$
$$y_2 = p_{21} x_1 + p_{22} x_2$$
$$\cdots$$
$$y_m = p_{m1} x_1 + p_{m2} x_2 + \cdots + p_{mm} x_m,$$

where $p_{ij} \in \mathbb{Z}$, $p_{ii} > 0$, and $0 \le p_{ij} < p_{ii}$ if $j < i$ (*Hermite's normal form*).

It is easily seen that in Proposition 7 we can choose $x_i = y_i$ $(1 \le i \le m)$ if and only if, for any $x \in \Lambda$ and any positive integer h, x is an integral linear combination of $y_1, \ldots, y_m$ whenever hx is.

By combining Propositions 6 and 7 we obtain

Proposition 8 *For a set $\Lambda \subseteq \mathbb{R}^n$ the following two conditions are equivalent:*

(i) *Λ is a discrete subgroup of $\mathbb{R}^n$ and there exists $R > 0$ such that, for each $y \in \mathbb{R}^n$, there is some $x \in \Lambda$ with $|y - x| < R$;*

(ii) *there exist n linearly independent vectors $x_1, \ldots, x_n$ in $\mathbb{R}^n$ such that*

$$\Lambda = \{\zeta_1 x_1 + \cdots + \zeta_n x_n : \zeta_1, \ldots, \zeta_n \in \mathbb{Z}\}.$$

Proof If (i) holds, then in the statement of Proposition 7 we must have $m = n$, i.e. (ii) holds. On the other hand, if (ii) holds then Λ is a discrete subgroup of $\mathbb{R}^n$, by Proposition 6. Moreover, for any $y \in \mathbb{R}^n$ we can choose $x \in \Lambda$ so that

$$y - x = \theta_1 x_1 + \cdots + \theta_n x_n,$$

where $0 \le \theta_j < 1 (j = 1, \ldots, n)$, and hence

$$|y - x| < |x_1| + \cdots + |x_n|.$$ □

A set $\Lambda \subseteq \mathbb{R}^n$ satisfying either of the two equivalent conditions of Proposition 8 will be called a *lattice* and any element of Λ a *lattice point*. The vectors $x_1, \ldots, x_n$ in (ii) will be said to be a *basis* for the lattice.

A lattice is sometimes defined to be any discrete subgroup of $\mathbb{R}^n$, and what we have called a lattice is then called a 'nondegenerate' lattice. Our definition is chosen simply to avoid repetition of the word 'nondegenerate'. We may occasionally use the more general definition and, with this warning, believe it will be clear from the context when this occurs.

The basis of a lattice is not uniquely determined. In fact $y_1, \ldots, y_n$ is also a basis if

$$y_j = \sum_{k=1}^{n} \alpha_{jk} x_k \quad (j = 1, \ldots, n),$$

where $A = (\alpha_{jk})$ is an $n \times n$ matrix of integers such that $\det A = \pm 1$, since A^{-1} is then also a matrix of integers. Moreover, every basis $y_1, \ldots, y_n$ is obtained in this way. For if

$$y_j = \sum_{k=1}^{n} \alpha_{jk} x_k, \quad x_i = \sum_{j=1}^{n} \beta_{ij} y_j, \quad (i, j = 1, \ldots, n),$$

where $A = (\alpha_{jk})$ and $B = (\beta_{ij})$ are $n \times n$ matrices of integers, then $BA = I$ and hence $(\det B)(\det A) = 1$. Since $\det A$ and $\det B$ are integers, it follows that $\det A = \pm 1$.

Let $x_1, \ldots, x_n$ be a basis for a lattice $\Lambda \subseteq \mathbb{R}^n$. If

$$x_k = \sum_{j=1}^{n} \gamma_{jk} e_j \quad (k = 1, \ldots, n),$$

where $e_1, \ldots, e_n$ is the canonical basis for $\mathbb{R}^n$ then, in terms of the nonsingular matrix $T = (\gamma_{jk})$, the lattice Λ is just the set of all vectors Tz with $z \in \mathbb{Z}^n$. The absolute value of the determinant of the matrix T does not depend on the choice of basis. For if $x'_1, \ldots, x'_n$ is any other basis, then

$$x'_i = \sum_{j=1}^{n} \alpha_{ij} x_j \quad (i = 1, \ldots, n),$$

where $A = (\alpha_{ij})$ is an $n \times n$ matrix of integers with $\det A = \pm 1$. Thus

$$x'_k = \sum_{j=1}^{n} \gamma'_{jk} e_j \quad (k = 1, \ldots, n),$$

where $T' = (\gamma'_{jk})$ satisfies $T' = T A^t$ and hence

$$|\det T'| = |\det T|.$$

The uniquely determined quantity $|\det T|$ will be called the *determinant* of the lattice Λ and denoted by $d(\Lambda)$. (Some authors, e.g. Conway and Sloane [14], call $|\det T|^2$ the determinant of Λ, but others prefer to call this the *discriminant* of Λ.)

The determinant $d(\Lambda)$ has a simple geometrical interpretation. In fact it is the volume of the parallelotope Π, consisting of all points $y \in \mathbb{R}^n$ such that

$$y = \theta_1 x_1 + \cdots + \theta_n x_n,$$

where $0 \le \theta_k \le 1$ $(k = 1, \ldots, n)$. The interior of Π is a *fundamental domain* for the subgroup Λ, since

$$\mathbb{R}^n = \bigcup_{x \in \Lambda} (\Pi + x),$$

$$\text{int}(\Pi + x) \cap \text{int}(\Pi + x') = \emptyset \quad \text{if } x, x' \in \Lambda \text{ and } x \neq x'.$$

For any lattice $\Lambda \subseteq \mathbb{R}^n$, the set Λ^* of all vectors $y \in \mathbb{R}^n$ such that $y^t x \in \mathbb{Z}$ for every $x \in \Lambda$ is again a lattice, the *dual* (or 'polar' or 'reciprocal') of Λ. In fact,

if $\Lambda = \{Tz : z \in \mathbb{Z}^n\}$, then $\Lambda^* \qquad = \{(T^t)^{-1}w : w \in \mathbb{Z}^n\}$.

Hence Λ is the dual of Λ^* and $\text{d}(\Lambda)\text{d}(\Lambda^*) = 1$. A lattice Λ is *self-dual* if $\Lambda^* = \Lambda$.

3 Proof of the Lattice Point Theorem; Other Results

In this section we take up the proof of Minkowski's lattice point theorem. The proof will be based on a very general result, due to Blichfeldt (1914), which is not restricted to convex sets.

Proposition 9 *Let S be a Lebesgue measurable subset of $\mathbb{R}^n$, Λ a lattice in $\mathbb{R}^n$ with determinant $\text{d}(\Lambda)$ and m a positive integer.*

If $\lambda(S) > m\,\text{d}(\Lambda)$, or if S is compact and $\lambda(S) = m\,\text{d}(\Lambda)$, then there exist $m + 1$ distinct points $x_1, \ldots, x_{m+1}$ of S such that the differences $x_j - x_k$ ($1 \leq j, k \leq m+1$) all lie in Λ.

Proof Let $b_1, \ldots, b_n$ be a basis for Λ and let P be the half-open parallelotope consisting of all points $x = \theta_1 b_1 + \cdots + \theta_n b_n$, where $0 \leq \theta_i < 1$ ($i = 1, \ldots, n$). Then $\lambda(P) = \text{d}(\Lambda)$ and

$$\mathbb{R}^n = \bigcup_{z \in \Lambda} (P + z), \quad (P + z) \cap (P + z') = \emptyset \quad \text{if } z \neq z'.$$

Suppose first that $\lambda(S) > m\,\text{d}(\Lambda)$. If we put

$$S_z = S \cap (P + z), \quad T_z = S_z - z,$$

then $T_z \subseteq P$, $\lambda(T_z) = \lambda(S_z)$ and

$$\lambda(S) = \sum_{z \in \Lambda} \lambda(S_z).$$

Hence

$$\sum_{z \in \Lambda} \lambda(T_z) = \lambda(S) > m\,\text{d}(\Lambda) = m\lambda(P).$$

Since $T_z \subseteq P$ for every z, it follows that some point $y \in P$ is contained in at least $m + 1$ sets T_z. (In fact this must hold for all y in a subset of P of positive measure.) Thus there exist $m + 1$ distinct points $z_1, \ldots, z_{m+1}$ of Λ and points $x_1, \ldots, x_{m+1}$ of S such that $y = x_j - z_j$ ($j = 1, \ldots, m + 1$). Then $x_1, \ldots, x_{m+1}$ are distinct and

$$x_j - x_k = z_j - z_k \in \Lambda \quad (1 \le j, k \le m+1).$$

Suppose next that S is compact and $\lambda(S) = m \, d(\Lambda)$. Let $\{\varepsilon_\nu\}$ be a decreasing sequence of positive numbers such that $\varepsilon_\nu \to 0$ as $\nu \to \infty$, and let S_ν denote the set of all points of $\mathbb{R}^n$ distant at most ε_ν from S. Then S_ν is compact, $\lambda(S_\nu) > \lambda(S)$ and

$$S_1 \supset S_2 \supset \cdots, \quad S = \bigcap_\nu S_\nu.$$

By what we have already proved, there exist $m+1$ distinct points $x_1^{(\nu)}, \ldots, x_{m+1}^{(\nu)}$ of S_ν such that $x_j^{(\nu)} - x_k^{(\nu)} \in \Lambda$ for all j, k. Since $S_\nu \subseteq S_1$ and S_1 is compact, by restricting attention to a subsequence we may assume that $x_j^{(\nu)} \to x_j$ as $\nu \to \infty$ ($j = 1, \ldots, m+1$). Then $x_j \in S$ and $x_j^{(\nu)} - x_k^{(\nu)} \to x_j - x_k$. Since $x_j^{(\nu)} - x_k^{(\nu)} \in \Lambda$, this is only possible if $x_j - x_k = x_j^{(\nu)} - x_k^{(\nu)}$ for all large ν. Hence $x_1, \ldots, x_{m+1}$ are distinct. $\qquad\square$

Siegel (1935) has given an analytic formula which underlies Proposition 9 and enables it to be generalized. Although we will make no use of it, this formula will now be established. For notational simplicity we restrict attention to the (self-dual) lattice $\Lambda = \mathbb{Z}^n$.

Proposition 10 *If $\Psi : \mathbb{R}^n \to \mathbb{C}$ is a bounded measurable function which vanishes outside some compact set, then*

$$\int_{\mathbb{R}^n} \Psi(x) \overline{\phi(x)} dx = \sum_{w \in \mathbb{Z}^n} \left| \int_{\mathbb{R}^n} \Psi(x) e^{-2\pi i w^t x} dx \right|^2,$$

where

$$\phi(x) = \sum_{z \in \mathbb{Z}^n} \Psi(x+z).$$

Proof Since Ψ vanishes outside a compact set, there exists a finite set $T \subseteq \mathbb{Z}^n$ such that $\Psi(x+z) = 0$ for all $x \in \mathbb{R}^n$ if $z \in \mathbb{Z}^n \backslash T$. Thus the sum defining $\phi(x)$ has only finitely many nonzero terms and ϕ also is a bounded measurable function which vanishes outside some compact set.

If we write

$$x = (\xi_1, \ldots, \xi_n), \quad z = (\zeta_1, \ldots, \zeta_n),$$

then the sum defining $\phi(x)$ is unaltered by the substitution $\zeta_j \to \zeta_j + 1$ and hence ϕ has period 1 in each of the variables ξ_j ($j = 1, \ldots, n$). Let Π denote the fundamental parallelotope

$$\Pi = \{x = (\xi_1, \ldots, \xi_n) \in \mathbb{R}^n : 0 \le \xi_j \le 1 \text{ for } j = 1, \ldots, n\}.$$

Since the functions $e^{2\pi i w^t x}$ ($w \in \mathbb{Z}^n$) are an orthogonal basis for $L^2(\Pi)$, Parseval's equality (Chapter I, §10) holds:

$$\int_\Pi |\phi(x)|^2 dx = \sum_{w \in \mathbb{Z}^n} |c_w|^2,$$

where

$$c_w = \int_\Pi \phi(x)e^{-2\pi i w^t x} dx.$$

But

$$c_w = \int_\Pi \sum_{z \in \mathbb{Z}^n} \Psi(x+z)e^{-2\pi i w^t x} dx$$

$$= \int_\Pi \sum_{z \in \mathbb{Z}^n} \Psi(x+z)e^{-2\pi i w^t(x+z)} dx,$$

since $e^{2k\pi i} = 1$ for any integer k. Hence

$$c_w = \int_{\mathbb{R}^n} \Psi(y)e^{-2\pi i w^t y} dy.$$

On the other hand,

$$\int_\Pi |\phi(x)|^2 dx = \int_\Pi \sum_{z',z'' \in \mathbb{Z}^n} \Psi(x+z')\overline{\Psi(x+z'')} dx$$

$$= \int_\Pi \sum_{z,z' \in \mathbb{Z}^n} \Psi(x+z')\overline{\Psi(x+z'+z)} dx$$

$$= \int_{\mathbb{R}^n} \sum_{z \in \mathbb{Z}^n} \Psi(y)\overline{\Psi(y+z)} dy = \int_{\mathbb{R}^n} \Psi(y)\overline{\phi(y)} dy.$$

Substituting these expressions in Parseval's equality, we obtain the result. $\qquad\square$

Suppose, in particular, that Ψ takes only real nonnegative values. Then so also does ϕ and

$$\int_{\mathbb{R}^n} \Psi(x)\phi(x) dx \le \sup_{x \in \mathbb{R}^n} \phi(x) \int_{\mathbb{R}^n} \Psi(x) dx.$$

On the other hand, omitting all terms with $w \ne 0$ we obtain

$$\sum_{w \in \mathbb{Z}^n} \left| \int_{\mathbb{R}^n} \Psi(x)e^{-2\pi i w^t x} dx \right|^2 \ge \left(\int_{\mathbb{R}^n} \Psi(x) dx \right)^2.$$

Hence, by Proposition 10,

$$\sup_{x \in \mathbb{R}^n} \phi(x) \geq \int_{\mathbb{R}^n} \Psi(x) dx.$$

For example, let $S \subseteq \mathbb{R}^n$ be a measurable set with $\lambda(S) > m$. Then there exists a *bounded* measurable set $S' \subseteq S$ with $\lambda(S') > m$. If we take Ψ to be the indicator function of S', then

$$\int_{\mathbb{R}^n} \Psi(x) dx = \lambda(S') > m$$

and we conclude that there exists $y \in \mathbb{R}^n$ such that

$$\sum_{z \in \mathbb{Z}^n} \Psi(y + z) = \phi(y) > m.$$

Since the only possible values of the summands on the left are 0 and 1, it follows that there exist $m + 1$ distinct points $z_1, \ldots, z_{m+1} \in \mathbb{Z}^n = \Lambda$ such that $y + z_j \in S$ for all j. The proof of Proposition 9 can now be completed in the same way as before.

Let $\{K_\alpha\}$ be a family of subsets of $\mathbb{R}^n$, where each K_α is the *closure* of a nonempty open set G_α, i.e. K_α is the intersection of all closed sets containing G_α. The family $\{K_\alpha\}$ is said to be a *packing* of $\mathbb{R}^n$ if $\alpha \neq \alpha'$ implies $G_\alpha \cap G_{\alpha'} = \emptyset$ and is said to be a *covering* of $\mathbb{R}^n$ if $\mathbb{R}^n = \bigcup_\alpha K_\alpha$. It is said to be a *tiling* of $\mathbb{R}^n$ if it is both a packing and a covering.

For example, if Π is a fundamental parallelotope of a lattice Λ, then the family $\{\Pi + a : a \in \Lambda\}$ is a tiling of $\mathbb{R}^n$. More generally, if G is a nonempty open subset of $\mathbb{R}^n$ with closure K, we may ask whether the family $\{K + a : a \in \Lambda\}$ of all Λ-translates of K is either a packing or a covering of $\mathbb{R}^n$. Some necessary conditions may be derived with the aid of Proposition 9:

Proposition 11 *Let K be the closure of a bounded nonempty open set $G \subseteq \mathbb{R}^n$ and let Λ be a lattice in $\mathbb{R}^n$.*

If the Λ-translates of K are a covering of $\mathbb{R}^n$ then $\lambda(K) \geq d(\Lambda)$, and the inequality is strict if they are not also a packing.

If the Λ-translates of K are a packing of $\mathbb{R}^n$ then $\lambda(K) \leq d(\Lambda)$, and the inequality is strict if they are not also a covering.

Proof Suppose first that the Λ-translates of K cover $\mathbb{R}^n$. Then every point of a fundamental parallelotope Π of Λ has the form $x - a$, where $x \in K$ and $a \in \Lambda$. Hence

$$\lambda(K) = \sum_{a \in \Lambda} \lambda(K \cap (\Pi + a))$$

$$= \sum_{a \in \Lambda} \lambda((K - a) \cap \Pi) \geq \lambda(\Pi) = d(\Lambda).$$

Suppose, in addition, that the Λ-translates of K are not a packing of $\mathbb{R}^n$. Then there exist distinct points x_1, x_2 in the interior G of K such that $a = x_1 - x_2 \in \Lambda$. Let

$$B_\varepsilon = \{x \in \mathbb{R}^n : |x| \leq \varepsilon\}.$$

We can choose $\varepsilon > 0$ so small that the balls $B_\varepsilon + x_1$ and $B_\varepsilon + x_2$ are disjoint and contained in G. Then $G' = G \setminus (B_\varepsilon + x_1)$ is a bounded nonempty open set with closure $K' = K \setminus (\text{int} B_\varepsilon + x_1)$. Since

$$B_\varepsilon + x_1 = B_\varepsilon + x_2 + a \subseteq K' + a,$$

the $\varLambda$-translates of K' contain K and therefore also cover $\mathbb{R}^n$. Hence, by what we have already proved, $\lambda(K') \geq \mathrm{d}(\varLambda)$. Since $\lambda(K) > \lambda(K')$, it follows that $\lambda(K) > \mathrm{d}(\varLambda)$.

Suppose now that the $\varLambda$-translates of K are a packing of $\mathbb{R}^n$. Then $\varLambda$ does not contain the difference of two distinct points in the interior G of K, since $G + a$ and $G + b$ are disjoint if a, b are distinct points of $\varLambda$. It follows from Proposition 9 that

$$\lambda(K) = \lambda(G) \leq \mathrm{d}(\varLambda).$$

Suppose, in addition, that the $\varLambda$-translates of K do not cover $\mathbb{R}^n$. Thus there exists a point $y \in \mathbb{R}^n$ which is not in any $\varLambda$-translate of K. We will show that we can choose $\varepsilon > 0$ so small that y is not in any $\varLambda$-translate of $K + B_\varepsilon$.

If this is not the case then, for any positive integer ν, there exists $a_\nu \in \varLambda$ such that

$$y \in K + B_{1/\nu} + a_\nu.$$

Evidently the sequence a_ν is bounded and hence there exists $a \in \varLambda$ such that $a_\nu = a$ for infinitely many ν. But then $y \in K + a$, which is contrary to hypothesis.

We may in addition assume ε chosen so small that $|x| > 2\varepsilon$ for every nonzero $x \in \varLambda$. Then the set $S = G \cup (B_\varepsilon + y)$ has the property that $\varLambda$ does not contain the difference of any two distinct points of S. Hence, by Proposition 9, $\lambda(S) \leq \mathrm{d}(\varLambda)$. Since

$$\lambda(K) = \lambda(G) < \lambda(S),$$

it follows that $\lambda(K) < \mathrm{d}(\varLambda)$. $\square$

We next apply Proposition 9 to convex sets. Minkowski's lattice point theorem (Theorem 1) is the special case $m = 1$ (and $\varLambda = \mathbb{Z}^n$) of the following generalization, due to van der Corput (1936):

Proposition 12 *Let C be a symmetric convex subset of $\mathbb{R}^n$, $\varLambda$ a lattice in $\mathbb{R}^n$ with determinant $\mathrm{d}(\varLambda)$, and m a positive integer.*

If $\lambda(C) > 2^n m\, \mathrm{d}(\varLambda)$, or if C is compact and $\lambda(C) = 2^n m\, \mathrm{d}(\varLambda)$, then there exist $2m$ distinct nonzero points $\pm y_1, \ldots, \pm y_m$ of $\varLambda$ such that

$$y_j \in C \quad (1 \leq j \leq m),$$
$$y_j - y_k \in C \quad (1 \leq j, k \leq m).$$

Proof The set $S = \{x/2 : x \in C\}$ has measure $\lambda(S) = \lambda(C)/2^n$. Hence, by Proposition 9, there exist $m + 1$ distinct points $x_1, \ldots, x_{m+1} \in C$ such that $(x_j - x_k)/2 \in \varLambda$ for all j, k.

The vectors of $\mathbb{R}^n$ may be totally ordered by writing $x > x'$ if $x - x'$ has its first nonzero coordinate positive. We assume the points $x_1, \ldots, x_{m+1} \in C$ numbered so that

$$x_1 > x_2 > \cdots > x_{m+1}.$$

Put

$$y_j = (x_j - x_{m+1})/2 \quad (j = 1, \ldots, m).$$

Then, by construction, $y_j \in \Lambda (j = 1, \ldots, m)$. Moreover $y_j \in C$, since $x_1, \ldots, x_{m+1} \in C$ and C is symmetric, and similarly $y_j - y_k = (x_j - x_k)/2 \in C$. Finally, since

$$y_1 > y_2 > \cdots > y_m > O,$$

we have $y_j \neq O$ and $y_j \neq \pm y_k$ if $j \neq k$. $\qquad \square$

The conclusion of Proposition 12 need no longer hold if C is not compact and $\lambda(C) = 2^n m \, d(\Lambda)$. For example, take $\Lambda = \mathbb{Z}^n$ and let C be the symmetric convex set

$$C = \{x = (\xi_1, \ldots, \xi_n) \in \mathbb{R}^n : |\xi_1| < m, |\xi_j| < 1 \text{ for } 2 \leq j \leq n\}.$$

Then $d(\Lambda) = 1$ and $\lambda(C) = 2^n m$. However, the only nonzero points of Λ in C are the $2(m-1)$ points $(\pm k, 0, \ldots, 0)$ $(1 \leq k \leq m-1)$.

To provide a broader view of the geometry of numbers we now mention without proof some further results. A different generalization of Minkowski's lattice point theorem was already proved by Minkowski himself. Let Λ be a lattice in $\mathbb{R}^n$ and let K be a compact symmetric convex subset of $\mathbb{R}^n$ with nonempty interior. Then ρK contains no nonzero point of Λ for small $\rho > 0$ and contains n linearly independent points of Λ for large $\rho > 0$. Let μ_i denote the infimum of all $\rho > 0$ such that ρK contains at least i linearly independent points of Λ $(i = 1, \ldots, n)$. Clearly the *successive minima* $\mu_i = \mu_i(K, \Lambda)$ satisfy the inequalities

$$0 < \mu_1 \leq \mu_2 \leq \cdots \leq \mu_n < \infty.$$

Minkowski's lattice point theorem says that

$$\mu_1^n \lambda(K) \leq 2^n d(\Lambda).$$

Minkowski's *theorem on successive minima* strengthens this to

$$2^n \, d(\Lambda)/n! \leq \mu_1 \mu_2 \cdots \mu_n \lambda(K) \leq 2^n d(\Lambda).$$

The lower bound is quite easy to prove, but the upper bound is more deep-lying — notwithstanding simplifications of Minkowski's original proof. If $\Lambda = \mathbb{Z}^n$, then equality holds in the upper bound for the *cube* $K = \{(\xi_1, \ldots, \xi_n) \in \mathbb{R}^n : |\xi_i| \leq 1 \text{ for all } i\}$ and in the lower bound for the *cross-polytope* $K = \{(\xi_1, \ldots, \xi_n) \in \mathbb{R}^n : \sum_{i=1}^n |\xi_i| \leq 1\}$.

If K is a compact symmetric convex subset of $\mathbb{R}^n$ with nonempty interior, we define its *critical determinant* $\Delta(K)$ to be the infimum, over all lattices Λ with no nonzero point in the interior of K, of their determinants $d(\Lambda)$. A lattice Λ for which $d(\Lambda) = \Delta(K)$ is called a *critical lattice* for K. It will be shown in §6 that a critical lattice always exists.

It follows from Proposition 12 that $\Delta(K) \geq 2^{-n}\lambda(K)$. A conjectured sharpening of Minkowski's theorem on successive minima, which has been proved by Minkowski (1896) himself for $n = 2$ and for n-dimensional ellipsoids, and by Woods (1956) for $n = 3$, claims that

$$\mu_1\mu_2\cdots\mu_n\Delta(K) \leq \mathrm{d}(\Lambda).$$

The successive minima of a convex body are connected with those of its dual body. If K is a compact symmetric convex subset of $\mathbb{R}^n$ with nonempty interior, then its *dual*

$$K^* = \{y \in \mathbb{R}^n : y^t x \leq 1 \text{ for all } x \in K\}$$

has the same properties, and K is the dual of K^*. Mahler (1939) showed that the successive minima of the dual body K^* with respect to the dual lattice Λ^* are related to the successive minima of K with respect to Λ by the inequalities

$$1 \leq \mu_i(K, \Lambda)\mu_{n-i+1}(K^*, \Lambda^*) \quad (i = 1, \ldots, n),$$

and hence, by applying Minkowski's theorem on successive minima also to K^* and Λ^*, he obtained inequalities in the opposite direction:

$$\mu_i(K, \Lambda)\mu_{n-i+1}(K^*, \Lambda^*) \leq 4^n/\lambda(K)\lambda(K^*) \quad (i = 1, \ldots, n).$$

By further proving that $\lambda(K)\lambda(K^*) \geq 4^n(n!)^{-2}$, he deduced that

$$\mu_i(K, \Lambda)\mu_{n-i+1}(K^*, \Lambda^*) \leq (n!)^2 \quad (i = 1, \ldots, n).$$

Dramatic improvements of these bounds have recently been obtained. Banaszczyk (1996), with the aid of techniques from harmonic analysis, has shown that there is a numerical constant $C > 0$ such that, for all $n \geq 1$ and all $i \in \{1, \ldots, n\}$,

$$\mu_i(K, \Lambda)\mu_{n-i+1}(K^*, \Lambda^*) \leq Cn(1 + \log n).$$

He had shown already (1993) that if $K = B_1$ is the n-dimensional closed unit ball, which is self-dual, then for all $n \geq 1$ and all $i \in \{1, \ldots, n\}$,

$$\mu_i(B_1, \Lambda)\mu_{n-i+1}(B_1, \Lambda^*) \leq n.$$

This result is close to being best possible, since there exists a numerical constant $C' > 0$ and self-dual lattices $\Lambda_n \subseteq \mathbb{R}^n$ such that

$$\mu_1(B_1, \Lambda_n)\mu_n(B_1, \Lambda_n) \geq \mu_1(B_1, \Lambda_n)^2 \geq C'n.$$

Two other applications of Minkowski's theorem on successive minima will be mentioned here. The first is a sharp form, due to Bombieri and Vaaler (1983), of 'Siegel's lemma'. In his investigations on transcendental numbers Siegel (1929) used Dirichlet's pigeonhole principle to prove that if $A = (\alpha_{jk})$ is an $m \times n$ matrix of integers, where

$m < n$, such that $|\alpha_{jk}| \leq \beta$ for all j, k, then the system of homogeneous linear equations

$$Ax = 0$$

has a solution $x = (\xi_k)$ in integers, not all 0, such that $|\xi_k| \leq 1 + (n\beta)^{m/(n-m)}$ for all k. Bombieri and Vaaler show that, if A has rank m and if $g > 0$ is the greatest common divisor of all $m \times m$ subdeterminants of A, then there are $n - m$ linearly independent integral solutions $x_j = (\xi_{jk})$ $(j = 1, \ldots, n - m)$ such that

$$\prod_{j=1}^{n-m} \|x_j\| \leq [\det(AA^t)]^{1/2}/g,$$

where $\|x_j\| = \max_k |\xi_{jk}|$.

The second application, due to Gillet and Soulé (1991), may be regarded as an arithmetic analogue of the Riemann–Roch theorem for function fields. Again let K be a compact symmetric convex subset of $\mathbb{R}^n$ with nonempty interior and let μ_i denote the infimum of all $\rho > 0$ such that ρK contains at least i linearly independent points of $\mathbb{Z}^n$ ($i = 1, \ldots, n$). If $M(K)$ is the number of points of $\mathbb{Z}^n$ in K, and if h is the maximum number of linearly independent points of $\mathbb{Z}^n$ in the interior of K, then Gillet and Soulé show that $\mu_1 \cdots \mu_h/M(K)$ is bounded above and below by positive constants, which depend on n but not on K.

A number of results in this section have dealt with compact symmetric convex sets with nonempty interior. Since such sets may appear rather special, it should be pointed out that they arise very naturally in connection with normed vector spaces.

The vector space $\mathbb{R}^n$ is said to be *normed* if with each $x \in \mathbb{R}^n$ there is associated a real number $|x|$ with the properties

(i) $|x| \geq 0$, with equality if and only if $x = O$,
(ii) $|x + y| \leq |x| + |y|$ for all $x, y \in \mathbb{R}^n$,
(iii) $|\alpha x| = |\alpha||x|$ for all $x \in \mathbb{R}^n$ and all $\alpha \in \mathbb{R}$.

Let K denote the set of all $x \in \mathbb{R}^n$ such that $|x| \leq 1$. Then K is bounded, since all norms on a finite-dimensional vector space are equivalent. In fact K is compact, since it follows from (ii) that K is closed. Moreover K is convex and symmetric, by (ii) and (iii). Furthermore, by (i) and (iii), $x/|x| \in K$ for each nonzero $x \in \mathbb{R}^n$. Hence the interior of K is nonempty and is actually the set of all $x \in \mathbb{R}^n$ such that $|x| < 1$.

Conversely, let K be a compact symmetric convex subset of $\mathbb{R}^n$ with nonempty interior. Then the origin is an interior point of K and for each nonzero $x \in \mathbb{R}^n$ there is a unique $\rho > 0$ such that ρx is on the boundary of K. If we put $|x| = \rho^{-1}$, and $|O| = 0$, then (i) obviously holds. Furthermore, since $|-x| = |x|$, it is easily seen that (iii) holds. Finally, if $y \in \mathbb{R}^n$ and $|y| = \sigma^{-1}$, then $\rho x, \sigma y \in K$ and hence, since K is convex,

$$\rho\sigma(\rho + \sigma)^{-1}(x + y) = \sigma(\rho + \sigma)^{-1}\rho x + \rho(\rho + \sigma)^{-1}\sigma y \in K.$$

Hence

$$|x + y| \leq (\rho + \sigma)/\rho\sigma = |x| + |y|.$$

Thus $\mathbb{R}^n$ is a normed vector space and K the set of all $x \in \mathbb{R}^n$ such that $|x| \leq 1$.

4 Voronoi Cells

Throughout this section we suppose $\mathbb{R}^n$ equipped with the *Euclidean metric*:

$$d(y, z) = \|y - z\|,$$

where $\|x\| = (x^t x)^{1/2}$. We call $\|x\|^2 = x^t x$ the *square-norm* of x and we denote the scalar product $y^t z$ by (y, z).

Fix some point $x_0 \in \mathbb{R}^n$. For any point $x \neq x_0$, the set of all points which are equidistant from x_0 and x is the hyperplane H_x which passes through the midpoint of the segment joining x_0 and x and is orthogonal to this segment. Analytically, H_x is the set of all $y \in \mathbb{R}^n$ such that

$$(x - x_0, y) = (x - x_0, x + x_0)/2,$$

which simplifies to

$$2(x - x_0, y) = \|x\|^2 - \|x_0\|^2.$$

The set of all points which are closer to x_0 than to x is the open half-space G_x consisting of all points $y \in \mathbb{R}^n$ such that

$$2(x - x_0, y) < \|x\|^2 - \|x_0\|^2.$$

The closed half-space $\bar{G}_x = H_x \cup G_x$ is the set of all points at least as close to x_0 as to x.

Let X be a subset of $\mathbb{R}^n$ containing more than one point which is *discrete*, i.e. for each $y \in \mathbb{R}^n$ there exists an open set containing y which contains at most one point of X. It follows that each bounded subset of $\mathbb{R}^n$ contains only finitely many points of X since, if there were infinitely many, they would have an accumulation point. Hence for each $y \in \mathbb{R}^n$ there exists an $x_0 \in X$ whose distance from y is minimal:

$$d(x_0, y) \leq d(x, y) \quad \text{for every } x \in X. \tag{1}$$

For each $x_0 \in X$ we define its *Voronoi cell* $V(x_0)$ to be the set of all $y \in \mathbb{R}^n$ for which (1) holds. Voronoi cells are also called 'Dirichlet domains', since they were used by Dirichlet (1850) in $\mathbb{R}^2$ before Voronoi (1908) used them in $\mathbb{R}^n$.

If we choose $r > 0$ so that the open ball

$$\beta_r(x_0) := \{y \in \mathbb{R}^n : d(x_0, y) < r\}$$

contains no point of X except x_0, then $\beta_{r/2}(x_0) \subseteq V(x_0)$. Thus x_0 is an interior point of $V(x_0)$.

Since

$$\bar{G}_x = \{y \in \mathbb{R}^n : d(x_0, y) \leq d(x, y)\},$$

we have $V(x_0) \subseteq \bar{G}_x$ and actually

$$V(x_0) = \bigcap_{x \in X \setminus x_0} \bar{G}_x. \tag{2}$$

It follows at once from (2) that $V(x_0)$ is closed and convex. Hence $V(x_0)$ is the closure of its nonempty interior.

According to the definitions of §3, the Voronoi cells form a tiling of $\mathbb{R}^n$, since

$$\mathbb{R}^n = \bigcup_{x \in X} V(x),$$

$$\mathrm{int}\, V(x) \cap \mathrm{int}\, V(x') = \emptyset \quad \text{if } x, x' \in X \text{ and } x \neq x'.$$

A subset A of a convex set C is said to be a *face* of C if A is convex and, for any $c, c' \in C$, $(c, c') \cap A \neq \emptyset$ implies $c, c' \in A$. The tiling by Voronoi cells has the additional property that $V(x) \cap V(x')$ is a face of both $V(x)$ and $V(x')$ if $x, x' \in X$ and $x \neq x'$. We will prove this by showing that if y_1, y_2 are distinct points of $V(x)$ and if $z \in (y_1, y_2) \cap V(x')$, then $y_1 \in V(x')$.

Since $z \in V(x) \cap V(x')$, we have $d(x, z) = d(x', z)$. Thus z lies on the hyperplane H which passes through the midpoint of the segment joining x and x' and is orthogonal to this segment. If $y_1 \notin V(x')$, then $d(x, y_1) < d(x', y_1)$. Thus y_1 lies in the open half-space G associated with the hyperplane H which contains x. But then y_2 lies in the open half-space G' which contains x', i.e. $d(x', y_2) < d(x, y_2)$, which contradicts $y_2 \in V(x)$.

We now assume that the set X is not only discrete, but also *relatively dense*, i.e.

(†) there exists $R > 0$ such that, for each $y \in \mathbb{R}^n$, there is some $x \in X$ with $d(x, y) < R$.

It follows at once that $V(x_0) \subseteq \beta_R(x_0)$. Thus $V(x_0)$ is bounded and, since it is closed, even compact. The ball $\beta_{2R}(x_0)$ contains only finitely many points $x_1, \ldots, x_m$ of X apart from x_0. We are going to show that

$$V(x_0) = \bigcap_{i=1}^{m} \bar{G}_{x_i}. \tag{3}$$

By (2) we need only show that if $y \in \bigcap_{i=1}^{m} \bar{G}_{x_i}$, then $y \in \bar{G}_x$ for every $x \in X$.

Assume that $d(x_0, y) \geq R$ and choose z on the segment joining x_0 and y so that $d(x_0, z) = R$. For some $x \in X$ we have $d(x, z) < R$ and hence $0 < d(x, x_0) < 2R$. Consequently $x = x_i$ for some $i \in \{1, \ldots, m\}$. Since $d(x_i, z) < R = d(x_0, z)$, we have $z \notin \bar{G}_{x_i}$. But this is a contradiction, since $x_0, y \in \bar{G}_{x_i}$ and z is on the segment joining them.

We conclude that $d(x_0, y) < R$. If $x \in X$ and $x \neq x_0, x_1, \ldots, x_m$, then

$$d(x, y) \geq d(x_0, x) - d(x_0, y)$$
$$\geq 2R - R = R > d(x_0, y).$$

Consequently $y \in \bar{G}_x$ for every $x \in X$.

It follows from (3) that $V(x_0)$ is a polyhedron. Since $V(x_0)$ is bounded and has a nonempty interior, it is actually an n-*dimensional polytope*.

The faces of a polytope are an important part of its structure. An $(n-1)$-dimensional face of an n-dimensional polytope is said to be a *facet* and a 0-dimensional face is said to be a *vertex*. We now apply to $V(x_0)$ some properties common to all polytopes.

In the representation (3) it may be possible to omit some closed half-spaces $\bar{G}_{x_i}$ without affecting the validity of the representation. By omitting as many half-spaces as possible we obtain an *irredundant representation*, which by suitable choice of notation we may take to be

$$V(x_0) = \bigcap_{i=1}^{l} \bar{G}_{x_i}$$

for some $l \leq m$. The intersections $V(x_0) \cap H_{x_i} (1 \leq i \leq l)$ are then the distinct facets of $V(x_0)$. Any nonempty proper face of $V(x_0)$ is contained in a facet and is the intersection of those facets which contain it. Furthermore, any nonempty face of $V(x_0)$ is the convex hull of those vertices of $V(x_0)$ which it contains.

It follows that for each x_i $(1 \leq i \leq l)$ there is a vertex v_i of $V(x_0)$ such that

$$d(x_0, v_i) = d(x_i, v_i).$$

For $d(x_0, v) \leq d(x_i, v)$ for every vertex v of $V(x_0)$. Assume that $d(x_0, v) < d(x_i, v)$ for every vertex v of $V(x_0)$. Then the open half-space G_{x_i} contains all vertices v and hence also their convex hull $V(x_0)$. But this is a contradiction, since $V(x_0) \cap H_{x_i}$ is a facet of $V(x_0)$.

To illustrate these results take $X = \mathbb{Z}^n$ and $x_0 = O$. Then the Voronoi cell $V(O)$ is the cube consisting of all points $y = (\eta_1, \ldots, \eta_n) \in \mathbb{R}^n$ with $|\eta_i| \leq 1/2$ $(i = 1, \ldots, n)$. It has the minimal number $2n$ of facets.

In fact any lattice Λ in $\mathbb{R}^n$ is discrete and has the property (†). *For a lattice Λ we can restrict attention to the Voronoi cell $V(\Lambda) := V(O)$*, since an arbitrary Voronoi cell is obtained from it by a translation: $V(x_0) = V(O) + x_0$. The Voronoi cell of a lattice has extra properties. Since $x \in \Lambda$ implies $-x \in \Lambda$, $y \in V(\Lambda)$ implies $-y \in V(\Lambda)$. Furthermore, if x_i is a lattice vector determining a facet of $V(\Lambda)$ and if $y \in V(\Lambda) \cap H_{x_i}$, then $\|y\| = \|y - x_i\|$. Since $x \in \Lambda$ implies $x_i - x \in \Lambda$, it follows that $y \in V(\Lambda) \cap H_{x_i}$ implies $x_i - y \in V(\Lambda) \cap H_{x_i}$. Thus *the Voronoi cell $V(\Lambda)$ and all its facets are centrosymmetric.*

In addition, any orthogonal transformation of $\mathbb{R}^n$ which maps onto itself the lattice Λ also maps onto itself the Voronoi cell $V(\Lambda)$. Furthermore the Voronoi cell $V(\Lambda)$ has volume $d(\Lambda)$, by Proposition 11, since the lattice translates of $V(\Lambda)$ form a tiling of $\mathbb{R}^n$.

We define a *facet vector* or 'relevant vector' of a lattice Λ to be a vector $x_i \in \Lambda$ such that $V(\Lambda) \cap H_{x_i}$ is a facet of the Voronoi cell $V(\Lambda)$. If $V(\Lambda)$ is contained in the closed ball $B_R = \{x \in \mathbb{R}^n : \|x\| \leq R\}$, then every facet vector x_i satisfies $\|x_i\| \leq 2R$. For, if $y \in V(\Lambda) \cap H_{x_i}$ then, by Schwarz's inequality (Chapter I, §4),

$$\|x_i\|^2 = 2(x_i, y) \leq 2\|x_i\|\|y\|.$$

The facet vectors were characterized by Voronoi (1908) in the following way:

Proposition 13 *A nonzero vector $x \in \Lambda$ is a facet vector of the lattice $\Lambda \subseteq \mathbb{R}^n$ if and only if every vector $x' \in x + 2\Lambda$, except $\pm x$, satisfies $\|x'\| > \|x\|$.*

Proof Suppose first that $\|x\| < \|x'\|$ for all $x' \neq \pm x$ such that $(x' - x)/2 \in \Lambda$. If $z \in \Lambda$ and $x' = 2z - x$, then $(x' - x)/2 \in \Lambda$. Hence if $z \neq O$, then

$$\|x/2\| < \|z - x/2\|,$$

i.e. $x/2 \in G_z$. Since $\|x/2\| = \|x - x/2\|$, it follows that $x/2 \in V(\Lambda)$ and x is a facet vector.

Suppose next that there exists $x' \neq \pm x$ such that $w = (x' - x)/2 \in \Lambda$ and $\|x'\| \leq \|x\|$. Then also $z = (x' + x)/2 \in \Lambda$ and $z, w \neq O$. If $y \in \bar{G}_z \cap \bar{G}_{-w}$, then

$$2(z, y) \leq \|z\|^2, \quad -2(w, y) \leq \|w\|^2.$$

Hence, by the parallelogram law (Chapter I, §10),

$$2(x, y) = 2(z, y) - 2(w, y) \leq \|z\|^2 + \|w\|^2$$
$$= \|x\|^2/2 + \|x'\|^2/2 \leq \|x\|^2.$$

That is, $y \in \bar{G}_x$. Thus $\bar{G}_x$ is not needed to define $V(\Lambda)$ and x is not a facet vector. □

Any lattice Λ contains a nonzero vector with minimal square-norm. Such a vector will be called a *minimal vector*. Its square-norm will be called the *minimum* of Λ and will be denoted by $m(\Lambda)$.

Proposition 14 *If $\Lambda \subseteq \mathbb{R}^n$ is a lattice with minimum $m(\Lambda)$, then any nonzero vector in Λ with square-norm $< 2m(\Lambda)$ is a facet vector. In particular, any minimal vector is a facet vector.*

Proof Put $r = m(\Lambda)$ and let x be a nonzero vector in Λ with $\|x\|^2 < 2r$. If x is not a facet vector, there exists $y \neq \pm x$ with $(y - x)/2 \in \Lambda$ such that $\|y\| \leq \|x\|$. Since $(y \pm x)/2 \in \Lambda$, $\|x \pm y\|^2 \geq 4r$. Thus

$$4r \leq \|x\|^2 + \|y\|^2 \pm 2(x, y) < 4r \pm 2(x, y),$$

which is impossible. □

Proposition 15 *For any lattice $\Lambda \subseteq \mathbb{R}^n$, the number of facets of its Voronoi cell $V(\Lambda)$ is at most $2(2^n - 1)$.*

Proof Let $x_1, \ldots, x_n$ be a basis for Λ. Then any vector $x \in \Lambda$ has a unique representation $x = x' + x''$, where $x' \in 2\Lambda$ and

$$x'' = \alpha_1 x_1 + \cdots + \alpha_n x_n,$$

with $\alpha_j \in \{0, 1\}$ for $j = 1, \ldots, n$. Thus the number of cosets of 2Λ in Λ is 2^n. But, by Proposition 13, each coset contains at most one pair $\pm y$ of facet vectors. Since 2Λ itself does not contain any facet vectors, the total number of facet vectors is at most $2(2^n - 1)$. □

There exist lattices $\Lambda \subseteq \mathbb{R}^n$ for which the upper bound of Proposition 15 is attained, e.g. the lattice $\Lambda = \{Tz : z \in \mathbb{Z}^n\}$ with $T = I + \beta J$, where J denotes the $n \times n$ matrix every element of which is 1 and $\beta = \{(1 + n)^{1/2} - 1\}/n$.

Proposition 16 *Every vector of a lattice $\Lambda \subseteq \mathbb{R}^n$ is an integral linear combination of facet vectors.*

Proof Let $b_1, \ldots, b_m$ be the facet vectors of Λ and put

$$\Lambda' = \{x = \beta_1 b_1 + \cdots + \beta_m b_m : \beta_1, \ldots, \beta_m \in \mathbb{Z}\}.$$

Evidently Λ' is a subgroup of $\mathbb{R}^n$ and actually a discrete subgroup, since $\Lambda' \subseteq \Lambda$. If Λ' were contained in a hyperplane of $\mathbb{R}^n$ any point on the line through the origin orthogonal to this hyperplane would belong to the Voronoi cell V of Λ, which is impossible because V is bounded. Hence Λ' contains n linearly independent vectors.

Thus Λ' is a sublattice of Λ. It follows that the Voronoi cell V of Λ is contained in the Voronoi cell V' of Λ'. But if $y \in V'$, then

$$\|y\| \leq \|b_i - y\|, \quad (i = 1, \ldots, m)$$

and hence $y \in V$. Thus $V' = V$. Hence the Λ'-translates of V and the Λ-translates of V are both tilings of $\mathbb{R}^n$. Since $\Lambda' \subseteq \Lambda$, this is possible only if $\Lambda' = \Lambda$. $\square$

Since every integral linear combination of facet vectors is in the lattice, Proposition 16 implies

Corollary 17 *Distinct lattices in $\mathbb{R}^n$ have distinct Voronoi cells.*

Proposition 16 does not say that the lattice has a basis of facet vectors. It is known that every lattice in $\mathbb{R}^n$ has a basis of facet vectors if $n \leq 6$, but if $n > 6$ this is still an open question. It is known also that every lattice in $\mathbb{R}^n$ has a basis of minimal vectors when $n \leq 4$ but, when $n > 4$, there are lattices with no such basis. In fact a lattice may have no basis of minimal vectors, even though every lattice vector is an integral linear combination of minimal vectors.

Lattices and their Voronoi cells have long been used in crystallography. An n-dimensional *crystal* may be defined mathematically to be a subset of $\mathbb{R}^n$ of the form

$$F + \Lambda = \{x + y : x \in F, y \in \Lambda\},$$

where F is a finite set and Λ a lattice. Crystals may be studied by means of their symmetry groups.

An *isometry* of $\mathbb{R}^n$ is an invertible affine transformation which leaves unaltered the Euclidean distance between any two points. For example, any orthogonal transformation is an isometry and so is a translation by an arbitrary vector v. Any isometry is the composite of a translation and an orthogonal transformation. The *symmetry group* of a set $X \subseteq \mathbb{R}^n$ is the group of all isometries of $\mathbb{R}^n$ which map X to itself.

We define an n-dimensional *crystallographic group* to be a group G of isometries of $\mathbb{R}^n$ such that the vectors corresponding to translations in G form an n-dimensional lattice. It is not difficult to show that a subset of $\mathbb{R}^n$ is an n-dimensional crystal if and only if it is discrete and its symmetry group is an n-dimensional crystallographic group.

It was shown by Bieberbach (1911) that a group G of isometries of $\mathbb{R}^n$ is a crystallographic group if and only if it is discrete and has a compact fundamental domain D, i.e. the sets $\{g(D) : g \in G\}$ form a tiling of $\mathbb{R}^n$. He could then show that the translations in a crystallographic group form a torsion-free abelian normal subgroup of finite index. He showed later (1912) that two crystallographic groups G_1, G_2 are isomorphic if and only if there exists an invertible affine transformation A such that

$G_2 = A^{-1}G_1A$. With the aid of results of Minkowski and Jordan it follows that, for a given dimension n, there are only finitely many non-isomorphic crystallographic groups. These results provided a positive answer to the first part of the 18th Problem of Hilbert (1900).

The structure of physical crystals is analysed by means of the corresponding 3-dimensional crystallographic groups. A stronger concept than isomorphism is useful for such applications. Two crystallographic groups G_1, G_2 may be said to be *properly isomorphic* if there exists an orientation-preserving invertible affine transformation A such that $G_2 = A^{-1}G_1A$. An isomorphism class of crystallographic groups either coincides with a proper isomorphism class or splits into two distinct proper isomorphism classes.

Fedorov (1891) showed that there are 17 isomorphism classes of 2-dimensional crystallographic groups, none of which splits. Collating earlier work of Sohncke (1879), Schoenflies (1889) and himself, Fedorov (1892) also showed that there are 219 isomorphism classes of 3-dimensional crystallographic groups, 11 of which split. More recently, Brown *et al.* (1978) have shown that there are 4783 isomorphism classes of 4-dimensional crystallographic groups, 112 of which split.

5 Densest Packings

The result of Hermite, mentioned at the beginning of the chapter, can be formulated in terms of lattices instead of quadratic forms. For any real non-singular matrix T, the matrix

$$A = T^tT$$

is a real positive definite symmetric matrix. Conversely, by a principal axes transformation, or more simply by induction, it may be seen that any real positive definite symmetric matrix A may be represented in this way.

Let Λ be the lattice

$$\Lambda = \{y = Tx \in \mathbb{R}^n : x \in \mathbb{Z}^n\}$$

and put

$$\gamma(\Lambda) = m(\Lambda)/d(\Lambda)^{2/n},$$

where $d(\Lambda)$ is the determinant and $m(\Lambda)$ the minimum of Λ. Then $\gamma(\rho\Lambda) = \gamma(\Lambda)$ for any $\rho > 0$. Hermite's result that there exists a positive constant c_n, depending only on n, such that $0 < x^tAx \leq c_n(\det A)^{1/n}$ for some $x \in \mathbb{Z}^n$ may be restated in the form

$$\gamma(\Lambda) \leq c_n.$$

Hermite's constant γ_n is defined to be the least positive constant c_n such that this inequality holds for all $\Lambda \subseteq \mathbb{R}^n$.

It may be shown that γ_n^n is a rational number for each n. It follows from Proposition 2 that $\overline{\lim}_{n\to\infty}\gamma_n/n \leq 2/\pi e$. Minkowski (1905) showed also that

$$\underline{\lim}_{n\to\infty} \gamma_n/n \geq 1/2\pi e = 0.0585\ldots,$$

and it is possible that actually $\lim_{n\to\infty} \gamma_n/n = 1/2\pi e$. The significance of Hermite's constant derives from its connection with lattice packings of balls, as we now explain.

Let Λ be a lattice in $\mathbb{R}^n$ and K a subset of $\mathbb{R}^n$ which is the closure of a nonempty open set G. We say that Λ gives a *lattice packing* for K if the family of translates $K + x$ ($x \in \Lambda$) is a packing of $\mathbb{R}^n$, i.e. if for any two distinct points $x, y \in \Lambda$ the interiors $G + x$ and $G + y$ are disjoint. This is the same as saying that Λ does not contain the difference of any two distinct points of the interior of K, since $g + x = g' + y$ if and only if $g' - g = x - y$. If K is a compact symmetric convex set with nonempty interior G, it is the same as saying that the interior of the set $2K$ contains no nonzero point of Λ, since in this case $g, g' \in G$ implies $(g' - g)/2 \in G$ and $2g = g - (-g)$.

The *density* of the lattice packing, i.e. the fraction of the total space which is occupied by translates of K, is clearly $\lambda(K)/d(\Lambda)$. Hence the maximum density of any lattice packing for K is

$$\delta(K) = \lambda(K)/\Delta(2K) = 2^{-n}\lambda(K)/\Delta(K),$$

where $\Delta(K)$ is the critical determinant of K, as defined in §3. The use of the word 'maximum' is justified, since it will be shown in §6 that the infimum involved in the definition of critical determinant is attained.

Our interest is in the special case of a closed ball: $K = B_\rho = \{x \in \mathbb{R}^n : \|x\| \leq \rho\}$. By what we have said, Λ gives a lattice packing for B_ρ if and only if the interior of $B_{2\rho}$ contains no nonzero point of Λ, i.e. if and only if $m(\Lambda)^{1/2} \geq 2\rho$. Hence

$$\delta(B_\rho) = \sup\{\lambda(B_\rho)/d(\Lambda) : m(\Lambda)^{1/2} = 2\rho\}$$
$$= \kappa_n \rho^n \sup\{d(\Lambda)^{-1} : m(\Lambda)^{1/2} = 2\rho\},$$

where $\kappa_n = \pi^{n/2}/(n/2)!$ again denotes the volume of the unit ball in $\mathbb{R}^n$. By virtue of homogeneity it follows that

$$\delta_n := \delta(B_\rho) = 2^{-n}\kappa_n \sup_\Lambda \gamma(\Lambda)^{n/2},$$

where the supremum is now over all lattices $\Lambda \subseteq \mathbb{R}^n$; that is, in terms of Hermite's constant γ_n,

$$\delta_n = 2^{-n}\kappa_n \gamma_n^{n/2}.$$

Thus γ_n, like δ_n, measures the densest lattice packing of balls. A lattice $\Lambda \subseteq \mathbb{R}^n$ for which $\gamma(\Lambda) = \gamma_n$, i.e. a critical lattice for a ball, will be called simply a *densest lattice*.

The densest lattice in $\mathbb{R}^n$ is known for each $n \leq 8$, and is uniquely determined apart from isometries and scalar multiples. In fact these densest lattices are all examples of indecomposable root lattices. These terms will now be defined.

A lattice Λ is said to be *decomposable* if there exist additive subgroups Λ_1, Λ_2 of Λ, each containing a nonzero vector, such that $(x_1, x_2) = 0$ for all $x_1 \in \Lambda_1$ and $x_2 \in \Lambda_2$, and every vector in Λ is the sum of a vector in Λ_1 and a vector in Λ_2. Since Λ_1 and Λ_2 are necessarily discrete, they are lattices in the wide sense (i.e. they are not full-dimensional). We say also that $\varLambda$ is the *orthogonal sum* of the lattices $\varLambda_1$ and $\varLambda_2$. The orthogonal sum of any finite number of lattices is defined similarly. A lattice is *indecomposable* if it is not decomposable.

The following result was first proved by Eichler (1952).

Proposition 18 *Any lattice $\varLambda$ is an orthogonal sum of finitely many indecomposable lattices, which are uniquely determined apart from order.*

Proof (i) Define a vector $x \in \varLambda$ to be 'decomposable' if there exist nonzero vectors $x_1, x_2 \in \varLambda$ such that $x = x_1 + x_2$ and $(x_1, x_2) = 0$. We show first that every nonzero $x \in \varLambda$ is a sum of finitely many indecomposable vectors.

By definition, x is either indecomposable or is the sum of two nonzero orthogonal vectors in $\varLambda$. Both these vectors have square-norm less than the square-norm of x, and for each of them the same alternative presents itself. Continuing in this way, we must eventually arrive at indecomposable vectors, since there are only finitely many vectors in $\varLambda$ with square-norm less than that of x.

(ii) If $\varLambda$ is the orthogonal sum of finitely many lattices L_v then, by the definition of an orthogonal sum, every indecomposable vector of $\varLambda$ lies in one of the sublattices L_v. Hence if two indecomposable vectors are not orthogonal, they lie in the same sublattice L_v.

(iii) Call two indecomposable vectors x, x' 'equivalent' if there exist indecomposable vectors $x = x_0, x_1, \ldots, x_{k-1}, x_k = x'$ such that $(x_j, x_{j+1}) \neq 0$ for $0 \leq j < k$. Clearly 'equivalence' is indeed an equivalence relation and thus the set of all indecomposable vectors is partitioned into equivalence classes $\mathscr{C}_\mu$. Two vectors from different equivalence classes are orthogonal and, if $\varLambda$ is an orthogonal sum of lattices L_v as in (ii), then two vectors from the same equivalence class lie in the same sublattice L_v.

(iv) Let $\varLambda_\mu$ be the subgroup of $\varLambda$ generated by the vectors in the equivalence class $\mathscr{C}_\mu$. Then, by (i), $\varLambda$ is generated by the sublattices $\varLambda_\mu$. Since, by (iii), $\varLambda_\mu$ is orthogonal to $\varLambda_{\mu'}$ if $\mu \neq \mu'$, $\varLambda$ is actually the orthogonal sum of the sublattices $\varLambda_\mu$. If $\varLambda$ is an orthogonal sum of lattices L_v as in (ii), then each $\varLambda_\mu$ is contained in some L_v. It follows that each $\varLambda_\mu$ is indecomposable and that these indecomposable sublattices are uniquely determined apart from order. $\square$

Let $\varLambda$ be a lattice in $\mathbb{R}^n$. If $\varLambda \subseteq \varLambda^*$, i.e. if $(x, y) \in \mathbb{Z}$ for all $x, y \in \varLambda$, then $\varLambda$ is said to be *integral*. If (x, x) is an even integer for every $x \in \varLambda$, then $\varLambda$ is said to be *even*. (It follows that an even lattice is also integral.) If $\varLambda$ is even and every vector in $\varLambda$ is an integral linear combination of vectors in $\varLambda$ with square-norm 2, then $\varLambda$ is said to be a *root lattice*.

Thus in a root lattice the minimal vectors have square-norm 2. It may be shown by a long, but elementary, argument that any root lattice has a basis of minimal vectors such that every minimal vector is an integral linear combination of the basis vectors with coefficients which are all nonnegative or all nonpositive. Such a basis will be called a *simple* basis. The facet vectors of a root lattice are precisely the minimal vectors, and hence its Voronoi cell is the set of all $y \in \mathbb{R}^n$ such that $(y, x) \leq 1$ for every minimal vector x.

Any root lattice is an orthogonal sum of indecomposable root lattices. It was shown by Witt (1941) that the indecomposable root lattices can be completely enumerated;

Table 1. Indecomposable root lattices

$$A_n = \{x = (\xi_0, \xi_1, \dots, \xi_n) \in \mathbb{Z}^{n+1} : \xi_0 + \xi_1 + \cdots + \xi_n = 0\} \ (n \geq 1);$$
$$D_n = \{x = (\xi_1, \dots, \xi_n) \in \mathbb{Z}^n : \xi_1 + \cdots + \xi_n \text{ even}\} \ (n \geq 3);$$
$$E_8 = D_8 \cup D_8^\dagger, \text{ where } D_8^\dagger = (1/2, 1/2, \dots, 1/2) + D_8;$$
$$E_7 = \{x = (\xi_1, \dots, \xi_8) \in E_8 : \xi_7 = -\xi_8\};$$
$$E_6 = \{x = (\xi_1, \dots, \xi_8) \in E_8 : \xi_6 = \xi_7 = -\xi_8\}.$$

they are all listed in Table 1. We give also their minimal vectors in terms of the canonical basis $e_1, \dots, e_n$ of $\mathbb{R}^n$.

The lattice A_n has $n(n + 1)$ minimal vectors, namely the vectors $\pm(e_j - e_k)$ $(0 \leq j < k \leq n)$, and the vectors $e_0 - e_1, e_1 - e_2, \dots, e_{n-1} - e_n$ form a simple basis. By calculating the determinant of $B^t B$, where B is the $(n+1) \times n$ matrix whose columns are the vectors of this simple basis, it may be seen that the determinant of the lattice A_n is $(n + 1)^{1/2}$.

The lattice D_n has $2n(n - 1)$ minimal vectors, namely the vectors $\pm e_j \pm e_k$ $(1 \leq j < k \leq n)$. The vectors $e_1 - e_2, e_2 - e_3, \dots, e_{n-1} - e_n, e_{n-1} + e_n$ form a simple basis and hence the lattice D_n has determinant 2.

The lattice E_8 has 240 minimal vectors, namely the 112 vectors $\pm e_j \pm e_k$ $(1 \leq j < k \leq 8)$ and the 128 vectors $(\pm e_1 \pm \cdots \pm e_8)/2$ with an even number of minus signs. The vectors

$$v_1 = (e_1 - e_2 - \cdots - e_7 + e_8)/2, \quad v_2 = e_1 + e_2,$$
$$v_3 = e_2 - e_1, \quad v_4 = e_3 - e_2, \dots, \quad v_8 = e_7 - e_6,$$

form a simple basis and hence the lattice has determinant 1.

The lattice E_7 has 126 minimal vectors, namely the 60 vectors $\pm e_j \pm e_k$ $(1 \leq j < k \leq 6)$, the vectors $\pm(e_7 - e_8)$ and the 64 vectors $\pm\left(\sum_{i=1}^6 (\pm e_i) - e_7 + e_8\right)/2$ with an odd number of minus signs in the sum. The vectors $v_1, \dots, v_7$ form a simple basis and the lattice has determinant $\sqrt{2}$.

The lattice E_6 has 72 minimal vectors, namely the 40 vectors $\pm e_j \pm e_k$ $(1 \leq j < k \leq 5)$ and the 32 vectors $\pm\left(\sum_{i=1}^5 (\pm e_i) - e_6 - e_7 + e_8\right)/2$ with an even number of minus signs in the sum. The vectors $v_1, \dots, v_6$ form a simple basis and the lattice has determinant $\sqrt{3}$.

We now return to lattice packings of balls. The densest lattices for $n \leq 8$ are given in Table 2. These lattices were shown to be densest by Lagrange (1773) for $n = 2$, by Gauss (1831) for $n = 3$, by Korkine and Zolotareff (1872,1877) for $n = 4, 5$ and by Blichfeldt (1925,1926,1934) for $n = 6, 7, 8$.

Although the densest lattice in $\mathbb{R}^n$ is unknown for every $n > 8$, there are plausible candidates in some dimensions. In particular, a lattice discovered by Leech (1967) is believed to be densest in 24 dimensions. This lattice may be constructed in the following way. Let p be a prime such that $p \equiv 3 \bmod 4$ and let H_n be the Hadamard matrix of order $n = p + 1$ constructed by Paley's method (see Chapter V, §2). The columns of the matrix

Table 2. Densest lattices in $\mathbb{R}^n$

n	Λ	γ_n	δ_n
1	A_1	1	1
2	A_2	$(4/3)^{1/2} = 1.1547\ldots$	$3^{1/2}\pi/6 = 0.9068\ldots$
3	D_3	$2^{1/3} = 1.2599\ldots$	$2^{1/2}\pi/6 = 0.7404\ldots$
4	D_4	$2^{1/2} = 1.4142\ldots$	$\pi^2/16 = 0.6168\ldots$
5	D_5	$8^{1/5} = 1.5157\ldots$	$2^{1/2}\pi^2/30 = 0.4652\ldots$
6	E_6	$(64/3)^{1/6} = 1.6653\ldots$	$3^{1/2}\pi^3/144 = 0.3729\ldots$
7	E_7	$(64)^{1/7} = 1.8114\ldots$	$\pi^3/105 = 0.2952\ldots$
8	E_8	2	$\pi^4/384 = 0.2536\ldots$

$$T = (n/4 + 1)^{-1/2} \begin{pmatrix} (n/4 + 1)I_n & H_n - I_n \\ 0_n & I_n \end{pmatrix}$$

generate a lattice in $\mathbb{R}^{2n}$. For $p = 3$ we obtain the root lattice E_8 and for $p = 11$ the Leech lattice Λ_{24}.

Leech's lattice may be characterized as the unique even lattice Λ in $\mathbb{R}^{24}$ with $d(\Lambda) = 1$ and $m(\Lambda) > 2$. It was shown by Conway (1969) that, if G is the group of all orthogonal transformations of $\mathbb{R}^{24}$ which map the Leech lattice Λ_{24} onto itself, then the factor group $G/\{\pm I_{24}\}$ is a finite simple group, and two more finite simple groups are easily obtained as (stabilizer) subgroups. These are three of the 26 sporadic simple groups which were mentioned in §7 of Chapter V.

Leech's lattice has 196560 minimal vectors of square-norm 4. Thus the packing of unit balls associated with Λ_{24} is such that each ball touches 196560 other balls. It has been shown that 196560 is the maximal number of nonoverlapping unit balls in $\mathbb{R}^{24}$ which can touch another unit ball and that, up to isometry, there is only one possible arrangement.

Similarly, since E_8 has 240 minimal vectors of square-norm 2, the packing of balls of radius $2^{-1/2}$ associated with E_8 is such that each ball touches 240 other balls. It has been shown that 240 is the maximal number of nonoverlapping balls of fixed radius in $\mathbb{R}^8$ which can touch another ball of the same radius and that, up to isometry, there is only one possible arrangement.

In general, one may ask what is the *kissing number* of $\mathbb{R}^n$, i.e. the maximal number of nonoverlapping unit balls in $\mathbb{R}^n$ which can touch another unit ball? The question, for $n = 3$, first arose in 1694 in a discussion between Newton, who claimed that the answer was 12, and Gregory, who said 13. It was first shown by Hoppe (1874) that Newton was right, but in this case the arrangement of the 12 balls in $\mathbb{R}^3$ is *not* unique up to isometry. One possibility is to take the centres of the 12 balls to be the vertices of a regular icosahedron, the centre of which is the centre of the unit ball they touch.

The kissing number of $\mathbb{R}^1$ is clearly 2. It is not difficult to show that the kissing number of $\mathbb{R}^2$ is 6 and that the centres of the six unit balls must be the vertices of a regular hexagon, the centre of which is the centre of the unit ball they touch. For $n > 3$ the kissing number of $\mathbb{R}^n$ is unknown, except for the two cases $n = 8$ and $n = 24$ already mentioned.

6 Mahler's Compactness Theorem

It is useful to study not only individual lattices, but also the family $\mathscr{L}_n$ of all lattices in $\mathbb{R}^n$. A sequence of lattices $\Lambda_k \in \mathscr{L}_n$ will be said to *converge* to a lattice $\Lambda \in \mathscr{L}_n$, in symbols $\Lambda_k \to \Lambda$, if there exist bases $b_{k1}, \ldots, b_{kn}$ of $\Lambda_k (k = 1, 2, \ldots)$ and a basis $b_1, \ldots, b_n$ of Λ such that

$$b_{kj} \to b_j \text{ as } k \to \infty \quad (j = 1, \ldots, n).$$

Evidently this implies that $d(\Lambda_k) \to d(\Lambda)$ as $k \to \infty$. Also, for any $x \in \Lambda$ there exist $x_k \in \Lambda_k$ such that $x_k \to x$ as $k \to \infty$. In fact if $x = \alpha_1 b_1 + \cdots + \alpha_n b_n$, where $\alpha_i \in \mathbb{Z} \ (i = 1, \ldots, n)$, we can take $x_k = \alpha_1 b_{k1} + \cdots + \alpha_n b_{kn}$.

It is not obvious from the definition that the limit of a sequence of lattices is uniquely determined, but this follows at once from the next result.

Proposition 19 *Let Λ be a lattice in $\mathbb{R}^n$ and let $\{\Lambda_k\}$ be a sequence of lattices in $\mathbb{R}^n$ such that $\Lambda_k \to \Lambda$ as $k \to \infty$. If $x_k \in \Lambda_k$ and $x_k \to x$ as $k \to \infty$, then $x \in \Lambda$.*

Proof With the above notation,

$$x = \alpha_1 b_1 + \cdots + \alpha_n b_n,$$

where $\alpha_i \in \mathbb{R} \ (i = 1, \ldots, n)$, and similarly

$$x_k = \alpha_{k1} b_1 + \cdots + \alpha_{kn} b_n,$$

where $\alpha_{ki} \in \mathbb{R}$ and $\alpha_{ki} \to \alpha_i$ as $k \to \infty \ (i = 1, \ldots, n)$.

The linear transformation T_k of $\mathbb{R}^n$ which maps b_i to $b_{ki}(i = 1, \ldots, n)$ can be written in the form

$$T_k = I - A_k,$$

where $A_k \to O$ as $k \to \infty$. It follows that

$$T_k^{-1} = (I - A_k)^{-1} = I + A_k + A_k^2 + \cdots = I + C_k,$$

where also $C_k \to O$ as $k \to \infty$. Hence

$$\begin{aligned} x_k &= T_k^{-1}(\alpha_{k1} b_{k1} + \cdots + \alpha_{kn} b_{kn}) \\ &= (\alpha_{k1} + \eta_{k1}) b_{k1} + \cdots + (\alpha_{kn} + \eta_{kn}) b_{kn}, \end{aligned}$$

where $\eta_{ki} \to 0$ as $k \to \infty \ (i = 1, \ldots, n)$. But $\alpha_{ki} + \eta_{ki} \in \mathbb{Z}$ for every k. Letting $k \to \infty$, we obtain $\alpha_i \in \mathbb{Z}$. That is, $x \in \Lambda$. $\square$

It is natural to ask if the Voronoi cells of a convergent sequence of lattices also converge in some sense. The required notion of convergence is in fact older than the notion of convergence of lattices and applies to arbitrary compact subsets of $\mathbb{R}^n$.

The *Hausdorff distance* $h(K, K')$ between two compact subsets K, K' of $\mathbb{R}^n$ is defined to be the infimum of all $\rho > 0$ such that every point of K is distant at most ρ from some point of K' and every point of K' is distant at most ρ from some point of K. We will show that this defines a metric, the *Hausdorff metric*, on the space of all compact subsets of $\mathbb{R}^n$.

Evidently

$$0 \le h(K, K') = h(K', K) < \infty.$$

Moreover $h(K, K') = 0$ implies $K = K'$. For if $x' \in K'$, there exist $x_k \in K$ such that $x_k \to x'$ and hence $x' \in K$, since K is closed. Thus $K' \subseteq K$, and similarly $K \subseteq K'$.

Finally we prove the triangle inequality

$$h(K, K'') \le h(K, K') + h(K', K'').$$

To simplify writing, put $\rho = h(K, K')$ and $\rho' = h(K', K'')$. For any $\varepsilon > 0$, if $x \in K$ there exist $x' \in K'$ such that $\|x - x'\| < \rho + \varepsilon$ and then $x'' \in K''$ such that $\|x' - x''\| < \rho' + \varepsilon$. Hence

$$\|x - x''\| < \rho + \rho' + 2\varepsilon.$$

Similarly, if $x'' \in K''$ there exists $x \in K$ for which the same inequality holds. Since ε can be arbitrarily small, this completes the proof.

The definition of Hausdorff distance can also be expressed in the form

$$h(K, K') = \inf\{\rho \ge 0 : K \subseteq K' + B_\rho, K' \subseteq K + B_\rho\},$$

where $B_\rho = \{x \in \mathbb{R}^n : \|x\| \le \rho\}$. A sequence K_j of compact subsets of $\mathbb{R}^n$ *converges* to a compact subset K of $\mathbb{R}^n$ if $h(K_j, K) \to 0$ as $j \to \infty$.

It was shown by Hausdorff (1927) that any uniformly bounded sequence of compact subsets of $\mathbb{R}^n$ has a convergent subsequence. In particular, any uniformly bounded sequence of compact convex subsets of $\mathbb{R}^n$ has a subsequence which converges to a compact convex set. This special case of Hausdorff's result, which is all that we will later require, had already been established by Blaschke (1916) and is known as *Blaschke's selection principle*.

Proposition 20 *Let $\{\Lambda_k\}$ be a sequence of lattices in $\mathbb{R}^n$ and let V_k be the Voronoi cell of Λ_k. If there exists a compact convex set V with nonempty interior such that $V_k \to V$ in the Hausdorff metric as $k \to \infty$, then V is the Voronoi cell of a lattice Λ and $\Lambda_k \to \Lambda$ as $k \to \infty$.*

Proof Since every Voronoi cell V_k is symmetric, so also is the limit V. Since V has nonempty interior, it follows that the origin is itself an interior point of V. Thus there exists $\delta > 0$ such that the ball $B_\delta = \{x \in \mathbb{R}^n : \|x\| \le \delta\}$ is contained in V.

It follows that $B_{\delta/2} \subseteq V_k$ for all large k. The quickest way to see this is to use *Rådström's cancellation law*, which says that if A, B, C are nonempty compact convex subsets of $\mathbb{R}^n$ such that $A + C \subseteq B + C$, then $A \subseteq B$. In the present case we have

$$B_{\delta/2} + B_{\delta/2} \subseteq B_\delta \subseteq V \subseteq V_k + B_{\delta/2} \text{ for } k \ge k_0,$$

and hence $B_{\delta/2} \subseteq V_k$ for $k \ge k_0$. Since also $V_k \subseteq V + B_{\delta/2}$ for all large k, there exists $R > 0$ such that $V_k \subseteq B_R$ for all k.

The lattice Λ_k has at most $2(2^n - 1)$ facet vectors, by Proposition 15. Hence, by restriction to a subsequence, we may assume that all Λ_k have the same number m of facet vectors. Let $x_{k1}, \ldots, x_{km}$ be the facet vectors of Λ_k and choose the notation so that $x_{k1}, \ldots, x_{kn}$ are linearly independent. Since they all lie in the ball B_{2R}, by restriction to a further subsequence we may assume that

$$x_{kj} \to x_j \quad \text{as } k \to \infty \quad (j = 1, \ldots, m).$$

Evidently $\|x_j\| \geq \delta$ $(j = 1, \ldots, m)$ since, for $k \geq k_0$, all nonzero $x \in \Lambda_k$ have $\|x\| \geq \delta$.

The set Λ of all integral linear combinations of $x_1, \ldots, x_m$ is certainly an additive subgroup of $\mathbb{R}^n$. Moreover Λ is discrete. For suppose $y \in \Lambda$ and $\|y\| < \delta$. We have

$$y = \alpha_1 x_1 + \cdots + \alpha_m x_m,$$

where $\alpha_j \in \mathbb{Z}$ $(j = 1, \ldots, m)$. If

$$y_k = \alpha_1 x_{k1} + \cdots + \alpha_m x_{km},$$

then $y_k \to y$ as $k \to \infty$ and hence $\|y_k\| < \delta$ for all large k. Since $y_k \in \Lambda_k$, it follows that $y_k = O$ for all large k and hence $y = O$.

Since the lattice Λ'_k with basis $x_{k1}, \ldots, x_{kn}$ is a sublattice of Λ_k, we have

$$\mathrm{d}(\Lambda'_k) \geq \mathrm{d}(\Lambda_k) = \lambda(V_k) \geq \lambda(B_{\delta/2}).$$

Since $\mathrm{d}(\Lambda'_k) = |\det(x_{k1}, \ldots, x_{kn})|$, it follows that also

$$|\det(x_1, \ldots, x_n)| \geq \lambda(B_{\delta/2}) > 0.$$

Thus the vectors $x_1, \ldots, x_n$ are linearly independent. Hence Λ is a lattice.

Let $b_1, \ldots, b_n$ be a basis of Λ. Then, by the definition of Λ,

$$b_i = \alpha_{i1} x_1 + \cdots + \alpha_{im} x_m,$$

where $\alpha_{ij} \in \mathbb{Z}$ $(1 \leq i \leq n, 1 \leq j \leq m)$. Put

$$b_{ki} = \alpha_{i1} x_{k1} + \cdots + \alpha_{im} x_{km}.$$

Then $b_{ki} \in \Lambda_k$ and $b_{ki} \to b_i$ as $k \to \infty$ $(i = 1, \ldots, n)$. Hence, for all large k, the vectors $b_{k1}, \ldots, b_{kn}$ are linearly independent. We are going to show that $b_{k1}, \ldots, b_{kn}$ is a basis of Λ_k for all large k.

Since $b_1, \ldots, b_n$ is a basis of Λ, we have

$$x_j = \gamma_{j1} b_1 + \cdots + \gamma_{jn} b_n,$$

where $\gamma_{ji} \in \mathbb{Z}$ $(1 \leq i \leq n, 1 \leq j \leq m)$. Hence, if

$$y_{kj} = \gamma_{j1} b_{k1} + \cdots + \gamma_{jn} b_{kn},$$

then $y_{kj} \in \Lambda_k$ and $y_{kj} \to x_j$ as $k \to \infty$ $(j = 1, \ldots, m)$. Thus, for all large k,

$$\|y_{kj} - x_{kj}\| < \delta \quad (j = 1, \ldots, m).$$

Since $y_{kj} - x_{kj} \in \Lambda_k$, this implies that, for all large k, $y_{kj} = x_{kj}$ $(j = 1, \ldots, m)$. Thus every facet vector of Λ_k is an integral linear combination of $b_{k1}, \ldots, b_{kn}$ and hence, by Proposition 16, every vector of Λ_k is an integral linear combination of $b_{k1}, \ldots, b_{kn}$. Since $b_{k1}, \ldots, b_{kn}$ are linearly independent, this shows that they are a basis of Λ_k.

Let W be the Voronoi cell of Λ. We wish to show that $V = W$. If $v \in V$, then there exist $v_k \in V_k$ such that $v_k \to v$. Assume $v \notin W$. Then $\|v\| > \|z - v\|$ for some $z \in \Lambda$, and so

$$\|v\| = \|z - v\| + \rho,$$

where $\rho > 0$. There exist $z_k \in \Lambda_k$ such that $z_k \to z$. Then, for all large k,

$$\|v\| > \|z_k - v\| + \rho/2$$

and hence, for all large k,

$$\|v_k\| > \|z_k - v_k\|.$$

But this contradicts $v_k \in V_k$.

This proves that $V \subseteq W$. On the other hand, V has volume

$$\begin{aligned}
\lambda(V) &= \lim_{k \to \infty} \lambda(V_k) = \lim_{k \to \infty} d(\Lambda_k) \\
&= \lim_{k \to \infty} |\det(b_{k1}, \ldots, b_{kn})| \\
&= |\det(b_1, \ldots, b_n)| = d(\Lambda) = \lambda(W).
\end{aligned}$$

It follows that every interior point of W is in V, and hence $W = V$. Corollary 17 now shows that the same lattice Λ would have been obtained if we had restricted attention to some other subsequence of $\{\Lambda_k\}$.

Let $a_1, \ldots, a_n$ be any basis of Λ. We are going to show that, for the sequence $\{\Lambda_k\}$ originally given, there exist $a_{ki} \in \Lambda_k$ such that

$$a_{ki} \to a_i \text{ as } k \to \infty \quad (i = 1, \ldots, n).$$

If this is not the case then, for some $i \in \{1, \ldots, n\}$ and some $\varepsilon > 0$, there exist infinitely many k such that

$$\|x - a_i\| > \varepsilon \quad \text{for all } x \in \Lambda_k.$$

From this subsequence we could as before pick a further subsequence $\Lambda_{k_v} \to \Lambda$. Then every $y \in \Lambda$ is the limit of a sequence $y_v \in \Lambda_{k_v}$. Taking $y = a_i$, we obtain a contradiction.

It only remains to show that $a_{k1}, \ldots, a_{kn}$ is a basis of Λ_k for all large k. Since

$$\lim_{k \to \infty} |\det(a_{k1}, \ldots, a_{kn})| = |\det(a_1, \ldots, a_n)|$$

$$= d(\Lambda) = \lambda(V) = \lim_{k \to \infty} \lambda(V_k),$$

for all large k we must have

$$0 < |\det(a_{k1}, \ldots, a_{kn})| < 2\lambda(V_k).$$

But if $a_{k1}, \ldots, a_{kn}$ were not a basis of Λ_k for all large k, then for infinitely many k we would have

$$|\det(a_{k1}, \ldots, a_{kn})| \geq 2d(\Lambda_k) = 2\lambda(V_k). \qquad \square$$

Proposition 20 has the following counterpart:

Proposition 21 *Let $\{\Lambda_k\}$ be a sequence of lattices in $\mathbb{R}^n$ and let V_k be the Voronoi cell of Λ_k. If there exists a lattice Λ such that $\Lambda_k \to \Lambda$ as $k \to \infty$, and if V is the Voronoi cell of Λ, then $V_k \to V$ in the Hausdorff metric as $k \to \infty$.*

Proof By hypothesis, there exists a basis $b_1, \ldots, b_n$ of Λ and a basis $b_{k1}, \ldots, b_{kn}$ of each Λ_k such that $b_{kj} \to b_j$ as $k \to \infty$ ($j = 1, \ldots, n$). Choose $R > 0$ so that the fundamental parallelotope of Λ is contained in the ball $B_R = \{x \in \mathbb{R}^n : \|x\| \leq R\}$. Then, for all $k \geq k_0$, the fundamental parallelotope of Λ_k is contained in the ball B_{2R}. It follows that, for all $k \geq k_0$, every point of $\mathbb{R}^n$ is distant at most $2R$ from some point of Λ_k and hence $V_k \subseteq B_{2R}$.

Consequently, by Blaschke's selection principle, the sequence $\{V_k\}$ has a subsequence $\{V_{k_\nu}\}$ which converges in the Hausdorff metric to a compact convex set W. Moreover,

$$\lambda(W) = \lim_{\nu \to \infty} \lambda(V_{k_\nu}) = \lim_{\nu \to \infty} d(\Lambda_{k_\nu}) = d(\Lambda) > 0.$$

Consequently, since W is convex, it has nonempty interior. It now follows from Proposition 20 that $W = V$.

Thus any convergent subsequence of $\{V_k\}$ has the same limit V. If the whole sequence $\{V_k\}$ did not converge to V, there would exist $\rho > 0$ and a subsequence $\{V_{k_\nu}\}$ such that

$$h(V_{k_\nu}, V) \geq \rho \quad \text{for all } \nu.$$

By the Blaschke selection principle again, this subsequence would itself have a convergent subsequence. Since its limit must be V, this yields a contradiction. $\qquad \square$

Suppose $\Lambda_k \in \mathcal{L}_n$ and $\Lambda_k \to \Lambda$ as $k \to \infty$. We will show that not only $d(\Lambda_k) \to d(\Lambda)$, but also $m(\Lambda_k) \to m(\Lambda)$ as $k \to \infty$. Since every $x \in \Lambda$ is the limit of a sequence $x_k \in \Lambda_k$, we must have $\overline{\lim}_{k \to \infty} m(\Lambda_k) \leq m(\Lambda)$. On the other hand, by Proposition 19, if $x_k \in \Lambda_k$ and $x_k \to x$, then $x \in \Lambda$. Hence $\underline{\lim}_{k \to \infty} m(\Lambda) \geq m(\Lambda)$, since $x \neq 0$ if $x_k \neq 0$ for large k.

Suppose now that a subset $\mathcal{F}$ of $\mathcal{L}_n$ has the property that any infinite sequence Λ_k of lattices in $\mathcal{F}$ has a convergent subsequence. Then there exist positive constants ρ, σ such that

$$m(\Lambda) \geq \rho^2, \quad d(\Lambda) \leq \sigma \quad \text{for all } \Lambda \in \mathcal{F}.$$

For otherwise there would exist a sequence Λ_k of lattices in $\mathcal{F}$ such that either $m(\Lambda_k) \to 0$ or $d(\Lambda_k) \to \infty$, and clearly this sequence could have no convergent subsequence.

We now prove the fundamental *compactness theorem* of Mahler (1946), which says that this necessary condition on $\mathcal{F}$ is also sufficient.

Proposition 22 *If* $\{\Lambda_k\}$ *is a sequence of lattices in* $\mathbb{R}^n$ *such that*

$$m(\Lambda_k) \geq \rho^2, \ d(\Lambda_k) \leq \sigma \quad \text{for all } k,$$

where ρ, σ *are positive constants, then the sequence* $\{\Lambda_k\}$ *certainly has a convergent subsequence.*

Proof Let V_k denote the Voronoi cell of Λ_k. We show first that the ball $B_{\rho/2} = \{x \in \mathbb{R}^n : \|x\| \leq \rho/2\}$ is contained in every Voronoi cell V_k. In fact if $\|x\| \leq \rho/2$ then, for every nonzero $y \in \Lambda_k$,

$$\|x - y\| \geq \|y\| - \|x\| \geq \rho - \rho/2 = \rho/2 \geq \|x\|,$$

and hence $x \in V_k$.

Let v_k be a point of V_k which is furthest from the origin. Then V_k contains the convex hull C_k of the set $v_k \cup B_{\rho/2}$. Since the volume of V_k is bounded above by σ, so also is the volume of C_k. But this implies that the sequence v_k is bounded. Thus there exists $R > 0$ such that the ball B_R contains every Voronoi cell V_k.

By Blaschke's selection principle, the sequence $\{V_k\}$ has a subsequence $\{V_{k_v}\}$ which converges in the Hausdorff metric to a compact convex set V. Since $B_{\rho/2} \subseteq V$, it follows from Proposition 20 that $\Lambda_{k_v} \to \Lambda$, where Λ is a lattice with Voronoi cell V. $\qquad\square$

To illustrate the utility of Mahler's compactness theorem, we now show that, as stated in Section 3, any compact symmetric convex set K with nonempty interior has a critical lattice.

By the definition of the critical determinant $\Delta(K)$, there exists a sequence Λ_k of lattices with no nonzero points in the interior of K such that $d(\Lambda_k) \to \Delta(K)$ as $k \to \infty$. Since K contains a ball B_ρ with radius $\rho > 0$, we have $m(\Lambda_k) \geq \rho^2$ for all k. Hence, by Proposition 22, there is a subsequence Λ_{k_v} which converges to a lattice Λ as $v \to \infty$. Since every point of Λ is a limit of points of Λ_{k_v}, no nonzero point of Λ lies in the interior of K. Furthermore,

$$d(\Lambda) = \lim_{v \to \infty} d(\Lambda_{k_v}) = \Delta(K),$$

and hence Λ is a critical lattice for K.

7 Further Remarks

The geometry of numbers is treated more extensively in Cassels [11], Erdős *et al.* [22] and Gruber and Lekkerkerker [27]. Minkowski's own account is available in [42].

Numerous references to the earlier literature are given in Keller [34]. Lagarias [36] gives an overview of lattice theory. For a simple proof that the indicator function of a convex set is Riemann integrable, see Szabo [57].

Diophantine approximation is studied in Cassels [12], Koksma [35] and Schmidt [50]. Minkowski's result that the discriminant of an algebraic number field other than $\mathbb{Q}$ has absolute value greater than 1 is proved in Narkiewicz [44], for example.

Minkowski's theorem on successive minima is proved in Bambah *et al.* [3]. For the results of Banaszczyk mentioned in §3, see [4] and [5]. Sharp forms of Siegel's lemma are proved not only in Bombieri and Vaaler [7], but also in Matveev [40]. The result of Gillet and Soulé appeared in [25]. Some interesting results and conjectures concerning the product $\lambda(K)\lambda(K^*)$ are described on pp. 425–427 of Schneider [51].

An algorithm of Lovász, which first appeared in Lenstra, Lenstra and Lovász [38], produces in finitely many steps a basis for a lattice Λ in $\mathbb{R}^n$ which is 'reduced'. Although the first vector of a reduced basis is in general not a minimal vector, it has square-norm at most $2^{n-1} m(\Lambda)$. This suffices for many applications and the algorithm has been used to solve a number of apparently unrelated computational problems, such as factoring polynomials in $\mathbb{Q}[t]$, integer linear programming and simultaneous Diophantine approximation. There is an account of the basis reduction algorithm in Schrijver [52]. The algorithmic geometry of numbers is surveyed in Kannan [33].

Mahler [39] has established an analogue of the geometry of numbers for formal Laurent series with coefficients from an arbitrary field F, the roles of $\mathbb{Z}, \mathbb{Q}$ and $\mathbb{R}$ being taken by $F[t], F(t)$ and $F((t))$. In particular, Eichler [19] has shown that the Riemann–Roch theorem for algebraic functions may be thus derived by geometry of numbers arguments.

There is also a generalization of Minkowski's lattice point theorem to locally compact groups, with Haar measure taking the place of volume; see Chapter 2 (Lemma 1) of Weil [60].

Voronoi *diagrams* and their uses are surveyed in Aurenhammer [1]. Proofs of the basic properties of polytopes referred to in §4 may be found in Brøndsted [9] and Coppel [15]. Planar tilings are studied in detail in Grünbaum and Shephard [28].

Mathematical crystallography is treated in Schwarzenberger [53] and Engel [21]. For the physicist's point of view, see Burckhardt [10], Janssen [32] and Birman [6]. There is much theoretical information, in addition to tables, in [31].

For Bieberbach's theorems, see Vince [59], Charlap [13] and Milnor [41]. Various equivalent forms for the definitions of crystal and crystallographic group are given in Dolbilin *et al.* [17]. It is shown in Charlap [13] that crystallographic groups may be abstractly characterized as groups containing a finitely generated maximal abelian torsion-free subgroup of finite index. (An abelian group is *torsion-free* if only the identity element has finite order.) The fundamental group of a compact flat Riemannian manifold is a torsion-free crystallographic group and all torsion-free crystallographic groups may be obtained in this way. For these connections with differential geometry, see Wolf [61] and Charlap [13].

In more than 4 dimensions the complete enumeration of all crystallographic groups is no longer practicable. However, algorithms for deciding if two crystallographic groups are equivalent in some sense have been developed by Opgenorth *et al.* [45].

An interesting subset of all crystallographic groups consists of those generated by reflections in hyperplanes, since Stiefel (1941/2) showed that they are in 1-1 correspondence with the compact simply-connected semi-simple Lie groups. See the 'Note historique' in Bourbaki [8].

There has recently been considerable interest in tilings of $\mathbb{R}^n$ which, although not lattice tilings, consist of translates of finitely many n-dimensional polytopes. The first example, in $\mathbb{R}^2$, due to Penrose (1974), was explained more algebraically by de Bruijn (1981). A substantial generalization of de Bruijn's construction was given by Katz and Duneau (1986), who showed that many such 'quasiperiodic' tilings may be obtained by a method of cut and projection from ordinary lattices in a higher-dimensional space. The subject gained practical significance with the discovery by Shechtman *et al.* (1984) that the diffraction pattern of an alloy of aluminium and magnesium has icosahedral symmetry, which is impossible for a crystal. Many other 'quasicrystals' have since been found. The papers referred to are reproduced, with others, in Steinhardt and Ostlund [56]. The mathematical theory of quasicrystals is surveyed in Le *et al.* [37].

Skubenko [54] has given an upper bound for Hermite's constant γ_n. Somewhat sharper bounds are known, but they have the same asymptotic behaviour and the proofs are much more complicated. A lower bound for γ_n was obtained with a new method by Ball [2].

For the densest lattices in $\mathbb{R}^n (n \leq 8)$, see Ryshkov and Baranovskii [49]. The enumeration of all root lattices is carried out in Ebeling [18]. (A more general problem is treated in Chap. 3 of Humphreys [30] and in Chap. 6 of Bourbaki [8].) For the Voronoi cells of root lattices, see Chap. 21 of Conway and Sloane [14] and Moody and Patera [43]. For the *Dynkin diagrams* associated with root lattices, see also Reiten [47].

Rajan and Shende [46] characterize root lattices as those lattices for which every facet vector is a minimal vector, but their definition of root lattice is not that adopted here. Their argument shows that if every facet vector of a lattice is a minimal vector then, after scaling to make the minimal vectors have square-norm 2, it is a root lattice in our sense.

There is a fund of information about lattice packings of balls in Conway and Sloane [14]. See also Thompson [58] for the Leech lattice and Coxeter [16] for the kissing number problem.

We have restricted attention to lattice packings and, in particular, to lattice packings of balls. Lattice packings of other convex bodies are discussed in the books on geometry of numbers cited above. Non-lattice packings have also been much studied. The notion of density is not so intuitive in this case and it should be realized that the density is unaltered if finitely many sets are removed from the packing.

Packings and coverings are discussed in the texts of Rogers [48] and Fejes Tóth [23], [24]. For packings of balls, see also Zong [62]. Sloane [55] and Elkies [20] provide introductions to the connections between lattice packings of balls and coding theory.

The third part of Hilbert's 18th problem, which is surveyed in Milnor [41], deals with the densest lattice or non-lattice packing of balls in $\mathbb{R}^n$. It is known that, for $n = 2$, the densest lattice packing is also a densest packing. The original proof by Thue (1882/1910) was incomplete, but a complete proof was given by L. Fejes Tóth (1940). The famous *Kepler conjecture* asserts that, also for $n = 3$, the densest lattice packing is a densest packing. A computer-aided proof has recently been announced by Hales [29]. It is unknown if the same holds for any $n > 3$.

Propositions 20 and 21 are due to Groemer [26], and are of interest quite apart from the application to Mahler's compactness theorem. Other proofs of the latter are given in Cassels [11] and Gruber and Lekkerkerker [27]. Blaschke's selection principle and Rådström's cancellation law are proved in [15] and [51], for example.

8 Selected References

[1] F. Aurenhammer, Voronoi diagrams – a survey of a fundamental geometric data structure, *ACM Computing Surveys* **23** (1991), 345–405.

[2] K. Ball, A lower bound for the optimal density of lattice packings, *Internat. Math. Res. Notices* 1992, no. 10, 217–221.

[3] R.P. Bambah, A.C. Woods and H. Zassenhaus, Three proofs of Minkowski's second inequality in the geometry of numbers, *J. Austral. Math. Soc.* **5** (1965), 453–462.

[4] W. Banaszczyk, New bounds in some transference theorems in the geometry of numbers, *Math. Ann.* **296** (1993), 625–635.

[5] W. Banaszczyk, Inequalities for convex bodies and polar reciprocal lattices in $\mathbb{R}^n$. II. Application of K-convexity, *Discrete Comput. Geom.* **16** (1996), 305–311.

[6] J.L. Birman, *Theory of crystal space groups and lattice dynamics,* Springer-Verlag, Berlin, 1984. [Original edition in *Handbuch der Physik,* 1974]

[7] E. Bombieri and J. Vaaler, On Siegel's lemma, *Invent. Math.* **73** (1983), 11–32.

[8] N. Bourbaki, *Groupes et algèbres de Lie, Chapitres 4,5 et 6,* Masson, Paris, 1981.

[9] A. Brøndsted, *An introduction to convex polytopes,* Springer-Verlag, New York, 1983.

[10] J.J. Burckhardt, *Die Bewegungsgruppen der Kristallographie,* 2nd ed., Birkhäuser, Basel, 1966.

[11] J.W.S. Cassels, *An introduction to the geometry of numbers,* corrected reprint, Springer-Verlag, Berlin, 1997. [Original edition, 1959]

[12] J.W.S. Cassels, *An introduction to Diophantine approximation,* Cambridge University Press, 1957.

[13] L.S. Charlap, *Bieberbach groups and flat manifolds,* Springer-Verlag, New York, 1986.

[14] J.H. Conway and N.J.A. Sloane, *Sphere packings, lattices and groups,* 3rd ed., Springer-Verlag, New York, 1999.

[15] W.A. Coppel, *Foundations of convex geometry,* Cambridge University Press, 1998.

[16] H.S.M. Coxeter, An upper bound for the number of equal nonoverlapping spheres that can touch another of the same size, *Convexity* (ed. V. Klee), pp. 53–71, Proc. Symp. Pure Math. **7**, Amer. Math. Soc., Providence, Rhode Island, 1963.

[17] N.P. Dolbilin, J.C. Lagarias and M. Senechal, Multiregular point systems, *Discrete Comput. Geom.* **20** (1998), 477–498.

[18] W. Ebeling, *Lattices and codes,* Vieweg, Braunschweig, 1994.

[19] M. Eichler, Ein Satz über Linearformen in Polynombereichen, *Arch. Math.* **10** (1959), 81–84.

[20] N.D. Elkies, Lattices, linear codes, and invariants, *Notices Amer. Math. Soc.* **47** (2000), 1238–1245 and 1382–1391.

[21] P. Engel, Geometric crystallography, *Handbook of convex geometry* (ed. P.M. Gruber and J.M. Wills), Volume B, pp. 989–1041, North-Holland, Amsterdam, 1993. (The same volume contains several other useful survey articles relevant to this chapter.)

[22] P. Erdős, P.M. Gruber and J. Hammer, *Lattice points,* Longman, Harlow, Essex, 1989.

[23] L. Fejes Tóth, *Regular Figures,* Pergamon, Oxford, 1964.

[24] L. Fejes Tóth, *Lagerungen in der Ebene auf der Kugel und im Raum*, 2nd ed., Springer-Verlag, Berlin, 1972.

[25] H. Gillet and C. Soulé, On the number of lattice points in convex symmetric bodies and their duals, *Israel J. Math.* **74** (1991), 347–357.

[26] H. Groemer, Continuity properties of Voronoi domains, *Monatsh. Math.* **75** (1971), 423–431.

[27] P.M. Gruber and C.G. Lekkerkerker, *Geometry of numbers*, 2nd ed., North-Holland, Amsterdam, 1987.

[28] B. Grünbaum and G.C. Shephard, *Tilings and patterns*, Freeman, New York, 1987.

[29] T.C. Hales, Cannonballs and honeycombs, *Notices Amer. Math. Soc.* **47** (2000), 440–449.

[30] J.E. Humphreys, *Introduction to Lie algebras and representation theory*, Springer-Verlag, New York, 1972.

[31] *International tables for crystallography, Vols. A-C*, Kluwer, Dordrecht, 1983–1993.

[32] T. Janssen, *Crystallographic groups*, North-Holland, Amsterdam, 1973.

[33] R. Kannan, Algorithmic geometry of numbers, *Annual review of computer science* **2** (1987), 231–267.

[34] O.-H. Keller, *Geometrie der Zahlen*, Enzyklopädie der mathematischen Wissenschaften I-2, 27, Teubner, Leipzig, 1954.

[35] J.F. Koksma, *Diophantische Approximationen*, Springer-Verlag, Berlin, 1936. [Reprinted Chelsea, New York, 1950]

[36] J.C. Lagarias, Point lattices, *Handbook of Combinatorics* (ed. R. Graham, M. Grötschel and L. Lovász), Vol. I, pp. 919–966, Elsevier, Amsterdam, 1995.

[37] T.Q.T. Le, S.A. Piunikhin and V.A. Sadov, The geometry of quasicrystals, *Russian Math. Surveys* **48** (1993), no. 1, 37–100.

[38] A.K. Lenstra, H.W. Lenstra and L. Lovász, Factoring polynomials with rational coefficients, *Math. Ann.* **261** (1982), 515–534.

[39] K. Mahler, An analogue to Minkowski's geometry of numbers in a field of series, *Ann. of Math.* **42** (1941), 488–522.

[40] E.M. Matveev, On linear and multiplicative relations, *Math. USSR-Sb.* **78** (1994), 411–425.

[41] J. Milnor, Hilbert's Problem 18: On crystallographic groups, fundamental domains, and on sphere packing, *Mathematical developments arising from Hilbert problems* (ed. F.E. Browder), pp. 491–506, Proc. Symp. Pure Math. **28**, Part 2, Amer. Math. Soc., Providence, Rhode Island, 1976.

[42] H. Minkowski, *Geometrie der Zahlen*, Teubner, Leipzig, 1896. [Reprinted Chelsea, New York, 1953]

[43] R.V. Moody and J. Patera, Voronoi and Delaunay cells of root lattices: classification of their faces and facets by Coxeter–Dynkin diagrams, *J. Phys. A* **25** (1992), 5089–5134.

[44] W. Narkiewicz, *Elementary and analytic theory of algebraic numbers*, 2nd ed., Springer-Verlag, Berlin, 1990.

[45] J. Opgenorth, W. Plesken and T. Schulz, Crystallographic algorithms and tables, *Acta Cryst. A* **54** (1998), 517–531.

[46] D.S. Rajan and A.M. Shende, A characterization of root lattices, *Discrete Math.* **161** (1996), 309–314.

[47] I. Reiten, Dynkin diagrams and the representation theory of Lie algebras, *Notices Amer. Math. Soc.* **44** (1997), 546–556.

[48] C.A. Rogers, *Packing and covering*, Cambridge University Press, 1964.

[49] S.S. Ryshkov and E.P. Baranovskii, Classical methods in the theory of lattice packings, *Russian Math. Surveys* **34** (1979), no. 4, 1–68.

[50] W.M. Schmidt, *Diophantine approximation*, Lecture Notes in Mathematics **785**, Springer-Verlag, Berlin, 1980.

[51] R. Schneider, *Convex bodies: the Brunn–Minkowski theory*, Cambridge University Press, 1993.

[52] A. Schrijver, *Theory of linear and integer programming*, corrected reprint, Wiley, Chichester, 1989.

[53] R.L.E. Schwarzenberger, *N-dimensional crystallography*, Pitman, London, 1980.

[54] B.F. Skubenko, A remark on an upper bound on the Hermite constant for the densest lattice packings of spheres, *J. Soviet Math.* **18** (1982), 960–961.

[55] N.J.A. Sloane, The packing of spheres, *Scientific American* **250** (1984), 92–101.

[56] P.J. Steinhardt and S. Ostlund (ed.), *The physics of quasicrystals*, World Scientific, Singapore, 1987.

[57] L. Szabo, A simple proof for the Jordan measurability of convex sets, *Elem. Math.* **52** (1997), 84–86.

[58] T.M. Thompson, *From error-correcting codes through sphere packings to simple groups*, Carus Mathematical Monograph No. 21, Mathematical Association of America, 1983.

[59] A. Vince, Periodicity, quasiperiodicity and Bieberbach's theorem, *Amer. Math. Monthly* **104** (1997), 27–35.

[60] A. Weil, *Basic number theory*, 2nd ed., Springer-Verlag, Berlin, 1973.

[61] J.A. Wolf, *Spaces of constant curvature*, 3rd ed., Publish or Perish, Boston, Mass., 1974.

[62] C. Zong, *Sphere packings*, Springer-Verlag, New York, 1999.

Additional References

F. Pfender and G. Ziegler, Kissing numbers, sphere packings and some unexpected proofs, *Notices Amer. Math. Soc.* **51** (2004), 873–883. [The Leech lattice is indeed the densest lattice in $\mathbb{R}^{24}$.]

O.R. Musin, The problem of the twenty-five spheres, *Russian Math. Surveys* **58** (2003), 794–795. [The kissing number of $\mathbb{R}^4$ is 24.]

G. Muraz and J.-L. Verger-Gaugry, On a generalization of the selection theorem of Mahler, *Journal de Théorie des Nombres de Bordeaux* **17** (2005), 237–269. [Extends Mahler's compactness theorem for lattices to sets which are uniformly discrete and uniformly relatively dense.]

The Number of Prime Numbers

1 Finding the Problem

It was already shown in Euclid's *Elements* (Book IX, Proposition 20) that there are infinitely many prime numbers. The proof is a model of simplicity: let $p_1, \ldots, p_n$ be any finite set of primes and consider the integer $N = p_1 \cdots p_n + 1$. Then $N > 1$ and each prime divisor p of N is distinct from $p_1, \ldots, p_n$, since $p = p_j$ would imply that p divides $N - p_1 \cdots p_n = 1$. It is worth noting that the same argument applies if we take $N = p_1^{\alpha_1} \cdots p_n^{\alpha_n} + 1$, with any positive integers $\alpha_1, \ldots, \alpha_n$.

Euler (1737) gave an analytic proof of Euclid's result, which provides also quantitative information about the distribution of primes:

Proposition 1 *The series* $\sum_p 1/p$, *where* p *runs through all primes, is divergent.*

Proof For any prime p we have

$$(1 - 1/p)^{-1} = 1 + p^{-1} + p^{-2} + \cdots$$

and hence

$$\prod_{p \leq x}(1 - 1/p)^{-1} = \prod_{p \leq x}(1 + p^{-1} + p^{-2} + \cdots) > \sum_{n \leq x} 1/n,$$

since any positive integer $n \leq x$ is a product of powers of primes $p \leq x$. Since

$$\sum_{n \leq x} 1/n > \sum_{n \leq x} \int_n^{n+1} dt/t > \log x,$$

it follows that

$$\prod_{p \leq x}(1 - 1/p)^{-1} > \log x.$$

On the other hand, since the representation of any positive integer as a product of prime powers is *unique*,

$$\prod_{p \leq x}(1 - 1/p^2)^{-1} = \prod_{p \leq x}(1 + p^{-2} + p^{-4} + \cdots) \leq \sum_{n=1}^{\infty} 1/n^2 =: S,$$

W.A. Coppel, *Number Theory: An Introduction to Mathematics*, Universitext,
DOI: 10.1007/978-0-387-89486-7_9, © Springer Science + Business Media, LLC 2009

and

$$S = 1 + \sum_{n=1}^{\infty} 1/(n+1)^2 < 1 + \sum_{n=1}^{\infty} \int_n^{n+1} dt/t^2 = 1 + \int_1^{\infty} dt/t^2 = 2.$$

(In fact $S = \pi^2/6$, as Euler (1735) also showed.) Since $1 - 1/p^2 = (1-1/p)(1+1/p)$, and since $1 + x \le e^x$, it follows that

$$\prod_{p \le x} (1 - 1/p)^{-1} \le S \prod_{p \le x} (1 + 1/p) < S \, e^{\sum_{p \le x} 1/p}.$$

Combining this with the inequality of the previous paragraph, we obtain

$$\sum_{p \le x} 1/p > \log\log x - \log S. \qquad \qquad \square$$

Since the series $\sum_{n=1}^{\infty} 1/n^2$ is convergent, Proposition 1 says that 'there are more primes than squares'. Proposition 1 can be made more precise. It was shown by Mertens (1874) that

$$\sum_{p \le x} 1/p = \log\log x + c + O(1/\log x),$$

where c is a constant ($c = 0.261497\ldots$).

Let $\pi(x)$ denote the number of primes $\le x$:

$$\pi(x) = \sum_{p \le x} 1.$$

It may be asked whether $\pi(x)$ has some simple asymptotic behaviour as $x \to \infty$. It is not obvious that this is a sensible question. The behaviour of $\pi(x)$ for small values of x is quite irregular. Moreover the sequence of positive integers contains arbitrarily large blocks without primes; for example, none of the integers

$$n! + 2, n! + 3, \ldots, n! + n$$

is a prime. Indeed Euler (1751) expressed the view that "there reigns neither order nor rule" in the sequence of prime numbers.

From an analysis of tables of primes Legendre (1798) was led to conjecture that, for large values of x, $\pi(x)$ is given approximately by the formula

$$x/(A \log x - B),$$

where A, B are constants and $\log x$ again denotes the natural logarithm of x (i.e., to the base e). In 1808 he proposed the specific values $A = 1$, $B = 1.08366$.

The first significant results on the asymptotic behaviour of $\pi(x)$ were obtained by Chebyshev (1849). He proved that, for each positive integer n,

$$\varliminf_{x \to \infty} \left(\pi(x) - \int_2^x dt/\log t \right) \log^n x / x \le 0$$

$$\le \varlimsup_{x \to \infty} \left(\pi(x) - \int_2^x dt/\log t \right) \log^n x / x,$$

where $\log^n x = (\log x)^n$. By repeatedly integrating by parts it may be seen that, for each positive integer n,

$$\int_2^x dt/\log t = \{1 + 1!/\log x + 2!/\log^2 x + \cdots + (n-1)!/\log^{n-1} x\}x/\log x$$

$$+ n! \int_2^x dt/\log^{n+1} t + c_n,$$

where c_n is a constant. Moreover, using the *Landau order symbol* defined under 'Notations',

$$\int_2^x dt/\log^{n+1} t = O(x/\log^{n+1} x),$$

since

$$\int_2^{x^{1/2}} dt/\log^{n+1} t < x^{1/2}/\log^{n+1} 2, \qquad \int_{x^{1/2}}^x dt/\log^{n+1} t < 2^{n+1} x/\log^{n+1} x.$$

Thus Chebyshev's result shows that $A = B = 1$ are the best possible values for a formula of Legendre's type and suggests that

$$Li(x) = \int_2^x dt/\log t$$

is a better approximation to $\pi(x)$.

If we interpret this approximation as an asymptotic formula, then it implies that $\pi(x) \log x/x \to 1$ as $x \to \infty$, i.e., using another *Landau order symbol*,

$$\pi(x) \sim x/\log x. \tag{1}$$

The validity of the relation (1) is now known as the *prime number theorem*. If the n-th prime is denoted by p_n, then the prime number theorem can also be stated in the form $p_n \sim n \log n$:

Proposition 2 $\pi(x) \sim x/\log x$ *if and only if* $p_n \sim n \log n$.

Proof If $\pi(x) \log x/x \to 1$, then

$$\log \pi(x) + \log \log x - \log x \to 0$$

and hence

$$\log \pi(x)/\log x \to 1.$$

Consequently

$$\pi(x) \log \pi(x)/x = \pi(x) \log x/x \cdot \log \pi(x)/\log x \to 1.$$

Since $\pi(p_n) = n$, this shows that $p_n \sim n \log n$.

Conversely, suppose $p_n/n \log n \to 1$. Since

$$(n + 1) \log(n + 1)/n \log n = (1 + 1/n)\{1 + \log(1 + 1/n)/\log n\} \to 1,$$

it follows that $p_{n+1}/p_n \to 1$. Furthermore

$$\log p_n - \log n - \log \log n \to 0,$$

and hence

$$\log p_n/\log n \to 1.$$

If $p_n \le x < p_{n+1}$, then $\pi(x) = n$ and

$$n \log p_n/p_{n+1} \le \pi(x) \log x/x \le n \log p_{n+1}/p_n.$$

Since

$$n \log p_n/p_{n+1} = p_n/p_{n+1} \cdot n \log n/p_n \cdot \log p_n/\log n \to 1$$

and similarly $n \log p_{n+1}/p_n \to 1$, it follows that also $\pi(x) \log x/x \to 1$.   □

Numerical evidence, both for the prime number theorem and for the fact that $Li(x)$ is a better approximation than $x/\log x$ to $\pi(x)$, is provided by Table 1.

In a second paper Chebyshev (1852) made some progress towards proving the prime number theorem by showing that

$$a \le \varliminf_{x \to \infty} \pi(x) \log x/x \le \varlimsup_{x \to \infty} \pi(x) \log x/x \le 6a/5,$$

where $a = 0.92129$. He used his results to give the first proof of *Bertrand's postulate*: for every real $x > 1$, there is a prime between x and $2x$.

New ideas were introduced by Riemann (1859), who linked the asymptotic behaviour of $\pi(x)$ with the behaviour of the function

$$\zeta(s) = \sum_{n=1}^{\infty} 1/n^s$$

Table 1.

x	$\pi(x)$	$x/\log x$	$Li(x)$	$\pi(x) \log x/x$	$\pi(x)/Li(x)$
10^3	168	144.	177.	1.16	0.94
10^4	1 229	1 085.	1 245.	1.132	0.987
10^5	9 592	8 685.	9 629.	1.1043	0.9961
10^6	78 498	72 382.	78 627.	1.08449	0.99835
10^7	664 579	620 420.	664 917.	1.07117	0.99949
10^8	5 761 455	5 428 681.	5 762 208.	1.06130	0.99987
10^9	50 847 534	48 254 942.	50 849 234.	1.05373	0.999966
10^{10}	455 052 511	434 294 481.	455 055 614.	1.04780	0.999993

for complex values of s. By developing these ideas, and by showing especially that $\zeta(s)$ has no zeros on the line $\mathcal{R}s = 1$, Hadamard and de la Vallée Poussin proved the prime number theorem (independently) in 1896. Shortly afterwards de la Vallée Poussin (1899) confirmed that $Li(x)$ was a better approximation than $x/\log x$ to $\pi(x)$ by proving (in particular) that

$$\pi(x) = Li(x) + O(x/\log^\alpha x) \quad \text{for every } \alpha > 0. \tag{2}$$

Better error bounds than de la Vallée Poussin's have since been obtained, but they still fall far short of what is believed to be true.

Another approach to the prime number theorem was found by Wiener (1927–1933), as an application of his general theory of Tauberian theorems. A convenient form for this application was given by Ikehara (1931), and Bochner (1933) showed that in this case Wiener's general theory could be avoided.

It came as a great surprise to the mathematical community when in 1949 Selberg, assisted by Erdős, found a new proof of the prime number theorem which uses only the simplest facts of real analysis. Though elementary in a technical sense, this proof was still quite complicated. As a result of several subsequent simplifications it can now be given quite a clear and simple form. Nevertheless the Wiener–Ikehara proof will be presented here on account of its greater versatility. The error bound (2) can be obtained by both the Wiener and Selberg approaches, in the latter case at the cost of considerable complication.

2 Chebyshev's Functions

In his second paper Chebyshev introduced two functions

$$\theta(x) = \sum_{p \le x} \log p, \quad \psi(x) = \sum_{p^\alpha \le x} \log p,$$

which have since played a major role. Although $\psi(x)$ has the most complicated definition, it is easier to treat analytically than either $\theta(x)$ or $\pi(x)$. As we will show, the asymptotic behaviour of $\theta(x)$ is essentially the same as that of $\psi(x)$, and the asymptotic behaviour of $\pi(x)$ may be deduced without difficulty from that of $\theta(x)$.

Evidently

$$\theta(x) = \psi(x) = 0 \quad \text{for } x < 2$$

and

$$0 < \theta(x) \le \psi(x) \quad \text{for } x \ge 2.$$

Lemma 3 *The asymptotic behaviours of $\psi(x)$ and $\theta(x)$ are connected by*

(i) $\psi(x) - \theta(x) = O(x^{1/2} \log^2 x)$;
(ii) $\psi(x) = O(x)$ *if and only if* $\theta(x) = O(x)$, *and in this case* $\psi(x) - \theta(x) = O(x^{1/2} \log x)$.

Proof Since

$$\psi(x) = \sum_{p \le x} \log p + \sum_{p^2 \le x} \log p + \cdots$$

and $k > \log x / \log 2$ implies $x^{1/k} < 2$, we have

$$\psi(x) = \theta(x) + \theta(x^{1/2}) + \cdots + \theta(x^{1/m}),$$

where $m = \lfloor \log x / \log 2 \rfloor$. (As is now usual, we denote by $\lfloor y \rfloor$ the greatest integer $\le y$.) But it is obvious from the definition of $\theta(x)$ that $\theta(x) = O(x \log x)$. Hence

$$\psi(x) - \theta(x) = O\left(\sum_{2 \le k \le m} x^{1/k} \log x \right) = O(x^{1/2} \log^2 x).$$

If $\theta(x) = O(x)$ the same argument yields $\psi(x) - \theta(x) = O(x^{1/2} \log x)$ and thus $\psi(x) = O(x)$. It is trivial that $\psi(x) = O(x)$ implies $\theta(x) = O(x)$.   □

The proof of Lemma 3 shows also that

$$\psi(x) = \theta(x) + \theta(x^{1/2}) + O(x^{1/3} \log^2 x).$$

Lemma 4 $\psi(x) = O(x)$ *if and only if* $\pi(x) = O(x / \log x)$, *and then*

$$\pi(x) \log x / x = \psi(x)/x + O(1/ \log x).$$

Proof Although their use can easily be avoided, it is more suggestive to use Stieltjes integrals. Suppose first that $\psi(x) = O(x)$. For any $x > 2$ we have

$$\pi(x) = \int_{2-}^{x+} 1/ \log t \, d\theta(t)$$

and hence, on integrating by parts,

$$\pi(x) = \theta(x)/ \log x + \int_{2}^{x} \theta(t)/t \log^2 t \, dt.$$

But

$$\int_{2}^{x} \theta(t)/t \log^2 t \, dt = O(x / \log^2 x),$$

since $\theta(t) = O(t)$ and, as we saw in §1,

$$\int_{2}^{x} dt / \log^2 t = O(x / \log^2 x).$$

Since

$$\theta(x)/ \log x = \psi(x)/ \log x + O(x^{1/2}),$$

by Lemma 3, it follows that

$$\pi(x) = \psi(x)/\log x + O(x/\log^2 x).$$

Suppose next that $\pi(x) = O(x/\log x)$. For any $x > 2$ we have

$$\theta(x) = \int_{2-}^{x+} \log t \, d\pi(t)$$

$$= \pi(x) \log x - \int_2^x \pi(t)/t \, dt = O(x),$$

and hence also $\psi(x) = O(x)$, by Lemma 3. $\qquad\square$

It follows at once from Lemma 4 that *the prime number theorem*, $\pi(x) \sim x/\log x$, *is equivalent to* $\psi(x) \sim x$.

The method of argument used in Lemma 4 can be carried further. Put

$$\theta(x) = x + R(x), \quad \pi(x) = \int_2^x dt/\log t + Q(x).$$

Subtracting

$$\int_2^x dt/\log t = x/\log x - 2/\log 2 + \int_2^x dt/\log^2 t$$

from

$$\pi(x) = \theta(x)/\log x + \int_2^x \theta(t)/t \log^2 t \, dt,$$

we obtain

$$Q(x) = R(x)/\log x + \int_2^x R(t)/t \log^2 t \, dt + 2/\log 2. \tag{3}_1$$

Also, adding

$$\int_2^x \left(\int_2^t du/\log u \right) dt/t = \int_2^x \left(\int_u^x dt/t \right) du/\log u$$

$$= \int_2^x (\log x - \log u) du/\log u$$

$$= \log x \int_2^x dt/\log t - x + 2$$

to

$$\theta(x) = \pi(x) \log x - \int_2^x \pi(t)/t \, dt$$

we obtain

$$R(x) = Q(x) \log x - \int_2^x Q(t)/t \, dt - 2. \tag{3}_2$$

It follows from $(3)_1$–$(3)_2$ that $R(x) = O(x/\log^\alpha x)$ for some $\alpha > 0$ if and only if $Q(x) = O(x/\log^{\alpha+1} x)$. Consequently, by Lemma 3,

$$\psi(x) = x + O(x/\log^\alpha x) \quad \text{for every } \alpha > 0$$

if and only if

$$\pi(x) = \int_2^x dt/\log t + O(x/\log^\alpha x) \quad \text{for every } \alpha > 0,$$

and $\pi(x)$ then has the asymptotic expansion

$$\pi(x) \sim \{1 + 1!/\log x + 2!/\log^2 x + \cdots\}x/\log x,$$

the error in breaking off the series after any finite number of terms having the order of magnitude of the first term omitted.

It follows from $(3)_1$–$(3)_2$ also that, *for a given α such that $1/2 \le \alpha < 1$,*

$$\psi(x) = x + O(x^\alpha \log^2 x),$$

if and only if

$$\pi(x) = \int_2^x dt/\log t + O(x^\alpha \log x).$$

The definition of $\psi(x)$ can be put in the form

$$\psi(x) = \sum_{n \le x} \Lambda(n),$$

where the *von Mangoldt function* $\Lambda(n)$ is defined by

$$\Lambda(n) = \log p \text{ if } n = p^\alpha \text{ for some prime } p \text{ and some } \alpha > 0,$$
$$= 0 \text{ otherwise.}$$

For any positive integer n we have

$$\log n = \sum_{d \mid n} \Lambda(d), \qquad (4)$$

since if $n = p_1^{\alpha_1} \cdots p_s^{\alpha_s}$ is the factorization of n into powers of distinct primes, then

$$\log n = \sum_{j=1}^{s} \alpha_j \log p_j.$$

3 Proof of the Prime Number Theorem

The *Riemann zeta-function* is defined by

$$\zeta(s) = \sum_{n=1}^{\infty} 1/n^s. \qquad (5)$$

This infinite series had already been considered by Euler, Dirichlet and Chebyshev, but Riemann was the first to study it for complex values of s. As customary, we write $s = \sigma + it$, where σ and t are real, and n^{-s} is defined for complex values of s by

$$n^{-s} = e^{-s \log n} = n^{-\sigma} (\cos(t \log n) - i \sin(t \log n)).$$

To show that the series (5) converges in the half-plane $\sigma > 1$ we compare as in §1 the sum with an integral. If $\lfloor x \rfloor$ denotes again the greatest integer $\leq x$, then on integrating by parts we obtain

$$\int_1^N x^{-s} dx - \sum_{n=1}^N n^{-s} = \int_{1-}^{N+} x^{-s} d\{x - \lfloor x \rfloor\}$$

$$= -1 + s \int_1^N x^{-s-1} \{x - \lfloor x \rfloor\} dx.$$

Since

$$\int_1^N x^{-s} dx = (1 - N^{1-s})/(s - 1),$$

by letting $N \to \infty$ we see that $\zeta(s)$ is defined for $\sigma > 1$ and

$$\zeta(s) = 1/(s - 1) + 1 - s \int_1^\infty x^{-s-1} \{x - \lfloor x \rfloor\} dx.$$

But, since $x - \lfloor x \rfloor$ is bounded, the integral on the right is uniformly convergent in any half-plane $\sigma \geq \delta > 0$. It follows that the definition of $\zeta(s)$ can be extended to the half-plane $\sigma > 0$, so that it is holomorphic there except for a simple pole with residue 1 at $s = 1$.

The connection between the zeta-function and prime numbers is provided by *Euler's product formula*, which may be viewed as an analytic version of the fundamental theorem of arithmetic:

Proposition 5 $\zeta(s) = \prod_p (1 - p^{-s})^{-1}$ *for* $\sigma > 1$, *where the product is taken over all primes p.*

Proof For $\sigma > 0$ we have

$$(1 - p^{-s})^{-1} = 1 + p^{-s} + p^{-2s} + \cdots .$$

Since each positive integer can be uniquely expressed as a product of powers of distinct primes, it follows that

$$\prod_{p \leq x} (1 - p^{-s})^{-1} = \sum_{n \leq N_x} n^{-s},$$

where N_x is the set of all positive integers, including 1, whose prime factors are all $\leq x$. But N_x contains all positive integers $\leq x$. Hence

$$\left| \zeta(s) - \prod_{p \leq x} (1 - p^{-s})^{-1} \right| \leq \sum_{n > x} n^{-\sigma} \quad \text{for } \sigma > 1,$$

and the sum on the right tends to zero as $x \to \infty$. □

It follows at once from Proposition 5 that $\zeta(s) \neq 0$ for $\sigma > 1$, since the infinite product is convergent and each factor is nonzero.

Proposition 6 $-\zeta'(s)/\zeta(s) = \sum_{n=1}^{\infty} \Lambda(n)/n^s$ *for* $\sigma > 1$, *where* $\Lambda(n)$ *denotes von Mangoldt's function.*

Proof The series $\omega(s) = \sum_{n=1}^{\infty} \Lambda(n)n^{-s}$ converges absolutely and uniformly in any half-plane $\sigma \geq 1 + \varepsilon$, where $\varepsilon > 0$, since

$$0 \leq \Lambda(n) \leq \log n < n^{\varepsilon/2} \quad \text{for all large } n.$$

Hence

$$\zeta(s)\omega(s) = \sum_{m=1}^{\infty} m^{-s} \sum_{k=1}^{\infty} \Lambda(k)k^{-s}$$

$$= \sum_{n=1}^{\infty} n^{-s} \sum_{d|n} \Lambda(d).$$

Since $\sum_{d|n} \Lambda(d) = \log n$, by (4), it follows that

$$\zeta(s)\omega(s) = \sum_{n=1}^{\infty} n^{-s} \log n = -\zeta'(s).$$

Since $\zeta(s) \neq 0$ for $\sigma > 1$, the result follows. However, we can also prove directly that $\zeta(s) \neq 0$ for $\sigma > 1$, and thus make the proof of the prime number theorem independent of Proposition 5.

Obviously if $\zeta(s_0) = 0$ for some s_0 with $\mathcal{R}s_0 > 1$ then $\zeta'(s_0) = 0$, and it follows by induction from Leibniz' formula for derivatives of a product that $\zeta^{(n)}(s_0) = 0$ for all $n \geq 0$. Since $\zeta(s)$ is holomorphic for $\sigma > 1$ and not identically zero, this is a contradiction. $\qquad\square$

Proposition 6 may be restated in terms of Chebyshev's ψ-function:

$$-\zeta'(s)/\zeta(s) = \int_1^{\infty} u^{-s} d\psi(u) = \int_0^{\infty} e^{-sx} d\psi(e^x) \quad \text{for } \sigma > 1. \qquad (6)$$

We are going to deduce from (6) that the function $\zeta(s)$ has no zeros on the line $\mathcal{R}s = 1$. Actually we will prove a more general result:

Proposition 7 *Let* $f(s)$ *be holomorphic in the closed half-plane* $\mathcal{R}s \geq 1$, *except for a simple pole at* $s = 1$. *If, for* $\mathcal{R}s > 1$, $f(s) \neq 0$ *and*

$$-f'(s)/f(s) = \int_0^{\infty} e^{-sx} d\phi(x),$$

where $\phi(x)$ *is a nondecreasing function for* $x \geq 0$, *then*

$$f(1 + it) \neq 0 \quad \text{for every real } t \neq 0.$$

Proof Put $s = \sigma + it$, where σ and t are real, and let

$$g(\sigma, t) = -\mathscr{R}\{f'(s)/f(s)\}.$$

Thus

$$g(\sigma, t) = \int_0^\infty e^{-\sigma x} \cos(tx)\, d\phi(x) \quad \text{for } \sigma > 1.$$

Hence, by Schwarz's inequality (Chapter I, §10),

$$g(\sigma, t)^2 \leq \int_0^\infty e^{-\sigma x}\, d\phi(x) \int_0^\infty e^{-\sigma x} \cos^2(tx)\, d\phi(x)$$

$$= g(\sigma, 0) \int_0^\infty e^{-\sigma x}\{1 + \cos(2tx)\}\, d\phi(x)/2$$

$$= g(\sigma, 0)\{g(\sigma, 0) + g(\sigma, 2t)\}/2.$$

Since $f(s)$ has a simple pole at $s = 1$, by comparing the Laurent series of $f(s)$ and $f'(s)$ at $s = 1$ (see Chapter I, §5) we see that

$$(\sigma - 1)g(\sigma, 0) \to 1 \quad \text{as } \sigma \to 1+ .$$

Similarly if $f(s)$ has a zero of multiplicity $m(t) \geq 0$ at $1 + it$, where $t \neq 0$, then by comparing the Taylor series of $f(s)$ and $f'(s)$ at $s = 1 + it$ we see that

$$(\sigma - 1)g(\sigma, t) \to -m(t) \quad \text{as } \sigma \to 1+ .$$

Thus if we multiply the inequality for $g(\sigma, t)^2$ by $(\sigma - 1)^2$ and let $\sigma \to 1+$, we obtain

$$m(t)^2 \leq \{1 - m(2t)\}/2 \leq 1/2.$$

Therefore, since $m(t)$ is an integer, $m(t) = 0$. $\qquad\qquad\square$

For $f(s) = \zeta(s)$, Proposition 7 gives the result of Hadamard and de la Vallée Poussin:

Corollary 8 $\zeta(1 + it) \neq 0$ *for every real $t \neq 0$.*

The use of Schwarz's inequality to prove Corollary 8 seems more natural than the usual proof by means of the inequality $3 + 4\cos\theta + \cos 2\theta \geq 0$. It follows from Corollary 8 that $-\zeta'(s)/\zeta(s) - 1/(s - 1)$ is holomorphic in the closed half-plane $\sigma \geq 1$. Hence, by (6), the hypotheses of the following theorem, due to Ikehara (1931), are satisfied with

$$F(s) = -\zeta'(s)/\zeta(s), \quad \phi(x) = \psi(e^x), \quad h = A = 1.$$

Theorem 9 *Let $\phi(x)$ be a nondecreasing function for $x \geq 0$ such that the Laplace transform*

$$F(s) = \int_0^\infty e^{-sx}\, d\phi(x)$$

is defined for $\mathscr{R}s > h$, where $h > 0$. If there exists a constant A and a function $G(s)$, which is continuous in the closed half-plane $\mathscr{R}s \geq h$, such that

$$G(s) = F(s) - Ah/(s-h) \quad for \; \mathscr{R}s > h,$$

then

$$\phi(x) \sim Ae^{hx} \quad for \; x \to +\infty.$$

Proof For each $X > 0$ we have

$$\int_0^X e^{-sx} d\phi(x) = e^{-sX}\{\phi(X) - \phi(0)\} + s\int_0^X e^{-sx}\{\phi(x) - \phi(0)\}\, dx.$$

For real $s = \rho > h$ both terms on the right are nonnegative and the integral on the left has a finite limit as $X \to \infty$. Hence $e^{-\rho X}\phi(X)$ is a bounded function of X for each $\rho > h$. It follows that if $\mathscr{R}s > h$ we can let $X \to \infty$ in the last displayed equation, obtaining

$$F(s) = s\int_0^\infty e^{-sx}\{\phi(x) - \phi(0)\}\, dx \quad for \; \mathscr{R}s > h.$$

Hence

$$[G(s) - A]/s = F(s)/s - A/(s-h) = \int_0^\infty e^{-(s-h)x}\{a(x) - A\}\, dx,$$

where $a(x) = e^{-hx}\{\phi(x) - \phi(0)\}$. Thus we will prove the theorem if we prove the following statement:

Let $a(x)$ be a nonnegative function for $x \geq 0$ such that

$$g(s) = \int_0^\infty e^{-sx}\{a(x) - A\}\, dx,$$

where $s = \sigma + it$, is defined for every $\sigma > 0$ and the limit

$$\gamma(t) = \lim_{\sigma \to +0} g(s)$$

exists uniformly on any finite interval $-T \leq t \leq T$. If, for some $h > 0$, $e^{hx}a(x)$ is a nondecreasing function, then

$$\lim_{x \to \infty} a(x) = A.$$

In the proof of this statement we will use the fact that the Fourier transform

$$\hat{k}(u) = \int_{-\infty}^\infty e^{iut} k(t)\, dt$$

of the function

$$k(t) = 1 - |t| \text{ for } |t| \le 1, \ = 0 \text{ for } |t| \ge 1,$$

has the properties

$$\hat{k}(u) \ge 0 \quad \text{for} - \infty < u < \infty, \quad C := \int_{-\infty}^{\infty} \hat{k}(u) du < \infty.$$

Indeed

$$\hat{k}(u) = \int_{-1}^{1} e^{iut} (1 - |t|) \, dt$$

$$= 2 \int_{0}^{1} (1 - t) \cos ut \, dt$$

$$= 2(1 - \cos u)/u^2.$$

Let ε, λ, y be arbitrary positive numbers. If $s = \varepsilon + i\lambda t$, then

$$\lambda \int_{-1}^{1} e^{i\lambda ty} k(t) g(s) \, dt = \lambda \int_{-1}^{1} e^{i\lambda ty} k(t) \int_{0}^{\infty} e^{-\varepsilon x} e^{-i\lambda tx} \{a(x) - A\} \, dx dt$$

$$= \lambda \int_{0}^{\infty} e^{-\varepsilon x} \{a(x) - A\} \int_{-1}^{1} e^{i\lambda t(y-x)} k(t) \, dt dx$$

$$= \lambda \int_{0}^{\infty} e^{-\varepsilon x} a(x) \hat{k}(\lambda(y - x)) \, dx$$

$$- \lambda A \int_{0}^{\infty} e^{-\varepsilon x} \hat{k}(\lambda(y - x)) \, dx.$$

When $\varepsilon \to +0$ the left side has the limit

$$\chi(y) := \lambda \int_{-1}^{1} e^{i\lambda ty} k(t) \gamma(\lambda t) \, dt$$

and the second term on the right has the limit

$$\lambda A \int_{0}^{\infty} \hat{k}(\lambda(y - x)) \, dx.$$

Consequently the first term on the right also has a finite limit. It follows that

$$\lambda \int_{0}^{\infty} a(x) \hat{k}(\lambda(y - x)) \, dx$$

is finite and is the limit of the first term on the right. Thus

$$\chi(y) = \lambda \int_{0}^{\infty} \{a(x) - A\} \hat{k}(\lambda(y - x)) \, dx$$

$$= \int_{-\infty}^{\lambda y} \{a(y - v/\lambda) - A\} \hat{k}(v) \, dv.$$

By the 'Riemann–Lebesgue lemma', $\chi(y) \to 0$ as $y \to \infty$. In fact this may be proved in the following way. We have

$$\chi(y) = \int_{-\infty}^{\infty} e^{i\lambda t y} \omega(t)\, dt$$

where

$$\omega(t) = \lambda k(t) \gamma(\lambda t).$$

Changing the variable of integration to $t + \pi/\lambda y$, we obtain

$$\chi(y) = -\int_{-\infty}^{\infty} e^{i\lambda t y} \omega(t + \pi/\lambda y)\, dt.$$

Hence

$$2\chi(y) = \int_{-\infty}^{\infty} e^{i\lambda t y} \{\omega(t) - \omega(t + \pi/\lambda y)\}\, dt$$

and

$$2|\chi(y)| \le \int_{-\infty}^{\infty} |\omega(t) - \omega(t + \pi/\lambda y)|\, dt.$$

Since $\omega(t)$ is continuous and vanishes outside a finite interval, it follows that $\chi(y) \to 0$ as $y \to \infty$.

Since

$$\int_{-\infty}^{\lambda y} \hat{k}(v)\, dv \to C \quad \text{as } y \to \infty,$$

we deduce that

$$\lim_{y \to \infty} = \int_{-\infty}^{\lambda y} \alpha(y - v/\lambda)\, \hat{k}(v)\, dv = AC \quad \text{for every } \lambda > 0.$$

We now make use of the fact that $e^{hx}\alpha(x)$ is a nondecreasing function. Choose any $\delta \in (0, 1)$. If $y = x + \delta$, where $x \ge 0$, then for $|v| \le \lambda\delta$

$$\alpha(y - v/\lambda) \ge e^{-h(\delta - v/\lambda)}\alpha(x) \ge e^{-2h\delta}\alpha(x)$$

and hence

$$\int_{-\infty}^{\lambda y} \alpha(y - v/\lambda)\hat{k}(v)\, dv \ge e^{-2h\delta}\alpha(x) \int_{-\lambda\delta}^{\lambda\delta} \hat{k}(v)\, dv.$$

We can choose $\lambda = \lambda(\delta)$ so large that the integral on the right exceeds $(1 - \delta)C$. Then, letting $x \to \infty$ we obtain

$$AC \ge e^{-2h\delta}(1 - \delta)C \varlimsup_{x \to \infty} \alpha(x).$$

Since this holds for arbitrarily small $\delta > 0$, it follows that

$$\varlimsup_{x \to \infty} \alpha(x) \le A.$$

Thus there exists a positive constant M such that

$$0 \le \alpha(x) \le M \quad \text{for all } x \ge 0.$$

On the other hand, if $y = x - \delta$, where $x \ge \delta$, then for $|v| \le \lambda\delta$

$$\alpha(y - v/\lambda) \le e^{h(\delta+v/\lambda)}\alpha(x) \le e^{2h\delta}\alpha(x)$$

and hence

$$\int_{-\infty}^{\lambda y} \alpha(y - v/\lambda)\hat{k}(v)\,dv \le e^{2h\delta}\alpha(x) \int_{-\lambda\delta}^{\lambda\delta} \hat{k}(v)\,dv + M \int_{|v|\ge\lambda\delta} \hat{k}(v)\,dv.$$

We can choose $\lambda = \lambda(\delta)$ so large that the second term on the right is less than δC. Then, letting $x \to \infty$ we obtain

$$AC \le e^{2h\delta}C \varlimsup_{x \to \infty} \alpha(x) + \delta C.$$

Since this holds for arbitrarily small $\delta > 0$, it follows that

$$A \le \varlimsup_{x \to \infty} \alpha(x).$$

Combining this with the inequality of the previous paragraph, we conclude that $\lim_{x \to \infty} \alpha(x) = A$. $\qquad\square$

Applying Theorem 9 to the special case mentioned before the statement of the theorem, we obtain $\psi(e^x) \sim e^x$. As we have already seen in §2, this is equivalent to the prime number theorem.

4 The Riemann Hypothesis

In his celebrated paper on the distribution of prime numbers Riemann (1859) proved only two results. He showed that the definition of $\zeta(s)$ can be extended to the whole complex plane, so that $\zeta(s) - 1/(s - 1)$ is everywhere holomorphic, and he proved that the values of $\zeta(s)$ and $\zeta(1 - s)$ are connected by a certain functional equation. This functional equation will now be derived by one of the two methods which Riemann himself used. It is based on a remarkable identity which Jacobi (1829) used in his treatise on elliptic functions.

Proposition 10 *For any $t, y \in \mathbb{R}$ with $y > 0$,*

$$\sum_{n=-\infty}^{\infty} e^{-(t+n)^2\pi y} = y^{-1/2} \sum_{n=-\infty}^{\infty} e^{-n^2\pi/y} e^{2\pi i n t}. \tag{7}$$

In particular,

$$\sum_{n=-\infty}^{\infty} e^{-n^2\pi y} = y^{-1/2} \sum_{n=-\infty}^{\infty} e^{-n^2\pi/y}. \tag{8}$$

Proof Put $f(v) = e^{-v^2 \pi y}$ and let

$$g(u) = \int_{-\infty}^{\infty} f(v) e^{-2\pi i u v} \, dv$$

be the Fourier transform of $f(v)$. We are going to show that

$$\sum_{n=-\infty}^{\infty} f(v+n) = \sum_{n=-\infty}^{\infty} g(n) e^{2\pi i n v}.$$

Let

$$F(v) = \sum_{n=-\infty}^{\infty} f(v+n).$$

This infinite series is uniformly convergent for $0 \le v \le 1$, and so also is the series obtained by term by term differentiation. Hence $F(v)$ is a continuously differentiable function. Consequently, since it is periodic with period 1, it is the sum of its own Fourier series:

$$F(v) = \sum_{m=-\infty}^{\infty} c_m e^{2\pi i m v},$$

where

$$c_m = \int_0^1 F(v) e^{-2\pi i m v} \, dv.$$

We can evaluate c_m by term by term integration:

$$c_m = \sum_{n=-\infty}^{\infty} \int_0^1 f(v+n) e^{-2\pi i m v} \, dv = \sum_{n=-\infty}^{\infty} \int_n^{n+1} f(v) e^{-2\pi i m v} \, dv$$

$$= \int_{-\infty}^{\infty} f(v) e^{-2\pi i m v} \, dv = g(m).$$

The argument up to this point is an instance of *Poisson's summation formula*. To evaluate $g(u)$ in the case $f(v) = e^{-v^2 \pi y}$ we differentiate with respect to u and integrate by parts, obtaining

$$g'(u) = -2\pi i \int_{-\infty}^{\infty} e^{-v^2 \pi y} v e^{-2\pi i u v} \, dv$$

$$= (i/y) \int_{-\infty}^{\infty} e^{-2\pi i u v} \, d e^{-v^2 \pi y}$$

$$= -(i/y) \int_{-\infty}^{\infty} e^{-v^2 \pi y} \, d e^{-2\pi i u v}$$

$$= -(2\pi u/y) g(u).$$

The solution of this first order linear differential equation is

$$g(u) = g(0)e^{-\pi u^2/y}.$$

Moreover

$$g(0) = \int_{-\infty}^{\infty} e^{-v^2 \pi y} dv = (\pi y)^{-1/2} J,$$

where

$$J = \int_{-\infty}^{\infty} e^{-v^2} dv.$$

Thus we have proved that

$$\sum_{n=-\infty}^{\infty} e^{-(v+n)^2 \pi y} = (\pi y)^{-1/2} J \sum_{n=-\infty}^{\infty} e^{-n^2 \pi/y} e^{2\pi i n v}.$$

Substituting $v = 0$, $y = 1$, we obtain $J = \pi^{1/2}$. $\square$

The *theta function*

$$\vartheta(x) = \sum_{n=-\infty}^{\infty} e^{-n^2 \pi x} \quad (x > 0)$$

arises not only in the theory of elliptic functions, as we will see in Chapter XII, but also in problems of heat conduction and statistical mechanics. The transformation law

$$\vartheta(x) = x^{-1/2}\vartheta(1/x)$$

is very useful for computational purposes since, when x is small, the series for $\vartheta(x)$ converges extremely slowly but the series for $\vartheta(1/x)$ converges extremely rapidly.

Since the functional equation of Riemann's zeta function involves Euler's *gamma function*, we summarize here the main properties of the latter. Euler (1729) defined his function $\Gamma(z)$ by

$$1/\Gamma(z) = \lim_{n \to \infty} z(z+1) \cdots (z+n)/n! n^z,$$

where $n^z = \exp(z \log n)$ and the limit exists for every $z \in \mathbb{C}$. It follows from the definition that $1/\Gamma(z)$ is everywhere holomorphic and that its only zeros are simple zeros at the points $z = 0, -1, -2, \ldots$. Moreover $\Gamma(1) = 1$ and

$$\Gamma(z+1) = z\Gamma(z).$$

Hence $\Gamma(n+1) = n!$ for any positive integer n. By putting $\Gamma(z+1) = z!$ the definition of the factorial function may be extended to any $z \in \mathbb{C}$ which is not a negative integer. Wielandt (1939) has characterized $\Gamma(z)$ as the only solution of the functional equation

$$F(z+1) = zF(z)$$

with $F(1) = 1$ which is holomorphic in the half-plane $\mathscr{R}z > 0$ and bounded for $1 < \mathscr{R}z < 2$.

It follows from the definition of $\Gamma(z)$ and the product formula for the sine function that

$$\Gamma(z)\Gamma(1-z) = \pi / \sin \pi z.$$

Many definite integrals may be evaluated in terms of the gamma function. By repeated integration by parts it may be seen that, if $\mathscr{R}z > 0$ and $n \in \mathbb{N}$, then

$$n!n^z/z(z+1)\cdots(z+n) = \int_0^n (1 - t/n)^n t^{z-1}dt,$$

where $t^{z-1} = \exp\{(z-1)\log t\}$. Letting $n \to \infty$, we obtain the integral representation

$$\Gamma(z) = \int_0^\infty e^{-t}t^{z-1}dt \quad \text{for } \mathscr{R}z > 0. \tag{9}$$

It follows that $\Gamma(1/2) = \pi^{1/2}$, since

$$\int_0^\infty e^{-t}t^{-1/2}dt = \int_{-\infty}^\infty e^{-v^2}dv = \pi^{1/2},$$

by the proof of Proposition 10. It was already shown by Euler (1730) that

$$B(x, y) := \int_0^1 t^{x-1}(1-t)^{y-1}dt = \Gamma(x)\Gamma(y)/\Gamma(x+y),$$

the relation holding for $\mathscr{R}x > 0$ and $\mathscr{R}y > 0$. The unit ball in $\mathbb{R}^n$ has volume $\kappa_n := \pi^{n/2}/(n/2)!$ and surface content $n\kappa_n$. Stirling's formula, $n! \approx (n/e)^n\sqrt{2\pi n}$, follows at once from the integral representation

$$\log \Gamma(z) = (z - 1/2)\log z - z + (1/2)\log 2\pi - \int_0^\infty (t - \lfloor t\rfloor - 1/2)(z+t)^{-1}dt,$$

valid for any $z \in \mathbb{C}$ which is not zero or a negative integer. Euler's constant

$$\gamma = \lim_{n\to\infty} (1 + 1/2 + 1/3 + \cdots + 1/n - \log n) \approx 0.5772157$$

may also be defined by $\gamma = -\Gamma'(1)$.

We now return to the Riemann zeta function.

Proposition 11 *The function* $Z(s) = \pi^{-s/2}\Gamma(s/2)\zeta(s)$ *satisfies the functional equation*

$$Z(s) = Z(1 - s) \text{ for } 0 < \sigma < 1.$$

Proof From the representation (9) of the gamma function we obtain, for $\sigma > 0$ and $n \geq 1$,

$$\int_0^\infty x^{s/2-1}e^{-n^2\pi x}dx = \pi^{-s/2}\Gamma(s/2)n^{-s}.$$

Hence, if $\sigma > 1$,

$$Z(s) = \sum_{n=1}^\infty \int_0^\infty x^{s/2-1}e^{-n^2\pi x}dx$$

$$= \int_0^\infty x^{s/2-1}\phi(x)dx,$$

where

$$\phi(x) = \sum_{n=1}^\infty e^{-n^2\pi x}.$$

By Proposition 10,

$$2\phi(x) + 1 = x^{-1/2}[2\phi(1/x) + 1].$$

Hence

$$Z(s) = \int_1^\infty x^{s/2-1}\phi(x)\,dx + \int_0^1 x^{s/2-1}\{x^{-1/2}\phi(1/x) + (1/2)x^{-1/2} - 1/2\}\,dx$$

$$= \int_1^\infty x^{s/2-1}\phi(x)\,dx + \int_0^1 x^{s/2-3/2}\phi(1/x)\,dx + 1/(s-1) - 1/s$$

$$= \int_1^\infty (x^{s/2-1} + x^{-s/2-1/2})\phi(x)\,dx + 1/s(s-1).$$

The integral on the right is convergent for all s and thus provides the analytic continuation of $Z(s)$ to the whole plane. Moreover the right side is unchanged if s is replaced by $1 - s$. $\quad\square$

The function $Z(s)$ in Proposition 11 is occasionally called the *completed* zeta function. In its product representation

$$Z(s) = \pi^{-s/2}\Gamma(s/2)\Pi_p(1 - p^{-s})^{-1}$$

it makes sense to regard $\pi^{-s/2}\Gamma(s/2)$ as an Euler factor at ∞, complementing the Euler factors $(1 - p^{-s})^{-1}$ at the primes p.

It follows from Proposition 11 and the previously stated properties of the gamma function that the definition of $\zeta(s)$ may be extended to the whole complex plane, so that $\zeta(s) - 1/(s - 1)$ is everywhere holomorphic and $\zeta(s) = 0$ if $s = -2, -4, -6, \ldots$. Since $\zeta(s) \neq 0$ for $\sigma \geq 1$ and $\zeta(0) = -1/2$, the functional equation shows that these 'trivial' zeros of $\zeta(s)$ are its only zeros in the half-plane $\sigma \leq 0$. Hence all 'nontrivial' zeros of $\zeta(s)$ lie in the strip $0 < \sigma < 1$ and are symmetrically situated with respect to the line $\sigma = 1/2$. The famous *Riemann hypothesis* asserts that all zeros in this strip actually lie on the line $\sigma = 1/2$.

Since $\zeta(\bar{s}) = \overline{\zeta(s)}$, the zeros of $\zeta(s)$ are also symmetric with respect to the real axis. Furthermore $\zeta(s)$ has no real zeros in the strip $0 < \sigma < 1$, since

$$(1 - 2^{-1-\sigma})\zeta(\sigma) = (1 - 2^{-\sigma}) + (3^{-\sigma} - 4^{-\sigma}) + \cdots > 0 \quad \text{for } 0 < \sigma < 1.$$

It has been verified by van de Lune *et al.* (1986), with the aid of a supercomputer, that the 1.5×10^9 zeros of $\zeta(s)$ in the rectangle $0 < \sigma < 1, 0 < t < T$, where $T = 545439823.215$, are all simple and lie on the line $\sigma = 1/2$.

The location of the zeros of $\zeta(s)$ is intimately connected with the asymptotic behaviour of $\pi(x)$. Let α^* denote the least upper bound of the real parts of all zeros of $\zeta(s)$. Then $1/2 \le \alpha^* \le 1$, since it is known that $\zeta(s)$ does have zeros in the strip $0 < \sigma < 1$, and the Riemann hypothesis is equivalent to $\alpha^* = 1/2$. It was shown by von Koch (1901) that

$$\psi(x) = x + O(x^{\alpha^*} \log^2 x)$$

and hence

$$\pi(x) = Li(x) + O(x^{\alpha^*} \log x).$$

(Actually von Koch assumed $\alpha^* = 1/2$, but his argument can be extended without difficulty.) It should be noted that these estimates are of interest only if $\alpha^* < 1$.

On the other hand if, for some α such that $0 < \alpha < 1$,

$$\pi(x) = Li(x) + O(x^{\alpha} \log x),$$

then

$$\theta(x) = x + O(x^{\alpha} \log^2 x).$$

By the remark after the proof of Lemma 3, it follows that

$$\psi(x) = x + x^{1/2} + O(x^{\alpha} \log^2 x) + O(x^{1/3} \log^2 x).$$

But for $\sigma > 1$ we have

$$-\zeta'(s)/\zeta(s) = \int_1^\infty x^{-s} d\psi(x) = s \int_1^\infty \psi(x) x^{-s-1} dx$$

and hence

$$-\zeta'(s)/\zeta(s) - s/(s-1) - s/(s-1/2) = s \int_1^\infty \{\psi(x) - x - x^{1/2}\} x^{-s-1} dx.$$

The integral on the right is uniformly convergent in the half-plane $\sigma \ge \varepsilon + \max(\alpha, 1/3)$, for any $\varepsilon > 0$, and represents there a holomorphic function. It follows that $1/2 \le \alpha^* \le \max(\alpha, 1/3)$. Consequently $\alpha^* \le \alpha$ and $\psi(x) = x + O(x^{\alpha} \log^2 x)$.

Combining this with von Koch's result, we see that the Riemann hypothesis is equivalent to

$$\pi(x) = Li(x) + O(x^{1/2} \log x)$$

and to

$$\psi(x) = x + O(x^{1/2} \log^2 x).$$

Since it is still not known if $\alpha^* < 1$, the error terms here are substantially smaller than any that have actually been established.

It has been shown by Cramér (1922) that

$$(\log x)^{-1} \int_2^x (\psi(t)/t - 1)^2 dt$$

has a finite limit as $x \to \infty$ if the Riemann hypothesis holds, and is unbounded if it does not. Similarly, for each $\alpha < 1$,

$$x^{-2(1-\alpha)} \int_2^x (\psi(t) - t)^2 t^{-2\alpha} dt$$

is bounded but does not have a finite limit as $x \to \infty$ if the Riemann hypothesis holds, and is unbounded otherwise.

For all values of x listed in Table 1 we have $\pi(x) < Li(x)$, and at one time it was conjectured that this inequality holds for all $x > 0$. However, Littlewood (1914) disproved the conjecture by showing that there exists a constant $c > 0$ such that

$$\pi(x_n) - Li(x_n) > cx_n^{1/2} \log \log \log x_n / \log x_n$$

for some sequence $x_n \to \infty$ and

$$\pi(\xi_n) - Li(\xi_n) < -c\xi_n^{1/2} \log \log \log \xi_n / \log \xi_n$$

for some sequence $\xi_n \to \infty$. This is a quite remarkable result, since no actual value of x is known for which $\pi(x) > Li(x)$. However, it is known that $\pi(x) > Li(x)$ for some x between 1.398201×10^{316} and 1.398244×10^{316}.

In this connection it may be noted that Rosser and Schoenfeld (1962) have shown that $\pi(x) > x/\log x$ for all $x \geq 17$. It had previously been shown by Rosser (1939) that $p_n > n \log n$ for all $n \geq 1$.

Not content with not being able to prove the Riemann hypothesis, Montgomery (1973) has assumed it and made a further conjecture. For given $\beta > 0$, let $N_T(\beta)$ be the number of zeros $1/2 + i\gamma$, $1/2 + i\gamma'$ of $\zeta(s)$ with $0 < \gamma' < \gamma \leq T$ such that

$$\gamma - \gamma' \leq 2\pi\beta/\log T.$$

Montgomery's conjecture is that, for each fixed $\beta > 0$,

$$N_T(\beta) \sim (T/2\pi) \log T \int_0^\beta \{1 - (\sin \pi u/\pi u)^2\} du \quad \text{as } T \to \infty.$$

Goldston (1988) has shown that this is equivalent to

$$\int_1^{T^\beta} \{\psi(x + x/T) - \psi(x) - x/T\}^2 x^{-2} dx \sim (\beta - 1/2) \log^2 T/T \quad \text{as } T \to \infty,$$

for each fixed $\beta \geq 1$, where $\psi(x)$ is Chebyshev's function.

In the language of physics Montgomery's conjecture says that $1 - (\sin \pi u/\pi u)^2$ is the *pair correlation function* of the zeros of $\zeta(s)$. Dyson pointed out that this is also the pair correlation function of the normalized eigenvalues of a random $N \times N$ Hermitian matrix in the limit $N \to \infty$. A great deal more is known about this so-called *Gaussian unitary ensemble*, which Wigner (1955) used to model the statistical properties of the spectra of complex nuclei. For example, if the eigenvalues are normalized so that the average difference between consecutive eigenvalues is 1, then the probability that the difference between an eigenvalue and the least eigenvalue greater than it does not exceed β converges as $N \to \infty$ to

$$\int_0^\beta p(u)\,du,$$

where the density function $p(u)$ can be explicitly specified.

It has been further conjectured that the spacings of the normalized zeros of the zeta-function have the same distribution. To make this precise, let the zeros $1/2 + i\gamma_n$ of $\zeta(s)$ with $\gamma_n > 0$ be numbered so that

$$\gamma_1 \leq \gamma_2 \leq \cdots.$$

Since it is known that the number of γ's in an interval $[T, T + 1]$ is asymptotic to $(\log T)/2\pi$ as $T \to \infty$, we put

$$\tilde{\gamma}_n = (\gamma_n \log \gamma_n)/2\pi,$$

so that the average difference between consecutive $\tilde{\gamma}_n$ is 1. If $\delta_n = \tilde{\gamma}_{n+1} - \tilde{\gamma}_n$, and if $\upsilon_N(\beta)$ is the number of $\delta_n \leq \beta$ with $n \leq N$, then the conjecture is that for each $\beta > 0$

$$\upsilon_N(\beta)/N \to \int_0^\beta p(u)\,du \quad \text{as } N \to \infty.$$

This nearest neighbour conjecture and the Montgomery pair correlation conjecture have been extensively tested by Odlyzko (1987/9) with the aid of a supercomputer. There is good agreement between the conjectures and the numerical results.

5 Generalizations and Analogues

The prime number theorem may be generalized to any algebraic number field in the following way. Let K be an algebraic number field, i.e. a finite extension of the field $\mathbb{Q}$ of rational numbers. Let R be the ring of all algebraic integers in K, $\mathscr{I}$ the set of all nonzero ideals of R, and $\mathscr{P}$ the subset of prime ideals. For any $A \in \mathscr{I}$, the quotient ring R/A is finite; its cardinality will be denoted by $|A|$ and called the *norm* of A.

It may be shown that the *Dedekind zeta-function*

$$\zeta_K(s) = \sum_{A \in \mathscr{I}} |A|^{-s}$$

is defined for $\mathscr{R}s > 1$ and that the product formula

$$\zeta_K(s) = \prod_{P \in \mathscr{P}} (1 - |P|^{-s})^{-1}$$

holds in this open half-plane. Furthermore the definition of $\zeta_K(s)$ may be extended so that it is nonzero and holomorphic in the closed half-plane $\mathscr{R}s \geq 1$, except for a simple pole at $s = 1$. By applying Ikehara's theorem we can then obtain the *prime ideal theorem*, which was first proved by Landau (1903):

$$\pi_K(x) \sim x/\log x,$$

where $\pi_K(x)$ denotes the number of prime ideals of R with norm $\leq x$.

It was shown by Hecke (1917) that the definition of the Dedekind zeta-function $\zeta_K(s)$ may also be extended so that it is holomorphic in the whole complex plane, except for the simple pole at $s = 1$, and so that, for some constant $A > 0$ and nonnegative integers r_1, r_2 (which can all be explicitly described in terms of the structure of the algebraic number field K),

$$Z_K(s) = A\Gamma(s/2)^{r_1} \Gamma(s)^{r_2} \zeta_K(s)$$

satisfies the functional equation

$$Z_K(s) = Z_K(1 - s).$$

The *extended Riemann hypothesis* asserts that, for every algebraic number field K,

$$\zeta_K(s) \neq 0 \quad \text{for } \mathscr{R}s > 1/2.$$

The numerical evidence for the extended Riemann hypothesis is favourable, although in the nature of things it cannot be tested as extensively as the ordinary Riemann hypothesis. The extended Riemann hypothesis implies error bounds for the prime ideal theorem of the same order as those which the ordinary Riemann hypothesis implies for the prime number theorem. However, it also has many other consequences. We mention only two.

It has been shown by Bach (1990), making precise an earlier result of Ankeny (1952), that if the extended Riemann hypothesis holds then, for each prime p, there is a quadratic non-residue $a \bmod p$ with $a < 2\log^2 p$. Thus we do not have to search far in order to find a quadratic non-residue, or to disprove the extended Riemann hypothesis.

It will be recalled from Chapter II that if p is a prime and a an integer not divisible by p, then $a^{p-1} \equiv 1 \bmod p$. For each prime p there exists a *primitive root*, i.e. an integer a such that $a^k \not\equiv 1 \bmod p$ for $1 \leq k < p - 1$. It is easily seen that an even square is never a primitive root, that an odd square (including 1) is a primitive root only for the prime $p = 2$, and that -1 is a primitive root only for the primes $p = 2, 3$.

Assuming the extended Riemann hypothesis, Hooley (1967) has proved a famous conjecture of Artin (1927): if the integer a is not a square or -1, then there exist infinitely many primes p for which a is a primitive root. Moreover, if $N_a(x)$ denotes the number of primes $p \leq x$ for which a is a primitive root, then

$$N_a(x) \sim A_a x/\log x \quad \text{for } x \to \infty,$$

where A_a is a positive constant which can be explicitly described. (The expression for A_a which Artin conjectured requires modification in some cases.)

There are also analogues for function fields of these results for number fields. Let K be an arbitrary field. A *field of algebraic functions of one variable over* K is a field L which satisfies the following conditions:

(i) $K \subseteq L$,
(ii) L contains an element v which is *transcendental* over K, i.e. v satisfies no monic polynomial equation

$$u^n + a_1 u^{n-1} + \cdots + a_n = 0$$

with coefficients $a_j \in K$,
(iii) L is a *finite extension* of the field $K(v)$ of rational functions of v with coeffients from K, i.e. L is finite-dimensional as a vector space over $K(v)$.

Let R be a ring with $K \subseteq R \subset L$ such that $x \in L \backslash R$ implies $x^{-1} \in R$. Then the set P of all $a \in R$ such that $a = 0$ or $a^{-1} \notin R$ is an ideal of R, and actually the unique maximal ideal of R. Hence the quotient ring R/P is a field. Since R is the set of all $x \in L$ such that $xP \subseteq P$, it is uniquely determined by P. The ideal P will be called a *prime divisor* of the field L and R/P its *residue field*. It may be shown that the residue field R/P is a finite extension of (a field isomorphic to) K.

An arbitrary *divisor* of the field L is a formal product $A = \prod_P P^{v_P}$ over all prime divisors P of L, where the exponents v_P are integers only finitely many of which are nonzero. The divisor is *integral* if $v_P \geq 0$ for all P.

The set K' of all elements of L which satisfy monic polynomial equations with coefficients from K is a subfield containing K, and L is also a field of algebraic functions of one variable over K'. It is easily shown that no element of $L \backslash R$ satisfies a monic polynomial equation with coefficients from R. Consequently $K' \subseteq R$ and the notion of prime divisor is the same whether we consider L to be over K or over K'. Since $(K')' = K'$, we may assume from the outset that $K' = K$. The elements of K will then be called *constants* and the elements of L *functions*.

Suppose now that the field of constants K is a finite field $\mathbb{F}_q$ containing q elements. We define the *norm* $N(P)$ of a prime divisor P to be the cardinality of the associated residue field R/P and the norm of an integral divisor $A = \prod_P P^{v_P}$ to be

$$N(A) = \prod_P N(P)^{v_P}.$$

It may be shown that, for each positive integer m, there exist only finitely many prime divisors of norm q^m. Moreover, for $\mathscr{R}s > 1$ the zeta-function of L can be defined by

$$\zeta_L(s) = \sum_A N(A)^{-s},$$

where the sum is over all integral divisors of L, and then

$$\zeta_L(s) = \prod_P (1 - N(P)^{-s})^{-1},$$

where the product is over all prime divisors of L.

This seems quite similar to the number field case, but the function field case is actually simpler. F.K. Schmidt (1931) deduced from the Riemann–Roch theorem that there exists a polynomial $p(u)$ of even degree $2g$, with integer coefficients and constant term 1, such that

$$\zeta_L(s) = p(q^{-s})/(1 - q^{-s})(1 - q^{1-s}),$$

and that the zeta-function satisfies the functional equation

$$q^{(g-1)s}\zeta_L(s) = q^{(g-1)(1-s)}\zeta_L(1 - s).$$

The non-negative integer g is the *genus* of the field of algebraic functions.

The analogue of the Riemann hypothesis, that all zeros of $\zeta_L(s)$ lie on the line $\mathscr{R}s = 1/2$, is equivalent to the statement that all zeros of the polynomial $p(u)$ have absolute value $q^{-1/2}$, or that the number N of prime divisors with norm q satisfies the inequality

$$|N - (q + 1)| \le 2gq^{1/2}.$$

This analogue has been *proved* by Weil (1948). A simpler proof has been given by Bombieri (1974), using ideas of Stepanov (1969).

The theory of function fields can also be given a geometric formulation. The prime divisors of a function field L with field of constants K can be regarded as the points of a non-singular projective curve over K, and vice versa. Weil (1949) conjectured far-reaching generalizations of the preceding results for curves over a finite field to algebraic varieties of higher dimension.

Let V be a nonsingular projective variety of dimension d, defined by homogeneous polynomials with coefficients in $\mathbb{Z}$. For any prime p, let V_p be the (possibly singular) variety defined by reducing the coefficients mod p and consider the formal power series

$$Z_p(T) := \exp\left(\sum_{n\ge 1} N_n(p)T^n/n\right),$$

where $N_n(p)$ denotes the number of points of V_p defined over the finite field $\mathbb{F}_{p^n}$. Weil conjectured that, if V_p is a nonsingular projective variety of dimension d over $\mathbb{F}_p$, then

(i) $Z_p(T)$ is a rational function of T,
(ii) $Z_p(1/p^d T) = \pm p^{de/2}T^e Z_p(T)$ for some integer e,
(iii) $Z_p(T)$ has a factorization of the form

$$Z_p(T) = P_1(T)\cdots P_{2d-1}(T)/P_0(T)P_2(T)\cdots P_{2d}(T),$$

where $P_0(T) = 1 - T$, $P_{2d}(T) = 1 - p^d T$ and $P_j(T) \in \mathbb{Z}[T]$ $(0 < j < 2d)$,
(iv) $P_j(T) = \prod_{k=1}^{b_j}(1 - \alpha_{jk}T)$, where $|\alpha_{jk}| = p^{j/2}$ for $1 \le k \le b_j$, $(0 < j < 2d)$.

The Weil conjectures have a topological significance, since the integer e in (ii) is the Euler characteristic of the original variety V, regarded as a complex manifold, and b_j in (iv) is its j-th Betti number.

Conjecture (i) was proved by Dwork (1960). The remaining conjectures were proved by Deligne (1974), using ideas of Grothendieck. The most difficult part is the proof that $|\alpha_{jk}| = p^{j/2}$ (the Riemann hypothesis for varieties over finite fields). Deligne's proof is a major achievement of 20th century mathematics, but unfortunately of a different order of difficulty than anything which will be proved here.

An analogue for function fields of Artin's primitive root conjecture was already proved by Bilharz (1937), assuming the Riemann hypothesis for this case. Function fields have been used by Goppa (1981) to construct linear codes. Good codes are obtained when the number of prime divisors is large compared to the genus, and this can be guaranteed by means of the Riemann 'hypothesis'.

Carlitz and Uchiyama (1957) used the Riemann hypothesis for function fields to obtain useful estimates for exponential sums in one variable, and Deligne (1977) showed that these estimates could be extended to exponential sums in several variables. Let $\mathbb{F}_p$ be the field of p elements, where p is a prime, and let $f \in \mathbb{F}_p[u_1, \ldots, u_n]$ be a polynomial in n variables of degree $d \geq 1$ with coefficients from $\mathbb{F}_p$ which is not of the form $g^p - g + b$, where $b \in \mathbb{F}_p$ and $g \in \mathbb{F}_p[u_1, \ldots, u_n]$. (This condition is certainly satisfied if $d < p$.) Then

$$\left| \sum_{x_1, \ldots, x_n \in \mathbb{F}_p} e^{2\pi i f(x_1, \ldots, x_n)/p} \right| \leq (d-1)p^{n-1/2}.$$

We mention one more application of the Weil conjectures. *Ramanujan's tau-function* is defined by

$$q \prod_{n=1}^{\infty} (1 - q^n)^{24} = \sum_{n=1}^{\infty} \tau(n) q^n.$$

It was conjectured by Ramanujan (1916), and proved by Mordell (1920), that

$$\sum_{n=1}^{\infty} \tau(n)/n^s = \prod_p (1 - \tau(p)p^{-s} + p^{11-2s})^{-1},$$

where the product is over all primes p. Ramanujan additionally conjectured that $|\tau(p)| \leq 2p^{11/2}$ for all p, and Deligne (1968/9) showed that this was a consequence of the (at that time unproven) Weil conjectures.

The prime number theorem also has an interesting analogue in the theory of dynamical systems. Let M be a compact Riemannian manifold with negative sectional curvatures, and let $N(T)$ denote the number of different (oriented) closed geodesics on M of length $\leq T$. It was first shown by Margulis (1970) that

$$N(T) \sim e^{hT}/hT \quad \text{as } T \to \infty,$$

where the positive constant h is the topological entropy of the associated geodesic flow.

Although much of the detail is specific to the problem, a proof may be given which has the same structure as the proof in §3 of the prime number theorem. If P is an arbitrary closed orbit of the geodesic flow and $\lambda(P)$ its least period, one shows that the zeta-function

$$\zeta_M(s) = \prod_P (1 - e^{-s\lambda(P)})^{-1}$$

is nonzero and holomorphic for $\mathscr{R}s \geq h$, except for a simple pole at $s = h$, and then applies Ikehara's theorem. The study of geodesics on a surface of negative curvature was initiated by Hadamard (1898), but it is unlikely that he realized there was a connection with the prime number theorem which he had proved two years earlier!

6 Alternative Formulations

There is an intimate connection between the *Dirichlet products* considered in §3 of Chapter III and *Dirichlet series*. It is easily seen that if the Dirichlet series

$$f(s) = \sum_{n=1}^{\infty} a(n)/n^s, \quad g(s) = \sum_{n=1}^{\infty} b(n)/n^s,$$

are absolutely convergent for $\mathscr{R}s > \alpha$, then the product $h(s) = f(s)g(s)$ may also be represented by an absolutely convergent Dirichlet series for $\mathscr{R}s > \alpha$:

$$h(s) = \sum_{n=1}^{\infty} c(n)/n^s,$$

where $c = a * b$, i.e.

$$c(n) = \sum_{d|n} a(d)b(n/d) = \sum_{d|n} a(n/d)b(d).$$

This implies, in particular, that for $\mathscr{R}s > 1$

$$\zeta^2(s) = \sum_{n=1}^{\infty} \tau(n)/n^s, \quad \zeta(s-1)\zeta(s) = \sum_{n=1}^{\infty} \sigma(n)/n^s,$$

where as in Chapter III (not as in §5),

$$\tau(n) = \sum_{d|n} 1, \quad \sigma(n) = \sum_{d|n} d,$$

denote respectively the number of positive divisors of n and the sum of the positive divisors of n. The relation for Euler's phi-function,

$$\sigma(n) = \sum_{d|n} \tau(n/d)\varphi(d),$$

which was proved in Chapter III, now yields for $\mathscr{R}s > 1$

$$\zeta(s - 1)/\zeta(s) = \sum_{n=1}^{\infty} \varphi(n)/n^s.$$

From the property by which we defined the Möbius function we obtain also, for $\mathcal{R}s > 1$,

$$1/\zeta(s) = \sum_{n=1}^{\infty} \mu(n)/n^s.$$

In view of this relation it is not surprising that the distribution of prime numbers is closely connected with the behaviour of the Möbius function. Put

$$M(x) = \sum_{n \leq x} \mu(n).$$

Since $|\mu(n)| \leq 1$, it is obvious that $|M(x)| \leq \lfloor x \rfloor$ for $x > 0$. The next result is not so obvious:

Proposition 12 $M(x)/x \to 0$ as $x \to \infty$.

Proof The function $f(s) := \zeta(s) + 1/\zeta(s)$ is holomorphic for $\sigma \geq 1$, except for a simple pole with residue 1 at $s = 1$. Moreover

$$f(s) = \sum_{n=1}^{\infty} \{1 + \mu(n)\}/n^s = \int_{1-}^{\infty} x^{-s} d\phi(x) \text{ for } \sigma > 1,$$

where $\phi(x) = \lfloor x \rfloor + M(x)$ is a nondecreasing function. Since

$$f(s) = \int_{0-}^{\infty} e^{-su} d\phi(e^u),$$

it follows from Ikehara's Theorem 9 that $\phi(x) \sim x$.   □

Proposition 12 is equivalent to the prime number theorem in the sense that either of the relations $M(x) = o(x)$, $\psi(x) \sim x$ may be deduced from the other by elementary (but not trivial) arguments.

The Riemann hypothesis also has an equivalent formulation in terms of the function $M(x)$. Suppose

$$M(x) = O(x^{\alpha}) \quad \text{as } x \to \infty,$$

for some α such that $0 < \alpha < 1$. For $\sigma > 1$ we have

$$1/\zeta(s) = \int_{1-}^{\infty} x^{-s} dM(x) = s \int_{1}^{\infty} x^{-s-1} M(x) \, dx.$$

But for $\sigma > \alpha$ the integral on the right is convergent and defines a holomorphic function. Consequently it is the analytic continuation of $1/\zeta(s)$. Thus if α^* again denotes the least upper bound of all zeros of $\zeta(s)$, then $\alpha \geq \alpha^* \geq 1/2$. On the other hand, Littlewood (1912) showed that

$$M(x) = O(x^{a^*+\varepsilon}) \quad \text{for every } \varepsilon > 0.$$

It follows that *the Riemann hypothesis holds if and only if* $M(x) = O(x^a)$ *for every* $a > 1/2$.

It has already been mentioned that the first 1.5×10^9 zeros of $\zeta(s)$ on the line $\sigma = 1/2$ are all simple. It is likely that the Riemann hypothesis does not tell the whole story and that all zeros of $\zeta(s)$ on the line $\sigma = 1/2$ are simple. Thus it is of interest that this is guaranteed by a sufficiently sharp bound for $M(x)$. We will show that *if*

$$M(x) = O(x^{1/2} \log^a x) \quad as \ x \to \infty,$$

for some $a < 1$, *then not only do all nontrivial zeros of* $\zeta(s)$ *lie on the line* $\sigma = 1/2$ *but they are all simple.*

Let $\rho = 1/2 + i\gamma$ be a zero of $\zeta(s)$ of multiplicity $m \geq 1$ and take $s = \rho + h$, where $h > 0$. Then $\sigma = 1/2 + h$ and, since

$$1/\zeta(s) = s \int_1^\infty x^{-s-1} M(x) \, dx \text{ for } \sigma > 1/2,$$

we have

$$|1/\zeta(s)| \leq |s| \int_1^\infty x^{-\sigma-1} |M(x)| \, dx = O(|s|) \int_1^\infty x^{-h-1} \log^a x \, dx$$

$$= O(|s|) \int_0^\infty e^{-hu} u^a \, du = O(|s|) \Gamma(a+1)/h^{a+1}.$$

Thus $h^{a+1} |1/\zeta(s)|$ is bounded for $h \to +0$ and hence $m \leq a+1$. Since m is an integer and $a < 1$, this implies $m = 1$ and $a \geq 0$.

The prime number theorem, in the form $M(x) = o(x)$, says that asymptotically $\mu(n)$ takes the values $+1$ and -1 with equal probability. By assuming that actually the values $\mu(n)$ asymptotically behave like independent random variables Good and Churchhouse (1968) have been led to two striking conjectures, analogous to the central limit theorem and the law of the iterated logarithm in the theory of probability:

Conjecture A *If* $N(n) \to \infty$ *and* $\log N / \log n \to 0$, *then*

$$P_n \left\{ \frac{M(m+N) - M(m)}{(6N/\pi^2)^{1/2}} < t \right\} \to (2\pi)^{-1/2} \int_{-\infty}^t e^{-u^2/2} \, du,$$

where

$$P_n\{f(m) < t\} = \#\{m \leq n : f(m) < t\}/n.$$

Conjecture B

$$\overline{\lim_{x \to \infty}} \, M(x)(2x \log \log x)^{-1/2} = \sqrt{6}/\pi$$

$$= - \lim_{x \to \infty} \, M(x)(2x \log \log x)^{-1/2}.$$

By what has been said, Conjecture B implies not only the Riemann hypothesis, but also that the zeros of $\zeta(s)$ are all simple. These probabilistic conjectures provide a more interesting reason than symmetry for believing in the validity of the Riemann hypothesis, but no progress has so far been made towards proving them.

7 Some Further Problems

A prime p is said to be a *twin prime* if $p + 2$ is also a prime. For example, 41 is a twin prime since both 41 and 43 are primes. It is still not known if there are infinitely many twin primes. However Brun (1919), using the sieve method which he devised for the purpose, showed that, if infinite, the sum of the reciprocals of all twin primes converges. Since the sum of the reciprocals of all primes diverges, this means that few primes are twin primes.

By a formal application of their circle method Hardy and Littlewood (1923) were led to conjecture that

$$\pi_2(x) \sim L_2(x) \quad \text{for } x \to \infty,$$

where $\pi_2(x)$ denotes the number of twin primes $\leq x$,

$$L_2(x) = 2C_2 \int_2^x dt / \log^2 t$$

and

$$C_2 = \prod_{p \geq 3} (1 - 1/(p - 1)^2) = 0.66016181 \ldots .$$

This implies that $\pi_2(x)/\pi(x) \sim 2C_2/\log x$. Table 2, adapted from Brent (1975), shows that Hardy and Littlewood's formula agrees well with the facts. Brent also calculates

$$\sum_{\text{twin} \, p \leq 10^{10}} (1/p + 1/(p + 2)) = 1.78748 \ldots$$

and, using the Hardy–Littlewood formula for the tail, obtains the estimate

$$\sum_{\text{all twin} \, p} (1/p + 1/(p + 2)) = 1.90216 \ldots .$$

His calculations have been considerably extended by Nicely (1995).

Table 2.

x	$\pi_2(x)$	$L_2(x)$	$\pi_2(x)/L_2(x)$
10^3	35	46	0.76
10^4	205	214	0.96
10^5	1224	1249	0.980
10^6	8169	8248	0.9904
10^7	58980	58754	1.0038
10^8	440312	440368	0.99987
10^9	3424506	3425308	0.99977
10^{10}	27412679	27411417	1.000046

Besides the twin prime formula many other asymptotic formulae were conjectured by Hardy and Littlewood. Most of them are contained in a general conjecture, which will now be described.

Let $f(t)$ be a polynomial in t of positive degree with integer coefficients. If $f(n)$ is prime for infinitely many positive integers n, then f has positive leading coefficient, f is irreducible over the field $\mathbb{Q}$ of rational numbers and, for each prime p, there is a positive integer n for which $f(n)$ is not divisible by p. It was conjectured by Bouniakowsky (1857) that conversely, if these three conditions are satisfied, then $f(n)$ is prime for infinitely many positive integers n. Schinzel (1958) extended the conjecture to several polynomials and Bateman and Horn (1962) gave Schinzel's conjecture the following quantitative form.

Let $f_j(t)$ be a polynomial in t of degree $d_j \geq 1$, with integer coefficients and positive leading coefficient, which is irreducible over the field $\mathbb{Q}$ of rational numbers ($j = 1, \ldots, m$). Suppose also that the polynomials $f_1(t), \ldots, f_m(t)$ are distinct and that, for each prime p, there is a positive integer n for which the product $f_1(n) \cdots f_m(n)$ is not divisible by p. Bateman and Horn's conjecture states that, if $N(x)$ is the number of positive integers $n \leq x$ for which $f_1(n), \ldots, f_m(n)$ are all primes, then

$$N(x) \sim (d_1 \cdots d_m)^{-1} C(f_1, \ldots, f_m) \int_2^x dt / \log^m t,$$

where

$$C(f_1, \ldots, f_m) = \prod_p \{(1 - 1/p)^{-m}(1 - \omega(p)/p)\},$$

the product being taken over all primes p and $\omega(p)$ denoting the number of $u \in \mathbb{F}_p$ (the field of p elements) such that $f_1(u) \cdots f_m(u) = 0$. (The convergence of the infinite product when the primes are taken in their natural order follows from the prime ideal theorem.)

The twin prime formula is obtained by taking $m = 2$ and $f_1(t) = t$, $f_2(t) = t+2$. By taking instead $f_1(t) = t$, $f_2(t) = 2t+1$, the Bateman–Horn conjecture gives the same asymptotic formula $\pi_G(x) \sim L_2(x)$ for the number $\pi_G(x)$ of primes $p \leq x$ for which $2p+1$ is also a prime ('Sophie Germain' primes). By taking $m = 1$ and $f_1(t) = t^2 + 1$ one obtains an asymptotic formula for the number of primes of the form $n^2 + 1$.

Bateman and Horn gave a heuristic derivation of their formula. However, the only case in which the formula has actually been proved is $m = 1, n_1 = 1$. This is the case of primes in an arithmetic progression which will be considered in the next chapter. When one considers the vast output of mathematical papers today compared with previous eras, it is salutary to recall that we still do not know as much about twin primes as Euclid knew about primes.

8 Further Remarks

The historical development of the prime number theorem is traced in Landau [33]. The original papers of Chebyshev are available in [56]. Pintz [48] has given a simple proof of Chebyshev's result that $\pi(x) = x/(A \log x - B + o(1))$ implies $A = B = 1$.

There is an English translation of Riemann's memoir in Edwards [20]. Complex variable proofs of the prime number theorem, with error term, are contained in the books of Ayoub [4], Ellison and Ellison [21], and Patterson [47]. For a simple complex variable proof without error term, due to Newman (1980), see Zagier [63].

A proof with error term by the Wiener–Ikehara method is given in Čižek [12]. Wiener's general Tauberian theorem is proved in Rudin [52]. For its algebraic interpretation, see the resumé of Fourier analysis in [13]. The development of Selberg's method is surveyed in Diamond [18]. An elementary proof of the prime number theorem which is quite different from that of Selberg and Erdős has been given by Daboussi [15].

A clear account of Stieltjes integrals is given in Widder [62]. However, we do not use Stieltjes integrals in any essential way, but only for the formal convenience of treating integration by parts and summation by parts in the same manner. Widder's book also contains the Wiener–Ikehara proof of the prime number theorem.

By a theorem of S. Bernstein (1928), proved in Widder's book and also in Mattner [38], the hypotheses of Proposition 7 can be stated without reference to the function $\phi(x)$. Bernstein's theorem says that a real-valued function $F(\sigma)$ can be represented in the form

$$F(\sigma) = \int_0^\infty e^{-\sigma x} d\phi(x),$$

where $\phi(x)$ is a nondecreasing function for $x \geq 0$ and the integral is convergent for every $\sigma > 1$, if and only if $F(\sigma)$ has derivatives of all orders and

$$(-1)^k F^{(k)}(\sigma) \geq 0 \quad \text{for every } \sigma > 1 \quad (k = 0, 1, 2, \ldots).$$

For the Poisson summation formula see, for example, Lasser [34] and Durán et al. [19]. There is a useful n-dimensional generalization, discussed more fully in §7 of Chapter XII, in which a sum over all points of a lattice is related to a sum over all points of the dual lattice. Further generalizations are mentioned in Chapter X.

More extended treatments of the gamma function are given in Andrews et al. [3] and Remmert [49].

More information about the Riemann zeta-function is given in the books of Patterson [47], Titchmarsh [57], and Karatsuba and Voronin [30]. For numerical data, see Rosser and Schoenfeld [50], van de Lune et al. [37] and Rumely [53].

For a proof that $\pi(x) - Li(x)$ changes sign infinitely often, see Diamond [17].
Estimates for values of x such that $\pi(x) > Li(x)$ are obtained by a technique due to
Lehman [35]; for the most recent estimate, see Bays and Hudson [8].

For the pair correlation conjecture, see Montgomery [40], Goldston [24] and
Odlyzko [45]. Random matrices are thoroughly discussed by Mehta [39]; for a nice
introduction, see Tracy and Widom [58].

For Dedekind zeta functions see Stark [54], besides the books on algebraic number
theory referred to in Chapter III. The prime ideal theorem is proved in Narkiewicz [44],
for example. For consequences of the extended Riemann hypothesis, see Bach [5],
Goldstein [23] and M.R. Murty [41]. Many other generalizations of the zeta function
are discussed in the article on zeta functions in [22].

Function fields are treated in the books of Chevalley [11] and Deuring [16]. The
lengthy review of Chevalley's book by Weil in *Bull. Amer. Math. Soc.* **57** (1951),
384–398, is useful but over-critical. Even if geometric methods are better adapted for
algebraic varieties of higher dimension, the algebraic methods available for curves
are essentially simpler. Moreover it was the close analogy with number fields that
suggested the possibility of a Riemann hypothesis for function fields. For a proof of
the latter, see Bombieri [9]. For the Weil conjectures, see Weil [61] and Katz [32].

Stichtenoth [55] gives a good account of the theory of function fields with special emphasis on its applications to coding theory. For these applications, see also
Goppa [26], Tsfasman *et al.* [60], and Tsfasman and Vladut [59]. Curves with a given
genus which have the maximal number of $\mathbb{F}_q$-points are discussed by Cossidente
et al. [14].

For introductions to Ramanujan's tau-function, see V.K. Murty [42] and Rankin's
article (pp. 245–268) in Andrews *et al.* [2]. For analogues of the prime number theorem in the theory of dynamical systems, see Katok and Hasselblatt [31] and Parry
and Pollicott [46]. Hadamard's pioneering study of geodesics on a surface of negative
curvature and his proof of the prime number theorem are both reproduced in [27].

The 'equivalence' of Proposition 12 with the prime number theorem is proved in
Ayoub [4]. A proof that the Riemann hypothesis is equivalent to $M(x) = O(x^{\alpha})$ for
every $\alpha > 1/2$ is contained in the book of Titchmarsh [57]. Good and Churchhouse's
probabilistic conjectures appeared in [25]. For the central limit theorem and the law of
the iterated logarithm see, for example, Adams [1], Kac [29], Bauer [7] and Loève [36].

Brun's theorem on twin primes is proved in Narkiewicz [43]. For numerical results,
see Brent [10]. For conjectural asymptotic formulas, see Hardy and Littlewood [28]
and Bateman and Horn [6]. There are several heuristic derivations of the twin prime
formula, the most recent being Rubenstein [51]. It would be useful to try to analyse
these heuristic derivations, so that the conclusion is seen as a consequence of precisely
stated assumptions.

9 Selected References

[1] W.J. Adams, *The life and times of the central limit theorem*, Kaedmon, New York, 1974.
[2] G.E. Andrews, R.A. Askey, B.C. Berndt, K.G. Ramanathan and R.A. Rankin (ed.),
Ramanujan revisited, Academic Press, London, 1988.
[3] G.E. Andrews, R. Askey and R. Roy, *Special functions*, Cambridge University Press, 1999.

[4] R. Ayoub, *An introduction to the analytic theory of numbers*, Math. Surveys no. 10, Amer. Math. Soc., Providence, 1963.

[5] E. Bach, Explicit bounds for primality testing and related problems, *Math. Comp.* **55** (1990), 353–380.

[6] P.T. Bateman and R.A. Horn, A heuristic asymptotic formula concerning the distribution of prime numbers, *Math. Comp.* **16** (1962), 363–367.

[7] H. Bauer, *Probability theory*, English transl. by R.B. Burckel, de Gruyter, Berlin, 1996.

[8] C. Bays and R.H. Hudson, A new bound for the smallest x with $\pi(x) > li(x)$, *Math. Comp.* **69** (1999), 1285–1296.

[9] E. Bombieri, Counting points on curves over finite fields (d'après S.A. Stepanov), *Séminaire Bourbaki vol. 1972/3, Exposés 418–435*, pp. 234–241, Lecture Notes in Mathematics **383** (1974), Springer-Verlag, Berlin.

[10] R.P. Brent, Irregularities in the distribution of primes and twin primes, *Math. Comp.* **29** (1975), 43–56.

[11] C. Chevalley, *Introduction to the theory of algebraic functions of one variable*, Math. Surveys no. 6, Amer. Math. Soc., New York, 1951.

[12] J. Čižek , On the proof of the prime number theorem, *Časopis Pěst. Mat.* **106** (1981), 395–401.

[13] W.A. Coppel, J.B. Fourier–On the occasion of his two hundredth birthday, *Amer. Math. Monthly* **76** (1969), 468–483.

[14] A. Cossidente, J.W.P. Hirschfeld, G. Korchmáros and F. Torres, On plane maximal curves, *Compositio Math.* **121** (2000), 163–181.

[15] H. Daboussi, Sur le théorème des nombres premiers, *C.R. Acad. Sci. Paris Sér. I* **298** (1984), 161–164.

[16] M. Deuring, *Lectures on the theory of algebraic functions of one variable*, Lecture Notes in Mathematics **314** (1973), Springer-Verlag, Berlin.

[17] H.G. Diamond, Changes of sign of $\pi(x) - li(x)$, *Enseign. Math.* **21** (1975), 1–14.

[18] H.G. Diamond, Elementary methods in the study of the distribution of prime numbers, *Bull. Amer. Math. Soc. (N.S.)* **7** (1982), 553–589.

[19] A.L. Durán, R. Estrada and R.P. Kanwal, Extensions of the Poisson summation formula, *J. Math. Anal. Appl.* **218** (1998), 581–606.

[20] H.M. Edwards, *Riemann's zeta function*, Academic Press, New York, 1974.

[21] W. Ellison and F. Ellison, *Prime numbers*, Wiley, New York, 1985.

[22] *Encyclopedic dictionary of mathematics* (ed. K. Ito), 2nd ed., Mathematical Society of Japan, MIT Press, Cambridge, Mass., 1987.

[23] L.J. Goldstein, Density questions in algebraic number theory, *Amer. Math. Monthly* **78** (1971), 342–351.

[24] D.A. Goldston, On the pair correlation conjecture for zeros of the Riemann zeta-function, *J. Reine Angew. Math.* **385** (1988), 24–40.

[25] I.J. Good and R.F. Churchhouse, The Riemann hypothesis and pseudorandom features of the Möbius sequence, *Math. Comp.* **22** (1968), 857–861.

[26] V.D. Goppa, Codes on algebraic curves, *Soviet Math. Dokl.* **24** (1981), 170–172.

[27] J. Hadamard, *Selecta*, Gauthier-Villars, Paris, 1935.

[28] G.H. Hardy and J.E. Littlewood, Some problems of partitio numerorum III, On the expression of a number as a sum of primes, *Acta Math.* **44** (1923), 1–70.

[29] M. Kac, *Statistical independence in probability, analysis and number theory*, Carus Mathematical Monograph **12**, Math. Assoc. of America, 1959.

[30] A.A. Karatsuba and S.M. Voronin, *The Riemann zeta-function*, English transl. by N. Koblitz, de Gruyter, Berlin, 1992.

[31] A. Katok and B. Hasselblatt, *Introduction to the modern theory of dynamical systems*, Cambridge University Press, Cambridge, 1995.

[32] N. Katz, An overview of Deligne's proof of the Riemann hypothesis for varieties over finite fields, *Mathematical developments arising from Hilbert problems*, Proc. Sympos. Pure Math. **28**, pp. 275–305, Amer. Math. Soc., Providence, 1976.

[33] E. Landau, *Handbuch der Lehre von der Verteilung der Primzahlen* (2 vols.), 2nd ed., Chelsea, New York, 1953.

[34] R. Lasser, *Introduction to Fourier series*, M. Dekker, New York, 1996.

[35] R.S. Lehman, On the difference $\pi(x) - li(x)$, *Acta Arith.* **11** (1966), 397–410.

[36] M. Loève, *Probability theory*, 4th ed. in 2 vols., Springer-Verlag, New York, 1978.

[37] J. van de Lune *et al.*, On the zeros of the Riemann zeta function in the critical strip IV, *Math. Comp.* **46** (1986), 667–681.

[38] L. Mattner, Bernstein's theorem, inversion formula of Post and Widder, and the uniqueness theorem for Laplace transforms, *Exposition. Math.* **11** (1993), 137–140.

[39] M.L. Mehta, *Random matrices*, 2nd ed., Academic Press, New York, 1991.

[40] H.L. Montgomery, The pair correlation of zeros of the zeta function, *Proc. Sympos. Pure Math.* **24**, pp. 181–193, Amer. Math. Soc., Providence, 1973.

[41] M. R. Murty, Artin's conjecture for primitive roots, *Math. Intelligencer* **10** (1988), no. 4, 59–67.

[42] V.K. Murty, Ramanujan and Harish-Chandra, *Math. Intelligencer* **15** (1993), no.2, 33–39.

[43] W. Narkiewicz, *Number theory*, World Scientific, Singapore, 1983.

[44] W. Narkiewicz, *Elementary and analytic theory of algebraic numbers*, 2nd ed., Springer-Verlag, Berlin, 1990.

[45] A.M. Odlyzko, On the distribution of spacings between zeros of the zeta function, *Math. Comp.* **48** (1987), 273–308.

[46] W. Parry and M. Pollicott, An analogue of the prime number theorem for closed orbits of Axiom A flows, *Ann. of Math.* **118** (1983), 573–591.

[47] S.J. Patterson, *An introduction to the theory of the Riemann zeta-function*, Cambridge University Press, Cambridge, 1988.

[48] J. Pintz, On Legendre's prime number formula, *Amer. Math. Monthly* **87** (1980), 733–735.

[49] R. Remmert, *Classical topics in complex function theory*, English transl. by L. Kay, Springer-Verlag, New York, 1998.

[50] J.B. Rosser and L. Schoenfeld, Approximate formulas for some functions of prime numbers, *Illinois J. Math.* **6** (1962), 64–94.

[51] M. Rubinstein, A simple heuristic proof of Hardy and Littlewood's conjecture B, *Amer. Math. Monthly* **100** (1993), 456–460.

[52] W. Rudin, *Functional analysis*, McGraw-Hill, New York, 1973.

[53] R. Rumely, Numerical computations concerning the ERH, *Math. Comp.* **61** (1993), 415–440.

[54] H.M. Stark, The analytic theory of algebraic numbers, *Bull. Amer. Math. Soc.* **81** (1975), 961–972.

[55] H. Stichtenoth, *Algebraic function fields and codes*, Springer-Verlag, Berlin, 1993.

[56] P.L. Tchebychef, *Oeuvres* (2 vols.), reprinted Chelsea, New York, 1962.

[57] E.C. Titchmarsh, *The theory of the Riemann zeta-function*, 2nd ed. revised by D.R. Heath-Brown, Clarendon Press, Oxford, 1986.

[58] C.A. Tracy and H. Widom, Introduction to random matrices, *Geometric and quantum aspects of integrable systems* (ed. G.F. Helminck), pp. 103–130, Lecture Notes in Physics **424**, Springer-Verlag, Berlin, 1993.

[59] M.A. Tsfasman and S.G. Vladut, *Algebraic-geometric codes*, Kluwer, Dordrecht, 1991.

[60] M.A. Tsfasman, S.G. Vladut and Th. Zink, Modular curves, Shimura curves, and Goppa codes, *Math. Nachr.* **109** (1982), 21–28.

[61] A. Weil, Number of solutions of equations in finite fields, *Bull. Amer. Math. Soc.* **55** (1949), 497–508.

[62] D.V. Widder, *The Laplace transform*, Princeton University Press, Princeton, 1941.
[63] D. Zagier, Newman's short proof of the prime number theorem, *Amer. Math. Monthly* **104** (1997), 705–708.

Additional References

J.B. Conrey, The Riemann hypothesis, *Notices Amer. Math. Soc.* **50** (2003), 341–353.
N.M. Katz and P. Sarnak, Zeroes of zeta functions and symmetry, *Bull. Amer. Math. Soc. (N.S.)* **36** (1999), 1–26.

X

A Character Study

1 Primes in Arithmetic Progressions

Let a and m be integers with $1 \le a < m$. If a and m have a common divisor $d > 1$, then no term after the first of the arithmetic progression

$$a, a + m, a + 2m, \ldots \qquad (*)$$

is a prime. Legendre (1788) conjectured, and later (1808) attempted a proof, that *if a and m are relatively prime, then the arithmetic progression* (*) *contains infinitely many primes.*

If $a_1, \ldots, a_h$ are the positive integers less than m and relatively prime to m, and if $\pi_j(x)$ denotes the number of primes $\le x$ in the arithmetic progression

$$a_j, a_j + m, a_j + 2m, \ldots,$$

then Legendre's conjecture can be stated in the form

$$\pi_j(x) \to \infty \quad \text{as } x \to \infty \quad (j = 1, \ldots, h).$$

Legendre (1830) subsequently conjectured, and again gave a faulty proof, that

$$\pi_j(x)/\pi_k(x) \to 1 \quad \text{as } x \to \infty \quad \text{for all } j, k.$$

Since the total number $\pi(x)$ of primes $\le x$ satisfies

$$\pi(x) = \pi_1(x) + \cdots + \pi_h(x) + c,$$

where c is the number of different primes dividing m, Legendre's second conjecture is equivalent to

$$\pi_j(x)/\pi(x) \to 1/h \quad \text{as } x \to \infty \quad (j = 1, \ldots, h).$$

Here $h = \varphi(m)$ is the number of positive integers less than m and relatively prime to m. If one assumes the truth of the prime number theorem, then the second conjecture is

W.A. Coppel, *Number Theory: An Introduction to Mathematics*, Universitext,
DOI: 10.1007/978-0-387-89486-7_10, © Springer Science + Business Media, LLC 2009

also equivalent to

$$\pi_j(x) \sim x/\varphi(m) \log x \quad (j = 1, \ldots, \varphi(m)).$$

The validity of the second conjecture in this form is known as the *prime number theorem for arithmetic progressions*.

Legendre's first conjecture was proved by Dirichlet (1837) in an outstanding paper which combined number theory, algebra and analysis. His algebraic innovation was the use of *characters* to isolate the primes belonging to a particular residue class mod m. Legendre's second conjecture, which implies the first, was proved by de la Vallée Poussin (1896), again using characters, at the same time that he proved the ordinary prime number theorem.

Selberg (1949), (1950) has given proofs of both conjectures which avoid the use of complex analysis, but they are not very illuminating. The prime number theorem for arithmetic progressions will be proved here by an extension of the method used in the previous chapter to prove the ordinary prime number theorem.

For any integer a, with $1 \le a < m$ and $(a, m) = 1$, let

$$\pi(x; m, a) = \sum_{p \le x, \, p \equiv a \bmod m} 1.$$

Also, generalizing the definition of Chebyshev's functions in the previous chapter, put

$$\theta(x; m, a) = \sum_{p \le x, \, p \equiv a \bmod m} \log p, \quad \psi(x; m, a) = \sum_{n \le x, \, n \equiv a \bmod m} \Lambda(n).$$

Exactly as in the last chapter, we can show that the prime number theorem for arithmetic progressions,

$$\pi(x; m, a) \sim x/\varphi(m) \log x \quad \text{as } x \to \infty,$$

is equivalent to

$$\psi(x; m, a) \sim x/\varphi(m) \quad \text{as } x \to \infty.$$

It is in this form that the theorem will be proved.

2 Characters of Finite Abelian Groups

Let G be an abelian group with identity element e. A *character* of G is a function $\chi : G \to \mathbb{C}$ such that

(i) $\chi(ab) = \chi(a)\chi(b)$ for all $a, b \in G$,
(ii) $\chi(c) \ne 0$ for some $c \in G$.

Since $\chi(c) = \chi(ca^{-1})\chi(a)$, by (i), it follows from (ii) that $\chi(a) \ne 0$ for every $a \in G$. (Thus χ is a *homomorphism* of G into the multiplicative group $\mathbb{C}^\times$ of nonzero complex numbers.) Moreover, since $\chi(a) = \chi(a)\chi(e)$, we must have $\chi(e) = 1$. Since $\chi(a)\chi(a^{-1}) = \chi(e)$, it follows that $\chi(a^{-1}) = \chi(a)^{-1}$.

The function $\chi_1 : G \to \mathbb{C}$ defined by $\chi_1(a) = 1$ for every $a \in G$ is obviously a character of G, the *trivial character* (also called the *principal* character!). Moreover, for any character χ of G, the function $\chi^{-1} : G \to \mathbb{C}$ defined by $\chi^{-1}(a) = \chi(a)^{-1}$ is also a character of G. Furthermore, if χ' and χ'' are characters of G, then the function $\chi'\chi'' : G \to \mathbb{C}$ defined by $\chi'\chi''(a) = \chi'(a)\chi''(a)$ is a character of G. Since

$$\chi_1\chi = \chi, \quad \chi'\chi'' = \chi''\chi', \quad \chi(\chi'\chi'') = (\chi\chi')\chi'',$$

it follows that the set $\hat{G}$ of all characters of G is itself an abelian group, the *dual group* of G, with the trivial character as identity element.

Suppose now that the group G is finite, of order g say. Then $\chi(a)$ is a g-th root of unity for every $a \in G$, since $a^g = e$ and hence

$$\chi(a)^g = \chi(a^g) = \chi(e) = 1.$$

It follows that $|\chi(a)| = 1$ and $\chi^{-1}(a) = \overline{\chi(a)}$. Thus we will sometimes write $\bar{\chi}$ instead of χ^{-1}.

Proposition 1 *The dual group $\hat{G}$ of a finite abelian group G is a finite abelian group of the same order. Moreover, if $a \in G$ and $a \neq e$, then $\chi(a) \neq 1$ for some $\chi \in \hat{G}$.*

Proof Let g denote the order of G. Suppose first that G is a cyclic group, generated by the element c. Then any character χ of G is uniquely determined by the value $\chi(c)$, which is a g-th root of unity. Conversely if $\omega_j = e^{2\pi i j/g} (0 \leq j < g)$ is a g-th root of unity, then the functions $\chi^{(j)} : G \to \mathbb{C}$ defined by $\chi^{(j)}(c^k) = \omega_j^k$ are distinct characters of G and $\chi^{(1)}(c^k) \neq 1$ for $1 \leq k < g$. It follows that the proposition is true when G is cyclic. The general case can be reduced to this by using the fact (see §4 of Chapter III) that any finite abelian group is a direct product of cyclic groups. However, it can also be treated directly in the following way.

We use induction on g and suppose that G is not cyclic. Let H be a maximal proper subgroup of G and let h be the order of H. Let $a \in G\backslash H$ and let r be the least positive integer such that $b = a^r \in H$. Since G is generated by H and a, and $a^n \in H$ if and only if r divides n, each $x \in G$ can be uniquely expressed in the form

$$x = a^k y,$$

where $y \in H$ and $0 \leq k < r$. Hence $g = rh$.

If χ is any character of G, its restriction to H is a character ψ of H. Moreover χ is uniquely determined by ψ and the value $\chi(a)$, since

$$\chi(a^k y) = \chi(a)^k \psi(y).$$

Since $\chi(a)^r = \psi(b)$ is a root of unity, $\omega = \chi(a)$ is a root of unity such that $\omega^r = \psi(b)$.

Conversely, it is easily verified that, for each character ψ of H and for each of the r roots of unity ω such that $\omega^r = \psi(b)$, the function $\chi : G \to \mathbb{C}$ defined by $\chi(a^k y) = \omega^k \psi(y)$ is a character of G. Since H has exactly h characters by the induction hypothesis, it follows that G has exactly $rh = g$ characters. It remains to show that if $a^k y \neq e$, then $\chi(a^k y) \neq 1$ for some χ. But if $\omega^k \psi(y) = 1$ for all ω, then $k = 0$; hence $y \neq e$ and $\chi(y) = \psi(y) \neq 1$ for some ψ, by the induction hypothesis. $\square$

Proposition 2 *Let G be a finite abelian group of order g and $\hat{G}$ its dual group. Then*

(i)

$$\sum_{a \in G} \chi(a) = \begin{cases} g & \text{if } \chi = \chi_1, \\ 0 & \text{if } \chi \neq \chi_1. \end{cases}$$

(ii)

$$\sum_{\chi \in \hat{G}} \chi(a) = \begin{cases} g & \text{if } a = e, \\ 0 & \text{if } a \neq e. \end{cases}$$

Proof Put

$$S = \sum_{a \in G} \chi(a).$$

Since it is obvious that $S = g$ if $\chi = \chi_1$, we assume $\chi \neq \chi_1$. Then $\chi(b) \neq 1$ for some $b \in G$. Since ab runs through all elements of G at the same time as a,

$$\chi(b)S = \sum_{a \in G} \chi(a)\chi(b) = \sum_{a \in G} \chi(ab) = S.$$

Since $\chi(b) \neq 1$, it follows that $S = 0$.

Now put

$$T = \sum_{\chi \in \hat{G}} \chi(a).$$

Evidently $T = g$ if $a = e$ since, by Proposition 1, $\hat{G}$ also has order g. Thus we now assume $a \neq e$. By Proposition 1 also, for some $\psi \in \hat{G}$ we have $\psi(a) \neq 1$. Since $\chi\psi$ runs through all elements of $\hat{G}$ at the same time as χ,

$$\psi(a)T = \sum_{\chi \in \hat{G}} \chi(a)\psi(a) = \sum_{\chi \in \hat{G}} \chi\psi(a) = T.$$

Since $\psi(a) \neq 1$, it follows that $T = 0$. $\square$

Since the product of two characters is again a character, and since $\bar{\psi}$ is the inverse of the character ψ, Proposition 2(i) can be stated in the apparently more general form

(i)′

$$\sum_{a \in G} \chi(a)\bar{\psi}(a) = \begin{cases} g & \text{if } \chi = \psi, \\ 0 & \text{if } \chi \neq \psi. \end{cases}$$

Similarly, since $\bar{\chi}(b) = \chi(b^{-1})$, Proposition 2(ii) can be stated in the form

(ii)′

$$\sum_{\chi \in \hat{G}} \chi(a)\bar{\chi}(b) = \begin{cases} g & \text{if } a = b, \\ 0 & \text{if } a \neq b. \end{cases}$$

The relations (i)′ and (ii)′ are known as the *orthogonality relations*, for the characters and elements respectively, of a finite abelian group.

3 Proof of the Prime Number Theorem for Arithmetic Progressions

The finite abelian group in which we are interested is the multiplicative group $\mathbb{Z}^\times_{(m)}$ of integers relatively prime to m, where $m > 1$ will be fixed from now on. The group $G_m = \mathbb{Z}^\times_{(m)}$ has order $\varphi(m)$, where $\varphi(m)$ denotes as usual the number of positive integers less than m and relatively prime to m.

A *Dirichlet character* mod m is defined to be a function $\chi : \mathbb{Z} \to \mathbb{C}$ with the properties

(i) $\chi(ab) = \chi(a)\chi(b)$ for all $a, b \in \mathbb{Z}$,
(ii) $\chi(a) = \chi(b)$ if $a \equiv b \bmod m$,
(iii) $\chi(a) \neq 0$ if and only if $(a, m) = 1$.

Any character χ of G_m can be extended to a Dirichlet character mod m by putting $\chi(a) = 0$ if $a \in \mathbb{Z}$ and $(a, m) \neq 1$. Conversely, on account of (ii), any Dirichlet character mod m uniquely determines a character of G_m.

To illustrate the definition, here are some examples of Dirichlet characters. In each case we set $\chi(a) = 0$ if $(a, m) \neq 1$.

(I) $m = p$ is an odd prime and $\chi(a) = (a/p)$ if $p \nmid a$, where (a/p) is the Legendre symbol;
(II) $m = 4$ and $\chi(a) = 1$ or -1 according as $a \equiv 1$ or $-1 \bmod 4$;
(III) $m = 8$ and $\chi(a) = 1$ or -1 according as $a \equiv \pm 1$ or $\pm 3 \bmod 8$.

We now return to the general case. By the results of the previous section we have

$$\sum_{n=1}^{m} \chi(n) \equiv \begin{cases} \varphi(m) & \text{if } \chi = \chi_1, \\ 0 & \text{if } \chi \neq \chi_1, \end{cases}$$

and

$$\sum_{\chi} \chi(a) = \begin{cases} \varphi(m) & \text{if } a \equiv 1 \bmod m, \\ 0 & \text{otherwise}, \end{cases}$$

where χ runs through all Dirichlet characters mod m. Furthermore

$$\sum_{n=1}^{m} \chi(n) \bar{\psi}(n) = \begin{cases} \varphi(m) & \text{if } \chi = \psi, \\ 0 & \text{if } \chi \neq \psi, \end{cases}$$

and

$$\sum_{\chi} \chi(a) \bar{\chi}(b) = \begin{cases} \varphi(m) & \text{if } (a, m) = 1 \text{ and } a \equiv b \bmod m, \\ 0 & \text{otherwise}. \end{cases}$$

Lemma 3 *If* $\chi \neq \chi_1$ *is a Dirichlet character* $\bmod m$ *then, for any positive integer* N,

$$\left| \sum_{n=1}^{N} \chi(n) \right| \leq \varphi(m)/2.$$

Proof Any positive integer N can be written in the form $N = qm + r$, where $q \geq 0$ and $1 \leq r \leq m$. Since $\chi(a) = \chi(b)$ if $a \equiv b \bmod m$, we have

$$\sum_{n=1}^{N} \chi(n) = \left(\sum_{n=1}^{m} + \sum_{n=m+1}^{2m} + \cdots + \sum_{n=(q-1)m+1}^{qm} \right) \chi(n) + \sum_{n=qm+1}^{qm+r} \chi(n)$$

$$= q \sum_{n=1}^{m} \chi(n) + \sum_{n=1}^{r} \chi(n).$$

But $\sum_{n=1}^{m} \chi(n) = 0$, since $\chi \neq \chi_1$. Hence

$$\sum_{n=1}^{N} \chi(n) = \sum_{n=1}^{r} \chi(n) = - \sum_{n=r+1}^{m} \chi(n).$$

Since $|\chi(n)| = 1$ or 0 according as $(n, m) = 1$ or $(n, m) \neq 1$, and since $\varphi(m)$ is the number of positive integers $n \leq m$ such that $(n, m) = 1$, the result follows. $\qquad\square$

With each Dirichlet character χ, there is associated a *Dirichlet L-function*

$$L(s, \chi) = \sum_{n=1}^{\infty} \chi(n)/n^s.$$

Since $|\chi(n)| \leq 1$ for all n, the series is absolutely convergent for $\sigma := \mathscr{R}s > 1$. We are going to show that if $\chi \neq \chi_1$, then the series is also convergent for $\sigma > 0$. (It does not converge if $\sigma \leq 0$, since then $|\chi(n)/n^s| \geq 1$ for infinitely many n.)

Put

$$H(x) = \sum_{n \leq x} \chi(n).$$

Then

$$\sum_{n \leq x} \chi(n) n^{-s} = \int_{1-}^{x+} t^{-s} dH(t)$$

$$= H(x) x^{-s} + s \int_{1}^{x} H(t) t^{-s-1} dt.$$

Since $H(x)$ is bounded, by Lemma 3, on letting $x \to \infty$ we obtain

$$L(s, \chi) = s \int_{1}^{\infty} H(t) t^{-s-1} dt \quad \text{for } \sigma > 0.$$

Moreover the integral on the right is uniformly convergent in any half-plane $\sigma \geq \delta$, where $\delta > 0$, and hence $L(s, \chi)$ is a holomorphic function for $\sigma > 0$.

The following discussion of Dirichlet L-functions and the prime number theorem for arithmetic progressions runs parallel to that of the Riemann ζ-function and the ordinary prime number theorem in the previous chapter. Consequently we will be more brief.

Proposition 4 $L(s, \chi) = \Pi_p(1 - \chi(p)p^{-s})^{-1}$ *for* $\sigma > 1$, *where the product is taken over all primes* p.

Proof The property $\chi(ab) = \chi(a)\chi(b)$ for all $a, b \in \mathbb{N}$ enables the proof of Euler's product formula for $\zeta(s)$ to be carried over to the present case. For $\sigma > 0$ we have

$$(1 - \chi(p)p^{-s})^{-1} = 1 + \chi(p)p^{-s} + \chi(p^2)p^{-2s} + \chi(p^3)p^{-3s} + \cdots$$

and hence for $\sigma > 1$

$$\prod_{p \leq x}(1 - \chi(p)p^{-s})^{-1} = \sum_{n \leq N_x} \chi(n)n^{-s},$$

where N_x is the set of all positive integers whose prime factors are all $\leq x$. Letting $x \to \infty$, we obtain the result. □

It follows at once that

$$L(s, \chi_1) = \zeta(s) \prod_{p|m}(1 - p^{-s})$$

and that, for any Dirichlet character χ, $L(s, \chi) \neq 0$ for $\sigma > 1$.

Proposition 5 $-L'(s, \chi)/L(s, \chi) = \sum_{n=1}^{\infty} \chi(n)\Lambda(n)/n^s$ *for* $\sigma > 1$.

Proof The series $\omega(s, \chi) = \sum_{n=1}^{\infty} \chi(n)\Lambda(n)n^{-s}$ converges absolutely and uniformly in any half-plane $\sigma \geq 1 + \varepsilon$, where $\varepsilon > 0$. Moreover, as in the proof of Proposition IX.6,

$$L(s, \chi)\omega(s, \chi) = \sum_{j=1}^{\infty} \chi(j)j^{-s} \sum_{k=1}^{\infty} \chi(k)\Lambda(k)k^{-s} = \sum_{n=1}^{\infty} n^{-s} \sum_{jk=n} \chi(j)\chi(k)\Lambda(k)$$

$$= \sum_{n=1}^{\infty} n^{-s}\chi(n) \sum_{d|n} \Lambda(d) = \sum_{n=1}^{\infty} n^{-s}\chi(n) \log n = -L'(s, \chi). \quad □$$

As in the proof of Proposition IX.6, we can also prove directly that $L(s, \chi) \neq 0$ for $\sigma > 1$, and thus make the proof of the prime number theorem for arithmetic progressions independent of Proposition 4.

The following general result, due to Landau (1905), considerably simplifies the subsequent argument (and has other applications).

Proposition 6 *Let* $\phi(x)$ *be a nondecreasing function for* $x \geq 0$ *such that the integral*

$$f(s) = \int_0^{\infty} e^{-sx}d\phi(x) \tag{†}$$

is convergent for $\mathcal{R}s > \beta$. *Thus* f *is holomorphic in this half-plane. If the definition of* f *can be extended so that it is holomorphic on the real segment* $(\alpha, \beta]$, *then the integral in* (†) *is convergent also for* $\mathcal{R}s > \alpha$. *Thus* f *is actually holomorphic, and* (†) *holds, in this larger half-plane.*

Proof Since f is holomorphic at β, we can choose $\delta > 0$ so that f is holomorphic in the disc $|s - (\beta + \delta)| < 2\delta$. Thus its Taylor series converges in this disc. But for $\mathscr{R}s > \beta$ the n-th derivative of f is given by

$$f^{(n)}(s) = (-1)^n \int_0^\infty e^{-sx} x^n d\phi(x).$$

Hence, for any σ such that $\beta - \delta < \sigma < \beta + \delta$,

$$f(\sigma) = \sum_{n=0}^\infty (\sigma - \beta - \delta)^n f^{(n)}(\beta + \delta)/n!$$

$$= \sum_{n=0}^\infty (\sigma - \beta - \delta)^n (-1)^n \int_0^\infty e^{-(\beta+\delta)x} x^n d\phi(x)/n!$$

$$= \sum_{n=0}^\infty \int_0^\infty e^{-(\beta+\delta)x} (\beta + \delta - \sigma)^n x^n /n!\, d\phi(x).$$

Since the integrands are non-negative, we can interchange the orders of summation and integration, obtaining

$$f(\sigma) = \int_0^\infty e^{-(\beta+\delta)x} \sum_{n=0}^\infty (\beta + \delta - \sigma)^n x^n /n!\, d\phi(x)$$

$$= \int_0^\infty e^{-(\beta+\delta)x} e^{(\beta+\delta-\sigma)x} d\phi(x)$$

$$= \int_0^\infty e^{-\sigma x} d\phi(x).$$

Thus the integral in (†) converges for real $s > \beta - \delta$.

Let γ be the greatest lower bound of all real $s \in (\alpha, \beta)$ for which the integral in (†) converges. Then the integral in (†) is also convergent for $\mathscr{R}s > \gamma$ and defines there a holomorphic function. Since this holomorphic function coincides with $f(s)$ for $\mathscr{R}s > \beta$, it follows that (†) holds for $\mathscr{R}s > \gamma$. Moreover $\gamma = \alpha$, since if $\gamma > \alpha$ we could replace β by γ in the preceding argument and thus obtain a contradiction to the definition of γ. □

The punch-line is the following proposition:

Proposition 7 $L(1 + it, \chi) \neq 0$ *for every real t and every $\chi \neq \chi_1$.*

Proof Assume on the contrary that $L(1 + i\alpha, \chi) = 0$ for some real α and some $\chi \neq \chi_1$. Then also $L(1 - i\alpha, \bar{\chi}) = 0$. If we put

$$f(s) = \zeta^2(s) L(s + i\alpha, \chi) L(s - i\alpha, \bar{\chi}),$$

then f is holomorphic and nonzero for $\sigma > 1$. Furthermore f is holomorphic on the real segment $[1/2, 1]$, since the double pole of $\zeta^2(s)$ at $s = 1$ is cancelled by the zeros of the other two factors. By logarithmic differentiation we obtain, for $\sigma > 1$,

$$-f'(s)/f(s)$$
$$= -2\zeta'(s)/\zeta(s) - L'(s+i\alpha, \chi)/L(s+i\alpha, \chi) - L'(s-i\alpha, \bar{\chi})/L(s-i\alpha, \bar{\chi})$$
$$= 2\sum_{n=1}^{\infty} \Lambda(n)n^{-s} + \sum_{n=1}^{\infty} \chi(n)\Lambda(n)n^{-s-i\alpha} + \sum_{n=1}^{\infty} \bar{\chi}(n)\Lambda(n)n^{-s+i\alpha}$$
$$= \sum_{n=2}^{\infty} c_n n^{-s},$$

where

$$c_n = \{2 + \chi(n)n^{-i\alpha} + \bar{\chi}(n)n^{i\alpha}\}\Lambda(n) = 2\{1 + \mathscr{R}(\chi(n)n^{-i\alpha})\}\Lambda(n).$$

Since $|\chi(n)| \le 1$ and $|n^{-i\alpha}| = 1$, it follows that $c_n \ge 0$ for all $n \ge 2$. If we put

$$g(s) = \sum_{n=2}^{\infty} c_n n^{-s}/\log n,$$

then $g'(s) = f'(s)/f(s)$ for $\sigma > 1$ and so the derivative of $e^{-g(s)}f(s)$ is

$$\{f'(s) - g'(s)f(s)\}e^{-g(s)} = 0.$$

Thus $f(s) = Ce^{g(s)}$, where C is a constant. In fact $C = 1$, since $g(\sigma) \to 0$ and $f(\sigma) \to 1$ as $\sigma \to +\infty$. Since $g(s)$ is the sum of an absolutely convergent Dirichlet series with nonnegative coefficients, so also are the powers $g^k(s)$ ($k = 2, 3, \ldots$). Hence also

$$f(s) = e^{g(s)} = 1 + g(s) + g^2(s)/2! + \cdots = \sum_{n=1}^{\infty} a_n n^{-s} \quad \text{for } \sigma > 1,$$

where $a_n \ge 0$ for every n. It follows from Proposition 6 that the series $\sum_{n=1}^{\infty} a_n n^{-\sigma}$ must actually converge with sum $f(\sigma)$ for $\sigma \ge 1/2$. We will show that this leads to a contradiction.

Take $n = p^2$, where p is a prime. Then, by the manner of its formation,

$$a_n \ge c_n/\log n + c_p^2/2\log^2 p$$
$$= \{2 + \chi(p)^2 p^{-2i\alpha} + \bar{\chi}(p)^2 p^{2i\alpha}\}/2 + \{2 + \chi(p)p^{-i\alpha} + \bar{\chi}(p)p^{i\alpha}\}^2/2$$
$$= 2 - \chi(p)\bar{\chi}(p) + \{1 + \chi(p)p^{-i\alpha} + \bar{\chi}(p)p^{i\alpha}\}^2 \ge 1,$$

since $|\chi(p)| \le 1$. Hence

$$f(1/2) = \sum_{n=1}^{\infty} a_n/n^{1/2} \ge \sum_{n=p^2} a_n/n^{1/2} \ge \sum_p 1/p.$$

Since $\sum_p 1/p$ diverges, this is a contradiction. $\qquad\square$

Proposition 8 $\sum_{n\leq x}\chi_1(n)\Lambda(n)\sim x$, $\sum_{n\leq x}\chi(n)\Lambda(n)=o(x)$ if $\chi\neq\chi_1$.

Proof For any Dirichlet character χ, put

$$g(s) = -\zeta'(s)/\zeta(s) - L'(s,\chi)/2L(s,\chi) - L'(s,\bar{\chi})/2L(s,\bar{\chi}),$$
$$h(s) = -\zeta'(s)/\zeta(s) - L'(s,\chi)/2iL(s,\chi) + L'(s,\bar{\chi})/2iL(s,\bar{\chi}).$$

For $\sigma = \mathscr{R}s > 1$ we have

$$g(s) = \sum_{n=1}^{\infty}\{1 + \mathscr{R}\chi(n)\}\Lambda(n)n^{-s},$$

$$h(s) = \sum_{n=1}^{\infty}\{1 + \mathscr{I}\chi(n)\}\Lambda(n)n^{-s}.$$

If $\chi \neq \chi_1$ then, by Proposition 7, $g(s) - 1/(s-1)$ and $h(s) - 1/(s-1)$ are holomorphic for $\mathscr{R}s \geq 1$. Since the coefficients of the Dirichlet series for $g(s)$ and $h(s)$ are nonnegative, it follows from Ikehara's theorem (Theorem IX.9) that

$$\sum_{n\leq x}\{1 + \mathscr{R}\chi(n)\}\Lambda(n) \sim x,$$

$$\sum_{n\leq x}\{1 + \mathscr{I}\chi(n)\}\Lambda(n) \sim x.$$

On the other hand, if $\chi = \chi_1$ then $g(s) - 2/(s-1)$ and $h(s) - 1/(s-1)$ are holomorphic for $\mathscr{R}s \geq 1$, from which we obtain in the same way

$$\sum_{n\leq x}\{1 + \chi_1(n)\}\Lambda(n) \sim 2x,$$

$$\sum_{n\leq x}\Lambda(n) \sim x.$$

The result follows.   □

The prime number theorem for arithmetic progressions can now be deduced immediately. For, by the orthogonality relations and Proposition 8, if $1 \leq a < m$ and $(a, m) = 1$, then

$$\psi(x;m,a) = \sum_{n\leq x, n\equiv a \bmod m}\Lambda(n)$$

$$= \sum_{\chi}\bar{\chi}(a)\sum_{n\leq x}\chi(n)\Lambda(n)/\varphi(m)$$

$$\sim x/\varphi(m).$$

It is also possible to obtain error bounds in the prime number theorem for arithmetic progressions of the same type as those in the ordinary prime number theorem. For example, it may be shown that for each $\alpha > 0$,

$$\psi(x; m, a) = x/\varphi(m) + O(x/\log^\alpha x),$$
$$\pi(x; m, a) = Li(x)/\varphi(m) + O(x/\log^\alpha x),$$

where the constants implied by the O-symbols depend on α, but not on m or a.

In the same manner as for the Riemann zeta-function $\zeta(s)$ it may be shown that the Dirichlet L-function $L(s, \chi)$ satisfies a functional equation, provided χ is a primitive character. (Here a Dirichlet character χ mod m is *primitive* if for each proper divisor d of m there exists an integer $a \equiv 1 \bmod d$ with $(a, m) = 1$ and $\chi(a) \neq 1$.) Explicitly, if χ is a primitive character mod m and if one puts

$$\Lambda(s, \chi) = (m/\pi)^{s/2} \Gamma((s+\delta)/2) L(s, \chi),$$

where $\delta = 0$ or 1 according as $\chi(-1) = 1$ or -1, then

$$\Lambda(1 - s, \bar\chi) = \varepsilon_\chi \Lambda(s, \chi),$$

where

$$\varepsilon_\chi = i^{-\delta} m^{-1/2} \sum_{k=1}^{m} \bar\chi(k) e^{2\pi i k/m}.$$

It follows from the functional equation that $|\varepsilon_\chi| = 1$. Indeed, by taking complex conjugates we obtain, for real s,

$$\Lambda(1 - s, \chi) = \bar\varepsilon_\chi \Lambda(s, \bar\chi)$$

and hence, on replacing s by $1 - s$,

$$\Lambda(s, \chi) = \bar\varepsilon_\chi \Lambda(1 - s, \bar\chi) = \varepsilon_\chi \bar\varepsilon_\chi \Lambda(s, \chi).$$

The extended Riemann hypothesis implies that no Dirichlet L-function $L(s, \chi)$ has a zero in the half-plane $\mathscr{R}s > 1/2$, since $f(s) = \prod_\chi L(s, \chi)$ is the Dedekind zeta-function of the algebraic number field $K = \mathbb{Q}(e^{2\pi i/m})$. Hence it may be shown that if the extended Riemann hypothesis holds, then

$$\psi(x; m, a) = x/\varphi(m) + O(x^{1/2} \log^2 x)$$

and

$$\pi(x; m, a) = Li(x)/\varphi(m) + O(x^{1/2} \log x),$$

where the constants implied by the O-symbols are independent of m and a. Assuming the extended Riemann hypothesis, Bach and Sorenson (1996) have shown that, for any a, m with $1 \leq a < m$ and $(a, m) = 1$, the least prime $p \equiv a \bmod m$ satisfies $p < 2(m \log m)^2$.

Without any hypothesis, Linnik (1944) proved that there exists an absolute constant L such that the least prime in any arithmetic progression $a, a+m, a+2m, \ldots$, where $1 \leq a < m$ and $(a, m) = 1$, does not exceed m^L if m is sufficiently large. Heath-Brown (1992) has shown that one can take any $L > 11/2$.

4 Representations of Arbitrary Finite Groups

The problem of extending the character theory of finite abelian groups to arbitrary finite groups was proposed by Dedekind and solved by Frobenius (1896). Simplifications were afterwards found by Frobenius himself, Burnside and Schur (1905). We will follow Schur's treatment, which is distinguished by its simplicity. It turns out that for nonabelian groups the concept of 'representation' is more fundamental than that of 'character'.

A *representation* of a group G is a mapping ρ of G into the set of all linear transformations of a finite-dimensional vector space V over the field $\mathbb{C}$ of complex numbers which preserves products, i.e.

$$\rho(st) = \rho(s)\rho(t) \quad \text{for all } s, t \in G, \tag{1}$$

and maps the identity element of G into the identity transformation of V : $\rho(e) = I$. The dimension of the vector space V is called the *degree* of the representation (although 'dimension' would be more natural).

It follows at once from (1) that

$$\rho(s)\rho(s^{-1}) = \rho(s^{-1})\rho(s) = I.$$

Thus, for every $s \in G$, $\rho(s)$ is an invertible linear transformation of V and $\rho(s^{-1}) = \rho(s)^{-1}$. (Hence a representation of G is a *homomorphism* of G into the group $GL(V)$ of all invertible linear transformations of V.)

Any group has a *trivial representation* of degree 1 in which every element of the group is mapped into the scalar 1.

Also, with any group G of finite order g a representation of degree g may be defined in the following way. Let $s_1, \ldots, s_g$ be an enumeration of the elements of G and let $e_1, \ldots, e_g$ be a basis for a g-dimensional vector space V over $\mathbb{C}$. We define a linear transformation $A(s_i)$ of V by its action on the basis elements:

$$A(s_i)e_j = e_k \quad \text{if } s_i s_j = s_k.$$

Then, for all $s, t \in G$,

$$A(s^{-1})A(s) = I, \quad A(st) = A(s)A(t).$$

Thus the mapping $\rho_R : s_i \to A(s_i)$ is a representation of G, known as the *regular representation*.

By choosing a basis for the vector space we can reformulate the preceding definitions in terms of matrices. A representation of a group G is then a product-preserving map $s \to A(s)$ of G into the group of all $n \times n$ non-singular matrices of complex numbers. The positive integer n is the degree of the representation. However, we must regard two matrix representations $s \to A(s)$ and $s \to B(s)$ as *equivalent* if one is obtained from the other simply by changing the basis of the vector space, i.e. if there exists a non-singular matrix T such that

$$T^{-1}A(s)T = B(s) \quad \text{for every } s \in G.$$

It is easily verified that if $s \to A(s)$ is a matrix representation of degree n of a group G, then $s \to A(s^{-1})^t$ (the transpose of $A(s^{-1})$) is a representation of the same degree, the *contragredient representation*. Furthermore, $s \to \det A(s)$ is a representation of degree 1.

Again, if $\rho : s \to A(s)$ and $\sigma : s \to B(s)$ are matrix representations of a group G, of degrees m and n respectively, then the Kronecker product mapping

$$s \to A(s) \otimes B(s)$$

is also a representation of G, of degree mn, since

$$(A(s) \otimes B(s))(A(t) \otimes B(t)) = A(st) \otimes B(st).$$

We will call this representation simply the *product* of the representations ρ and σ, and denote it by $\rho \otimes \sigma$.

The basic problem of representation theory is to determine all possible representations of a given group. As we will see, all representations may in fact be built up from certain 'irreducible' ones.

Let ρ be a representation of a group G by linear transformations of a vector space V. If a subspace U of V is *invariant* under G, i.e. if

$$\rho(s)U \subseteq U \quad \text{for every } s \in G,$$

then the restrictions to U of the given linear transformations provide a representation ρ_U of G by linear transformations of the vector space U. If it happens that there exists another subspace W invariant under G such that V is the direct sum of U and W, i.e. $V = U + W$ and $U \cap W = \{0\}$, then the representation ρ is completely determined by the representations ρ_U and ρ_W and will be said simply to be their *sum*.

A representation ρ of a group G by linear transformations of a vector space V is said to be *irreducible* if no nontrivial proper subspace of V is invariant under G, and *reducible* otherwise. Evidently any representation of degree 1 is irreducible.

A matrix representation $s \to A(s)$, of degree n, of a group G is reducible if it is equivalent to a representation in which all matrices have the block form

$$\begin{pmatrix} P(s) & Q(s) \\ 0 & R(s) \end{pmatrix},$$

where $P(s)$ is a square matrix of order m, $0 < m < n$. Then $s \to P(s)$ and $s \to R(s)$ are representations of G of degrees m and $n - m$ respectively. The given representation is the sum of these representations if there exists a non-singular matrix T such that

$$T^{-1}A(s)T = \begin{pmatrix} P(s) & 0 \\ 0 & R(s) \end{pmatrix} \quad \text{for every } s \in G.$$

The following theorem of Maschke (1899) reduces the problem of finding all representations of a *finite* group to that of finding all irreducible representations.

Proposition 9 *Every representation of a finite group is (equivalent to) a sum of irreducible representations.*

Proof We give a constructive proof due to Schur. Let $s \to A(s)$, where

$$A(s) = \begin{pmatrix} P(s) & Q(s) \\ 0 & R(s) \end{pmatrix},$$

be a reducible representation of a group G of finite order g. Since the mapping $s \to A(s)$ preserves products, we have

$$P(st) = P(s)P(t), \quad R(st) = R(s)R(t), \quad Q(st) = P(s)Q(t) + Q(s)R(t). \quad (2)$$

The non-singular matrix

$$T = \begin{pmatrix} I & M \\ 0 & I \end{pmatrix}$$

satisfies

$$\begin{pmatrix} P(t) & Q(t) \\ 0 & R(t) \end{pmatrix} T = T \begin{pmatrix} P(t) & 0 \\ 0 & R(t) \end{pmatrix} \quad (3)$$

if and only if

$$MR(t) = P(t)M + Q(t).$$

Take

$$M = g^{-1} \sum_{s \in G} Q(s)R(s^{-1}).$$

Then, by (2),

$$P(t)M = g^{-1} \sum_{s \in G} \{Q(ts) - Q(t)R(s)\}R(s^{-1})$$

$$= g^{-1} \sum_{s \in G} Q(ts)R(s^{-1}t^{-1})R(t) - Q(t) = MR(t) - Q(t),$$

and hence (3) holds.

Thus the given reducible representation $s \to A(s)$ is the sum of two representations $s \to P(s)$ and $s \to R(s)$ of lower degree. The result follows by induction on the degree. $\qquad \square$

Maschke's original proof of Proposition 9 depended on showing that every representation of a finite group is equivalent to a representation by *unitary* matrices. We briefly sketch the argument. Let $\rho : s \to A(s)$ be a representation of a finite group G by linear transformations of a finite-dimensional vector space V. We may suppose V equipped with a positive definite inner product (u, v). It is easily verified that

$$(u, v)_G = g^{-1} \sum_{t \in G} (A(t)u, A(t)v)$$

is also a positive definite inner product on V and that it is invariant under G, i.e.

$$(A(s)u, A(s)v)_G = (u, v)_G \quad \text{for every } s \in G.$$

If U is a subspace of V which is invariant under G, and if $U^\perp$ is the subspace consisting of all vectors $v \in V$ such that $(u, v)_G = 0$ for every $u \in U$, then $U^\perp$ is also invariant under G and V is the direct sum of U and $U^\perp$. Thus ρ is the sum of its restrictions to U and $U^\perp$.

The basic result for irreducible representations is *Schur's lemma*, which comes in two parts:

Proposition 10 (i) *Let* $s \to A_1(s)$ *and* $s \to A_2(s)$ *be irreducible representations of a group* G *by linear transformations of the vector spaces* V_1 *and* V_2. *If there exists a linear transformation* $T \neq 0$ *of* V_1 *into* V_2 *such that*

$$T A_1(s) = A_2(s)T \text{ for every } s \in G,$$

then the spaces V_1 *and* V_2 *have the same dimension and* T *is invertible, so that the representations are equivalent.*

(ii) *Let* $s \to A(s)$ *be an irreducible representation of a group* G *by linear transformations of a vector space* V. *A linear transformation* T *of* V *has the property*

$$T A(s) = A(s)T \quad \text{for every } s \in G \tag{4}$$

if and only if $T = \lambda I$ *for some* $\lambda \in \mathbb{C}$.

Proof (i) The image of V_1 under T is a subspace of V_2 which is invariant under the second representation. Since $T \neq 0$ and the representation is irreducible, it must be the whole space: $TV_1 = V_2$. On the other hand, those vectors in V_1 whose image under T is 0 form a subspace of V_1 which is invariant under the first representation. Since $T \neq 0$ and the representation is irreducible, it must contain only the zero vector. Hence distinct vectors of V_1 have distinct images in V_2 under T. Thus T is a one-to-one mapping of V_1 onto V_2.

(ii) By the fundamental theorem of algebra, there exists a complex number λ such that $\det(\lambda I - T) = 0$. Hence $T - \lambda I$ is not invertible. But if T has the property (4), so does $T - \lambda I$. Therefore $T - \lambda I = 0$, by (i) with $A_1 = A_2$. It is obvious that, conversely, (4) holds if $T = \lambda I$. $\square$

Corollary 11 *Every irreducible representation of an abelian group is of degree 1.*

Proof By Proposition 10 (ii) all elements of the group must be represented by scalar multiples of the identity transformation. But such a representation is irreducible only if its degree is 1. $\square$

5 Characters of Arbitrary Finite Groups

By definition, the *trace* of an $n \times n$ matrix $A = (\alpha_{ij})$ is the sum of its main diagonal elements:

$$\mathrm{tr}A = \sum_{i=1}^{n} \alpha_{ii}.$$

It is easily verified that, for any $n \times n$ matrices A, B and any scalars λ, μ, we have

$$\mathrm{tr}(\lambda A + \mu B) = \lambda \mathrm{tr}\, A + \mu \mathrm{tr}\, B,$$
$$\mathrm{tr}(AB) = \mathrm{tr}(BA), \quad \mathrm{tr}(A \otimes B) = (\mathrm{tr}\, A)(\mathrm{tr}\, B).$$

Let $\rho : s \to A(s)$ be a matrix representation of a group G. By the *character* of the representation ρ we mean the mapping $\chi : G \to \mathbb{C}$ defined by

$$\chi(s) = \mathrm{tr}A(s).$$

Since $\mathrm{tr}(T^{-1}AT) = \mathrm{tr}(ATT^{-1}) = \mathrm{tr}A$, equivalent representations have the same character. The significance of characters stems from the converse, which will be proved below.

Clearly the character χ of a representation ρ is a *class function*, i.e.

$$\chi(st) = \chi(ts) \quad \text{for all } s, t \in G.$$

The degree n of the representation ρ is determined by its character χ, since $A(e) = I_n$ and hence $\chi(e) = n$.

If the representation ρ is the sum of two representations ρ' and ρ'', the corresponding characters χ, χ', χ'' evidently satisfy

$$\chi(s) = \chi'(s) + \chi''(s) \quad \text{for every } s \in G.$$

On the other hand, if the representation ρ is the product of the representations ρ' and ρ'', then

$$\chi(s) = \chi'(s)\chi''(s) \quad \text{for every } s \in G.$$

Thus the set of all characters of a group is closed under addition and multiplication. The character of an irreducible representation will be called simply an *irreducible character*.

Let G be a group and ρ a representation of G of degree n with character χ. If s is an element of G of finite order m, then by restriction ρ defines a representation of the cyclic group generated by s. By Proposition 9 and Corollary 11, this representation is equivalent to a sum of representations of degree 1. Thus if S is the matrix representing s, there exists an invertible matrix T such that

$$T^{-1}ST = \mathrm{diag}[\omega_1, \ldots, \omega_n]$$

is a diagonal matrix. Moreover, since

$$T^{-1}S^kT = \text{diag}[\omega_1^k, \ldots, \omega_n^k],$$

$\omega_1, \ldots, \omega_n$ are all m-th roots of unity. Thus

$$\chi(s) = \omega_1 + \cdots + \omega_n$$

is a sum of n m-th roots of unity. Since the inverse of a root of unity ω is its complex conjugate $\bar{\omega}$, it follows that

$$\chi(s^{-1}) = \omega_1^{-1} + \cdots + \omega_n^{-1} = \overline{\chi(s)}.$$

Now let G be a group of finite order g, and let $\rho : s \to A(s)$ and $\sigma : s \to B(s)$ be irreducible matrix representations of G of degrees n and m respectively. For any $n \times m$ matrix C, form the matrix

$$T = \sum_{s \in G} A(s)CB(s^{-1}).$$

Since ts runs through the elements of G at the same time as s,

$$A(t)T = TB(t) \quad \text{for every } t \in G.$$

Therefore, by Schur's lemma, $T = O$ if ρ is not equivalent to σ and $T = \lambda I$ if $\rho = \sigma$. In particular, take C to be any one of the mn matrices which have a single entry 1 and all other entries 0. Then if $A = (\alpha_{ij})$, $B = (\beta_{kl})$, we get

$$\sum_{s \in G} \alpha_{ij}(s)\beta_{kl}(s^{-1}) = \begin{cases} 0 & \text{if } \rho, \sigma \text{ are inequivalent,} \\ \lambda_{jk}\delta_{il} & \text{if } \rho = \sigma, \end{cases}$$

where $\delta_{il} = 1$ or 0 according as $i = l$ or $i \neq l$ ('Kronecker delta'). Since for $(\alpha_{ij}) = (\beta_{ij})$ the left side is unchanged when i is interchanged with k and j with l, we must have $\lambda_{jk} = \lambda\delta_{jk}$. To determine λ set $i = l$, $j = k$ and sum with respect to k. Since the matrices representing s and s^{-1} are inverse, we get $g1 = n\lambda$. Thus

$$\sum_{s \in G} \alpha_{ij}(s)\alpha_{kl}(s^{-1}) = \begin{cases} g/n & \text{if } j = k \text{ and } i = l, \\ 0 & \text{otherwise.} \end{cases}$$

If μ, ν run through an index set for the inequivalent irreducible representations of G, then the relations which have been obtained can be rewritten in the form

$$\sum_{s \in G} \alpha_{ij}^{(\mu)}(s)\alpha_{kl}^{(\nu)}(s^{-1}) = \begin{cases} g/n_\mu & \text{if } \mu = \nu, j = k, i = l, \\ 0 & \text{otherwise.} \end{cases} \tag{5}$$

The *orthogonality relations* (5) for the irreducible matrix elements have several corollaries:

(i) *The functions* $a_{ij}^{(\mu)} : G \to \mathbb{C}$ *are linearly independent.*

For suppose there exist $\lambda_{ij}^{(\mu)} \in \mathbb{C}$ such that

$$\sum_{i,j,\mu} \lambda_{ij}^{(\mu)} a_{ij}^{(\mu)}(s) = 0 \quad \text{for every } s \in G.$$

Multiplying by $a_{kl}^{(v)}(s^{-1})$ and summing over all $s \in G$, we get $(g/n_v)\lambda_{lk}^{(v)} = 0$. Hence every coefficient $\lambda_{lk}^{(v)}$ vanishes.

(ii)

$$\sum_{s \in G} \chi_\mu(s)\chi_v(s^{-1}) = g\delta_{\mu v}. \tag{6}$$

This follows from (5) by setting $i = j, k = l$ and summing over j, l.

(iii) *The irreducible characters χ_μ are linearly independent.*

In fact (iii) follows from (6) in the same way that (i) follows from (5).

The *orthogonality relations* (6) for the irreducible characters enable us to decompose a given representation ρ into irreducible representations. For if $\rho = \oplus m_\mu \rho_\mu$ is a direct sum decomposition of ρ into irreducible components ρ_μ, where the coefficients m_μ are non-negative integers, and if ρ has character χ, then

$$\chi(s) = \sum_\mu m_\mu \chi_\mu(s).$$

Multiplying by $\chi_v(s^{-1})$ and summing over all $s \in G$, we deduce from (6) that

$$g^{-1} \sum_{s \in G} \chi(s)\chi_v(s^{-1}) = m_v. \tag{7}$$

Thus the multiplicities m_v are uniquely determined by the character χ of the representation ρ. It follows that *two representations are equivalent if and only if they have the same character.*

In the same way we find

$$g^{-1} \sum_{s \in G} \chi(s)\chi(s^{-1}) = \sum_\mu m_\mu^2. \tag{8}$$

Hence *a representation ρ with character χ is irreducible if and only if*

$$g^{-1} \sum_{s \in G} \chi(s)\chi(s^{-1}) = 1.$$

The procedure for decomposing a representation into its irreducible components may be applied, in particular, to the regular representation. Evidently the $g \times g$ matrix representing an element s has all its main diagonal elements 0 if $s \neq e$ and all its main diagonal elements 1 if $s = e$. Thus the character χ_R of the regular representation ρ_R is given by

$$\chi_R(e) = g, \quad \chi_R(s) = 0 \quad \text{if } s \neq e.$$

Since $\chi_v(e) = n_v$ is the degree of the v-th irreducible representation, it follows from (7) that $m_v = n_v$. Thus *every irreducible representation is contained in the direct sum decomposition of the regular representation, and moreover each occurs as often as its degree.*

It follows that

$$\sum_\mu n_\mu^2 = g, \quad \sum_\mu n_\mu \chi_\mu(s) = 0 \quad \text{if } s \neq e. \tag{9}$$

Thus the total number of functions $a_{ij}^{(\mu)}$ is $\sum_\mu n_\mu^2 = g$. Therefore, since they are linearly independent, *every function $\phi : G \to \mathbb{C}$ is a linear combination of functions $a_{ij}^{(\mu)}$ occurring in irreducible matrix representations.*

We show next that *every class function $\phi : G \to \mathbb{C}$ is a linear combination of irreducible characters χ_μ.* By what we have just proved $\phi = \sum_\mu \phi_\mu$, where

$$\phi_\mu = \sum_{i,j=1}^{n_\mu} \lambda_{ij}^{(\mu)} a_{ij}^{(\mu)}$$

and $\lambda_{ij}^{(\mu)} \in \mathbb{C}$. But $\phi(st) = \phi(ts)$ and

$$\phi_\mu(st) = \sum_{i,j,k} \lambda_{ik}^{(\mu)} a_{ij}^{(\mu)}(s) a_{jk}^{(\mu)}(t), \quad \phi_\mu(ts) = \sum_{i,j,k} \lambda_{kj}^{(\mu)} a_{ki}^{(\mu)}(t) a_{ij}^{(\mu)}(s).$$

Since the functions $a_{ij}^{(\mu)}$ are linearly independent, we must have

$$\sum_k \lambda_{ik}^{(\mu)} a_{jk}^{(\mu)}(t) = \sum_k \lambda_{kj}^{(\mu)} a_{ki}^{(\mu)}(t).$$

If we denote by $T^{(\mu)}$ the transpose of the matrix $(\lambda_{ik}^{(\mu)})$, we can rewrite this in the form

$$A^{(\mu)}(t) T^{(\mu)} = T^{(\mu)} A^{(\mu)}(t).$$

Consequently, by Schur's lemma, $T^{(\mu)} = \lambda_\mu I_{n_\mu}$ and hence $\phi_\mu = \lambda_\mu \chi_\mu$. Thus $\phi = \sum_\mu \lambda_\mu \chi_\mu$.

Two elements u, v of a group G are said to be *conjugate* if $v = s^{-1}us$ for some $s \in G$. It is easily verified that conjugacy is an equivalence relation. Consequently G is the union of pairwise disjoint subsets, called *conjugacy classes*, such that two elements belong to the same subset if and only if they are conjugate. The inverses of all elements in a conjugacy class again form a conjugacy class, the *inverse class.*

In this terminology a function $\phi : G \to \mathbb{C}$ is a class function if and only if $\phi(u) = \phi(v)$ whenever u and v belong to the same conjugacy class. Thus the number of linearly independent class functions is just the number of conjugacy classes in G. Since the characters χ_μ form a basis for the class functions, it follows that *the number of inequivalent irreducible representations is equal to the number of conjugacy classes in the group.*

If a group of order g has r conjugacy classes then, by (9), $g = n_1^2 + \cdots + n_r^2$. Since it is abelian if and only if every conjugacy class contains exactly one element, i.e. if and only if $r = g$, it follows that *a finite group is abelian if and only if every irreducible representation has degree* 1.

Let $\mathscr{C}_1, \ldots, \mathscr{C}_r$ be the conjugacy classes of the group G and let h_k be the number of elements in $\mathscr{C}_k$ ($k = 1, \ldots, r$). Changing notation, we will now denote by χ_{ik} the common value of the character of all elements in the k-th conjugacy class in the i-th irreducible representation. Then, since $\chi(s^{-1}) = \overline{\chi(s)}$, the orthogonality relations (6) can be rewritten in the form

$$g^{-1} \sum_{j=1}^{r} h_j \chi_{ij} \overline{\chi_{kj}} = \begin{cases} 1 & \text{if } i = k, \\ 0 & \text{if } i \neq k. \end{cases} \tag{10}$$

Thus the $r \times r$ matrices $A = (\chi_{ik})$, $B = (g^{-1} h_i \overline{\chi_{ki}})$ satisfy $AB = I$. Therefore also $BA = I$, i.e.

$$\sum_{j=1}^{r} \overline{\chi_{ji}} \chi_{jk} = \begin{cases} g/h_k & \text{if } i = k, \\ 0 & \text{if } i \neq k. \end{cases} \tag{11}$$

It may be noted that h_k divides g since, for any $s_k \in \mathscr{C}_k$, g/h_k is the order of the subgroup formed by all elements of G which commute with s_k. We are going to show finally that *the degree of any irreducible representation divides the order of the group.*

Any representation $\rho : s \to A(s)$ of a finite group G may be extended by linearity to the set of all linear combinations of elements of G:

$$\rho \left(\sum_{s \in G} \alpha_s s \right) = \sum_{s \in G} \alpha_s A(s).$$

In particular, let C_k denote the sum of all elements in the k-th conjugacy class $\mathscr{C}_k$ of G. For any $t, u \in G$,

$$u^{-1} s_k u t = t(t^{-1} u^{-1} s_k u t)$$

and hence

$$\rho(C_k) A(t) = \sum_{s \in \mathscr{C}_k} A(st) = \sum_{s \in \mathscr{C}_k} A(ts) = A(t) \rho(C_k).$$

If $\rho = \rho_i$ is an irreducible representation, it follows from Schur's lemma that $\rho_i(C_k) = \lambda_{ik} I_{n_i}$. Moreover, since

$$\operatorname{tr} \rho_i(C_k) = h_k \chi_{ik},$$

where h_k again denotes the number of elements in $\mathscr{C}_k$, we must have $\lambda_{ik} = h_k \chi_{ik}/n_i$. Now let

$$C = \sum_{k=1}^{r} (g/h_k) C_k C_{k'},$$

where $\mathscr{C}_{k'}$ is the conjugacy class inverse to $\mathscr{C}_k$. (Otherwise expressed, $C = \sum_{s,t \in G} sts^{-1} t^{-1}$.) Then $\rho_i(C) = \gamma_i I_{n_i}$, where

$$\gamma_i = \sum_{k=1}^{r} (g/h_k) \lambda_{ik} \overline{\lambda_{ik}} = (g/n_i^2) \sum_{k=1}^{r} h_k \chi_{ik} \overline{\chi_{ik}} = (g/n_i)^2,$$

by (10). If $\rho_R(C)$ is the matrix representing C in the regular representation, it follows that there exists an invertible matrix T such that $T^{-1}\rho_R(C)T$ is a diagonal matrix, consisting of the matrices $(g/n_i)^2 I_{n_i}$, repeated n_i times, for every i. In particular, $(g/n_i)^2$ is a root of the characteristic polynomial $\phi(\lambda) = \det(\lambda I_g - \rho_R(C))$ for every i. But $\rho_R(C)$ is a matrix with integer entries and hence the polynomial $\phi(\lambda) = \lambda^g + a_1\lambda^{g-1} + \cdots + a_g$ has integer coefficients $a_1, \ldots, a_g$. The following lemma, already proved in Proposition II.16 but reproved for convenience of reference here, now implies that $(g/n_i)^2$ is an integer and hence that n_i divides g.

Lemma 12 *If $\phi(\lambda) = \lambda^n + a_1\lambda^{n-1} + \cdots + a_n$ is a monic polynomial with integer coefficients $a_1, \ldots, a_n$ and r a rational number such that $\phi(r) = 0$, then r is an integer.*

Proof We can write $r = b/c$, where b and c are relatively prime integers and $c > 0$. Then

$$b^n + a_1 b^{n-1}c + \cdots + a_n c^n = 0$$

and hence c divides b^n. Since c and b have no common prime factor, this implies $c = 1$. $\square$

If we apply the preceding argument to C_k, rather than to C, we see that there exists an invertible matrix T_k such that $T_k^{-1}\rho_R(C_k)T_k$ is a diagonal matrix, consisting of the matrices $(h_k\chi_{ik}/n_i)I_{n_i}$ repeated n_i times, for every i. Thus $h_k\chi_{ik}/n_i$ is a root of the characteristic polynomial $\phi_k(\lambda) = \det(\lambda I_g - \rho_R(C_k))$. Since this is a monic polynomial with integer coefficients, it follows that $h_k\chi_{ik}/n_i$ *is an algebraic integer.*

6 Induced Representations and Examples

Let H be a subgroup of finite index n of a group G, i.e. G is the disjoint union of n left cosets of H:

$$G = s_1 H \cup \cdots \cup s_n H.$$

Also, let there be given a representation $\sigma : t \to A(t)$ of H by linear transformations of a vector space V. The representation $\tilde{\sigma} : s \to \tilde{A}(s)$ of G *induced* by the given representation σ of H is defined in the following way:

Take the vector space $\tilde{V}$ to be the direct sum of n subspaces V_i, where V_i consists of all formal products $s_i \cdot v$ ($v \in V$) with the rules of combination

$$s_i \cdot (v + v') = s_i \cdot v + s_i \cdot v', \qquad s_i \cdot (\lambda v) = \lambda(s_i \cdot v).$$

Then we set

$$\tilde{A}(s)s_i \cdot v = s_j \cdot A(t)v,$$

where t and s_j are determined from s and s_i by requiring that $t = s_j^{-1}ss_i \in H$. The degree of the induced representation of G is thus n times the degree of the original representation of H.

With respect to a given basis of V let $A(t)$ now denote the matrix representing $t \in H$ and put $A(s) = O$ if $s \in G \backslash H$. If one adopts corresponding bases for each of the subspaces V_i, then the matrix $\tilde{A}(s)$ representing $s \in G$ in the induced representation is the block matrix

$$
\tilde{A}(s) = \begin{pmatrix} A(s_1^{-1}ss_1) & A(s_1^{-1}ss_2) & \cdots & A(s_1^{-1}ss_n) \\ A(s_2^{-1}ss_1) & A(s_2^{-1}ss_2) & \cdots & A(s_2^{-1}ss_n) \\ \cdots & \cdots & \cdots & \cdots \\ A(s_n^{-1}ss_1) & A(s_n^{-1}ss_2) & \cdots & A(s_n^{-1}ss_n) \end{pmatrix}.
$$

Evidently each row and each column contains exactly one nonzero block. It should be noted also that a different choice of coset representatives $s_i' = s_i t_i$, where $t_i \in H (i = 1, \ldots, n)$, yields an equivalent representation, since

$$
\begin{pmatrix} A(t_1)^{-1} & \cdots & 0 \\ \cdots & \cdots & \cdots \\ 0 & \cdots & A(t_n)^{-1} \end{pmatrix} \tilde{A}(s) \begin{pmatrix} A(t_1) & \cdots & 0 \\ \cdots & \cdots & \cdots \\ 0 & \cdots & A(t_n) \end{pmatrix}
$$

$$
= \begin{pmatrix} A(s_1'^{-1}ss_1') & \cdots & A(s_1'^{-1}ss_n') \\ \cdots & \cdots & \cdots \\ A(s_n'^{-1}ss_1') & \cdots & A(s_n'^{-1}ss_n') \end{pmatrix}.
$$

Furthermore, changing the order of the cosets corresponds to performing the same permutation on the rows and columns of $\tilde{A}(s)$, and thus also yields an equivalent representation.

It follows that if ψ is the character of the original representation σ of H, then the character $\tilde{\psi}$ of the induced representation $\tilde{\sigma}$ of G is given by

$$
\tilde{\psi}(s) = \sum_{i=1}^{n} \psi(s_i^{-1}ss_i),
$$

where we set $\psi(s) = 0$ if $s \notin H$. If H is of finite order h, this can be rewritten in the form

$$
\tilde{\psi}(s) = h^{-1} \sum_{u \in G} \psi(u^{-1}su), \tag{12}
$$

since $\psi(t^{-1}s_i^{-1}ss_it) = \psi(s_i^{-1}ss_i)$ if $t \in H$.

From any representation of a group G we can also obtain a representation of a subgroup H simply by restricting the given representation to H. We will say that the representation of H is *deduced* from that of G. There is a remarkable reciprocity between induced and deduced representations, discovered by Frobenius (1898):

Proposition 13 *Let* $\rho: s \to A(s)$ *be an irreducible representation of the finite group* G *and* $\sigma: t \to B(t)$ *an irreducible representation of the subgroup* H. *Then the number of times that* σ *occurs in the representation of* H *deduced from the representation* ρ *of* G *is equal to the number of times that* ρ *occurs in the representation of* G *induced by the representation* σ *of* H.

Proof Let χ denote the character of the representation ρ of G and ψ the character of the representation σ of H. By (7), the number of times that ρ occurs in the complete reduction of the induced representation $\tilde{\sigma}$ is

$$g^{-1} \sum_{s \in G} \tilde{\psi}(s)\chi(s^{-1}) = (gh)^{-1} \sum_{s,u \in G} \psi(u^{-1}su)\chi(s^{-1}).$$

If we put $u^{-1}s^{-1}u = t$, $u^{-1} = v$, then $s^{-1} = v^{-1}tv$ and (t, v) runs through all elements of $G \times G$ at the same time as (s, u). Therefore

$$g^{-1} \sum_{s \in G} \tilde{\psi}(s)\chi(s^{-1}) = (gh)^{-1} \sum_{t,v \in G} \chi(v^{-1}tv)\psi(t^{-1})$$

$$= h^{-1} \sum_{t \in G} \chi(t)\psi(t^{-1}) = h^{-1} \sum_{t \in H} \chi(t)\psi(t^{-1}),$$

which is the number of times that σ occurs in the complete reduction of the restriction of ρ to H. $\square$

Corollary 14 *Each irreducible representation of a finite group G is contained in a representation induced by some irreducible representation of a given subgroup H.*

A simple, but still significant, application of these results is to the case where the order of the subgroup H is half that of the whole group G. The subgroup H is then necessarily *normal* (as defined in Chapter I, §7) since, for any $v \in G \backslash H$, the elements of $G \backslash H$ form both a single left coset vH and a single right coset Hv. Hence if $s \to A(s)$ is a representation of H, then so also is $s \to A(v^{-1}sv)$, its *conjugate representation*. Since $v^2 \in H$, the conjugate of the conjugate is equivalent to the original representation. Evidently a representation is irreducible if and only if its conjugate representation is irreducible.

On the other hand G has a nontrivial character λ of degree 1, defined by

$$\lambda(s) = 1 \text{ or } -1 \text{ according as } s \in H \text{ or } s \notin H.$$

If χ is an irreducible character of G, then the character $\chi\lambda$ of the product representation is also irreducible, since

$$1 = g^{-1} \sum_{s \in G} \chi(s)\chi(s^{-1}) = \sum_{s \in G} \chi(s)\lambda(s)\chi(s^{-1})\lambda(s^{-1}).$$

Evidently χ and $\chi\lambda$ have the same degree.

If ψ_i is the character of an irreducible representation of H, we will denote by ψ_i^v the character of its conjugate representation. Thus

$$\psi_i^v(s) = \psi_i(v^{-1}sv).$$

The representation and its conjugate are equivalent if and only if $\psi_i^v(s) = \psi_i(s)$ for every $s \in H$.

Consider now the induced representation $\tilde{\psi}_i$ of G. Since H is a normal subgroup, it follows from (12) that

$$\tilde{\psi}_i(s) = \tilde{\psi}_i^v(s) = 0 \quad \text{if } s \in G\backslash H,$$
$$\tilde{\psi}_i(s) = \tilde{\psi}_i^v(s) = \psi_i(s) + \psi_i^v(s) \quad \text{if } s \in H.$$

Hence $\tilde{\psi}_i = \tilde{\psi}_i^v$ and

$$\sum_{s \in G} \tilde{\psi}_i(s)\tilde{\psi}_i(s^{-1}) = \sum_{s \in H}\{\psi_i(s) + \psi_i^v(s)\}\{\psi_i(s^{-1}) + \psi_i^v(s^{-1})\}$$
$$= \sum_{s \in H} \psi_i(s)\psi_i(s^{-1}) + \sum_{s \in H} \psi_i^v(s)\psi_i^v(s^{-1})$$
$$+ \sum_{s \in H}\{\psi_i(s)\psi_i^v(s^{-1}) + \psi_i(s^{-1})\psi_i^v(s)\}.$$

Consequently, by the orthogonality relations for H,

$$\sum_{s \in G} \tilde{\psi}_i(s)\tilde{\psi}_i(s^{-1}) = 2h + 2\sum_{s \in H} \psi_i(s)\psi_i^v(s^{-1}).$$

If ψ_i and ψ_i^v are inequivalent, the second term on the right vanishes and we obtain

$$\sum_{s \in G} \tilde{\psi}_i(s)\tilde{\psi}_i(s^{-1}) = g.$$

Thus the induced representation $\tilde{\psi}_i$ of G is irreducible, its degree being twice that of ψ_i.

On the other hand, if ψ_i and ψ_i^v are equivalent, then

$$\sum_{s \in G} \tilde{\psi}_i(s)\tilde{\psi}_i(s^{-1}) = 2g.$$

If $\tilde{\psi}_i = \sum_j m_j \chi_j$ is the decomposition of $\tilde{\psi}_i$ into irreducible characters χ_j of G, it follows from (8) that $\sum_j m_j^2 = 2$. This implies that $\tilde{\psi}_i$ decomposes into two inequivalent irreducible characters of G, say $\tilde{\psi}_i = \chi_k + \chi_l$. We will show that in fact $\chi_l = \chi_k\lambda$.

If $\chi_k(s) = 0$ for all $s \notin H$, then

$$\sum_{s \in H} \chi_k(s)\chi_k(s^{-1}) = \sum_{s \in G} \chi_k(s)\chi_k(s^{-1}) = g = 2h$$

and hence, by the same argument as that just used, the restriction of χ_k to H decomposes into two inequivalent irreducible characters of H. Since the restriction of $\tilde{\psi}_i$ to H is $2\psi_i$, this is a contradiction. We conclude that $\chi_k(s) \neq 0$ for some $s \notin H$, i.e. $\chi_k\lambda \neq \chi_k$. Since χ_k occurs once in the decomposition of $\tilde{\psi}_i$, and $\tilde{\psi}_i(s) = 0$ if $s \notin H$,

$$1 = g^{-1} \sum_{s \in G} \tilde{\psi}_i(s) \chi_k(s^{-1})$$

$$= g^{-1} \sum_{s \in H} \tilde{\psi}_i(s) \chi_k(s^{-1})$$

$$= g^{-1} \sum_{s \in H} \tilde{\psi}_i(s) \chi_k(s^{-1}) \lambda(s^{-1})$$

$$= g^{-1} \sum_{s \in G} \tilde{\psi}_i(s) \chi_k(s^{-1}) \lambda(s^{-1}).$$

Thus $\chi_k \lambda$ also occurs once in the decomposition of $\tilde{\psi}_i$, and since $\chi_k \lambda \neq \chi_k$ we must have $\chi_k \lambda = \chi_l$.

In the relation $\sum_i \psi_i(1)^2 = h$, partition the sum into a sum over pairs of distinct conjugate characters and a sum over self-conjugate characters:

$$\Sigma' \{ \psi_i(1)^2 + \psi_i^v(1)^2 \} + \Sigma'' \psi_i(1)^2 = h.$$

Then for the corresponding characters of G we have

$$\Sigma' \tilde{\psi}_i(1)^2 + \Sigma'' \{ \chi_k(1)^2 + \chi_l(1)^2 \} = 2\Sigma' \{ \psi_i(1)^2 + \psi_i^v(1)^2 \} + 2\Sigma'' \psi_i(1)^2 = 2h = g.$$

Since, by Corollary 14, each irreducible character of G appears in the sum on the left, it follows from (9) that each occurs exactly once. Thus we have proved

Proposition 15 *Let the finite group G have a subgroup H of half its order. Then each pair of distinct conjugate characters of H yields by induction a single irreducible character of G of twice the degree, whereas each self-conjugate character of H yields by induction two distinct irreducible characters of G of the same degree, which coincide on H and differ in sign on $G \setminus H$. The irreducible characters of G thus obtained are all distinct, and every irreducible character of G is obtained in this way.*

We will now use Proposition 15 to determine the irreducible characters of several groups of mathematical and physical interest. Let $\mathscr{S}_n$ denote the *symmetric* group consisting of all permutations of the set $\{1, 2, \ldots, n\}$, $\mathscr{A}_n$ the *alternating* group consisting of all even permutations, and C_n the *cyclic* group consisting of all cyclic permutations. Thus $\mathscr{S}_n$ has order $n!$, $\mathscr{A}_n$ has order $n!/2$ and C_n has order n.

The irreducible characters of the abelian group $\mathscr{A}_3 = C_3$ are all of degree 1 and can be arranged as a table in the following way, where ω is a primitive cube root of unity, say $\omega = e^{2\pi i/3} = (-1 + i\sqrt{3})/2$.

		$\mathscr{A}_3$	
	e	(123)	(132)
ψ_1	1	1	1
ψ_2	1	ω	ω^2
ψ_3	1	ω^2	ω

The group $\mathscr{S}_3$ contains $\mathscr{A}_3$ as a subgroup of index 2. The elements of $\mathscr{S}_3$ form three conjugacy classes: $\mathscr{C}_1$ containing only the identity element e, $\mathscr{C}_2$ containing the three elements $(12),(13),(23)$ of order 2, and $\mathscr{C}_3$ containing the two elements $(123),(132)$ of order 3. The irreducible character ψ_1 of $\mathscr{A}_3$ is self-conjugate and yields two irreducible characters of $\mathscr{S}_3$ of degree 1, the trivial character χ_1 and the sign character $\chi_2 = \chi_1\lambda$. The irreducible characters ψ_2, ψ_3 of $\mathscr{A}_3$ are conjugate and yield a single irreducible character χ_3 of $\mathscr{S}_3$ of degree 2. Thus we obtain the character table:

$\mathscr{S}_3$			
	$\mathscr{C}_1$	$\mathscr{C}_2$	$\mathscr{C}_3$
χ_1	1	1	1
χ_2	1	-1	1
χ_3	2	0	-1

The elements of $\mathscr{A}_4$ form four conjugacy classes: $\mathscr{C}_1$ containing only the identity element e, $\mathscr{C}_2$ containing the three elements $t_1 = (12)(34)$, $t_2 = (13)(24)$, $t_3 = (14)(23)$ of order 2, $\mathscr{C}_3$ containing four elements of order 3, namely c, ct_1, ct_2, ct_3, where $c = (123)$, and $\mathscr{C}_4$ containing the remaining four elements of order 3, namely $c^2, c^2t_1, c^2t_2, c^2t_3$. Moreover $N = \mathscr{C}_1 \cup \mathscr{C}_2$ is a normal subgroup of order 4, $H = \{e, c, c^2\}$ is a cyclic subgroup of order 3, and

$$\mathscr{A}_4 = HN, \quad H \cap N = \{e\}.$$

If χ is a character of degree 1 of H, then a character ψ of degree 1 of $\mathscr{A}_4$ is defined by

$$\psi(hn) = \chi(h) \quad \text{for all } h \in H, n \in N.$$

Since H is isomorphic to $\mathscr{A}_3$, we obtain in this way three characters ψ_1, ψ_2, ψ_3 of $\mathscr{A}_4$ of degree 1. Since $\mathscr{A}_4$ has order 12, and $12 = 1 + 1 + 1 + 9$, the remaining irreducible character ψ_4 of $\mathscr{A}_4$ has degree 3. The character table of $\mathscr{A}_4$ can be completed by means of the orthogonality relations (11) and has the following form, where again $\omega = (-1 + i\sqrt{3})/2$.

$\mathscr{A}_4$				
$\|\mathscr{C}\|$	1	3	4	4
$\mathscr{C}$	$\mathscr{C}_1$	$\mathscr{C}_2$	$\mathscr{C}_3$	$\mathscr{C}_4$
ψ_1	1	1	1	1
ψ_2	1	1	ω	ω^2
ψ_3	1	1	ω^2	ω
ψ_4	3	-1	0	0

The group $\mathscr{S}_4$ contains $\mathscr{A}_4$ as a subgroup of index 2 and $v = (12) \in \mathscr{S}_4\backslash\mathscr{A}_4$. The elements of $\mathscr{S}_4$ form five conjugacy classes: $\mathscr{C}_1$ containing only the identity element e, $\mathscr{C}_2$ containing six transpositions (jk) $(1 \le j < k \le 4)$, $\mathscr{C}_3$ containing the three elements of order 2 in $\mathscr{A}_4$, $\mathscr{C}_4$ containing eight elements of order 3, and $\mathscr{C}_5$ containing six elements of order 4.

The self-conjugate character ψ_1 of $\mathscr{A}_4$ yields two characters of $\mathscr{S}_4$ of degree 1, the trivial character χ_1 and the sign character $\chi_2 = \chi_1\lambda$; the pair of conjugate characters ψ_2, ψ_3 of $\mathscr{A}_4$ yields an irreducible character χ_3 of $\mathscr{S}_4$ of degree 2; and the self-conjugate character ψ_4 of $\mathscr{A}_4$ yields two irreducible characters χ_4, χ_5 of $\mathscr{S}_4$ of degree 3. The rows of the character table corresponding to χ_4, χ_5 must have the form

$$
\begin{array}{ccccc}
3 & x & z & w & y \\
3 & -x & z & w & -y
\end{array}
$$

and from the orthogonality relations (11) we obtain $z = -1$, $w = 0$, $xy = -1$. From the orthogonality relations (10) we further obtain $x + y = 0$. Hence $x^2 = 1$ and the complete character table is

			$\mathscr{S}_4$		
$|\mathscr{C}|$	1	6	3	8	6
$\mathscr{C}$	$\mathscr{C}_1$	$\mathscr{C}_2$	$\mathscr{C}_3$	$\mathscr{C}_4$	$\mathscr{C}_5$
χ_1	1	1	1	1	1
χ_2	1	-1	1	1	-1
χ_3	2	0	2	-1	0
χ_4	3	1	-1	0	-1
χ_5	3	-1	-1	0	1

   The physical significance of these groups derives from the fact that $\mathscr{A}_4$ (resp. $\mathscr{S}_4$) is isomorphic to the group of all rotations (resp. orthogonal transformations) of $\mathbb{R}^3$ which map a regular tetrahedron onto itself. Similarly $\mathscr{A}_3$ (resp. $\mathscr{S}_3$) is isomorphic to the group of all plane rotations (resp. plane rotations and reflections) which map an equilateral triangle onto itself.
   An important property of induced representations was proved by R. Brauer (1953): each character of a finite group is a linear combination with integer coefficients (not necessarily non-negative) of characters induced from characters of elementary subgroups. Here a group is said to be *elementary* if it is the direct product of a group whose order is a power of a prime and a cyclic group whose order is not divisible by that prime.
   It may be deduced without difficulty from Brauer's theorem that, if G is a finite group and m the least common multiple of the orders of its elements, then (as had long been conjectured) any irreducible representation of G is equivalent to a representation in the field $\mathbb{Q}(e^{2\pi i/m})$. Green (1955) has shown that Brauer's theorem is actually best possible: if each character of a finite group G is a linear combination with integer coefficients of characters induced from characters of subgroups belonging to some family $\mathscr{F}$, then each elementary subgroup of G is contained in a conjugate of some subgroup in $\mathscr{F}$.

7 Applications

Character theory has turned out to be an invaluable tool in the study of abstract groups. We illustrate this by two results of Burnside (1904) and Frobenius (1901). It is remarkable, first that these applications were found so soon after the development of character theory and secondly that, one century later, there are still no proofs known which do not use character theory.

Lemma 16 *If $\rho: s \to A(s)$ is a representation of degree n of a finite group G, then the character χ of ρ satisfies*

$$|\chi(s)| \leq n \quad \text{for any } s \in G.$$

Moreover, equality holds for some s if and only if $A(s) = \omega I_n$, where $\omega \in \mathbb{C}$.

Proof If $s \in G$ has order m, there exists an invertible matrix T such that

$$T^{-1}A(s)T = \text{diag}[\omega_1, \ldots, \omega_n],$$

where $\omega_1, \ldots, \omega_n$ are m-th roots of unity. Hence $\chi(s) = \omega_1 + \cdots + \omega_n$ and

$$|\chi(s)| \leq |\omega_1| + \cdots + |\omega_n| = n.$$

Moreover $|\chi(s)| = n$ only if $\omega_1, \ldots, \omega_n$ all lie on the same ray through the origin and hence only if they are all equal, since they lie on the unit circle. But then $A(s) = \omega I_n$. $\qquad\square$

The *kernel* of the representation ρ is the set K_ρ of all $s \in G$ for which $\rho(s) = I_n$. Evidently K_ρ is a normal subgroup of G. By Lemma 16, K_ρ may be characterized as the set of all $s \in G$ such that $\chi(s) = n$.

Lemma 17 *Let $\rho: s \to A(s)$ be an irreducible representation of degree n of a finite group G, with character χ, and let $\mathscr{C}$ be a conjugacy class of G containing h elements. If h and n are relatively prime then, for any $s \in \mathscr{C}$, either $\chi(s) = 0$ or $A(s) = \omega I_n$ for some $\omega \in \mathbb{C}$.*

Proof Since h and n are relatively prime, there exist integers a, b such that $ah + bn = 1$. Then

$$\chi(s)/n = ah\chi(s)/n + b\chi(s).$$

Since $h\chi(s)/n$ and $\chi(s)$ are algebraic integers, it follows that $\chi(s)/n$ is an algebraic integer. We may assume that $|\chi(s)| < n$, since otherwise the result follows from Lemma 16.

Suppose s has order m. If $(k, m) = 1$, then the conjugacy class containing s^k also has cardinality h and thus $\chi(s^k)/n$ is an algebraic integer, by what we have already proved. Hence

$$\alpha = \prod_k \chi(s^k)/n,$$

where k runs through all positive integers less than m and relatively prime to m, is also an algebraic integer. But $\chi(s^k) = f(\omega^k)$, where ω is a primitive m-th root of unity and

$$f(x) = x^{r_1} + \cdots + x^{r_n}$$

for some non-negative integers $r_1, \ldots, r_n$ less than m. Thus α is a symmetric function of the primitive roots ω^k. Since the *cyclotomic polynomial*

$$\Phi_n(x) = \prod_k (x - \omega^k)$$

has integer coefficients, it follows that $\alpha \in \mathbb{Q}$. Consequently, by Lemma 12, $\alpha \in \mathbb{Z}$.

But $|\alpha| < 1$, since $|\chi(s)| < n$ and $|\chi(s^k)| \leq n$ for every k. Hence $\alpha = 0$, and thus $\chi(s^k) = 0$ for some k with $(k, m) = 1$. If $g(x)$ is the monic polynomial in $\mathbb{Q}[x]$ of least positive degree such that $g(\omega^k) = 0$, then any polynomial in $\mathbb{Q}[x]$ with ω^k as a root must be divisible by $g(x)$. Since we showed in Chapter II, §5 that the cyclotomic polynomial $\Phi_n(x)$ is irreducible over the field $\mathbb{Q}$, it follows that $g(x) = \Phi_n(x)$ and that $\Phi_n(x)$ divides $f(x)$. Hence also $\chi(s) = f(\omega) = 0$.   □

Before stating the next result we recall from Chapter I, §7 that a group is said to be *simple* if it contains more than one element and has no nontrivial proper normal subgroup.

Proposition 18 *If a finite group G has a conjugacy class $\mathscr{C}$ of cardinality p^a, for some prime p and positive integer a, then G is not a simple group.*

Proof If $s \in \mathscr{C}$ then, by (9),

$$\sum_\mu n_\mu \chi_\mu(s) = 0.$$

Assume the notation chosen so that χ_1 is the character of the trivial representation. If $\chi_\mu(s) = 0$ for every $\mu > 1$ for which p does not divide n_μ, then the displayed equation has the form $1 + p\zeta = 0$, where ζ is an algebraic integer. Since $-1/p$ is not an integer, this contradicts Lemma 12. Consequently, by Lemma 17, for some $v > 1$ we must have $A^{(v)}(s) = \omega I_{n_v}$, where $\omega \in \mathbb{C}$. The set K_v of all elements of G which are represented by the identity transformation in the v-th irreducible representation is a normal subgroup of G. Moreover $K_v \neq \{e\}$, since K_v contains all elements $u^{-1}s^{-1}us$, and $K_v \neq G$, since $v > 1$. Thus G is not simple.   □

Corollary 19 *If G is a group of order $p^a q^b$, where p, q are distinct primes and a, b non-negative integers such that $a + b > 1$, then G is not simple.*

Proof Let $\mathscr{C}_1, \ldots, \mathscr{C}_r$ be the conjugacy classes of G, with $\mathscr{C}_1 = \{e\}$, and let h_k be the cardinality of $\mathscr{C}_k$ $(k = 1, \ldots, r)$. Then h_k divides the order g of G and

$$g = h_1 + \cdots + h_r.$$

Suppose first that $h_j = 1$ for some $j > 1$. Then $\mathscr{C}_j = \{s_j\}$, where s_j commutes with every element of G. Thus the cyclic group H generated by s_j is a normal subgroup of G. Then G is not simple even if $H = G$, since $a + b > 1$ and any proper subgroup of a cyclic group is normal.

Suppose next that $h_k \neq 1$ for every $k > 1$. If G is simple then, by Proposition 18, q divides h_k for every $k > 1$. Since q divides g, it follows that q divides $h_1 = 1$, which is a contradiction.   □

It has been shown by Kazarin (1990) that the normal subgroup generated by the elements of the conjugacy class $\mathscr{C}$ in Proposition 18 is *solvable*. Although no proof of Burnside's Proposition 18 is known which does not use character theory, Goldschmidt (1970) and Matsuyama (1973) have given a rather intricate proof of the important Corollary 19 which is purely group theoretic.

The restriction to *two* distinct primes in the statement of Corollary 19 is essential, since the alternating group $\mathscr{A}_5$ of order $60 = 2^2 \cdot 3 \cdot 5$ is simple. It follows at once from Corollary 19, by induction on the order, that any finite group whose order is divisible by at most two distinct primes is *solvable*. P. Hall (1928/1937) has used Corollary 19 to show that a finite group G of order g is solvable if and only if G has a subgroup H of order h for every factorization $g = p^a h$, where $a > 0$ and p is a prime not dividing h.

The second application of group characters, due to Frobenius, has the following statement:

Proposition 20 *If the finite group G has a nontrivial proper subgroup H such that*

$$x^{-1} H x \cap H = \{e\} \quad \text{for every } x \in G \backslash H,$$

then G contains a normal subgroup N such that G is the semidirect product of H and N, i.e.

$$G = NH, \quad H \cap N = \{e\}.$$

Proof Obviously $x^{-1} H x = y^{-1} H y$ if $y \in Hx$ and the hypotheses imply that $x^{-1} H x \cap y^{-1} H y = \{e\}$ if $y \notin Hx$. If g, h are the orders of G, H respectively, it follows that the number of distinct conjugate subgroups $x^{-1} H x$ (including H itself) is $n = g/h$. Furthermore the number of elements of G which belong to some conjugate subgroup is $n(h-1) + 1 = g - (n-1)$. Thus the set S of elements of G which do not belong to any conjugate subgroup has cardinality $n - 1$.

Let ψ_μ be the character of an irreducible representation of H and $\tilde{\psi}_\mu$ the character of the induced representation of G. By (12) and the hypotheses,

$$\tilde{\psi}_\mu(e) = n\psi_\mu(e), \quad \tilde{\psi}_\mu(s) = 0 \quad \text{if } s \in S, \quad \tilde{\psi}_\mu(s) = \psi_\mu(s) \quad \text{if } s \in H \backslash e.$$

For any fixed μ, form the class function

$$\chi = \tilde{\psi}_\mu - \psi_\mu(e)\{\tilde{\psi}_1 - \chi_1\},$$

where ψ_1 and χ_1 are the characters of the trivial representations of H and G respectively. Then χ is a *generalized character* of G, i.e. $\chi = \sum_\nu m_\nu \chi_\nu$ is a linear combination of irreducible characters χ_ν with integral, but not necessarily non-negative, coefficients m_ν. Moreover

$$\chi(e) = \psi_\mu(e), \quad \chi(s) = \psi_\mu(e) \quad \text{if } s \in S, \quad \chi(s) = \psi_\mu(s) \quad \text{if } s \in H \backslash e.$$

Hence

$$\sum_{s \in H \backslash e} \chi(s)\chi(s^{-1}) = \sum_{s \in H \backslash e} \psi_\mu(s)\psi_\mu(s^{-1}) = h - \psi_\mu(e)^2.$$

Since S has cardinality $n - 1$, it follows that

$$\sum_{s \in G} \chi(s)\chi(s^{-1}) = n\{h - \psi_\mu(e)^2\} + \psi_\mu(e)^2 + (n - 1)\psi_\mu(e)^2 = g.$$

But the formula (8) holds also for generalized characters. Since $\chi(e) > 0$, we conclude that χ is in fact an irreducible character of G. Thus we have an irreducible representation of degree $\chi(e)$ in which the matrices representing elements of S have trace $\chi(e)$. The elements of S must therefore be represented by the unit matrix, i.e. they belong to the kernel K_μ of the representation.

On the other hand, for any $t \in H \backslash e$ we have

$$\sum_\mu \psi_\mu(e)\psi_\mu(t) = 0$$

and hence $\psi_\mu(t) \neq \psi_\mu(e)$ for some μ. Thus the intersection of the kernels K_μ for varying μ contains just the elements of S and e. Since K_μ is a normal subgroup of G, it follows that $N = S \cup \{e\}$ is also a normal subgroup. Furthermore, since $H \cap N = \{e\}$, HN has cardinality $hn = g$ and hence $HN = G$. $\qquad\square$

A finite group G which satisfies the hypotheses of Proposition 20 is said to be a *Frobenius group*. The subgroup H is said to be a *Frobenius complement* and the normal subgroup N a *Frobenius kernel*. It is readily shown that a finite permutation group is a Frobenius group if and only if it is transitive and no element except the identity fixes more than one symbol. Another characterization follows from Proposition 20: a finite group G is a Frobenius group if and only if it has a nontrivial proper normal subgroup N such that, if $x \in N$ and $x \neq e$, then $xy \neq yx$ for all $y \in G \backslash N$.

Frobenius groups are of some general significance and much is known about their structure. It is easily seen that h divides $n - 1$, so that the subgroups H and N have relatively prime orders. It has been shown by Thompson (1959) that the normal subgroup N is a direct product of groups of prime power order. The structure of H is known even more precisely through the work of Burnside (1901) and others.

Applications of group characters of quite a different kind arise in the study of molecular vibrations. We describe one such application within classical mechanics, due to Wigner (1930). However, there are further applications within quantum mechanics, e.g. to the determination of the possible spectral lines in the Raman scattering of light by a substance whose molecules have a particular symmetry group.

A basic problem of classical mechanics deals with the *small oscillations* of a system of particles about an equilibrium configuration. The equations of motion have the form

$$B\ddot{x} + Cx = 0, \tag{13}$$

where $x \in \mathbb{R}^n$ is a vector of generalized coordinates and B, C are positive definite real symmetric matrices. In fact the kinetic energy is $(1/2)\dot{x}^t B\dot{x}$ and, as a first approximation for x near 0, the potential energy is $(1/2)x^t Cx$.

Since B and C are positive definite, there exists (see Chapter V, §4) a non-singular matrix T such that

$$T^t BT = I, \quad T^t CT = D,$$

where D is a diagonal matrix with positive diagonal elements. By the linear transformation $x = Ty$ the equations of motion are brought to the form

$$\ddot{y} + Dy = 0.$$

These 'decoupled' equations can be solved immediately: if

$$y = (\eta_1, \ldots, \eta_n)^t, \quad D = \operatorname{diag}[\omega_1^2, \ldots, \omega_n^2],$$

with $\omega_k > 0$ $(k = 1, \ldots, n)$, then

$$\eta_k = \alpha_k \cos \omega_k t + \beta_k \sin \omega_k t,$$

where $\alpha_k, \beta_k (k = 1, \ldots, n)$ are arbitrary constants of integration. Hence there exist vectors $a_k, b_k \in \mathbb{R}^n$ such that every solution of (13) is a linear combination of solutions of the form

$$a_k \cos \omega_k t, \quad b_k \sin \omega_k t \quad (k = 1, \ldots, n),$$

the so-called *normal modes* of oscillation. The eigenvalues of the matrix $B^{-1}C$ are the squares of the *normal frequencies* $\omega_1, \ldots, \omega_n$.

An important example is the system of particles formed by a molecule of N atoms. Since the displacement of each atom from its equilibrium position is specified by three coordinates, the internal configuration of the molecule without regard to its position and orientation in space may be specified by $n = 3N - 6$ internal coordinates. The determination of the corresponding normal frequencies $\omega_1, \ldots, \omega_n$ may be a formidable task even for moderate values of N. However, the problem is considerably reduced by taking advantage of the symmetry of the molecule.

A *symmetry operation* is an isometry of $\mathbb{R}^3$ which sends the equilibrium position of any atom into the equilibrium position of an atom of the same type. The set of all symmetry operations is clearly a group under composition, the *symmetry group* of the molecule.

For example, the methane molecule CH_4 has four hydrogen atoms at the vertices of a regular tetrahedron and a carbon atom at the centre, from which it follows that the symmetry group of CH_4 is isomorphic to $\mathscr{S}_4$. Similarly, the ammonia molecule NH_3 has three hydrogen atoms and a nitrogen atom at the four vertices of a regular tetrahedron, and hence the symmetry group of NH_3 is isomorphic to $\mathscr{S}_3$.

We return now to the general case. If G is the symmetry group of the molecule, then to each $s \in G$ there corresponds a linear transformation $A(s)$ of the configuration space $\mathbb{R}^n$. Moreover the map $\rho: s \to A(s)$ is a representation of G. Since the kinetic and potential energies are unchanged by a symmetry operation, we have

$$A(s)^t B A(s) = B, \quad A(s)^t C A(s) = C \quad \text{for every } s \in G.$$

It follows that

$$B^{-1}C A(s) = A(s)B^{-1}C \quad \text{for every } s \in G.$$

Assume the notation chosen so that the distinct ω's are $\omega_1, \ldots, \omega_p$ and ω_k occurs m_k times in the sequence $\omega_1, \ldots, \omega_n$ $(k = 1, \ldots, p)$. Thus $n = m_1 + \cdots + m_p$. If V_k is the set of all $v \in \mathbb{R}^n$ such that

$$B^{-1}Cv = \omega_k^2 v,$$

then V_k is an m_k-dimensional subspace of $\mathbb{R}^n (k = 1, \ldots, p)$ and $\mathbb{R}^n$ is the direct sum of $V_1, \ldots, V_p$. Moreover each eigenspace V_k is invariant under $A(s)$ for every $s \in G$. Hence, by Maschke's theorem (which holds also for representations in a real vector space), V_k is a direct sum of real-irreducible invariant subspaces. It follows that there exists a real non-singular matrix T such that, for every $s \in G$,

$$T^{-1}A(s)T = \begin{pmatrix} A_1(s) & 0 & \cdots & 0 \\ 0 & A_2(s) & \cdots & 0 \\ \cdots & \cdots & \cdots & \cdots \\ 0 & 0 & \cdots & A_q(s) \end{pmatrix},$$

where $s \to A_k(s)$ is a real-irreducible representation of G, of degree n_k say ($k = 1, \ldots, q$), and

$$T^{-1}B^{-1}CT = \begin{pmatrix} \lambda_1 I_{n_1} & 0 & \cdots & 0 \\ 0 & \lambda_2 I_{n_2} & \cdots & 0 \\ \cdots & \cdots & \cdots & \cdots \\ 0 & 0 & \cdots & \lambda_q I_{n_q} \end{pmatrix}.$$

If the real-irreducible representations $s \to A_k(s)$ ($k = 1, \ldots, q$) are also complex-irreducible, then their degrees and multiplicities can be found by character theory. Thus by decomposing the representation ρ of G into its irreducible components we can determine the degeneracy of the normal frequencies.

We will not consider here the modifications needed when some real-irreducible component is not also complex-irreducible. Also, it should be noted that it may happen 'accidentally' that $\lambda_j = \lambda_k$ for some $j \neq k$.

As a simple illustration of the preceding discussion we consider the ammonia molecule NH_3. Its internal configuration may be described by the six internal coordinates r_1, r_2, r_3 and $\alpha_{23}, \alpha_{31}, \alpha_{12}$, where r_j is the change from its equilibrium value of the distance from the nitrogen atom to the j-th hydrogen atom, and α_{jk} is the change from its equilibrium value of the angle between the rays joining the nitrogen atom to the j-th and k-th hydrogen atoms.

We will determine the character χ of the corresponding representation ρ of the symmetry group $\mathscr{S}_3$. In the notation of the character table previously given for $\mathscr{S}_3$, there is an element $s \in \mathscr{C}_3$ for which the symmetry operation $A(s)$ cyclically permutes r_1, r_2, r_3 and $\alpha_{23}, \alpha_{31}, \alpha_{12}$. Consequently $\chi(s) = 0$ if $s \in \mathscr{C}_3$. Also, there is an element $t \in \mathscr{C}_2$ for which the symmetry operation $A(t)$ interchanges r_1 with r_2 and α_{23} with α_{31}, but fixes r_3 and α_{12}. Consequently $\chi(t) = 2$ if $t \in \mathscr{C}_2$. Since it is obvious that $\chi(e) = 6$, this determines χ and we adjoin it to the character table of $\mathscr{S}_3$:

$|\mathscr{C}|$	1	3	2
$\mathscr{C}$	$\mathscr{C}_1$	$\mathscr{C}_2$	$\mathscr{C}_3$
χ_1	1	1	1
χ_2	1	-1	1
χ_3	2	0	-1
χ	6	2	0

Decomposing the character χ into its irreducible components by means of (7), we obtain $\chi = 2\chi_1 + 2\chi_3$. Since the irreducible representations of $\mathscr{S}_3$ are all real, this means that the configuration space $\mathbb{R}^6$ is the direct sum of four irreducible invariant subspaces, two of dimension 1 and two of dimension 2. Knowing what to look for, we may verify that the one-dimensional subspaces spanned by $r_1 + r_2 + r_3$ and $\alpha_{23} + \alpha_{31} + \alpha_{12}$ are invariant. Also, the two-dimensional subspace formed by all vectors $\mu_1 r_1 + \mu_2 r_2 + \mu_3 r_3$ with $\mu_1 + \mu_2 + \mu_3 = 0$ is invariant and irreducible, and so is the two-dimensional subspace formed by all vectors $v_1 \alpha_{23} + v_2 \alpha_{31} + v_3 \alpha_{12}$ with $v_1 + v_2 + v_3 = 0$. Hence we can find a real non-singular matrix T such that

$$T^{-1} B^{-1} C T = \begin{pmatrix} \lambda_1 I_1 & 0 & 0 & 0 \\ 0 & \lambda_2 I_1 & 0 & 0 \\ 0 & 0 & \lambda_3 I_2 & 0 \\ 0 & 0 & 0 & \lambda_4 I_2 \end{pmatrix}.$$

This shows that the ammonia molecule NH_3 has two nondegenerate normal frequencies and two doubly degenerate normal frequencies.

8 Generalizations

During the past century the character theory of finite groups has been extensively generalized to infinite groups with a topological structure. It may be helpful to give an overview here, without proofs, of this vast development. The reader wishing to pursue some particular topic may consult the references at the end of the chapter.

A *topological group* is a group G with a topology such that the map $(s, t) \to st^{-1}$ of $G \times G$ into G is continuous. Throughout the following discussion we will assume that G is a topological group which, as a topological space, is *locally compact and Hausdorff*, i.e. any two distinct points are contained in open sets whose closures are disjoint compact sets. (A closed set E in a topological space is *compact* if each open cover of E has a finite subcover. In a metric space this is consistent with the definition of sequential compactness in Chapter I, §4.)

Let $\mathscr{C}_0(G)$ denote the set of all continuous functions $f: G \to \mathbb{C}$ such that $f(s) = 0$ for all s outside some compact subset of G (which may depend on f). A map $M: \mathscr{C}_0(G) \to \mathbb{C}$ is said to be a *nonnegative linear functional* if

(i) $M(f_1 + f_2) = M(f_1) + M(f_2)$ for all $f_1, f_2 \in \mathscr{C}_0(G)$,
(ii) $M(\lambda f) = \lambda M(f)$ for all $\lambda \in \mathbb{C}$ and $f \in \mathscr{C}_0(G)$,
(iii) $M(f) \geq 0$ if $f(s) \geq 0$ for every $s \in G$.

It is said to be a *left* (resp. *right*) *Haar integral* if, in addition, it is nontrivial, i.e. $M(f) \neq 0$ for some $f \in \mathscr{C}_0(G)$, and left (resp. right) invariant, i.e.

(iv) $M(_t f) = M(f)$ for every $t \in G$ and $f \in \mathscr{C}_0(G)$, where $_t f(s) = f(t^{-1}s)$, (resp. $M(f_t) = M(f)$ for every $t \in G$ and $f \in \mathscr{C}_0(G)$, where $f_t(s) = f(st)$).

It was shown by Haar (1933) that a left Haar integral exists on any locally compact group; it was later shown to be uniquely determined apart from a positive multiplicative constant. By defining $M^*(f) = M(f^*)$, where $f^*(s) = f(s^{-1})$ for every $s \in G$, it follows that a right Haar integral also exists and is uniquely determined apart from a positive multiplicative constant.

The notions of left and right Haar integral obviously coincide if the group G is abelian, and it may be shown that they also coincide if G is compact or is a semi-simple Lie group.

We now restrict attention to the case of a left Haar integral. It is easily seen that

$$M(\bar{f}) = \overline{M(f)},$$

where $\bar{f}(s) = \overline{f(s)}$ for every $s \in G$. If we set $(f, g) = M(f\bar{g})$, then the usual inner product properties hold:

$$(f_1 + f_2, g) = (f_1, g) + (f_2, g),$$
$$(\lambda f, g) = \lambda(f, g),$$
$$(f, g) = \overline{(g, f)},$$
$$(f, f) \geq 0, \text{ with equality only if } f \equiv 0.$$

By the *Riesz representation theorem*, there is a unique *positive measure* μ on the σ-algebra $\mathcal{M}$ generated by the compact subsets of G (cf. Chapter XI, §3) such that $\mu(K)$ is finite for every compact set $K \subseteq G$, $\mu(E)$ is the supremum of $\mu(K)$ over all compact $K \subseteq E$ for each $E \in \mathcal{M}$, and

$$M(f) = \int_G f d\mu \quad \text{for every } f \in \mathcal{C}_0(G).$$

The measure μ is necessarily left invariant:

$$\mu(E) = \mu(sE) \quad \text{for all } E \in \mathcal{M} \text{ and } s \in G,$$

where $sE = \{sx : x \in E\}$.

For $p = 1$ or 2, let $L^p(G)$ denote the set of all μ-measurable functions $f : G \to \mathbb{C}$ such that

$$\int_G |f|^p d\mu < \infty.$$

The definition of M can be extended to $L^1(G)$ by setting

$$M(f) = \int_G f d\mu,$$

and the inner product can be extended to $L^2(G)$ by setting

$$(f, g) = \int_G f\bar{g} \, d\mu.$$

Moreover, with this inner product $L^2(G)$ is a *Hilbert space*. If we define the *convolution product* $f * g$ of $f, g \in L^1(G)$ by

$$f * g(s) = \int_G f(st)g(t^{-1}) \, d\mu(t),$$

then $L^1(G)$ is a *Banach algebra* and

$$M(f * g) = M(f)M(g) \quad \text{for all } f, g \in L^1(G).$$

A *unitary representation* of G in a Hilbert space $\mathcal{H}$ is a map ρ of G into the set of all linear transformations of $\mathcal{H}$ which maps the identity element e of G into the identity transformation of $\mathcal{H}$:

$$\rho(e) = I,$$

which preserves not only products in G:

$$\rho(st) = \rho(s)\rho(t) \quad \text{for all } s, t \in G,$$

but also inner products in $\mathcal{H}$:

$$(\rho(s)u, \rho(s)v) = (u, v) \quad \text{for all } s \in G \text{ and all } u, v \in \mathcal{H},$$

and for which the map $(s, v) \rightarrow \rho(s)v$ of $G \times \mathcal{H}$ into $\mathcal{H}$ is continuous (or, equivalently, for which the map $s \rightarrow (\rho(s)v, v)$ of G into $\mathbb{C}$ is continuous at e for every $v \in \mathcal{H}$).

For example, any locally compact group G has a unitary representation ρ in $L^2(G)$, its *regular representation*, defined by

$$(\rho(t)f)(s) = f(t^{-1}s) \quad \text{for all } f \in L^2(G) \text{ and all } s, t \in G.$$

If ρ is a unitary representation of G in a Hilbert space $\mathcal{H}$, and if a closed subspace V of $\mathcal{H}$ is invariant under $\rho(s)$ for every $s \in G$, then so also is its orthogonal complement $V^\perp$. The representation ρ is said to be *irreducible* if the only closed subspaces of $\mathcal{H}$ which are invariant under $\rho(s)$ for every $s \in G$ are $\mathcal{H}$ and $\{0\}$. It has been shown by Gelfand and Raikov (1943) that, for any locally compact group G and any $s \in G \backslash e$, there is an irreducible unitary representation ρ of G with $\rho(s) \neq I$.

Consider now the case in which the locally compact group G is abelian. Then any irreducible unitary representation of G is one-dimensional. Hence if we define a *character* of G to be a continuous function $\chi : G \rightarrow \mathbb{C}$ such that

(i) $\chi(st) = \chi(s)\chi(t)$ for all $s, t \in G$,
(ii) $|\chi(s)| = 1$ for every $s \in G$,

then every irreducible unitary representation is a character, and vice versa.

If multiplication and inversion of characters are defined pointwise, then the set $\hat{G}$ of all characters of G is again an abelian group, the *dual group* of G. Moreover, we can put a topology on $\hat{G}$ by defining a subset of $\hat{G}$ to be open if it is a union of sets of the form

$$N(\psi, \varepsilon, K) = \{\chi \in \hat{G} : |\chi(s)/\psi(s) - 1| < \varepsilon \text{ for all } s \in K\},$$

where $\psi \in \hat{G}$, $\varepsilon > 0$ and K is a compact subset of G. Then $\hat{G}$ is not only abelian, but also a locally compact topological group.

For each fixed $s \in G$, the map $\hat{s}: \chi \to \chi(s)$ is a character of $\hat{G}$. Moreover the map $s \to \hat{s}$ is one-to-one, by the theorem of Gelfand and Raikov, and every character of $\hat{G}$ is obtained in this way. In fact the *duality theorem* of Pontryagin and van Kampen (1934/5) states that G is isomorphic and homeomorphic to the dual group of $\hat{G}$.

The *Fourier transform* of a function $f \in L^1(G)$ is the function $\hat{f}: \hat{G} \to \mathbb{C}$ defined by

$$\hat{f}(\chi) = \int_G f(s)\overline{\chi(s)}\,d\mu(s),$$

where μ is the Haar measure on G. If $f_1, f_2 \in L^1(G) \cap L^2(G)$, then $\hat{f}_1, \hat{f}_2 \in L^2(\hat{G})$ and, with a suitable fixed normalization of the Haar measure $\hat{\mu}$ on $\hat{G}$,

$$(f_1, f_2)_G = (\hat{f}_1, \hat{f}_2)_{\hat{G}}.$$

Furthermore, the map $f \to \hat{f}$ can be uniquely extended to a unitary map of $L^2(G)$ onto $L^2(\hat{G})$. This generalizes *Plancherel's theorem* for Fourier integrals on the real line.

If $f = g * h$, where $g, h \in L^1(G)$, then $f \in L^1(G)$ and

$$\hat{f}(\chi) = \hat{g}(\chi)\hat{h}(\chi) \quad \text{for every } \chi \in \hat{G}.$$

If, in addition, $g, h \in L^2(G)$, then $\hat{f} \in L^1(\hat{G})$ and, with the same choice as before for the Haar measure $\hat{\mu}$ on $\hat{G}$, the *Fourier inversion formula* holds:

$$f(s) = \int_{\hat{G}} \hat{f}(\chi)\chi(s)\,d\hat{\mu}(\chi).$$

The *Poisson summation formula* can also be extended to this general setting. Let H be a closed subgroup of G and let K denote the factor group G/H. If the Haar measures $\mu, \hat{v}$ on $H, \hat{K}$ are suitably chosen then, with appropriate hypotheses on $f \in L^1(G)$,

$$\int_H f(t)\,d\mu(t) = \int_{\hat{K}} \hat{f}(\psi)\,d\hat{v}(\psi).$$

We now give some examples (without spelling out the topologies). If $G = \mathbb{R}$ is the additive group of all real numbers, then its characters are the functions $\chi_t: \mathbb{R} \to \mathbb{C}$, with $t \in \mathbb{R}$, defined by

$$\chi_t(s) = e^{its}.$$

In this case G is isomorphic and homeomorphic to $\hat{G}$ itself under the map $t \to \chi_t$. The Haar integral of $f \in L^1(G)$ is the ordinary Lebesgue integral

$$M(f) = \int_{-\infty}^{\infty} f(s)\,ds,$$

the Fourier transform of f is

$$\hat{f}(t) = \int_{-\infty}^{\infty} f(s)e^{-its}\,ds,$$

and the Fourier inversion formula has the form

$$f(s) = (1/2\pi)\int_{-\infty}^{\infty} \hat{f}(t)e^{its}\,dt.$$

If $G = \mathbb{Z}$ is the additive group of all integers, then its characters are the functions $\chi_z \colon \mathbb{Z} \to \mathbb{C}$, with $z \in \mathbb{C}$ and $|z| = 1$, defined by

$$\chi_z(n) = z^n.$$

Thus $\hat{G}$ is the multiplicative group of all complex numbers of absolute value 1. The Haar integral of $f \in L^1(G)$ is

$$M(f) = \sum_{n=-\infty}^{\infty} f(n),$$

the Fourier transform of f is

$$\hat{f}(e^{i\phi}) = \sum_{n=-\infty}^{\infty} f(n)e^{-in\phi},$$

and the Fourier inversion formula has the form

$$f(n) = (1/2\pi)\int_0^{2\pi} \hat{f}(e^{i\phi})e^{in\phi}\,d\phi.$$

Thus the classical theories of Fourier integrals and Fourier series are just special cases. As another example, let $G = \mathbb{Q}_p$ be the additive group of all p-adic numbers. The characters in this case are the functions $\chi_t \colon \mathbb{Q}_p \to \mathbb{C}$, with $t \in \mathbb{Q}_p$, defined by

$$\chi_t(s) = e^{2\pi i\lambda(st)},$$

where $\lambda(x) = \sum_{j<0} x_j p^j$ if $x \in \mathbb{Q}_p$ is given by $x = \sum_{j=-\infty}^{\infty} x_j p^j$, $x_j \in \{0, 1, \dots, p-1\}$ and $x_j = 0$ for all large $j < 0$. Also in this case $\hat{G}$ is isomorphic and homeomorphic to $\hat{G}$ itself under the map $t \to \chi_t$. If we choose the Haar measure on G so that the measure of the compact set $\mathbb{Z}_p$ of all p-adic integers is 1, then the same choice for $\hat{G}$ is the appropriate one for Plancherel's theorem and the Fourier inversion formula.

Consider next the case in which the group G is compact, but not necessarily abelian. In this case $\mathscr{C}_0(G)$ coincides with the set $\mathscr{C}(G)$ of all continuous functions $f \colon G \to \mathbb{C}$. The Haar integral is both left and right invariant, and we suppose it normalized so that the integral of the constant 1 has the value 1. Then the integral $M(f)$ of any $f \in \mathscr{C}(G)$, or $L^1(G)$, may be called the *invariant mean* of f.

It may be shown that if ρ is a unitary representation of a compact group G in a Hilbert space $\mathscr{H}$, then $\mathscr{H}$ may be represented as a direct sum $\mathscr{H} = \oplus_\alpha \mathscr{H}_\alpha$ of mutually orthogonal finite-dimensional invariant subspaces $\mathscr{H}_\alpha$ such that, for every α, the restriction of ρ to $\mathscr{H}_\alpha$ is irreducible.

In particular, any irreducible unitary representation of a compact group is finite-dimensional. Consequently it is possible to talk about matrix elements and traces, i.e. characters, of irreducible unitary representations. The orthogonality relations for matrix elements and for characters of irreducible representations of finite groups remain valid for irreducible unitary representations of compact groups if one replaces $g^{-1} \sum_{s \in G} f(s)$ by the invariant mean $M(f)$.

Furthermore, any function $f \in \mathscr{C}(G)$ can be uniformly approximated by finite linear combinations of matrix elements of irreducible unitary representations, and any class function $f \in \mathscr{C}(G)$ can be uniformly approximated by finite linear combinations of characters of irreducible unitary representations. Finally, in the direct sum decomposition of the regular representation into finite-dimensional irreducible unitary representations, each irreducible representation occurs as often as its dimension.

Thus the representation theory of compact groups is completely analogous to that of finite groups. Indeed we may regard the representation theory of finite groups as a special case, since any finite group is compact with the discrete topology and any representation is equivalent to a unitary representation.

An example of a compact group which is neither finite nor abelian is the group $G = SU(2)$ of all 2×2 unitary matrices with determinant 1. The elements of G have the form

$$
g = \begin{bmatrix} \gamma & \delta \\ -\bar{\delta} & \bar{\gamma} \end{bmatrix},
$$

where γ, δ are complex numbers such that $|\gamma|^2 + |\delta|^2 = 1$. Writing $\gamma = \xi_0 + i\xi_3$, $\delta = \xi_1 + i\xi_2$, we see that topologically $SU(2)$ is homeomorphic to the sphere

$$
S^3 = \{x = (\xi_0, \xi_1, \xi_2, \xi_3) \in \mathbb{R}^4 : \xi_0^2 + \xi_1^2 + \xi_2^2 + \xi_3^2 = 1\}
$$

and hence is compact and *simply-connected* (i.e. it is path-connected and any closed path can be continuously deformed to a point).

For any integer $n \geq 0$, let V_n denote the vector space of all polynomials $f(z_1, z_2)$ with complex coefficients which are homogeneous of degree n. Writing $z = (z_1, z_2)$, we have

$$
zg = (\gamma z_1 - \bar{\delta} z_2, \delta z_1 + \bar{\gamma} z_2).
$$

Hence if we define a linear transformation T_g of V_n by $(T_g f)(z) = f(zg)$, then $\rho_n : g \to T_g$ is a representation of $SU(2)$ in V_n. It may be shown that this representation is irreducible and is unitary with respect to the inner product

$$
\left(\sum_{k=0}^n \alpha_k z_1^k z_2^{n-k}, \sum_{k=0}^n \beta_k z_1^k z_2^{n-k} \right) = \sum_{k=0}^n k!(n-k)! \alpha_k \bar{\beta}_k.
$$

Moreover, every irreducible representation of $SU(2)$ is equivalent to ρ_n for some $n \geq 0$.

To determine the character χ_n of ρ_n we observe that any $g \in G$ is conjugate in G to a diagonal matrix

$$t = \begin{pmatrix} e^{i\theta} & 0 \\ 0 & e^{-i\theta} \end{pmatrix},$$

where $\theta \in \mathbb{R}$. If $f_k(z_1, z_2) = z_1^k z_2^{n-k} (0 \leq k \leq n)$, then

$$(T_t f_k)(z_1, z_2) = (e^{i\theta} z_1)^k (e^{-i\theta} z_2)^{n-k} = e^{i(2k-n)\theta} f_k(z_1, z_2).$$

Since the polynomials $f_0, \ldots, f_n$ are a basis for V_n it follows that

$$\chi_n(g) = \chi_n(t) = \sum_{k=0}^{n} e^{i(2k-n)\theta}.$$

Thus $\chi_n(I) = n + 1$, $\chi_n(-I) = (-1)^n(n+1)$ and

$$\chi_n(g) = \{e^{i(n+1)\theta} - e^{-i(n+1)\theta}\}/\{e^{i\theta} - e^{-i\theta}\} = \sin(n+1)\theta / \sin\theta \text{ if } g \neq I, -I.$$

From this formula we can easily deduce the decomposition of the product representation $\rho_m \otimes \rho_n$ into irreducible components. Since

$$\chi_m(g)\chi_n(g) = (e^{in\theta} + e^{i(n-2)\theta} + \cdots + e^{-in\theta})\{e^{i(m+1)\theta} - e^{-i(m+1)\theta}\}/\{e^{i\theta} - e^{-i\theta}\}$$
$$= \chi_{m+n}(g) + \chi_{m+n-2}(g) + \cdots + \chi_{|m-n|}(g),$$

we have the *Clebsch–Gordan formula*

$$\rho_m \otimes \rho_n = \rho_{m+n} + \rho_{m+n-2} + \cdots + \rho_{|m-n|}.$$

This formula is the group-theoretical basis for the rule in atomic physics which determines the possible values of the angular momentum when two systems with given angular momenta are coupled.

The complex numbers γ, δ with $|\gamma|^2 + |\delta|^2 = 1$ which specify the matrix $g \in SU(2)$ can be uniquely expressed in the form

$$\gamma = e^{i(\psi+\varphi)/2} \cos\theta/2, \qquad \delta = e^{i(\psi-\varphi)/2} \sin\theta/2,$$

where $0 \leq \theta \leq \pi, 0 \leq \varphi < 2\pi, -2\pi \leq \psi < 2\pi$. Then the invariant mean of any continuous function $f: SU(2) \to \mathbb{C}$ is given by

$$M(f) = (1/16\pi^2) \int_{-2\pi}^{2\pi} \int_0^{2\pi} \int_0^{\pi} f(\theta, \varphi, \psi) \sin\theta \, d\theta \, d\varphi \, d\psi.$$

Another example of a compact group which is neither finite nor abelian is the group $SO(3)$ of all 3×3 real orthogonal matrices with determinant 1. The representations of $SO(3)$ may actually be obtained from those of $SU(2)$, since the two groups are intimately related. This was already shown in §6 of Chapter I, but another version of the proof will now be given.

The set V of all 2×2 matrices v which are skew-Hermitian and have zero trace,

$$v = \begin{pmatrix} \alpha & \beta \\ -\bar{\beta} & \bar{\alpha} \end{pmatrix}, \quad \text{where } \mathscr{R}\alpha = 0,$$

is a three-dimensional real vector space which may be identified with $\mathbb{R}^3$ by writing $\alpha = i\xi_3$, $\beta = \xi_1 + i\xi_2$. Any $g \in G = SU(2)$ defines a linear transformation $T_g : v \to gvg^{-1}$ of $\mathbb{R}^3$. Moreover T_g is an orthogonal transformation, since if

$$T_g v = v_1 = \begin{pmatrix} \alpha_1 & \beta_1 \\ -\bar{\beta}_1 & \bar{\alpha}_1 \end{pmatrix}$$

then, by the product rule for determinants,

$$|\alpha_1|^2 + |\beta_1|^2 = |\alpha|^2 + |\beta|^2.$$

Hence $\det T_g = \pm 1$. In fact, since T_g is a continuous function of g and $SU(2)$ is connected, we must have $\det T_g = \det T_e = 1$ for every $g \in G$. Thus $T_g \in SO(3)$. Since $T_{gh} = T_g T_h$, the map $g \to T_g$ is a representation of G.

Every element of $SO(3)$ is represented in this way, since

$$\text{if } g_\varphi = \begin{pmatrix} e^{-i\varphi/2} & 0 \\ 0 & e^{i\varphi/2} \end{pmatrix} \quad \text{then } T_{g_\varphi} = B_\varphi = \begin{pmatrix} \cos\varphi & \sin\varphi & 0 \\ -\sin\varphi & \cos\varphi & 0 \\ 0 & 0 & 1 \end{pmatrix},$$

$$\text{if } h_\theta = \begin{pmatrix} \cos\theta/2 & -\sin\theta/2 \\ \sin\theta/2 & \cos\theta/2 \end{pmatrix} \quad \text{then } T_{h_\theta} = C_\theta = \begin{pmatrix} 1 & 0 & 0 \\ 0 & \cos\theta & \sin\theta \\ 0 & -\sin\theta & \cos\theta \end{pmatrix},$$

and every $A \in SO(3)$ can be expressed as a product $A = B_\psi C_\theta B_\varphi$, where φ, θ, ψ are *Euler's angles*.

If $T_g = I_3$ is the identity matrix, i.e. if $gv = vg$ for every $v \in V$, then $g = \pm I_2$, since any 2×2 matrix which commutes with both the matrices

$$\begin{pmatrix} 0 & 1 \\ -1 & 0 \end{pmatrix}, \begin{pmatrix} i & 0 \\ 0 & -i \end{pmatrix}$$

must be a scalar multiple of the identity matrix. It follows that $SO(3)$ *is isomorphic to the factor group* $SU(2)/\{\pm I_2\}$.

These examples, and higher-dimensional generalizations, can be treated systematically by the theory of Lie groups. A *Lie group* is a group G with the structure of a finite-dimensional real analytic manifold such that the map $(x, y) \to xy^{-1}$ of $G \times G$ into G is real analytic.

Some examples of Lie groups are

(i) a *Euclidean space* $\mathbb{R}^n$ under vector addition;
(ii) an *n-dimensional torus* (or *n-torus*) $\mathbb{T}^n$, i.e. the direct product of n copies of the multiplicative group $\mathbb{T}^1$ of all complex numbers of absolute value 1;

(iii) the *general linear group* $GL(n)$ of all real nonsingular $n \times n$ matrices under matrix multiplication;

(iv) the *orthogonal group* $O(n)$ of all matrices $X \in GL(n)$ such that $X^t X = I_n$;

(v) the *unitary group* $U(n)$ of all complex $n \times n$ matrices X such that $X^* X = I_n$, where X^* is the conjugate transpose of X; ($U(n)$ may be viewed as a subgroup of $GL(2n)$)

(vi) the *unitary symplectic group* $Sp(n)$ of all quaternion $n \times n$ matrices X such that $X^* X = I_n$, where X^* is the conjugate transpose of X. ($Sp(n)$ may be viewed as a subgroup of $GL(4n)$)

The definition implies that any Lie group is a locally compact topological group. The fifth Paris problem of Hilbert (1900) asks for a characterization of Lie groups among all topological groups. A complete solution was finally given by Gleason, Montgomery and Zippin (1953): a topological group can be given the structure of a Lie group if and only if it is *locally Euclidean*, i.e. there is a neighbourhood of the identity which is homeomorphic to $\mathbb{R}^n$ for some n.

The advantage of Lie groups over arbitrary topological groups is that, by replacing them by their Lie algebras, they can be studied by the methods of *linear* analysis.

A real (resp. complex) *Lie algebra* is a finite-dimensional real (resp. complex) vector space L with a map $(u, v) \to [u, v]$ of $L \times L$ into L, which is linear in u and in v and has the properties

(i) $[v, v] = 0$ for every $v \in L$,

(ii) $[u, [v, w]] + [v, [w, u]] + [w, [u, v]] = 0$ for all $u, v, w \in L$. (Jacobi identity)

It follows from (i) and the linearity of the bracket product that

$$[u, v] + [v, u] = 0 \quad \text{for all } u, v \in L.$$

An example of a real (resp. complex) Lie algebra is the vector space $\mathfrak{gl}(n, \mathbb{R})$ (resp. $\mathfrak{gl}(n, \mathbb{C})$) of all $n \times n$ real (resp. complex) matrices X with $[X, Y] = XY - YX$. Other examples are easily constructed as subalgebras.

A *Lie subalgebra* of a Lie algebra L is a vector subspace M of L such that $u \in M$ and $v \in M$ imply $[u, v] \in M$. Some Lie subalgebras of $\mathfrak{gl}(n, \mathbb{C})$ are

(i) the set A_n of all $X \in \mathfrak{gl}(n + 1, \mathbb{C})$ with tr $X = 0$,

(ii) the set B_n of all $X \in \mathfrak{gl}(2n + 1, \mathbb{C})$ such that $X^t + X = 0$,

(iii) the set C_n of all $X \in \mathfrak{gl}(2n, \mathbb{C})$ such that $X^t J + J X = 0$, where

$$J = \begin{pmatrix} 0 & I_n \\ -I_n & 0 \end{pmatrix},$$

(iv) the set D_n of all $X \in \mathfrak{gl}(2n, \mathbb{C})$ such that $X^t + X = 0$.

The manifold structure of a Lie group G implies that with each $s \in G$ there is associated a real vector space, the *tangent space* at s. The group structure of the Lie group G implies that the tangent space at the identity e of G is a real Lie algebra, which will be denoted by $L(G)$. For example, if $G = GL(n)$ then $L(G) = \mathfrak{gl}(n, \mathbb{R})$. The properties of Lie groups are mirrored by those of their Lie algebras in the following way.

For every real Lie algebra L, there is a simply-connected Lie group $\tilde{G}$ such that $L(\tilde{G}) = L$. Moreover, $\tilde{G}$ is uniquely determined up to isomorphism by L. A connected Lie group G has $L(G) = L$ if and only if G is isomorphic to a factor group $\tilde{G}/D$, where D is a discrete subgroup of the centre of $\tilde{G}$.

A *Lie subgroup* of a Lie group G is a real analytic submanifold H of G which is also a Lie group under the restriction to H of the group structure on G. It may be shown that a subgroup H of a Lie group G is a Lie subgroup if it is a closed subset of G, and is a connected Lie subgroup if and only if it is path-connected. Thus any closed subgroup of $GL(n)$ is a Lie group.

If H is a Lie subgroup of the Lie group G, then $L(H)$ is a Lie subalgebra of $L(G)$. Moreover, if M is a Lie subalgebra of $L(G)$, there is a unique connected Lie subgroup H of G such that $L(H) = M$.

If G_1, G_2 are Lie groups, then a map $f : G_1 \to G_2$ is a *Lie group homomorphism* if it is an analytic map, regarding G_1, G_2 as manifolds, and a homomorphism, regarding G_1, G_2 as groups. It may be shown that any continuous map $f : G_1 \to G_2$ which is a group homomorphism is actually a Lie group homomorphism. (It follows that a locally Euclidean topological group can be given the structure of a Lie group in only one way.)

If L_1, L_2 are Lie algebras, then a map $T : L_1 \to L_2$ is a *Lie algebra homomorphism* if it is linear and $T[u, v] = [Tu, Tv]$ for all $u, v \in L_1$. If G_1, G_2 are Lie groups and if $f : G_1 \to G_2$ is a Lie group homomorphism, then the derivative of f at the identity, $f'(e) : L(G_1) \to L(G_2)$, is a Lie algebra homomorphism. Moreover, if G_1 is connected then distinct Lie group homomorphisms give rise to distinct Lie algebra homomorphisms, and if G_1 is simply-connected then every Lie algebra homomorphism $L(G_1) \to L(G_2)$ arises from some Lie group homomorphism. (In particular, the representations of a connected Lie group are determined by the representations of its Lie algebra.)

A Lie algebra L is *abelian* if $[u, v] = 0$ for all $u, v \in L$. A connected Lie group is abelian if and only if its Lie algebra is abelian. Since the Euclidean space $\mathbb{R}^n$ is a simply-connected Lie group with an n-dimensional abelian Lie algebra, it follows that any n-dimensional connected abelian Lie group is isomorphic to a direct product $\mathbb{R}^{n-k} \times \mathbb{T}^k$ (where $\mathbb{T}^k$ is a k-torus) for some k such that $0 \le k \le n$.

An *ideal* of a Lie algebra L is a vector subspace M of L such that $u \in L$ and $v \in M$ imply $[u, v] \in M$. A connected Lie subgroup H of a connected Lie group G is a normal subgroup if and only if $L(H)$ is an ideal of $L(G)$.

A Lie algebra L is *simple* if it has no ideals except $\{0\}$ and L and is not one-dimensional, and *semisimple* if it has no abelian ideal except $\{0\}$. It may be shown that a Lie algebra is semisimple if and only if it is the direct sum of finitely many ideals, each of which is a simple Lie algebra.

A Lie group is *semisimple* if it is connected and has no connected abelian normal Lie subgroup except $\{e\}$. It follows that a connected Lie group G is semisimple if and only if its Lie algebra $L(G)$ is semisimple.

We turn our attention now to compact Lie groups. It may be shown that a compact topological group can be given the structure of a Lie group if and only if it is finite-dimensional and locally connected. Furthermore, a compact Lie group is isomorphic to a closed subgroup of $GL(n)$ for some n. Other basic results are:

(i) a compact Lie group, and even any compact topological group, has only finitely many connected components;

(ii) a connected compact Lie group is abelian if and only if it is an n-torus $\mathbb{T}^n$ for some n;

(iii) a semisimple connected compact Lie group G has a finite centre. Moreover the simply-connected Lie group $\tilde{G}$ such that $L(\tilde{G}) = L(G)$ is not only semisimple but also compact;

(iv) an arbitrary connected compact Lie group G has the form $G = ZH$, where Z, H are connected compact Lie subgroups, H is semisimple and Z is the component of the centre of G which contains the identity e.

These results essentially reduce the classification of arbitrary compact Lie groups to the classification of those which are semisimple and simply-connected. It may be shown that the latter are in one-to-one correspondence with the semisimple *complex* Lie algebras. Since a semisimple Lie algebra is a direct sum of finitely many simple Lie algebras, we are thus reduced to the classification of the simple complex Lie algebras. The miracle is that these can be completely enumerated: the non-isomorphic simple complex Lie algebras consist of the four infinite families $A_n(n \geq 1)$, $B_n(n \geq 2)$, $C_n(n \geq 3)$, $D_n(n \geq 4)$, of dimensions $n(n+2)$, $n(2n+1)$, $n(2n+1)$, $n(2n-1)$ respectively, and five *exceptional* Lie algebras G_2, F_4, E_6, E_7, E_8 of dimensions $14, 52, 78, 133, 248$ respectively.

To the simple complex Lie algebra A_n corresponds the compact Lie group $SU(n+1)$ of all matrices in $U(n+1)$ with determinant 1; to B_n corresponds the compact Lie group $SO(2n+1)$ of all matrices in $O(2n+1)$ with determinant 1; to C_n corresponds the compact Lie group $Sp(n)$ (whose matrices all have determinant 1), and to D_n corresponds the compact Lie group $SO(2n)$ of all matrices in $O(2n)$ with determinant 1. The groups $SU(n)$ and $Sp(n)$ are simply-connected if $n \geq 2$, whereas $SO(n)$ is connected but has index 2 in its simply-connected covering group $Spin(n)$ if $n \geq 5$. The compact Lie groups corresponding to the five exceptional simple complex Lie algebras are all related to the algebra of *octonions* or Cayley numbers.

Space does not permit consideration here of the methods by which this classification has been obtained, although the methods are just as significant as the result. Indeed they provide a uniform approach to many problems involving the classical groups, giving explicit formulas for the invariant mean and for the characters of all irreducible representations. There is also a notable connection with *groups generated by reflections*.

The classification of arbitrary semisimple Lie groups reduces similarly to the classification of simple *real* Lie algebras, which have also been completely enumerated. The irreducible unitary representations of non-compact semisimple Lie groups have been extensively studied, notably by Harish-Chandra. However, the non-compact case is essentially more difficult than the compact, since any nontrivial representation is infinite-dimensional, and the results are still incomplete. Much of the motivation for this work has come from elementary particle physics where, in the original formulation of Wigner (1939), a particle (specified by its mass and spin) corresponds to an irreducible unitary representation of the inhomogeneous Lorentz group.

9 Further Remarks

The history of Legendre's conjectures on primes in arithmetic progressions is described in Vol. I of Dickson [13]. Dirichlet's original proof is contained in [33], pp. 313–342. Although no simple general proof of Dirichlet's theorem is known, simple proofs have been given for the existence of infinitely many primes congruent to 1 mod m; see Sedrakian and Steinig [41].

If all arithmetic progressions $a, a + m, \ldots$ with $(a, m) = 1$ contain a prime, then they all contain infinitely many, since for any $k > 1$ the arithmetic progression $a + m^k, a + 2m^k, \ldots$ contains a prime.

It may be shown that any finite abelian group G is *isomorphic* to its dual group $\hat{G}$ (although not in a canonical way) by expressing G as a direct product of cyclic groups; see, for example, W. & F. Ellison [15].

In the final step of the proof of Proposition 7 we have followed Bateman [3]. Other proofs that $L(1, \chi) \neq 0$ for every $\chi \neq \chi_1$, which do not use Proposition 6, are given in Hasse [21]. The functional equation for Dirichlet L-functions was first proved by Hurwitz (1882). For proofs of some of the results stated at the end of §3, see Bach and Sorenson [1], Davenport [12], W. & F. Ellison [15] and Prachar [40]. Funakura [18] characterizes Dirichlet L-functions by means of their analytic properties.

The history of the theory of group representations and group characters is described in Curtis [10]. More complete expositions of the subject than ours are given by Serre [42], Feit [16], Huppert [27], and Curtis and Reiner [11]. The proof given here that the degree of an irreducible representation divides the order of the group is not Frobenius' original proof. It first appeared in a footnote of a paper by Schur (1904) on projective representations, where it is attributed to Frobenius. Zassenhaus [50] gives an interpretation in terms of *Casimir operators*.

A character-free proof of Corollary 19 is given in Gagen [19]. P. Hall's theorem is proved in Feit [16], for example. Frobenius groups are studied further in Feit [16] and Huppert [27].

For physical and chemical applications of group representations, see Cornwell [9], Janssen [29], Meijer [36], Birman [4] and Wilson *et al.* [48].

Dym and McKean [14] give an outward-looking introduction to the classical theory of Fourier series and integrals. The formal definition of a topological group is due to Schreier (1926). The Haar integral is discussed by Nachbin [37]. General introductions to abstract harmonic analysis are given by Weil [46], Loomis [34] and Folland [17]. More detailed information on topological groups and their representations is contained in Pontryagin [39], Hewitt and Ross [23] and Gurarii [20]. A simple proof that the additive group $\mathbb{Q}_p$ of all p-adic numbers is isomorphic to its dual group is given by Washington [45]. In the adelic approach to algebraic number theory this isomorphism lies behind the functional equation of the Riemann zeta function; see, for example, Lang [31].

For Hilbert's fifth problem, see Yang [49] and Hirschfeld [24]. The correspondence between Lie groups and Lie algebras was set up by Sophus Lie (1873–1893) in a purely local way, i.e. between neighbourhoods of the identity in the Lie group and of zero in the Lie algebra. Over half a century elapsed before the correspondence was made global by Cartan, Pontryagin and Chevalley. A basic property of solvable Lie algebras was established by Lie, but we owe to Killing (1888–1890) the remarkable classification of simple complex Lie algebras. Some gaps and inaccuracies in Killing's pioneering work were filled and corrected in the thesis of Cartan (1894). The classification of simple real Lie algebras is due to Cartan (1914). The representation theory of semisimple Lie algebras and compact semisimple Lie groups is the creation of Cartan (1913) and Weyl (1925–7). The introduction of groups generated by reflections is due to Weyl.

For the theory of Lie groups, see Chevalley [7], Warner [44], Varadarajan [43], Helgason [22] and Barut and Raczka [2]. The last reference also has information on representations of noncompact Lie groups and applications to quantum theory. The purely algebraic theory of Lie algebras is discussed by Jacobson [28] and Humphreys [25]. Niederle [38] gives a survey of the applications of the exceptional Lie algebras and Lie superalgebras in particle physics. Groups generated by reflections are treated by Humphreys [26], Bourbaki [5] and Kac [30], while Cohen [8] gives a useful overview.

The character theory of locally compact abelian groups, whose roots lie in Dirichlet's theorem on primes in arithmetic progressions, has given something back to number theory in the adelic approach to algebraic number fields; see the thesis of Tate, reproduced (pp. 305–347) in Cassels and Fröhlich [6], Lang [31] and Weil [47]. For a broad historical perspective and future plans, see Mackey [35] and Langlands [32].

10 Selected References

[1] E. Bach and J. Sorenson, Explicit bounds for primes in residue classes, *Math. Comp.* **65** (1996), 1717–1735.

[2] A.O. Barut and R. Raczka, *Theory of group representations and applications*, 2nd ed., Polish Scientific Publishers, Warsaw, 1986.

[3] P.T. Bateman, A theorem of Ingham implying that Dirichlet's L-functions have no zeros with real part one, *Enseign. Math.* **43** (1997), 281–284.

[4] J.L. Birman, *Theory of crystal space groups and lattice dynamics*, Springer-Verlag, Berlin, 1984.

[5] N. Bourbaki, *Groupes et algèbres de Lie: Chapitres 4,5 et 6*, Masson, Paris, 1981.

[6] J.W.S. Cassels and A. Fröhlich (ed.), *Algebraic number theory*, Academic Press, London, 1967.

[7] C. Chevalley, *Theory of Lie groups I*, Princeton University Press, Princeton, 1946. [Reprinted, 1999]

[8] A.M. Cohen, Coxeter groups and three related topics, *Generators and relations in groups and geometries* (ed. A. Barlotti *et al.*), pp. 235–278, Kluwer, Dordrecht, 1991.

[9] J.F. Cornwell, *Group theory in physics*, 3 vols., Academic Press, London, 1984–1989.

[10] C.W. Curtis, *Pioneers of representation theory: Frobenius, Burnside, Schur, and Brauer*, American Mathematical Society, Providence, R.I., 1999.

[11] C.W. Curtis and I. Reiner, *Methods of representation theory*, 2 vols., Wiley, New York, 1990.

[12] H. Davenport, *Multiplicative number theory*, 3rd ed. revised by H.L. Montgomery, Springer-Verlag, New York, 2000.

[13] L.E. Dickson, *History of the theory of numbers*, 3 vols., reprinted Chelsea, New York, 1966.

[14] H. Dym and H.P. McKean, *Fourier series and integrals*, Academic Press, Orlando, FL, 1972.

[15] W. Ellison and F. Ellison, *Prime numbers*, Wiley, New York, 1985.

[16] W. Feit, *Characters of finite groups*, Benjamin, New York, 1967.

[17] G.B. Folland, *A course in abstract harmonic analysis*, CRC Press, Boca Raton, FL, 1995.

[18] T. Funakura, On characterization of Dirichlet L-functions, *Acta Arith.* **76** (1996), 305–315.

[19] T.M. Gagen, *Topics in finite groups*, London Mathematical Society Lecture Note Series **16**, Cambridge University Press, 1976.

[20] V.P. Gurarii, *Group methods in commutative harmonic analysis*, English transl. by D. and S. Dynin, Encyclopaedia of Mathematical Sciences **25**, Springer-Verlag, Berlin, 1998.

[21] H. Hasse, *Vorlesungen über Zahlentheorie*, 2nd ed., Springer-Verlag, Berlin, 1964.

[22] S. Helgason, *Differential geometry, Lie groups and symmetric spaces*, Academic Press, New York, 1978.

[23] E. Hewitt and K.A. Ross, *Abstract harmonic analysis*, 2 vols., Springer-Verlag, Berlin, 1963/1970. [Corrected reprint of Vol. I, 1979]

[24] J. Hirschfeld, The nonstandard treatment of Hilbert's fifth problem, *Trans. Amer. Math. Soc.* **321** (1990), 379–400.

[25] J.E. Humphreys, *Introduction to Lie algebras and representation theory*, Springer-Verlag, New York, 1972.

[26] J.E. Humphreys, *Reflection groups and Coxeter groups*, Cambridge University Press, Cambridge, 1990.

[27] B. Huppert, *Character theory of finite groups,* de Gruyter, Berlin, 1998.

[28] N. Jacobson, *Lie algebras*, Interscience, New York, 1962.

[29] T. Janssen, *Crystallographic groups*, North-Holland, Amsterdam, 1973.

[30] V.G. Kac, *Infinite dimensional Lie Algebras*, corrected reprint of 3rd ed., Cambridge University Press, Cambridge, 1995.

[31] S. Lang, *Algebraic number theory*, 2nd ed., Springer-Verlag, New York, 1994.

[32] R.P. Langlands, Representation theory: its rise and its role in number theory, *Proceedings of the Gibbs symposium* (ed. D.G. Caldi and G.D. Mostow), pp. 181–210, Amer. Math. Soc., Providence, Rhode Island, 1990.

[33] G. Lejeune-Dirichlet, *Werke*, reprinted in one volume, Chelsea, New York, 1969.

[34] L.H. Loomis, *An introduction to abstract harmonic analysis*, Van Nostrand, New York, 1953.

[35] G.W. Mackey, Harmonic analysis as the exploitation of symmetry - a historical survey, *Bull. Amer. Math. Soc. (N.S.)* **3** (1980), 543–698. [Reprinted, with related articles, in G.W. Mackey, *The scope and history of commutative and noncommutative harmonic analysis*, American Mathematical Society, Providence, R.I., 1992]

[36] P.H. Meijer (ed.), *Group theory and solid state physics: a selection of papers*, Vol. 1, Gordon and Breach, New York, 1964.

[37] L. Nachbin, *The Haar integral*, reprinted, Krieger, Huntington, New York, 1976.

[38] J. Niederle, The unusual algebras and their applications in particle physics, *Czechoslovak J. Phys. B* **30** (1980), 1–22.

[39] L.S. Pontryagin, *Topological groups*, English transl. of 2nd ed. by A. Brown, Gordon and Breach, New York, 1966. [Russian original, 1954]

[40] K. Prachar, *Primzahlverteilung*, Springer-Verlag, Berlin, 1957.

[41] N. Sedrakian and J. Steinig, A particular case of Dirichlet's theorem on arithmetic progressions, *Enseign. Math.* **44** (1998), 3–7.

[42] J.-P. Serre, *Linear representations of finite groups*, Springer-Verlag, New York, 1977.

[43] V.S. Varadarajan, *Lie groups, Lie algebras and their representations*, corrected reprint, Springer-Verlag, New York, 1984.

[44] F.W. Warner, *Foundations of differentiable manifolds and Lie groups*, corrected reprint, Springer-Verlag, New York, 1983.

[45] L. Washington, On the self-duality of Q_p, *Amer. Math. Monthly* **81** (1974), 369–371.

[46] A. Weil, *L'integration dans les groupes topologiques et ses applications*, 2nd ed., Hermann, Paris, 1953.

[47] A. Weil, *Basic number theory*, 2nd ed., Springer-Verlag, Berlin, 1973.

[48] E.B. Wilson, J.C. Decius and P.C. Cross, *Molecular vibrations*, McGraw-Hill, New York, 1955.

[49] C.T. Yang, Hilbert's fifth problem and related problems on transformation groups, *Mathematical developments arising from Hilbert problems* (ed. F.E. Browder), pp. 142–146, Amer. Math. Soc., Providence, R.I., 1976.

[50] H. Zassenhaus, An equation for the degrees of the absolutely irreducible representations of a group of finite order, *Canad. J. Math.* **2** (1950), 166–167.

XI

Uniform Distribution and Ergodic Theory

A trajectory of a system which is evolving with time may be said to be 'recurrent' if it keeps returning to any neighbourhood, however small, of its initial point, and 'dense' if it passes arbitrarily near to every point. It may be said to be 'uniformly distributed' if the proportion of time it spends in any region tends asymptotically to the ratio of the volume of that region to the volume of the whole space. In the present chapter these notions will be made precise and some fundamental properties derived. The subject of dynamical systems has its roots in mechanics, but we will be particularly concerned with its applications in number theory.

1 Uniform Distribution

Before introducing our subject, we establish the following interesting result:

Lemma 0 *Let $J = [a, b]$ be a compact interval and $f_n : J \to \mathbb{R}$ a sequence of nondecreasing functions. If $f_n(t) \to f(t)$ for every $t \in J$ as $n \to \infty$, where $f : J \to \mathbb{R}$ is a continuous function, then $f_n(t) \to f(t)$ uniformly on J.*

Proof Evidently f is also nondecreasing. Furthermore, since J is compact, f is uniformly continuous on J. It follows that, for any $\varepsilon > 0$, there is a subdivision $a = t_0 < t_1 < \cdots < t_m = b$ such that

$$f(t_k) - f(t_{k-1}) < \varepsilon \ (k = 1, \ldots, m).$$

We can choose a positive integer p so that, for all $n > p$,

$$|f_n(t_k) - f(t_k)| < \varepsilon \ (k = 0, 1, \ldots, m).$$

If $t \in J$, then $t \in [t_{k-1}, t_k]$ for some $k \in \{1, \ldots, m\}$. Hence

$$f_n(t) - f(t) \leq f_n(t_k) - f(t_k) + f(t_k) - f(t_{k-1}) < 2\varepsilon$$

and similarly

$$f_n(t) - f(t) \geq f_n(t_{k-1}) - f(t_{k-1}) + f(t_{k-1}) - f(t_k) > -2\varepsilon.$$

Thus $|f_n(t) - f(t)| < 2\varepsilon$ for every $t \in J$ if $n > p$. $\qquad\square$

W.A. Coppel, *Number Theory: An Introduction to Mathematics*, Universitext,
DOI: 10.1007/978-0-387-89486-7_11, © Springer Science + Business Media, LLC 2009

For any real number ξ, let $\lfloor \xi \rfloor$ denote again the greatest integer $\leq \xi$ and let

$$\{\xi\} = \xi - \lfloor \xi \rfloor$$

denote the *fractional part* of ξ. We are going to prove that, if ξ is irrational, then the sequence $(\{n\xi\})$ of the fractional parts of the multiples of ξ is *dense* in the unit interval $I = [0, 1]$, i.e. every point of I is a limit point of the sequence.

It is sufficient to show that the points $z_n = e^{2\pi i n \xi}$ $(n = 1, 2, \ldots)$ are dense on the unit circle. Since ξ is irrational, the points z_n are all distinct and $z_n \neq \pm 1$. Consequently they have a limit point on the unit circle. Thus, for any given $\varepsilon > 0$, there exist positive integers m, r such that

$$|z_{m+r} - z_m| < \varepsilon.$$

But

$$|z_{m+r} - z_m| = |z_r - 1| = |z_{n+r} - z_n| \quad \text{for every } n \in \mathbb{N}.$$

If we write $z_r = e^{2\pi i \theta}$, where $0 < \theta < 1$, then $z_{kr} = e^{2\pi i k \theta}$ $(k = 1, 2, \ldots)$. Define the positive integer N by $1/(N + 1) < \theta < 1/N$. Then the points $z_r, z_{2r}, \ldots, z_{Nr}$ follow one another in order on the unit circle and every point of the unit circle is distant less than ε from one of these points.

It may be asked if the sequence $(\{n\xi\})$ is not only dense in I, but also spends 'the right amount of time' in each subinterval of I. To make the question precise we introduce the following definition:

A sequence (ξ_n) of real numbers is said to be *uniformly distributed mod* 1 if, for all α, β with $0 \leq \alpha < \beta \leq 1$,

$$\varphi_{\alpha,\beta}(N)/N \to \beta - \alpha \quad \text{as } N \to \infty,$$

where $\varphi_{\alpha,\beta}(N)$ is the number of positive integers $n \leq N$ such that $\alpha \leq \{\xi_n\} < \beta$.

In this definition we need only require that $\varphi_{0,\alpha}(N)/N \to \alpha$ for every $\alpha \in (0, 1)$, since

$$\varphi_{\alpha,\beta}(N) = \varphi_{0,\beta}(N) - \varphi_{0,\alpha}(N)$$

and hence

$$|\varphi_{\alpha,\beta}(N)/N - (\beta - \alpha)| \leq |\varphi_{0,\beta}(N)/N - \beta| + |\varphi_{0,\alpha}(N)/N - \alpha|.$$

It follows from Lemma 0, with $f_n(t) = \varphi_{0,t}(n)/n$ and $f(t) = t$, that the sequence (ξ_n) is uniformly distributed mod 1 if and only if

$$\varphi_{\alpha,\beta}(N)/N \to \beta - \alpha \quad \text{as } N \to \infty$$

uniformly for all α, β with $0 \leq \alpha < \beta \leq 1$.

It was first shown by Bohl (1909) that, if ξ is irrational, the sequence $(n\xi)$ is uniformly distributed mod 1 in the sense of our definition. Later Weyl (1914,1916) established this result by a less elementary, but much more general argument, which was equally applicable to multi-dimensional problems. The following two theorems, due to Weyl, replace the problem of showing that a sequence is uniformly distributed mod 1 by a more tractable analytic problem.

Theorem 1 *A real sequence (ξ_n) is uniformly distributed mod 1 if and only if, for every function $f : I \to \mathbb{C}$ which is Riemann integrable,*

$$N^{-1} \sum_{n=1}^{N} f(\{\xi_n\}) \to \int_I f(t)\, dt \quad as\ N \to \infty. \tag{1}$$

Proof For any $\alpha, \beta \in I$ with $\alpha < \beta$, let $\chi_{\alpha,\beta}$ denote the *indicator function* of the interval $[\alpha, \beta)$, i.e.

$$\chi_{\alpha,\beta}(t) = 1 \quad \text{for } \alpha \le t < \beta,$$
$$= 0 \quad \text{otherwise.}$$

Since

$$\int_I \chi_{\alpha,\beta}(t)\, dt = \beta - \alpha,$$

the definition of uniform distribution can be rephrased by saying that the sequence (ξ_n) is uniformly distributed mod 1 if and only if, for all choices of α and β,

$$N^{-1} \sum_{n=1}^{N} \chi_{\alpha,\beta}(\{\xi_n\}) \to \int_I \chi_{\alpha,\beta}(t)\, dt \quad as\ N \to \infty.$$

Thus the sequence (ξ_n) is certainly uniformly distributed mod 1 if (1) holds for every Riemann integrable function f.

Suppose now that the sequence (ξ_n) is uniformly distributed mod 1. Then (1) holds not only for every function $f = \chi_{\alpha,\beta}$, but also for every finite linear combination of such functions, i.e. for every *step-function* f. But, for any real-valued Riemann integrable function f and any $\varepsilon > 0$, there exist step-functions f_1, f_2 such that

$$f_1(t) \le f(t) \le f_2(t) \quad \text{for every } t \in I$$

and

$$\int_I (f_2(t) - f_1(t))\, dt < \varepsilon.$$

Hence

$$N^{-1} \sum_{n=1}^{N} f(\{\xi_n\}) - \int_I f(t)\, dt \le N^{-1} \sum_{n=1}^{N} f_2(\{\xi_n\}) - \int_I f_2(t)\, dt + \varepsilon$$
$$< 2\varepsilon \quad \text{for all large } N,$$

and similarly

$$N^{-1} \sum_{n=1}^{N} f(\{\xi_n\}) - \int_I f(t)\, dt > -2\varepsilon \quad \text{for all large } N.$$

Thus (1) holds when the Riemann integrable function f is real-valued and also, by linearity, when it is complex-valued. □

A converse of Theorem 1 has been proved by de Bruijn and Post (1968): if a function $f : I \to \mathbb{C}$ has the property that

$$\lim_{N \to \infty} N^{-1} \sum_{n=1}^{N} f(\{\xi_n\})$$

exists for every sequence (ξ_n) which is uniformly distributed mod 1, then f is Riemann integrable.

In the statement of the next result, and throughout the rest of the chapter, we use the abbreviation

$$e(t) = e^{2\pi i t}.$$

In the proof of the next result we use the *Weierstrass approximation theorem*: any continuous function $f : I \to \mathbb{C}$ of period 1 is the uniform limit of a sequence (f_n) of trigonometric polynomials. In fact, as Fejér (1904) showed, one can take f_n to be the arithmetic mean $(S_0 + \cdots + S_{n-1})/n$, where

$$S_m = S_m(x) := \sum_{h=-m}^{m} c_h e(hx)$$

is the m-th partial sum of the Fourier series for f. This yields the explicit formula

$$f_n(x) = \int_I K_n(x - t) f(t) \, dt,$$

where

$$K_n(u) = (\sin^2 n\pi u)/(n \sin^2 \pi u).$$

Theorem 2 *A real sequence (ξ_n) is uniformly distributed mod 1 if and only if, for every integer $h \neq 0$,*

$$N^{-1} \sum_{n=1}^{N} e(h\xi_n) \to 0 \quad as \ N \to \infty. \tag{2}$$

Proof If the sequence (ξ_n) is uniformly distributed mod 1 then, by taking $f(t) = e(ht)$ in Theorem 1 we obtain (2) since, for every integer $h \neq 0$,

$$\int_I e(ht) \, dt = 0.$$

Conversely, suppose (2) holds for every nonzero integer h. Then, by linearity, for any trigonometric polynomial

$$g(t) = \sum_{h=-m}^{m} b_h e(ht)$$

we have

$$N^{-1} \sum_{n=1}^{N} g(\{\xi_n\}) \to b_0 = \int_I g(t) \, dt \quad as \ N \to \infty.$$

If f is a continuous function of period 1 then, by the Weierstrass approximation theorem, for any $\varepsilon > 0$ there exists a trigonometric polynomial $g(t)$ such that $|f(t) - g(t)| < \varepsilon$ for every $t \in I$. Hence

$$\left| N^{-1} \sum_{n=1}^{N} f(\{\xi_n\}) - \int_I f(t)\, dt \right|$$

$$\leq \left| N^{-1} \sum_{n=1}^{N} (f(\{\xi_n\}) - g(\{\xi_n\})) \right| + \left| N^{-1} \sum_{n=1}^{N} g(\{\xi_n\}) - \int_I g(t)\, dt \right|$$

$$+ \left| \int_I (g(t) - f(t))\, dt \right|$$

$$< 2\varepsilon + \left| N^{-1} \sum_{n=1}^{N} g(\{\xi_n\}) - \int_I g(t)\, dt \right|$$

$$< 3\varepsilon \quad \text{for all large } N.$$

Thus (1) holds for every continuous function f of period 1.

Finally, if $\chi_{\alpha,\beta}$ is the function defined in the proof of Theorem 1 then, for any $\varepsilon > 0$, there exist continuous functions f_1, f_2 of period 1 such that

$$f_1(t) \leq \chi_{\alpha,\beta}(t) \leq f_2(t) \quad \text{for every } t \in I$$

and

$$\int_I (f_2(t) - f_1(t))\, dt < \varepsilon,$$

from which it follows similarly that

$$N^{-1} \sum_{n=1}^{N} \chi_{\alpha,\beta}(\{\xi_n\}) \to \int_I \chi_{\alpha,\beta}(t)\, dt \quad \text{as } N \to \infty.$$

Thus the sequence (ξ_n) is uniformly distributed mod 1. $\qquad \square$

Weyl's criterion, as Theorem 2 is usually called, immediately implies Bohl's result:

Proposition 3 *If ξ is irrational, the sequence $(n\xi)$ is uniformly distributed mod 1.*

Proof For any nonzero integer h,

$$e(h\xi) + e(2h\xi) + \cdots + e(Nh\xi) = (e((N+1)h\xi) - e(h\xi))/(e(h\xi) - 1).$$

Hence

$$\left| N^{-1} \sum_{n=1}^{N} e(hn\xi) \right| \leq 2|e(h\xi) - 1|^{-1} N^{-1},$$

and the result follows from Theorem 2. $\qquad \square$

These results can be immediately extended to higher dimensions. A sequence (x_n) of vectors in $\mathbb{R}^d$ is said to be *uniformly distributed* mod 1 if, for all vectors $a = (\alpha_1, \ldots, \alpha_d)$ and $b = (\beta_1, \ldots, \beta_d)$ with $0 \le \alpha_k < \beta_k \le 1$ ($k = 1, \ldots, d$),

$$\varphi_{a,b}(N)/N \to \prod_{k=1}^{d}(\beta_k - \alpha_k) \quad \text{as } N \to \infty,$$

where $x_n = (\xi_n^{(1)}, \ldots, \xi_n^{(d)})$ and $\varphi_{a,b}(N)$ is the number of positive integers $n \le N$ such that $\alpha_k \le \{\xi_n^{(k)}\} < \beta_k$ for every $k \in \{1, \ldots, d\}$. Let I^d be the set of all $x = (\xi^{(1)}, \ldots, \xi^{(d)})$ such that $0 \le \xi^{(k)} \le 1$ ($k = 1, \ldots, d$) and, for an arbitrary vector $x = (\xi^{(1)}, \ldots, \xi^{(d)})$, put

$$\{x\} = (\{\xi^{(1)}\}, \ldots, \{\xi^{(d)}\}).$$

Then Theorems 1 and 2 have the following generalizations:

Theorem 1′ *A sequence (x_n) of vectors in $\mathbb{R}^d$ is uniformly distributed mod 1 if and only if, for every function $f : I^d \to \mathbb{C}$ which is Riemann integrable,*

$$N^{-1} \sum_{n=1}^{N} f(\{x_n\}) \to \int_I \cdots \int_I f(t_1, \ldots, t_d)\, dt_1 \cdots dt_d \quad \text{as } N \to \infty.$$

Theorem 2′ *A sequence (x_n) of vectors in $\mathbb{R}^d$ is uniformly distributed mod 1 if and only if, for every nonzero vector $m = (\mu_1, \ldots, \mu_d) \in \mathbb{Z}^d$,*

$$N^{-1} \sum_{n=1}^{N} e(m \cdot x_n) \to 0 \quad \text{as } N \to \infty,$$

where $m \cdot x_n = \mu_1 \xi_n^{(1)} + \cdots + \mu_d \xi_n^{(d)}$.

Proposition 3 can also be generalized in the following way:

Proposition 3′ *If $x = (\xi^{(1)}, \ldots, \xi^{(d)})$ is any vector in $\mathbb{R}^d$ such that $1, \xi^{(1)}, \ldots, \xi^{(d)}$ are linearly independent over the field $\mathbb{Q}$ of rational numbers, then the sequence (nx) is uniformly distributed mod 1.*

In particular, the sequence $(\{nx\}) = (\{n\xi^{(1)}\}, \ldots, \{n\xi^{(d)}\})$ is dense in the d-dimensional unit cube if $1, \xi^{(1)}, \ldots, \xi^{(d)}$ are linearly independent over the field $\mathbb{Q}$ of rational numbers. This much weaker assertion had already been proved before Weyl by Kronecker (1884).

It is easily seen that the linear independence of $1, \xi^{(1)}, \ldots, \xi^{(d)}$ over the field $\mathbb{Q}$ of rational numbers is also necessary for the sequence $(\{nx\})$ to be dense in the d-dimensional unit cube and, *a fortiori*, for the sequence (nx) to be uniformly distributed mod 1. For if $1, \xi^{(1)}, \ldots, \xi^{(d)}$ are linearly dependent over $\mathbb{Q}$ there exists a nonzero vector $m = (\mu_1, \ldots, \mu_d) \in \mathbb{Z}^d$ such that

$$m \cdot x = \mu_1 \xi^{(1)} + \cdots + \mu_d \xi^{(d)} \in \mathbb{Z}.$$

It follows that each point of the sequence (nx) lies on some hyperplane $m \cdot y = h$, where $h \in \mathbb{Z}$. Without loss of generality, suppose $\mu_1 \neq 0$. Then no point of the d-dimensional unit cube which is sufficiently close to the point $(|2\mu_1|^{-1}, 0, \ldots, 0)$ lies on such a hyperplane.

We now return to the one-dimensional case. Weyl used Theorem 2 to prove, not only Proposition 3, but also a deeper result concerning the uniform distribution of the sequence $(f(n))$, where f is a polynomial of any positive degree. We will derive Weyl's result by a more general argument due to van der Corput (1931), based on the following inequality:

Lemma 4 *If $\zeta_1, \ldots, \zeta_N$ are arbitrary complex numbers then, for any positive integer $M \leq N$,*

$$M^2 \left| \sum_{n=1}^{N} \zeta_n \right|^2 \leq M(M + N - 1) \sum_{n=1}^{N} |\zeta_n|^2 + 2(M + N - 1) \sum_{m=1}^{M-1} (M - m) \left| \sum_{n=1}^{N-m} \overline{\zeta_n} \zeta_{n+m} \right|.$$

Proof Put $\zeta_n = 0$ if $n \leq 0$ or $n > N$. Then it is easily verified that

$$M \sum_{n=1}^{N} \zeta_n = \sum_{h=1}^{M+N-1} \left(\sum_{k=0}^{M-1} \zeta_{h-k} \right).$$

Applying Schwarz's inequality (Chapter I, §4), we get

$$M^2 \left| \sum_{n=1}^{N} \zeta_n \right|^2 \leq (M + N - 1) \sum_{h=1}^{M+N-1} \left| \sum_{k=0}^{M-1} \zeta_{h-k} \right|^2$$

$$= (M + N - 1) \sum_{h=1}^{M+N-1} \sum_{j,k=0}^{M-1} \zeta_{h-k} \overline{\zeta_{h-j}}.$$

On the right side any term $|\zeta_n|^2$ occurs exactly M times, namely for $h - k = h - j = n$. A term $\overline{\zeta_n}\zeta_{n+m}$ or $\zeta_n\overline{\zeta_{n+m}}$, where $m > 0$, occurs only if $m < M$ and then it occurs exactly $M - m$ times. Thus the right side is equal to

$$M(M + N - 1) \sum_{n=1}^{N} |\zeta_n|^2 + (M + N - 1) \sum_{m=1}^{M-1} (M - m) \sum_{n=1}^{N-m} (\overline{\zeta_n}\zeta_{n+m} + \zeta_n\overline{\zeta_{n+m}}).$$

The lemma follows. □

Corollary 5 *If (ξ_n) is a real sequence such that, for each positive integer m,*

$$N^{-1} \sum_{n=1}^{N} e(\xi_{n+m} - \xi_n) \to 0 \quad \text{as } N \to \infty,$$

then

$$N^{-1} \sum_{n=1}^{N} e(\xi_n) \to 0 \quad \text{as } N \to \infty.$$

Proof By taking $\zeta_n = e(\xi_n)$ in Lemma 4 we obtain, for $1 \leq M \leq N$,

$$N^{-2} \left| \sum_{n=1}^{N} e(\xi_n) \right|^2 \leq 2(M+N-1)M^{-2}N^{-2} \sum_{m=1}^{M-1} (M-m) \left| \sum_{n=1}^{N-m} e(\xi_{n+m} - \xi_n) \right|$$
$$+ (M+N-1)M^{-1}N^{-1}.$$

Keeping M fixed and letting $N \to \infty$, we get

$$\varlimsup_{N \to \infty} N^{-2} \left| \sum_{n=1}^{N} e(\xi_n) \right|^2 \leq M^{-1}.$$

But M can be chosen as large as we please.   □

An immediate consequence is van der Corput's *difference theorem*:

Proposition 6 *The real sequence (ξ_n) is uniformly distributed mod 1 if, for each positive integer m, the sequence $(\xi_{n+m} - \xi_n)$ is uniformly distributed mod 1.*

Proof If the sequences $(\xi_{n+m} - \xi_n)$ are uniformly distributed mod 1 then, by Theorem 2,

$$N^{-1} \sum_{n=1}^{N} e(h(\xi_{n+m} - \xi_n)) \to 0 \quad \text{as } N \to \infty$$

for all integers $h \neq 0, m > 0$. Replacing ξ_n by $h\xi_n$ in Corollary 5 we obtain, for all integers $h \neq 0$,

$$N^{-1} \sum_{n=1}^{N} e(h\xi_n) \to 0 \quad \text{as } N \to \infty.$$

Hence, by Theorem 2 again, the sequence (ξ_n) is uniformly distributed mod 1.   □

The sequence $(n\xi)$, with ξ irrational, shows that we cannot replace 'if' by 'if and only if' in the statement of Proposition 6. Weyl's result will now be derived from Proposition 6:

Proposition 7 *If*

$$f(t) = \alpha_r t^r + \alpha_{r-1} t^{r-1} + \cdots + \alpha_0$$

is any polynomial with real coefficients α_k such that α_k is irrational for at least one $k > 0$, then the sequence $(f(n))$ is uniformly distributed mod 1.

Proof If $r = 1$, then the result holds by the same argument as in Proposition 3. We assume that $r > 1, \alpha_r \neq 0$ and the result holds for polynomials of degree less than r.
For any positive integer m,

$$g_m(t) = f(t+m) - f(t)$$

is a polynomial of degree $r - 1$ with leading coefficient rma_r. If α_r is irrational, then rma_r is also irrational and hence, by the induction hypothesis, the sequence $(g_m(n))$ is uniformly distributed mod 1. Consequently, by Proposition 6, the sequence $(f(n))$ is also uniformly distributed mod 1.

Suppose next that the leading coefficient α_r is rational, and let $\alpha_s (1 \le s < r)$ be the coefficient nearest to it which is irrational. Then the coefficients of $t^{r-1}, \ldots, t^s$ of the polynomial $g_m(t)$ are rational, but the coefficient of t^{s-1} is irrational. If $s > 1$, it follows again from the induction hypothesis and Proposition 6 that the sequence $(f(n))$ is uniformly distributed mod 1.

Suppose finally that $s = 1$ and put

$$F(t) = \alpha_r t^r + \alpha_{r-1} t^{r-1} + \cdots + \alpha_2 t^2.$$

If $q > 0$ is a common denominator for the rational numbers $\alpha_2, \ldots, \alpha_r$ then, for any integer $h \ne 0$ and any nonnegative integers j, k,

$$e(hF(jq + k)) = e(hF(k)).$$

Write $N = \ell q + k$, where $\ell = \lfloor N/q \rfloor$ and $0 \le k < q$. Since $f(t) = F(t) + \alpha_1 t + \alpha_0$, we obtain

$$N^{-1} \sum_{n=0}^{N-1} e(hf(n)) = N^{-1} \sum_{k=0}^{q-1} \sum_{j=0}^{\ell-1} e(hf(jq + k)) + N^{-1} \sum_{n=\ell q}^{N} e(hf(n))$$

$$= N^{-1} \lfloor N/q \rfloor \sum_{k=0}^{q-1} e(hF(k)) \sum_{j=0}^{\ell-1} \ell^{-1} e(h(jq\alpha_1 + k\alpha_1 + \alpha_0))$$

$$+ N^{-1} \sum_{n=\ell q}^{N} e(hf(n)).$$

The last term tends to zero as $N \to \infty$, since the sum contains at most q terms, each of absolute value 1. By Theorem 2, each of the q inner sums in the first term also tends to zero as $N \to \infty$, because the result holds for $r = 1$. Hence, by Theorem 2 again, the sequence $(f(n))$ is uniformly distributed mod 1. $\square$

An interesting extension of Proposition 6 was derived by Korobov and Postnikov (1952):

Proposition 8 *If, for every positive integer m, the sequence $(\xi_{n+m} - \xi_n)$ is uniformly distributed mod 1 then, for all integers $q > 0$ and $r \ge 0$, the sequence (ξ_{qn+r}) is uniformly distributed mod 1.*

Proof We may suppose $q > 1$, since the assertion follows at once from Proposition 6 if $q = 1$. By Theorem 2 it is enough to show that, for every integer $m \ne 0$,

$$S := N^{-1} \sum_{n=1}^{N} e(m\xi_{qn+r}) \to 0 \quad \text{as } N \to \infty.$$

Since

$$q^{-1} \sum_{k=1}^{q} e(nk/q) = 1 \quad \text{if } n \equiv 0 \bmod q,$$

$$= 0 \quad \text{if } n \not\equiv 0 \bmod q,$$

we can write

$$S = (qN)^{-1} \sum_{n=1}^{qN} e(m\xi_{n+r}) \sum_{k=1}^{q} e(nk/q)$$

$$= (qN)^{-1} \sum_{k=1}^{q} \sum_{n=1}^{qN} e(m\eta_n^{(k)}),$$

where we have put

$$\eta_n^{(k)} = \xi_{n+r} + nk/mq.$$

By hypothesis, for every positive integer h, the sequence

$$\eta_{n+h}^{(k)} - \eta_n^{(k)} = \xi_{n+h+r} - \xi_{n+r} - hk/mq$$

is uniformly distributed mod 1. Hence $\eta_n^{(k)}$ is uniformly distributed mod 1, by Proposition 6. Thus, for each $k \in \{1, \ldots, q\}$,

$$(qN)^{-1} \sum_{n=1}^{qN} e(m\eta_n^{(k)}) \to 0 \quad \text{as } N \to \infty,$$

and consequently also $S \to 0$ as $N \to \infty$.   □

As an application of Proposition 8 we prove

Proposition 9 *Let A be a $d \times d$ matrix of integers, no eigenvalue of which is a root of unity. If, for some $x \in \mathbb{R}^d$, the sequence $(A^n x)$ is uniformly distributed mod 1 then, for any integers $q > 0$ and $r \geq 0$, the sequence $(A^{qn+r} x)$ is also uniformly distributed mod 1.*

Proof It follows from Theorem 2′ that, for any nonzero vector $m \in \mathbb{Z}^d$, the scalar sequence $\xi_n = m \cdot A^n x$ is uniformly distributed mod 1. For any positive integer h, the sequence

$$\xi_{n+h} - \xi_n = m \cdot (A^h - I)A^n x = (A^h - I)^t m \cdot A^n x$$

has the same form as the sequence ξ_n, since the hypotheses ensure that $(A^h - I)^t m$ is a nonzero vector in $\mathbb{Z}^d$. Hence the sequence $\xi_{n+h} - \xi_n$ is uniformly distributed mod 1. Therefore, by Proposition 8, the sequence $\xi_{qn+r} = m \cdot A^{qn+r} x$ is uniformly distributed mod 1, and thus the sequence $A^{qn+r} x$ is uniformly distributed mod 1.   □

It may be noted that the matrix A in Proposition 9 is necessarily non-singular. For if $\det A = 0$, there exists a nonzero vector $z \in \mathbb{Z}^d$ such that $A^t z = 0$. Then, for any $x \in \mathbb{R}^d$ and any positive integer n, $e(z \cdot A^n x) = e((A^t)^n z \cdot x) = 1$. Thus $N^{-1} \sum_{n=1}^{N} e(z \cdot A^n x) = 1$ and therefore, by Theorem 2′, the sequence $A^n x$ is not uniformly distributed mod 1.

Further examples of uniformly distributed sequences are provided by the following result, which is due to Fejér (c. 1924):

Proposition 10 *Let (ξ_n) be a sequence of real numbers such that $\eta_n := \xi_{n+1} - \xi_n$ tends to zero monotonically as $n \to \infty$. Then (ξ_n) is uniformly distributed mod 1 if $n|\eta_n| \to \infty$ as $n \to \infty$.*

Proof By changing the signs of all ξ_n we may restrict attention to the case where the sequence (η_n) is strictly decreasing. For any real numbers α, β we have

$$|e(\alpha) - e(\beta) - 2\pi i(\alpha - \beta)e(\beta)| = |e(\alpha - \beta) - 1 - 2\pi i(\alpha - \beta)|$$

$$= 4\pi^2 \left| \int_0^{\alpha-\beta} (\alpha - \beta - t)e(t)\,dt \right|$$

$$\leq 4\pi^2 \left| \int_0^{\alpha-\beta} (\alpha - \beta - t)\,dt \right|$$

$$= 2\pi^2(\alpha - \beta)^2.$$

If we take $\alpha = h\xi_{n+1}$ and $\beta = h\xi_n$, where h is any nonzero integer, this yields

$$|e(h\xi_{n+1})/\eta_n - e(h\xi_n)/\eta_n - 2\pi i h e(h\xi_n)| \leq 2\pi^2 h^2 \eta_n$$

and hence

$$|e(h\xi_{n+1})/\eta_{n+1} - e(h\xi_n)/\eta_n - 2\pi i h e(h\xi_n)| \leq 1/\eta_{n+1} - 1/\eta_n + 2\pi^2 h^2 \eta_n.$$

Taking $n = 1, \ldots, N$ and adding, we obtain

$$\left| 2\pi h \sum_{n=1}^{N} e(h\xi_n) \right| \leq 1/\eta_{N+1} + 1/\eta_1 + \sum_{n=1}^{N}(1/\eta_{n+1} - 1/\eta_n) + 2\pi^2 h^2 \sum_{n=1}^{N} \eta_n$$

$$= 2/\eta_{N+1} + 2\pi^2 h^2 \sum_{n=1}^{N} \eta_n.$$

Thus

$$N^{-1} \left| \sum_{n=1}^{N} e(h\xi_n) \right| \leq (\pi|h|N\eta_{N+1})^{-1} + \pi|h|N^{-1} \sum_{n=1}^{N} \eta_n.$$

But the right side of this inequality tends to zero as $N \to \infty$, since $N\eta_N \to \infty$ and $\eta_N \to 0$. $\qquad\square$

By the mean value theorem, the hypotheses of Proposition 10 are certainly satisfied if $\xi_n = f(n)$, where f is a differentiable function such that $f'(t) \to 0$ monotonically as $t \to \infty$ and $t|f'(t)| \to \infty$ as $t \to \infty$. Consequently the sequence (an^α) is uniformly distributed mod 1 if $a \neq 0$ and $0 < \alpha < 1$, and the sequence $(a(\log n)^\alpha)$ is uniformly distributed mod 1 if $a \neq 0$ and $\alpha > 1$. By using van der Corput's difference theorem and an inductive argument starting from Proposition 10, it may be further shown that the sequence (an^α) is uniformly distributed mod 1 for any $a \neq 0$ and any $\alpha > 0$ which is not an integer.

It has been shown by Kemperman (1973) that 'if' may be replaced by 'if and only if' in the statement of Proposition 10. Consequently the sequence $(a(\log n)^\alpha)$ is not uniformly distributed mod 1 if $0 < \alpha \leq 1$.

The theory of uniform distribution has an application, and its origin, in astronomy. In his investigations on the secular perturbations of planetary orbits Lagrange (1782) was led to the problem of *mean motion*: if

$$z(t) = \sum_{k=1}^{n} \rho_k e(\omega_k t + \alpha_k),$$

where $\rho_k > 0$ and $\alpha_k, \omega_k \in \mathbb{R}$ $(k = 1, \ldots, n)$, does $t^{-1} \arg z(t)$ have a finite limit as $t \to +\infty$? It is assumed that $z(t)$ never vanishes and $\arg z(t)$ is then defined by continuity. (Zeros of $z(t)$ can be admitted by writing $z(t) = \rho(t)e(\phi(t))$, where $\rho(t)$ and $\phi(t)$ are continuous real-valued functions and $\rho(t)$ is required to change sign at a zero of $z(t)$ of odd multiplicity.)

In the astronomical application $\arg z(t)$ measures the longitude of the perihelion of the planetary orbit. Lagrange showed that the limit

$$\mu = \lim_{t \to +\infty} t^{-1} \arg z(t)$$

does exist when $n = 2$ and also, for arbitrary n, when some ρ_k exceeds the sum of all the others. The only planets which do not satisfy this second condition are Venus and Earth. Lagrange went on to say that, when neither of the two conditions was satisfied, the problem was "very difficult and perhaps impossible".

There was no further progress until the work of Bohl (1909), who took $n = 3$ and considered the non-Lagrangian case when there exists a triangle with sidelengths ρ_1, ρ_2, ρ_3. He showed that the limit μ exists if $\omega_1, \omega_2, \omega_3$ are linearly independent over the rational field $\mathbb{Q}$ and then $\mu = \lambda_1 \omega_1 + \lambda_2 \omega_2 + \lambda_3 \omega_3$, where $\pi\lambda_1, \pi\lambda_2, \pi\lambda_3$ are the angles of the triangle with sidelengths ρ_1, ρ_2, ρ_3. In the course of the proof he stated and proved Proposition 3 (without formulating the general concept of uniform distribution).

Using his earlier results on uniform distribution, Weyl (1938) showed that the limit μ exists if $\omega_1, \ldots, \omega_n$ are linearly independent over the rational field $\mathbb{Q}$ and then

$$\mu = \lambda_1 \omega_1 + \cdots + \lambda_n \omega_n,$$

where $\lambda_k \geq 0$ $(k = 1, \ldots, n)$ and $\sum_{k=1}^{n} \lambda_k = 1$. The coefficients λ_k depend only on the ρ's, not on the α's or ω's, and there is even an explicit expression for λ_k, involving Bessel functions, which is derived from the theory of random walks.

Finally, it was shown by Jessen and Tornehave (1945) that the limit μ exists for arbitrary $\omega_k \in \mathbb{R}$.

2 Discrepancy

The *star discrepancy* of a finite set of points $\xi_1, \ldots, \xi_N$ in the unit interval $I = [0, 1]$ is defined to be

$$D_N^* = D_N^*(\xi_1, \ldots, \xi_N) = \sup_{0 < \alpha \leq 1} |\varphi_\alpha(N)/N - \alpha|,$$

where $\varphi_\alpha(N) = \varphi_{0,\alpha}(N)$ denotes the number of positive integers $n \leq N$ such that $0 \leq \xi_n < \alpha$. Here we will omit the qualifier 'star', since we will not be concerned with any other type of discrepancy and the notation D_N^* should provide adequate warning.

It was discovered only in 1972, by Niederreiter, that the preceding definition may be reformulated in the following simple way:

Proposition 11 *If* $\xi_1, \ldots, \xi_N$ *are real numbers such that* $0 \leq \xi_1 \leq \cdots \leq \xi_N \leq 1$, *then*

$$D_N^* = D_N^*(\xi_1, \ldots, \xi_N) = \max_{1 \leq k \leq N} \max(|\xi_k - k/N|, |\xi_k - (k-1)/N|)$$

$$= (2N)^{-1} + \max_{1 \leq k \leq N} |\xi_k - (2k-1)/2N|.$$

Proof Put $\xi_0 = 0, \xi_{N+1} = 1$. Since the distinct ξ_k with $0 \leq k \leq N+1$ define a subdivision of the unit interval I, we have

$$D_N^* = \max_{k:\xi_k < \xi_{k+1}} \sup_{\xi_k \leq \alpha < \xi_{k+1}} |\varphi_\alpha(N)/N - \alpha|$$

$$= \max_{k:\xi_k < \xi_{k+1}} \sup_{\xi_k \leq \alpha < \xi_{k+1}} |k/N - \alpha|.$$

But the function $f_k(t) = |k/N - t|$ attains its maximum in the interval $\xi_k \leq t \leq \xi_{k+1}$ at one of the endpoints of this interval. Consequently

$$D_N^* = \max_{k:\xi_k < \xi_{k+1}} \max(|k/N - \xi_k|, |k/N - \xi_{k+1}|).$$

We are going to show that in fact

$$D_N^* = \max_{0 \leq k \leq N} \max(|k/N - \xi_k|, |k/N - \xi_{k+1}|).$$

Suppose $\xi_k < \xi_{k+1} = \xi_{k+2} = \cdots = \xi_{k+r} < \xi_{k+r+1}$ for some $r \geq 2$. By applying the same reasoning as before to the function $g_k(t) = |t - \xi_{k+1}|$ we obtain, for $1 \leq j < r$,

$$|(k+j)/N - \xi_{k+j}| = |(k+j)/N - \xi_{k+j+1}| = |(k+j)/N - \xi_{k+1}|$$
$$< \max(|k/N - \xi_{k+1}|, |(k+r)/N - \xi_{k+1}|)$$
$$= \max(|k/N - \xi_{k+1}|, |(k+r)/N - \xi_{k+r}|).$$

Since both terms in the last maximum appear in the expression already obtained for D_N^*, it follows that this expression is not altered by dropping the restriction to those k for which $\xi_k < \xi_{k+1}$.

Since $|0/N - \xi_0| = |N/N - \xi_{N+1}| = 0$, we can now also write

$$D_N^* = \max_{1 \leq k \leq N} \max(|k/N - \xi_k|, |(k-1)/N - \xi_k|).$$

The second expression for D_N^* follows immediately, since

$$\max(|k/N - \alpha|, |(k-1)/N - \alpha|) = |(k-1/2)/N - \alpha| + 1/2N. \qquad \square$$

Corollary 12 *If $\xi_1, \ldots, \xi_N$ are real numbers such that $0 \leq \xi_1 \leq \cdots \leq \xi_N \leq 1$, then $D_N^* \geq (2N)^{-1}$. Moreover, equality holds if and only if $\xi_k = (2k-1)/N$ for $k = 1, \ldots, N$.*

Thus Proposition 11 says that the discrepancy of any set of N points of I is obtained by adding to its minimal value $1/2N$ the maximum deviation of the set from the unique minimizing set, when both sets are arranged in order of magnitude.

The next result shows that the discrepancy $D_N^*(\xi_1, \ldots, \xi_N)$ is a continuous function of $\xi_1, \ldots, \xi_N$.

Proposition 13 *If $\xi_1, \ldots, \xi_N$ and $\eta_1, \ldots, \eta_N$ are two sets of N points of I, with the discrepancies D_N^* and E_N^* respectively, then*

$$|D_N^* - E_N^*| \leq \max_{1 \leq k \leq N} |\xi_k - \eta_k|.$$

Proof Let $x_1 \leq \cdots \leq x_N$ and $y_1 \leq \cdots \leq y_N$ be the two given sets rearranged in order of magnitude. It is enough to show that

$$\max_{1 \leq k \leq N} |x_k - y_k| \leq \delta := \max_{1 \leq k \leq N} |\xi_k - \eta_k|,$$

since it then follows from Proposition 11 that

$$D_N^* \leq \delta + E_N^*, \quad E_N^* \leq \delta + D_N^*.$$

Assume, on the contrary, that $|x_k - y_k| > \delta$ for some k. Then either $x_k > y_k + \delta$ or $y_k > x_k + \delta$. Without loss of generality we restrict attention to the first case. By hypothesis, for each y_i with $1 \leq i \leq k$ there exists an x_{j_i} with $1 \leq j_i \leq N$ such that $|y_i - x_{j_i}| \leq \delta$ and such that the subscripts j_i are distinct. Since $y_1 \leq \cdots \leq y_k$, it follows that

$$x_{j_i} \leq y_i + \delta \leq y_k + \delta < x_k.$$

But this is a contradiction, since there are at most $k - 1$ x's less than x_k. $\qquad \square$

We now show how the notion of discrepancy makes it possible to obtain estimates for the accuracy of various methods of numerical integration.

Proposition 14 *If the function f satisfies the 'Lipschitz condition'*

$$|f(t_2) - f(t_1)| \leq L|t_2 - t_1| \quad \text{for all } t_1, t_2 \in I,$$

then for any finite set $\xi_1, \ldots, \xi_N \in I$ with discrepancy D_N^*,

$$\left| N^{-1} \sum_{n=1}^{N} f(\xi_n) - \int_I f(t)\,dt \right| \leq L D_N^*.$$

Proof Without loss of generality we may assume $\xi_1 \leq \cdots \leq \xi_N$. Writing

$$\int_I f(t)\,dt = \sum_{n=1}^N \int_{(n-1)/N}^{n/N} f(t)\,dt,$$

we obtain

$$\left| N^{-1} \sum_{n=1}^N f(\xi_n) - \int_I f(t)\,dt \right| \leq \sum_{n=1}^N \int_{(n-1)/N}^{n/N} |f(\xi_n) - f(t)|\,dt$$

$$\leq L \sum_{n=1}^N \int_{(n-1)/N}^{n/N} |\xi_n - t|\,dt.$$

But for $(n-1)/N \leq t \leq n/N$ we have

$$|\xi_n - t| \leq \max(|\xi_n - n/N|, |\xi_n - (n-1)/N|) \leq D_N^*,$$

by Proposition 11. The result follows. $\qquad\square$

As Koksma (1942) first showed, Proposition 14 can be sharpened in the following way:

Proposition 15 *If the function* f *has bounded variation on the unit interval* I, *with total variation* V, *then for any finite set* $\xi_1, \ldots, \xi_N \in I$ *with discrepancy* D_N^*,

$$\left| N^{-1} \sum_{n=1}^N f(\xi_n) - \int_I f(t)\,dt \right| \leq V D_N^*.$$

Proof Without loss of generality we may assume $\xi_1 \leq \cdots \leq \xi_N$ and we put $\xi_0 = 0$, $\xi_{N+1} = 1$. By integration and summation by parts we obtain

$$\sum_{n=0}^N \int_{\xi_n}^{\xi_{n+1}} (t - n/N)\,df(t) = \int_I t\,df(t) - N^{-1} \sum_{n=0}^N n(f(\xi_{n+1}) - f(\xi_n))$$

$$= [tf(t)]_0^1 - \int_I f(t)\,dt - f(1) + N^{-1} \sum_{n=0}^{N-1} f(\xi_{n+1})$$

$$= N^{-1} \sum_{n=1}^N f(\xi_n) - \int_I f(t)\,dt.$$

The result follows, since for $\xi_n \leq t \leq \xi_{n+1}$ we have

$$|t - n/N| \leq \max(|\xi_n - n/N|, |\xi_{n+1} - n/N|) \leq D_N^*. \qquad\square$$

As an application of Proposition 15 we prove

Proposition 16 *If $\xi_1, \ldots, \xi_N$ are points of the unit interval I with discrepancy D_N^* then, for any integer $h \neq 0$,*

$$\left| N^{-1} \sum_{n=1}^{N} e(h\xi_n) \right| \leq 4|h| D_N^*.$$

Proof We can write

$$N^{-1} \sum_{n=1}^{N} e(h\xi_n) = \rho e(\alpha),$$

where $\rho \geq 0$ and $\alpha \in I$. Thus

$$\rho = N^{-1} \sum_{n=1}^{N} e(h\xi_n - \alpha).$$

Adding this relation to its complex conjugate, we obtain

$$\rho = N^{-1} \sum_{n=1}^{N} \cos 2\pi (h\xi_n - \alpha).$$

The result follows by applying Proposition 15 to the function $f(t) = \cos 2\pi (ht - \alpha)$, which has bounded variation on I with total variation $\int_I |f'(t)| \, dt = 4|h|$. $\square$

An inequality in the opposite direction to Proposition 16 was obtained by Erdős and Turán (1948) who showed that, for any positive integer m,

$$D_N^* \leq C \left(m^{-1} + \sum_{h=1}^{m} h^{-1} \left| N^{-1} \sum_{n=1}^{N} e(h\xi_n) \right| \right),$$

where the positive constant C is independent of m, N and the ξ's. Niederreiter and Philipp (1973) showed that one can take $C = 4$. Furthermore they generalized the result and simplified the proof.

The connection between these results and the theory of uniform distribution is close at hand. Let (ξ_n) be an arbitrary sequence of real numbers and let δ_N denote the discrepancy of the fractional parts $\{\xi_1\}, \ldots, \{\xi_N\}$. By the remark after the definition of uniform distribution in §1, *the sequence (ξ_n) is uniformly distributed mod 1 if and only if $\delta_N \to 0$ as $N \to \infty$.* It follows from Proposition 16 and the inequality of Erdős and Turán (in which m may be arbitrarily large) that $\delta_N \to 0$ as $N \to \infty$ if and only if, for every integer $h \neq 0$,

$$N^{-1} \sum_{n=1}^{N} e(h\xi_n) \to 0 \quad \text{as } N \to \infty.$$

This provides a new proof of Theorem 2. Furthermore, from bounds for the exponential sums we can obtain estimates for the rapidity with which δ_N tends to zero.

Propositions 14 and 15 show that in a formula for numerical integration the nodes (ξ_n) should be chosen to have as small a discrepancy as possible. For a given finite number N of nodes Corollary 12 shows how this can be achieved. In practice, however, one does not know in advance an appropriate choice of N, since universal error bounds may grossly overestimate the error in a specific case. Consequently it is also of interest to consider the problem of choosing an infinite sequence (ξ_n) of nodes so that the discrepancy δ_N of $\xi_1, \ldots, \xi_N$ tends to zero as rapidly as possible when $N \to \infty$. There is a limit to what can be achieved in this way. W. Schmidt (1972), improving earlier results of van Aardenne-Ehrenfest (1949) and Roth (1954), showed that there exists an absolute constant $C > 0$ such that

$$\varlimsup_{N \to \infty} N\delta_N / \log N \geq C$$

for *every* infinite sequence (ξ_n). Kuipers and Niederreiter (1974) showed that a possible value for C was $(132 \log 2)^{-1} = 0.0109\ldots$, which Bejian (1979) improved to $(24 \log 2)^{-1} = 0.0601\ldots$.

Schmidt's result is best possible, apart from the value of the constant. Ostrowski (1922) had already shown that for the sequence $(\{n\alpha\})$, where $\alpha \in (0, 1)$ is irrational, one has

$$s^*(\alpha) := \varlimsup_{N \to \infty} N\delta_N / \log N < \infty$$

if in the continued fraction expansion

$$\alpha = [0; a_1, a_2, \ldots] = \cfrac{1}{a_1 + \cfrac{1}{a_2 + \cdots}}$$

the partial quotients a_k are bounded. Dupain and Sós (1984) have shown that the minimum value of $s^*(\alpha)$, for all such α, is $(4 \log(1 + \sqrt{2}))^{-1} = 0.283\ldots$ and the minimum is attained for $\alpha = \sqrt{2} - 1 = [0; 2, 2, \ldots]$. Schoessengeier (1984) has proved that, for any irrational $\alpha \in (0, 1)$, one has $N\delta_N = O(\log N)$ if and only if the partial quotients a_k satisfy $\sum_{k=1}^{n} a_k = O(n)$.

There are other low discrepancy sequences. Haber (1966) showed that, for a sequence (ξ_n) constructed by van der Corput (1935),

$$\varlimsup_{N \to \infty} N\delta_N / \log N = (3 \log 2)^{-1} = 0.481\ldots.$$

van der Corput's sequence is defined as follows: if $n - 1 = a_m 2^m + \cdots + a_1 2^1 + a_0$, where $a_k \in \{0, 1\}$, then $\xi_n = a_0 2^{-1} + a_1 2^{-2} + \cdots + a_m 2^{-m-1}$. In other words, the expression for ξ_n in the base 2 is obtained from that for $n - 1$ by reflection in the 'decimal' point, a construction which is easily implemented on a computer. Various generalizations of this construction have been given, and Faure (1981) defined in this way a sequence (ξ_n) for which

$$\varlimsup_{N \to \infty} N\delta_N / \log N = (1919)(3454 \log 12)^{-1} = 0.223\ldots.$$

Thus if C^* is the least upper bound for all admissible values of C in Schmidt's result then, by what has been said, $0.060\ldots \leq C^* \leq 0.223\ldots$. It is natural to ask: what is the exact value of C^*, and is there a sequence (ξ_n) for which it is attained?

The notion of discrepancy is easily extended to higher dimensions by defining the discrepancy of a finite set of vectors $x_1, \ldots, x_N$ in the d-dimensional unit cube $I^d = I \times \cdots \times I$ to be

$$D_N^*(x_1, \ldots, x_N) = \sup_{\substack{0 < a_k \leq 1 \ (k=1,\ldots,d)}} |\varphi_a(N)/N - a_1 \cdots a_d|,$$

where $x_n = (\xi_n^{(1)}, \ldots, \xi_n^{(d)})$, $a = (a_1, \ldots, a_d)$ and $\varphi_a(N)$ is the number of positive integers $n \leq N$ such that $0 \leq \xi_n^{(k)} < a_k$ for every $k \in \{1, \ldots, d\}$.

For $d > 1$ there is no simple reformulation of the definition analogous to Proposition 11, but many results do carry over. In particular, Proposition 15 was generalized and applied to the numerical evaluation of multiple integrals by Hlawka (1961/62). Indeed this application has greater value in higher dimensions, where other methods perform poorly.

For the application one requires a set of vectors $x_1, \ldots, x_N \in I^d$ whose discrepancy $D_N^*(x_1, \ldots, x_N)$ is small. A simple procedure for obtaining such a set, which is most useful when the integrand is smooth and has period 1 in each of its variables, is the method of 'good lattice points' introduced by Korobov (1959). Here, for a suitably chosen $g \in \mathbb{Z}^d$, one takes $x_n = \{(n-1)g/N\}$ $(n = 1, \ldots, N)$. A result of Niederreiter (1986) implies that, for every $d \geq 2$ and every $N \geq 2$, one can choose g so that

$$ND_N^* \leq (1 + \log N)^d + d2^d.$$

The van der Corput sequence has also been generalized to any finite number of dimensions by Halton (1960). He defined an infinite sequence (x_n) of vectors in $\mathbb{R}^d$ for which

$$\varlimsup_{N \to \infty} N\delta_N/(\log N)^d < \infty.$$

It is conjectured that for each $d > 1$ (as for $d = 1$) there exists an absolute constant $C_d > 0$ such that

$$\varlimsup_{N \to \infty} N\delta_N/(\log N)^d \geq C_d$$

for every infinite sequence (x_n) of vectors in $\mathbb{R}^d$. However, the best known result remains that of Roth (1954), in which the exponent d is replaced by $d/2$.

3 Birkhoff's Ergodic Theorem

In statistical mechanics there is a procedure for calculating the physical properties of a system by simply averaging over all possible states of the system. To justify this procedure Boltzmann (1871) introduced what he later called the 'ergodic hypothesis'. In the formulation of Maxwell (1879) this says that "the system, if left to itself in its actual state of motion, will, sooner or later, pass through every phase which is consistent with the equation of energy". The word *ergodic*, coined by Boltzmann (1884), was a composite of the Greek words for 'energy' and 'path'. It was recognized by Poincaré (1894) that it was too much to ask that a path pass through every state on the same energy surface as its initial state, and he suggested instead that it pass arbitrarily close to every such state. Moreover, he observed that it would still be necessary to exclude certain exceptional initial states.

A breakthrough came with the work of G.D. Birkhoff (1931), who showed that Lebesgue measure was the appropriate tool for treating the problem. He established a deep and general result which says that, apart from a set of initial states of measure zero, there is a definite limiting value for the proportion of time which a path spends in any given measurable subset B of an energy surface X. The proper formulation for the ergodic hypothesis was then that this limiting value should coincide with the ratio of the measure of B to that of X, i.e. that 'the paths through almost all initial states should be uniformly distributed over arbitrary measurable sets'. It was not difficult to deduce that this was the case if and only if 'any invariant measurable subset of X either had measure zero or had the same measure as X'.

Birkhoff proved his theorem in the framework of classical mechanics and for *flows* with continuous time. We will prove his theorem in the abstract setting of probability spaces and for *cascades* with discrete time. The abstract formulation makes possible other applications, for which continuous time is not appropriate.

Let $\mathscr{B}$ be a σ-*algebra* of subsets of a given set X, i.e. a nonempty family of subsets of X such that

(B1) the complement of any set in $\mathscr{B}$ is again a set in $\mathscr{B}$,
(B2) the union of any finite or countable collection of sets in $\mathscr{B}$ is again a set in $\mathscr{B}$.

It follows that $X \in \mathscr{B}$, since $B \in \mathscr{B}$ implies $B^c := X \backslash B \in \mathscr{B}$ and $X = B \cup B^c$. Hence also $\emptyset = X^c \in \mathscr{B}$. Furthermore, the intersection of any finite or countable collection of sets in $\mathscr{B}$ is again a set in $\mathscr{B}$, since $\bigcap_n B_n = X \backslash (\bigcup_n B_n^c)$. Hence if $A, B \in \mathscr{B}$, then

$$B \backslash A = B \cap A^c \in \mathscr{B}$$

and the *symmetric difference*

$$A \, \varDelta \, B := (B \backslash A) \cup (A \backslash B) \in \mathscr{B}.$$

The family of all subsets of X is certainly a σ-algebra. Furthermore, the intersection of any collection of σ-algebras is again a σ-algebra. It follows that, for any family $\mathscr{A}$ of subsets of X, there is a σ-algebra $\sigma(\mathscr{A})$ which contains $\mathscr{A}$ and is contained in every σ-algebra which contains $\mathscr{A}$. We call $\sigma(\mathscr{A})$ the σ-algebra of subsets of X *generated* by $\mathscr{A}$.

Suppose $\mathscr{B}$ is a σ-algebra of subsets of X and a function $\mu : \mathscr{B} \to \mathbb{R}$ is defined such that

(Pr1) $\mu(B) \geq 0$ for every $B \in \mathscr{B}$,
(Pr2) $\mu(X) = 1$,
(Pr3) if (B_n) is a sequence of pairwise disjoint sets in $\mathscr{B}$, then $\mu(\bigcup_n B_n) = \sum_n \mu(B_n)$.

Then μ is said to be a *probability measure* and the triple $(X, \mathcal{B}, \mu)$ is said to be a *probability space*.

It is easily seen that the definition implies

 (i) $\mu(\emptyset) = 0$,
 (ii) $\mu(B^c) = 1 - \mu(B)$,
 (iii) $\mu(A) \le \mu(B)$ if $A, B \in \mathcal{B}$ and $A \subseteq B$,
 (iv) $\mu(B_n) \to \mu(B)$ if (B_n) is a sequence of sets in $\mathcal{B}$ such that $B_1 \supseteq B_2 \supseteq \cdots$ and $B = \bigcap_n B_n$.

If a property of points in a probability space $(X, \mathcal{B}, \mu)$ holds for all $x \in B$, where $B \in \mathcal{B}$ and $\mu(B) = 1$, then the property is said to hold for $(\mu\text{-})$ *almost all* $x \in X$, or simply *almost everywhere* (a.e.).

A function $f : X \to \mathbb{R}$ is *measurable* if, for every $\alpha \in \mathbb{R}$, the set $\{x \in X : f(x) < \alpha\}$ is in $\mathcal{B}$. Let $f : X \to [0, \infty)$ be measurable and for any partition $\mathscr{P}$ of X into finitely many pairwise disjoint sets $B_1, \ldots, B_n \in \mathcal{B}$, put

$$L_{\mathscr{P}}(f) = \sum_{k=1}^{n} f_k \, \mu(B_k),$$

where $f_k = \inf\{f(x) : x \in B_k\}$. We say that f is *integrable* if

$$\int_X f \, d\mu := \sup_{\mathscr{P}} L_{\mathscr{P}}(f) < \infty.$$

The set of all measurable functions $f : X \to \mathbb{R}$ such that $|f|$ is integrable is denoted by $L(X, \mathcal{B}, \mu)$.

A map $T : X \to X$ is said to be a *measure-preserving transformation* of the probability space $(X, \mathcal{B}, \mu)$ if, for every $B \in \mathcal{B}$, the set $T^{-1}B = \{x \in X : Tx \in B\}$ is again in $\mathcal{B}$ and $\mu(T^{-1}B) = \mu(B)$. This is equivalent to $\mu(TB) = \mu(B)$ for every $B \in \mathcal{B}$ if the measure-preserving transformation T is *invertible*, i.e. if T is bijective and $TB \in \mathcal{B}$ for every $B \in \mathcal{B}$. However, we do not wish to restrict attention to the invertible case. Several important examples of measure-preserving transformations of probability spaces will be given in the next section.

Birkhoff's ergodic theorem, which is also known as the 'individual' or 'pointwise' ergodic theorem, has the following statement:

Theorem 17 *Let T be a measure-preserving transformation of the probability space $(X, \mathcal{B}, \mu)$. If $f \in L(X, \mathcal{B}, \mu)$ then, for almost all $x \in X$, the limit*

$$f^*(x) = \lim_{n \to \infty} n^{-1} \sum_{k=0}^{n-1} f(T^k x)$$

exists and $f^*(Tx) = f^*(x)$. Moreover, $f^* \in L(X, \mathcal{B}, \mu)$ and $\int_X f^* d\mu = \int_X f \, d\mu$.

Proof It is sufficient to prove the theorem for nonnegative functions, since we can write $f = f_+ - f_-$, where

$$f_+(x) = \max\{f(x), 0\}, \quad f_-(x) = \max\{-f(x), 0\},$$

and $f_+, f_- \in L(X, \mathcal{B}, \mu)$.

Put

$$\bar{f}(x) = \varlimsup_{n\to\infty} n^{-1} \sum_{k=0}^{n-1} f(T^k x), \quad \underline{f}(x) = \varliminf_{n\to\infty} n^{-1} \sum_{k=0}^{n-1} f(T^k x).$$

Then $\bar{f}$ and $\underline{f}$ are μ-measurable functions since, for any sequence (g_n),

$$\varlimsup_{n\to\infty} g_n(x) = \inf_m (\sup_{n\ge m} g_n(x)), \quad \varliminf_{n\to\infty} g_n(x) = \sup_m (\inf_{n\ge m} g_n(x)).$$

Moreover $\bar{f}(x) = \bar{f}(Tx)$, $\underline{f}(x) = \underline{f}(Tx)$ for every $x \in X$, since

$$(n+1)^{-1} \sum_{k=0}^{n} f(T^k x) = (n+1)^{-1} f(x) + (1+1/n)^{-1} n^{-1} \sum_{k=0}^{n-1} f(T^{k+1} x).$$

It is sufficient to show that

$$\int_X \bar{f} \, d\mu \le \int_X f \, d\mu \le \int_X \underline{f} \, d\mu.$$

For then, since $\underline{f} \le \bar{f}$, it follows that $\bar{f}(x) = \underline{f}(x) = f^*(x)$ for μ-almost all $x \in X$ and

$$\int_X f^* \, d\mu = \int_X f \, d\mu.$$

Fix some $M > 0$ and define the 'cut-off' function $\bar{f}_M$ by

$$\bar{f}_M(x) = \min\{M, \bar{f}(x)\}.$$

Then $\bar{f}_M$ is bounded and $\bar{f}_M(Tx) = \bar{f}_M(x)$ for every $x \in X$. Fix also any $\varepsilon > 0$. By the definition of $\bar{f}(x)$, for each $x \in X$ there exists a positive integer n such that

$$\bar{f}_M(x) \le n^{-1} \sum_{k=0}^{n-1} f(T^k x) + \varepsilon. \tag{$*$}$$

Thus if F_n is the set of all $x \in X$ for which $(*)$ holds and if $E_n = \bigcup_{k=1}^{n} F_k$, then $E_1 \subseteq E_2 \subseteq \cdots$ and $X = \bigcup_{n\ge 1} E_n$. Since the sets E_n are μ-measurable, we can choose N so large that $\mu(E_N) > 1 - \varepsilon/M$.

Put

$$\tilde{f}(x) = f(x) \quad \text{if } x \in E_N,$$
$$= \max\{f(x), M\} \quad \text{if } x \notin E_N.$$

Also, let $\tau(x)$ be the least positive integer $n \le N$ for which $(*)$ holds if $x \in E_N$, and let $\tau(x) = 1$ if $x \notin E_N$. Since $\bar{f}_M$ is T-invariant, $(*)$ implies

$$\sum_{k=0}^{n-1} \bar{f}_M(T^k x) \le \sum_{k=0}^{n-1} f(T^k x) + n\varepsilon$$

and hence

$$\sum_{k=0}^{\tau(x)-1} \bar{f}_M(T^k x) \le \sum_{k=0}^{\tau(x)-1} \tilde{f}(T^k x) + \tau(x)\varepsilon.$$

To estimate the sum $\sum_{k=0}^{L-1} \bar{f}_M(T^k x)$ for any $L > N$, we partition it into blocks of the form

$$\sum_{k=0}^{\tau(y)-1} \bar{f}_M(T^k y)$$

and a remainder block. More precisely, define inductively

$$n_0(x) = 0, \quad n_k(x) = n_{k-1}(x) + \tau(T^{n_{k-1}} x) \quad (k = 1, 2, \ldots)$$

and define h by $n_h(x) < L \le n_{h+1}(x)$. Then

$$\sum_{k=0}^{n_1(x)-1} \bar{f}_M(T^k x) \le \sum_{k=0}^{n_1(x)-1} \tilde{f}(T^k x) + \tau(x)\varepsilon,$$

$$\sum_{k=n_1(x)}^{n_2(x)-1} \bar{f}_M(T^k x) \le \sum_{k=n_1(x)}^{n_2(x)-1} \tilde{f}(T^k x) + \tau(T^{n_1} x)\varepsilon,$$

$$\cdots$$

$$\sum_{k=n_{h-1}(x)}^{n_h(x)-1} \bar{f}_M(T^k x) \le \sum_{k=n_{h-1}(x)}^{n_h(x)-1} \tilde{f}(T^k x) + \tau(T^{n_{h-1}} x)\varepsilon.$$

Since $n_h(x) < L$, we obtain by addition

$$\sum_{k=0}^{n_h(x)-1} \bar{f}_M(T^k x) \le \sum_{k=0}^{n_h(x)-1} \tilde{f}(T^k x) + L\varepsilon.$$

On the other hand, since $L \le n_{h+1}(x) \le n_h(x) + N$, we have

$$\sum_{k=n_h(x)}^{L-1} \bar{f}_M(T^k x) \le NM.$$

Since $\tilde{f} \ge 0$, it follows that

$$\sum_{k=0}^{L-1} \bar{f}_M(T^k x) \le \sum_{k=0}^{L-1} \tilde{f}(T^k x) + L\varepsilon + NM.$$

Dividing by L and integrating over X, we obtain

$$\int_X \bar{f}_M \, d\mu \le \int_X \tilde{f} \, d\mu + \varepsilon + NM/L,$$

since the measure-preserving nature of T implies that, for any $g \in L(X, \mathscr{B}, \mu)$,

$$\int_X g(Tx)\, d\mu(x) = \int_X g(x)\, d\mu(x).$$

Since

$$\int_X \tilde{f}\, d\mu \le \int_X f\, d\mu + \int_{X \setminus E_N} M\, d\mu \le \int_X f\, d\mu + \varepsilon,$$

it follows that

$$\int_X \bar{f}_M\, d\mu \le \int_X f\, d\mu + 2\varepsilon + NM/L.$$

Since L may be chosen arbitrarily large and then ε arbitrarily small, we conclude that

$$\int_X \bar{f}_M\, d\mu \le \int_X f\, d\mu.$$

Now letting $M \to \infty$, we obtain

$$\int_X \bar{f}\, d\mu \le \int_X f\, d\mu.$$

The proof that

$$\int_X f\, d\mu \le \int_X \underline{f}\, d\mu$$

is similar. Given $\varepsilon > 0$, there exists for each $x \in X$ a positive integer n such that

$$n^{-1} \sum_{k=0}^{n-1} f(T^k x) \le \underline{f}(x) + \varepsilon. \qquad (**)$$

If F_n is the set of all $x \in X$ for which $(**)$ holds and if $E_n = \bigcup_{k=1}^{n} F_k$, we can choose N so large that

$$\int_{X \setminus E_N} f\, d\mu < \varepsilon.$$

Put

$$\tilde{f}(x) = f(x) \quad \text{if } x \in E_N,$$
$$= 0 \quad \text{if } x \notin E_N.$$

Let $\tau(x)$ be the least positive integer n for which $(**)$ holds if $x \in E_N$, and $\tau(x) = 1$ otherwise. The proof now goes through in the same way as before. $\qquad \square$

It should be noticed that the preceding proof simplifies if the function f is bounded. In Birkhoff's original formulation the function f was the indicator function χ_B of an arbitrary set $B \in \mathcal{B}$. In this case the theorem says that, if $v_n(x)$ is the number of $k < n$ for which $T^k x \in B$, then $\lim_{n \to \infty} v_n(x)/n$ exists for almost all $x \in X$. That is, 'almost every point has an average sojourn time in any measurable set'.

A measure-preserving transformation T of the probability space $(X, \mathcal{B}, \mu)$ is said to be *ergodic* if, for every $B \in \mathcal{B}$ with $T^{-1} B = B$, either $\mu(B) = 0$ or $\mu(B) = 1$. Part (ii) of the next proposition says that this is the case if and only if 'time means and space means are equal'.

Proposition 18 *Let T be a measure-preserving transformation of the probability space $(X, \mathcal{B}, \mu)$. Then T is ergodic if and only if one of the following equivalent properties holds:*

(i) *if $f \in L(X, \mathcal{B}, \mu)$ satisfies $f(Tx) = f(x)$ almost everywhere, then f is constant almost everywhere;*
(ii) *if $f \in L(X, \mathcal{B}, \mu)$ then, for almost all $x \in X$,*

$$n^{-1} \sum_{k=0}^{n-1} f(T^k x) \to \int_X f \, d\mu \quad as \ n \to \infty;$$

(iii) *if $A, B \in \mathcal{B}$, then*

$$n^{-1} \sum_{k=0}^{n-1} \mu(T^{-k} A \cap B) \to \mu(A)\mu(B) \quad as \ n \to \infty;$$

(iv) *if $C \in \mathcal{B}$ and $\mu(C) > 0$, then $\mu(\bigcup_{n \geq 1} T^{-n} C) = 1$;*
(v) *if $A, B \in \mathcal{B}$ and $\mu(A) > 0$, $\mu(B) > 0$, then $\mu(T^{-n} A \cap B) > 0$ for some $n > 0$.*

Proof Suppose first that T is ergodic and let $f \in L(X, \mathcal{B}, \mu)$ satisfy $f(Tx) = f(x)$ a.e. Put

$$\bar{f}(x) = \varlimsup_{n \to \infty} n^{-1} \sum_{k=0}^{n-1} f(T^k x).$$

Then $\bar{f}(Tx) = \bar{f}(x)$ for every $x \in X$ and $\bar{f}(x) = f(x)$ a.e. For any $\alpha \in \mathbb{R}$, let

$$A_\alpha = \{x \in X : \bar{f}(x) < \alpha\}.$$

Then $\mu(A_\alpha) = 0$ or 1, since $T^{-1} A_\alpha = A_\alpha$ and T is ergodic. Since $\mu(A_\alpha)$ is a nondecreasing function of α and $\mu(A_\alpha) \to 0$ as $\alpha \to -\infty$, $\mu(A_\alpha) \to 1$ as $\alpha \to +\infty$, there exists $\beta \in \mathbb{R}$ such that $\mu(A_\alpha) = 0$ for $\alpha < \beta$ and $\mu(A_\alpha) = 1$ for $\alpha > \beta$. It follows that $\mu(A_\beta) = 0$ and $\mu(B_\beta) = 1$, where

$$B_\beta = \{x \in X : \bar{f}(x) \leq \beta\}.$$

Hence $f(x) = \beta$ a.e. and (i) holds.

Suppose now that (i) holds and let $f \in L(X, \mathcal{B}, \mu)$. Then the function f^* in the statement of Theorem 17 must be constant a.e. Moreover, if γ is its constant value, we must have

$$\gamma = \int_X f^* \, d\mu = \int_X f \, d\mu.$$

Thus (i) implies (ii).

Suppose next that (ii) holds and let $A, B \in \mathcal{B}$. Then, for almost all $x \in X$,

$$\lim_{n \to \infty} n^{-1} \sum_{k=0}^{n-1} \chi_A(T^k x) = \int_X \chi_A \, d\mu = \mu(A).$$

Hence, for almost all $x \in X$,

$$\lim_{n \to \infty} n^{-1} \sum_{k=0}^{n-1} \chi_A(T^k x) \chi_B(x) = \mu(A) \chi_B(x)$$

and so, by the dominated convergence theorem,

$$\mu(A)\mu(B) = \int_X \lim_{n \to \infty} n^{-1} \sum_{k=0}^{n-1} \chi_A(T^k x) \chi_B(x) \, d\mu(x)$$

$$= \lim_{n \to \infty} n^{-1} \sum_{k=0}^{n-1} \int_X \chi_A(T^k x) \chi_B(x) \, d\mu(x)$$

$$= \lim_{n \to \infty} n^{-1} \sum_{k=0}^{n-1} \mu(T^{-k} A \cap B).$$

Thus (ii) implies (iii).

Suppose now that (iii) holds and choose $C \in \mathcal{B}$ with $\mu(C) > 0$. Put $A = \bigcup_{n \geq 0} T^{-n} C$ and $B = (\bigcup_{n \geq 1} T^{-n} C)^c$. Then, for every $k \geq 1$, $T^{-k} A \subseteq \bigcup_{n \geq 1} T^{-n} C$ and hence $\mu(T^{-k} A \cap B) = 0$. Thus

$$n^{-1} \sum_{k=0}^{n-1} \mu(T^{-k} A \cap B) = \mu(A \cap B)/n \to 0 \quad \text{as } n \to \infty.$$

Since $\mu(A) \geq \mu(C) > 0$, it follows from (iii) that $\mu(B) = 0$. Thus (iii) implies (iv).

Next choose any $A, B \in \mathcal{B}$ such that $\mu(A) > 0$, $\mu(B) > 0$. If (iv) holds, then $\mu(\bigcup_{n \geq 1} T^{-n} A) = 1$ and hence

$$\mu(B) = \mu\left(B \cap \bigcup_{n \geq 1} T^{-n} A\right) = \mu\left(\bigcup_{n \geq 1} (B \cap T^{-n} A)\right).$$

Since $\mu(B) > 0$, it follows that $\mu(B \cap T^{-n} A) > 0$ for some $n > 0$. Thus (iv) implies (v).

Finally choose $A \in \mathcal{B}$ with $T^{-1} A = A$ and put $B = A^c$. Then, for every $n \geq 1$, we have $\mu(T^{-n} A \cap B) = \mu(A \cap B) = 0$. If (v) holds, it follows that either $\mu(A) = 0$ or $\mu(B) = 0$. Hence (v) implies that T is ergodic. $\qquad\square$

4 Applications

We now give some examples to illustrate the general concepts and results of the previous section.

(i) Suppose $X = \mathbb{R}^d/\mathbb{Z}^d$ is a d-dimensional torus, $\mathscr{B}$ is the family of *Borel subsets* of X (i.e., the σ-algebra of subsets generated by the family of open sets), and $\mu = \lambda$ is Lebesgue measure, i.e. $\mu(B) = \int_X \chi_B(x)\,dx$ for any $B \in \mathscr{B}$, where χ_B is the indicator function of B. Every $x \in X$ is represented by a unique vector $(\xi_1, \ldots, \xi_d)$, where $0 \le \xi_k < 1$ $(k = 1, \ldots, d)$, and X is an abelian group with addition $z = x + y$ defined by $\zeta_k \equiv \xi_k + \eta_k \bmod 1$ $(k = 1, \ldots, d)$.

For any $a \in X$, the *translation* $T_a : X \to X$ defined by $T_a x = x + a$ is a measure-preserving transformation of the probability space $(X, \mathscr{B}, \lambda)$.

Proposition 19 *The translation $T_a : X \to X$ of the d-dimensional torus $X = \mathbb{R}^d/\mathbb{Z}^d$ is ergodic if and only if $1, \alpha_1, \ldots, \alpha_d$ are linearly independent over the rational field $\mathbb{Q}$, where $(\alpha_1, \ldots, \alpha_d)$ is the vector which represents a.*

Proof Suppose first that $1, \alpha_1, \ldots, \alpha_d$ are not linearly independent over $\mathbb{Q}$. Then there exists a nonzero vector $n \in \mathbb{Z}^d$ such that

$$n \cdot a = v_1 \alpha_1 + \cdots + v_d \alpha_d \in \mathbb{Z}.$$

Hence if $f(x) = e(n \cdot x)$, then $f(T_a x) = f(x)$ for all x. Since f is not constant a.e., it follows from part (i) of Proposition 18 that T_a is not ergodic.

Suppose on the other hand that $1, \alpha_1, \ldots, \alpha_d$ are linearly independent over $\mathbb{Q}$ and let f be an integrable function such that $f(T_a x) = f(x)$ a.e. Then $f(T_a x)$ and $f(x)$ have the same Fourier coefficients:

$$\int_X f(x)e(-n \cdot x)\,dx = \int_X f(x + a)e(-n \cdot x)\,dx = e(n \cdot a)\int_X f(x)e(-n \cdot x)\,dx.$$

Since $e(n \cdot a) \ne 1$ for all $n \ne 0$, it follows that

$$\int_X f(x)e(-n \cdot x)\,dx = 0 \quad \text{for all } n \ne 0.$$

Since integrable functions with the same Fourier coefficients must agree almost everywhere, this proves that f is constant a.e. Hence, by Proposition 18 again, T_a is ergodic. $\square$

If we compare Proposition 3′ and the remarks after its proof with Proposition 19, then we see from Theorems 1′-2′ and Proposition 18 that the following five statements are equivalent for $X = \mathbb{R}^d/\mathbb{Z}^d$ and any $a \in X$:

(α) the sequence $(\{na\})$ is dense in X;
(β) for every $x \in X$, the sequence $(x + na)$ is uniformly distributed in X;
(γ) the translation $T_a : X \to X$ is ergodic;
(δ) for each continuous function $f : X \to \mathbb{C}$, $\lim_{n\to\infty} n^{-1}\sum_{k=0}^{n-1} f(T_a^k x) = \int_X f\,d\lambda$
    for all $x \in X$;

(ε) for each function $f \in L(X, \mathscr{B}, \lambda)$, $\lim_{n\to\infty} n^{-1} \sum_{k=0}^{n-1} f(T_a^k x) = \int_X f \, d\lambda$ for almost all $x \in X$.

(ii) Again suppose $X = \mathbb{R}^d / \mathbb{Z}^d$ is a d-dimensional torus, $\mathscr{B}$ is the family of Borel subsets of X and $\mu = \lambda$ is Lebesgue measure. For any $d \times d$ matrix $A = (\alpha_{jk})$ of integers, let $R_A : X \to X$ be the map defined by $R_A x = x'$, where

$$\xi_j' \equiv \sum_{k=1}^{d} \alpha_{jk} \xi_k \mod 1 \; (j = 1, \ldots, d).$$

If $\det A = 0$ then R_A is not measure-preserving, since the image of $\mathbb{R}^d$ under A is contained in a hyperplane of $\mathbb{R}^d$. However, if $\det A \neq 0$ then R_A is measure-preserving, since each point of X is the image under R_A of $|\det A|$ distinct points of X, and a small region B of X is the image under R_A of $|\det A|$ disjoint regions, each with volume $|\det A|^{-1}$ times that of B. (This argument is certainly valid if A is a diagonal matrix, and the general case may be reduced to this by Proposition III.41.) Thus R_A is an *endomorphism* of the torus $\mathbb{R}^d / \mathbb{Z}^d$ if and only if A is nonsingular, and an *automorphism* if and only if $\det A = \pm 1$.

Proposition 20 *The endomorphism $R_A : X \to X$ of the d-dimensional torus $X = \mathbb{R}^d / \mathbb{Z}^d$ is ergodic if and only if no eigenvalue of the nonsingular matrix A is a root of unity.*

Proof For any $n \in \mathbb{Z}^d$ we have

$$e(n \cdot R_A x) = e(n \cdot Ax) = e(Dn \cdot x),$$

where $D = A^t$ is the transpose of A.

Suppose first that A, and hence also D, has an eigenvalue ω which is a root of unity: $\omega^p = 1$ for some positive integer p. Then $(D^p - I)z = 0$ for some nonzero vector z. Moreover, since D is a matrix of integers, we may assume that $z = m \in \mathbb{Z}^d$. We may further assume that $D^i m \neq D^j m$ for $0 \le i < j < p$, by choosing p to have its least possible value. If we put

$$f(x) = e(m \cdot x) + e(m \cdot Ax) + \cdots + e(m \cdot A^{p-1}x),$$

then $f(R_A x) = f(x)$, but f is not constant a.e. Hence R_A is not ergodic, by Proposition 18.

Suppose next that R_A is not ergodic. Then, by Proposition 18 again, there exists a function $f \in L(X, \mathscr{B}, \lambda)$ such that $f(R_A x) = f(x)$ a.e., but $f(x)$ is not constant a.e. If the Fourier series of $f(x)$ is

$$\sum_{n \in \mathbb{Z}^d} c_n e(n \cdot x),$$

then the Fourier series of $f(R_A x)$ is

$$\sum_{n \in \mathbb{Z}^d} c_n e(n \cdot Ax) = \sum_{n \in \mathbb{Z}^d} c_n e(Dn \cdot x) = \sum_{n \in \mathbb{Z}^d} c_{D^{-1}n} e(n \cdot x)$$

and hence

$$c_n = c_{D^{-1}n} \quad \text{for every } n \in \mathbb{Z}^d.$$

But $c_m \neq 0$ for some nonzero $m \in \mathbb{Z}^d$, since f is not constant a.e., and $|c_n| \to 0$ as $|n| \to \infty$, since $f \in L(X, \mathcal{B}, \lambda)$. Since $c_{D^{-k}m} = c_m$ for every positive integer k, it follows that the subscripts $D^{-k}m$ are not all distinct. Hence $D^p m = m$ for some positive integer p and A has an eigenvalue which is a root of unity.   □

(There are generalizations of Propositions 19 and 20 to translations and endomorphisms of any compact abelian group X, with Haar measure in place of Lebesgue measure.)

The preceding results have an application to the theory of 'normal numbers'. In fact, without any extra effort, we will consider also higher-dimensional generalizations. A vector $x \in \mathbb{R}^d$ is said to be *normal with respect to the matrix A*, where A is a $d \times d$ matrix of integers, if the sequence $(A^n x)$ is uniformly distributed mod 1.

Proposition 21 *Let A be a $d \times d$ matrix of integers. Then (λ-) almost all vectors $x \in \mathbb{R}^d$ are normal with respect to A if and only if A is nonsingular and no eigenvalue of A is a root of unity.*

Proof If A is nonsingular and no eigenvalue of A is a root of unity then, by Proposition 20, R_A is an ergodic measure-preserving transformation of the torus $X = \mathbb{R}^d / \mathbb{Z}^d$. Hence, by Proposition 18(ii), for each nonzero $m \in \mathbb{Z}^d$,

$$n^{-1} \sum_{k=0}^{n-1} e(m \cdot A^n x) \to 0 \quad \text{as } n \to \infty \quad \text{for almost all } x \in \mathbb{R}^d.$$

Since $\mathbb{Z}^d$ is countable, and the union of a countable number of sets of measure zero is again a set of measure zero, it follows that, for almost all $x \in \mathbb{R}^d$,

$$n^{-1} \sum_{k=0}^{n-1} e(m \cdot A^n x) \to 0 \quad \text{as } n \to \infty \quad \text{for every nonzero } m \in \mathbb{Z}^d.$$

Hence, by Theorem 2′, almost all $x \in \mathbb{R}^d$ are normal with respect to A.

If A is singular then, by the remark following the proof of Proposition 9, no $x \in \mathbb{R}^d$ is normal with respect to A. Suppose finally that some eigenvalue of A is a root of unity. Then there exists a positive integer p and a nonzero vector $z \in \mathbb{Z}^d$ such that $D^p z = z$, where $D = A^t$. If

$$f(x) = e(z \cdot x) + e(z \cdot Ax) + \cdots + e(z \cdot A^{p-1}x),$$

then $f(Ax) = f(x)$ and hence

$$n^{-1} \sum_{k=0}^{n-1} f(A^k x) = f(x).$$

But if x is normal with respect to A then, by Theorem 1′,

$$n^{-1} \sum_{k=0}^{n-1} f(A^k x) \to \int_X f \, d\lambda = 0.$$

Since f is not zero a.e., it follows that the set of all x which are normal with respect to A does not have full measure. $\qquad\square$

We consider next when normality with respect to one matrix coincides with normality with respect to another matrix.

Proposition 22 *Let A be a $d \times d$ nonsingular matrix of integers, no eigenvalue of which is a root of unity. Then, for any positive integer q, the vector $x \in \mathbb{R}^d$ is normal with respect to A^q if and only if it is normal with respect to A.*

Proof It follows at once from Proposition 9 that if x is normal with respect to A, then it is also normal with respect to A^q.

Suppose, on the other hand, that x is normal with respect to A^q. Then, by Theorem $2'$, for every nonzero vector $m \in \mathbb{Z}^d$,

$$N^{-1} \sum_{n=0}^{N-1} e(m \cdot A^{nq} x) \to 0 \quad \text{as } N \to \infty.$$

Put $D = A^t$. Since D is a nonsingular matrix of integers, $D^j m$ is a nonzero vector in $\mathbb{Z}^d$ for any integer $j \geq 0$ and hence

$$N^{-1} \sum_{n=0}^{N-1} e(m \cdot A^{nq+j} x) = N^{-1} \sum_{n=0}^{N-1} e(D^j m \cdot A^{nq} x) \to 0 \quad \text{as } N \to \infty.$$

Adding these relations for $j = 0, 1, \ldots, q - 1$ and dividing by q, we obtain

$$(Nq)^{-1} \sum_{n=0}^{Nq-1} e(m \cdot A^n x) \to 0 \quad \text{as } N \to \infty.$$

Since the sum of at most q terms $e(m \cdot A^n x)$ has absolute value at most q it follows that, also without restricting N to be a multiple of q,

$$N^{-1} \sum_{n=0}^{N-1} e(m \cdot A^n x) \to 0 \quad \text{as } N \to \infty.$$

Hence, by Theorem $2'$, x is normal with respect to A. $\qquad\square$

Corollary 23 *Let A be a $d \times d$ nonsingular integer matrix, no eigenvalue of which is a root of unity, and let B be a $d \times d$ integer matrix such that $A^p = B^q$ for some positive integers p, q. Then $x \in \mathbb{R}^d$ is normal with respect to A if and only if x is normal with respect to B.*

Proof This follows at once from Proposition 22, since the hypotheses imply that also B is nonsingular and has no eigenvalue which is a root of unity. $\qquad\square$

Brown and Moran (1993) have shown, conversely, that if A, B are *commuting* $d \times d$ nonsingular integer matrices, no eigenvalues of which are roots of unity, such that the set of all vectors normal with respect to A coincides with the set of all vectors normal with respect to B, then $A^p = B^q$ for some positive integers p, q.

These results will now be specialized to the scalar case. A real number x is said to be *normal to the base* a, where a is an integer ≥ 2, if the sequence $(a^n x)$ is uniformly distributed mod 1. It is readily shown that x is normal to the base a if and only if, in the expansion of x to the base a:

$$x = \lfloor x \rfloor + x_1/a + x_2/a^2 + \cdots,$$

where $x_i \in \{0, 1, \ldots, a - 1\}$ for all $i \geq 1$ and $x_i = a - 1$ for at most finitely many i, every block of digits occurs with the proper frequency; i.e., for any positive integer k and any $a_1, \ldots, a_k \in \{0, 1, \ldots, a-1\}$, the number $v(N)$ of i with $1 \leq i \leq N$ such that

$$x_i = a_1, x_{i+1} = a_2, \ldots, x_{i+k-1} = a_k,$$

satisfies $v(N)/N \to a^{-k}$ as $N \to \infty$. By Proposition 21, almost all real numbers x are normal to a given base a. The original proof of this by Borel (1909) was a forerunner of Birkhoff's ergodic theorem. (In fact Borel's proof was faulty, but his paper was influential. Borel used a different definition of normal number, but Wall (1949) showed that it was equivalent to the definition in terms of uniform distribution adopted here.)

The first published proof of the scalar case of Corollary 23 was given by Schmidt (1960), who also proved the scalar version of the result of Brown and Moran: the set of all numbers normal to the base a coincides with the set of all numbers normal to the base b, where a and b are integers ≥ 2, if and only if $a^p = b^q$ for some positive integers p, q.

Although almost all real numbers are normal to *every* base a, it is still not known if such familiar irrational numbers as $\sqrt{2}, e$ or π are normal to some base. There are, however, various explicit constructions of normal numbers. In particular, Champernowne (1933) showed that the real number θ whose expansion to the base 10 is composed of the positive integers in their natural order, in other words, $\theta = 0.123456789101112\ldots$, is itself normal to the base 10.

(iii) Let A be a set of finite cardinality r, which for definiteness we take to be the set $\{1, \ldots, r\}$, and let $p_1, \ldots, p_r$ be positive real numbers with sum 1. If $\mathscr{B}_0$ is the family of all subsets of the finite set A and if, for any $B_0 \in \mathscr{B}_0$, we put $\mu_0(B_0) = \sum_{a \in B_0} p_a$, then μ_0 is a probability measure and $(A, \mathscr{B}_0, \mu_0)$ is a probability space.

Now let X be the set of all bi-infinite sequences $x = (\ldots, x_{-2}, x_{-1}, x_0, x_1, x_2, \ldots)$ with $x_i \in A$ for every $i \in \mathbb{Z}$. Thus X is the product of infinitely many copies of A. We construct a *product measure* on X in the following way.

For any finite sequence $(a_{-m}, \ldots, a_0, \ldots, a_m)$ with $a_i \in A$ for $-m \leq i \leq m$, define the (special) *cylinder set* $[a_{-m}, \ldots, a_m]$ of *order* m to be the set of all $x \in X$ such that $x_i = a_i$ for $-m \leq i \leq m$. There are r^{2m+1} distinct cylinder sets of order m, distinct cylinder sets are disjoint and X is the union of them all.

Let $\mathscr{C}_m$ denote the collection of all unions of distinct cylinder sets of order m. Thus $X \in \mathscr{C}_m$ and, if $B \in \mathscr{C}_m$, then $B^c = X \backslash B \in \mathscr{C}_m$. Moreover, $B, C \in \mathscr{C}_m$ implies $B \cup C \in \mathscr{C}_m$ and $B \cap C \in \mathscr{C}_m$. If $B \in \mathscr{C}_m$, say

$$B = [a_{-m}, \ldots, a_m] \cup \cdots \cup [a'_{-m}, \ldots, a'_m],$$

we define

$$\mu_m(B) = p_{a_{-m}} \cdots p_{a_m} + \cdots + p_{a'_{-m}} \cdots p_{a'_m}.$$

Then $\mu_m(X) = 1$, $\mu_m(B) \geq 0$ for every $B \in \mathscr{C}_m$, and

$$\mu_m(B \cup C) = \mu_m(B) + \mu_m(C) \text{ if } B, C \in \mathscr{C}_m \text{ and } B \cap C = \emptyset.$$

Every union of cylinder sets of order m is also a union of cylinder sets of order $m + 1$, since

$$[a_{-m}, \ldots, a_m] = \bigcup_{a, a' \in A} [a, a_{-m}, \ldots, a_m, a'].$$

Thus $\mathscr{C}_m \subseteq \mathscr{C}_{m+1}$. Moreover μ_{m+1} continues μ_m, since

$$\mu_{m+1}([a_{-m}, \ldots, a_m]) = \sum_{j,j'=1}^{r} p_j p_{j'} p_{a_{-m}} \cdots p_{a_m}$$

$$= \mu_m([a_{-m}, \ldots, a_m]) \left(\sum_{j=1}^{r} p_j \right) \left(\sum_{j'=1}^{r} p_{j'} \right)$$

$$= \mu_m([a_{-m}, \ldots, a_m]).$$

Let μ denote the continuation of all μ_m to $\mathscr{C} = \mathscr{C}_0 \cup \mathscr{C}_1 \cup \ldots$. If $B, C \in \mathscr{C}$, then $B, C \in \mathscr{C}_m$ for some m. Hence, for given $C \in \mathscr{C}$, there are only finitely many distinct $B \in \mathscr{C}$ such that $B \subseteq C$. Consequently, if C is the union of a sequence of disjoint sets $C_n \in \mathscr{C}$ $(n = 1, 2, \ldots)$, then $C_n = \emptyset$ for all large n and $\mu(C) = \sum_{n \geq 1} \mu(C_n)$. It follows, by a construction due to Carathéodory (1914), that μ can be uniquely extended to the σ-algebra $\mathscr{B}$ of subsets of X generated by $\mathscr{C}$ so that $(X, \mathscr{B}, \mu)$ is a probability space. For any $\varepsilon > 0$ there exists, for each $B \in \mathscr{B}$, some $C \in \mathscr{C}$ such that $\mu(B \triangle C) < \varepsilon$.

The *two-sided Bernoulli shift* $B_{p_1,\ldots,p_r}$ is the map $\sigma : X \to X$ defined by $\sigma x = x'$, where $x'_i = x_{i+1}$ for every $i \in \mathbb{Z}$. It is a measure-preserving transformation of the probability space $(X, \mathscr{B}, \mu)$, since

$$\sigma^{-1}[a_{-m}, \ldots, a_m] = \bigcup_{a, a' \in A} [a, a', a_{-m}, \ldots, a_m]$$

and hence

$$\mu(\sigma^{-1}[a_{-m}, \ldots, a_m]) = \sum_{j,j'=1}^{r} p_j p_{j'} p_{a_{-m}} \cdots p_{a_m}$$

$$= \sum_{j,j'=1}^{r} p_j p_{j'} \mu([a_{-m}, \ldots, a_m]) = \mu([a_{-m}, \ldots, a_m]).$$

The Bernoulli shift $B_{1/2,1/2}$ is a model for the random process consisting of bi-infinite sequences of coin-tossings.

We may define the *general cylinder set* $C_{i_1 \ldots i_k}^{a_1 \ldots a_k}$, where $i_1, \ldots, i_k$ are distinct integers, to be the set of all $x \in X$ such that

$$x_{i_1} = a_1, \ldots, x_{i_k} = a_k.$$

In particular, $C_i^a = \sigma^{-i}[a]$ and hence $\mu(C_i^a) = p_a$. It follows by induction on k that

$$\mu(C_{i_1 \ldots i_k}^{a_1 \ldots a_k}) = p_{a_1} \cdots p_{a_k}.$$

Proposition 24 *For any given positive numbers $p_1, \ldots, p_r$ with sum 1, the two-sided Bernoulli shift $B_{p_1, \ldots, p_r}$ is ergodic.*

Proof Suppose $B \in \mathscr{B}$ and $\sigma^{-1} B = B$. For any $\varepsilon > 0$ there exists a set $C \in \mathscr{C}$ such that

$$\mu(B \, \Delta C) = \mu(B \backslash C) + \mu(C \backslash B) < \varepsilon.$$

Then

$$|\mu(B) - \mu(C)| = |\mu(C \cap B) + \mu(B \backslash C) - \mu(C \cap B) - \mu(C \backslash B)|$$
$$\leq \mu(B \backslash C) + \mu(C \backslash B) < \varepsilon$$

and hence

$$|\mu(B)^2 - \mu(C)^2| = \{\mu(C) + \mu(B)\} \, |\mu(B) - \mu(C)| < 2\varepsilon.$$

We may suppose that C is the union of finitely many special cylinder sets of order m. Since

$$\sigma^{-n}[a_{-m}, \ldots, a_m] = C_{-m+n, \ldots, m+n}^{a_{-m}, \ldots, a_m},$$

for $n > 2m$ we have

$$[a'_{-m}, \ldots, a'_m] \cap \sigma^{-n}[a_{-m}, \ldots, a_m] = C_{-m, \ldots, m, -m+n, \ldots, m+n}^{a'_{-m}, \ldots, a'_m, a_{-m}, \ldots, a_m},$$

and hence

$$\mu([a'_{-m}, \ldots, a'_m] \cap \sigma^{-n}[a_{-m}, \ldots, a_m]) = p_{a'_{-m}} \cdots p_{a'_m} p_{a_{-m}} \cdots p_{a_m},$$
$$= \mu([a'_{-m}, \ldots, a'_m]) \mu([a_{-m}, \ldots, a_m]).$$

It follows that if $n > 2m$, then

$$\mu(C \cap \sigma^{-n} C) = \mu(C)^2.$$

But

$$\mu(B \backslash (C \cap \sigma^{-n} C)) \leq 2\mu(B \backslash C),$$

since

$$B\backslash(C \cap \sigma^{-n}C) \subseteq (B\backslash C) \cup (B\backslash\sigma^{-n}C) \subseteq (B\backslash C) \cup \sigma^{-n}(B\backslash C),$$

and similarly

$$\mu((C \cap \sigma^{-n}C)\backslash B) \leq 2\mu(C\backslash B).$$

Hence

$$|\mu(B) - \mu(C \cap \sigma^{-n}C)| \leq \mu(B\backslash(C \cap \sigma^{-n}C)) + \mu((C \cap \sigma^{-n}C)\backslash B) < 2\varepsilon.$$

Thus

$$0 \leq \mu(B) - \mu(B)^2 = \mu(B) - \mu(C \cap \sigma^{-n}C) + \mu(C \cap \sigma^{-n}C) - \mu(B)^2$$
$$< 2\varepsilon + \mu(C)^2 - \mu(B)^2 < 4\varepsilon.$$

Since ε is arbitrary, we conclude that $\mu(B) = \mu(B)^2$. Hence $\mu(B) = 0$ or 1, and σ is ergodic. $\square$

Similarly, if Y is the set of all infinite sequences $y = (y_1, y_2, y_3, \ldots)$ with $y_i \in A$ for every $i \in \mathbb{N}$, then the *one-sided Bernoulli shift* $B^+_{p_1,\ldots,p_r}$, i.e. the map $\tau : Y \to Y$ defined by $\tau y = y'$, where $y'_i = y_{i+1}$ for every $i \in \mathbb{N}$, is a measure-preserving transformation of the analogously constructed probability space $(Y, \mathscr{B}, \mu)$. It should be noted that, although $\tau Y = Y$, τ is not invertible. In the same way as for the two-sided shift, it may be shown that the one-sided Bernoulli shift $B^+_{p_1,\ldots,p_r}$ is always ergodic.

(iv) An example of some historical interest is the 'continued fraction' or *Gauss* map. Let $X = [0, 1]$ be the unit interval and $T : X \to X$ the map defined (in the notation of §1) by

$$T\xi = \{\xi^{-1}\} \quad \text{if } \xi \in (0, 1),$$
$$= 0 \quad \text{if } \xi = 0 \text{ or } 1.$$

Thus T acts as the shift operator on the continued fraction expansion of ξ: if

$$\xi = [0; a_1, a_2, \ldots] = \cfrac{1}{a_1 + \cfrac{1}{a_2 + \cdots}},$$

then $T\xi = [0; a_2, a_3, \ldots]$. (In the terminology of Chapter IV, the complete quotients of ξ are $\xi_{n+1} = 1/T^n\xi$.)

It is not difficult to show that T is a measure-preserving transformation of the probability space $(X, \mathscr{B}, \mu)$, where $\mathscr{B}$ is the family of Borel subsets of $X = [0, 1]$ and μ is the 'Gauss' measure defined by

$$\mu(B) = (\log 2)^{-1} \int_B (1 + x)^{-1} \, dx.$$

It may further be shown that T is ergodic. Hence, by Birkhoff's ergodic theorem, if f is an integrable function on the interval X then, for almost all $\xi \in X$,

$$\lim_{n \to \infty} n^{-1} \sum_{k=0}^{n-1} f(T^k\xi) = (\log 2)^{-1} \int_X f(x)(1 + x)^{-1} \, dx.$$

Here it makes no difference if 'integrable' and 'almost all' refer to the invariant measure μ or to Lebesgue measure, since $1/2 \le (1+x)^{-1} \le 1$.

Taking f to be the indicator function of the set $\{\xi \in X : a_1 = m\}$, we see that the asymptotic relative frequency of the positive integer m among the partial quotients $a_1, a_2, \ldots$ is almost always

$$(\log 2)^{-1} \int_{(m+1)^{-1}}^{m^{-1}} (1+x)^{-1}\, dx = (\log 2)^{-1} \log((m+1)^2/(m(m+2))).$$

It follows, in particular, that almost all $\xi \in X$ have unbounded partial quotients.

Again, by taking $f(\xi) = \log \xi$ it may be shown that, for almost all $\xi \in X$,

$$\lim_{n \to \infty} (1/n) \log q_n(\xi) = \pi^2/(12 \log 2),$$

where $q_n(\xi)$ is the denominator of the n-th convergent p_n/q_n of ξ. This was first proved by Lévy (1929).

In a letter to Laplace, Gauss (1812) stated that, for each $x \in (0, 1)$, the proportion of $\xi \in X$ for which $T^n \xi < x$ converges as $n \to \infty$ to $\log(1 + x)/(\log 2)$ and he asked if Laplace could provide an estimate for the rapidity of convergence. If one writes

$$r_n(x) = m_n(x) - \log(1 + x)/(\log 2),$$

where $m_n(x)$ is the Lebesgue measure of the set of all $\xi \in X$ such that $T^n \xi < x$, then Gauss's statement is that $r_n(x) \to 0$ as $n \to \infty$ and his question is, how fast?

Gauss's statement was first proved by Kuz'min (1928), who also gave an estimate for the rapidity of convergence. If one regards Gauss's statement as a proposition in ergodic theory, then one needs to know that T is not only ergodic but even *mixing*, i.e. for all $A, B \in \mathcal{B}$,

$$\mu(T^{-n} A \cap B) \to \mu(A)\mu(B) \quad \text{as } n \to \infty.$$

Kuz'min's estimate $r_n(x) = O(q^{\sqrt{n}})$ for some $q \in (0, 1)$ was improved by Lévy (1929) and Szüsz (1961) to $r_n(x) = O(q^n)$ with $q = 0.7$ and $q = 0.485$ respectively. A substantial advance was made by Wirsing (1974). By means of an infinite-dimensional generalization of a theorem of Perron (1907) and Frobenius (1908) on positive matrices, he showed that

$$r_n(x) = (-\lambda)^n \psi(x) + O(x(1 - x)\mu^n),$$

where ψ is a twice continuously differentiable function with $\psi(0) = \psi(1) = 0$, $0 < \mu < \lambda$ and $\lambda = 0.303663 \ldots$. Wirsing's analysis has been extended by Babenko (1978) and Mayer (1990).

(v) Suppose we are given a system of ordinary differential equations

$$dx/dt = f(x), \tag{$\dagger$}$$

where $x \in \mathbb{R}^d$ and $f : \mathbb{R}^d \to \mathbb{R}^d$ is a continuously differentiable function. Then, for any $x \in \mathbb{R}^d$, there is a unique solution $\varphi_t(x)$ of ($\dagger$) such that $\varphi_0(x) = x$.

Suppose further that there exists an *invariant region* $X \subseteq \mathbb{R}^d$. That is, X is the closure of a bounded connected open set and $x \in X$ implies $\varphi_t(x) \in X$. Then the map $T_t \colon X \to X$ given by $T_t x = \varphi_t(x)$ is defined for every $t \in \mathbb{R}$ and satisfies $T_{t+s}x = T_t(T_s x)$.

Suppose finally that $\operatorname{div} f = 0$ for every $x \in \mathbb{R}^d$, where $x = (x_1, \ldots, x_d)$, $f = (f_1, \ldots, f_d)$ and

$$\operatorname{div} f := \sum_{k=1}^{d} \partial f_k / \partial x_k.$$

Then, by a theorem due to Liouville, the map T_t sends an arbitrary region into a region of the same volume. (For the statement and proof of Liouville's theorem see, for example, V.I. Arnold, *Mathematical methods of classical mechanics*, Springer-Verlag, New York, 1978.) It follows that if $\mathscr{B}$ is the family of Borel subsets of X and μ Lebesgue measure, normalized so that $\mu(X) = 1$, then T_t is a measure-preserving transformation of the probability space $(X, \mathscr{B}, \mu)$.

An important special case is the Hamiltonian system of ordinary differential equations

$$dp_i/dt = -\partial H/\partial q_i, \quad dq_i/dt = \partial H/\partial p_i \quad (i = 1, \ldots, n),$$

where $H(p_1, \ldots, p_n, q_1, \ldots, q_n)$ is a twice continuously differentiable real-valued function. The divergence does indeed vanish identically in this case, since

$$-\sum_{i=1}^{n} \partial^2 H/\partial p_i \partial q_i + \sum_{i=1}^{n} \partial^2 H/\partial q_i \partial p_i = 0.$$

Furthermore, for any $h \in \mathbb{R}$, the energy surface $X \colon H(p, q) = h$ is invariant, since

$$dH[p(t), q(t)]/dt = \sum_{i=1}^{n} \partial H/\partial p_i (-\partial H/\partial q_i) + \sum_{i=1}^{n} \partial H/\partial q_i \partial H/\partial p_i = 0.$$

It is not difficult to show that if σ is the volume element on X induced by the Euclidean metric $\| \ \|$ on $\mathbb{R}^{2n}$, and if

$$\nabla H = (\partial H/\partial p_1, \ldots, \partial H/\partial p_n, \partial H/\partial q_1, \ldots, \partial H/\partial q_n)$$

is the gradient of H, then the maps T_t preserve the measure μ on X defined by

$$\mu(B) = \int_B d\sigma / \|\nabla H\|.$$

If X is compact, this measure can be normalized and we obtain a family of measure-preserving transformations $T_t (t \in \mathbb{R})$ of the corresponding probability space.

(vi) Many problems arising in mechanics may be reduced by a change of variables to the geometric problem of *geodesic flow*. If M is a smooth Riemannian manifold then the set of all pairs (x, v), where $x \in M$ and v is a unit vector in the tangent space to

M at x, can be given the structure of a Riemannian manifold, the *unit tangent bundle* T_1M. Evidently T_1M is a $(2n-1)$-dimensional manifold if M is n-dimensional. There is a natural measure μ on T_1M such that $d\mu = dv_q\, d\omega_q$, where dv_q is the volume element at q of the Riemannian manifold M and ω_q is Lebesgue measure on the unit sphere S^{n-1} in the tangent space to M at x. If M is compact, then the measure μ can be normalized so that $\mu(T_1M) = 1$.

A *geodesic* on M is a curve $\gamma \subseteq M$ such that the length of every curve in M joining a point $x \in \gamma$ to any sufficiently close point $y \in \gamma$ is not less than the length of the arc of γ which joins x and y. Given any point $(x, v) \in T_1M$, there is a unique geodesic passing through x in the direction of v. The geodesic flow on T_1M is the flow $\varphi_t: T_1M \to T_1M$ defined by $\varphi_t(x, v) = (x_t, v_t)$, where x_t is the point of M reached from x after time t by travelling with unit speed along the geodesic determined by (x, v) and v_t is the unit tangent vector to this geodesic at x_t. If M is compact then, for every real t, φ_t is defined and is a measure-preserving transformation of the corresponding probability space $(T_1M, \mathscr{B}, \mu)$.

The geodesics on a compact 2-dimensional manifold M whose curvature at each point is negative were profoundly studied by Hadamard (1898). It was first shown by E. Hopf (1939) that in this case φ_t is ergodic for every $t > 0$. (We must exclude $t = 0$, since φ_0 is the identity map.) This result has been considerably generalized by Anosov (1967) and others. In particular, the geodesic flow on a compact n-dimensional Riemannian manifold is ergodic if at each point the curvature of every 2-dimensional section is negative.

Although the preceding examples look quite different, some of them are not 'really' different, i.e. apart from sets of measure zero. More precisely, if $(X_1, \mathscr{B}_1, \mu_1)$ and $(X_2, \mathscr{B}_2, \mu_2)$ are probability spaces with measure-preserving transformations $T_1: X_1 \to X_1$ and $T_2: X_2 \to X_2$, we say that T_1 is *isomorphic* to T_2 if there exist sets $X_1' \in \mathscr{B}_1$, $X_2' \in \mathscr{B}_2$ with $\mu_1(X_1') = 1$, $\mu_2(X_2') = 1$ and $T_1X_1' \subseteq X_1'$, $T_2X_2' \subseteq X_2'$, and a bijective map φ of X_1' onto X_2' such that

(i) for any $B_1 \subseteq X_1'$, $B_1 \in \mathscr{B}_1$ if and only if $\varphi(B_1) \in \mathscr{B}_2$ and then $\mu_1(B_1) = \mu_2(\varphi(B_1))$;

(ii) $\varphi(T_1x) = T_2\varphi(x)$ for every $x \in X_1'$.

For example, it is easily shown that the Bernoulli shift $B_{p_1,\ldots,p_r}$ is isomorphic to the following transformation of the unit square, equipped with Lebesgue measure. Divide the square into r vertical strips of width $p_1, \ldots, p_r$; then contract the height of the i-th strip and expand its width so that it has height p_i and width 1; finally combine these rectangles to form the unit square again by regarding them as horizontal strips of height $p_1, \ldots, p_r$. (For $r = 2$ and $p_1 = p_2 = 1/2$, this transformation of the unit square is allegedly used by bakers when kneading dough.)

It is easily shown also that isomorphism is an equivalence relation and that it preserves ergodicity. However, it is usually quite difficult to show that two measure-preserving transformations are indeed isomorphic. A period of rapid growth was initiated with the definition by Kolmogorov (1958), and its practical implementation by Sinai (1959), of a new numerical isomorphism invariant, the *entropy* of a measure-preserving transformation. For the formal definition of entropy we refer to the texts on ergodic theory cited at the end of the chapter. Here we merely state its value for some of the preceding examples.

Any translation T_a of the torus $\mathbb{R}^d/\mathbb{Z}^d$ has entropy zero, whereas the endomorphism R_A of $\mathbb{R}^d/\mathbb{Z}^d$ has entropy

$$\sum_{i:\,|\lambda_i|>1} \log|\lambda_i|,$$

where $\lambda_1, \ldots, \lambda_d$ are the eigenvalues of the matrix A and the summation is over those of them which lie outside the unit circle.

The two-sided Bernoulli shift $B_{p_1,\ldots,p_r}$ has entropy

$$-\sum_{j=1}^{r} p_j \log p_j,$$

and the entropy of the one-sided Bernoulli shift $B^+_{p_1,\ldots,p_r}$ is given by the same formula. It follows that $B_{1/2,1/2}$ is not isomorphic to $B_{1/3,1/3,1/3}$, since the first has entropy $\log 2$ and the second has entropy $\log 3$. Ornstein (1970) established the remarkable result that two-sided Bernoulli shifts are completely classified by their entropy: $B_{p_1,\ldots,p_r}$ is isomorphic to $B_{q_1,\ldots,q_s}$ if and only if

$$-\sum_{j=1}^{r} p_j \log p_j = -\sum_{k=1}^{s} q_k \log q_k.$$

This is no longer true for one-sided Bernoulli shifts. Walters (1973) has shown that $B^+_{p_1,\ldots,p_r}$ is isomorphic to $B^+_{q_1,\ldots,q_s}$ if and only if $r = s$ and $q_1, \ldots, q_s$ is a permutation of $p_1, \ldots, p_r$.

The Gauss map $Tx = \{x^{-1}\}$ has entropy $\pi^2/6\log 2$. Although it is mixing, it is not isomorphic to a Bernoulli shift.

Katznelson (1971) showed that any ergodic automorphism of the torus $\mathbb{R}^d/\mathbb{Z}^d$ is isomorphic to a two-sided Bernoulli shift, and Lind (1977) has extended this result to ergodic automorphisms of any compact abelian group.

Ornstein and Weiss (1973) showed that, if φ_t is the geodesic flow on a smooth (of class C^3) compact two-dimensional Riemannian manifold whose curvature at each point is negative, then φ_t is isomorphic to a two-sided Bernoulli shift for every $t > 0$. Although, as Hilbert showed, a compact surface of negative curvature cannot be imbedded in $\mathbb{R}^3$, the geodesic flow on a surface of negative curvature can be realized as the motion of a particle constrained to move on a surface in $\mathbb{R}^3$ subject to centres of attraction and repulsion in the ambient space. The isomorphism with a Bernoulli shift shows that a deterministic mechanical system can generate a random process. Thus philosophical objections to 'Laplacian determinism' or to 'God playing dice' do not seem to have much point.

5 Recurrence

It was shown by Poincaré (1890) that the paths of a Hamiltonian system of differential equations almost always return to any neighbourhood, however small, of their initial

points. Poincaré's proof was inevitably incomplete, since at the time measure theory
did not exist. However, Carathéodory (1919) showed that his argument could be made
rigorous with the aid of Lebesgue measure:

Proposition 25 *Let $T: X \to X$ be a measure-preserving transformation of the probability space $(X, \mathscr{B}, \mu)$. Then almost all points of any $B \in \mathscr{B}$ return to B infinitely
often, i.e. for each $x \in B$, apart from a set of μ-measure zero, there exists an increasing
sequence (n_k) of positive integers such that $T^{n_k} x \in B$ $(k = 1, 2, \ldots)$.*
Furthermore, if $\mu(B) > 0$, then $\mu(B \cap T^{-n}B) > 0$ for infinitely many $n \geq 1$.

Proof For any $N \geq 0$, put $B_N = \bigcup_{n \geq N} T^{-n}B$. Then

$$A := \bigcap_{N \geq 0} B_N$$

is the set of all points $x \in X$ such that $T^n x \in B$ for infinitely many positive integers
n. Since $B_{N+1} = T^{-1}B_N$, we have $\mu(B_{N+1}) = \mu(B_N)$ and hence $\mu(B_N) = \mu(B_0)$
for all $N \geq 1$. Since $B_{N+1} \subseteq B_N$, it follows that

$$\mu(A) = \lim_{N \to \infty} \mu(B_N) = \mu(B_0).$$

Since $A \subseteq B_0$, this implies

$$\mu(B_0 \backslash A) = \mu(B_0) - \mu(A) = 0$$

and hence, since $B \subseteq B_0$, $\mu(B \backslash A) = 0$.
   This proves the first statement of the proposition. If $\mu(B \cap T^{-n}B) = 0$ for all
$n \geq m$, then $\mu(B \cap B_N) = 0$ for all $N \geq m$ and hence

$$\mu(B \cap A) = \lim_{N \to \infty} \mu(B \cap B_N) = 0.$$

Consequently

$$\mu(B) = \mu(B \backslash A) + \mu(B \cap A) = 0,$$

which proves the second statement of the proposition. $\square$

   Furstenberg (1977) extended Proposition 25 in the following way:
   *Let T be a measure-preserving transformation of the probability space $(X, \mathscr{B}, \mu)$.
If $B \in \mathscr{B}$ with $\mu(B) > 0$ and if $p \geq 2$, then $\mu(B \cap T^{-n}B \cap \cdots \cap T^{-(p-1)n}B) > 0$
for some $n \geq 1$.*
   His proof of this theorem made heavy use of ergodic theory and, in particular,
of a new structure theory for measure-preserving transformations. From his theorem
he was able to deduce quite easily a result for which Szemeredi (1975) had given a
complicated combinatorial proof:
   *Let S be a subset of the set $\mathbb{N}$ of positive integers which has positive upper density;
i.e., for some $\alpha \in (0, 1)$, there exist arbitrarily long intervals $I \subseteq \mathbb{N}$ containing at least
$\alpha|I|$ elements of S. Then S contains arithmetic progressions of arbitrary finite length.*

Furstenberg's approach to this result is not really shorter than Szemeredi's, but it is much more systematic. In fact the following generalization of Furstenberg's theorem was given soon afterwards by Furstenberg and Katznelson (1978):

If $T_1, \ldots, T_p$ are commuting measure-preserving transformations of the probability space $(X, \mathcal{B}, \mu)$ and if $B \in \mathcal{B}$ with $\mu(B) > 0$, then $\mu(B \cap T_1^{-n} B \cap \cdots \cap T_p^{-n} B) > 0$ for infinitely many $n \geq 1$.

Furstenberg and Katznelson could then deduce quite easily a multi-dimensional extension of Szemeredi's theorem which is still beyond the reach of combinatorial methods. Szemeredi's theorem was itself a far-reaching generalization of a famous theorem of van der Waerden (1927):

If $\mathbb{N} = S_1 \cup \cdots \cup S_r$ is a partition of the set of all positive integers into finitely many subsets, then one of the subsets S_j contains arithmetic progressions of arbitrary finite length.

Szemeredi's result further indicates how the subset S_j should be chosen.

Poincaré's measure-theoretic recurrence theorem has a topological counterpart due to Birkhoff (1912):

If X is a compact metric space and $T : X \to X$ a continuous map, then there exists a point $z \in X$ and an increasing sequence (n_k) of positive integers such that $T^{n_k} z \to z$ as $k \to \infty$.

Before Furstenberg and Katznelson proved their measure-theoretic theorem, Furstenberg and Weiss (1978) had already proved its topological counterpart:

If X is a compact metric space and $T_1, \ldots, T_p$ commuting continuous maps of X into itself, then there exists a point $z \in X$ and an increasing sequence (n_k) of positive integers such that $T_i^{n_k} z \to z$ as $k \to \infty$ $(i = 1, \ldots, p)$.

From their theorem Furstenberg and Weiss were able to deduce quite easily both van der Waerden's theorem and a known multi-dimensional generalization of it, due to Grünwald. It would take too long to prove here Szemeredi's theorem by the method of Furstenberg and Katznelson, but we will prove van der Waerden's theorem by the method of Furstenberg and Weiss. The proof illustrates how results in one area of mathematics can find application in another area which is apparently unrelated.

Proposition 26 *Let (X, d) be a compact metric space and $T : X \to X$ a continuous map. Then, for any real $\varepsilon > 0$ and any $p \in \mathbb{N}$, there exists some $z \in X$ and $n \in \mathbb{N}$ such that*

$$\mathrm{d}(T^n z, z) < \varepsilon, \quad \mathrm{d}(T^{2n} z, z) < \varepsilon, \ldots, \mathrm{d}(T^{pn} z, z) < \varepsilon.$$

Proof (i) A subset A of X is said to be *invariant* under T if $T A \subseteq A$. The closure $\bar{A}$ of an invariant set A is again invariant since, by the continuity of T, $T \bar{A} \subseteq \overline{T A}$. Let $\mathcal{F}$ be the collection of all nonempty closed invariant subsets of X. Clearly $\mathcal{F}$ is not empty, since $X \in \mathcal{F}$. If we regard $\mathcal{F}$ as partially ordered by inclusion then, by *Hausdorff's maximality theorem*, $\mathcal{F}$ contains a maximal totally ordered subcollection $\mathcal{T}$. The intersection Z of all the subsets in $\mathcal{T}$ is both closed and invariant. It is also nonempty, since X is compact. Hence $Z \in \mathcal{T}$ and, by construction, no nonempty proper closed subset of Z is invariant.

By replacing X by its compact subset Z we may now assume that the only closed invariant subsets of X itself are X and $\emptyset$.

(ii) For any given $z \in X$, the closure of the set $(T^n z)_{n \geq 1}$ is a nonempty closed invariant subset of X and therefore coincides with X. Thus for every $\varepsilon > 0$ there exists $n = n(\varepsilon) \geq 1$ such that $d(T^n z, z) < \varepsilon$. This proves the proposition for $p = 1$.

We suppose now that $p > 1$ and the proposition holds with p replaced by $p - 1$.

(iii) We show next that, for any $\varepsilon > 0$, there exists a finite set K of positive integers such that, for all $x, x' \in X$,

$$d(T^k x', x) < \varepsilon/2 \quad \text{for some } k \in K.$$

If B is a nonempty open subset of X, then for every $z \in X$ there exists some $n \geq 1$ such that $T^n z \in B$. Hence $X = \bigcup_{n \geq 1} T^{-n} B$. Since X is compact and the sets $T^{-n} B$ are open, there is a finite set $K(B)$ of positive integers such that

$$X = \bigcup_{k \in K(B)} T^{-k} B.$$

Since X is compact again, there exist finitely many open balls $B_1, \ldots, B_r$ with radius $\varepsilon/4$ such that $X = B_1 \cup \cdots \cup B_r$. If $x, x' \in X$, then $x \in B_i$ for some $i \in \{1, \ldots, r\}$ and $x' \in T^{-k} B_i$ for some $k \in K(B_i)$. Thus we can take $K = K(B_1) \cup \cdots \cup K(B_r)$.

(iv) We now show that, for any $\varepsilon > 0$ and any $x \in X$, there exists $y \in X$ and $n \geq 1$ such that

$$d(T^n y, x) < \varepsilon, \quad d(T^{2n} y, x) < \varepsilon, \ldots, \quad d(T^{pn} y, x) < \varepsilon.$$

In fact, since each $T^k (k \in K)$ is uniformly continuous on X, we can choose $\rho > 0$ so that $d(x_1, x_2) < \rho$ implies $d(T^k x_1, T^k x_2) < \varepsilon/2$ for all $x_1, x_2 \in X$ and all $k \in K$. By the induction hypothesis, there exist $x' \in X$ and $n \geq 1$ such that

$$d(T^n x', x') < \rho, \ldots, d(T^{(p-1)n} x', x') < \rho.$$

But the invariant set TX is closed, since X is compact, and so $TX = X$. Hence $T^n X = X$ and we can choose $y' \in X$ so that $T^n y' = x'$. Thus

$$d(T^n y', x') = 0, \quad d(T^{2n} y', x') < \rho, \ldots, \quad d(T^{pn} y', x') < \rho.$$

It follows that, for all $k \in K$,

$$d(T^{n+k} y', T^k x') < \varepsilon/2, \ldots, d(T^{pn+k} y', T^k x') < \varepsilon/2.$$

For each $x \in X$ there is a $k \in K$ such that $d(T^k x', x) < \varepsilon/2$. Thus if $y = T^k y'$, then

$$d(T^n y, x) < \varepsilon, \ldots, d(T^{pn} y, x) < \varepsilon.$$

(v) Let $\varepsilon_0 > 0$ and $x_0 \in X$ be given. By (iv) there exist $x_1 \in X$ and $n_1 \geq 1$ such that

$$d(T^{n_1} x_1, x_0) < \varepsilon_0, \ldots, d(T^{pn_1} x_1, x_0) < \varepsilon_0.$$

We can now choose $\varepsilon_1 \in (0, \varepsilon_0)$ so that $d(x, x_1) < \varepsilon_1$ implies

$$d(T^{n_1} x, x_0) < \varepsilon_0, \ldots, d(T^{pn_1} x, x_0) < \varepsilon_0.$$

Suppose we have defined points $x_1, \ldots, x_k$, positive integers $n_1, \ldots, n_k$, and $\varepsilon_1, \ldots, \varepsilon_k \in (0, \varepsilon_0)$ such that, for $i = 1, \ldots, k$,

$$d(T^{n_i} x_i, x_{i-1}) < \varepsilon_{i-1}, \ldots, d(T^{pn_i} x_i, x_{i-1}) < \varepsilon_{i-1},$$

and $d(x, x_i) < \varepsilon_i$ implies

$$d(T^{n_i} x, x_{i-1}) < \varepsilon_{i-1}, \ldots, d(T^{pn_i} x, x_{i-1}) < \varepsilon_{i-1}.$$

By (iv) there exist $x_{k+1} \in X$ and $n_{k+1} \geq 1$ such that

$$d(T^{n_{k+1}} x_{k+1}, x_k) < \varepsilon_k, \ldots, d(T^{pn_{k+1}} x_{k+1}, x_k) < \varepsilon_k,$$

and we can then choose $\varepsilon_{k+1} \in (0, \varepsilon_0)$ so that $d(x, x_{k+1}) < \varepsilon_{k+1}$ implies

$$d(T^{n_{k+1}} x, x_k) < \varepsilon_k, \ldots, d(T^{pn_{k+1}} x, x_k) < \varepsilon_k.$$

Thus the process can be continued indefinitely.

By taking successively $i = j - 1, j - 2, \ldots$ we see that, if $i < j$, then

$$d(T^{n_{i+1} + \cdots + n_{j-1} + n_j} x_j, x_i) < \varepsilon_i, \ldots, d(T^{p(n_{i+1} + \cdots + n_{j-1} + n_j)} x_j, x_i) < \varepsilon_i.$$

Since X is compact, it is covered by a finite number r of open balls with radius $\varepsilon_0/2$. Hence there exist i, j with $0 \leq i < j \leq r$ such that $d(x_i, x_j) < \varepsilon_0$. If we put $n = n_{i+1} + \cdots + n_{j-1} + n_j$ then, since $\varepsilon_i < \varepsilon_0$, we obtain from the triangle inequality

$$d(T^n x_j, x_j) < 2\varepsilon_0, \ldots, d(T^{pn} x_j, x_j) < 2\varepsilon_0.$$

But $\varepsilon_0 > 0$ was arbitrary. $\qquad\square$

It may be deduced from Proposition 26, by means of *Baire's category theorem*, that under the same hypotheses there exists a point $z \in X$ and an increasing sequence (n_k) of positive integers such that $T^{in_k} z \to z$ as $k \to \infty$ ($i = 1, \ldots, p$). However, as we now show, Proposition 26 already suffices to proves van der Waerden's theorem.

The set X^* of all infinite sequences $x = (x_1, x_2, \ldots)$, where $x_i \in \{1, 2, \ldots, r\}$ for every $i \geq 1$, can be given the structure of a compact metric space by defining $d(x, x) = 0$ and $d(x, y) = 2^{-k}$ if $x \neq y$ and k is the least positive integer such that $x_k \neq y_k$. The shift map $\tau: X^* \to X^*$, defined by $\tau((x_1, x_2, \ldots)) = (x_2, x_3, \ldots)$, is continuous, since

$$d(\tau(x), \tau(y)) \leq 2 \, d(x, y).$$

With the partition $\mathbb{N} = S_1 \cup \cdots \cup S_r$ in the statement of van der Waerden's theorem we associate the infinite sequence $x \in X^*$ defined by $x_i = j$ if $i \in S_j$.

Let X denote the closure of the set $(\tau^n x)_{n \geq 1}$. Then X is a closed subset of X^* which is invariant under τ. By Proposition 26, there exists a point $z \in X$ and a positive integer n such that

$$d(\tau^n z, z) < 1/2, \quad d(\tau^{2n} z, z) < 1/2, \ldots, \quad d(\tau^{pn} z, z) < 1/2;$$

i.e. $z_1 = z_{n+1} = z_{2n+1} = \cdots = z_{pn+1}$. Since $z \in X$, there is a positive integer m such that $d(\tau^m x, z) < 2^{-pn-1}$, i.e. $x_{m+i} = z_i$ for $1 \leq i \leq pn + 1$. It follows that

$$x_{m+1} = x_{m+n+1} = \cdots = x_{m+pn+1}.$$

Thus for every positive integer p there is a set $S_{j(p)}$ which contains an arithmetic progression of length p. Since there are only r possible values for $j(p)$, one of the sets S_j must contain arithmetic progressions of arbitrary finite length.

A far-reaching generalization of van der Waerden's theorem has been given by Hales and Jewett (1963). Let $A = \{a_1, \ldots, a_q\}$ be a finite set and let A^n be the set of all n-tuples with elements from A. A set $W = \{w^1, \ldots, w^q\} \subseteq A^n$ of q n-tuples $w^k = (w_1^k, \ldots, w_n^k)$ is said to be a *combinatorial line* if there exists a partition

$$\{1, \ldots, n\} = I \cup J, \ I \cap J = \emptyset,$$

such that

$$w_i^k = a_k \ (k = 1, \ldots, q) \quad \text{for } i \in I; \quad w_j^1 = \cdots = w_j^q \quad \text{for } j \in J.$$

The Hales–Jewett theorem says that, for any positive integer r, there exists a positive integer $N = N(q, r)$ such that, if A^N is partitioned into r classes, then at least one of these classes contains a combinatorial line.

If one takes $A = \{0, 1, \ldots, q-1\}$ and interprets A^n as the set of expansions to base q of all non-negative integers less than q^n, then a combinatorial line is an arithmetic progression. On the other hand, if one takes $A = \mathbb{F}_q$ to be a finite field with q elements and interprets A^n as the n-dimensional vector space $\mathbb{F}_q^n$, then a combinatorial line is an affine line. The interesting feature of the Hales–Jewett theorem is that it is purely combinatorial and does not involve any notion of addition.

6 Further Remarks

Uniform distribution and discrepancy are thoroughly discussed in Kuipers and Niederreiter [30]. For later results, see Drmota and Tichy [13]. Since these two books have extensive bibliographies, we will be sparing with references. However, it would be remiss not to recommend the great paper of Weyl [52], which remains as fresh as when it was written.

Lemma 0 is often attributed to Polya (1920), but it was already proved by Buchanan and Hildebrandt [9].

Fejér's proof that continuous periodic functions can be uniformly approximated by trigonometric polynomials is given in Dym and McKean [15]. The theorem also follows directly from the the theorem of Weierstrass (1885) on the uniform approximation of continuous functions by ordinary polynomials. A remarkable generalization of both results was given by Stone (1937); see Stone [49]. The 'Stone–Weierstrass theorem' is also proved in Rudin [44], for example.

Chen [11] gives a quantitative version of Kronecker's theorem of a different type from Proposition 3′.

The converse of Proposition 10 is proved by Kemperman [27]. For the history of the problem of mean motion, and generalizations to almost periodic functions, see Jessen and Tornehave [24]. Methods for estimating exponential sums were developed in connection with the theory of uniform distribution, but then found other applications. See Chandrasekharan [10] and Graham and Kolesnik [21].

For applications of discrepancy to numerical integration, see Niederreiter [36, 37]. For the basic properties of functions of bounded variation and the definition of total variation see, for example, Riesz and Sz.-Nagy [42].

Sharper versions of the original Erdős–Turan inequality are proved by Niederreiter and Philipp [38] and in Montgomery [35]. The discrepancy of the sequence ($\{n\alpha\}$), where α is an irrational number whose continued fraction expansion has bounded partial quotients (i.e., is *badly approximable*), is discussed by Dupain and Sós [14]. The discrepancy of the sequence ($\{n\alpha\}$), where $\alpha \in \mathbb{R}^d$, has been deeply studied by Beck [3]. The work of Roth, Schmidt and others is treated in Beck and Chen [4].

For accounts of measure theory, see Billingsley [6], Halmos [22], Loève [32] and Saks [46]. More detailed treatments of ergodic theory are given in the books of Petersen [39], Walters [51] and Cornfeld *et al.* [12]. The prehistory of ergodic theory is described by the Ehrenfests [16]. However, they do not refer to the paper of Poincaré (1894), which is reproduced in [41].

The proof of Birkhoff's ergodic theorem given here follows Katznelson and Weiss [26]. A different proof is given in the book of Walters.

Many other ergodic theorems besides Birkhoff's are discussed in Krengel [29]. We mention only the *subadditive ergodic theorem* of Kingman (1968): if T is a measure-preserving transformation of the probability space $(X, \mathcal{B}, \mu)$ and if (g_n) is a sequence of functions in $L(X, \mathcal{B}, \mu)$ such that $\inf_n n^{-1} \int_X g_n \, d\mu > -\infty$ and, for all $m, n \geq 1$,

$$g_{n+m}(x) \leq g_n(x) + g_m(T^n x) \text{ a.e.,}$$

then $n^{-1} g_n(x) \to g^*(x)$ a.e., where $g^*(Tx) = g^*(x)$ a.e., $g^* \in L(X, \mathcal{B}, \mu)$ and

$$\int_X g^* \, d\mu = \lim_{n \to \infty} n^{-1} \int_X g_n \, d\mu = \inf_n n^{-1} \int_X g_n \, d\mu.$$

Birkhoff's ergodic theorem may be regarded as a special case by taking $g_n(x) = \sum_{k=0}^{n-1} f(T^k x)$. A simple proof of Kingman's theorem is given by Steele [48]. For applications of Kingman's theorem to percolation processes and products of random matrices, see Kingman [28]. The multiplicative ergodic theorem of Oseledets is derived from Kingman's theorem by Ruelle [45].

The book of Kuipers and Niederreiter cited above has an extensive discussion of normal numbers. For normality with respect to a matrix, see also Brown and Moran [8].

Proofs of Gauss's statement on the continued fraction map are contained in the books by Billingsley [7] and Rockett and Szusz [43]. For more recent work, see Wirsing [53], Babenko [2] and Mayer [33]. For the deviation of $(1/n) \log q_n(\xi)$ from its (a.e.) limiting value $\pi^2/(12 \log 2)$ there are analogues of the central limit theorem and the law of the iterated logarithm; see Philipp and Stackelberg [40]. For higher-dimensional generalizations of Gauss's invariant measure, see Hardcastle and Khanin [23].

Applications of ergodic theory to classical mechanics are discussed in the books of Arnold and Avez [1] and Katok and Hasselblatt [25]. For connections between ergodic theory and the '$3x + 1$ problem', see Lagarias [31].

Ergodic theory has been used to generalize considerably some of the results on lattices in Chapter VIII. A *lattice* in a locally compact group G is a discrete subgroup Γ such that the G-invariant measure of the quotient space G/Γ is finite. (In Chapter VIII, $G = \mathbb{R}^n$ and $\Gamma = \mathbb{Z}^n$.) Zimmer [54] gives a good introduction to the results which have been obtained in this area.

An attractive account of the work of Furstenberg and his collaborators is given in Furstenberg [17]. See also Graham *et al.* [20] and the book of Petersen cited above. The discovery of van der Waerden's theorem is described in van der Waerden [50]. For a recent direct proof, see Mills [34].

The direct proofs reduce the theorem to an equivalent finite form: *for any positive integer p, there exists a positive integer N such that, whenever the set $\{1, 2, \ldots, N\}$ is partitioned into two subsets, at least one subset contains an arithmetic progression of length p.* The original proofs provided an upper bound for the least possible value $N(p)$ of N, but it was unreasonably large. Some progress towards obtaining reasonable upper bounds has recently been made by Shelah [47] and Gowers [19].

The Hales–Jewett theorem is proved, and then extensively generalized, in Bergelson and Leibman [5]. Furstenberg and Katznelson [18] prove a density version of the Hales–Jewett theorem, analogous to Szemeredi's density version of van der Waerden's theorem.

7 Selected References

[1] V.I. Arnold and A. Avez, *Ergodic problems of classical mechanics*, Benjamin, New York, 1968.

[2] K.I. Babenko, On a problem of Gauss, *Soviet Math. Dokl.* **19** (1978), 136–140.

[3] J. Beck, Probabilistic diophantine approximation, I. Kronecker sequences, *Ann. of Math.* **140** (1994), 451–502.

[4] J. Beck and W.W.L. Chen, *Irregularities of distribution*, Cambridge University Press, 1987.

[5] V. Bergelson and A. Leibman, Set polynomials and polynomial extension of the Hales–Jewett theorem, *Ann. of Math.* **150** (1999), 33–75.

[6] P. Billingsley, *Probability and measure*, 3rd ed., Wiley, New York, 1995.

[7] P. Billingsley, *Ergodic theory and information*, reprinted, Krieger, Huntington, N.Y., 1978.

[8] G. Brown and W. Moran, Schmidt's conjecture on normality for commuting matrices, *Invent. Math.* **111** (1993), 449–463.

[9] H.E. Buchanan and H.T. Hildebrandt, Note on the convergence of a sequence of functions of a certain type, *Ann. of Math.* **9** (1908), 123–126.

[10] K. Chandrasekharan, Exponential sums in the development of number theory, *Proc. Steklov Inst. Math.* **132** (1973), 3–24.

[11] Y.-G. Chen, The best quantitative Kronecker's theorem, *J. London Math. Soc.* (2) **61** (2000), 691–705.

[12] I.P. Cornfeld, S.V. Fomin and Ya. G. Sinai, *Ergodic theory*, Springer-Verlag, New York, 1982.

[13] M. Drmota and R.F. Tichy, *Sequences, discrepancies and applications*, Lecture Notes in Mathematics **1651**, Springer, Berlin, 1997.

[14] I. Dupain and V.T. Sós, On the discrepancy of $(n\alpha)$ sequences, *Topics in classical number theory* (ed. G. Halász), Vol. I, pp. 355–387, North-Holland, Amsterdam, 1984.

[15] H. Dym and H.P. McKean, *Fourier series and integrals*, Academic Press, Orlando, FL, 1972.

[16] P. and T. Ehrenfest, *The conceptual foundations of the statistical approach in mechanics*, English translation by M.J. Moravcsik, Cornell University Press, Ithaca, 1959. [German original, 1912]

[17] H. Furstenberg, *Recurrence in ergodic theory and combinatorial number theory*, Princeton University Press, 1981.

[18] H. Furstenberg and Y. Katznelson, A density version of the Hales–Jewett theorem, *J. Analyse Math.* **57** (1991), 64–119.

[19] W.T. Gowers, A new proof of Szemeredi's theorem, *Geom. Funct. Anal.* **11** (2001), 465–588.

[20] R.L. Graham, B.L. Rothschild and J.H. Spencer, *Ramsey theory*, 2nd ed., Wiley, New York, 1990.

[21] S.W. Graham and G. Kolesnik, *Van der Corput's method of exponential sums*, London Math. Soc. Lecture Notes **126**, Cambridge University Press, 1991.

[22] P.R. Halmos, *Measure theory*, 2nd printing, Springer-Verlag, New York, 1974.

[23] D.M. Hardcastle and K. Khanin, Continued fractions and the d-dimensional Gauss transformation, *Comm. Math. Phys.* **215** (2001), 487–515.

[24] B. Jessen and H. Tornehave, Mean motion and zeros of almost periodic functions, *Acta Math.* **77** (1945), 137–279.

[25] A. Katok and B. Hasselblatt, *Introduction to the modern theory of dynamical systems*, Cambridge University Press, 1995.

[26] Y. Katznelson and B. Weiss, A simple proof of some ergodic theorems, *Israel J. Math.* **42** (1982), 291–296.

[27] J.H.B. Kemperman, Distributions modulo 1 of slowly changing sequences, *Nieuw Arch. Wisk.* (3) **21** (1973), 138–163.

[28] J.F.C. Kingman, Subadditive processes, *Ecole d'Eté de Probabilités de Saint-Flour* V-1975 (ed. A. Badrikian), pp. 167–223, Lecture Notes in Mathematics **539**, Springer-Verlag, 1976.

[29] U. Krengel, *Ergodic theorems*, de Gruyter, Berlin, 1985.

[30] L. Kuipers and H. Niederreiter, *Uniform distribution of sequences*, Wiley, New York, 1974.

[31] J.C. Lagarias, The $3x + 1$ problem and its generalizations, *Amer. Math. Monthly* **92** (1985), 3–23.

[32] M. Loève, *Probability theory*, 4th ed. in 2 vols., Springer-Verlag, New York, 1978.

[33] D.H. Mayer, On the thermodynamic formalism for the Gauss map, *Comm. Math. Phys.* **130** (1990), 311–333.

[34] G. Mills, A quintessential proof of van der Waerden's theorem on arithmetic progressions, *Discrete Math.* **47** (1983), 117–120.

[35] H.L. Montgomery, *Ten lectures on the interface between analytic number theory and harmonic analysis*, CBMS Regional Conference Series in Mathematics **84**, American Mathematical Society, Providence, R.I., 1994.

[36] H. Niederreiter, Quasi-Monte Carlo methods and pseudo-random numbers, *Bull. Amer. Math. Soc.* **84** (1978), 957–1041.

[37] H. Niederreiter, *Random number generation and quasi-Monte Carlo methods*, CBMS–NSF Regional Conference Series in Applied Mathematics **63**, SIAM, Philadelphia, 1992.

[38] H. Niederreiter and W. Philipp, Berry–Esseen bounds and a theorem of Erdös and Turán on uniform distribution mod 1, *Duke Math. J.* **40** (1973), 633–649.

[39] K. Petersen, *Ergodic theory*, Cambridge University Press, 1983.

[40] W. Philipp and O.P. Stackelberg, Zwei Grenzwertsätze für Kettenbrüche, *Math. Ann.* **181** (1969), 152–156.

[41] H. Poincaré, Sur la théorie cinétique des gaz, *Oeuvres*, t. X, pp. 246–263, Gauthier-Villars, Paris, 1954.

[42] F. Riesz and B. Sz.-Nagy, *Functional analysis*, English transl. by L.F. Boron, Ungar, New York, 1955.

[43] A. Rockett and P. Szusz, *Continued fractions*, World Scientific, Singapore, 1992.

[44] W. Rudin, *Principles of mathematical analysis*, 3rd ed., McGraw-Hill, New York, 1976.

[45] D. Ruelle, Ergodic theory of differentiable dynamical systems, *Inst. Hautes Études Sci. Publ. Math.* **50** (1979), 27–58.

[46] S. Saks, *Theory of the integral*, 2nd revised ed., English transl. by L.C. Young, reprinted, Dover, New York, 1964.

[47] S. Shelah, Primitive recursion bounds for van der Waerden numbers, *J. Amer. Math. Soc.* **1** (1988), 683–697.

[48] J.M. Steele, Kingman's subadditive ergodic theorem, *Ann. Inst. H. Poincaré Sect. B* **25** (1989), 93–98.

[49] M.H. Stone, A generalized Weierstrass approximation theorem, *Studies in modern analysis* (ed. R.C. Buck), pp. 30–87, Mathematical Association of America, 1962.

[50] B.L. van der Waerden, How the proof of Baudet's conjecture was found, *Studies in Pure Mathematics* (ed. L. Mirsky), pp. 251–260, Academic Press, London, 1971.

[51] P. Walters, *An introduction to ergodic theory*, Springer-Verlag, New York, 1982.

[52] H. Weyl, Über die Gleichverteilung von Zahlen mod Eins, *Math. Ann.* **77** (1916), 313–352. [Reprinted in *Selecta Hermann Weyl*, pp. 111–147, Birkhäuser, Basel, 1956 and in *Hermann Weyl, Gesammelte Abhandlungen* (ed. K. Chandrasekharan), *Band I*, pp. 563–599, Springer-Verlag, Berlin, 1968]

[53] E. Wirsing, On the theorem of Gauss–Kusmin–Lévy and a Frobenius type theorem for function spaces, *Acta Arith.* **24** (1974), 507–528.

[54] R.J. Zimmer, *Ergodic theory and semi-simple groups*, Birkhäuser, Boston, 1984.

Additional Reference

B. Kra, The Green-Tao theorem on arithmetic progressions in the primes: an ergodic point of view, *Bull. Amer. Math. Soc. (N.S.)* **43** (2006), 3–23.

XII

Elliptic Functions

Our discussion of elliptic functions may be regarded as an essay in revisionism, since we do not use Liouville's theorem, Riemann surfaces or the Weierstrassian functions. We wish to show that the methods used by the founding fathers of the subject provide a natural and rigorous approach, which is very well suited for applications.

The work is arranged so that the initial sections are mutually independent, although motivation for each section is provided by those which precede it. To some extent we have also separated the discussion for real and for complex parameters, so that those interested only in the real case may skip the complex one.

1 Elliptic Integrals

After the development of the integral calculus in the second half of the 17th century, it was natural to apply it to the determination of the arc length of an ellipse since, by Kepler's first law, the planets move in elliptical orbits with the sun at one focus.

An ellipse is described in rectangular coordinates by an equation

$$x^2/a^2 + y^2/b^2 = 1,$$

where a and b are the *semi-axes* of the ellipse $(a > b > 0)$. It is also given parametrically by

$$x = a \sin\theta, \ y = b \cos\theta \quad (0 \le \theta \le 2\pi).$$

The arc length $s(\Theta)$ from $\theta = 0$ to $\theta = \Theta$ is given by

$$s(\Theta) = \int_0^\Theta [(dx/d\theta)^2 + (dy/d\theta)^2]^{1/2} d\theta$$

$$= \int_0^\Theta (a^2 \cos^2\theta + b^2 \sin^2\theta)^{1/2} d\theta$$

$$= \int_0^\Theta [a^2 - (a^2 - b^2)\sin^2\theta]^{1/2} d\theta.$$

W.A. Coppel, *Number Theory: An Introduction to Mathematics*, Universitext,
DOI: 10.1007/978-0-387-89486-7_12, © Springer Science + Business Media, LLC 2009

If we put $b^2 = a^2(1 - k^2)$, where k $(0 < k < 1)$ is the *eccentricity* of the ellipse, this takes the form

$$s(\Theta) = a \int_0^\Theta (1 - k^2 \sin^2 \theta)^{1/2} d\theta.$$

If we further put $z = \sin\theta = x/a$ and restrict attention to the first quadrant, this assumes the algebraic form

$$a \int_0^Z [(1 - k^2 z^2)/(1 - z^2)]^{1/2} dz.$$

Since the arc length of the whole quadrant is obtained by taking $Z = 1$, the arc length of the whole ellipse is

$$L = 4a \int_0^1 [(1 - k^2 z^2)/(1 - z^2)]^{1/2} dz.$$

Consider next Galileo's problem of the simple pendulum. If θ is the angle of deflection from the downward vertical, the equation of motion of the pendulum is

$$d^2\theta/dt^2 + (g/l) \sin\theta = 0,$$

where l is the length of the pendulum and g is the gravitational constant. This differential equation has the first integral

$$(d\theta/dt)^2 = (2g/l)(\cos\theta - a),$$

where a is a constant. In fact $a < 1$ for a real motion, and for oscillatory motion we must also have $a > -1$. We can then put $a = \cos\alpha$ $(0 < \alpha < \pi)$, where α is the maximum value of θ, and integrate again to obtain

$$t = (l/2g)^{1/2} \int_0^\Theta (\cos\theta - \cos\alpha)^{-1/2} d\theta$$

$$= (l/4g)^{1/2} \int_0^\Theta (\sin^2 \alpha/2 - \sin^2 \theta/2)^{-1/2} d\theta.$$

Putting $k = \sin\alpha/2$ and $kx = \sin\theta/2$, we can rewrite this in the form

$$t = (l/g)^{1/2} \int_0^X [(1 - k^2 x^2)(1 - x^2)]^{-1/2} dx.$$

The angle of deflection θ attains its maximum value α when $X = 1$, and the motion is periodic with period

$$T = 4(l/g)^{1/2} \int_0^1 [(1 - k^2 x^2)(1 - x^2)]^{-1/2} dx.$$

Attempts to evaluate the integrals in both these problems in terms of algebraic and elementary transcendental functions proved fruitless. Thus the idea arose of treating them as fundamental entities in terms of which other integrals could be expressed.

An example is the determination of the arc length of a *lemniscate*. This curve, which was studied by Jacob Bernoulli (1694), has the form of a figure of eight and is the locus of all points $z \in \mathbb{C}$ such that $|2z^2 - 1| = 1$ or, in polar coordinates,

$$r^2 = \cos 2\theta \quad (-\pi/4 \le \theta \le \pi/4 \cup 3\pi/4 \le \theta \le 5\pi/4).$$

If $-\pi/4 \le \Theta \le 0$, the arc length $s(\Theta)$ from $\theta = -\pi/4$ to $\theta = \Theta$ is given by

$$
\begin{aligned}
s(\Theta) &= \int_{-\pi/4}^{\Theta} [r^2 + (dr/d\theta)^2]^{1/2} d\theta \\
&= \int_{-\pi/4}^{\Theta} [r^2 + (1 - r^4)/r^2]^{1/2} d\theta \\
&= \int_0^R (1 - r^4)^{-1/2} dr.
\end{aligned}
$$

If we make the change of variables $x = \sqrt{2}r/(1 + r^2)^{1/2}$, then on account of $dx/dr = \sqrt{2}/(1 + r^2)^{3/2}$ we obtain

$$s(\Theta) = 2^{-1/2} \int_0^X [(1 - x^2/2)(1 - x^2)]^{-1/2} dx.$$

Another example is the determination of the surface area of an ellipsoid. Suppose the ellipsoid is described in rectangular coordinates by the equation

$$x^2/a^2 + y^2/b^2 + z^2/c^2 = 1,$$

where $a > b > c > 0$. The total surface area is $8S$, where S is the surface area of the part contained in the positive octant. In this octant we have

$$z = c[1 - (x/a)^2 - (y/b)^2]^{1/2}$$

and hence

$$1 + (\partial z/\partial x)^2 + (\partial z/\partial y)^2 = [1 - (\alpha x/a)^2 - (\beta y/b)^2]/[1 - (x/a)^2 - (y/b)^2],$$

where

$$\alpha = (a^2 - c^2)^{1/2}/a, \quad \beta = (b^2 - c^2)^{1/2}/b.$$

Consequently

$$S = \int_0^a \int_0^{b(1-(x/a)^2)^{1/2}} [1 - (\alpha x/a)^2 - (\beta y/b)^2]^{1/2}[1 - (x/a)^2 - (y/b)^2]^{-1/2} dy dx.$$

If we make the change of variables

$$x = ar\cos\theta, \quad y = br\sin\theta,$$

with Jacobian $J = abr$, we obtain

$$S = ab \int_0^{\pi/2} d\theta \int_0^1 (1 - \sigma r^2)^{1/2}(1 - r^2)^{-1/2} r \, dr,$$

where

$$\sigma = \alpha^2 \cos^2 \theta + \beta^2 \sin^2 \theta.$$

If we now put

$$u^2 = (1 - r^2)/(1 - \sigma r^2),$$

then $r^2 = (1 - u^2)/(1 - \sigma u^2)$ and

$$r\, dr/du = -(1 - \sigma)u/(1 - \sigma u^2)^2.$$

Hence

$$S = ab \int_0^{\pi/2} d\theta \int_0^1 (1 - \sigma)(1 - \sigma u^2)^{-2} du.$$

Inverting the order of integration and giving σ its value, we obtain

$$S = ab \int_0^1 du \int_0^{\pi/2} [(1 - \alpha^2) \cos^2 \theta + (1 - \beta^2) \sin^2 \theta]$$
$$\times [(1 - \alpha^2 u^2) \cos^2 \theta + (1 - \beta^2 u^2) \sin^2 \theta]^{-2} d\theta.$$

It is readily verified that

$$\int_0^{\pi/2} \cos^2 \theta (m \cos^2 \theta + n \sin^2 \theta)^{-2} d\theta = \pi/4m(mn)^{1/2},$$
$$\int_0^{\pi/2} \sin^2 \theta (m \cos^2 \theta + n \sin^2 \theta)^{-2} d\theta = \pi/4n(mn)^{1/2}.$$

Thus we obtain finally

$$S = (\pi ab/4) \int_0^1 [(1 - \alpha^2)/(1 - \alpha^2 u^2) + (1 - \beta^2)/(1 - \beta^2 u^2)]$$
$$\times [(1 - \alpha^2 u^2)(1 - \beta^2 u^2)]^{-1/2} du.$$

By an *elliptic integral* one understands today any integral of the form

$$\int R(x, w)\, dx,$$

where $R(x, w)$ is a rational function of x and w, and where $w^2 = g(x)$ is a polynomial in x of degree 3 or 4 without repeated roots. The elliptic integral is said to be *complete* if it is a definite integral in which the limits of integration are distinct roots of $g(x)$.

The case of a quartic is easily reduced to that of a cubic. In the preceding examples we can simply put $y = x^2$. Thus, for the lemniscate,

$$s(\Theta) = 2^{-1/2} \int_0^Y [4y(1 - y)(1 - y/2)]^{-1/2} dy.$$

In general, suppose $g(x) = (x - \alpha)h(x)$, where h is a cubic. If

$$h(x) = h_0(x - \alpha)^3 + h_1(x - \alpha)^2 + h_2(x - \alpha) + h_3$$

and we make the change of variables $x = \alpha + 1/y$, then $g(x) = g^*(y)/y^4$, where

$$g^*(y) = h_0 + h_1 y + h_2 y^2 + h_3 y^3,$$

and

$$\int R(x, w)\,dx = \int R^*(y, v)\,dy,$$

where $R^*(y, v)$ is a rational function of y and v, and $v^2 = g^*(y)$.

Since any even power of w is a polynomial in x, the integrand can be written in the form $R(x, w) = (A + Bw)/(C + Dw)$, where A, B, C, D are polynomials in x. Multiplying numerator and denominator by $(C - Dw)w$, we obtain

$$R(x, w) = N/L + M/Lw,$$

where L, M, N are polynomials in x. By decomposing the rational function N/L into partial fractions its integral can be evaluated in terms of rational functions and (real or complex) logarithms. By similarly decomposing the rational function M/L into partial fractions, we are reduced to evaluating the integrals

$$I_0 = \int dx/w, \quad I_n = \int x^n dx/w, \quad J_n(\gamma) = \int (x - \gamma)^{-n} dx/w,$$

where $n \in \mathbb{N}$ and $\gamma \in \mathbb{C}$.

The argument of the preceding paragraph is actually valid if $w^2 = g$ is any polynomial. Suppose now that g is a cubic without repeated roots, say

$$g(x) = a_0 x^3 + a_1 x^2 + a_2 x + a_3.$$

By differentiation we obtain, for any integer $m \geq 0$,

$$(x^m w)' = mx^{m-1} w + x^m g'/2w = (2mx^{m-1} g + x^m g')/2w.$$

Since the numerator on the right is the polynomial

$$(2m + 3)a_0 x^{m+2} + (2m + 2)a_1 x^{m+1} + (2m + 1)a_2 x^m + 2ma_3 x^{m-1},$$

it follows on integration that

$$2x^m w = (2m + 3)a_0 I_{m+2} + (2m + 2)a_1 I_{m+1} + (2m + 1)a_2 I_m + 2ma_3 I_{m-1}.$$

It follows by induction that, for each integer $n > 1$,

$$I_n = p_n(x)w + c_n I_0 + c_n' I_1,$$

where $p_n(x)$ is a polynomial of degree $n - 2$ and c_n, c_n' are constants. Thus the evaluation of I_n for $n > 1$ reduces to the evaluation of I_0 and I_1.

Consider now the integral $J_n(\gamma)$. In the same way as before, for any integer $m \geq 1$,

$$d\{(x-\gamma)^{-m}w\}/dx = -m(x-\gamma)^{-m-1}w + (x-\gamma)^{-m}g'/2w$$
$$= \{-2mg + (x-\gamma)g'\}/2w(x-\gamma)^{m+1}.$$

We can write

$$g(x) = b_0 + b_1(x-\gamma) + b_2(x-\gamma)^2 + b_3(x-\gamma)^3$$

and the numerator on the right of the previous equation is then

$$-2mb_0 + (1-2m)b_1(x-\gamma) + (2-2m)b_2(x-\gamma)^2 + (3-2m)b_3(x-\gamma)^3.$$

It follows on integration that

$$2(x-\gamma)^{-m}w = -2mb_0 J_{m+1}(\gamma) + (1-2m)b_1 J_m(\gamma)$$
$$+ (2-2m)b_2 J_{m-1}(\gamma) + (3-2m)b_3 J_{m-2}(\gamma),$$

where $J_{-1}(\gamma) = \int(x-\gamma)\,dx/w$ is a constant linear combination of I_0 and I_1. Since g does not have repeated roots, $b_1 \neq 0$ if $b_0 = 0$.

It follows by induction that if $g(\gamma) = b_0 \neq 0$ then, for any $n > 1$,

$$J_n(\gamma) = q_n((x-\gamma)^{-1})w + d_n J_1(\gamma) + d'_n I_0 + d''_n I_1,$$

where $q_n(t)$ is a polynomial of degree $n-1$ and d_n, d'_n, d''_n are constants. On the other hand, if $g(\gamma) = 0$ then $g'(\gamma) = b_1 \neq 0$ and, for any $n \geq 1$,

$$J_n(\gamma) = r_n((x-\gamma)^{-1})w + e_n I_0 + e'_n I_1,$$

where $r_n(t)$ is a polynomial of degree n and e_n, e'_n are constants.

Thus the evaluation of an arbitrary elliptic integral can be reduced to the evaluation of

$$I_0 = \int dx/w, \quad I_1 = \int x\,dx/w, \quad J_1(\gamma) = \int(x-\gamma)^{-1}dx/w,$$

where $w^2 = g$ is a cubic without repeated roots, $\gamma \in \mathbb{C}$ and $g(\gamma) \neq 0$. Following Legendre (1793), to whom this reduction is due, integrals of these types are called respectively *elliptic integrals of the first, second and third kinds*.

The cubic g can itself be simplified. If α is a root of g then, by replacing x by $x - \alpha$, we may assume that $g(0) = 0$. If β is now another root of g then, by replacing x by x/β, we may further assume that $g(1) = 0$. Thus the evaluation of an arbitrary elliptic integral may be reduced to one for which g has the form

$$g_\lambda(x) := 4x(1-x)(1-\lambda x),$$

where $\lambda \in \mathbb{C}$ and $\lambda \neq 0, 1$. This normal form, which was used by Riemann (1858) in lectures, is obtained from the normal form of Legendre by the change of variables $x = \sin^2\theta$. To draw attention to the difference, it is convenient to call it *Riemann's normal form*.

The range of λ can be further restricted by linear changes of variables. The transformation $y = (1 - \lambda x)/(1 - \lambda)$ replaces Riemann's normal form by one of the same type with λ replaced by $U\lambda = 1 - \lambda$. Similarly, the transformation $y = 1 - \lambda x$ replaces Riemann's normal form by one of the same type with λ replaced by $V\lambda = 1/(1 - \lambda)$. The transformations U and V together generate a group $\mathscr{G}$ of order 6 (isomorphic to the symmetric group $\mathscr{S}_3$ of all permutations of three letters), since

$$U^2 = V^3 = (UV)^2 = I.$$

The values of λ corresponding to the elements I, V, V^2, U, UV, UV^2 of $\mathscr{G}$ are

$$\lambda, \quad 1/(1 - \lambda), \quad (\lambda - 1)/\lambda, \quad 1 - \lambda, \quad \lambda/(\lambda - 1), \quad 1/\lambda.$$

The region $\mathscr{F}$ of the complex plane $\mathbb{C}$ defined by the inequalities

$$|\lambda - 1| < 1, \quad 0 < \mathscr{R}\lambda < 1/2,$$

is a *fundamental domain* for the group $\mathscr{G}$; i.e., no point of $\mathscr{F}$ is mapped to a different point of $\mathscr{F}$ by an element of $\mathscr{G}$ and each point of $\mathbb{C}$ is mapped to a point of $\mathscr{F}$ or its boundary $\partial\mathscr{F}$ by some element of $\mathscr{G}$. Consequently the sets $\{G(\mathscr{F}) : G \in \mathscr{G}\}$ form a *tiling* of $\mathbb{C}$; i.e.,

$$\mathbb{C} = \bigcup_{G\in\mathscr{G}} G(\mathscr{F} \cup \partial\mathscr{F}), \quad G(\mathscr{F}) \cap G'(\mathscr{F}) = \emptyset \quad \text{if } G, G' \in \mathscr{G} \text{ and } G \neq G'.$$

This is illustrated in Figure 1, where the set $G(\mathscr{F})$ is represented simply by the group element G and, in particular, $\mathscr{F}$ is represented by I. It follows that in Riemann's normal form we may suppose $\lambda \in \mathscr{F} \cup \partial\mathscr{F}$.

The changes of variable in the preceding reduction to Riemann's normal form may be complex, even though the original integrand was real. It will now be shown that any real elliptic integral can be reduced by a real change of variables to one in Riemann's normal form, where $0 < \lambda < 1$ and the independent variable is restricted to the interval $0 \le x \le 1$.

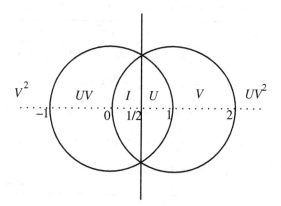

Fig. 1. Fundamental domain for λ.

If g is a cubic or quartic with only real roots, this can be achieved by a linear fractional transformation, mapping roots of g to roots of g_λ. Appropriate transformations are listed in Tables 1 and 2. It should be noted that λ is always a *cross-ratio* of the four roots of g in Table 2, and that λ is always a cross-ratio of the three roots of g and the point '∞' in Table 1.

Table 1. Reduction to Riemann's normal form, g a cubic with all roots real

$$dx/g(x)^{1/2} = dy/\mu g_\lambda(y)^{1/2}$$

$g(x) = A(x - \alpha_1)(x - \alpha_2)(x - \alpha_3)$, where $\alpha_1 > \alpha_2 > \alpha_3$; $\alpha_{jk} = \alpha_j - \alpha_k$

$g_\lambda(y) = 4y(1 - y)(1 - \lambda y)$, where $0 < \lambda < 1$, $y \in (0, 1)$

$\mu = (\alpha_{13})^{1/2}/2$, $\lambda_0 = \alpha_{23}/\alpha_{13}$, $1 - \lambda_0 = \alpha_{12}/\alpha_{13}$.

A	λ	Range	Transformation	Corresponding values	
$+1$	λ_0	$x \geq \alpha_1$	$y = (x - \alpha_1)/(x - \alpha_2)$	$x = \infty$	$y = 1$
				α_1	0
-1	$1 - \lambda_0$	$\alpha_2 \leq x \leq \alpha_1$	$= \alpha_{13}(x - \alpha_2)/\alpha_{12}(x - \alpha_3)$	α_1	1
				α_2	0
$+1$	λ_0	$\alpha_3 \leq x \leq \alpha_2$	$= (x - \alpha_3)/\alpha_{23}$	α_2	1
				α_3	0
-1	$1 - \lambda_0$	$x \leq \alpha_3$	$= \alpha_{13}/(\alpha_1 - x)$	α_3	1
				$-\infty$	0

Table 2. Reduction to Riemann's normal form, g a quartic with all roots real

$$dx/g(x)^{1/2} = dy/\mu g_\lambda(y)^{1/2}$$

$g(x) = A(x - \alpha_1)(x - \alpha_2)(x - \alpha_3)(x - \alpha_4)$, where $\alpha_1 > \alpha_2 > \alpha_3 > \alpha_4$; $\alpha_{jk} = \alpha_j - \alpha_k$

$g_\lambda(y) = 4y(1 - y)(1 - \lambda y)$, where $0 < \lambda < 1$, $y \in (0, 1)$

$\mu = (\alpha_{13}\alpha_{24})^{1/2}/2$, $\lambda_0 = \alpha_{23}\alpha_{14}/\alpha_{13}\alpha_{24}$, $1 - \lambda_0 = \alpha_{12}\alpha_{34}/\alpha_{13}\alpha_{24}$.

A	λ	Range	Transformation	Corresponding values	
$+1$	λ_0	$x \geq \alpha_1$	$y = \alpha_{24}(x - \alpha_1)/\alpha_{14}(x - \alpha_2)$	$x = \infty$	$y = \alpha_{24}/\alpha_{14}$
				α_1	0
-1	$1 - \lambda_0$	$\alpha_2 \leq x \leq \alpha_1$	$= \alpha_{13}(x - \alpha_2)/\alpha_{12}(x - \alpha_3)$	α_1	1
				α_2	0
$+1$	λ_0	$\alpha_3 \leq x \leq \alpha_2$	$= \alpha_{24}(x - \alpha_3)/\alpha_{23}(x - \alpha_4)$	α_2	1
				α_3	0
-1	$1 - \lambda_0$	$\alpha_4 \leq x \leq \alpha_3$	$= \alpha_{13}(x - \alpha_4)/\alpha_{34}(\alpha_1 - x)$	α_3	1
				α_4	0
$+1$	λ_0	$x \leq \alpha_4$	$= \alpha_{24}(x - \alpha_1)/\alpha_{14}(x - \alpha_2)$	α_4	1
				$-\infty$	α_{24}/α_{14}

Suppose now that g is a real cubic or quartic with a pair of conjugate complex roots. Then we can write

$$g(x) = Q_1 Q_2 = (a_1 x^2 + 2b_1 x + c_1)(a_2 x^2 + 2b_2 x + c_2),$$

where the coefficients are real, $a_1 c_1 - b_1^2 > 0$ and $a_2 c_2 - b_2^2 \neq 0$, but a_2 may be zero. Consider first the case where $a_2 \neq 0$ and $b_1 = b_2 a_1/a_2$. Then

$$Q_1 = a_1 (x + b_1/a_1)^2 + b_1', \quad Q_2 = a_2 (x + b_1/a_1)^2 + b_2',$$

where

$$b_1' = (a_1 c_1 - b_1^2)/a_1, \quad b_2' = (a_2 c_2 - b_2^2)/a_2.$$

If we put $y = (x + b_1/a_1)^2$, then

$$R(x) = R_1(y) + R_2(y) y^{1/2},$$

where the rational functions R_1, R_2 are determined by the rational function R, and

$$dx/g(x)^{1/2} = \pm dy/2[y(a_1 y + b_1')(a_2 y + b_2')]^{1/2}.$$

Thus we are reduced to the case of a cubic with 3 distinct real roots.

In the remaining cases there exist distinct real values s_1, s_2 of s such that the polynomial $Q_1 + s Q_2$ is proportional to a perfect square. For $Q_1 + s Q_2$ is proportional to a perfect square if

$$D(s) := (a_1 + s a_2)(c_1 + s c_2) - (b_1 + s b_2)^2 = 0.$$

We have $D(0) = a_1 c_1 - b_1^2 > 0$. If $a_2 = 0$, then $b_2 \neq 0$ and $D(\pm\infty) = -\infty$. On the other hand, if $a_2 \neq 0$, then $D(-a_1/a_2) < 0$, since $b_1 \neq b_2 a_1/a_2$, and $D(s)$ has the sign of $a_2 c_2 - b_2^2$ for both large positive and large negative s. Thus the quadratic $D(s)$ has distinct real roots s_1, s_2. Hence

$$Q_1 + s_1 Q_2 = (a_1 + s_1 a_2)(x + d_1)^2, \quad Q_1 + s_2 Q_2 = (a_1 + s_2 a_2)(x + d_2)^2,$$

where $a_1 + s_j a_2 \neq 0$ $(j = 1, 2)$ and

$$d_1 = (b_1 + s_1 b_2)/(a_1 + s_1 a_2), \quad d_2 = (b_1 + s_2 b_2)/(a_1 + s_2 a_2).$$

Consequently

$$Q_1 = A_1 (x + d_1)^2 + B_1 (x + d_2)^2, \quad Q_2 = A_2 (x + d_1)^2 + B_2 (x + d_2)^2,$$

where

$$A_1 = -s_2 (a_1 + s_1 a_2)/(s_1 - s_2), \quad B_1 = s_1 (a_1 + s_2 a_2)/(s_1 - s_2),$$
$$A_2 = (a_1 + s_1 a_2)/(s_1 - s_2), \quad B_2 = -(a_1 + s_2 a_2)/(s_1 - s_2).$$

If we put $y = \{(x + d_1)/(x + d_2)\}^2$, then

$$R(x) = R_1(y) + R_2(y)y^{1/2},$$

where again the rational functions R_1, R_2 are determined by the rational function R, and

$$dx/g(x)^{1/2} = \pm dy/2|d_2 - d_1|[y(A_1y + B_1)(A_2y + B_2)]^{1/2}.$$

Thus we are again reduced to the case of a cubic with 3 distinct real roots.

The preceding argument may be applied also when g has only real roots, provided the factors Q_1 and Q_2 are chosen so that their zeros do not interlace. Suppose (without loss of generality) that $g = g_\lambda$ is in Riemann's normal form and take

$$Q_1 = (1 - x)(1 - \lambda x), \quad Q_2 = 4x.$$

In this case we can write

$$Q_1 = \{(1 + \sqrt{\lambda})^2(x - 1/\sqrt{\lambda})^2 - (1 - \sqrt{\lambda})^2(x + 1/\sqrt{\lambda})^2\}\sqrt{\lambda}/4,$$
$$Q_2 = -\sqrt{\lambda}\{(x - 1/\sqrt{\lambda})^2 - (x + 1/\sqrt{\lambda})^2\}.$$

If we put

$$1 - 4\sqrt{\lambda}y/(1 + \sqrt{\lambda})^2 = \{(x - 1/\sqrt{\lambda})/(x + 1/\sqrt{\lambda})\}^2,$$

we obtain

$$dx/g_\lambda(x)^{1/2} = dy/\mu g_\rho(y)^{1/2},$$

where

$$\mu = 1 + \sqrt{\lambda}, \quad \rho = 4\sqrt{\lambda}/(1 + \sqrt{\lambda})^2.$$

The usefulness of this change of variables will be seen in the next section.

2 The Arithmetic-Geometric Mean

Let a and b be positive real numbers, with $a > b$, and let

$$a_1 = (a + b)/2, \quad b_1 = (ab)^{1/2}$$

be respectively their arithmetic and geometric means. Then

$$a_1 < (a + a)/2 = a, \quad b_1 > (bb)^{1/2} = b,$$

and

$$a_1 - b_1 = (a^{1/2} - b^{1/2})^2/2 > 0.$$

Thus a_1, b_1 satisfy the same hypotheses as a, b and the procedure can be repeated. If we define sequences $\{a_n\}, \{b_n\}$ inductively by

$$a_0 = a, \quad b_0 = b,$$
$$a_{n+1} = (a_n + b_n)/2, \quad b_{n+1} = (a_n b_n)^{1/2} \quad (n = 0, 1, \ldots),$$

then

$$0 < b_0 < b_1 < b_2 < \cdots < a_2 < a_1 < a_0.$$

It follows that $a_n \to \lambda$ and $b_n \to \mu$ as $n \to \infty$, where $\lambda \geq \mu > 0$. In fact $\lambda = \mu$, as one sees by letting $n \to \infty$ in the relation $a_{n+1} = (a_n + b_n)/2$. The convergence of the sequences $\{a_n\}$ and $\{b_n\}$ to their common limit is extremely rapid, since

$$a_n - b_n = (a_{n-1} - b_{n-1})^2/8a_{n+1}.$$

(As an example, if $a = \sqrt{2}$ and $b = 1$, calculation shows that a_4 and b_4 differ by only one unit in the 20th decimal place.)

The common limit of the sequences $\{a_n\}$ and $\{b_n\}$ will be denoted by $M(a, b)$. The definition can be extended to arbitrary positive real numbers a, b by putting

$$M(a, a) = a, \quad M(b, a) = M(a, b).$$

Following Gauss (1818), $M(a, b)$ is known as the *arithmetic-geometric mean* of a and b. However, the preceding algorithm, which we will call the *AGM algorithm*, was first introduced by Lagrange (1784/5), who showed that it had a remarkable application to the numerical calculation of arbitrary elliptic integrals. The first tables of elliptic integrals, which made them as accessible as logarithms, were constructed in this way under the supervision of Legendre (1826). Today the algorithm can be used directly by electronic computers.

By putting $1 - \lambda x = t^2/a^2$ in Riemann's normal form, it may be seen that any real elliptic integral may be brought to the form

$$\int \varphi(t)[(a^2 - t^2)(t^2 - b^2)]^{-1/2}dt,$$

where $\varphi(t)$ is a rational function of t^2 with real coefficients, $a > b > 0$ and $t \in [b, a]$. We will restrict attention here to the *complete* elliptic integral

$$J = \int_b^a \varphi(t)[(a^2 - t^2)(t^2 - b^2)]^{-1/2}dt,$$

but at the cost of some complication the discussion may be extended to *incomplete* elliptic integrals (where the interval of integration is a proper subinterval of $[b, a]$).

If we make the change of variables

$$t^2 = a^2 \sin^2\theta + b^2 \cos^2\theta \quad (0 \leq \theta \leq \pi/2),$$

then

$$tdt/d\theta = (a^2 - b^2)\sin\theta\cos\theta = [(a^2 - t^2)(t^2 - b^2)]^{1/2}$$

and

$$J = \int_0^{\pi/2} \varphi((a^2\sin^2\theta + b^2\cos^2\theta)^{1/2})d\theta/(a^2\sin^2\theta + b^2\cos^2\theta)^{1/2}.$$

Now put

$$t_1 = (1/2)(t + ab/t)$$

and, as before,

$$a_1 = (a + b)/2, \quad b_1 = (ab)^{1/2}.$$

Then

$$a_1^2 - t_1^2 = (a^2 - t^2)(t^2 - b^2)/4t^2,$$
$$t_1^2 - b_1^2 = (t^2 - ab)^2/4t^2,$$
$$dt_1/dt = (t^2 - ab)/2t^2.$$

As t increases from b to b_1, t_1 decreases from a_1 to b_1, and as t further increases from b_1 to a, t_1 increases from b_1 back to a_1. Since

$$t = t_1 \pm (t_1^2 - b_1^2)^{1/2},$$

it follows from these observations that

$$\int_b^a \varphi(t)[(a^2 - t^2)(t^2 - b^2)]^{-1/2}dt = \int_{b_1}^{a_1} \psi(t_1)[(a_1^2 - t_1^2)(t_1^2 - b_1^2)]^{-1/2}dt_1,$$

where

$$\psi(t_1) = (1/2)\{\varphi[(t_1 + (t_1^2 - b_1^2)^{1/2}] + \varphi[(t_1 - (t_1^2 - b_1^2)^{1/2}]\}.$$

In particular, if we take $\varphi(t) = 1$ and put

$$\mathscr{K}(a, b) := \int_b^a [(a^2 - t^2)(t^2 - b^2)]^{-1/2}dt,$$

we obtain

$$\mathscr{K}(a, b) = \mathscr{K}(a_1, b_1).$$

Hence, by repeating the process, $\mathscr{K}(a, b) = \mathscr{K}(a_n, b_n)$. But

$$\mathscr{K}(a_n, b_n) = \int_0^{\pi/2} (a_n^2\sin^2\theta + b_n^2\cos^2\theta)^{-1/2}d\theta$$

and

$$b_n \leq (a_n^2 \sin^2 \theta + b_n^2 \cos^2 \theta)^{1/2} \leq a_n.$$

Consequently, by letting $n \to \infty$ we obtain

$$\mathscr{K}(a, b) = \pi/2M(a, b). \tag{1}$$

Now take $\varphi(t) = a^2 - t^2$ and put

$$\mathscr{E}(a, b) := \int_b^a [(a^2 - t^2)/(t^2 - b^2)]^{1/2} dt.$$

In this case

$$\psi(t_1) = (a^2 - b^2)/2 + 2(a_1^2 - t_1^2)$$

and hence

$$\mathscr{E}(a, b) = (a^2 - b^2)\mathscr{K}(a, b)/2 + 2\mathscr{E}(a_1, b_1).$$

If we write

$$e_n = 2^n(a_n^2 - b_n^2)$$

then, since $\mathscr{K}(a, b) = \mathscr{K}(a_n, b_n)$, by repeating the process we obtain

$$\mathscr{E}(a, b)/\mathscr{K}(a, b) = (e_0 + e_1 + \cdots + e_{n-1})/2 + 2^n \mathscr{E}(a_n, b_n)/\mathscr{K}(a_n, b_n).$$

But

$$2^n \mathscr{E}(a_n, b_n) = e_n \int_0^{\pi/2} \cos^2 \theta (a_n^2 \sin^2 \theta + b_n^2 \cos^2 \theta)^{-1/2} d\theta$$

and $e_n \to 0$ (rapidly) as $n \to \infty$, since

$$e_n = 2^n(a_{n-1} - b_{n-1})^2/4 = e_{n-1}(a_{n-1} - b_{n-1})/4a_n.$$

Hence

$$\mathscr{E}(a, b)/\mathscr{K}(a, b) = (e_0 + e_1 + e_2 + \cdots)/2. \tag{2}$$

To avoid taking differences of nearly equal quantities, the constants e_n may be calculated by means of the recurrence relations

$$e_n = e_{n-1}^2/2^{n+2}a_n^2 \quad (n = 1, 2, \ldots).$$

Next take

$$\varphi(t) = p[(p^2 - a^2)(p^2 - b^2)]^{1/2}/(p^2 - t^2),$$

where either $p > a$ or $0 < p < b$, and put

$$\mathscr{P}(a, b, p) := \int_b^a p[(p^2 - a^2)(p^2 - b^2)]^{1/2} dt/(p^2 - t^2)[(a^2 - t^2)(t^2 - b^2)]^{1/2}.$$

In this case

$$\psi(t_1) = q_1 \pm p_1[(p_1^2 - a_1^2)(p_1^2 - b_1^2)]^{1/2}/(p_1^2 - t_1^2),$$

where

$$p_1 = (1/2)(p + ab/p),$$
$$q_1 = (p_1^2 - a_1^2)^{1/2} = [(p^2 - a^2)(p^2 - b^2)]^{1/2}/2p,$$

and the $+$ or $-$ sign is taken according as $p > a$ or $0 < p < b$. Since $p_1 > a_1$ in either event, without loss of generality *we now assume that* $p > a$. Then also $p_1 < p$ and

$$\mathscr{P}(a, b, p) = q_1 \mathscr{K}(a, b) + \mathscr{P}(a_1, b_1, p_1).$$

Define the sequence $\{p_n\}$ inductively by

$$p_0 = p, \quad p_{n+1} = (1/2)(p_n + a_n b_n/p_n) \quad (n = 0, 1, \ldots),$$

and put

$$q_{n+1} = (p_{n+1}^2 - a_{n+1}^2)^{1/2} = [(p_n^2 - a_n^2)(p_n^2 - b_n^2)]^{1/2}/2p_n.$$

Then $p_n \to v \geq M(a, b)$ as $n \to \infty$, since $a_n < p_n < p_{n-1}$. In fact $v = M(a, b)$, as one sees by letting $n \to \infty$ in the recurrence relation defining the sequence $\{p_n\}$. Moreover

$$\delta_n := (a_n^2 - b_n^2)/(p_n^2 - a_n^2) \to 0 \text{ as } n \to \infty,$$

since

$$\delta_{n+1} = \delta_n \left(\frac{p_n^2}{4a_{n+1}^2}\right)\left(\frac{a_n^2 - b_n^2}{p_n^2 - b_n^2}\right) < \delta_n p_n^2/4a_{n+1}^2.$$

Hence

$$(p_n^2 - b_n^2)/(p_n^2 - a_n^2) = 1 + \delta_n \to 1.$$

Since $\mathscr{P}(a_n, b_n, p_n)$

$$= p_n[(p_n^2 - a_n^2)(p_n^2 - b_n^2)]^{1/2} \int_0^{\pi/2} \frac{(a_n^2 \sin^2 \theta + b_n^2 \cos^2 \theta)^{-1/2} d\theta}{(p_n^2 - a_n^2) \sin^2 \theta + (p_n^2 - b_n^2) \cos^2 \theta},$$

it follows that $\mathscr{P}(a_n, b_n, p_n) \to \pi/2$ as $n \to \infty$. Hence

$$\mathscr{P}(a, b, p) = (q_1 + q_2 + \cdots) \mathscr{K}(a, b) + \pi/2. \tag{3}$$

To avoid taking differences of nearly equal quantities, the constants q_n may be calculated by means of the recurrence relations

$$\delta_{n+1} = \delta_n^2 p_n^2/4a_{n+1}^2(1 + \delta_n), \quad q_{n+1} = (1 + \delta_n)^{1/2} q_n^2/2p_n \quad (n = 1, 2, \ldots).$$

Using (1)–(3), complete elliptic integrals of all three kinds can be calculated by the *AGM* algorithm. We now consider another application, the utility of which will be seen in §6.

By putting $t_1 = (1/2)(t + ab/t)$ again, one sees that

$$\int_a^\infty [(t^2 - a^2)(t^2 - b^2)]^{-1/2}dt = (1/2)\int_{a_1}^\infty [(t_1^2 - a_1^2)(t_1^2 - b_1^2)]^{-1/2}dt_1.$$

But the change of variables $u = a(1 - b^2/t^2)^{1/2}$ shows that

$$\int_a^\infty [(t^2 - a^2)(t^2 - b^2)]^{-1/2}dt = \mathscr{K}(a, c),$$

where $c = (a^2 - b^2)^{1/2}$. It follows that

$$\mathscr{K}(a, c) = \mathscr{K}(a_1, c_1)/2 = \cdots = \mathscr{K}(a_n, c_n)/2^n,$$

where $c_n = (a_n^2 - b_n^2)^{1/2}$. The asymptotic behaviour of $\mathscr{K}(a_n, c_n)$ may be determined in the following way.

If we put $s = ac/t$, then s decreases from a to c as t increases from c to a, and

$$ds/dt = -[(a^2 - s^2)(s^2 - c^2)]^{1/2}/[(a^2 - t^2)(t^2 - c^2)]^{1/2}.$$

Since $s = t$ when $t = h := (ac)^{1/2}$, it follows that

$$\mathscr{K}(a, c) = 2\int_c^h [(a^2 - t^2)(t^2 - c^2)]^{-1/2}dt.$$

But, for $c \leq t \leq h$,

$$b^{-1} = (a^2 - c^2)^{-1/2} \leq (a^2 - t^2)^{-1/2} \leq (a^2 - h^2)^{-1/2} = a^{-1}(1 - c/a)^{-1/2}.$$

Hence

$$2b^{-1}L \leq \mathscr{K}(a, c) \leq 2a^{-1}(1 - c/a)^{-1/2}L,$$

where

$$L := \int_c^h (t^2 - c^2)^{-1/2}dt = \log\{(a/c)^{1/2} + (a/c - 1)^{1/2}\}.$$

If we now replace a, b, c by a_n, b_n, c_n then, since $a_n/c_n \to \infty$ and moreover $a_n, b_n \to M(a, b)$, we deduce that

$$2^n \mathscr{K}(a, c)/\log(4a_n/c_n) \to 1/M(a, b) = 2\mathscr{K}(a, b)/\pi.$$

But $4a_n/c_n = (4a_n/c_{n-1})^2$, since $c_n = (a_{n-1} - b_{n-1})/2$, and hence

$$2^{-n}\log(4a_n/c_n)$$
$$= 2^{1-n}\log(4a_{n-1}/c_{n-1}) - 2^{1-n}\log(a_{n-1}/a_n)$$
$$= \cdots$$
$$= \log(4a_0/c_0) - \log(a_0/a_1) - 2^{-1}\log(a_1/a_2) - \cdots - 2^{1-n}\log(a_{n-1}/a_n).$$

It follows that

$$\pi \mathcal{K}(a, c)/2\mathcal{K}(a, b) = \log(4a_1/c_0) - \sum_{n=1}^{\infty} 2^{-n} \log(a_n/a_{n+1}). \qquad (4)$$

Finally, to determine $\mathscr{E}(a, c)$ we can use the relation

$$\mathcal{K}(a, b)\mathscr{E}(a, c) + \mathcal{K}(a, c)\mathscr{E}(a, b) - a^2 \mathcal{K}(a, b)\mathcal{K}(a, c) = \pi/2.$$

By homogeneity we need only establish this relation for $a = 1$. Since

$$\mathcal{K}(1, (1 - \lambda)^{1/2}) = \int_0^1 [4x(1 - x)(1 - \lambda x)]^{-1/2} dx,$$

$$\mathscr{E}(1, (1 - \lambda)^{1/2}) = \int_0^1 [(1 - \lambda x)/4x(1 - x)]^{1/2} dx,$$

it is in fact equivalent to the following relation, due to Legendre, between the complete elliptic integrals of the first and second kinds:

Proposition 1 *If*

$$K(\lambda) = \int_0^1 [4x(1 - x)(1 - \lambda x)]^{-1/2} dx, \quad E(\lambda) = \int_0^1 [(1 - \lambda x)/4x(1 - x)]^{1/2} dx,$$

then

$$K(\lambda)E(1 - \lambda) + K(1 - \lambda)E(\lambda) - K(\lambda)K(1 - \lambda) = \pi/2 \quad for \ 0 < \lambda < 1. \qquad (5)$$

Proof We show first that the derivative of the left side of (5) is zero. Evidently

$$dE/d\lambda = -(1/2) \int_0^1 x[4x(1 - x)(1 - \lambda x)]^{-1/2} dx = [E(\lambda) - K(\lambda)]/2\lambda.$$

Similarly,

$$dK/d\lambda = (1/2) \int_0^1 x(1 - \lambda x)^{-1}[4x(1 - x)(1 - \lambda x)]^{-1/2} dx.$$

Substituting $x = (1 - u)/(1 - \lambda u)$ and writing $\lambda' = 1 - \lambda$, we obtain

$$dK/d\lambda = (1/2\lambda') \int_0^1 [(1 - u)/4u(1 - \lambda u)]^{1/2} du$$
$$= [E(\lambda) - \lambda' K(\lambda)]/2\lambda\lambda'.$$

It follows that

$$d(\lambda\lambda' dK/d\lambda)/d\lambda = K/4.$$

Thus $y_1(\lambda) = K(\lambda)$ is a solution of the second order linear differential equation

$$d(\lambda\lambda' dy/d\lambda)/d\lambda - y/4 = 0. \qquad (6)$$

By symmetry, $y_2(\lambda) = K(\lambda')$ is also a solution. It follows that the 'Wronskian'

$$W = \lambda\lambda'(y_2 dy_1/d\lambda - y_1 dy_2/d\lambda)$$

has derivative zero and so is constant. But, writing

$$K'(\lambda) = K(1 - \lambda), \quad E'(\lambda) = E(1 - \lambda),$$

we have

$$2W = K'(E - \lambda'K) + K(E' - \lambda K') = KE' + K'(E - K).$$

To evaluate this constant we let $\lambda \to 0$. Putting $x = \sin^2\theta$, we obtain

$$K(\lambda) = \int_0^{\pi/2} (1 - \lambda\sin^2\theta)^{-1/2} d\theta, \quad E(\lambda) = \int_0^{\pi/2} (1 - \lambda\sin^2\theta)^{1/2} d\theta$$

and hence, as $\lambda \to 0$,

$$K(\lambda) \to \pi/2, E(\lambda) \to \pi/2, E(\lambda') \to 1.$$

Moreover

$$K(\lambda')[E(\lambda) - K(\lambda)] \to 0,$$

since

$$K(\lambda) - E(\lambda) = \lambda \int_0^1 x[4x(1 - x)(1 - \lambda x)]^{-1/2} dx = O(\lambda)$$

and

$$0 \le K(\lambda') \le \int_0^{\pi/2} [1 - (1 - \lambda)]^{-1/2} d\theta = O(\lambda^{-1/2}).$$

It follows that $2W = \pi/2$.   □

If $\lambda = 1/2$, then $\lambda' = \lambda$ and (5) takes the simple form

$$K(1/2)[2E(1/2) - K(1/2)] = \pi/2.$$

By the remarks preceding the statement of Proposition 1, the left side can be evaluated by the *AGM* algorithm. In this way π has recently been calculated to millions of decimal places. (It will be recalled that the value $\lambda = 1/2$ occurred in the rectification of the lemniscate.)

3 Elliptic Functions

According to Jacobi, the theory of elliptic functions was conceived on 23 December 1751, the day on which the Berlin Academy asked Euler to report on the *Produzioni*

Matematiche of Count Fagnano, a copy of which had been sent them by the author. The papers which aroused Euler's interest had in fact already appeared in an obscure Italian journal between 1715 and 1720. Fagnano had shown first how a quadrant of a lemniscate could be halved, then how it could be divided algebraically into 2^m, $3 \cdot 2^m$ or $5 \cdot 2^m$ equal parts. He had also established an algebraic relation between the length of an elliptic arc, the length of another suitably chosen arc and the length of a quadrant. By analysing and extending his arguments, Euler was led ultimately (1761) to a general addition theorem for elliptic integrals. An elegant proof of Euler's theorem was given by Lagrange (1768/9), using differential equations. We follow this approach here.

Let

$$g_\lambda(x) = 4x(1-x)(1-\lambda x) = 4\lambda x^3 - 4(1+\lambda)x^2 + 4x$$

be Riemann's normal form and let $2 f_\lambda(x)$ be its derivative:

$$f_\lambda(x) = 6\lambda x^2 - 4(1+\lambda)x + 2.$$

By the fundamental existence and uniqueness theorem for ordinary differential equations, the second order differential equation

$$x'' = f_\lambda(x) \tag{7}$$

has a unique solution $S(t) = S(t, \lambda)$, defined (and holomorphic) for $|t|$ sufficiently small, which satisfies the initial conditions

$$S(0) = S'(0) = 0. \tag{8}$$

The solution $S(t, \lambda)$ is an elementary function if $\lambda = 0$ or 1:

$$S(t, 0) = \sin^2 t, \quad S(t, 1) = \tanh^2 t.$$

(For other values of λ, $S(t)$ coincides with the Jacobian elliptic function $\mathrm{sn}^2 t$.)

Evidently $S(t)$ is an even function of t, since $S(-t)$ is also a solution of (7) and satisfies the same initial conditions (8).

For any solution $x(t)$ of (7), the function $x'(t)^2 - g_\lambda[x(t)]$ is a constant, since its derivative is zero. In particular,

$$S'(t)^2 = g_\lambda[S(t)], \tag{9}$$

since both sides vanish for $t = 0$.

If $|\tau|$ is sufficiently small, then $x_1(t) = S(t+\tau)$ and $x_2(t) = S(t-\tau)$ are solutions of (7) near $t = 0$. Moreover,

$$x_j'(t)^2 = g_\lambda[x_j(t)] \quad (j = 1, 2),$$

since these relations hold for $t = 0$. From

$$(x_1 x_2' + x_1' x_2)' = x_1 f_\lambda(x_2) + x_2 f_\lambda(x_1) + 2x_1' x_2'$$

and

$$(x_1 x_2' + x_1' x_2)^2 = x_1^2 g_\lambda(x_2) + x_2^2 g_\lambda(x_1) + 2x_1 x_2 x_1' x_2'$$

we obtain

$$2x_1x_2(x_1x_2' + x_1'x_2)' - (x_1x_2' + x_1'x_2)^2 - 2x_1x_2x_1'x_2'$$
$$= x_1^2\{2x_2 f_\lambda(x_2) - g_\lambda(x_2)\} + x_2^2\{2x_1 f_\lambda(x_1) - g_\lambda(x_1)\}.$$

But if $g_\lambda(x) = \alpha x^3 + \beta x^2 + \gamma x$ and $f_\lambda(x) = g_\lambda'(x)/2$, then

$$2x f_\lambda(x) - g_\lambda(x) = x^2(2\alpha x + \beta).$$

Hence

$$2x_1x_2(x_1x_2' + x_1'x_2)' - (x_1x_2' + x_1'x_2)^2 = 2x_1^2 x_2^2 \{\alpha(x_1 + x_2) + \beta\} + 2x_1x_2x_1'x_2'.$$

On the other hand,

$$(x_1' - x_2')(x_1x_2' + x_1'x_2) = x_2 g_\lambda(x_1) - x_1 g_\lambda(x_2) + (x_1 - x_2)x_1'x_2'$$
$$= x_1x_2(x_1 - x_2)\{\alpha(x_1 + x_2) + \beta\} + (x_1 - x_2)x_1'x_2'.$$

Comparing these two relations, we obtain

$$\{2x_1x_2(x_1x_2' + x_1'x_2)' - (x_1x_2' + x_1'x_2)^2\}(x_1 - x_2) = 2x_1x_2(x_1' - x_2')(x_1x_2' + x_1'x_2).$$

If we divide by $2x_1x_2(x_1 - x_2)(x_1x_2' + x_1'x_2)$, this takes the form

$$\frac{(x_1x_2' + x_1'x_2)'}{x_1x_2' + x_1'x_2} - \frac{x_1x_2' + x_1'x_2}{2x_1x_2} = \frac{x_1' - x_2'}{x_1 - x_2},$$

which can be integrated to give

$$(x_1x_2' + x_1'x_2)^2 = C(\tau)x_1x_2(x_1 - x_2)^2,$$

where the constant $C(\tau)$ depends on τ. Equivalently,

$$[S(u)S'(v) - S'(u)S(v)]^2 = C((u + v)/2)S(u)S(v)[S(u) - S(v)]^2.$$

To evaluate the constant, we divide throughout by $S(v)$ and let $v \to 0$. By (9), this yields $C(u/2) = \gamma/S(u)$. Since $\gamma = 4$ (for Riemann's normal form), we obtain finally

$$S(u + v) = 4S(u)S(v)[S(u) - S(v)]^2/[S(u)S'(v) - S'(u)S(v)]^2. \tag{10}$$

Thus $S(u + v)$ is a rational function of $S(u)$, $S(v)$, $S'(u)$, $S'(v)$. Moreover, since $(S')^2 = g_\lambda(S)$, there exists a polynomial $p(x, y, z)$, not identically zero and with coefficients independent of u and v, such that $p[S(u+v), S(u), S(v)] = 0$. In other words, the function $S(u)$ has an *algebraic addition theorem*.

The relation (10) can also be written in the form

$$S(u + v) = [S(u)S'(v) + S'(u)S(v)]^2/4S(u)S(v)[1 - \lambda S(u)S(v)]^2, \tag{11}$$

since

$$S(u)^2 S'(v)^2 - S'(u)^2 S(v)^2 = S(u)^2 g_\lambda[S(v)] - S(v)^2 g_\lambda[S(u)]$$
$$= 4S(u)S(v)[S(u) - S(v)][1 - \lambda S(u)S(v)].$$

Replacing v by $-v$ in (11) and subtracting the result from (11), we obtain

$$S(u + v) - S(u - v) = S'(u)S'(v)/[1 - \lambda S(u)S(v)]^2. \tag{12}$$

In particular, for $v = u$,

$$S(2u) = g_\lambda[S(u)]/[1 - \lambda S^2(u)]^2. \tag{13}$$

We recall that a function is *meromorphic* in a connected open set D if it is holomorphic throughout D, except for isolated singularities which are poles. Since, by (13), $S(2t)$ is a rational function of $S(t)$, it follows that if $S(t)$ is meromorphic and a solution (wherever it is finite) of the differential equation (7) in an open disc $|t| < R$, then its definition can be extended so that it is meromorphic and a solution (wherever it is finite) of the differential equation (7) also in the disc $|t| < 2R$. But the fundamental existence and uniqueness theorem guarantees that $S(t)$ is holomorphic in a neighbourhood of the origin. Consequently we can extend its definition so that it is meromorphic and a solution of (7) in the whole complex plane $\mathbb{C}$.

Further properties of the function $S(t)$ may be derived from the differential equation (7). For any constants α, β, if $y(t) = \alpha S(\beta t)$, then $y(0) = y'(0) = 0$. It is readily seen that $y(t)$ satisfies a differential equation of the form (7) if and only if either $\alpha = 1$, $\beta = \pm 1$ or $\alpha = \lambda$, $\lambda\beta^2 = 1$, and in the latter case with λ replaced by $1/\lambda$ in (7). It follows that, for any $\lambda \neq 0$,

$$S(t, 1/\lambda) = \lambda S(\lambda^{-1/2}t, \lambda). \tag{14}$$

By differentiation it may be shown also that $S(it, \lambda)/[S(it, \lambda)-1]$, where $i^2 = -1$, is a solution of the differential equation (7) with λ replaced by $1 - \lambda$. It follows that

$$S(t, 1 - \lambda) = S(it, \lambda)/[S(it, \lambda) - 1]. \tag{15}$$

By combining (14) and (15) we obtain, for any $\lambda \neq 0, 1$, three more relations:

$$S(t, 1/(1 - \lambda)) = (1 - \lambda)S(i(1 - \lambda)^{-1/2}t, \lambda)/[S(i(1 - \lambda)^{-1/2}t, \lambda) - 1], \tag{16}$$

$$S(t, (\lambda - 1)/\lambda) = \lambda S(i\lambda^{-1/2}t, \lambda)/[\lambda S(i\lambda^{-1/2}t, \lambda) - 1], \tag{17}$$

$$S(t, \lambda/(\lambda - 1)) = (1 - \lambda)S((1 - \lambda)^{-1/2}t, \lambda)/[1 - \lambda S((1 - \lambda)^{-1/2}t, \lambda)]. \tag{18}$$

As in §1, it follows from (14)–(18) that the evaluation of $S(t, \lambda)$ for all $t, \lambda \in \mathbb{C}$ reduces to its evaluation for λ in the region $|\lambda - 1| \leq 1, 0 \leq \mathscr{R}\lambda \leq 1/2$. Similarly it follows from (14) and (18) that the evaluation of $S(t, \lambda)$ for all $t, \lambda \in \mathbb{R}$ reduces to its evaluation for λ in the interval $0 < \lambda < 1$. We now show that $S(t, \lambda)$ can then be calculated by the *AGM* algorithm.

It is easily verified that if

$$z(t) = (1 + \sqrt{\lambda})^2 S(t, \lambda)/[1 + \sqrt{\lambda}S(t, \lambda)]^2,$$

then

$$(dz/dt)^2 = (1 + \sqrt{\lambda})^2 \{4\lambda_0 z^3 - 4(1 + \lambda_0)z^2 + 4z\},$$

where

$$\lambda_0 = 4\sqrt{\lambda}/(1 + \sqrt{\lambda})^2. \tag{19}$$

Since $z(0) = z'(0) = 0$ and $z''(0) \neq 0$, it follows that $z(t) = S((1 + \sqrt{\lambda})t, \lambda_0)$. Thus

$$S((1 + \sqrt{\lambda})t, \lambda_0) = (1 + \sqrt{\lambda})^2 S(t, \lambda)/[1 + \sqrt{\lambda}S(t, \lambda)]^2. \tag{20}$$

The inequality $0 < \lambda < 1$ implies $\lambda < \lambda_0 < 1$. Hence, by regarding (19) as a quadratic equation for $\sqrt{\lambda}$, we obtain

$$\sqrt{\lambda} = [1 - (1 - \lambda_0)^{1/2}]^2/\lambda_0. \tag{21}$$

If we write $\sqrt{\lambda_0} = c_0/a_0$, where $c_0 = (a_0^2 - b_0^2)^{1/2}$ and $0 < b_0 < a_0$, then

$$\sqrt{\lambda} = (a_0 - b_0)/(a_0 + b_0) = c_1/a_1,$$

where

$$a_1 = (a_0 + b_0)/2, \quad b_1 = (a_0 b_0)^{1/2}, \quad c_1 = (a_1^2 - b_1^2)^{1/2}.$$

Since $1 + \sqrt{\lambda} = a_0/a_1$, we can rewrite (20) in the form

$$S(a_0 t, \lambda_0) = (1 + c_1/a_1)^2 S(a_1 t, \lambda_1)/[1 + (c_1/a_1)S(a_1 t, \lambda_1)]^2,$$

where $\lambda_1 = \lambda = (c_1/a_1)^2$. Repeating the process, we obtain

$$S(a_{n-1} t, \lambda_{n-1}) = (1 + c_n/a_n)^2 S(a_n t, \lambda_n)/[1 + (c_n/a_n)S(a_n t, \lambda_n)]^2,$$

where $\lambda_n = (c_n/a_n)^2$. As $n \to \infty$,

$$a_n \to \mu := M(a, b), \quad c_n \to 0, \quad \lambda_n \to 0.$$

Since $S(t, 0) = \sin^2 t$, for some (not very large) $n = N$ we have $S(a_N t, \lambda_N) \approx \sin^2 \mu t$, which may be considered as known. Then, by taking successively $n = N, N-1, \ldots, 1$ we can calculate $S(a_0 t, \lambda_0)$. Moreover, we can start the process by taking $a_0 = 1$, $b_0 = (1 - \lambda_0)^{1/2}$.

We now consider periodicity properties. If $\lambda \neq 1$ and $S(h) = 1$ for some nonzero $h \in \mathbb{C}$ then, by (13), $S(2h) = 0$. Furthermore $S'(2h) = 0$, by (9). It follows that $S(t)$ has period $2h$, since $S(t + 2h)$ is a solution of the differential equation (7) which satisfies the same initial conditions (8) as $S(t)$. It remains to show that there exists such an h.

Suppose first that $\lambda \in \mathbb{R}$ and $0 < \lambda < 1$. Since $S''(0) = 2$, we have $S'(t) > 0$ for small $t > 0$. If $S'(t) > 0$ for $0 < t < T$, then $S(t)$ is a positive increasing function for $0 < t < T$. Since $g_\lambda[S(t)] > 0$, we must also have $S(t) < 1$ for $0 < t < T$. From the relation

$$t = \int_0^{S(t)} dx/g_\lambda(x)^{1/2},$$

it follows that $T \leq K(\lambda)$, where

$$K(\lambda) := \int_0^1 dx / g_\lambda(x)^{1/2}.$$

Hence $S'(t)$ vanishes for some t such that $0 < t \leq K(\lambda)$ and we can now take T to be the least $t > 0$ for which $S'(t) = 0$. Then $S'(T) = 0$, $S(T) = 1$ and by letting $t \to T$ we obtain $T = K(\lambda)$.

This shows that $S(u)$ maps the interval $[0, K(\lambda)]$ bijectively onto $[0, 1]$, and if

$$u(\xi) = \int_0^\xi dx / g_\lambda(x)^{1/2} \quad (0 \leq \xi \leq 1),$$

then $S[u(\xi)] = \xi$. Thus, in the real domain, the elliptic integral of the first kind is *inverted* by the function $S(u)$.

Since $\lambda \neq 1$, it follows that $S(t) = S(t, \lambda)$ has period $2K(\lambda)$. Since $\lambda \neq 0$, it follows from (15) that $S(t, \lambda)$ also has period $2iK(1 - \lambda)$. Thus $S(t, \lambda)$ is a *doubly-periodic* function, with a real period and a pure imaginary period. We will show that all periods are given by

$$2mK(\lambda) + 2niK(1 - \lambda) \quad (m, n \in \mathbb{Z}).$$

The periods of a nonconstant meromorphic function f form a discrete additive subgroup of $\mathbb{C}$. If f has two periods whose ratio is not real then, by the simple case $n = 2$ of Proposition VIII.7, it has periods ω_1, ω_2 such that all periods are given by

$$m\omega_1 + n\omega_2 \quad (m, n \in \mathbb{Z}).$$

In the present case we can take $\omega_1 = 2K(\lambda)$, $\omega_2 = 2iK(1 - \lambda)$ since, by construction, $2K(\lambda)$ is the least positive period.

Suppose next that $\lambda \in \mathbb{R}$ and either $\lambda > 1$ or $\lambda < 0$. Then, by (14) and (15), $S(t, \lambda)$ is again a doubly-periodic function with a real period and a pure imaginary period.

Suppose finally that $\lambda \in \mathbb{C} \backslash \mathbb{R}$. Without loss of generality, *we assume $\mathscr{I}\lambda > 0$*. Then $g_\lambda(z)$ does not vanish in the upper half-plane $\mathscr{H}$. It follows that there exists a unique function $h_\lambda(z)$, holomorphic for $z \in \mathscr{H}$ with $\mathscr{R}h_\lambda(z) > 0$ for z near 0, such that

$$h_\lambda(z)^2 = g_\lambda(z). \tag{22}$$

Moreover, we may extend the definition so that $h_\lambda(z)$ is continuous and (22) continues to hold for $z \in \mathscr{H} \cup \mathbb{R}$.

We can write $S(t) = \psi(t^2)$, where

$$\psi(w) = w + a_2 w^2 + \cdots$$

is holomorphic at the origin. By inversion of series, there exists a function

$$\phi(z) = z + b_2 z^2 + \cdots,$$

which is holomorphic at the origin, such that $\psi[\phi(z)] = z$. For $z \in \mathscr{H}$ near 0, put

$$u(z) = \phi(z)^{1/2},$$

where the square root is chosen so that $\mathscr{R}u(z) > 0$. Then $S[u(z)] = z$. Differentiating and then squaring, we obtain

$$S'[u(z)]u'(z) = 1, \quad u'(z)^2 = 1/g_\lambda(z).$$

But $u'(z)$ also has positive real part, since $S'[u(z)] \sim 2u(z)$ for $z \to 0$. Consequently $u'(z) = 1/h_\lambda(z)$. Since $u(z) \to 0$ as $z \to 0$, we conclude that

$$u(z) = \int_0^z d\zeta/h_\lambda(\zeta), \tag{23}$$

where the path of integration is (say) a straight line segment. However, the function on the right is holomorphic for all $z \in \mathscr{H}$. Consequently, if we define $u(z)$ by (23) then, by analytic continuation, the relation $S[u(z)] = z$ continues to hold for all $z \in \mathscr{H}$. Letting $z \to 1$, we now obtain $S(h) = 1$ for $h = K(\lambda)$, where

$$K(\lambda) := \int_0^1 dx/g_\lambda(x)^{1/2}$$

and the square root is chosen so that $g_\lambda(x)^{1/2}$ is continuous and has positive real part for small $x > 0$ and actually, as we will see in a moment, for $0 < x < 1$. Hence $S(t)$ has period $2K(\lambda)$. Furthermore, by (15), $S(t)$ also has period $2iK(1-\lambda)$.

For $0 < x < 1$ we have

$$1/g_\lambda(x)^{1/2} = (1 - \bar{\lambda}x)^{1/2}/[4x(1-x)]^{1/2}|1 - \lambda x|.$$

If $\lambda = \mu + i\nu$, where $\nu > 0$, then $1 - \bar{\lambda}x = \gamma + i\delta$, where $\gamma = 1 - \mu x$ and $\delta = \nu x > 0$ for $0 < x < 1$. Hence

$$(1 - \bar{\lambda}x)^{1/2} = \alpha + i\beta,$$

where

$$\alpha = \{\gamma + (\gamma^2 + \delta^2)^{1/2}\}^{1/2}/\sqrt{2}, \quad 2\alpha\beta = \delta,$$

first for small $x > 0$ and then, by continuity, for $0 < x < 1$. Thus α and β are positive for $0 < x < 1$. Consequently $\mathscr{R}g_\lambda(x)^{1/2} > 0$ for $0 < x < 1$ and

$$K(\lambda) = A + iB,$$

where $A > 0, B > 0$.

Similarly, for $0 < y < 1$ we have

$$1/g_{1-\lambda}(y)^{1/2} = (1 - (1 - \bar{\lambda})y)^{1/2}/[4y(1-y)]^{1/2}|1 - (1 - \lambda)y|$$

and $1 - (1 - \bar{\lambda})y = \gamma' - i\delta'$, where $\gamma' = 1 - (1 - \mu)y$ and $\delta' = \nu y > 0$ for $0 < y < 1$. Hence

$$(1 - (1 - \bar{\lambda})y)^{1/2} = \alpha' - i\beta',$$

where

$$\alpha' = \{\gamma' + (\gamma'^2 + \delta'^2)^{1/2}\}^{1/2}/\sqrt{2}, \quad 2\alpha'\beta' = \delta'.$$

Thus α' and β' are positive for $0 < y < 1$, and

$$K(1 - \lambda) = A' - iB',$$

where $A' > 0$, $B' > 0$.

We will now show that the period ratio $iK(1 - \lambda)/K(\lambda)$ is not real by showing that the quotient $K(1 - \lambda)/K(\lambda)$ has positive real part. Since this is equivalent to showing that

$$AA' - BB' > 0,$$

it is sufficient to show that $\alpha\alpha' - \beta\beta' > 0$ for all $x, y \in (0, 1)$. The inequality is certainly satisfied for all x, y near 0, since $\alpha \to 1$, $\beta \to 0$ as $x \to 0$ and $\alpha' \to 1$, $\beta' \to 0$ as $y \to 0$. Thus we need only show that we never have $\alpha\alpha' = \beta\beta'$. But

$$2\alpha^2 = (\gamma^2 + \delta^2)^{1/2} + \gamma, \quad 2\beta^2 = (\gamma^2 + \delta^2)^{1/2} - \gamma,$$

with analogous expressions for $2\alpha'^2, 2\beta'^2$. Hence, if $\alpha\alpha' = \beta\beta'$, then by squaring we obtain

$$[(\gamma^2 + \delta^2)^{1/2} + \gamma][(\gamma'^2 + \delta'^2)^{1/2} + \gamma'] = [(\gamma^2 + \delta^2)^{1/2} - \gamma][(\gamma'^2 + \delta'^2)^{1/2} - \gamma'],$$

which reduces to

$$\gamma(\gamma'^2 + \delta'^2)^{1/2} = -\gamma'(\gamma^2 + \delta^2)^{1/2}.$$

Squaring again, we obtain $\gamma^2\delta'^2 = \gamma'^2\delta^2$. Since the previous equation shows that γ and γ' do not have the same sign, it follows that

$$\gamma\delta' + \gamma'\delta = 0.$$

Giving $\gamma, \delta, \gamma', \delta'$ their explicit expressions, this takes the form $v(x + y - xy) = 0$. Hence $x(1 - y) + y = 0$, which is impossible if $0 < y < 1$ and $x > 0$.

The relation $S[u(z)] = z$, where $u(z)$ is defined by (23), shows that the elliptic integral of the first kind is *inverted* by the elliptic function $S(u)$. We may use this to simplify other elliptic integrals. The change of variables $x = S(u)$ replaces the integral

$$\int R(x)\,dx/g_\lambda(x)^{1/2}$$

by $\int R[S(u)]du$. Following Jacobi, we take

$$E(u) := \int_0^u [1 - \lambda S(v)]dv \tag{24}$$

as the standard elliptic integral of the second kind, and

$$\Pi(u,a) := (\lambda/2) \int_0^u S'(a)S(v)dv/[1 - \lambda S(a)S(v)] \qquad (25)$$

as the standard elliptic integral of the third kind.

Many properties of these functions may be obtained by integration from corresponding properties of the function $S(u)$. By way of example, we show that

$$E(u+a) - E(u-a) - 2E(a) = -\lambda S'(a)S(u)/[1 - \lambda S(a)S(u)]. \qquad (26)$$

Indeed it is evident that both sides vanish when $u = 0$, and it follows from (12) that they have the same derivative with respect to u. Integrating (26) with respect to u, we further obtain

$$\Pi(u,a) = uE(a) - (1/2) \int_{u-a}^{u+a} E(v)dv. \qquad (27)$$

Thus the function $\Pi(u,a)$, which depends on two variables (as well as the parameter λ) can be expressed in terms of functions of only one variable. Furthermore, we have the *interchange property* (due, in other notation, to Legendre)

$$\Pi(u,a) - uE(a) = \Pi(a,u) - aE(u). \qquad (28)$$

If we take $u = 2K = 2K(\lambda)$, then $S'(u) = 0$ and hence $\Pi(a,u) = 0$. Thus

$$\Pi(2K,a) = 2KE(a) - aE(2K), \qquad (29)$$

which shows that the complete elliptic integral of the third kind can be expressed in terms of complete and incomplete elliptic integrals of the first and second kinds.

In order to justify taking $\Pi(u,a)$ as the standard elliptic integral of the third kind, we show finally that $S(a)$ takes all complex values. Otherwise, if $S(u) \neq c$ for all $u \in \mathbb{C}$, then $c \neq 0$ and

$$f(u) = S(u)/[S(u) - c]$$

is holomorphic in the whole complex plane. Furthermore, it is doubly-periodic with two periods ω_1, ω_2 whose ratio is not real. Since it is bounded in the parallelogram with vertices $0, \omega_1, \omega_2, \omega_1 + \omega_2$, it follows that it is bounded in $\mathbb{C}$. Hence, by Liouville's theorem, f is a constant. Since S is not constant and $c \neq 0$, this is a contradiction.

4 Theta Functions

Theta functions arise not only in connection with elliptic functions (as we will see), but also in problems of heat conduction, statistical mechanics and number theory.

Consider the bi-infinite series

$$\sum_{n=-\infty}^{\infty} q^{n^2} z^n = 1 + \sum_{n=1}^{\infty} q^{n^2} z^n + \sum_{n=1}^{\infty} q^{n^2} z^{-n},$$

where $q, z \in \mathbb{C}$ and $z \neq 0$. Both series on the right converge if $|q| < 1$, both diverge if $|q| > 1$, and at most one converges if $|q| = 1$. Thus we now assume $|q| < 1$.

A remarkable representation for the series on the left was given by Jacobi (1829), in §64 of his *Fundamenta Nova*, and is now generally known as *Jacobi's triple product formula*:

Proposition 2 *If $|q| < 1$ and $z \neq 0$, then*

$$\sum_{n=-\infty}^{\infty} q^{n^2} z^n = \prod_{n=1}^{\infty}(1 + q^{2n-1}z)(1 + q^{2n-1}z^{-1})(1 - q^{2n}). \tag{30}$$

Proof Put

$$f_N(z) = \prod_{n=1}^{N}(1 + q^{2n-1}z)(1 + q^{2n-1}z^{-1}).$$

Then we can write

$$f_N(z) = c_0^N + c_1^N(z + z^{-1}) + \cdots + c_N^N(z^N + z^{-N}). \tag{31}$$

To determine the coefficients c_n^N we use the functional relation

$$f_N(q^2 z) = (1 + q^{2N+1}z)(1 + q^{-1}z^{-1})f_N(z)/(1 + qz)(1 + q^{2N-1}z^{-1})$$
$$= (1 + q^{2N+1}z)f_N(z)/(qz + q^{2N}).$$

Multiplying both sides by $qz + q^{2N}$ and equating coefficients of z^{n+1} we get, for $n = 0, 1, \ldots, N - 1$,

$$q^{2n+1}c_n^N + q^{2N+2n+2}c_{n+1}^N = c_{n+1}^N + q^{2N+1}c_n^N,$$

i.e.,

$$q^{2n+1}(1 - q^{2N-2n})c_n^N = (1 - q^{2N+2n+2})c_{n+1}^N.$$

But, since $\sum_{n=1}^{N}(2n-1) = N^2$, it follows from the definition of $f_N(z)$ that $c_N^N = q^{N^2}$. Hence, for $0 \leq n \leq N$,

$$c_n^N = (1 - q^{2N+2n+2})(1 - q^{2N+2n+4}) \cdots (1 - q^{4N})q^{n^2}/D,$$

where $D = (1 - q^2)(1 - q^4) \cdots (1 - q^{2N-2n})$.

If $|q| < 1$ and $z \neq 0$, then the infinite products

$$\prod_{n=1}^{\infty}(1 + q^{2n-1}z), \quad \prod_{n=1}^{\infty}(1 + q^{2n-1}z^{-1}), \quad \prod_{n=1}^{\infty}(1 - q^{2n})$$

are all convergent. From the convergence of the last it follows that, for each fixed n,

$$\lim_{N \to \infty} c_n^N = q^{n^2} \Big/ \prod_{k=1}^{\infty}(1 - q^{2k}).$$

Moreover, there exists a constant $A > 0$, depending on q but not on n or N, such that

$$|c_n^N| \leq A|q|^{n^2}.$$

For we can choose $B > 0$ so that $|\prod_{k=1}^{m}(1 - q^{2k})| \geq B$ for all m, we can choose $C > 0$ so that $|\prod_{k=1}^{m}(1 - q^{2k})| \leq C$ for all m, and we can then take $A = C/B^2$. Since the series $\sum_{n=-\infty}^{\infty} q^{n^2} z^n$ is absolutely convergent, it follows that we can proceed to the limit term by term in (31) to obtain (30). $\qquad \square$

In the series $\sum_{n=-\infty}^{\infty} q^{n^2} z^n$ we now put

$$q = e^{\pi i \tau}, \quad z = e^{2\pi i v},$$

so that $|q| < 1$ corresponds to $\mathscr{I}\tau > 0$, and we define the *theta function*

$$\theta(v; \tau) = \sum_{n=-\infty}^{\infty} e^{\pi i \tau n^2} e^{2\pi i v n}.$$

The function $\theta(v; \tau)$ is holomorphic in v and τ for all $v \in \mathbb{C}$ and $\tau \in \mathscr{H}$ (the upper half-plane). Since initially we will be more interested in the dependence on v, with τ just a parameter, we will often write $\theta(v)$ in place of $\theta(v; \tau)$. Furthermore, we will still use q as an abbreviation for $e^{\pi i \tau}$.

Evidently

$$\theta(v + 1) = \theta(v) = \theta(-v).$$

Moreover,

$$\theta(v + \tau) = \sum_{n=-\infty}^{\infty} q^{n^2 + 2n} e^{2\pi i v n}$$

$$= q^{-1} e^{-2\pi i v} \sum_{n=-\infty}^{\infty} q^{(n+1)^2} e^{2\pi i v(n+1)}$$

$$= e^{-\pi i (2v + \tau)} \theta(v).$$

It may be immediately verified that

$$\partial^2 \theta / \partial v^2 = -4\pi^2 q \partial \theta / \partial q = 4\pi i \partial \theta / \partial \tau,$$

which becomes the partial differential equation of heat conduction in one dimension on putting $\tau = 4\pi i t$.

By Proposition 2, we have also the product representation

$$\theta(v) = \prod_{n=1}^{\infty}(1 + q^{2n-1} e^{2\pi i v})(1 + q^{2n-1} e^{-2\pi i v})(1 - q^{2n}).$$

It follows that the points

$$v = 1/2 + \tau/2 + m + n\tau \quad (m, n \in \mathbb{Z})$$

are simple zeros of $\theta(v)$, and that these are the only zeros.

One important property of the theta function is almost already known to us:

Proposition 3 *For all $v \in \mathbb{C}$ and $\tau \in \mathscr{H}$,*

$$\theta(v; -1/\tau) = (\tau/i)^{1/2} e^{\pi i \tau v^2} \theta(\tau v; \tau), \tag{32}$$

where the square root is chosen to have positive real part.

Proof Suppose first that $\tau = iy$, where $y > 0$. We wish to show that

$$\sum_{n=-\infty}^{\infty} e^{-n^2 \pi / y} e^{2n\pi i v} = y^{1/2} \sum_{n=-\infty}^{\infty} e^{-(v+n)^2 \pi y}.$$

But this was already proved in Proposition IX.10.

Thus (32) holds when τ is pure imaginary. Since, with the stated choice of square root, both sides of (32) are holomorphic functions for $v \in \mathbb{C}$ and $\tau \in \mathscr{H}$, the relation continues to hold throughout this extended domain, by analytic continuation.   □

Following Hermite (1858), for any integers α, β we now put

$$\theta_{\alpha,\beta}(v) = \theta_{\alpha,\beta}(v; \tau) = \sum_{n=-\infty}^{\infty} (-1)^{\beta n} e^{\pi i \tau (n+\alpha/2)^2} e^{2\pi i v (n+\alpha/2)}.$$

(The factor $(-1)^{\beta n}$ may be made less conspicuous by writing it as $e^{\pi i \beta n}$.) Since

$$\theta_{\alpha+2,\beta}(v) = (-1)^{\beta} \theta_{\alpha,\beta}(v), \quad \theta_{\alpha,\beta+2}(v) = \theta_{\alpha,\beta}(v),$$

there are only four essentially distinct functions, namely

$$\theta_{00}(v) = \sum_{n=-\infty}^{\infty} e^{\pi i \tau n^2} e^{2\pi i v n},$$

$$\theta_{01}(v) = \sum_{n=-\infty}^{\infty} (-1)^n e^{\pi i \tau n^2} e^{2\pi i v n},$$

$$\theta_{10}(v) = \sum_{n=-\infty}^{\infty} e^{\pi i \tau (n+1/2)^2} e^{\pi i v (2n+1)}, \tag{33}$$

$$\theta_{11}(v) = \sum_{n=-\infty}^{\infty} (-1)^n e^{\pi i \tau (n+1/2)^2} e^{\pi i v (2n+1)}.$$

Moreover,

$$\theta_{00}(v; \tau) = \theta(v; \tau), \quad \theta_{01}(v; \tau) = \theta(v + 1/2; \tau),$$

$$\theta_{10}(v; \tau) = e^{\pi i (v+\tau/4)} \theta(v + \tau/2; \tau), \quad \theta_{11}(v; \tau) = e^{\pi i (v+\tau/4)} \theta(v + 1/2 + \tau/2; \tau).$$

In fact, for all integers m, n,

$$\theta_{\alpha,\beta}(v + m\tau/2 + n/2) = \theta_{\alpha+m,\beta+n}(v) e^{-\pi i (mv + m^2 \tau/4 - \alpha n/2)}. \tag{34}$$

Since the zeros of $\theta(v; \tau)$ are the points $v = 1/2 + \tau/2 + m\tau + n$, the zeros of $\theta_{\alpha,\beta}(v)$ are the points

$$v = (\beta + 1)/2 + (\alpha + 1)\tau/2 + m\tau + n \ (m, n \in \mathbb{Z}).$$

The notation for theta functions is by no means standardized. Hermite's notation reflects the underlying symmetry, but for purposes of comparison we indicate its connection with the more commonly used notation in Whittaker and Watson [29]:

$$\theta_{00}(v; \tau) = \vartheta_3(\pi v, q), \quad \theta_{01}(v; \tau) = \vartheta_4(\pi v, q),$$
$$\theta_{10}(v; \tau) = \vartheta_2(\pi v, q), \quad \theta_{11}(v; \tau) = i\vartheta_1(\pi v, q).$$

It follows from the definitions that $\theta_{00}(v; \tau)$, $\theta_{01}(v; \tau)$ and $\theta_{10}(v; \tau)$ are even functions of v, whereas $\theta_{11}(v; \tau)$ is an odd function of v. Moreover $\theta_{00}(v; \tau)$ and $\theta_{01}(v; \tau)$ are periodic with period 1 in v, but $\theta_{10}(v; \tau)$ and $\theta_{11}(v; \tau)$ change sign when v is increased by 1.

All four theta functions satisfy the same partial differential equation as $\theta(v; \tau)$. From the product expansion of $\theta(v; \tau)$ we obtain the product expansions

$$\theta_{00}(v) = Q_0 \prod_{n=1}^{\infty}(1 + q^{2n-1}e^{2\pi iv})(1 + q^{2n-1}e^{-2\pi iv}),$$

$$\theta_{01}(v) = Q_0 \prod_{n=1}^{\infty}(1 - q^{2n-1}e^{2\pi iv})(1 - q^{2n-1}e^{-2\pi iv}),$$

$$\theta_{10}(v) = 2Q_0e^{\pi i\tau/4} \cos \pi v \prod_{n=1}^{\infty}(1 + q^{2n}e^{2\pi iv})(1 + q^{2n}e^{-2\pi iv}), \tag{35}$$

$$\theta_{11}(v) = 2i\, Q_0e^{\pi i\tau/4} \sin \pi v \prod_{n=1}^{\infty}(1 - q^{2n}e^{2\pi iv})(1 - q^{2n}e^{-2\pi iv}),$$

where $q = e^{\pi i\tau}$ and

$$Q_0 = \prod_{n=1}^{\infty}(1 - q^{2n}).$$

In particular,

$$\theta_{00}(0) = Q_0 \prod_{n=1}^{\infty}(1 + q^{2n-1})^2,$$

$$\theta_{01}(0) = Q_0 \prod_{n=1}^{\infty}(1 - q^{2n-1})^2,$$

$$\theta_{10}(0) = 2q^{1/4}Q_0 \prod_{n=1}^{\infty}(1 + q^{2n})^2.$$

By differentiating with respect to v and then putting $v = 0$, we obtain in addition $\theta'_{11}(0) = 2\pi i q^{1/4} Q_0^3$. But

$$Q_0 = \prod_{n=1}^{\infty} (1 - q^n)(1 + q^n)$$

$$= \prod_{n=1}^{\infty} (1 - q^{2n})(1 - q^{2n-1})(1 + q^{2n})(1 + q^{2n-1}),$$

which implies

$$\prod_{n=1}^{\infty} (1 - q^{2n-1})(1 + q^{2n})(1 + q^{2n-1}) = 1.$$

It follows that

$$\theta_{00}(0)\theta_{01}(0)\theta_{10}(0) = 2q^{1/4} Q_0^3$$

and hence

$$\theta'_{11}(0) = \pi i \theta_{00}(0)\theta_{01}(0)\theta_{10}(0). \tag{36}$$

It is evident from their series definitions that, when q is replaced by $-q$, the functions θ_{00} and θ_{01} are interchanged, whereas the functions $q^{-1/4}\theta_{10}$ and $q^{-1/4}\theta_{11}$ are unaltered. Hence

$$\begin{aligned}
\theta_{00}(v; \tau + 1) &= \theta_{01}(v; \tau), \quad \theta_{10}(v; \tau + 1) = e^{\pi i/4}\theta_{10}(v; \tau), \\
\theta_{01}(v; \tau + 1) &= \theta_{00}(v; \tau), \quad \theta_{11}(v; \tau + 1) = e^{\pi i/4}\theta_{11}(v; \tau).
\end{aligned} \tag{37}$$

From Proposition 3 we obtain also the transformation formulas

$$\begin{aligned}
\theta_{00}(v; -1/\tau) &= (\tau/i)^{1/2} e^{\pi i \tau v^2} \theta_{00}(\tau v; \tau), \\
\theta_{10}(v; -1/\tau) &= (\tau/i)^{1/2} e^{\pi i \tau v^2} \theta_{01}(\tau v; \tau), \\
\theta_{01}(v; -1/\tau) &= (\tau/i)^{1/2} e^{\pi i \tau v^2} \theta_{10}(\tau v; \tau), \\
\theta_{11}(v; -1/\tau) &= -i(\tau/i)^{1/2} e^{\pi i \tau v^2} \theta_{11}(\tau v; \tau).
\end{aligned} \tag{38}$$

Up to this point we have used Hermite's notation just to dress up old results in new clothes. The next result breaks fresh ground.

Proposition 4 *For all $v, w \in \mathbb{C}$ and $\tau \in \mathscr{H}$,*

$$\begin{aligned}
\theta_{00}(v; \tau)\theta_{00}(w; \tau) &= \theta_{00}(v + w; 2\tau)\theta_{00}(v - w; 2\tau) + \theta_{10}(v + w; 2\tau)\theta_{10}(v - w; 2\tau), \\
\theta_{10}(v; \tau)\theta_{10}(w; \tau) &= \theta_{10}(v + w; 2\tau)\theta_{00}(v - w; 2\tau) + \theta_{00}(v + w; 2\tau)\theta_{10}(v - w; 2\tau), \\
\theta_{00}(v; \tau)\theta_{01}(w; \tau) &= \theta_{01}(v + w; 2\tau)\theta_{01}(v - w; 2\tau) + \theta_{11}(v + w; 2\tau)\theta_{11}(v - w; 2\tau), \\
\theta_{01}(v; \tau)\theta_{01}(w; \tau) &= \theta_{00}(v + w; 2\tau)\theta_{00}(v - w; 2\tau) - \theta_{10}(v + w; 2\tau)\theta_{10}(v - w; 2\tau), \\
\theta_{10}(v; \tau)\theta_{11}(w; \tau) &= \theta_{11}(v + w; 2\tau)\theta_{01}(v - w; 2\tau) - \theta_{01}(v + w; 2\tau)\theta_{11}(v - w; 2\tau), \\
\theta_{11}(v; \tau)\theta_{11}(w; \tau) &= \theta_{10}(v + w; 2\tau)\theta_{00}(v - w; 2\tau) - \theta_{00}(v + w; 2\tau)\theta_{10}(v - w; 2\tau).
\end{aligned}$$

Proof From the definition of θ_{00},

$$\theta_{00}(v; \tau)\theta_{00}(w; \tau) = \sum_{j,k} e^{\pi i \tau (j^2 + k^2)} e^{2\pi i v j} e^{2\pi i w k} = \sum_{j+k \ even} + \sum_{j+k \ odd}.$$

In the first sum on the right we can write $j + k = 2m$, $j - k = 2n$. Then $j = m + n$, $k = m - n$ and

$$\sum_{j+k \ even} = \sum_{m,n \in \mathbb{Z}} e^{2\pi i \tau (m^2 + n^2)} e^{2\pi i (v+w)m} e^{2\pi i (v-w)n}$$

$$= \theta_{00}(v + w; 2\tau)\theta_{00}(v - w; 2\tau).$$

In the second sum we can write $j + k = 2m + 1$, $j - k = 2n + 1$. Then $j = m + n + 1$, $k = m - n$ and

$$\sum_{j+k \ odd} = \sum_{m,n \in \mathbb{Z}} e^{2\pi i \tau \{(m+1/2)^2 + (n+1/2)^2\}} e^{2\pi i v (m+n+1)} e^{2\pi i w (m-n)}$$

$$= \theta_{10}(v + w; 2\tau)\theta_{10}(v - w; 2\tau).$$

Adding, we obtain the first relation of the proposition.

We obtain the second relation from the first by replacing v by $v + \tau/2$ and w by $w + \tau/2$. The remaining relations are obtained from the first two by increasing v and/or w by $1/2$. $\qquad\square$

By taking $w = v$ in Proposition 4, and adding or subtracting pairs of equations whose right sides differ only in one sign, we obtain the *duplication formulas*:

Proposition 5 *For all $v \in \mathbb{C}$ and $\tau \in \mathscr{H}$,*

$$\theta_{00}(2v; 2\tau) = [\theta_{00}^2(v; \tau) + \theta_{01}^2(v; \tau)]/2\theta_{00}(0; 2\tau)$$

$$= [\theta_{10}^2(v; \tau) - \theta_{11}^2(v; \tau)]/2\theta_{10}(0; 2\tau),$$

$$\theta_{10}(2v; 2\tau) = [\theta_{00}^2(v; \tau) - \theta_{01}^2(v; \tau)]/2\theta_{10}(0; 2\tau)$$

$$= [\theta_{10}^2(v; \tau) + \theta_{11}^2(v; \tau)]/2\theta_{00}(0; 2\tau),$$

$$\theta_{01}(2v; 2\tau) = \theta_{00}(v; \tau)\theta_{01}(v; \tau)/\theta_{01}(0; 2\tau),$$

$$\theta_{11}(2v; 2\tau) = \theta_{10}(v; \tau)\theta_{11}(v; \tau)/\theta_{01}(0; 2\tau).$$

From Proposition 4 we can also derive the following *addition formulas*:

Proposition 6 *For all $v, w \in \mathbb{C}$ and $\tau \in \mathscr{H}$,*

$$\theta_{01}^2(0)\theta_{01}(v + w)\theta_{01}(v - w)$$

$$= \theta_{01}^2(v)\theta_{01}^2(w) - \theta_{11}^2(v)\theta_{11}^2(w) = \theta_{00}^2(v)\theta_{00}^2(w) - \theta_{10}^2(v)\theta_{10}^2(w),$$

$$\theta_{00}(0)\theta_{01}(0)\theta_{00}(v + w)\theta_{01}(v - w)$$

$$= \theta_{00}(v)\theta_{01}(v)\theta_{00}(w)\theta_{01}(w) + \theta_{10}(v)\theta_{11}(v)\theta_{10}(w)\theta_{11}(w),$$

$$\theta_{01}(0)\theta_{10}(0)\theta_{10}(v + w)\theta_{01}(v - w)$$

$$= \theta_{01}(v)\theta_{10}(v)\theta_{01}(w)\theta_{10}(w) + \theta_{00}(v)\theta_{11}(v)\theta_{00}(w)\theta_{11}(w),$$

$$\theta_{00}(0)\theta_{10}(0)\theta_{11}(v + w)\theta_{01}(v - w)$$

$$= \theta_{01}(v)\theta_{11}(v)\theta_{00}(w)\theta_{10}(w) + \theta_{00}(v)\theta_{10}(v)\theta_{01}(w)\theta_{11}(w),$$

where all theta functions have the same second argument τ.

Proof Consider the second relation. If we use the first and fourth relations of Proposition 4 to evaluate the products $\theta_{00}(v)\theta_{00}(w)$ and $\theta_{01}(v)\theta_{01}(w)$, we obtain

$$\theta_{00}(v)\theta_{01}(v)\theta_{00}(w)\theta_{01}(w) = \theta_{00}^2(v + w; 2\tau)\theta_{00}^2(v - w; 2\tau)$$
$$- \theta_{10}^2(v + w; 2\tau)\theta_{10}^2(v - w; 2\tau).$$

Similarly, if we use the second and sixth relations of Proposition 4 to evaluate the products $\theta_{10}(v)\theta_{10}(w)$ and $\theta_{11}(v)\theta_{11}(w)$, we obtain

$$\theta_{10}(v)\theta_{11}(v)\theta_{10}(w)\theta_{11}(w) = \theta_{10}^2(v + w; 2\tau)\theta_{00}^2(v - w; 2\tau)$$
$$- \theta_{00}^2(v + w; 2\tau)\theta_{10}^2(v - w; 2\tau).$$

Hence, in the second relation of the present proposition the right side is equal to

$$[\theta_{00}^2(v + w; 2\tau) + \theta_{10}^2(v + w; 2\tau)][\theta_{00}^2(v - w; 2\tau) - \theta_{10}^2(v - w; 2\tau)].$$

On the other hand, if we use the first and fourth relations of Proposition 4 to evaluate the products $\theta_{00}(0)\theta_{00}(v+w)$ and $\theta_{01}(0)\theta_{01}(v-w)$, we see that the left side is likewise equal to

$$[\theta_{00}^2(v + w; 2\tau) + \theta_{10}^2(v + w; 2\tau)][\theta_{00}^2(v - w; 2\tau) - \theta_{10}^2(v - w; 2\tau)].$$

This proves the second relation of the proposition, and the others may be proved similarly.   □

Corollary 7 *For all $v \in \mathbb{C}$ and $\tau \in \mathscr{H}$,*

$$\theta_{00}^2(0)\theta_{01}^2(v) + \theta_{10}^2(0)\theta_{11}^2(v) = \theta_{01}^2(0)\theta_{00}^2(v), \tag{39}$$
$$\theta_{10}^2(0)\theta_{01}^2(v) + \theta_{00}^2(0)\theta_{11}^2(v) = \theta_{01}^2(0)\theta_{10}^2(v). \tag{40}$$

Moreover, for all $\tau \in \mathscr{H}$,

$$\theta_{00}^4(0) = \theta_{01}^4(0) + \theta_{10}^4(0). \tag{41}$$

Proof We get (39) and (40) from the first relation of Proposition 6 by taking $w = 1/2$ and $w = (1 + \tau)/2$ respectively. We obtain (41) from (39) by taking $v = 1/2$.   □

If we regard (39) and (40) as a system of simultaneous linear equations for the unknowns $\theta_{01}^2(v), \theta_{11}^2(v)$, then the determinant of this system is $\theta_{00}^4(0) - \theta_{10}^4(0) = \theta_{01}^4(0) \neq 0$. It follows that the square of any theta function may be expressed as a linear combination of the squares of any other two theta functions.

By substituting for the theta functions their expansions as infinite products, the formula (41) may be given the following remarkable form:

$$\prod_{n=1}^{\infty}(1 + q^{2n-1})^8 = \prod_{n=1}^{\infty}(1 - q^{2n-1})^8 + 16q \prod_{n=1}^{\infty}(1 + q^{2n})^8.$$

Proposition 8 *For all $v \in \mathbb{C}$ and $\tau \in \mathscr{H}$,*

$$\{\theta_{00}(v)/\theta_{01}(v)\}' = \pi i \theta_{10}^2(0)\theta_{10}(v)\theta_{11}(v)/\theta_{01}^2(v), \tag{42}$$

$$\{\theta_{10}(v)/\theta_{01}(v)\}' = \pi i \theta_{00}^2(0)\theta_{00}(v)\theta_{11}(v)/\theta_{01}^2(v), \tag{43}$$

$$\{\theta_{11}(v)/\theta_{01}(v)\}' = \pi i \theta_{01}^2(0)\theta_{00}(v)\theta_{10}(v)/\theta_{01}^2(v), \tag{44}$$

$$\{\theta_{01}'(v)/\theta_{01}(v)\}' = \theta_{01}''(0)/\theta_{01}(0) + \pi^2\theta_{00}^2(0)\theta_{10}^2(0)\theta_{11}^2(v)/\theta_{01}^2(v). \tag{45}$$

Proof By differentiating the second relation of Proposition 6 with respect to w and then putting $w = 0$, we obtain

$$\theta_{00}(0)\theta_{01}(0)[\theta_{00}'(v)\theta_{01}(v) - \theta_{00}(v)\theta_{01}'(v)] = \theta_{10}(0)\theta_{11}'(0)\theta_{10}(v)\theta_{11}(v),$$

since not only $\theta_{11}(0) = 0$ but also $\theta_{00}'(0) = \theta_{01}'(0) = \theta_{10}'(0) = 0$. Dividing by $\theta_{01}^2(v)$ and recalling the expression (36) for $\theta_{11}'(0)$, we obtain (42). Similarly, from the third and fourth relations of Proposition 6 we obtain (43) and (44).

In the same way, if we differentiate the first relation of Proposition 6 twice with respect to w and then put $w = 0$, we obtain

$$\theta_{01}^2(0)[\theta_{01}''(v)\theta_{01}(v) - \theta_{01}'(v)^2] = \theta_{01}(0)\theta_{01}''(0)\theta_{01}^2(v) - \theta_{11}'(0)^2\theta_{11}^2(v).$$

Hence, using (36) again, we obtain (45). $\square$

We are now in a position to make the connection between theta functions and elliptic functions.

5 Jacobian Elliptic Functions

The behaviour of the theta functions when their argument is increased by 1 or τ makes it clear that doubly-periodic functions may be constructed from their quotients. We put

$$\operatorname{sn} u = \operatorname{sn}(u; \tau) := -i\theta_{00}(0)\theta_{11}(v)/\theta_{10}(0)\theta_{01}(v),$$

$$\operatorname{cn} u = \operatorname{cn}(u; \tau) := \theta_{01}(0)\theta_{10}(v)/\theta_{10}(0)\theta_{01}(v), \tag{46}$$

$$\operatorname{dn} u = \operatorname{dn}(u; \tau) := \theta_{01}(0)\theta_{00}(v)/\theta_{00}(0)\theta_{01}(v),$$

where $u = \pi\theta_{00}^2(0)v$.

The constant multiples are chosen so that, in addition to $\operatorname{sn} 0 = 0$, we have $\operatorname{cn} 0 = \operatorname{dn} 0 = 1$. The independent variable is scaled so that, by (42)–(44),

$$d(\operatorname{sn} u)/du = \operatorname{cn} u \operatorname{dn} u,$$

$$d(\operatorname{cn} u)/du = -\operatorname{sn} u \operatorname{dn} u, \tag{47}$$

$$d(\operatorname{dn} u)/du = -\lambda \operatorname{sn} u \operatorname{cn} u,$$

where

$$\lambda = \lambda(\tau) := \theta_{10}^4(0; \tau)/\theta_{00}^4(0; \tau). \tag{48}$$

It follows at once from the definitions that $\operatorname{sn} u$ is an odd function of u, whereas $\operatorname{cn} u$ and $\operatorname{dn} u$ are even functions of u. It follows from (41) that

$$1 - \lambda(\tau) = \theta_{01}^4(0; \tau)/\theta_{00}^4(0; \tau), \tag{49}$$

and from (39)–(40) that

$$\operatorname{cn}^2 u = 1 - \operatorname{sn}^2 u, \quad \operatorname{dn}^2 u = 1 - \lambda \operatorname{sn}^2 u. \tag{50}$$

Evidently (47) implies

$$d(\operatorname{sn}^2 u)/du = 2\operatorname{sn} u \operatorname{cn} u \operatorname{dn} u,$$

$$d^2(\operatorname{sn}^2 u)/du^2 = 2(\operatorname{cn}^2 u \operatorname{dn}^2 u - \operatorname{sn}^2 u \operatorname{dn}^2 u - \lambda \operatorname{sn}^2 u \operatorname{cn}^2 u).$$

If we write $S(u) = S(u; \tau) := \operatorname{sn}^2 u$ and use (50), we can rewrite this in the form

$$d^2 S/du^2 = 2[(1 - S)(1 - \lambda S) - S(1 - \lambda S) - \lambda S(1 - S)]$$

$$= 6\lambda S^2 - 4(1 + \lambda)S + 2.$$

Since $S(0) = S'(0) = 0$, we conclude that $S(u)$ coincides with the function denoted by the same symbol in §3. However, it should be noted that now λ is not given, but is determined by τ. Thus the question arises: can we choose $\tau \in \mathcal{H}$ (the upper half-plane) so that $\lambda(\tau)$ is any prescribed complex number other than 0 or 1?

For many applications it is sufficient to know that we can choose $\tau \in \mathcal{H}$ so that $\lambda(\tau)$ is any prescribed real number between 0 and 1. Since this case is much simpler, we will deal with it now and defer treatment of the general case until the next section. We have

$$\lambda(\tau) = 1 - \theta_{01}^4(0; \tau)/\theta_{00}^4(0; \tau) = 1 - \prod_{n=1}^{\infty} \{(1 - q^{2n-1})/(1 + q^{2n-1})\}^8,$$

where $q = e^{\pi i \tau}$. If $\tau = iy$, where $y > 0$, then $0 < q < 1$. Moreover, as y increases from 0 to ∞, q decreases from 1 to 0 and the infinite product increases from 0 to 1. Thus $\lambda(\tau)$ decreases continuously from 1 to 0 and, for each $w \in (0, 1)$, there is a unique pure imaginary $\tau \in \mathcal{H}$ such that $\lambda(\tau) = w$.

It should be mentioned that, also with our previous approach, $S(u)$ could have been recognized as the square of a meromorphic function by defining sn u, cn u, dn u to be the solution, for *given* $\lambda \in \mathbb{C}$, of the system of differential equations (47) which satisfies the initial condition sn $0 = 0$, cn $0 = $ dn $0 = 1$.

Elliptic functions were first defined by Abel (1827) as the inverses of elliptic integrals. His definitions were modified by Jacobi (1829) to accord with Legendre's normal form for elliptic integrals, and the functions sn u, cn u, dn u are generally known as the *Jacobian elliptic functions*. The actual notation is due to Gudermann (1838). The definition by means of theta functions was given later by Jacobi (1838) in lectures.

Several properties of the Jacobian elliptic functions are easy consequences of the later definition. In the first place, all three are meromorphic in the whole u-plane, since the theta functions are everywhere holomorphic. Their poles are determined by the zeros of $\theta_{01}(v)$ and are all simple. Similarly, the zeros of sn u, cn u and dn u are determined by the zeros of $\theta_{11}(v)$, $\theta_{10}(v)$ and $\theta_{00}(v)$ respectively and are all simple. If we put

$$\mathbf{K} = \mathbf{K}(\tau) := \pi \theta_{00}^2(0; \tau)/2, \quad \mathbf{K}' = \mathbf{K}'(\tau) := \tau \mathbf{K}(\tau)/i, \tag{51}$$

then we have

$$\textit{Poles of } \operatorname{sn} u, \operatorname{cn} u, \operatorname{dn} u: \quad u = 2m\mathbf{K} + (2n+1)i\mathbf{K}' \ (m, n \in \mathbb{Z}). \tag{52}$$

$$\textit{Zeros of } \operatorname{sn} u: \quad u = 2m\mathbf{K} + 2ni\mathbf{K}',$$

$$\operatorname{cn} u: \quad u = (2m+1)\mathbf{K} + 2ni\mathbf{K}', \ (m, n \in \mathbb{Z}) \tag{53}$$

$$\operatorname{dn} u: \quad u = (2m+1)\mathbf{K} + (2n+1)i\mathbf{K}'.$$

From the definitions (46) of the Jacobian elliptic functions and the behaviour of the theta functions when v is increased by 1 or τ we further obtain

$$\operatorname{sn} u = -\operatorname{sn}(u + 2\mathbf{K}) = \operatorname{sn}(u + 2i\mathbf{K}'),$$

$$\operatorname{cn} u = -\operatorname{cn}(u + 2\mathbf{K}) = -\operatorname{cn}(u + 2i\mathbf{K}'), \tag{54}$$

$$\operatorname{dn} u = \operatorname{dn}(u + 2\mathbf{K}) = -\operatorname{dn}(u + 2i\mathbf{K}').$$

It follows that all three functions are *doubly-periodic*. In fact sn u has periods $4\mathbf{K}$ and $2i\mathbf{K}'$, cn u has periods $4\mathbf{K}$ and $2\mathbf{K} + 2i\mathbf{K}'$, and dn u has periods $2\mathbf{K}$ and $4i\mathbf{K}'$. In each case the ratio of the two periods is not real, since $\tau \in \mathscr{H}$.

Since any period must equal a difference between two poles, it must have the form $2m\mathbf{K} + 2ni\mathbf{K}'$ for some $m, n \in \mathbb{Z}$. Since $4\mathbf{K}$ and $2i\mathbf{K}'$ are periods of sn u, but $2\mathbf{K}$ is not, and since any integral linear combination of periods is again a period, it follows that the periods of sn u are precisely the integral linear combinations of $4\mathbf{K}$ and $2i\mathbf{K}'$. Similarly the periods of cn u are the integral linear combinations of $4\mathbf{K}$ and $2\mathbf{K} + 2i\mathbf{K}'$, and the periods of dn u are the integral linear combinations of $2\mathbf{K}$ and $4i\mathbf{K}'$.

It was shown in §3 that, if $0 < \lambda < 1$, then $S(t, \lambda)$ has least positive period $2K(\lambda)$, where

$$K(\lambda) = \int_0^1 dx/g_\lambda(x)^{1/2}.$$

But, as we have seen, there is a unique pure imaginary $\tau \in \mathscr{H}$ such that $\lambda = \lambda(\tau)$, and $2K[\lambda(\tau)]$ is then the least positive period of $\operatorname{sn}^2(u; \tau)$. Since the periods of $\operatorname{sn}^2(u; \tau)$ are $2m\mathbf{K} + 2ni\mathbf{K}'(m, n \in \mathbb{Z})$, and since $\mathbf{K}, \mathbf{K}'$ are real and positive when τ is pure imaginary, it follows that

$$K[\lambda(\tau)] = \mathbf{K}(\tau).$$

The domain of validity of this relation may be extended by appealing to results which will be established in §6. In fact it holds, by analytic continuation, for all τ in the region $\mathscr{D}$ illustrated in Figure 3, since $\lambda(\tau) \in \mathscr{H}$ for $\tau \in \mathscr{D}$.

From the definitions (46) of the Jacobian elliptic functions, the addition formulas for the theta functions (Proposition 6) and the expression (48) for λ, we obtain *addition formulas* for the Jacobian functions:

$$\operatorname{sn}(u_1 + u_2) = (\operatorname{sn} u_1 \operatorname{cn} u_2 \operatorname{dn} u_2 + \operatorname{sn} u_2 \operatorname{cn} u_1 \operatorname{dn} u_1)/(1 - \lambda \operatorname{sn}^2 u_1 \operatorname{sn}^2 u_2),$$

$$\operatorname{cn}(u_1 + u_2) = (\operatorname{cn} u_1 \operatorname{cn} u_2 - \operatorname{sn} u_1 \operatorname{sn} u_2 \operatorname{dn} u_1 \operatorname{dn} u_2)/(1 - \lambda \operatorname{sn}^2 u_1 \operatorname{sn}^2 u_2), \tag{55}$$

$$\operatorname{dn}(u_1 + u_2) = (\operatorname{dn} u_1 \operatorname{dn} u_2 - \lambda \operatorname{sn} u_1 \operatorname{sn} u_2 \operatorname{cn} u_1 \operatorname{cn} u_2)/(1 - \lambda \operatorname{sn}^2 u_1 \operatorname{sn}^2 u_2).$$

The addition formulas show that the evaluation of the Jacobian elliptic functions for arbitrary complex argument may be reduced to their evaluation for real and pure imaginary arguments.

The usual addition formulas for the sine and cosine functions may be regarded as limiting cases of (55). For if $\tau = iy$ and $y \to \infty$, the product expansions (35) show that

$$\theta_{00}(v) \to 1, \quad \theta_{01}(v) \to 1,$$
$$\theta_{10}(v) \sim 2e^{\pi i \tau/4} \cos \pi v, \quad \theta_{11}(v) \sim 2ie^{\pi i \tau/4} \sin \pi v,$$

and hence

$$\lambda \to 0, \quad u \to \pi v,$$
$$\operatorname{sn} u \to \sin u, \quad \operatorname{cn} u \to \cos u, \quad \operatorname{dn} u \to 1.$$

The definitions (46) of the Jacobian elliptic functions and the transformation formulas (37)–(38) for the theta functions imply also *transformation formulas* for the Jacobian functions:

Proposition 9 *For all $u \in \mathbb{C}$ and $\tau \in \mathcal{H}$,*

$$\operatorname{sn}(u; \tau + 1) = (1 - \lambda(\tau))^{1/2} \operatorname{sn}(u'; \tau)/\operatorname{dn}(u'; \tau),$$
$$\operatorname{cn}(u; \tau + 1) = \operatorname{cn}(u'; \tau)/\operatorname{dn}(u'; \tau),$$
$$\operatorname{dn}(u; \tau + 1) = 1/\operatorname{dn}(u'; \tau),$$

where

$$u' = u/(1 - \lambda(\tau))^{1/2}$$

and

$$(1 - \lambda(\tau))^{1/2} = \theta_{01}^2(0; \tau)/\theta_{00}^2(0; \tau).$$

Furthermore,

$$\lambda(\tau + 1) = \lambda(\tau)/[\lambda(\tau) - 1],$$
$$\mathbf{K}(\tau + 1) = (1 - \lambda(\tau))^{1/2} \mathbf{K}(\tau).$$

Proof With $v = u/\pi \theta_{00}^2(0; \tau + 1)$ we have, by (37),

$$\operatorname{dn}(u; \tau + 1) = \theta_{00}(0; \tau)\theta_{01}(v; \tau)/\theta_{01}(0; \tau)\theta_{00}(v; \tau) = 1/\operatorname{dn}(u'; \tau),$$

where

$$u' = \pi \theta_{00}^2(0; \tau)v = \theta_{00}^2(0; \tau)u/\theta_{01}^2(0; \tau) = u/(1 - \lambda(\tau))^{1/2}.$$

Similarly, from (37) and (48)-(49), we obtain

$$\lambda(\tau + 1) = -\theta_{10}^4(0; \tau)/\theta_{01}^4(0; \tau) = \lambda(\tau)/[\lambda(\tau) - 1].$$

The other relations are established in the same way. □

Proposition 10 *For all* $u \in \mathbb{C}$ *and* $\tau \in \mathscr{H}$,

$$\operatorname{sn}(u; -1/\tau) = -i\operatorname{sn}(iu; \tau)/\operatorname{cn}(iu; \tau),$$
$$\operatorname{cn}(u; -1/\tau) = 1/\operatorname{cn}(iu; \tau),$$
$$\operatorname{dn}(u; -1/\tau) = \operatorname{dn}(iu; \tau)/\operatorname{cn}(iu; \tau),$$

Furthermore,

$$\lambda(-1/\tau) = 1 - \lambda(\tau),$$
$$\mathbf{K}(-1/\tau) = \mathbf{K}'(\tau).$$

Proof With $v = u/\pi \theta_{00}^2(0; -1/\tau)$ we have, by (38),

$$\operatorname{sn}(u; -1/\tau) = -i\theta_{00}(0; -1/\tau)\theta_{11}(v; -1/\tau)/\theta_{10}(0; -1/\tau)\theta_{01}(v; -1/\tau)$$
$$= -\theta_{00}(0; \tau)\theta_{11}(\tau v; \tau)/\theta_{01}(0; \tau)\theta_{10}(\tau v; \tau).$$

On the other hand, with $v' = iu/\pi \theta_{00}^2(0; \tau)$ we have

$$\operatorname{sn}(iu; \tau)/\operatorname{cn}(iu; \tau) = -i\theta_{00}(0; \tau)\theta_{11}(v'; \tau)/\theta_{01}(0; \tau)\theta_{10}(v'; \tau).$$

Since $\tau v = v'$, by comparing these two relations we obtain the first assertion of the proposition.

The next two assertions may be obtained in the same way. The final two assertions follow from (38), together with (48), (49) and (51). $\qquad\square$

It follows from Proposition 10 that the evaluation of the Jacobian elliptic functions for pure imaginary argument and parameter τ may be reduced to their evaluation for real argument and parameter $-1/\tau$.

From the definition (46) of the Jacobian elliptic functions and the duplication formulas for the theta functions we can also obtain formulas for the Jacobian functions when the parameter τ is doubled ('Landen's transformation'):

Proposition 11 *For all* $u \in \mathbb{C}$ *and* $\tau \in \mathscr{H}$,

$$\operatorname{sn}(u''; 2\tau) = [1 + (1 - \lambda(\tau))^{1/2}]\operatorname{sn}(u; \tau)\operatorname{cn}(u; \tau)/\operatorname{dn}(u; \tau),$$
$$\operatorname{cn}(u''; 2\tau) = \{1 - [1 + (1 - \lambda(\tau))^{1/2}]\operatorname{sn}^2(u; \tau)\}/\operatorname{dn}(u; \tau),$$
$$\operatorname{dn}(u''; 2\tau) = \{1 - [1 - (1 - \lambda(\tau))^{1/2}]\operatorname{sn}^2(u; \tau)\}/\operatorname{dn}(u; \tau),$$

where $u'' = [1 + (1 - \lambda(\tau))^{1/2}]u$ *and* $(1 - \lambda(\tau))^{1/2} = \theta_{01}^2(0; \tau)/\theta_{00}^2(0; \tau)$.
Furthermore,

$$\lambda(2\tau) = \lambda^2(\tau)/[1 + (1 - \lambda(\tau))^{1/2}]^4,$$
$$\mathbf{K}(2\tau) = [1 + (1 - \lambda(\tau))^{1/2}]\mathbf{K}(\tau)/2.$$

Proof If $u = \pi \theta_{00}^2(0; \tau)v$ and $u'' = \pi \theta_{00}^2(0; 2\tau)2v$ then, by Proposition 5,

$$u'' = 2\theta_{00}^2(0; 2\tau)u/\theta_{00}^2(0; \tau)$$
$$= [\theta_{00}^2(0; \tau) + \theta_{01}^2(0; \tau)]u/\theta_{00}^2(0; \tau).$$

Hence, by (49),

$$u'' = [1 + (1 - \lambda(\tau))^{1/2}]u.$$

By Proposition 5 also,

$$\text{sn}\,(u''; 2\tau) = -i\theta_{00}(0; 2\tau)\theta_{10}(v; \tau)\theta_{11}(v; \tau)/\theta_{10}(0; 2\tau)\theta_{00}(v; \tau)\theta_{01}(v; \tau).$$

On the other hand,

$$\text{sn}\,(u; \tau)\text{cn}\,(u; \tau)/\text{dn}\,(u; \tau) = -i\theta_{00}^2(0; \tau)\theta_{10}(v; \tau)\theta_{11}(v; \tau)/D,$$

where $D = \theta_{10}^2(0; \tau)\theta_{00}(v; \tau)\theta_{01}(v; \tau)$.

Since $2\theta_{00}(0; 2\tau)\theta_{10}(0; 2\tau) = \theta_{10}^2(0; \tau)$, it follows that

$$\text{sn}\,(u''; 2\tau) = 2\theta_{00}^2(0; 2\tau)\,\text{sn}\,(u; \tau)\,\text{cn}\,(u; \tau)/\theta_{00}^2(0; \tau)\,\text{dn}\,(u; \tau).$$

Since $2\theta_{00}^2(0; 2\tau)/\theta_{00}^2(0; \tau) = u''/u$, this proves the first assertion of the proposition. The remaining assertions may be proved similarly.   □

We show finally how the standard elliptic integrals of the second and third kinds, defined by (24) and (25), may be expressed in terms of theta functions. If we put

$$\Theta(u) = \theta_{01}(v), \tag{56}$$

where $u = \pi\theta_{00}^2(0)v$, then since

$$\lambda S(u) = \lambda\text{sn}^2 u = -\theta_{10}^2(0)\theta_{11}^2(v)/\theta_{00}^2(0)\theta_{01}^2(v),$$

we can rewrite (45) in the form

$$d\{\Theta'(u)/\Theta(u)\}/du = -\alpha + 1 - \lambda S(u),$$

where α is independent of u and the prime on the left denotes differentiation with respect to u. Since $\Theta'(0) = 0$, by integrating we obtain

$$E(u) = \Theta'(u)/\Theta(u) + \alpha u.$$

To determine α we take $u = \mathbf{K}$. Since $\theta_{01}'(1/2) = \theta_{00}'(1) = \theta_{00}'(0) = 0$, we obtain $\alpha = \mathbf{E}/\mathbf{K}$, where

$$\mathbf{E} = \mathbf{E}(\mathbf{K}) = \int_0^{\mathbf{K}} \{1 - \lambda S(u)\}\,du = \int_0^1 (1 - \lambda x)\,dx/g_\lambda(x)^{1/2}$$

is a complete elliptic integral of the second kind. Thus

$$E(u) = \Theta'(u)/\Theta(u) + u\mathbf{E}/\mathbf{K}. \tag{57}$$

Substituting this expression for $E(u)$ in (27), we further obtain

$$\Pi(u, a) = u\Theta'(a)/\Theta(a) + (1/2)\log\{\Theta(u - a)/\Theta(u + a)\}. \tag{58}$$

6 The Modular Function

The function

$$\lambda(\tau) := \theta_{10}^4(0; \tau)/\theta_{00}^4(0; \tau),$$

which was introduced in §5, is known as the *modular function*. In this section we study its remarkable properties. (The term 'modular function', without the definite article, is also used in a more general sense, which we do not consider here.)

The modular function is holomorphic in the upper half-plane $\mathcal{H}$. Furthermore, we have

Proposition 12 *For any $\tau \in \mathcal{H}$,*

$$\lambda(\tau + 1) = \lambda(\tau)/[\lambda(\tau) - 1],$$
$$\lambda(-1/\tau) = 1 - \lambda(\tau),$$
$$\lambda(-1/(\tau + 1)) = 1/[1 - \lambda(\tau)],$$
$$\lambda((\tau - 1)/\tau) = [\lambda(\tau) - 1]/\lambda(\tau),$$
$$\lambda(\tau/(\tau + 1)) = 1/\lambda(\tau).$$

Proof The first two relations have already been established in Propositions 9 and 10. If, as in §1, we put

$$U\lambda = 1 - \lambda, \quad V\lambda = 1/(1 - \lambda),$$

and if we also put $T\tau = \tau + 1$, $S\tau = -1/\tau$, then they may be written in the form

$$\lambda(T\tau) = UV\lambda(\tau), \quad \lambda(S\tau) = U\lambda(\tau).$$

It follows that

$$\lambda(-1/(\tau + 1)) = \lambda(ST\tau) = U\lambda(T\tau) = U^2V\lambda(\tau) = V\lambda(\tau) = 1/[1 - \lambda(\tau)].$$

Similarly,

$$\lambda((\tau - 1)/\tau) = \lambda(TS\tau) = V^2\lambda(\tau) = [\lambda(\tau) - 1]/\lambda(\tau),$$
$$\lambda(\tau/(\tau + 1)) = \lambda(TST\tau) = UV^2\lambda(\tau) = 1/\lambda(\tau). \qquad \square$$

As we saw in Proposition IV.12, together the transformations $S\tau = -1/\tau$ and $T\tau = \tau + 1$ generate the *modular group* Γ, consisting of all linear fractional transformations

$$\tau' = (a\tau + b)/(c\tau + d),$$

where $a, b, c, d \in \mathbb{Z}$ and $ad - bc = 1$. Consequently we can deduce the effect on $\lambda(\tau)$ of any modular transformation on τ. However, Proposition 12 contains the only cases which we require.

We will now study in some detail the behaviour of the modular function in the upper half-plane. We first observe that we need only consider the behaviour of $\lambda(\tau)$ in the right half of $\mathcal{H}$. For, from the definitions of the theta functions as infinite series,

$$\overline{\theta_{00}(0; \tau)} = \theta_{00}(0; -\bar{\tau}), \quad \overline{\theta_{01}(0; \tau)} = \theta_{01}(0; -\bar{\tau}),$$

where the bar denotes complex conjugation, and hence

$$\lambda(-\bar{\tau}) = \overline{\lambda(\tau)}. \tag{59}$$

We next note that, by taking $\tau = i$ in the relation $\lambda(-1/\tau) = 1 - \lambda(\tau)$, we obtain $\lambda(i) = 1/2$. We have already seen in §5 that $\lambda(\tau)$ is real on the imaginary axis $\tau = iy$ ($y > 0$), and decreases from 1 to 0 as y increases from 0 to ∞. Since $\lambda(\tau + 1) = \lambda(\tau)/[\lambda(\tau) - 1]$, it follows that $\lambda(\tau)$ is real also on the half-line $\tau = 1 + iy$ ($y > 0$), and increases from $-\infty$ to 0 as y increases from 0 to ∞. Moreover, $\lambda(1 + i) = -1$.

The linear fractional map $\tau = (\tau' - 1)/\tau'$ maps the half-line $\mathscr{R}\tau' = 1, \mathscr{I}\tau' > 0$ onto the semi-circle $|\tau - 1/2| = 1/2, \mathscr{I}\tau > 0$, and $\tau' = 1 + i$ is mapped to $\tau = (1 + i)/2$. Since

$$\lambda((\tau' - 1)/\tau') = [\lambda(\tau') - 1]/\lambda(\tau'),$$

it follows from what we have just proved that, as τ traverses this semi-circle from 0 to 1, $\lambda(\tau)$ is real and increases from 1 to ∞. Moreover, $\lambda((1 + i)/2) = 2$.

If $\mathscr{R}\tau = 1/2$, then $\bar{\tau} = 1 - \tau$ and hence, by (59),

$$\overline{\lambda(\tau)} = \lambda(\tau - 1) = \lambda(\tau)/[\lambda(\tau) - 1],$$

which implies

$$|\lambda(\tau) - 1|^2 = 1.$$

Thus $w = \lambda(\tau)$ maps the half-line $\mathscr{R}\tau = 1/2, \mathscr{I}\tau > 0$ into the circle $|w - 1| = 1$. Furthermore, the map is injective. For if $\lambda(\tau_1) = \lambda(\tau_2)$, then $\lambda(2\tau_1) = \lambda(2\tau_2)$, by Proposition 11, and the map is injective on the half-line $\mathscr{R}\tau = 1, \mathscr{I}\tau > 0$. If $\tau = 1/2 + iy$, where $y \to +\infty$, then

$$\theta_{00}(0; \tau) \to 1, \quad \theta_{10}(0; \tau) \sim 2e^{\pi i \tau/4}$$

and hence

$$\lambda(\tau) \sim 16i e^{-\pi y}.$$

In particular, $\lambda(\tau) \in \mathscr{H}$ and $\lambda(\tau) \to 0$. Since $\lambda((1 + i)/2) = 2$, it follows that $w = \lambda(\tau)$ maps the half-line $\tau = 1/2 + iy$ ($y > 1/2$) bijectively onto the semi-circle $|w - 1| = 1, \mathscr{I}w > 0$.

If $|\tau| = 1, \mathscr{I}\tau > 0$ and $\tau' = \tau/(1 + \tau)$, then $\mathscr{R}\tau' = 1/2, \mathscr{I}\tau' > 0$ and $\lambda(\tau') = 1/\lambda(\tau)$. Consequently, by what we have just proved, $w = \lambda(\tau)$ maps the semi-circle $|\tau| = 1, \mathscr{I}\tau > 0$ bijectively onto the half-line $\mathscr{R}w = 1/2$, $\mathscr{I}w > 0$.

The point $e^{\pi i/3} = (1 + i\sqrt{3})/2$ is in $\mathscr{H}$ and lies on both the line $\mathscr{R}\tau = 1/2$ and the circle $|\tau| = 1$. Hence $\lambda(e^{\pi i/3})$ lies on both the semi-circle $|w - 1| = 1, \mathscr{I}w > 0$ and the line $\mathscr{R}w = 1/2$, which implies that

$$\lambda(e^{\pi i/3}) = e^{\pi i/3}.$$

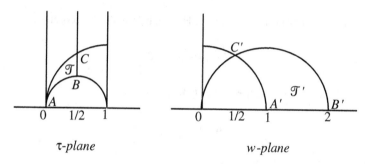

Fig. 2. $w = \lambda(\tau)$ maps $\mathscr{T}$ onto $\mathscr{T}'$.

Again, since $\lambda(\tau - 1) = \lambda(\tau)/[\lambda(\tau) - 1]$, $w = \lambda(\tau)$ maps the semi-circle $|\tau - 1| = 1$, $\mathscr{I}\tau > 0$ bijectively onto the semi-circle $|w| = 1$, $\mathscr{I}w > 0$.

In particular, we have the behaviour illustrated in Figure 2: $w = \lambda(\tau)$ maps the boundary of the (non-Euclidean) 'triangle' $\mathscr{T}$ with vertices $A = 0$, $B = (1 + i)/2$, $C = e^{\pi i/3}$ bijectively onto the boundary of the 'triangle' $\mathscr{T}'$ with vertices $A' = 1$, $B' = 2$, $C' = e^{\pi i/3}$. We are going to deduce from this that the region inside $\mathscr{T}$ is mapped bijectively onto the region inside $\mathscr{T}'$. The reasoning here does not depend on special properties of the function or the domain, but is quite general (the 'principle of the argument'). To emphasize this, we will temporarily denote the independent variable by z, instead of τ.

Choose any $w_0 \in \mathbb{C}$ which is either inside or outside the 'triangle' $\mathscr{T}'$, and let Δ denote the change in the argument of $w - w_0$ as w traverses $\mathscr{T}'$ in the direction $A'B'C'$. Thus $\Delta = 2\pi$ or 0 according as w_0 is inside or outside $\mathscr{T}'$. But Δ is also the change in the argument of $\lambda(z) - w_0$ as z traverses $\mathscr{T}$ in the direction ABC. Since $\lambda(z)$ is a nonconstant holomorphic function, the number of times that it assumes the value w_0 inside $\mathscr{T}$ is either zero or a positive integer p.

Suppose the latter, and let $z = \zeta_1, \ldots, \zeta_p$ be the points inside $\mathscr{T}$ for which $\lambda(z) = w_0$. In the neighbourhood of ζ_j we have, for some positive integer m_j and some $a_{0j} \neq 0$,

$$\lambda(z) - w_0 = a_{0j}(z - \zeta_j)^{m_j} + a_{1j}(z - \zeta_j)^{m_j+1} + \cdots$$

and

$$\lambda'(z) = m_j a_{0j}(z - \zeta_j)^{m_j-1} + (m_j + 1)a_{1j}(z - \zeta_j)^{m_j} + \cdots.$$

Hence

$$\lambda'(z)/[\lambda(z) - w_0] = m_j/(z - \zeta_j) + f_j(z),$$

where $f_j(z)$ is holomorphic at ζ_j. Consequently

$$f(z) := \lambda'(z)/[\lambda(z) - w_0] - \sum_{j=1}^{p} m_j/(z - \zeta_j)$$

is holomorphic at every point z inside $\mathscr{T}$. Hence, by Cauchy's theorem,

$$\int_{\mathscr{T}} f(z)\, dz = 0.$$

But, since $\log \lambda(z) = \log |\lambda(z)| + i \arg \lambda(z)$,

$$\int_{\mathscr{T}} \lambda'(z)\, dz / [\lambda(z) - w_0] = i\, \Delta.$$

Similarly, since ζ_j is inside $\mathscr{T}$,

$$\int_{\mathscr{T}} dz / (z - \zeta_j) = 2\pi i.$$

It follows that

$$\Delta = 2\pi \sum_{j=1}^{p} m_j.$$

If w_0 is outside $\mathscr{T}'$, then $\Delta = 0$ and we have a contradiction. Hence $\lambda(z)$ is never outside $\mathscr{T}'$ if z is inside $\mathscr{T}$. If w_0 is inside $\mathscr{T}'$, then $\Delta = 2\pi$. Hence $\lambda(z)$ assumes each value inside $\mathscr{T}'$ at exactly one point z inside $\mathscr{T}$, and at this point $\lambda'(z) \neq 0$.

Finally, if $\lambda(z)$ assumed a value w_0 *on* $\mathscr{T}'$ at a point z_0 inside $\mathscr{T}$, then it would assume all values near w_0 in the neighbourhood of z_0. In particular, it would assume values outside $\mathscr{T}'$, which we have shown to be impossible. It follows that $w = \lambda(z)$ maps the region inside $\mathscr{T}$ bijectively onto the region inside $\mathscr{T}'$, and $\lambda'(z) \neq 0$ for all z inside $\mathscr{T}$.

We must also have $\lambda'(z) \neq 0$ for all $z \neq 0$ on $\mathscr{T}$. Otherwise, if $\lambda(z_0) = w_0$ and $\lambda'(z_0) = 0$ for some $z_0 \in \mathscr{T} \cap \mathscr{H}$ then, for some $m > 1$ and $c \neq 0$,

$$\lambda(z) - w_0 \sim c(z - z_0)^m \text{ as } z \to z_0.$$

But this implies that $\lambda(z)$ takes values outside $\mathscr{T}'$ for some z near z_0 inside $\mathscr{T}$.

By putting together the preceding results we see that $w = \lambda(\tau)$ maps the domain

$$\mathscr{D} = \{\tau \in \mathscr{H} : 0 < \mathscr{R}\tau < 1, |\tau - 1/2| > 1/2\}$$

bijectively onto the upper half-plane $\mathscr{H}$, with the subdomain k of $\mathscr{D}$ mapped onto the subdomain k' of $\mathscr{H}$ ($k = 1, \ldots, 6$), as illustrated in Figure 3. Moreover, the boundary in $\mathscr{H}$ of $\mathscr{D}$ is mapped bijectively onto the real axis, with the points 0 and 1 omitted.

If we denote by $\bar{\mathscr{D}}$ the closure of $\mathscr{D}$ in $\mathscr{H}$ and by $\mathscr{D}^*$ the reflection of $\mathscr{D}$ in the imaginary axis, then it follows from (59) that $w = \lambda(\tau)$ maps the region

$$\bar{\mathscr{D}} \cup \mathscr{D}^* = \{\tau \in \mathscr{H} : 0 \leq \mathscr{R}\tau \leq 1, |\tau - 1/2| \geq 1/2\}$$
$$\cup \{\tau \in \mathscr{H} : -1 < \mathscr{R}\tau < 0, |\tau + 1/2| > 1/2\}$$

bijectively onto the whole complex plane $\mathbb{C}$, with the points 0 and 1 omitted. *This answers the question raised in §5.*

There remains the practical problem, for a given $w \in \mathbb{C}$, of determining $\tau \in \mathcal{H}$ such that $\lambda(\tau) = w$. If $0 < w < 1$, we can calculate τ by the *AGM* algorithm, using the formula (4), since $\tau = iK(1-w)/K(w)$. For complex w we can use an extension of the *AGM* algorithm, or proceed in the following way.

Since

$$(1 - \lambda(\tau))^{1/4} = \theta_{01}(0; \tau)/\theta_{00}(0; \tau)$$

and

$$\theta_{00}(0; \tau) = 1 + 2\sum_{n=1}^{\infty} q^{n^2}, \quad \theta_{01}(0; \tau) = 1 + 2\sum_{n=1}^{\infty}(-1)^n q^{n^2},$$

we have

$$[1 - (1 - \lambda(\tau))^{1/4}]/[1 + (1 - \lambda(\tau))^{1/4}]$$
$$= [\theta_{00}(0; \tau) - \theta_{01}(0; \tau)]/[\theta_{00}(0; \tau) + \theta_{01}(0; \tau)]$$
$$= 2(q + q^9 + q^{25} + \cdots)/(1 + 2q^4 + 2q^{16} + \cdots).$$

Thus if we put

$$\ell := [1 - (1 - w)^{1/4}]/[1 + (1 - w)^{1/4}],$$

we have to solve for q the equation

$$\ell/2 = (q + q^9 + q^{25} + \cdots)/(1 + 2q^4 + 2q^{16} + \cdots).$$

Expanding the right side as a power series in q and inverting the relationship, we obtain

$$q = \ell/2 + 2(\ell/2)^5 + 15(\ell/2)^9 + 150(\ell/2)^{13} + O(\ell/2)^{17}.$$

To ensure rapid convergence we may suppose that, in Figure 3, w is situated in the region $5'$ or on its boundary, since the general case may be reduced to this by a linear fractional transformation. It is not difficult to show that in this region $|\ell|$ takes its maximum value when $w = e^{\pi i/3}$, and then

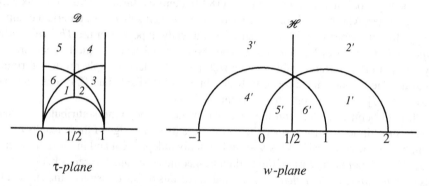

Fig. 3. $w = \lambda(\tau)$ maps $\mathcal{D}$ onto $\mathcal{H}$.

$$\ell = (1 - e^{-\pi i/12})/(1 + e^{-\pi i/12}) = i \tan \pi/24.$$

Thus $|\ell| \leq \tan \pi/24 < 2/15$ and $|\ell/2|^4 < 2 \times 10^{-5}$. Since $\mathscr{I}\tau \geq \sqrt{3}/2$ for τ in the region 5, for the solution q we have

$$|q| \leq e^{-\pi\sqrt{3}/2} < 1/15.$$

Having determined q, we may calculate $\mathbf{K}(\tau)$, sn u, ... from their representations by theta functions.

7 Further Remarks

Numerous references to the older literature on elliptic integrals and elliptic functions are given by Fricke [12]. The more important original contributions are readily available in Euler [10], Lagrange [21], Legendre [22], Gauss [13], Abel [1] and Jacobi [16], which includes his lecture course of 1838.

It was shown by Landen (1775) that the length of arc of a hyperbola could be expressed as the difference of the lengths of two elliptic arcs. The change of variables involved is equivalent to that used by Lagrange (1784/5) in his application of the *AGM* algorithm. However, Lagrange used the transformation in much greater generality, and it was his idea that elliptic integrals could be calculated numerically by iterating the transformation. The connection with the result of Landen was made explicit by Legendre (1786).

By bringing together his own results and those of others the treatise of Legendre [22], and his earlier *Exercices de calcul integral* (1811/19), contributed substantially to the discoveries of Abel and Jacobi. The supplementary third volume of his treatise, published in 1828 when he was 76, contains the first account of their work in book form.

The most important contribution of Abel (1827) was not the replacement of elliptic integrals by elliptic functions, but the study of the latter in the complex domain. In this way he established their double periodicity, determined their zeros and poles and (besides much else) showed that they could be represented as quotients of infinite products.

The triple product formula of Jacobi (1829) identified these infinite products with infinite series, whose rapid convergence made them well suited for numerical computation. Infinite series of this type had in fact already appeared in the *Théorie analytique de la Chaleur* of Fourier (1822), and Proposition 3 had essentially been proved by Poisson (1827). Remarkable generalizations of the Jacobi triple product formula to affine Lie algebras have recently been obtained by Macdonald [23] and Kac and Peterson [17]. For an introductory account, see Neher [24].

It is difficult to understand the glee with which some authors attribute to Gauss results on elliptic functions, since the world owes its knowledge of these results not to him, but to others. Gauss's work was undoubtedly independent and in most cases earlier, although not in the case of the arithmetic-geometric mean. The remark, in §335 of his *Disquisitiones Arithmeticae* (1801), that his results on the division of the circle into n equal parts applied also to the lemniscate, was one of the motivations for Abel, who carried out this extension. (For a modern account, see Rosen [25].) However, Gauss's claim in a letter to Schumacher of 30 May 1828, quoted in Krazer [20], that Abel had anticipated about a third of his own research is quite unjustified, and not only because of his inability to bring his work to a form in which it could be presented to the world.

It was *proved* by Liouville (1834) that elliptic integrals of the first and second kinds are always 'nonelementary'. For an introductory account of Liouville's theory, see Kasper [18]. (But elliptic integrals of the third kind may be 'elementary'; see Chapter IV, §7.)

The three kinds of elliptic integral may also be characterized function-theoretically. On the Riemann surface of the algebraic function $w^2 = g(z)$, where g is a cubic without repeated roots, the differential dz/w is everywhere holomorphic, the differential zdz/w is holomorphic except for a double pole at ∞ with zero residue, and the differential $[w(z) + w(a)]dz/2(z-a)w(z)$ is holomorphic except for two simple poles at a and ∞ with residues 1 and -1 respectively.

Many integrals which are not visibly elliptic may be reduced to elliptic integrals by a change of variables. A compilation is given by Byrd and Friedman [8], pp. 254–271.

The arithmetic-geometric mean may also be defined for pairs of complex numbers; a thorough discussion is given by Cox [9]. For the application of the *AGM* algorithm to integrals which are not strictly elliptic, see Bartky [4].

The differential equation (6) is a special case of the hypergeometric differential equation. In fact, if $|\lambda| < 1$, then by expanding $(1 - \lambda x)^{-1/2}$, resp. $(1 - \lambda x)^{1/2}$, by the binomial theorem and integrating term by term, the complete elliptic integrals

$$K(\lambda) = \int_0^1 [4x(1-x)(1-\lambda x)]^{-1/2}dx, \quad E(\lambda) = \int_0^1 [(1-\lambda x)/4x(1-x)]^{1/2}dx,$$

may be identified with the hypergeometric functions

$$(\pi/2)F(1/2, 1/2; 1; \lambda), \quad (\pi/2)F(-1/2, 1/2; 1; \lambda),$$

where

$$F(\alpha, \beta; \gamma; z) = 1 + \alpha\beta z/1 \cdot \gamma + \alpha(\alpha+1)\beta(\beta+1)z^2/1 \cdot 2 \cdot \gamma(\gamma+1) + \cdots.$$

Many transformation formulas for the complete elliptic integrals may be regarded as special cases of more general transformation formulas for the hypergeometric function.

The proof in §3 that $K(1 - \lambda)/K(\lambda)$ has positive real part is due to Falk [11].

It follows from (12)–(13) by induction that $S(nu)$ and $S'(nu)/S'(u)$ are rational functions of $S(u)$ for every integer n. The elliptic function $S(u)$ is said to admit *complex multiplication* if $S(\mu u)$ is a rational function of $S(u)$ for some complex number μ which is not an integer. It may be shown that $S(u)$ admits complex multiplication if and only if $\lambda \neq 0, 1$ and the period ratio $iK(1 - \lambda)/K(\lambda)$ is a quadratic irrational, in the sense of Chapter IV. This condition is obviously satisfied if $\lambda = 1/2$, the case of the lemniscate.

A function $f(u)$ is said to possess an *algebraic addition theorem* if there is a polynomial $p(x, y, z)$, not identically zero and with coefficients independent of u, v, such that

$$p(f(u+v), f(u), f(v)) = 0 \quad \text{for all } u, v.$$

It may be shown that a function f, which is meromorphic in the whole complex plane, has an algebraic addition theorem if and only if it is either a rational function or, when the independent variable is scaled by a constant factor, a rational function of $S(u, \lambda)$ and its derivative $S'(u, \lambda)$ for some $\lambda \in \mathbb{C}$. This result (in different notation) is due to Weierstrass and is proved in Akhiezer [3], for example. A generalization of Weierstrass' theorem, due to Myrberg, is proved in Belavin and Drinfeld [6].

The term 'elliptic function' is often used to denote any function which is meromorphic in the whole complex plane and has two periods whose ratio is not real. It may be shown that, if the independent variable is scaled by a constant factor, an elliptic function in this general sense is a rational function of $S(u, \lambda)$ and $S'(u, \lambda)$ for some $\lambda \neq 0, 1$.

The functions $f(v)$ which are holomorphic in the whole complex plane $\mathbb{C}$ and satisfy the functional equations

$$f(v + 1) = f(v), \quad f(v + \tau) = e^{-n\pi i(2v + \tau)} f(v),$$

where $n \in \mathbb{N}$ and $\tau \in \mathscr{H}$, form an n-dimensional complex vector space. It was shown by Hermite (1862) that this may be used to derive many relations between theta functions, such as Proposition 6.

Proposition 11 can be extended to give transformation formulas for the Jacobian functions when the parameter τ is multiplied by any positive integer n. See, for example, Tannery and Molk [27], vol. II.

The modular function was used by Picard (1879) to prove that a function $f(z)$, which is holomorphic for all $z \in \mathbb{C}$ and not a constant, assumes every complex value except perhaps one. The exponential function $\exp z$, which does not assume the value 0, illustrates that an exceptional value may exist. A careful proof of Picard's theorem is given in Ahlfors [2]. (There are also proofs which do not use the modular function.)

It was already observed by Lagrange (1813) that there is a correspondence between addition formulas for elliptic functions and the formulas of spherical trigonometry. This correspondence has been most intensively investigated by Study [26].

There is an n-dimensional generalization of theta functions, which has a useful application to the lattices studied in Chapter VIII. The theta function of an *integral lattice* Λ in $\mathbb{R}^n$ is defined by

$$\theta_\Lambda(\tau) = \sum_{u \in \Lambda} q^{(u,u)} = 1 + \sum_{m \geq 1} N_m q^m,$$

where $q = e^{\pi i \tau}$ and N_m is the number of vectors in Λ with square-norm m. If $n = 1$ and $\Lambda = \mathbb{Z}$, then

$$\theta_\mathbb{Z}(\tau) = 1 + 2q + 2q^4 + 2q^9 + \cdots = \theta(0; \tau).$$

It is easily seen that $\theta_\Lambda(\tau)$ is a holomorphic function of τ in the half-plane $\mathscr{I}\tau > 0$. It follows from Poisson's summation formula that the theta function of the *dual lattice* Λ^* is given by

$$\theta_{\Lambda^*}(\tau) = d(\Lambda)(i/\tau)^{n/2} \theta_\Lambda(-1/\tau) \quad \text{for } \mathscr{I}\tau > 0.$$

Many geometrical properties of a lattice are reflected in its theta function. However, a lattice is not uniquely determined by its theta function, since there are lattices in $\mathbb{R}^4$ (and in higher dimensions) which are not isometric but have the same theta function.

For applications of elliptic functions and theta functions to classical mechanics, conformal mapping, geometry, theoretical chemistry, statistical mechanics and approximation theory, see Halphen [15] (vol. 2), Kober [19], Bos *et al.* [7], Glasser and Zucker [14], Baxter [5] and Todd [28]. Applications to number theory will be considered in the next chapter.

8 Selected References

[1] N.H. Abel, *Oeuvres complètes,* Tome 1, 2nd ed. (ed. L. Sylow et S. Lie), Grondahl, Christiania, 1881. [Reprinted J. Gabay, Sceaux, 1992]

[2] L.V. Ahlfors, *Complex analysis*, 3rd ed., McGraw-Hill, New York, 1979.

[3] N.I. Akhiezer, *Elements of the theory of elliptic functions*, American Mathematical Society, Providence, R.I., 1990. [English transl. of 2nd Russian edition, 1970]

[4] W. Bartky, Numerical calculation of a generalized complete elliptic integral, *Rev. Modern Phys.* **10** (1938), 264–269.

[5] R.J. Baxter, *Exactly solved models in statistical mechanics*, Academic Press, London, 1982. [Reprinted, 1989]

[6] A.A. Belavin and V.G. Drinfeld, Triangle equations and simple Lie algebras, *Soviet Sci. Rev. Sect. C: Math. Phys.* **4** (1984), 93–165. [Reprinted, Harwood, Amsterdam, 1998]

[7] H.J.M. Bos, C. Kers, F. Oort and D.W. Raven, Poncelet's closure theorem, *Exposition. Math.* **5** (1987), 289–364.

[8] P.F. Byrd and M.D. Friedman, *Handbook of elliptic integrals for engineers and scientists*, 2nd ed., Springer, Berlin, 1971.

[9] D.A. Cox, The arithmetic-geometric mean of Gauss, *Enseign. Math.* **30** (1984), 275–330.

[10] L. Euler, *Opera omnia*, Ser. I, Vol. XX (ed. A. Krazer), Leipzig, 1912.

[11] M. Falk, Beweis eines Satzes aus der Theorie der elliptischen Functionen, *Acta Math.* **7** (1885/6), 197–200.

[12] R. Fricke, *Elliptische Funktionen*, Encyklopädie der Mathematischen Wissenschaften, Band II, Teil 2, pp. 177–348, Teubner, Leipzig, 1921.

[13] C.F. Gauss, *Werke*, Band III, Göttingen, 1866. [Reprinted G. Olms, Hildesheim, 1973]

[14] M.L. Glasser and I.J. Zucker, Lattice sums, *Theoretical chemistry: Advances and perspectives* **5** (1980), 67–139.

[15] G.H. Halphen, *Traité des fonctions elliptiques et de leurs applications,* 3 vols., Gauthier-Villars, Paris, 1886–1891.

[16] C.G.J. Jacobi, *Gesammelte Werke,* Band I (ed. C.W. Borchardt), Berlin, 1881. [Reprinted Chelsea, New York, 1969]

[17] V.G. Kac and D.H. Peterson, Infinite-dimensional Lie algebras, theta functions and modular forms, *Adv. in Math.* **53** (1984), 125–264.

[18] T. Kasper, Integration in finite terms: the Liouville theory, *Math. Mag.* **53** (1980), 195–201.

[19] H. Kober, *Dictionary of conformal representations*, Dover, New York, 1952.

[20] A. Krazer, Zur Geschichte des Umkehrproblems der Integral, *Jahresber. Deutsch. Math.-Verein.* **18** (1909), 44–75.

[21] J.L. Lagrange, *Oeuvres,* t. 2 (ed. J.-A. Serret), Gauthier-Villars, Paris, 1868. [Reprinted G. Olms, Hildesheim, 1973]

[22] A.M. Legendre, *Traité des fonctions elliptiques et des intégrales Eulériennes, avec des tables pour en faciliter le calcul numérique*, Paris, t.1 (1825), t.2 (1826), t.3 (1828). [Microform, Readex Microprint Corporation, New York, 1970]

[23] I.G. Macdonald, Affine root systems and Dedekind's η-function, *Invent. Math.* **15** (1972), 91–143.

[24] E. Neher, Jacobis Tripelprodukt-Identität und η-Identitäten in der Theorie affiner Lie-Algebren, *Jahresber. Deutsch. Math.-Verein.* **87** (1985), 164–181.

[25] M. Rosen, Abel's theorem on the lemniscate, *Amer. Math. Monthly* **88** (1981), 387–395.

[26] E. Study, *Sphärische Trigonometrie, orthogonale Substitutionen und elliptische Funktionen*, Leipzig, 1893.

[27] J. Tannery and J. Molk, *Éléments de la théorie des fonctions elliptiques*, 4 vols., Gauthier-Villars, Paris, 1893–1902. [Reprinted Chelsea, New York, 1972]

[28] J. Todd, Applications of transformation theory: a legacy from Zolotarev (1847-1878), *Approximation theory and spline functions* (ed. S.P. Singh et al.), pp. 207–245, Reidel, Dordrecht, 1984.

[29] E.T. Whittaker and G.N. Watson, *A course of modern analysis*, 4th ed., Cambridge University Press, 1927. [Reprinted, 1996]

XIII

Connections with Number Theory

1 Sums of Squares

In Proposition II.40 we proved Lagrange's theorem that every positive integer can be represented as a sum of 4 squares. Jacobi (1829), at the end of his *Fundamenta Nova*, gave a completely different proof of this theorem with the aid of theta functions. Moreover, his proof provided a formula for the number of different representations. Hurwitz (1896), by developing further the arithmetic of quaternions which was used in Chapter II, also derived this formula. Here we give Jacobi's argument preference since, although it is less elementary, it is more powerful.

Proposition 1 *The number of representations of a positive integer m as a sum of 4 squares of integers is equal to 8 times the sum of those positive divisors of m which are not divisible by* 4.

Proof From the series expansion

$$\theta_{00}(0) = \sum_{n \in \mathbb{Z}} q^{n^2}$$

we obtain

$$\theta_{00}^4(0) = \sum_{n_1,\dots,n_4 \in \mathbb{Z}} q^{n_1^2 + \dots + n_4^2} = 1 + \sum_{m \geq 1} r_4(m) q^m,$$

where $r_4(m)$ is the number of solutions in integers $n_1, \dots, n_4$ of the equation

$$n_1^2 + \dots + n_4^2 = m.$$

We will prove the result by comparing this with another expression for $\theta_{00}^4(0)$.
    We can write equation (43) of Chapter XII in the form

$$\theta_{10}'(v)/\theta_{10}(v) - \theta_{01}'(v)/\theta_{01}(v) = \pi i \theta_{00}^2(0)\theta_{00}(v)\theta_{11}(v)/\theta_{01}(v)\theta_{10}(v).$$

Differentiating with respect to v and then putting $v = 0$, we obtain

$$\theta_{10}''(0)/\theta_{10}(0) - \theta_{01}''(0)/\theta_{01}(0) = \pi i \theta_{00}^3(0)\theta_{11}'(0)/\theta_{01}(0)\theta_{10}(0) = -\pi^2 \theta_{00}^4(0),$$

W.A. Coppel, *Number Theory: An Introduction to Mathematics*, Universitext,
DOI: 10.1007/978-0-387-89486-7_13, © Springer Science + Business Media, LLC 2009

by (36) of Chapter XII. Since the theta functions are all solutions of the partial differential equation

$$\partial^2 y/\partial v^2 = -4\pi^2 q \partial y/\partial q,$$

the last relation can be written in the form

$$4q\partial/\partial q \log\{\theta_{10}(0)/\theta_{01}(0)\} = \theta_{00}^4(0).$$

On the other hand, the product expansions of the theta functions show that

$$\theta_{10}(0)/\theta_{01}(0) = 2q^{1/4} \prod_{n\geq 1}(1+q^{2n})^2 \Big/ \prod_{n\geq 1}(1-q^{2n-1})^2$$

$$= 2q^{1/4} \prod_{n\geq 1}(1-q^{4n})^2 \Big/ \prod_{n\geq 1}(1-q^{2n})^2(1-q^{2n-1})^2$$

$$= 2q^{1/4} \prod_{n\geq 1}(1-q^{4n})^2(1-q^n)^{-2}.$$

Differentiating logarithmically, we obtain

$$\theta_{00}^4(0) = 4q\partial/\partial q \log\{\theta_{10}(0)/\theta_{01}(0)\}$$

$$= 1 + 8\sum_{n\geq 1} nq^n/(1-q^n) - 8\sum_{n\geq 1} 4nq^{4n}/(1-q^{4n})$$

$$= 1 + 8\sum_{n\geq 1}\sum_{k\geq 1}(nq^{kn} - 4nq^{4kn})$$

$$= 1 + 8\sum_{m\geq 1}\{\sigma(m) - \sigma'(m)\}q^m,$$

where $\sigma(m)$ is the sum of all positive divisors of m and $\sigma'(m)$ is the sum of all positive divisors of m which are divisible by 4. Since the coefficients in a power series expansion are uniquely determined, it follows that

$$r_4(m) = 8\{\sigma(m) - \sigma'(m)\}. \qquad \square$$

Proposition 1 may also be restated in the form: the number of representations of m as a sum of 4 squares is equal to 8 times the sum of the odd positive divisors of m if m is odd, and 24 times this sum if m is even. For example,

$$r_4(10) = 24(1+5) = 144.$$

Since any positive integer has the odd positive divisor 1, Proposition 1 provides a new proof of Proposition II.40.

The number of representations of a positive integer as a sum of 2 squares may be treated in the same way, as Jacobi also showed (or, alternatively, by developing further the arithmetic of Gaussian integers):

Proposition 2 *The number of representations of a positive integer m as a sum of 2 squares of integers is equal to 4 times the excess of the number of positive divisors of m of the form $4h + 1$ over the number of positive divisors of the form $4h + 3$.*

Proof We have

$$\theta_{00}^2(0) = \sum_{n_1, n_2 \in \mathbb{Z}} q^{n_1^2 + n_2^2} = 1 + \sum_{m \geq 1} r_2(m) q^m,$$

where $r_2(m)$ is the number of solutions in integers n_1, n_2 of the equation

$$n_1^2 + n_2^2 = m.$$

To obtain another expression for $\theta_{00}^2(0)$ we use again the relation

$$\theta_{10}'(v)/\theta_{10}(v) - \theta_{01}'(v)/\theta_{01}(v) = \pi i \theta_{00}^2(0)\theta_{00}(v)\theta_{11}(v)/\theta_{01}(v)\theta_{10}(v),$$

but this time we simply take $v = 1/4$. Since

$$\theta_{01}(1/4) = \sum_{n \in \mathbb{Z}}(-i)^n q^{n^2} = \sum_{n \in \mathbb{Z}} i^{-n} q^{n^2} = \theta_{00}(1/4),$$

and similarly $\theta_{11}(1/4) = i\,\theta_{10}(1/4)$, we obtain

$$\pi \theta_{00}^2(0) = \theta_{01}'(1/4)/\theta_{01}(1/4) - \theta_{10}'(1/4)/\theta_{10}(1/4).$$

By differentiating logarithmically the product expansion for $\theta_{10}(v)$ and then putting $v = 1/4$, we get

$$\theta_{10}'(1/4)/\theta_{10}(1/4) = -\pi - 4\pi \sum_{n \geq 1} q^{2n}/(1 + q^{4n}).$$

Similarly, by differentiating logarithmically the product expansion for $\theta_{01}(v)$ and then putting $v = 1/4$, we get

$$\theta_{01}'(1/4)/\theta_{01}(1/4) = 4\pi \sum_{n \geq 1} q^{2n-1}/(1 + q^{4n-2}).$$

Thus

$$\theta_{01}'(1/4)/\theta_{01}(1/4) - \theta_{10}'(1/4)/\theta_{10}(1/4) = \pi + 4\pi \sum_{n \geq 1} q^n/(1 + q^{2n})$$

and hence

$$\theta_{00}^2(0) = 1 + 4 \sum_{n \geq 1} q^n/(1 + q^{2n}).$$

Since

$$q^n/(1 + q^{2n}) = q^n(1 - q^{2n})/(1 - q^{4n}) = (q^n - q^{3n}) \sum_{k \geq 0} q^{4kn},$$

it follows that

$$\theta_{00}^2(0) = 1 + 4 \sum_{n \geq 1} \sum_{k \geq 0} \{q^{(4k+1)n} - q^{(4k+3)n}\}$$

$$= 1 + 4 \sum_{m \geq 1} \{d_1(m) - d_3(m)\}q^m,$$

where $d_1(m)$ and $d_3(m)$ are respectively the number of positive divisors of m congruent to 1 and 3 mod 4. Hence

$$r_2(m) = 4\{d_1(m) - d_3(m)\}. \qquad \square$$

From Proposition 2 we immediately obtain again that any prime $p \equiv 1 \bmod 4$ may be represented as a sum of 2 squares and that the representation is essentially unique. Proposition II.39 may also be rederived.

The number $r_s(m)$ of representations of a positive integer m as a sum of s squares has been expressed by explicit formulas for many other values of s besides 2 and 4. Systematic ways of attacking the problem are provided by the theory of modular forms and the circle method of Hardy, Ramanujan and Littlewood.

2 Partitions

A *partition* of a positive integer n is a set of positive integers with sum n. For example, $\{2, 1, 1\}$ is a partition of 4. We denote the number of distinct partitions of n by $p(n)$. For example, $p(4) = 5$, since all partitions of 4 are given by

$$\{4\}, \{3, 1\}, \{2, 2\}, \{2, 1, 1\}, \{1, 1, 1, 1\}.$$

It was shown by Euler (1748) that the sequence $p(n)$ has a simple *generating function*:

Proposition 3 *If* $|x| < 1$, *then*

$$1/(1 - x)(1 - x^2)(1 - x^3) \cdots = 1 + \sum_{n \geq 1} p(n)x^n.$$

Proof If $|x| < 1$, then the infinite product $\prod_{m \geq 1}(1 - x^m)$ converges and its reciprocal has a convergent power series expansion. To determine the coefficients of this expansion note that, since

$$(1 - x^m)^{-1} = \sum_{k \geq 0} x^{km},$$

the coefficient of $x^n (n \geq 1)$ in the product $\prod_{m \geq 1}(1 - x^m)^{-1}$ is the number of representations of n in the form

$$n = 1k_1 + 2k_2 + \cdots,$$

where the k_j are non-negative integers. But this number is precisely $p(n)$, since any partition is determined by the number of 1's, 2's, ... that it contains. $\qquad \square$

For many purposes the discussion of convergence is superfluous and Proposition 3 may be regarded simply as a relation between formal products and formal power series.

Euler also obtained an interesting counterpart to Proposition 3, which we will derive from Jacobi's triple product formula.

Proposition 4 *If $|x| < 1$, then*

$$(1 - x)(1 - x^2)(1 - x^3) \cdots = \sum_{m \in \mathbb{Z}} (-1)^m x^{m(3m+1)/2}.$$

Proof If we take $q = x^{3/2}$ and $z = -x^{1/2}$ in Proposition XII.2, we obtain at once the result, since

$$\prod_{n \geq 1} (1 - x^{3n})(1 - x^{3n-1})(1 - x^{3n-2}) = \prod_{k \geq 1} (1 - x^k). \qquad \square$$

Proposition 4 also has a combinatorial interpretation. The coefficient of $x^n (n \geq 1)$ in the power series expansion of $\prod_{k \geq 1} (1 - x^k)$ is

$$s_n = \sum (-1)^v,$$

where the sum is over all partitions of n into *unequal* parts and v is the number of parts in the partition. In other words,

$$s_n = p_e^*(n) - p_o^*(n),$$

where $p_e^*(n)$, resp. $p_o^*(n)$, is the number of partitions of the positive integer n into an even, resp. odd, number of unequal parts. On the other hand,

$$\sum_{m \in \mathbb{Z}} (-1)^m x^{m(3m+1)/2} = 1 + \sum_{m \geq 1} (-1)^m \{x^{m(3m+1)/2} + x^{m(3m-1)/2}\}.$$

Thus Proposition 4 says that $p_e^*(n) = p_o^*(n)$ unless $n = m(3m \pm 1)/2$ for some $m \in \mathbb{N}$, in which case $p_e^*(n) - p_o^*(n) = (-1)^m$.

From Propositions 3 and 4 we obtain

$$\left[1 + \sum_{m \geq 1} (-1)^m \{x^{m(3m+1)/2} + x^{m(3m-1)/2}\}\right]\left[1 + \sum_{k \geq 1} p(k)x^k\right] = 1.$$

Multiplying out on the left side and equating to zero the coefficient of $x^n (n \geq 1)$, we obtain the recurrence relation:

$$\begin{aligned}
p(n) = \ &p(n - 1) + p(n - 2) - p(n - 5) - p(n - 7) \\
&+ \cdots + (-1)^{m-1} p(n - m(3m - 1)/2) \\
&+ (-1)^{m-1} p(n - m(3m + 1)/2) + \cdots,
\end{aligned}$$

where $p(0) = 1$ and $p(k) = 0$ for $k < 0$. This recurrence relation is quite an efficient way of calculating $p(n)$. It was used by MacMahon (1918) to calculate $p(n)$ for $n \leq 200$.

In the same way that we proved Proposition 3 we may show that, if $|x| < 1$, then

$$1/(1 - x)(1 - x^2) \cdots (1 - x^m) = 1 + \sum_{n \geq 1} p_m(n) x^n,$$

where $p_m(n)$ is the number of partitions of n into parts not exceeding m.

From the vast number of formulas involving partitions and their generating functions we select only one more pair, the celebrated *Rogers–Ramanujan identities*. The proof of these identities will be based on the following preliminary result:

Proposition 5 *If $|q| < 1$ and $|x| < |q|^{-1}$, then*

$$1 + \sum_{n \geq 1} x^n q^{n^2}/(q)_n = \sum_{n \geq 0} (-1)^n x^{2n} q^{5n(n+1)/2 - 2n} \{1 - x^2 q^{2(2n+1)}\}/(q)_n (xq^{n+1})_\infty,$$

where $(a)_0 = 1$,

$$(a)_n = (1 - a)(1 - aq) \cdots (1 - aq^{n-1}) \quad \text{if } n \geq 1, \quad \text{and}$$
$$(a)_\infty = (1 - a)(1 - aq)(1 - aq^2) \cdots .$$

Proof Consider the *q-difference equation*

$$f(x) = f(xq) + xq\, f(xq^2).$$

A formal power series $\sum_{n \geq 0} a_n x^n$ satisfies this equation if and only if

$$a_n(1 - q^n) = a_{n-1} q^{2n-1} \quad (n \geq 1).$$

Thus the only formal power series solution with $a_0 = 1$ is

$$f(x) = 1 + xq/(1 - q) + x^2 q^4/(1 - q)(1 - q^2)$$
$$+ x^3 q^9/(1 - q)(1 - q^2)(1 - q^3) + \cdots .$$

Moreover, if $|q| < 1$, this power series converges for all $x \in \mathbb{C}$.

If $|q| < 1$, the functions

$$F(x) = \sum_{n \geq 0} (-1)^n x^{2n} q^{5n(n+1)/2 - 2n} \{1 - x^2 q^{2(2n+1)}\}/(q)_n (xq^{n+1})_\infty,$$

$$G(x) = \sum_{n \geq 0} (-1)^n x^{2n} q^{5n(n+1)/2 - n} \{1 - xq^{2n+1}\}/(q)_n (xq^{n+1})_\infty$$

are holomorphic for $|x| < |q|^{-1}$.

We have

$$F(x) - G(x)$$

$$= \sum_{n \geq 0} (-1)^n x^{2n} q^{5n(n+1)/2} \{q^{-2n} - x^2 q^{2(n+1)} - q^{-n} + xq^{n+1}\}/(q)_n (xq^{n+1})_\infty$$

$$= \sum_{n \geq 0} (-1)^n x^{2n} q^{5n(n+1)/2} \{q^{-2n}(1 - q^n) + xq^{n+1}(1 - xq^{n+1})\}/(q)_n (xq^{n+1})_\infty$$

$$= \sum_{n \geq 1} (-1)^n x^{2n} q^{5n(n+1)/2 - 2n}/(q)_{n-1} (xq^{n+1})_\infty$$

$$\quad + xq \sum_{n \geq 0} (-1)^n x^{2n} q^{5n(n+1)/2 + n}/(q)_n (xq^{n+2})_\infty$$

$$= -x^2 q^3 \sum_{n \geq 0} (-1)^n x^{2n} q^{5n(n+1)/2 + 3n}/(q)_n (xq^{n+2})_\infty$$

$$\quad + xq \sum_{n \geq 0} (-1)^n x^{2n} q^{5n(n+1)/2 + n}/(q)_n (xq^{n+2})_\infty$$

$$= xq \sum_{n \geq 0} (-1)^n (xq)^{2n} q^{5n(n+1)/2 - n} \{1 - (xq)q^{2n+1}\}/(q)_n (xq^{n+2})_\infty$$

$$= xq G(xq).$$

Similarly,

$$G(x) = \sum_{n \geq 0} (-1)^n x^{2n} q^{5n(n+1)/2} \{q^{-n} - xq^{n+1}\}/(q)_n (xq^{n+1})_\infty$$

$$= \sum_{n \geq 0} (-1)^n x^{2n} q^{5n(n+1)/2} \{q^{-n}(1 - q^n) + 1 - xq^{n+1}\}/(q)_n (xq^{n+1})_\infty$$

$$= \sum_{n \geq 1} (-1)^n x^{2n} q^{5n(n+1)/2 - n}/(q)_{n-1} (xq^{n+1})_\infty$$

$$\quad + \sum_{n \geq 0} (-1)^n x^{2n} q^{5n(n+1)/2}/(q)_n (xq^{n+2})_\infty$$

$$= \sum_{n \geq 0} (-1)^n (xq)^{2n} q^{5n(n+1)/2 - 2n} \{1 - (xq)^2 q^{2(2n+1)}\}/(q)_n (xq^{n+2})_\infty$$

$$= F(xq).$$

Combining this with the previous relation, we obtain

$$F(x) = F(xq) + xq\, F(xq^2).$$

But we have seen that this q-difference equation has a unique holomorphic solution $f(x)$ such that $f(0) = 1$. Hence $F(x) = f(x)$. $\qquad\square$

The Rogers–Ramanujan identities may now be easily derived:

Proposition 6 *If $|q| < 1$, then*

$$\sum_{n \geq 0} q^{n^2}/(1 - q)(1 - q^2) \cdots (1 - q^n) = \prod_{m \geq 0} (1 - q^{5m+1})^{-1}(1 - q^{5m+4})^{-1},$$

$$\sum_{n \geq 0} q^{n(n+1)}/(1 - q)(1 - q^2) \cdots (1 - q^n) = \prod_{m \geq 0} (1 - q^{5m+2})^{-1}(1 - q^{5m+3})^{-1}.$$

Proof Put $P = \prod_{k \geq 1}(1 - q^k)$. By Proposition 5 and its proof we have

$$\sum_{n \geq 0} q^{n^2}/(1 - q)(1 - q^2) \cdots (1 - q^n) = F(1)$$

$$= \left[1 + \sum_{n \geq 1}(-1)^n\{q^{n(5n+1)/2} + q^{n(5n-1)/2}\}\right]\bigg/ P$$

and, since $F(q) = G(1)$,

$$\sum_{n \geq 0} q^{n(n+1)}/(1 - q)(1 - q^2) \cdots (1 - q^n) = F(q)$$

$$= \left[1 + \sum_{n \geq 1}(-1)^n\{q^{n(5n+3)/2} + q^{n(5n-3)/2}\}\right]\bigg/ P.$$

On the other hand, by replacing q by $q^{5/2}$ and z by $-q^{1/2}$, resp. $-q^{3/2}$, in Jacobi's triple product formula (Proposition XII.2), we obtain

$$\sum_{n \in \mathbb{Z}}(-1)^n q^{n(5n+1)/2} = \prod_{m \geq 1}(1 - q^{5m})(1 - q^{5m-2})(1 - q^{5m-3})$$

$$= P \bigg/ \prod_{m \geq 0}(1 - q^{5m+1})(1 - q^{5m+4})$$

and

$$\sum_{n \in \mathbb{Z}}(-1)^n q^{n(5n+3)/2} = \prod_{m \geq 1}(1 - q^{5m})(1 - q^{5m-1})(1 - q^{5m-4})$$

$$= P \bigg/ \prod_{m \geq 0}(1 - q^{5m+2})(1 - q^{5m+3}).$$

Combining these relations with the previous ones, we obtain the result. $\quad\square$

The combinatorial interpretation of the Rogers–Ramanujan identities was pointed out by MacMahon (1916). The first identity says that the number of partitions of a positive integer n into parts congruent to $\pm 1 \bmod 5$ is equal to the number of partitions of n into parts that differ by at least 2. The second identity says that the number of partitions of a positive integer n into parts congruent to $\pm 2 \bmod 5$ is equal to the number of partitions of n into parts greater than 1 that differ by at least 2.

A remarkable application of the Rogers–Ramanujan identities to the hard hexagon model of statistical mechanics was found by Baxter (1981). Many other models in statistical mechanics have been exactly solved with the aid of theta functions. A unifying principle is provided by the vast theory of infinite-dimensional Lie algebras which has been developed over the past 25 years.

The number $p(n)$ of partitions of n increases rapidly with n. It was first shown by Hardy and Ramanujan (1918) that

$$p(n) \sim e^{\pi \sqrt{2n/3}}/4n\sqrt{3} \quad \text{as } n \to \infty.$$

They further obtained an asymptotic series for $p(n)$, which was modified by Rademacher (1937) into a convergent series, from which it is even possible to calculate $p(n)$ exactly. A key role in the difficult proof is played by the behaviour under transformations of the modular group of *Dedekind's eta function*

$$\eta(\tau) = q^{1/12} \prod_{k \geq 1} (1 - q^{2k}),$$

where $q = e^{\pi i \tau}$ and $\tau \in \mathscr{H}$ (the upper half-plane).

The paper of Hardy and Ramanujan contained the first use of the 'circle method', which was subsequently applied by Hardy and Littlewood to a variety of problems in analytic number theory.

3 Cubic Curves

We define an *affine plane curve* over a field K to be a polynomial $f(X, Y)$ in two indeterminates with coefficients from K, but we regard two polynomials $f(X, Y)$ and $f^*(X, Y)$ as defining the same affine curve if $f^* = \lambda f$ for some nonzero $\lambda \in K$. The *degree* of the curve is defined without ambiguity to be the degree of the polynomial f.

If

$$f(X, Y) = aX + bY + c$$

is a polynomial of degree 1, the curve is said to be an *affine line*. If

$$f(X, Y) = aX^2 + bXY + cY^2 + lX + mY + n$$

is a polynomial of degree 2, the curve is said to be an *affine conic*. If $f(X, Y)$ is a polynomial of degree 3, the curve is said to be an *affine cubic*. It is the cubic case in which we will be most interested.

Let $\mathscr{C}$ be an affine plane curve over the field K, defined by the polynomial $f(X, Y)$. We say that $(x, y) \in K^2$ is a *point* or, more precisely, a *K-point* of the affine curve $\mathscr{C}$ if $f(x, y) = 0$. The K-point (x, y) is said to be *non-singular* if there exist $a, b \in K$, not both zero, such that

$$f(x + X, y + Y) = aX + bY + \cdots,$$

where all unwritten terms have degree > 1. Since a, b are uniquely determined by f, we can define the *tangent* to the affine curve $\mathscr{C}$ at the non-singular point (x, y) to be the affine line

$$\ell(X, Y) = aX + bY - (ax + by).$$

It is easily seen that these definitions do not depend on the choice of polynomial within an equivalence class $\{\lambda f : 0 \neq \lambda \in K\}$.

The study of the asymptotes of an affine plane curve leads one to consider also its 'points at infinity', the asymptotes being the tangents at these points. We will now make this precise.

If the polynomial $f(X, Y)$ has degree d, then

$$F(X, Y, Z) = Z^d f(X/Z, Y/Z)$$

is a homogeneous polynomial of degree d such that

$$f(X, Y) = F(X, Y, 1).$$

Furthermore, if $\mathscr{F}(X, Y, Z)$ is any homogeneous polynomial such that $f(X, Y) = \mathscr{F}(X, Y, 1)$, then $\mathscr{F}(X, Y, Z) = Z^m F(X, Y, Z)$ for some non-negative integer m.

We define a *projective plane curve* over a field K to be a homogeneous polynomial $F(X, Y, Z)$ of degree $d > 0$ in three indeterminates with coefficients from K, but we regard two homogeneous polynomials $F(X, Y, Z)$ and $F^*(X, Y, Z)$ as defining the same projective curve if $F^* = \lambda F$ for some nonzero $\lambda \in K$. The projective curve is said to be a *projective line, conic* or *cubic* if F has degree $1, 2$ or 3 respectively.

If $\mathscr{C}$ is an affine plane curve, defined by a polynomial $f(X, Y)$ of degree $d > 0$, the projective plane curve $\bar{\mathscr{C}}$, defined by the homogeneous polynomial $Z^d f(X/Z, Y/Z)$ of the same degree, is called the *projective completion* of $\mathscr{C}$. Thus the projective completion of an affine line, conic or cubic is respectively a projective line, conic or cubic.

Let $\bar{\mathscr{C}}$ be a projective plane curve over the field K, defined by the homogeneous polynomial $F(X, Y, Z)$. We say that $(x, y, z) \in K^3$ is a *point*, or *K-point*, of $\mathscr{C}$ if $(x, y, z) \neq (0, 0, 0)$ and $F(x, y, z) = 0$, but we regard two triples (x, y, z) and (x^*, y^*, z^*) as defining the same K-point if

$$x^* = \lambda x, y^* = \lambda y, z^* = \lambda z \quad \text{for some nonzero } \lambda \in K.$$

If $\bar{\mathscr{C}}$ is the projective completion of the affine plane curve $\mathscr{C}$, then a point (x, y, z) of $\bar{\mathscr{C}}$ with $z \neq 0$ corresponds to a point $(x/z, y/z)$ of $\mathscr{C}$, and a point $(x, y, 0)$ of $\bar{\mathscr{C}}$ corresponds to a *point at infinity* of $\mathscr{C}$.

The K-point (x, y, z) of the projective plane curve defined by the homogeneous polynomial $F(X, Y, Z)$ is said to be *non-singular* if there exist $a, b, c \in K$, not all zero, such that

$$F(x + X, y + Y, z + Z) = aX + bY + cZ + \cdots,$$

where all unwritten terms have degree > 1. Since a, b, c are uniquely determined by F, we can define the *tangent* to the projective curve at the non-singular point (x, y, z)

to be the projective line defined by $aX + bY + cZ$. It follows from Euler's theorem on homogeneous functions that (x, y, z) is itself a point of the tangent.

It is easily seen that if $\bar{\mathscr{C}}$ is the projective completion of an affine plane curve $\mathscr{C}$, and if $z \neq 0$, then (x, y, z) is a non-singular point of $\bar{\mathscr{C}}$ if and only if $(x/z, y/z)$ is a non-singular point of $\mathscr{C}$. Moreover, if the tangent to $\bar{\mathscr{C}}$ at (x, y, z) is the projective line

$$\bar{\ell}(X, Y, Z) = aX + bY + cZ,$$

then the tangent to $\mathscr{C}$ at $(x/z, y/z)$ is the affine line defined by

$$\ell(X, Y) = aX + bY + c.$$

Let $\mathscr{C}$ be an affine plane curve over the field K, defined by the polynomial $f(X, Y)$, and let (x, y) be a non-singular K-point of $\mathscr{C}$. Then we can write

$$f(x + X, y + Y) = aX + bY + f_2(X, Y) + \cdots,$$

where a, b are not both zero, $f_2(X, Y)$ is a homogeneous polynomial of degree 2, and all unwritten terms have degree > 2. The non-singular point (x, y) is said to be an *inflection point* or, more simply, a *flex* of $\mathscr{C}$ if $f_2(X, Y)$ is divisible by $aX + bY$.

Similarly we can define a flex for a projective plane curve. Let (x, y, z) be a non-singular point of the projective plane curve over the field K, defined by the homogeneous polynomial $F(X, Y, Z)$. Then we can write

$$F(x + X, y + Y, z + Z) = aX + bY + cZ + F_2(X, Y, Z) + \cdots,$$

where a, b, c are not all zero, $F_2(X, Y, Z)$ is a homogeneous polynomial of degree 2, and all unwritten terms have degree > 2. The non-singular point (x, y, z) is said to be a *flex* if $F_2(X, Y, Z)$ is divisible by $aX + bY + cZ$.

Two more definitions are required before we embark on our study of cubic curves. A projective curve over the field K, defined by the homogeneous polynomial $F(X, Y, Z)$ of degree $d > 0$, is said to be reducible over K if

$$F(X, Y, Z) = F_1(X, Y, Z)F_2(X, Y, Z),$$

where F_1 and F_2 are homogeneous polynomials of degree less than d with coefficients from K. The K-points of the curve defined by F are then just the K-points of the curve defined by F_1, together with the K-points of the curve defined by F_2. A curve is said to be *irreducible over K* if it is not reducible over K.

Two projective curves over the field K, defined by the homogeneous polynomials $F(X, Y, Z)$ and $G(X', Y', Z')$, are said to be *projectively equivalent* if there exists an invertible linear transformation

$$X = a_{11}X' + a_{12}Y' + a_{13}Z'$$
$$Y = a_{21}X' + a_{22}Y' + a_{23}Z'$$
$$Z = a_{31}X' + a_{32}Y' + a_{33}Z'$$

with coefficients $a_{ij} \in K$ such that

$$F(a_{11}X' + \cdots, a_{21}X' + \cdots, a_{31}X' + \cdots) = G(X', Y', Z').$$

It is clear that F and G necessarily have the same degree, and that projective equivalence is in fact an equivalence relation.

Consider now the affine cubic curve $\mathscr{C}$ defined by the polynomial

$$f(X, Y) = a_{30}X^3 + a_{21}X^2Y + a_{12}XY^2 + a_{03}Y^3 + a_{20}X^2 + a_{11}XY$$
$$+ a_{02}Y^2 + a_{10}X + a_{01}Y + a_{00}.$$

We assume that $\mathscr{C}$ has a non-singular K-point which is a flex. Without loss of generality, suppose that this is the origin. Then $a_{00} = 0$, a_{10} and a_{01} are not both zero, and

$$a_{20}X^2 + a_{11}XY + a_{02}Y^2 = (a_{10}X + a_{01}Y)(a'_{10}X + a'_{01}Y)$$

for some $a'_{10}, a'_{01} \in K$. By an invertible linear change of variables we may suppose that $a_{10} = 0$, $a_{01} = 1$. Then f has the form

$$f(X, Y) = Y + a_1XY + a_3Y^2 - a_0X^3 - a_2X^2Y - a_4XY^2 - a_6Y^3.$$

If $a_0 = 0$, then f is divisible by Y and the corresponding projective curve is reducible. Thus we now assume $a_0 \neq 0$. In fact we may assume $a_0 = 1$, by replacing f by a constant multiple and then scaling Y. The projective completion $\bar{\mathscr{C}}$ of $\mathscr{C}$ is now defined by the homogeneous polynomial

$$YZ^2 + a_1XYZ + a_3Y^2Z - X^3 - a_2X^2Y - a_4XY^2 - a_6Y^3.$$

If we interchange Y and Z, the flex becomes the unique point at infinity of the affine cubic curve defined by the polynomial

$$Y^2 + a_1XY + a_3Y - (X^3 + a_2X^2 + a_4X + a_6).$$

This can be further simplified by making mild restrictions on the field K. If K has characteristic $\neq 2$, i.e. if $1 + 1 \neq 0$, then by replacing Y by $(Y - a_1X - a_3)/2$ we obtain the cubic curve defined by the polynomial

$$Y^2 - (4X^3 + b_2X^2 + 2b_4X + b_6).$$

If K also has characteristic $\neq 3$, i.e. if $1 + 1 + 1 \neq 0$, then by replacing X by $(X - 3b_2)/6^2$ and Y by $2Y/6^3$, we obtain the cubic curve defined by the polynomial $Y^2 - (X^3 + aX + b)$. Thus we have proved:

Proposition 7 *If a projective cubic curve over the field K is irreducible and has a non-singular K-point which is a flex, then it is projectively equivalent to the projective completion $\mathscr{W} = \mathscr{W}(a_1, \ldots, a_6)$ of an affine curve of the form*

$$Y^2 + a_1XY + a_3Y - (X^3 + a_2X^2 + a_4X + a_6).$$

If K has characteristic $\neq 2, 3$, then it is projectively equivalent to the projective completion $\mathscr{C} = \mathscr{C}_{a,b}$ of an affine curve of the form

$$Y^2 - (X^3 + aX + b).$$

It is easily seen that, conversely, for any choice of $a_1, \ldots, a_6 \in K$ the curve $\mathscr{W}$, and in particular $\mathscr{C}_{a,b}$, is irreducible over K and that $\mathbf{0}$, the unique point at infinity, is a flex. For any $u, r, s, t \in K$ with $u \neq 0$, the invertible linear change of variables

$$X = u^2 X' + r,$$
$$Y = u^3 Y' + su^2 X' + t$$

replaces the curve $\mathscr{W} = \mathscr{W}(a_1, \ldots, a_6)$ by a curve $\mathscr{W}' = \mathscr{W}'(a_1', \ldots, a_6')$ of the same form. The numbering of the coefficients reflects the fact that if $r = s = t = 0$, then

$$a_1 = ua_1', \quad a_2 = u^2 a_2', \quad a_3 = u^3 a_3',$$
$$a_4 = u^4 a_4', \quad a_6 = u^6 a_6'.$$

In particular, for any nonzero $u \in K$, the invertible linear change of variables

$$X = u^2 X',$$
$$Y = u^3 Y'$$

replaces $\mathscr{C}_{a,b}$ by $\mathscr{C}_{a',b'}$, where

$$a = u^4 a',$$
$$b = u^6 b'.$$

By replacing X by $x + X$ and Y by $y + Y$, we see that if a K-point (x, y) of $\mathscr{C}_{a,b}$ is singular, then

$$3x^2 + a = y = 0,$$

which implies $4a^3 + 27b^2 = 0$. Thus the curve $\mathscr{C}_{a,b}$ has no singular points if $4a^3 + 27b^2 \neq 0$.

We will call

$$d := 4a^3 + 27b^2$$

the *discriminant* of the curve $\mathscr{C}_{a,b}$. It is not difficult to verify that if the cubic polynomial $X^3 + aX + b$ has roots e_1, e_2, e_3, then

$$d = -[(e_1 - e_2)(e_1 - e_3)(e_2 - e_3)]^2.$$

If $d = 0$, $a \neq 0$, then the polynomial $X^3 + aX + b$ has the repeated root $x_0 = -3b/2a$ and $P = (x_0, 0)$ is the unique singular point. If $d = a = 0$, then $b = 0$ and $P = (0, 0)$ is the unique singular point.

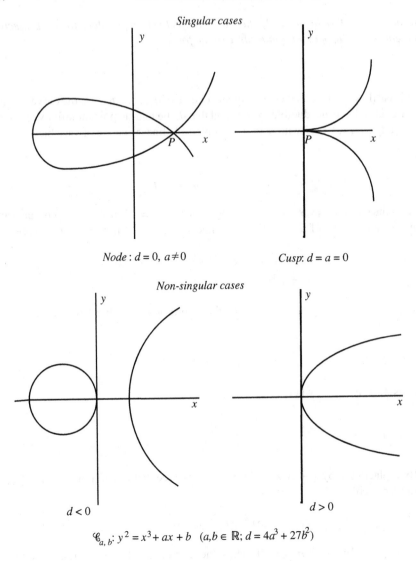

Singular cases

Node : $d = 0$, $a \neq 0$ Cusp: $d = a = 0$

Non-singular cases

$d < 0$ $d > 0$

$\mathscr{C}_{a,b}$: $y^2 = x^3 + ax + b$ $(a,b \in \mathbb{R};\ d = 4a^3 + 27b^2)$

Fig. 1. Cubic curves over $\mathbb{R}$.

The different types of curve which arise when $K = \mathbb{R}$ is the field of real numbers are illustrated in Figure 1. The unique point at infinity $\mathbf{0}$ may be thought of as being at both ends of the y-axis. (In the case of a node, Figure 1 illustrates the situation for $x_0 > 0$. For $x_0 < 0$ the singular point is an isolated point of the curve.)

Suppose now that K is any field of characteristic $\neq 2, 3$ and that the curve $\mathscr{C}_{a,b}$ has zero discriminant. Because of the geometrical interpretation when $K = \mathbb{R}$, the unique singular point of the curve $\mathscr{C}_{a,b}$ is said to be a *node* if $a \neq 0$ and a *cusp* if $a = 0$. In the cusp case, if we put $T = Y/X$, then the cubic curve has the parametrization

$X = T^2, Y = T^3$. In the node case, if we put $T = Y/(X + 3b/2a)$, then it has the parametrization

$$X = T^2 + 3b/a, Y = T^3 + 9bT/2a.$$

Thus in both cases the cubic curve is in fact elementary.

We now restrict attention to non-singular cubic curves, i.e. curves which do not have a singular point.

Two K-points of a projective cubic curve determine a projective line, which intersects the curve in a third K-point. This procedure for generating additional K-points was used implicitly by Diophantus and explicitly by Newton. There is also another procedure, which may be regarded as a limiting case: the tangent to a projective cubic curve at a K-point intersects the curve in another K-point. The combination of the two procedures is known as the 'chord and tangent' process. It will now be described analytically for the cubic curve $\mathscr{C}_{a,b}$.

If O is the unique point at infinity of the cubic curve $\mathscr{C}_{a,b}$ and if $P = (x, y)$ is any finite K-point, then the affine line determined by O and P is $X - x$ and its other point of intersection with $\mathscr{C}_{a,b}$ is $P^* = (x, -y)$.

Now let $P_1 = (x_1, y_1)$ and $P_2 = (x_2, y_2)$ be any two finite K-points. If $x_1 \neq x_2$, then the affine line determined by P_1 and P_2 is

$$Y - mX - c,$$

where

$$m = (y_2 - y_1)/(x_2 - x_1), \quad c = (y_1 x_2 - y_2 x_1)/(x_2 - x_1),$$

and its third point of intersection with $\mathscr{C}_{a,b}$ is $P_3 = (x_3, y_3)$, where

$$x_3 = m^2 - x_1 - x_2, \quad y_3 = mx_3 + c.$$

If $x_1 = x_2$, but $y_1 \neq y_2$, then the affine line determined by P_1 and P_2 is $X - x_1$ and its other point of intersection with $\mathscr{C}_{a,b}$ is O. Finally, if $P_1 = P_2$, it may be verified that the tangent to $\mathscr{C}_{a,b}$ at P_1 is the affine line

$$Y - mX - c,$$

where

$$m = (3x_1^2 + a)/2y_1, \quad c = (-x_1^3 + ax_1 + 2b)/2y_1,$$

and its other point of intersection with $\mathscr{C}_{a,b}$ is the point $P_3 = (x_3, y_3)$, where x_3 and y_3 are given by the same formulas as before, but with the new values of m and c (and with $x_2 = x_1$).

It is rather remarkable that the K-points of a non-singular projective cubic curve can be given the structure of an abelian group. That this is possible is suggested by the addition theorem for elliptic functions.

Suppose that $K = \mathbb{C}$ is the field of complex numbers and that the cubic curve is the projective completion $\mathscr{C}_\lambda$ of the affine curve

$$Y^2 - g_\lambda(X),$$

where

$$g_\lambda(X) = 4\lambda X^3 - 4(1 + \lambda)X^2 + 4X$$

is Riemann's normal form and $\lambda \neq 0, 1$. If $S(u)$ is the elliptic function defined in §3 of Chapter XII, then $P(u) = (S(u), S'(u))$ is a point of $\mathscr{C}_\lambda$ for any $u \in \mathbb{C}$. If we define the sum of $P(u)$ and $P(v)$ to be the point $P(u + v)$, then the set of all $\mathbb{C}$-points of $\mathscr{C}_\lambda$ becomes an abelian group, with $P(0) = (0, 0)$ as identity element and with $P(-u) = (S(u), -S'(u))$ as the inverse of $P(u)$. In order to carry this construction over to the cubic curve $\mathscr{C}_{a,b}$ and to other fields than $\mathbb{C}$, we interpret it geometrically.

It was shown in (10) of Chapter XII that

$$S(u + v) = 4S(u)S(v)[S(v) - S(u)]^2/[S'(u)S(v) - S'(v)S(u)]^2.$$

The points $(x_1, y_1) = (S(u), S'(u))$ and $(x_2, y_2) = (S(v), S'(v))$ determine the affine line

$$Y - mX - c,$$

where

$$m = [S'(v) - S'(u)]/[S(v) - S(u)],$$
$$c = [S'(u)S(v) - S'(v)S(u)]/[S(v) - S(u)].$$

The third point of intersection of this line with the cubic $\mathscr{C}_\lambda$ is the point (x_3, y_3), where

$$\begin{aligned}x_3 &= c^2/4\lambda x_1 x_2 \\ &= [S'(u)S(v) - S'(v)S(u)]^2/4\lambda S(u)S(v)[S(v) - S(u)]^2 \\ &= 1/\lambda S(u + v).\end{aligned}$$

On the other hand, the points $(0, 0) = (S(0), S'(0))$ and $(x_3^*, y_3^*) = (S(u + v), S'(u + v))$ determine the affine line $Y - (y_3^*/x_3^*)X$ and its third point of intersection with $\mathscr{C}_\lambda$ is the point (x_4, y_4), where $x_4 = 1/\lambda x_3^* = x_3$. Evidently $y_4^2 = y_3^2$, and it may be verified that actually $y_4 = y_3$. Thus (x_3^*, y_3^*) is the third point of intersection with $\mathscr{C}_\lambda$ of the line determined by the points $(0, 0)$ and (x_3, y_3).

The origin $(0, 0)$ may not be a point of the cubic curve $\mathscr{C}_{a,b}$ but O, the point at infinity, certainly is. Consequently, as illustrated in Figure 2, we now define the *sum* $P_1 + P_2$ of two K-points P_1, P_2 of $\mathscr{C}_{a,b}$ to be the K-point P_3^*, where P_3 is the third point of $\mathscr{C}_{a,b}$ on the line determined by P_1, P_2 and P_3^* is the third point of $\mathscr{C}_{a,b}$ on the line determined by O, P_3. If $P_1 = P_2$, the line determined by P_1, P_2 is understood to mean the tangent to $\mathscr{C}_{a,b}$ at P_1.

It is simply a matter of elementary algebra to deduce from the formulas previously given that, if addition is defined in this way, the set of all K-points of $\mathscr{C}_{a,b}$ becomes an abelian group, with O as identity element and with $-P = (x, -y)$ as the inverse of $P = (x, y)$. Since $-P = P$ if and only if $y = 0$, the elements of order 2 in this group are the points $(x_0, 0)$, where x_0 is a root of the polynomial $X^3 + aX + b$ (if it has any roots in K).

Throughout the preceding discussion of cubic curves we restricted attention to those with a flex. It will now be shown that in a sense this is no restriction.

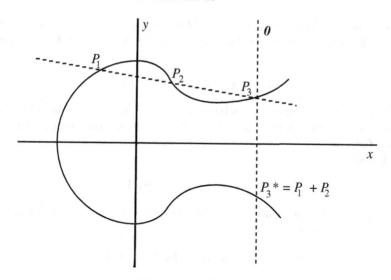

Fig. 2. Addition on $\mathscr{C}_{a,b}$.

Let $\mathscr{C}$ be a projective cubic curve over the field K, defined by the homogeneous polynomial $F_1(X, Y, Z)$, and suppose that $\mathscr{C}$ has a non-singular K-point P. Without loss of generality we assume that $P = (1, 0, 0)$ and that the tangent at P is the projective line Z. Then F_1 has no term in X^3 or in $X^2 Y$:

$$F_1(X, Y, Z) = aY^3 + bY^2 Z + cYZ^2 + dZ^3 + eX^2 Z + gXY^2 + hXZ^2.$$

Here $e \neq 0$, since P is non-singular, and we may suppose $g \neq 0$, since otherwise P is a flex. If we replace $gX + aY$ by X, this assumes the form

$$F_2(X, Y, Z) = XY^2 + bY^2 Z + cYZ^2 + dZ^3 + eX^2 Z + gXYZ + hXZ^2,$$

with new values for the coefficients. If we now replace $X + bZ$ by X, this assumes the form

$$F_3(X, Y, Z) = XY^2 + cYZ^2 + dZ^3 + eX^2 Z + gXYZ + hXZ^2,$$

again with new values for the coefficients. The projective cubic curve $\mathscr{D}$ over the field K, defined by the homogeneous polynomial

$$F_4(U, V, W) = VW^2 + cV^2 W + dUV^2 + eU^3 + gUVW + hU^2 V,$$

has a flex at the point $(0, 0, 1)$. Moreover,

$$F_3(U^2, VW, UV) = U^2 V F_4(U, V, W),$$
$$F_4(XZ, Z^2, XY) = XZ^2 F_3(X, Y, Z).$$

This shows that *any projective cubic curve over the field K with a non-singular K-point is* birationally equivalent *to one with a flex*.

Birational equivalence may be defined in the following way. A *rational transformation* of the projective plane with points $X = (X_1, X_2, X_3)$ is a map $X \to Y = \boldsymbol{\varphi}(X)$, where

$$\boldsymbol{\varphi}(X) = (\varphi_1(X), \varphi_2(X), \varphi_3(X))$$

and $\varphi_1, \varphi_2, \varphi_3$ are homogeneous polynomials without common factor of the same degree m, say. (In the corresponding affine plane the coordinates are transformed by *rational* functions.) The transformation is *birational* if there exists an inverse map $Y \to X = \boldsymbol{\psi}(Y)$, where

$$\boldsymbol{\psi}(Y) = (\psi_1(Y), \psi_2(Y), \psi_3(Y))$$

and ψ_1, ψ_2, ψ_3 are homogeneous polynomials without common factor of the same degree n, say, such that

$$\psi[\boldsymbol{\varphi}(X)] = \omega(X)X, \quad \varphi[\boldsymbol{\psi}(Y)] = \theta(Y)Y$$

for some scalar polynomials $\omega(X), \theta(Y)$. Two irreducible projective plane curves $\mathscr{C}$ and $\mathscr{D}$ over the field K, defined respectively by the homogeneous polynomials $F(X)$ and $G(Y)$ (not necessarily of the same degree), are *birationally equivalent* if there exists a birational transformation $Y = \varphi(X)$ with inverse $X = \psi(Y)$ such that $G[\varphi(X)]$ is divisible by $F(X)$ and $F[\psi(Y)]$ is divisible by $G[(Y)]$.

It is clear that birational equivalence is indeed an equivalence relation, and that irreducible projective curves which are projectively equivalent are also birationally equivalent. Birational transformations are often used to simplify the singular points of a curve. Indeed the theorem on *resolution of singularities* says that any irreducible curve is birationally equivalent to a non-singular curve, although it may be a curve in a higher-dimensional space rather than in the plane. The algebraic geometry of curves may be regarded as the study of those properties which are invariant under birational equivalence.

It was shown by Poincaré (1901) that any non-singular curve of *genus* 1 defined over the field $\mathbb{Q}$ of rational numbers and with at least one rational point is birationally equivalent over $\mathbb{Q}$ to a cubic curve. Such a curve is now said to be an *elliptic curve* (for the somewhat inadequate reason that it may be parametrized by elliptic functions over the field of complex numbers.) However, for our purposes it is sufficient to define an elliptic curve to be a non-singular cubic curve of the form $\mathscr{W}$, over a field K of arbitrary characteristic, or of the form $\mathscr{C}_{a,b}$, over a field K of characteristic $\neq 2, 3$.

4 Mordell's Theorem

We showed in the previous section that, for any field K of characteristic $\neq 2, 3$, the K-points of the elliptic curve $\mathscr{C}_{a,b}$ defined by the polynomial

$$Y^2 - X^3 - aX - b,$$

where $a, b \in K$ and $d := 4a^3 + 27b^2 \neq 0$, form an abelian group, $E(K)$ say. We now restrict our attention to the case when $K = \mathbb{Q}$ is the field of rational numbers, and we write simply $E := E(\mathbb{Q})$. This section is devoted to the basic theorem of Mordell (1922), which says that *the abelian group E is finitely generated.*

By replacing X by X/c^2 and Y by Y/c^3 for some nonzero $c \in \mathbb{Q}$, we may (and will) assume that a and b are both integers. Let $P = (x, y)$ be any finite rational point of $\mathscr{C}_{a,b}$ and write $x = p/q$, where p and q are coprime integers. The *height* $h(P)$ of P is uniquely defined by

$$h(P) = \log \max(|p|, |q|).$$

We also set $h(O) = 0$, where O is the unique point at infinity of $\mathscr{C}_{a,b}$.

Evidently $h(P) \geq 0$. Furthermore, $h(-P) = h(P)$, since $P = (x, y)$ implies $-P = (x, -y)$. Also, for any $r > 0$, there exist only finitely many elements $P = (x, y)$ of E with $h(P) \leq r$, since x determines y up to sign.

Proposition 8 *There exists a constant $C = C(a, b) > 0$ such that*

$$|h(2P) - 4h(P)| \leq C \quad \text{for all } P \in E.$$

Proof By the formulas given in §3, if $P = (x, y)$, then $2P = (x', y')$, where

$$x' = m^2 - 2x, \quad m = (3x^2 + a)/2y.$$

Since $y^2 = x^3 + ax + b$, it follows that

$$x' = (x^4 - 2ax^2 - 8bx + a^2)/4(x^3 + ax + b).$$

If $x = p/q$, where p and q are coprime integers, then $x' = p'/q'$, where

$$p' = p^4 - 2ap^2q^2 - 8bpq^3 + a^2q^4,$$
$$q' = 4q(p^3 + apq^2 + bq^3).$$

Evidently p' and q' are also integers, but they need not be coprime. However, since

$$p' = ep'', \quad q' = eq'',$$

where e, p'', q'' are integers and p'', q'' are coprime, we have

$$h(2P) = \log \max(|p''|, |q''|) \leq \log \max(|p'|, |q'|).$$

Since

$$\max(|p'|, |q'|) \leq \max(|p|, |q|)^4 \max\{1 + 2|a| + 8|b| + a^2, 4(1 + |a| + |b|)\},$$

it follows that

$$h(2P) \leq 4h(P) + C'$$

for some constant $C' = C'(a, b) > 0$.

The Euclidean algorithm may be used to derive the polynomial identity

$$(3X^2 + 4a)(X^4 - 2aX^2 - 8bX + a^2) - (3X^3 - 5aX - 27b)(X^3 + aX + b) = d,$$

where once again $d = 4a^3 + 27b^2$. Substituting p/q for X, we obtain

$$4dq^7 = 4(3p^2q + 4aq^3)p' - (3p^3 - 5apq^2 - 27bq^3)q'.$$

Similarly, the Euclidean algorithm may be used to derive the polynomial identity

$$f(X)(1 - 2aX^2 - 8bX^3 + a^2X^4) + g(X)X(1 + aX^2 + bX^3) = d,$$

where

$$f(X) = 4a^3 + 27b^2 - a^2bX + a(3a^3 + 22b^2)X^2 + 3b(a^3 + 8b^2)X^3,$$
$$g(X) = a^2b + a(5a^3 + 32b^2)X + 2b(13a^3 + 96b^2)X^2 - 3a^2(a^3 + 8b^2)X^3.$$

Substituting q/p for X, we obtain

$$4dp^7 = 4\{(4a^3 + 27b^2)p^3 - a^2bp^2q + (3a^4 + 22ab^2)pq^2 + 3(a^3b + 8b^3)q^3\}p'$$
$$+ \{a^2bp^3 + (5a^4 + 32ab^2)p^2q + (26a^3b + 192b^3)pq^2 - 3(a^5 + 8a^2b^2)q^3\}q'.$$

Since $d \neq 0$, it follows from these two relations that

$$\max(|p|, |q|)^7 \leq C_1 \max(|p|, |q|)^3 \max(|p'|, |q'|)$$

and hence

$$\max(|p|, |q|)^4 \leq C_1 \max(|p'|, |q'|).$$

But the two relations also show that the greatest common divisor e of p' and q' divides both $4dq^7$ and $4dp^7$, and hence also $4d$, since p and q are coprime. Consequently

$$\max(|p'|, |q'|) \leq 4|d| \max(|p''|, |q''|).$$

Combining this with the previous inequality, we obtain

$$4h(P) \leq h(2P) + C''$$

for some constant $C'' = C''(a, b) > 0$.
This proves the result, with $C = \max(C', C'')$.   □

Proposition 9 *There exists a unique function $\hat{h}: E \to \mathbb{R}$ such that*

(i) *$\hat{h} - h$ is bounded,*
(ii) *$\hat{h}(2P) = 4\hat{h}(P)$ for every $P \in E$.*

Furthermore, it is given by the formula $\hat{h}(P) = \lim_{n\to\infty} h(2^n P)/4^n$.

Proof Suppose $\hat{h}$ has the properties (i),(ii). Then, by (ii), $4^n \hat{h}(P) = \hat{h}(2^n P)$ and hence, by (i), $4^n \hat{h}(P) - h(2^n P)$ is bounded. Dividing by 4^n, we see that $h(2^n P)/4^n \to \hat{h}(P)$ as $n \to \infty$. This proves uniqueness.

To prove existence, choose C as in the statement of Proposition 8. Then, for any integers m, n with $n > m > 0$,

$$|4^{-n}h(2^n P) - 4^{-m}h(2^m P)| = \left| \sum_{j=m}^{n-1} \{4^{-j-1}h(2^{j+1}P) - 4^{-j}h(2^j P)\} \right|$$

$$\leq \sum_{j=m}^{n-1} 4^{-j-1} |h(2^{j+1}P) - 4h(2^j P)|$$

$$\leq \sum_{j=m}^{n-1} 4^{-j-1} C < 4^{-m} C/3.$$

Thus the sequence $\{4^{-n}h(2^n P)\}$ is a fundamental sequence and consequently convergent. If we denote its limit by $\hat{h}(P)$, then clearly $\hat{h}(2P) = 4\hat{h}(P)$. On the other hand, by taking $m = 0$ and letting $n \to \infty$ in the preceding inequality we obtain

$$|\hat{h}(P) - h(P)| \leq C/3.$$

Thus $\hat{h}$ has both the required properties. $\qquad\square$

The value $\hat{h}(P)$ is called the *canonical height* of the rational point P. The formula for $\hat{h}(P)$ shows that, for all $P \in E$,

$$\hat{h}(-P) = \hat{h}(P) \geq 0.$$

Moreover, by Proposition 9(i), for any $r > 0$ there exist only finitely many elements P of E with $\hat{h}(P) \leq r$.

It will now be shown that the canonical height satisfies the *parallelogram law*:

Proposition 10 *For all $P_1, P_2 \in E$,*

$$\hat{h}(P_1 + P_2) + \hat{h}(P_1 - P_2) = 2\hat{h}(P_1) + 2\hat{h}(P_2).$$

Proof It is sufficient to show that there exists a constant $C' > 0$ such that, for all $P_1, P_2 \in E$,

$$h(P_1 + P_2) + h(P_1 - P_2) \leq 2h(P_1) + 2h(P_2) + C'. \qquad (*)$$

For it then follows from the formula in Proposition 9 that, for all $P_1, P_2 \in E$,

$$\hat{h}(P_1 + P_2) + \hat{h}(P_1 - P_2) \leq 2\hat{h}(P_1) + 2\hat{h}(P_2).$$

But, replacing P_1 by $P_1 + P_2$ and P_2 by $P_1 - P_2$, we also have

$$\hat{h}(2P_1) + \hat{h}(2P_2) \leq 2\hat{h}(P_1 + P_2) + 2\hat{h}(P_1 - P_2)$$

and hence, by Proposition 9(ii),

$$2\hat{h}(P_1) + 2\hat{h}(P_2) \leq \hat{h}(P_1 + P_2) + \hat{h}(P_1 - P_2).$$

To prove (∗) we may evidently assume that $P_1 = (x_1, y_1)$ and $P_2 = (x_2, y_2)$ are both finite. Moreover, by Proposition 8, we may assume that $P_1 \neq P_2$. Then, by the formulas of §3,

$$P_1 + P_2 = (x_3, y_3), \quad P_1 - P_2 = (x_4, y_4),$$

where

$$x_3 = (y_2 - y_1)^2/(x_2 - x_1)^2 - (x_1 + x_2),$$
$$x_4 = (y_2 + y_1)^2/(x_2 - x_1)^2 - (x_1 + x_2).$$

Hence

$$x_3 + x_4 = 2[y_2^2 + y_1^2 - (x_2 - x_1)(x_2^2 - x_1^2)]/(x_2 - x_1)^2$$

and

$$x_3 x_4 = (y_2^2 - y_1^2)^2/(x_2 - x_1)^4 - 2(x_1 + x_2)(y_1^2 + y_2^2)/(x_2 - x_1)^2 + (x_1 + x_2)^2.$$

Since $y_j^2 = x_j^3 + a x_j + b$ $(j = 1, 2)$, these relations simplify to

$$x_3 + x_4 = 2[x_1 x_2(x_1 + x_2) + a(x_1 + x_2) + 2b]/(x_2 - x_1)^2$$

and

$$x_3 x_4 = N/(x_2 - x_1)^2,$$

where

$$N = (x_2^2 + x_1 x_2 + x_1^2 + a)^2 - 2(x_1 + x_2)^2(x_2^2 - x_1 x_2 + x_1^2 + a)$$
$$- 4b(x_1 + x_2) + (x_2^2 - x_1^2)^2$$
$$= (x_1 x_2 - a)^2 - 4b(x_1 + x_2).$$

Put $x_j = p_j/q_j$, where $(p_j, q_j) = 1$ $(1 \leq j \leq 4)$. Then x_3, x_4 are the roots of the quadratic polynomial

$$AX^2 + BX + C$$

with integer coefficients

$$A = (p_2 q_1 - p_1 q_2)^2,$$
$$B = (p_1 p_2 + a q_1 q_2)(p_1 q_2 + p_2 q_1) + 2b q_1^2 q_2^2,$$
$$C = (p_1 p_2 - a q_1 q_2)^2 - 4b q_1 q_2(p_1 q_2 + p_2 q_1).$$

Consequently

$$A p_3 p_4 = C q_3 q_4,$$
$$A(p_3 q_4 + p_4 q_3) = B q_3 q_4.$$

By Proposition II.16, q_3 and q_4 each divide A, and so their product divides A^2. Hence, for some integer $D \neq 0$,

$$A^2 = D q_3 q_4, \quad AC = D p_3 p_4, \quad AB = D(p_3 q_4 + p_4 q_3).$$

But it is easily seen that $q_3 q_4$, $p_3 p_4$ and $p_3 q_4 + p_4 q_3$ have no common prime divisor. It follows that A divides D.

Hence, if we put

$$\rho_j = \max(|p_j|, |q_j|) \ (1 \leq j \leq 4),$$

then

$$|q_3 q_4| \leq |A| \leq 4\rho_1^2 \rho_2^2,$$
$$|p_3 p_4| \leq |C| \leq [(1 + |a|)^2 + 8|b|]\rho_1^2 \rho_2^2,$$
$$|p_3 q_4 + p_4 q_3| \leq |B| \leq 2(1 + |a| + |b|)\rho_1^2 \rho_2^2.$$

But

$$\max(|p_3|, |q_3|) \max(|p_4|, |q_4|) \leq \max(|p_3 p_4|, |q_3 q_4| + |p_3 q_4 + p_4 q_3|),$$

since if $|q_3| \leq |p_3|$ and $|p_4| \leq |q_4|$, for example, then

$$|p_3 q_4| \leq |p_4 q_3| + |p_3 q_4 + p_4 q_3| \leq |q_3 q_4| + |p_3 q_4 + p_4 q_3|.$$

It follows that there exists a constant $C'' > 0$ such that

$$\rho_3 \rho_4 \leq C'' \rho_1^2 \rho_2^2,$$

which is equivalent to $(*)$ with $C' = \log C''$. □

Corollary 11 *For any $P \in E$ and any integer n,*

$$\hat{h}(nP) = n^2 \hat{h}(P).$$

Proof Since $\hat{h}(-P) = \hat{h}(P)$, we may assume $n > 0$. We may actually assume $n > 2$, since the result is trivial for $n = 1$ and it holds for $n = 2$ by Proposition 9. By Proposition 10 we have

$$\hat{h}(nP) + \hat{h}((n-2)P) = 2\hat{h}((n-1)P) + 2\hat{h}(P),$$

from which the general case follows by induction. □

It follows from Corollary 11 that if an element P of the group E has finite order, then $\hat{h}(P) = 0$. The converse is also true. In fact, by Proposition 10, the set of all $P \in E$ such that $\hat{h}(P) = 0$ is a subgroup of E, and this subgroup is finite since there are only finitely many points P such that $\hat{h}(P) < 1$.

We now deduce from Proposition 10 that a non-negative quadratic form can be constructed from the canonical height. If we put

$$(P, Q) = \hat{h}(P + Q) - \hat{h}(P) - \hat{h}(Q),$$

then evidently

$$(P, Q) = (Q, P), (P, P) = 2\hat{h}(P) \geq 0.$$

It remains to show that

$$(P, Q + R) = (P, Q) + (P, R),$$

and we do this by proving that

$$\hat{h}(P + Q + R) = \hat{h}(P + Q) + \hat{h}(P + R) + \hat{h}(Q + R) - \hat{h}(P) - \hat{h}(Q) - \hat{h}(R).$$

But, by the parallelogram law,

$$\hat{h}(P + Q + R + P) + \hat{h}(Q + R) = \hat{h}(P + Q + R + P) + \hat{h}(P + Q + R - P)$$
$$= 2\hat{h}(P + Q + R) + 2\hat{h}(P)$$

and

$$\hat{h}(P + Q + R + P) + \hat{h}(Q - R) = \hat{h}(P + Q + R + P) + \hat{h}(P + Q - P - R)$$
$$= 2\hat{h}(P + Q) + 2\hat{h}(P + R).$$

Subtracting the second relation from the first, we obtain

$$\hat{h}(Q + R) - \hat{h}(Q - R) = 2\hat{h}(P + Q + R) + 2\hat{h}(P) - 2\hat{h}(P + Q) - 2\hat{h}(P + R).$$

Since, by the parallelogram law again,

$$\hat{h}(Q + R) + \hat{h}(Q - R) = 2\hat{h}(Q) + 2\hat{h}(R),$$

this is equivalent to what we wished to prove.

Proposition 12 *The abelian group E is finitely generated if, for some integer $m > 1$, the factor group E/mE is finite.*

Proof Let S be a set of representatives of the cosets of the subgroup mE. Since S is finite, by hypothesis, we can choose $C > 0$ so that $\hat{h}(Q) \leq C$ for all $Q \in S$. The set

$$S' = \{Q' \in E : \hat{h}(Q') \leq C\}$$

contains S and is also finite. We will show that it generates E.

Let E' be the subgroup of E generated by the elements of S'. If $E' \neq E$, choose $P \in E \backslash E'$ so that $\hat{h}(P)$ is minimal. Then

$$P = mP_1 + Q_1 \quad \text{for some } P_1 \in E \text{ and } Q_1 \in S.$$

Since

$$\hat{h}(P + Q_1) + \hat{h}(P - Q_1) = 2\hat{h}(P) + 2\hat{h}(Q_1),$$

it follows that

$$\hat{h}(mP_1) = \hat{h}(P - Q_1) \leq 2\hat{h}(P) + 2C$$

and hence

$$\hat{h}(P_1) \leq 2[\hat{h}(P) + C]/m^2 \leq [\hat{h}(P) + C]/2.$$

But $P_1 \notin E'$, since $P \notin E'$, and hence $\hat{h}(P_1) \geq \hat{h}(P)$. It follows that $\hat{h}(P) \leq C$, which is a contradiction. Hence $E' = E$. □

Proposition 12 shows that to complete the proof of Mordell's theorem it is enough to show that the factor group $E/2E$ is finite. *We will prove this only for the case when E contains an element of order* 2. A similar proof may be given for the general case, but it requires some knowledge of algebraic number theory.

The assumption that E contains an element of order 2 means that there is a rational point $(x_0, 0)$, where x_0 is a root of the polynomial $X^3 + aX + b$. Since a and b are taken to be integers, and the polynomial has highest coefficient 1, x_0 must also be an integer. By changing variable from X to $x_0 + X$, we replace the cubic $\mathscr{C}_{a,b}$ by a cubic $C_{A,B}$ defined by a polynomial

$$Y^2 - (X^3 + AX^2 + BX),$$

where $A, B \in \mathbb{Z}$. The non-singularity condition $d := 4a^3 + 27b^2 \neq 0$ becomes

$$D := B^2(4B - A^2) \neq 0,$$

but this is the only restriction on A, B. The chord joining two rational points of $C_{A,B}$ is given by the same formulas as for $\mathscr{C}_{a,b}$ in §3, but the tangent to $C_{A,B}$ at the finite point $P_1 = (x_1, y_1)$ is now the affine line

$$Y - mX - c,$$

where

$$m = (3x_1^2 + 2Ax_1 + B)/2y_1, \quad c = -x_1(x_1^2 - B)/2y_1.$$

The geometrical interpretation of the group law remains the same as before. We will now denote by E the group of all rational points of $C_{A,B}$. Our change of variable has made the point $N = (0, 0)$ an element of E of order 2.

Let $P = (x, y)$ be a rational point of $C_{A,B}$ with $x \neq 0$. We are going to show that, in a sense which will become clear, there are only finitely many rational square classes to which x can belong.

Write $x = m/n$, $y = p/q$, where m, n, p, q are integers with $n, q > 0$ and $(m, n) = (p, q) = 1$. Then

$$p^2 n^3 = (m^3 + Am^2n + Bmn^2)q^2,$$

which implies both $q^2|n^3$ and $n^3|q^2$. Thus $n^3 = q^2$. From $n^2|q^2$ we obtain $n|q$. Hence $q = en$ for some integer e, and it follows that $n = e^2$, $q = e^3$. Thus

$$x = m/e^2, \quad y = p/e^3, \quad \text{where } e > 0 \quad \text{and} \quad (m, e) = (p, e) = 1.$$

Moreover,

$$p^2 = m(m^2 + Ame^2 + Be^4).$$

This shows that each prime which divides m, but not $m^2 + Ame^2 + Be^4$, must occur to an even power in m. On the other hand, each prime which divides both m and $m^2 + Ame^2 + Be^4$ must also divide B, since $(m, e) = 1$. Consequently we can write

$$x = \pm p_1^{\varepsilon_1} \cdots p_k^{\varepsilon_k} (u/e)^2,$$

where $u \in \mathbb{N}$, $p_1, \ldots, p_k$ are the distinct primes dividing B and $\varepsilon_j \in \{0, 1\}$ $(1 \le j \le k)$. Hence there are at most 2^{k+1} rational square classes to which x can belong.

Suppose now that $P_1 = (x_1, y_1)$ and $P_2 = (x_2, y_2)$ are distinct rational points of $C_{A,B}$ for which $x_1 x_2$ is a nonzero rational square, and let $P_3 = (x_3, y_3)$ be the third point of intersection with $C_{A,B}$ of the line through P_1 and P_2. Then x_1, x_2, x_3 are the three roots of a cubic equation

$$(mX + c)^2 = X^3 + AX^2 + BX.$$

From the constant term we see that $x_1 x_2 x_3 = c^2$. It follows that x_3 is a nonzero rational square if $c \ne 0$. If $c = 0$, then $P_3 = N$ and $x_1 x_2 = B$.

Suppose next that $P = (x, y)$ is any rational point of $C_{A,B}$ with $x \ne 0$, and let $2P = (\bar{x}, -\bar{y})$. Then $\bar{P} = (\bar{x}, \bar{y})$ is the other point of intersection with $C_{A,B}$ of the tangent to $C_{A,B}$ at P. By the same argument as before, $x^2 \bar{x} = c^2$. Hence $\bar{x}$ is a nonzero rational square if $c \ne 0$. If $c = 0$, then $2P = N$ and $x^2 = B$.

To deduce that $E/2E$ is finite from these observations we will use an arithmetic analogue of Landen's transformation. We saw in Chapter XII that, over the field $\mathbb{C}$ of complex numbers, the cubic curve $\mathscr{C}_\lambda$ defined by the polynomial $Y^2 - g_\lambda(X)$, where $g_\lambda(X) = 4X(1 - X)(1 - \lambda X)$, admits the parametrization

$$X = S(u, \lambda), \quad Y = S'(u, \lambda).$$

It follows from Proposition XII.11 that the cubic curve $\mathscr{C}_{\lambda'}$, where λ' is given by $\lambda' = \lambda^2 / [1 + (1 - \lambda)^{1/2}]^4$, admits the parametrization

$$X' = [1 + (1 - \lambda)^{1/2}] X (1 - X)/(1 - \lambda X),$$
$$Y' = [1 + (1 - \lambda)^{1/2}] Y (1 - 2X + \lambda X^2)/(1 - \lambda X)^2,$$

where again $X = S(u, \lambda)$, $Y = S'(u, \lambda)$ and where $(1 - 2X + \lambda X^2)/(1 - \lambda X)^2$ is the derivative with respect to X of $X(1 - X)/(1 - \lambda X)$. Since also $X' = S(u', \lambda')$, where $u' = [1 + (1 - \lambda)^{1/2}] u$, the map $(X, Y) \to (X', Y')$ defines a homomorphism of the group of complex points of $\mathscr{C}_\lambda$ into the group of complex points of $\mathscr{C}_{\lambda'}$.

We will simply state analogous results for the cubic curve $C_{A,B}$ over the field $\mathbb{Q}$ of rational numbers, since their verification is elementary. If (x, y) is a rational point of $C_{A,B}$ with $x \ne 0$ and if

$$x' = (x^2 + Ax + B)/x, \quad y' = y(x^2 - B)/x^2,$$

then (x', y') is a rational point of $C_{A',B'}$, where

$$A' = -2A, \quad B' = A^2 - 4B.$$

Moreover, if we define a map φ of the group E of all rational points of $C_{A,B}$ into the group E' of all rational points of $C_{A',B'}$ by putting

$$\varphi(x, y) = (x', y') \quad \text{if } x \neq 0, \quad \varphi(N) = \varphi(O) = O,$$

then φ is a homomorphism, i.e.

$$\varphi(P + Q) = \varphi(P) + \varphi(Q), \quad \varphi(-P) = -\varphi(P).$$

The range $\varphi(E)$ may not be the whole of E'. In fact, since

$$x' = (x^3 + Ax^2 + Bx)/x^2 = (y/x)^2,$$

the first coordinate of any finite point of $\varphi(E)$ must be a rational square. Furthermore, if $N = (0, 0)$ is a point of $\varphi(E)$, the integer $B' = A^2 - 4B$ must be a square. We will show that these conditions completely characterize $\varphi(E)$.

Evidently if $A^2 - 4B$ is a square, then the quadratic polynomial $X^2 + AX + B$ has a rational root $x_0 \neq 0$ and $\varphi(x_0, 0) = N$. Suppose now that (x', y') is a rational point of $C_{A',B'}$ and that $x' = t^2$ is a nonzero rational square. We will show that if

$$x_1 = (t^2 - A + y'/t)/2, \quad y_1 = tx_1,$$
$$x_2 = (t^2 - A - y'/t)/2, \quad y_2 = -tx_2,$$

then $(x_j, y_j) \in E$ and $\varphi(x_j, y_j) = (x', y')$ $(j = 1, 2)$. It is easily seen that $(x_j, y_j) \in E$ if and only if

$$t^2 = x_j + A + B/x_j.$$

But

$$
\begin{aligned}
x_1 x_2 &= [(t^2 - A)^2 - y'^2/t^2]/4 \\
&= [(x' - A)^2 - y'^2/x']/4 \\
&= (x'^3 - 2Ax'^2 + A^2 x' - y'^2)/4x'.
\end{aligned}
$$

Since

$$y'^2 = x'^3 - 2Ax'^2 + (A^2 - 4B)x',$$

it follows that $x_1 x_2 = B$. Hence (x_1, y_1) and (x_2, y_2) are both in E if $t^2 = x_1 + A + x_2$, and this condition is certainly satisfied by the definitions of x_1 and x_2.

In addition to

$$x_j + A + B/x_j = t^2 = x' \quad (j = 1, 2),$$

we have

$$y_1(x_1^2 - B)/x_1^2 = t(x_1^2 - x_1 x_2)/x_1 = t(x_1 - x_2) = y',$$

and similarly $y_2(x_2^2 - B)/x_2^2 = y'$. It follows that

$$\varphi(x_1, y_1) = \varphi(x_2, y_2) = (x', y').$$

Since φ is a homomorphism, the range $\varphi(E)$ is a subgroup of E'. We are going to show that this subgroup is of finite index in E'. By what we have already proved for E, there exists a finite (or empty) set $P_1' = (x_1', y_1'), \ldots, P_s' = (x_s', y_s')$ of points of E' such that x_i' is not a rational square $(1 \le i \le s)$ and such that, if $P' = (x', y')$ is any other point of E' for which x' is not a rational square, then $x'x_j'$ is a nonzero rational square for a unique $j \in \{1, \ldots, s\}$. Let $P'' = (x'', y'')$ be the third point of intersection with $C_{A', B'}$ of the line through P' and P_j', so that

$$P' + P_j' + P'' = O.$$

By what we have already proved, either x'' is a nonzero rational square or $P'' = N$ and $x'x_j' = B'$ is a square. In either case, $P'' \in \varphi(E)$. Furthermore, if $2P_j' = (\bar{x}, -\bar{y})$, then either $\bar{x}$ is a nonzero rational square or $2P_j' = N$ and $x_j'^2 = B'$. In either case again, $2P_j' \in \varphi(E)$. Since

$$P' = P_j' - (2P_j' + P''),$$

it follows that P' and P_j' are in the same coset of $\varphi(E)$. Consequently $P_1', \ldots, P_s'$, together with O, and also N if B' is not a square, form a complete set of representatives of the cosets of $\varphi(E)$ in E'.

The preceding discussion can be repeated with $C_{A', B'}$ in the place of $C_{A, B}$. It yields a homomorphism φ' of the group E' of all rational points of $C_{A', B'}$ into the group E'' of all rational points of $C_{A'', B''}$, where

$$A'' = -2A' = 4A, \quad B'' = A'^2 - 4B' = 16B.$$

But the simple transformation $(X, Y) \to (X/4, Y/8)$ replaces $C_{A'', B''}$ by $C_{A, B}$ and defines an isomorphism χ of E'' with E. Hence the composite map $\psi = \chi \circ \varphi'$ is a homomorphism of E' into E, and $\psi \circ \varphi$ is a homomorphism of E into itself.

We now show that the homomorphism $P \to \psi \circ \varphi(P)$ is just the doubling map $P \to 2P$. Since this is obvious if $P = O$ or N, we need only verify it for $P = (x, y)$ with $x \ne 0$.

For $P'' = \varphi' \circ \varphi(P)$ we have

$$x'' = (y'/x')^2 = [y(1 - B/x^2) \cdot x^2/y^2]^2 = (x^2 - B)^2/y^2$$

and

$$\begin{aligned}
y'' &= y'(1 - B'/x'^2) = y(1 - B/x^2)[1 - (A^2 - 4B)x^4/y^4] \\
&= (x^2 - B)[y^4 - (A^2 - 4B)x^4]/x^2 y^3 \\
&= (x^2 - B)[(x^2 + Ax + B)^2 - (A^2 - 4B)x^2]/y^3.
\end{aligned}$$

Hence for $\psi \circ \varphi(P) = P^* = (x^*, y^*)$ we have

$$x^* = (x^2 - B)^2/4y^2,$$
$$y^* = (x^2 - B)[(x^2 + Ax + B)^2 - (A^2 - 4B)x^2]/8y^3.$$

On the other hand, if the tangent to $C_{A,B}$ at P intersects $C_{A,B}$ again at $(\bar{x}, \bar{y})$, then $2P = (\bar{x}, -\bar{y})$. The cubic equation

$$(mx + c)^2 = X^3 + AX^2 + BX$$

has x as a double root and $\bar{x}$ as its third root. Hence $\bar{x} = (c/x)^2$. Using the formula for c given previously, we obtain

$$\bar{x} = (x^2 - B)^2/4y^2 = x^*.$$

Furthermore, using the formula for m given previously,

$$\bar{y} = m\bar{x} + c = [(3x^2 + 2Ax + B)\bar{x} - x(x^2 - B)]/2y$$
$$= (x^2 - B)[(3x^2 + 2Ax + B)(x^2 - B) - 4xy^2)]/8y^3.$$

Substituting $x^3 + Ax^2 + Bx$ for y^2, we obtain $\bar{y} = -y^*$. Thus $\psi \circ \varphi(P) = 2P$, as claimed.

Since $\varphi(E)$ has finite index in E', and likewise $\psi(E')$ has finite index in E, it follows that $2E = \psi \circ \varphi(E)$ has finite index in E. (The proof shows that the index is at most $2^{\alpha+\beta+2}$, where α is the number of distinct prime divisors of B and β is the number of distinct prime divisors of $A^2 - 4B$.)

By the remarks after the proof of Proposition 12, Mordell's theorem has now been completely proved in the case where E contains an element of order 2.

5 Further Results and Conjectures

Let $\mathscr{C}_{a,b}$ be the elliptic curve defined by the polynomial

$$Y^2 - (X^3 + aX + b),$$

where $a, b \in \mathbb{Z}$ and $d := 4a^3 + 27b^2 \neq 0$. By Mordell's theorem, the abelian group $E = E_{a,b}(\mathbb{Q})$ of all rational points of $\mathscr{C}_{a,b}$ is finitely generated. It follows from the structure theorem for finitely generated abelian groups (Chapter III, §4) that E is the direct sum of a finite abelian group E^t and a 'free' abelian group E^f, which is the direct sum of $r \geq 0$ infinite cyclic subgroups. The non-negative integer r is called the *rank* of the elliptic curve and E^t its *torsion group*.

The torsion group can, in principle, be determined by a finite amount of computation. A theorem of Nagell (1935) and Lutz (1937) says that if $P = (x, y)$ is a point of E of finite order, then x and y are integers and either $y = 0$ or y^2 divides d. Thus there are only finitely many possibilities to check.

A deep theorem of Mazur (1977) says that the torsion group must be one of the following:

(i) a cyclic group of order n ($1 \leq n \leq 10$ or $n = 12$),
(ii) the direct sum of a cyclic group of order 2 and a cyclic group of order $2n$ ($1 \leq n \leq 4$).

It was already known that each of these possibilities occurs. It is easy to check if the torsion group is of type (i) or type (ii), since in the latter case there are three elements of order 2, whereas in the former case there is at most one. Mazur's result shows that an element has infinite order, if it does not have order ≤ 12.

It is conjectured that there exist elliptic curves over $\mathbb{Q}$ with arbitrarily large rank. (Examples are known of elliptic curves with rank ≥ 22.) At present no infallible algorithm is known for determining the rank of an elliptic curve, let alone a basis for the torsion-free group E^f. However, Manin (1971) devised a conditional algorithm, based on the strong conjecture of Birch and Swinnerton-Dyer which will be mentioned later. This conjecture is still unproved, but is supported by much numerical evidence.

An important way of obtaining arithmetic information about an elliptic curve is by reduction modulo a prime p. We regard the coefficients not as integers, but as integers mod p, and we look not for $\mathbb{Q}$-points, but for $\mathbb{F}_p$-points. Since the normal form $\mathscr{C}_{a,b}$ was obtained by assuming that the field had characteristic $\neq 2, 3$, we now adopt a more general normal form.

Let $\mathscr{W} = \mathscr{W}(a_1, \ldots, a_6)$ be the projective completion of the affine cubic curve defined by the polynomial

$$Y^2 + a_1 XY + a_3 Y - (X^3 + a_2 X^2 + a_4 X + a_6),$$

where $a_j \in \mathbb{Q}$ ($j = 1, 2, 3, 4, 6$). It may be shown that $\mathscr{W}$ is non-singular if and only if the *discriminant* $\varDelta \neq 0$, where

$$\varDelta = -b_2^2 b_8 - 8b_4^3 - 27b_6^2 + 9b_2 b_4 b_6$$

and

$$
\begin{aligned}
b_2 &= a_1^2 + 4a_2, \\
b_4 &= a_1 a_3 + 2a_4, \\
b_6 &= a_3^2 + 4a_6, \\
b_8 &= a_1^2 a_6 - a_1 a_3 a_4 + 4a_2 a_6 + a_2 a_3^2 - a_4^2.
\end{aligned}
$$

(We retain the name 'discriminant', although $\varDelta = -16d$ for $\mathscr{W} = \mathscr{C}_{a,b}$.) The definition of addition on $\mathscr{W}$ has the same geometrical interpretation as on $\mathscr{C}_{a,b}$, although the corresponding algebraic formulas are different. They are written out in §7.

For any $u, r, s, t \in \mathbb{Q}$ with $u \neq 0$, the invertible linear change of variables

$$X = u^2 X' + r, \quad Y = u^3 Y' + su^2 X' + t$$

replaces $\mathscr{W}$ by a curve $\mathscr{W}'$ of the same form with discriminant $\varDelta' = u^{-12} \varDelta$. By means of such a transformation we may assume that the coefficients a_j are integers and that $\varDelta$, which is now an integer, has minimal absolute value. (It has been proved by Tate that we then have $|\Delta| > 1$.) The discussion which follows presupposes that $\mathcal{W}$ is chosen in this way so that, in particular, discriminant means 'minimal discriminant'. We say that such a $\mathcal{W}$ is a *minimal model* for the elliptic curve.

For any prime p, let $\mathcal{W}_p$ be the cubic curve defined over the finite field $\mathbb{F}_p$ by the polynomial

$$Y^2 + \tilde{a}_1 XY + \tilde{a}_3 Y - (X^3 + \tilde{a}_2 X^2 + \tilde{a}_4 X + \tilde{a}_6),$$

where $\tilde{a}_j \in a_j + p\mathbb{Z}$. If $p \nmid \Delta$ the cubic curve $\mathcal{W}_p$ is non-singular, but if $p \mid \Delta$ then $\mathcal{W}_p$ has a unique singular point. The singular point (x_0, y_0) of $\mathcal{W}_p$ is a *cusp* if, on replacing X and Y by $x_0 + X$ and $y_0 + Y$, we obtain a polynomial of the form

$$c(aX + bY)^2 + \cdots,$$

where $a, b, c \in \mathbb{F}_p$ and the unwritten terms are of degree > 2. Otherwise, the singular point is a *node*.

For any prime p, let N_p denote the number of $\mathbb{F}_p$-points of $\mathcal{W}_p$, including the point at infinity O, and put

$$c_p = p + 1 - N_p.$$

It was conjectured by Artin (1924), and proved by Hasse (1934), that

$$|c_p| \leq 2p^{1/2} \quad \text{if } p \nmid \Delta.$$

Since $2p^{1/2}$ is not an integer, this inequality says that the quadratic polynomial

$$1 - c_p T + p T^2$$

has conjugate complex roots $\gamma_p, \bar{\gamma}_p$ of absolute value $p^{-1/2}$ or, if we put $T = p^{-s}$, that the zeros of

$$1 - c_p p^{-s} + p^{1-2s}$$

lie on the line $\mathcal{R}s = 1/2$. Thus it is an analogue of the Riemann hypothesis on the zeros of $\zeta(s)$, but differs from it by having been proved. (As mentioned in §5 of Chapter IX, Hasse's result was considerably generalized by Weil (1948) and Deligne (1974).)

The *L-function* of the original elliptic curve $\mathcal{W}$ is defined by

$$L(s) = L(s, \mathcal{W}) := \prod_{p \mid \Delta} (1 - c_p p^{-s})^{-1} \prod_{p \nmid \Delta} (1 - c_p p^{-s} + p^{1-2s})^{-1}.$$

The first product on the right side has only finitely many factors. The infinite second product is convergent for $\mathcal{R}s > 3/2$, since

$$1 - c_p p^{-s} + p^{1-2s} = (p^{1/2-s} - p^{1/2}\gamma_p)(p^{1/2-s} - p^{1/2}\bar{\gamma}_p)$$

and $|\gamma_p| = |\bar{\gamma}_p| = p^{-1/2}$. Multiplying out the products, we obtain for $\mathscr{R}s > 3/2$ an absolutely convergent Dirichlet series

$$L(s) = \sum_{n \geq 1} c_n n^{-s}$$

with integer coefficients c_n. (If $n = p$ is prime, then c_n is the previously defined c_p.)

The *conductor* $N = N(\mathscr{W})$ of the elliptic curve $\mathscr{W}$ is defined by the singular reductions $\mathscr{W}_p$ of $\mathscr{W}$:

$$N = \prod_{p \mid \Delta} p^{f_p},$$

where $f_p = 1$ if $\mathscr{W}_p$ has a node, whereas $f_p = 2$ if $p > 3$ and $\mathscr{W}_p$ has a cusp. We will not define f_p if $p \in \{2, 3\}$ and $\mathscr{W}_p$ has a cusp, but we mention that f_p is then an integer ≥ 2 which can be calculated by an algorithm due to Tate (1975). (It may be shown that $f_2 \leq 8$ and $f_3 \leq 5$.)

The elliptic curve $\mathscr{W}$ is said to be *semi-stable* if $\mathscr{W}_p$ has a node for every $p \mid \Delta$. Thus, for a semi-stable elliptic curve, the conductor N is precisely the product of the distinct primes dividing the discriminant Δ. (The semi-stable case is the only one in which the conductor is square-free.)

Three important conjectures about elliptic curves, involving their L-functions and conductors, will now be described.

It was conjectured by Hasse (1954) that the function

$$\zeta(s, \mathscr{W}) := \zeta(s)\zeta(s - 1)/L(s, \mathscr{W})$$

may be analytically continued to a function which is meromorphic in the whole complex plane and that $\zeta(2 - s, \mathscr{W})$ is connected with $\zeta(s, \mathscr{W})$ by a functional equation similar to that satisfied by the Riemann zeta-function $\zeta(s)$. In terms of L-functions, Hasse's conjecture was given the following precise form by Weil (1967):

HW-Conjecture: *If the elliptic curve $\mathscr{W}$ has L-function $L(s)$ and conductor N, then $L(s)$ may be analytically continued, so that the function*

$$\Lambda(s) = (2\pi)^{-s} \Gamma(s) L(s),$$

where $\Gamma(s)$ denotes Euler's gamma-function, is holomorphic throughout the whole complex plane and satisfies the functional equation

$$\Lambda(s) = \pm N^{1-s} \Lambda(2 - s).$$

(In fact it is the functional equation which determines the precise definition of the conductor.)

The second conjecture, due to Birch and Swinnerton-Dyer (1965), connects the L-function with the group of rational points:

BSD-Conjecture: *The L-function $L(s)$ of the elliptic curve $\mathscr{W}$ has a zero at $s = 1$ of order exactly equal to the rank $r \geq 0$ of the group $E = E(\mathscr{W}, \mathbb{Q})$ of all rational points of $\mathscr{W}$.*

This is sometimes called the 'weak' conjecture of Birch and Swinnerton-Dyer, since they also gave a 'strong' version, in which the nonzero constant C such that

$$L(s) \sim C(s-1)^r \quad \text{for } s \to 1$$

is expressed by other arithmetic invariants of $\mathcal{W}$. The strong conjecture may be regarded as an analogue for elliptic curves of a known formula for the Dedekind zeta-function of an algebraic number field. An interesting reformulation of the strong form has been given by Bloch (1980).

The statement of the third conjecture requires some preparation. For any positive integer N, let $\Gamma_0(N)$ denote the multiplicative group of all matrices

$$A = \begin{pmatrix} a & b \\ c & d \end{pmatrix},$$

where a, b, c, d are integers such that $ad - bc = 1$ and $c \equiv 0 \bmod N$. A function $f(\tau)$ which is holomorphic for $\tau \in \mathcal{H}$ (the upper half-plane) is said to be a *modular form of weight 2 for* $\Gamma_0(N)$ if, for every such A,

$$f((a\tau + b)/(c\tau + d)) = (c\tau + d)^2 f(\tau).$$

An elliptic curve $\mathcal{W}$, with L-function

$$L(s) = \sum_{n \geq 1} c_n n^{-s}$$

and conductor N, is said to be *modular* if the function

$$f(\tau) = \sum_{n \geq 1} c_n e^{2\pi i n \tau},$$

which is certainly holomorphic in $\mathcal{H}$, is a modular form of weight 2 for $\Gamma_0(N)$. This actually implies that f is a 'cusp form' and satisfies a functional equation

$$f(-1/N\tau) = \mp N\tau^2 f(\tau).$$

It follows that the *Mellin transform*

$$\Lambda(s) = \int_0^\infty f(iy) y^{s-1} dy$$

may be analytically continued for all $s \in \mathbb{C}$ and satisfies the functional equation

$$\Lambda(s) = \pm N^{1-s} \Lambda(2-s).$$

(Note the reversal of sign.) But

$$\Lambda(s) = (2\pi)^{-s} \Gamma(s) L(s),$$

since, by (9) of Chapter IX,

$$\int_0^\infty e^{-2\pi n y} y^{s-1} dy = (2\pi n)^{-s} \Gamma(s).$$

Hence any modular elliptic curve satisfies the *HW*-conjecture.

It was shown by Weil (1967) that, conversely, an elliptic curve is modular if not only its L-function $L(s) = \sum_{n \geq 1} c_n n^{-s}$ has the properties required in the

HW-conjecture but also, for sufficiently many Dirichlet characters χ, the 'twisted' *L*-functions

$$L(s, \chi) = \sum_{n \geq 1} \chi(n) c_n n^{-s}$$

have analogous properties.

The definition of modular elliptic curve can be given a more intuitive form: the elliptic curve $\mathscr{C}_{a,b}$ is modular if there exist non-constant functions $X = f(\tau), Y = g(\tau)$ which are holomorphic in the upper half-plane, which are invariant under $\Gamma_0(N)$, i.e.

$$f((a\tau + b)/(c\tau + d)) = f(\tau), \quad g((a\tau + b)/(c\tau + d)) = g(\tau)$$

for every

$$A = \begin{pmatrix} a & b \\ c & d \end{pmatrix} \in \Gamma_0(N),$$

and which parametrize $\mathscr{C}_{a,b}$:

$$g^2(\tau) = f^3(\tau) + af(\tau) + b.$$

The significance of modular elliptic curves is that one can apply to them the extensive analytic theory of modular forms. For example, through the work of Kolyvagin (1990), together with results of Gross and Zagier (1986) and others, it is known that (as the *BSD*-conjecture predicts) a modular elliptic curve has rank 0 if its *L*-function does not vanish at $s = 1$, and has rank 1 if its *L*-function has a simple zero at $s = 1$.

The third conjecture, stated rather roughly by Taniyama (1955) and more precisely by Weil (1967), is simply this:

TW-Conjecture: *Every elliptic curve over the field $\mathbb{Q}$ of rational numbers is modular.*

The name of Shimura is often also attached to this conjecture, since he certainly contributed to its ultimate formulation. Shimura (1971) further showed that any elliptic curve which admits complex multiplication is modular. A big step forward was made by Wiles (1995) who, with assistance from Taylor, showed that any semi-stable elliptic curve is modular. A complete proof of the *TW*-conjecture, due to Diamond and others, has recently been announced by Darmon (1999). Thus all the results which had previously been established for modular elliptic curves actually hold for all elliptic curves over $\mathbb{Q}$.

It should be mentioned that there is also a 'Riemann hypothesis' for elliptic curves over $\mathbb{Q}$, namely that all zeros of the *L*-function in the critical strip $1/2 < \mathscr{R}s < 3/2$ lie on the line $\mathscr{R}s = 1$.

Mordell's theorem was extended from elliptic curves over $\mathbb{Q}$ to abelian varieties over any algebraic number field by Weil (1928). Many other results in the arithmetic of elliptic curves have been similarly extended. The topic is too vast to be considered here, but it should be said that our exposition for the prototype case is not always in the most appropriate form for such generalizations.

In the same paper in which he proved his theorem, Mordell (1922) conjectured that if a non-singular irreducible projective curve, defined by a homogeneous polynomial $F(x, y, z)$ with rational coefficients, has infinitely many rational points, then it is birationally equivalent to a line, a conic or a cubic. Mordell's conjecture was first proved by Faltings (1983). Actually Falting's result was not restricted to *plane* algebraic curves, and on the way he proved two other important conjectures of Tate and Shafarevich.

Falting's result implies that the Fermat equation $x^n + y^n = z^n$ has at most finitely many solutions in integers if $n > 3$. In the next section we will see that Wiles' result that semi-stable elliptic curves are modular implies that there are *no* solutions in nonzero integers.

6 Some Applications

The arithmetic of elliptic curves has an interesting application to the ancient problem of congruent numbers. A positive integer n is (confusingly) said to be *congruent* if it is the area of a right-angled triangle whose sides all have rational length, i.e. if there exist positive rational numbers u, v, w such that $u^2 + v^2 = w^2$, $uv = 2n$. For example, 6 is congruent, since it is the area of the right-angled triangle with sides of length $3, 4, 5$. Similarly, 5 is congruent, since it is the area of the right-angled triangle with sides of length $3/2, 20/3, 41/6$.

In the margin of his copy of Diophantus' *Arithmetica* Fermat (c. 1640) gave a complete proof that 1 is not congruent. The following is a paraphrase of his argument. Assume that 1 is congruent. Then there exist positive rational numbers u, v, w such that

$$u^2 + v^2 = w^2, \quad uv = 2.$$

Since an integer is a rational square only if it is an integral square, on clearing denominators it follows that there exist positive integers a, b, c, d such that

$$a^2 + b^2 = c^2, \quad 2ab = d^2.$$

Choose such a quadruple a, b, c, d for which c is minimal. Then $(a, b) = 1$. Since d is even, exactly one of a, b is even and we may suppose it to be a. Then

$$a = 2g^2, \quad b = h^2$$

for some positive integers g, h. Since b and c are both odd and $(b, c) = 1$,

$$(c - b, c + b) = 2.$$

Since

$$(c - b)(c + b) = a^2 = 4g^4,$$

it follows that

$$c + b = 2c_1^4, \quad c - b = 2d_1^4,$$

for some relatively prime positive integers c_1, d_1. Then

$$(c_1^2 - d_1^2)(c_1^2 + d_1^2) = c_1^4 - d_1^4 = b = h^2.$$

But

$$(c_1^2 - d_1^2, \ c_1^2 + d_1^2) = 1,$$

since $(c_1^2, d_1^2) = 1$ and b is odd. Hence

$$c_1^2 - d_1^2 = p^2, \quad c_1^2 + d_1^2 = q^2,$$

for some odd positive integers p, q. Thus

$$a_1 = (q + p)/2, \quad b_1 = (q - p)/2$$

are positive integers and

$$a_1^2 + b_1^2 = (q^2 + p^2)/2 = c_1^2,$$
$$2a_1 b_1 = (q^2 - p^2)/2 = d_1^2.$$

Since $c_1 \le c_1^4 < c$, this contradicts the minimality of c.

It follows that the Fermat equation

$$x^4 + y^4 = z^4$$

has no solutions in nonzero integers x, y, z. For if a solution existed and if we put

$$u = 2|yz|/x^2, \quad v = x^2/|yz|, \quad w = (y^4 + z^4)/x^2|yz|,$$

we would have $u^2 + v^2 = w^2, uv = 2$.

It is easily seen that a positive integer n is congruent if and only if there exists a rational number x such that $x, x + n$ and $x - n$ are all rational squares. For suppose

$$x = r^2, \quad x + n = s^2, \quad x - n = t^2,$$

and put

$$u = s - t, \quad v = s + t, \quad w = 2r.$$

Then

$$uv = s^2 - t^2 = 2n$$

and

$$u^2 + v^2 = 2(s^2 + t^2) = 4x = w^2.$$

Conversely, if u, v, w are rational numbers such that $uv = 2n$ and $u^2 + v^2 = w^2$, then

$$(u + v)^2 = w^2 + 4n, \quad (u - v)^2 = w^2 - 4n.$$

Thus, if we put $x = (w/2)^2$, then x, $x + n$ and $x - n$ are all rational squares.

It may be noted that if x is a rational number such that x, $x + n$ and $x - n$ are all rational squares, then $x \neq -n, 0, n$, since $n > 0$ and 2 is not a rational square.

The problem of determining which positive integers are congruent was considered by Arab mathematicians of the 10th century AD, and later by Fibonacci (1225) in his *Liber Quadratorum*. The connection with elliptic curves will now be revealed:

Proposition 13 *A positive integer n is congruent if and only if the cubic curve C_n defined by the polynomial*

$$Y^2 - (X^3 - n^2 X)$$

has a rational point $P = (x, y)$ with $y \neq 0$.

Proof Suppose first that n is congruent. Then there exists a rational number x such that x, $x + n$ and $x - n$ are all rational squares. Hence their product

$$x^3 - n^2 x = x(x - n)(x + n)$$

is also a rational square. Since $x \neq -n, 0, n$, it follows that $x^3 - n^2 x = y^2$, where y is a nonzero rational number.

Suppose now that $P = (x, y)$ is any rational point of the curve C_n with $y \neq 0$. If we put

$$u = |(x^2 - n^2)/y|, \quad v = |2nx/y|, \quad w = |(x^2 + n^2)/y|,$$

then u, v, w are positive rational numbers such that

$$u^2 + v^2 = w^2, \quad uv = 2n. \qquad \square$$

It is readily verified that $\lambda = 1/2$ in the Riemann normal form for C_n.

We now show that, for every positive integer n, the torsion group of C_n has order 4, consisting of the identity element O, and the three elements $(0, 0)$, $(n, 0)$, $(-n, 0)$ of order 2. Assume on the contrary that for some positive integer n the curve C_n has a rational point $P = (x, y)$ of finite order with $y \neq 0$ and take n to be the least positive integer with this property. Then $2P = (x', y')$ is also a rational point of C_n of finite order. The formula for the other point of intersection with C_n of the tangent to C_n at P shows that

$$x' = [(x^2 + n^2)/2y]^2.$$

It follows that

$$x' + n = [(x^2 - n^2 + 2nx)/2y]^2,$$
$$x' - n = [(x^2 - n^2 - 2nx)/2y]^2.$$

Moreover x', $x' + n$ and $x' - n$ are all *nonzero* rational squares. Since $2P$ is of finite order, the theorem of Nagell and Lutz mentioned in §5 implies that x' is an integer. Consequently

$$x' = r^2, \quad x' + n = s^2, \quad x' - n = t^2$$

for some positive *integers* r, s, t. Hence n is even, since

$$2n = s^2 - t^2 = (s - t)(s + t)$$

and if one of $s - t$ and $s + t$ is even, so also is the other. Since $n = s^2 - r^2$ and any integral square is congruent to 0 or 1 mod 4, we cannot have $n \equiv 2 \bmod 4$. Hence $n \equiv 0 \bmod 4$. But then $(x'/4, y'/8)$ is a rational point of finite order of $C_{n/4}$, which contradicts the minimality of n.

If n is congruent, then so also is $m^2 n$ for any positive integer m. Thus it is enough to determine which square-free positive integers are congruent. By what we have just proved and Proposition 13, a square-free positive integer n is congruent if and only if the elliptic curve C_n has positive rank. Since C_n admits complex multiplication, a result of Coates and Wiles (1977) shows that if C_n has positive rank, then its L-function vanishes at $s = 1$. (According to the *BSD*-conjecture, C_n has positive rank if and only if its L-function vanishes at $s = 1$.)

By means of the theory of modular forms, Tunnell (1983) has obtained a practical necessary and sufficient condition for the L-function $L(s, C_n)$ of C_n to vanish at $s = 1$: if n is a square-free positive integer, then $L(1, C_n) = 0$ if and only if $A_+(n) = A_-(n)$, where $A_+(n)$, resp. $A_-(n)$, is the number of triples $(x, y, z) \in \mathbb{Z}^3$ with z even, resp. z odd, such that

$$x^2 + 2y^2 + 8z^2 = n \quad \text{if } n \text{ is odd, or} \quad 2x^2 + 2y^2 + 16z^2 = n \quad \text{if } n \text{ is even.}$$

It is not difficult to show that $A_+(n) = A_-(n)$ when $n \equiv 5, 6$ or $7 \bmod 8$, but there seems to be no such simple criterion in other cases. With the aid of a computer it has been verified that, for every $n < 10000$, n is congruent if and only if $A_+(n) = A_-(n)$.

The arithmetic of elliptic curves also has a useful application to the class number problem of Gauss. For any square-free integer $d < 0$, let $h(d)$ be the *class number* of the quadratic field $\mathbb{Q}(\sqrt{d})$. As mentioned in §8 of Chapter IV, it was conjectured by Gauss (1801), and proved by Heilbronn (1934), that $h(d) \to \infty$ as $d \to -\infty$. However, the proof does not provide a method of determining an upper bound for the values of d for which the class number $h(d)$ has a given value. As mentioned in Chapter II, Stark (1967) showed that there are no other negative values of d for which $h(d) = 1$ besides the nine values already known to Gauss. Using methods developed by Baker (1966) for the theory of transcendental numbers, it was shown by Baker (1971) and Stark (1971) that there are exactly 18 negative values of d for which $h(d) = 2$. A simpler and more powerful method for attacking the problem was found by Goldfeld (1976). He obtained an effective lower bound for $h(d)$, provided that there exists a modular elliptic curve over $\mathbb{Q}$ whose L-function has a triple zero at $s = 1$. Gross and Zagier (1986) showed that such an elliptic curve does indeed exist. However, to show that this elliptic curve was modular required a considerable amount of computation. The proof of the *TW*-conjecture makes any computation unnecessary.

The most celebrated application of the arithmetic of elliptic curves has been the recent proof of Fermat's last theorem. In his copy of the translation by Bachet of Diophantus' *Arithmetica* Fermat also wrote "It is impossible to separate a cube into two cubes, or a fourth power into two fourth powers or, in general, any power higher than the second into two like powers. I have discovered a truly marvellous proof of this, which this margin is too narrow to contain."

In other words, Fermat asserted that, if $n > 2$, the equation

$$x^n + y^n = z^n$$

has no solutions in nonzero integers x, y, z. In §2 of Chapter III we pointed out that it was sufficient to prove his assertion when $n = 4$ and when $n = p$ is an odd prime, and we gave a proof there for $n = 3$.

A nice application to cubic curves of the case $n = 3$ was made by Kronecker (1859). If we make the change of variables

$$x = 2a/(3b - 1), \quad y = (3b + 1)/(3b - 1),$$

with inverse

$$a = x/(y - 1), \quad b = (y + 1)/3(y - 1),$$

then

$$x^3 + y^3 - 1 = 2(4a^3 + 27b^2 + 1)/(3b - 1)^3.$$

Since the equation $x^3 + y^3 = 1$ has no solution in nonzero rational numbers, the only solutions in rational numbers of the equation

$$4a^3 + 27b^2 = -1$$

are $a = -1, b = \pm 1/3$. Consequently the only cubic curves $\mathscr{C}_{a,b}$ with rational coefficients a, b and discriminant $d = -1$ are $Y^2 - X^3 + X \pm 1/3$.

We return now to Fermat's assertion. In the present section we have already given Fermat's own proof for $n = 4$. Suppose now that $p \geq 5$ is prime and assume, contrary to Fermat's assertion, that the equation

$$a^p + b^p + c^p = 0$$

does have a solution in nonzero integers a, b, c. By removing any common factor we may assume that $(a, b) = 1$, and then also $(a, c) = (b, c) = 1$. Since a, b, c cannot all be odd, we may assume that b is even. Then a and c are odd, and we may assume that $a \equiv -1 \bmod 4$.

We now consider the projective cubic curve $\mathscr{E}_{A,B}$ defined by the polynomial

$$Y^2 - X(X - A)(X + B),$$

where $A = a^p$ and $B = b^p$. By construction, $(A, B) = 1$ and

$$A \equiv -1 \bmod 4, \quad B \equiv 0 \bmod 32.$$

Moreover, if we put $C = -(A + B)$, then $C \neq 0$ and $(A, C) = (B, C) = 1$. The linear change of variables

$$X \to 4X, \quad Y \to 8Y + 4X$$

replaces $\mathscr{E}_{A,B}$ by the elliptic curve $\mathscr{W}_{A,B}$ defined by

$$Y^2 + XY - \{X^3 + (B - A - 1)X^2/4 - ABX/16\},$$

which has discriminant

$$\Delta = (ABC)^2/2^8.$$

Our hypotheses ensure that the coefficients of $\mathscr{W}_{A,B}$ are integers and that Δ is a nonzero integer. It may be shown that $\mathscr{W}_{A,B}$ is actually a minimal model for $\mathscr{E}_{A,B}$. Moreover, when we reduce modulo any prime ℓ which divides Δ, the singular point which arises is a node. Thus $\mathscr{W}_{A,B}$ is semi-stable and its conductor N is the product of the distinct primes dividing ABC.

Fermat's last theorem will be proved, for any prime $p \geq 5$, if we show that such an elliptic curve cannot exist if A, B, C are all p-th powers. If p is large, one reason for suspecting that such an elliptic curve cannot exist is that the discriminant is then very large compared with the conductor. Another reason, which does not depend on the size of p, was suggested by Frey (1986). Frey gave a heuristic argument that $\mathscr{W}_{A,B}$ could not then be modular, which would contradict the TW-conjecture.

Frey's intuition was made more precise by Serre (1987). Let G be the group of all automorphisms of the field of all algebraic numbers. With any modular form for $\Gamma_0(N)$ one can associate a 2-dimensional representation of G over a finite field. Serre showed that Fermat's last theorem would follow from the TW-conjecture, together with a conjecture about lowering the level of such 'Galois representations' associated with modular forms. The latter conjecture was called Serre's ε-conjecture, because it was a special case of a much more general conjecture which Serre made.

Serre's ε-conjecture was proved by Ribet (1990), although the proof might be described as being of order ε^{-1}. Now, for the first time, the falsity of Fermat's last theorem would have a significant consequence: the falsity of the TW-conjecture. Since $\mathscr{W}_{A,B}$ is semi-stable with the normalizations made above, to prove Fermat's last theorem it was actually enough to show that any semi-stable elliptic curve was modular. As stated in §5, this was accomplished by Wiles (1995) and Taylor and Wiles (1995). We will not attempt to describe the proof since, besides Fermat's classic excuse, it is beyond the scope of this work.

Fermat's last theorem contributed greatly to the development of mathematics, but Fermat was perhaps lucky that his assertion turned out to be correct. After proving Fermat's assertion for $n = 3$, that the cube of a positive integer could not be the sum of two cubes of positive integers, Euler asserted that, also for any $n \geq 4$, an n-th power of a positive integer could not be expressed as a sum of $n - 1$ n-th powers of positive integers. A counterexample to Euler's conjecture was first found, for $n = 5$, by Lander and Parkin (1966):

$$27^5 + 84^5 + 110^5 + 133^5 = 144^5.$$

Elkies (1988) used the arithmetic of elliptic curves to find infinitely many counterexamples for $n = 4$, the simplest being

$$95800^4 + 217519^4 + 414560^4 = 422481^4.$$

A prize has been offered by Beal (1997) for a proof or disproof of his conjecture that the equation

$$x^l + y^m = z^n$$

has no solution in coprime positive integers x, y, z if l, m, n are integers > 2. (The exponent 2 must be excluded since, for example, $2^5 + 7^2 = 3^4$ and $2^7 + 17^3 = 71^2$.) Will Beal's conjecture turn out to be like Fermat's or like Euler's?

7 Further Remarks

For sums of squares, see Grosswald [31], Rademacher [46], and Volume II, Chapter IX of Dickson [23]. A recent contribution is Milne [42].

A general reference for the theory of partitions is Andrews [2]. Proposition 4 is often referred to as *Euler's pentagonal number theorem*, since $m(3m - 1)/2$ $(m > 1)$ represents the number of dots needed to construct successively larger and larger pentagons. A direct proof of the combinatorial interpretation of Proposition 4 was given by Franklin (1881). It is reproduced in Andrews [2] and in van Lint and Wilson [41]. The replacement of proofs using generating functions by purely combinatorial proofs has become quite an industry; see, for example, Bressoud and Zeilberger [13], [14].

Besides the q-difference equations used in the proof of Proposition 5, there are also q-integrals:

$$\int_0^a f(x) d_q x := \sum_{n \geq 0} f(aq^n)(aq^n - aq^{n+1}).$$

The q-binomial coefficients (mentioned in §2 of Chapter II)

$$\begin{bmatrix} n \\ m \end{bmatrix} = \begin{bmatrix} n \\ m \end{bmatrix}_q := (q)_n/(q)_m(q)_{n-m} \quad (0 \leq m < n),$$

where $(a)_0 = 1$ and

$$(a)_n = (1 - a)(1 - aq) \cdots (1 - aq^{n-1}) \ (n \geq 1),$$

have recurrence properties similar to those of ordinary binomial coefficients:

$$\begin{bmatrix} n \\ m \end{bmatrix} = \begin{bmatrix} n - 1 \\ m - 1 \end{bmatrix} + q^m \begin{bmatrix} n - 1 \\ m \end{bmatrix} = \begin{bmatrix} n - 1 \\ m \end{bmatrix} + q^{n-m} \begin{bmatrix} n - 1 \\ m - 1 \end{bmatrix} \quad (0 < m < n).$$

The q-hypergeometric series

$$\sum_{n \geq 0} (a)_n (b)_n x^n / (c)_n (q)_n$$

was already studied by Heine (1847). There is indeed a whole world of q-analysis, which may be regarded as having the same relation to classical analysis as quantum mechanics has to classical mechanics. (The choice of the letter 'q' nearly a century before the advent of quantum mechanics showed remarkable foresight.) There are introductions to this world in Andrews et al. [4] and Vilenkin and Klimyk [58]. For Macdonald's conjectures concerning q-analogues of orthogonal polynomials, see Kirillov [36].

Although q-analysis always had its devotees, it remained outside the mainstream of mathematics until recently. Now it arises naturally in the study of *quantum groups*, which are not groups but q-deformations of the universal enveloping algebra of a Lie algebra.

The Rogers–Ramanujan identities were discovered independently by Rogers (1894), Ramanujan (1913) and Schur (1917). Their romantic history is retold in Andrews [2], which contains also generalizations. For the applications of the identities in statistical mechanics, see Baxter's article (pp. 69–84) in Andrews et al. [3]. (The same volume contains other interesting articles on mathematical developments arising from Ramanujan's work.)

The Jacobi triple product formula was derived in Chapter XII as the limit of a formula for polynomials. Andrews [1] has given a similar derivation of the Rogers–Ramanujan identities. This approach has found applications and generalizations in conformal field theory, with the two sides of the polynomial identity corresponding to fermionic and bosonic bases for Fock space; see Berkovich and McCoy [9].

These connections go much further than the Rogers–Ramanujan identities. There is now a vast interacting area which involves, besides the theory of partitions, solvable models of statistical mechanics, conformal field theory, integrable systems in classical and quantum mechanics, infinite-dimensional Lie algebras, quantum groups, knot theory and operator algebras. For introductory accounts, see [45], [10] and various articles in [24] and [27]. More detailed treatments of particular aspects are given in Baxter [8], Faddeev and Takhtajan [26], Jantzen [33], Jones [34], Kac [35] and Korepin et al. [38].

For the Hardy–Ramanujan–Rademacher expansion for $p(n)$, see Rademacher [46] and Andrews [2]. An interesting proof by means of probability theory for the first term of the expansion has been given by Báez-Duarte [5].

The definition of birational equivalence in §3 is adequate for our purposes, but has been superseded by a more general definition in the language of 'schemes', which is applicable to algebraic varieties of arbitrary dimension without any given embedding in a projective space. For the evolution of the modern concept, see Čižmár [18].

The history of the discovery of the group law on a cubic curve is described by Schappacher [48].

Several good accounts of the arithmetic of elliptic curves are now available; e.g., Knapp [37] and the trilogy [52], [50], [51]. Although the subject has been transformed in the past 25 years, the survey articles by Cassels [16], Tate [55] and Gelbart [28] are still of use. Tate gives a helpful introduction, Cassels has many references to the older literature, and Gelbart explains the connection with the Langlands program, for which see also Gelbart [29].

For reference, we give here the formulas for addition on an elliptic curve in the so-called Weierstrass's normal form. If $P_1 = (x_1, y_1)$ and $P_2 = (x_2, y_2)$ are points of the curve

$$Y^2 + a_1XY + a_3Y - (X^3 + a_2X^2 + a_4X + a_6),$$

then

$$-P_1 = (x_1, -y_1 - a_1x_1 - a_3), \quad P_1 + P_2 = P_3^* = (x_3, -y_3),$$

where

$$x_3 = \lambda(\lambda + a_1) - a_2 - x_1 - x_2, \quad y_3 = (\lambda + a_1)x_3 + \mu + a_3,$$

and

$$\lambda = (y_2 - y_1)/(x_2 - x_1), \quad \mu = (y_1x_2 - y_2x_1)/(x_2 - x_1) \quad \text{if } x_1 \neq x_2;$$
$$\lambda = (3x_1^2 + 2a_2x_1 + a_4 - a_1y_1)/N, \quad \mu = (-x_1^3 + a_4x_1 + 2a_6 - a_3y_1)/N,$$

with $N = 2y_1 + a_1x_1 + a_3$ if $x_1 = x_2$, $P_2 \neq -P_1$.

An algorithm for obtaining a minimal model of an elliptic curve is described in Laska [40]. Other algorithms connected with elliptic curves are given in Cremona [21].

The original conjecture of Birch and Swinnerton-Dyer was generalized by Tate [54] and Bloch [11]. For a first introduction to the theory of modular forms see Serre [49], and for a second see Lang [39].

Hasse actually showed that, if $\mathscr{E}$ is an elliptic curve over any finite field $\mathbb{F}_q$ containing q elements, then the number N_q of $\mathbb{F}_q$-points on $\mathscr{E}$ (including the point at infinity) satisfies the inequality

$$|N_q - (q + 1)| \leq 2q^{1/2}.$$

For an elementary proof, see Chahal [17]. Hasse's result is the special case, when the genus $g = 1$, of the Riemann hypothesis for function fields, which was mentioned in Chapter IX, §5.

It follows from the result of Siegel (1929), mentioned in §9 of Chapter IV, and even from the earlier work of Thue (1909), that an elliptic curve with integral coefficients has at most finitely many *integral* points. However, their method is not constructive. Baker [6], using the results on linear forms in the logarithms of algebraic numbers which he developed for the theory of transcendental numbers, obtained an explicit upper bound for the magnitude of any integral point in terms of an upper bound for the absolute values of all coefficients. Sharper bounds have since been obtained, e.g. by Bugeaud [15]. (For modern proofs of Baker's theorem on the linear independence of logarithms of algebraic numbers, see Waldschmidt [59]. The history of Baker's method is described in Baker [7].)

For information about the proof of Mordell's conjecture we refer to Bloch [12], Szpiro [53], and Cornell and Silverman [19]. The last includes an English translation of Faltings' original article. As mentioned in §9 of Chapter IV, Vojta (1991) has given a proof of the Mordell conjecture which is completely different from that of Faltings. There is an exposition of this proof, with simplifications due to Bombieri (1990), in Hindry and Silverman [32].

For congruent numbers, see Volume II, Chapter XVI of Dickson [23], Tunnell [57], and Noda and Wada [43]. The survey articles of Goldfeld [30] and Oesterlé [44] deal with Gauss's class number problem.

References for earlier work on Fermat's last theorem were given in Chapter III. Ribet [47] and Cornell *et al.* [20] provide some preparation for the original papers of Wiles [60] and Taylor and Wiles [56]. For the *TW*-conjecture, see also Darmon [22]. For Euler's conjecture, see Elkies [25].

8 Selected References

[1] G.E. Andrews, A polynomial identity which implies the Rogers-Ramanujan identities, *Scripta Math.* **28** (1970), 297–305.

[2] G.E. Andrews, *The theory of partitions*, Addison-Wesley, Reading, Mass., 1976. [Paperback edition, Cambridge University Press, 1998]

[3] G.E. Andrews, R.A. Askey, B.C. Berndt, K.G. Ramanathan and R.A. Rankin (ed.), *Ramanujan revisited*, Academic Press, London, 1988.

[4] G.E. Andrews, R. Askey and R. Roy, *Special functions*, Cambridge University Press, 1999.

[5] L. Báez-Duarte, Hardy-Ramanujan's asymptotic formula for partitions and the central limit theorem, *Adv. in Math.* **125** (1997), 114–120.

[6] A. Baker, The diophantine equation $y^2 = ax^3 + bx^2 + cx + d$, *J. London Math. Soc.* **43** (1968), 1–9.

[7] A. Baker, The theory of linear forms in logarithms, *Transcendence theory: advances and applications* (ed. A. Baker and D.W. Masser), pp. 1–27, Academic Press, London, 1977.

[8] R.J. Baxter, *Exactly solved models in statistical mechanics*, Academic Press, London, 1982. [Reprinted, 1989]

[9] A. Berkovich and B.M. McCoy, Rogers-Ramanujan identities: a century of progress from mathematics to physics, *Proceedings of the International Congress of Mathematicians*: *Berlin* 1998, Vol. III, pp. 163–172, Documenta Mathematica, Bielefeld, 1998.

[10] J.S. Birman, New points of view in knot theory, *Bull. Amer. Math. Soc. (N.S.)* **28** (1993), 253–287.

[11] S. Bloch, A note on height pairings, Tamagawa numbers, and the Birch and Swinnerton-Dyer conjecture, *Invent. Math.* **58** (1980), 65–76.

[12] S. Bloch, The proof of the Mordell conjecture, *Math. Intelligencer* **6** (1984), no. 2, 41–47.

[13] D.M. Bressoud and D. Zeilberger, A short Rogers-Ramanujan bijection, *Discrete Math.* **38** (1982), 313–315.

[14] D.M. Bressoud and D. Zeilberger, Bijecting Euler's partitions-recurrence, *Amer. Math. Monthly* **92** (1985), 54–55.

[15] Y. Bugeaud, On the size of integer solutions of elliptic equations, *Bull. Austral. Math. Soc.* **57** (1998), 199–206.

[16] J.W.S. Cassels, Diophantine equations with special reference to elliptic curves, *J. London Math. Soc.* **41** (1966), 193–291.

[17] J.S. Chahal, Manin's proof of the Hasse inequality revisited, *Nieuw Arch. Wisk. (4)* **13** (1995), 219–232.

[18] J. Čižmár, Birationale Transformationen (Ein historischer Überblick), *Period. Polytech. Mech. Engrg.* **39** (1995), 9–24.

[19] G. Cornell and J.H. Silverman (ed.), *Arithmetic geometry*, Springer-Verlag, New York, 1986.

[20] G. Cornell, J.H. Silverman and G. Stevens (ed.), *Modular forms and Fermat's last theorem*, Springer, New York, 1997.

[21] J.E. Cremona, *Algorithms for modular elliptic curves*, 2nd ed., Cambridge University Press, 1997.

[22] H. Darmon, A proof of the full Shimura–Taniyama–Weil conjecture is announced, *Notices Amer. Math. Soc.* **46** (1999), 1397–1401.

[23] L.E. Dickson, *History of the theory of numbers*, 3 vols., Carnegie Institute, Washington, D.C., 1919–1923. [Reprinted Chelsea, New York, 1992]

[24] L. Ehrenpreis and R.C. Gunning (ed.), *Theta functions: Bowdoin 1987*, Proc. Symp. Pure Math. **49**, Amer. Math. Soc., Providence, R.I., 1989.

[25] N.D. Elkies, On $A^4 + B^4 + C^4 = D^4$, *Math. Comp.* **51** (1988), 825–835.

[26] L.D. Faddeev and L.A. Takhtajan, *Hamiltonian methods in soliton theory*, Springer-Verlag, Berlin, 1987.

[27] A.S. Fokas and V.E. Zakharov (ed.), *Important developments in soliton theory*, Springer-Verlag, Berlin, 1993.

[28] S. Gelbart, Elliptic curves and automorphic representations, *Adv. in Math.* **21** (1976), 235–292.

[29] S. Gelbart, An elementary introduction to the Langlands program, *Bull. Amer. Math. Soc. (N.S.)* **10** (1984), 177–219.

[30] D. Goldfeld, Gauss' class number problem for imaginary quadratic fields, *Bull. Amer. Math. Soc. (N.S.)* **13** (1985), 23–37.

[31] E. Grosswald, *Representations of integers as sums of squares*, Springer-Verlag, New York, 1985.

[32] M. Hindry and J.H. Silverman, *Diophantine geometry*, Springer, New York, 2000.

[33] J.C. Jantzen, *Lectures on quantum groups*, American Mathematical Society, Providence, R.I., 1996.

[34] V.F.R. Jones, *Subfactors and knots*, CBMS Regional Conference Series in Mathematics **80**, Amer. Math. Soc., Providence, R.I., 1991.

[35] V.G. Kac, *Infinite-dimensional Lie algebras*, 3rd ed., Cambridge University Press, 1990.

[36] A.A. Kirillov, Jr., Lectures on affine Hecke algebras and Macdonald's conjectures, *Bull. Amer. Math. Soc. (N.S.)* **34** (1997), 251–292.

[37] A.W. Knapp, *Elliptic curves*, Princeton University Press, Princeton, N.J., 1992.

[38] V.E. Korepin, N.M. Bogoliubov and A.G. Izergin, *Quantum inverse scattering method and correlation functions*, Cambridge University Press, 1993.

[39] S. Lang, *Introduction to modular forms*, Springer-Verlag, Berlin, corr. reprint, 1995.

[40] M. Laska, An algorithm for finding a minimal Weierstrass equation for an elliptic curve, *Math. Comp.* **38** (1982), 257–260.

[41] J.H. van Lint and R.M. Wilson, *A course in combinatorics*, Cambridge University Press, 1992.

[42] S.C. Milne, New infinite families of exact sums of squares formulas, Jacobi elliptic functions and Ramanujan's tau function, *Proc. Nat. Acad. Sci. U.S.A.* **93** (1996), 15004–15008.

[43] K. Noda and H. Wada, All congruent numbers less than 10000, *Proc. Japan Acad. Ser. A Math. Sci.* **69** (1993), 175–178.

[44] J. Oesterlé, Le problème de Gauss sur le nombre de classes, *Enseign. Math.* **34** (1988), 43–67.

[45] M. Okado, M. Jimbo and T. Miwa, Solvable lattice models in two dimensions and modular functions, *Sugaku Exp.* **2** (1989), 29–54.

[46] H. Rademacher, *Topics in analytic number theory*, Springer-Verlag, Berlin, 1973.

[47] K.A. Ribet, Galois representations and modular forms, *Bull. Amer. Math. Soc. (N.S.)* **32** (1995), 375–402.

[48] N. Schappacher, Développement de la loi de groupe sur une cubique, *Séminaire de Théorie des Nombres, Paris 1988–89* (ed. C. Goldstein), pp. 159–184, Birkhäuser, Boston, 1990.

[49] J.-P. Serre, *A course in arithmetic*, Springer-Verlag, New York, 1973.

[50] J.H. Silverman, *The arithmetic of elliptic curves*, Springer-Verlag, New York, 1986.

[51] J.H. Silverman, *Advanced topics in the arithmetic of elliptic curves*, Springer-Verlag, New York, 1994.

[52] J.H. Silverman and J. Tate, *Rational points on elliptic curves*, Springer-Verlag, New York, 1992.

[53] L. Szpiro, La conjecture de Mordell [d'après G. Faltings], *Astérisque* **121–122** (1985), 83–103.

[54] J.T. Tate, On the conjectures of Birch and Swinnerton-Dyer and a geometric analog, *Séminaire Bourbaki: Vol. 1965/1966, Exposé no. 306*, Benjamin, New York, 1966.

[55] J.T. Tate, The arithmetic of elliptic curves, *Invent. Math.* **23** (1974), 179–206.

[56] R.L. Taylor and A. Wiles, Ring theoretic properties of certain Hecke algebras, *Ann. of Math.* **141** (1995), 553–572.

[57] J.B. Tunnell, A classical Diophantine problem and modular forms of weight 3/2, *Invent. Math.* **72** (1983), 323–334.

[58] N. Ja. Vilenkin and A.V. Klimyk, *Representation of Lie groups and special functions*, 4 vols., Kluwer, Dordrecht, 1991–1995.

[59] M. Waldschmidt, *Diophantine approximation on linear algebraic groups*, Springer, Berlin, 2000.

[60] A. Wiles, Modular elliptic curves and Fermat's last theorem, *Ann. of Math.* **141** (1995), 443–551.

Additional References

R.E. Borcherds, What is moonshine?, *Proceedings of the International Congress of Mathematicians: Berlin 1998*, Vol. I, pp. 607–615, Documenta Mathematica, Bielefeld, 1998.

C. Breuil, B. Conrad, F. Diamond and R. Taylor, On the modularity of elliptic curves over Q, *J. Amer. Math. Soc.* **14** (2001), 843–939.

Chandrasekhar Khare, Serre's modularity conjecture, *Preprint*.

Notations

The *Landau order symbols* are defined in the following way: if $I = [t_0, \infty)$ is a half-line and if $f, g: I \to \mathbb{R}$ are real-valued functions with $g(t) > 0$ for all $t \in I$, we write

$f = O(g)$ if there exists a constant $C > 0$ such that $|f(t)|/g(t) \le C$ for all $t \in I$;

$f = o(g)$ if $f(t)/g(t) \to 0$ as $t \to \infty$;

$f \sim g$ if $f(t)/g(t) \to 1$ as $t \to \infty$.

The *end of a proof* is denoted by $\square$.

Axioms

Index